ELEMENTS OF ELECTRICAL ENGINEERING

By

J.B. Gupta

B.Sc. (Elec. Engg.); F.I.E. (India);
C. Eng. (I); (I); M.I.S.T.E.

S.K. KATARIA & SONS®

Publisher of Engineering Books

4885/109, Prakash Mahal, Dr. Subhash Bhargav Lane,
Opposite Delhi Medical Association, Daryaganj, New Delhi–110002 (INDIA)
Phone: +91-11-43551243; ***Mobile:*** +91-9871775858, 7678425209
e-mail: katariabook@gmail.com, katariabooks@yahoo.com
Website: www.skkatariaandsons.com
(e-books, subscriptions and bulk buying also available on our website)

Published By:

S.K. Kataria & Sons®
4885/109, Prakash Mahal, Dr. Subhash Bhargav Lane,
Opposite Delhi Medical Association, Daryaganj, New Delhi-110002 (INDIA)
Phone: +91-11-43551243; *Mobile:* +91-9871775858, 7678425209
e-mail: katariabook@gmail.com, katariabooks@yahoo.com
Website: www.skkatariaandsons.com

***Edited By*:**
Rajeev Manglik
B.E. (Elec.) Pb. University
Rohit Manglik
B.E. (Elec.) University of Rookee

First Edition : 2009; Reprint: 2026

Laser Typesetting by : SARA Assignments

PREFACE

The primary aim of presenting this book is to meet the needs of first year students of degree engineering (all branches) studying in Engineering Colleges affiliated to the Gujarat Technological University. The subject matter has been presented in an easy comprehensive manner, logical sequence using simple language and to give a detailed exposition of the fundamentals.

The book consists of 15 chapters, divided in 9 units, providing a comprehensive coverage of course as prescribed by Gujarat Technological University for first year students. A large number of worked-out examples, theoretical exercises and numerical problems, all of degree standard, have been included in the text. The solutions to illustrative examples covering almost all topics and subtopics of various chapters make the entire presentation easy to follow. Exercises and numerical problems with answers given at the end of each chapter help the students in evaluating their understanding of the subject. Highlights given at the end of almost each chapter help the students in refreshing their knowledge while short answer type questions with answers help them in preparing for viva-voce/practical examinations and interviews. Objective type questions given at the end of book will also prove very useful to the students.

Every care has been taken to eliminate misprints, omissions and errors but it is too much to expect that no inaccuracy/misprint/error has crept in and the author would be grateful to the readers for bringing to his notice any such errors/misprints/omissions they may come across while going through the book.

The author shall feel satisfied if the book meets the needs of the students for whom it is meant. Suggestions and criticism for the improvement of the book are always welcomed.

The author will like to appreciate sincere efforts and contribution made by his sons, Rajeev Manglik and Rohit Manglik in presenting the book in this form.

The author will like to thank all the Scientists, Engineers, Authors and Publishers whose works and text have been the source of enlightenment, inspiration and guidance to him in the presentation.

Lastly, but not the least, the author is thankful to his publishers for bringing out the book in such a short period and for pricing it moderately despite heavy cost of paper and printing.

January 1, 2009 **J.B. Gupta**

CONTENTS

UNIT-I

DC Circuits

Contents in This Unit

DC Circuits: Effect of temperature upon resistance; Solution of series, parallel in brief; Star-delta combination of resistances; KVL & KCL.

Chapter

1 DC Circuits

1.1. NATURE OF ELECTRICITY

It is comparatively easy to describe electricity by its action than by its nature. Electricity is convenient form of energy and have got numerous applications, such as for lighting, transportation, communication, heating, elevators, and for driving of various types of machine tools owing to its inherent numerous advantages.

Although no one knows precisely what electricity is, it has been possible to develop theories about electricity through experiments and observations of its behaviour. As a result, it is now believed that all matters are essentially electrical in nature.

In order to appreciate the nature of electricity, it is necessary to consider the structure of matter. Matter has been defined any thing that occupies space and has weight such as copper, iron, water, air, smoke, aluminium, silver etc. Matter is made up of extremely small particles known as *molecules*. A molecule is the smallest particle into which a given substance may be divided without changing its identity. Molecules can further be subdivided chemically into still minute particles known as *atoms*. An atom may be regarded as the smallest particle of an element that can exist. The substances, whose molecules consist of similar atoms, are called the *elements* whereas the substances whose molecules consist of dissimilar atoms are called the *compounds*.

According to modern electron theory, the atoms of all elements are composed of two parts, namely, the central *nucleus* and the surrounding or *orbital electrons* (Fig. 1.1).

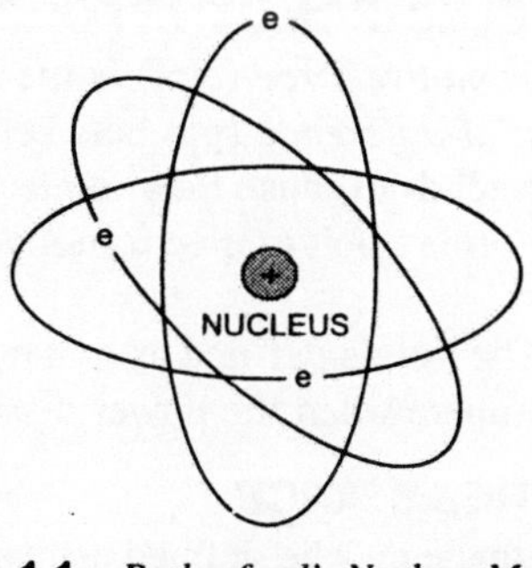

Fig. 1.1 *Rutherford's Nuclear Model of Atomic Structure*

The electrons each have a negative electric charge of 1.602×10^{-19} coulomb and some particles within the nucleus have a positive charge of the same magnitude. Since opposite charges attract each other, a force of attraction exists between the oppositely charged electrons and nucleus. Compared to the mass of nucleus, electrons are relatively tiny particles of almost negligible mass (each electron weighing 9.107×10^{-31} kg).

The nucleus of an atom is largely a cluster of two types of particles, called the *protons* and *neutrons*. Protons each have a positive electrical charge, equal in magnitude (but opposite in polarity) to the negative charge of an electron. A neutron has no charge at all. Protons and neutrons, each have mass about 1800 times the mass of an electron. For a given atom, the number of protons in the nucleus normally is equal to the number of orbiting electrons. Thus, an atom is electrically neutral. If an atom loses an electron, it loses the negative charge and becomes a *positive ion*. Similarly if an atom gains an additional electron, it becomes a *negative ion*. However, nearly all atoms have some electrons which are loosely bound to their nuclei. These electrons are known as *free electrons* and may be dislodged by one means or another and transferred from one atom to another. The body, that contains unequal number of electrons and protons is said to be *electrically charged.* If a body contains electrons more than its normal number then the body is said to be – *vely charged.* Similarly a body containing electrons less than its normal number is said to be + *vely charged.*

The atoms of some materials such as silver, copper, aluminium and zinc, known as *conductors,* have many free electrons. These free electrons of a metal move about haphazardly in all directions from atom to atom but when a certain electrical pressure or potential is applied to such metals at the two ends, the electrons move only in one direction. The drift of electrons in a conductor in one direction is known as the *electric current i.e.* the flow of electric current takes place by the movement of the electrons in a conductor. Since electrons are negatively charged, the direction of their motion is opposite to the direction of conventional current which is from higher potential to the lower potential. In non-metallic materials, such as glass, mica, slate and porcelain, the electrons are very closely bound to the nucleus and it is difficult to remove the electrons from the atoms. Such materials are known as *non-conductors* or *insulators.*

1.2. ELECTRIC CURRENT

Electric current may be defined as the time rate of net motion of an electric charge across a cross-sectional boundary (Fig. 1.2). A random motion of electrons in a metal does not constitute a current unless there is a net transfer of charge with time

i.e. electric current, i = Rate of transfer of electric charge

$$= \frac{\text{Quantity of electric charge transferred during a given time duration}}{\text{Time duration}}$$

$$= \frac{dQ}{dt} \qquad ...(1.1)$$

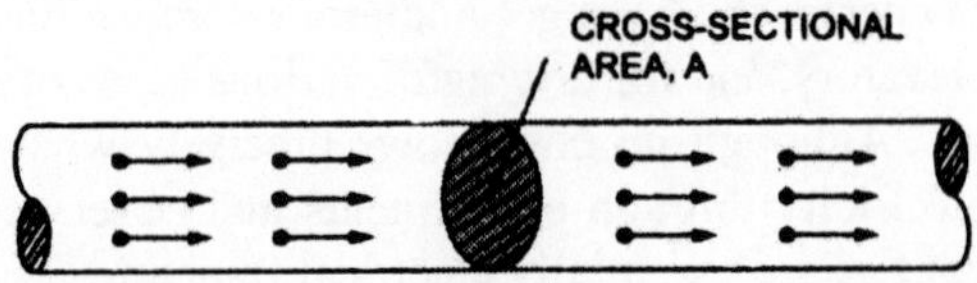

Fig. 1.2

Coulomb is the practical as well as SI unit for measurement of electric charge. One coulomb is approximately equal to 624 × 10^{16} electrons.

Since current is the rate of flow of electric charge through a conductor and coulomb is the unit of electric charge, the current may be specified in coulombs per second. In practice the term coulomb per second is seldom used, a shorter term, *ampere* is used instead.

1.3. ELECTROMOTIVE FORCE AND POTENTIAL DIFFERENCE

Electromotive force (emf) is the force that causes an electric current to flow in an electric circuit while the *potential difference* (pd) between two points in an electric circuit is that difference in their electrical state which tends to cause flow of electric current between them.

Volt is a unit of electromotive force as well as potential difference in practical as well as in SI system of units.

The volt is defined as that potential difference between two points of a conductor carrying a current of one ampere when the power dissipated between these points is equal to one watt.

1.4. RESISTANCE

Resistance may be defined as that property of a substance which opposes (or restricts) the flow of an electric current (or electrons) through it.

The practical as well as mks (or SI) unit of resistance is ohm (Ω), which is defined as that resistance between two points of a conductor when a potential difference of one volt, applied between these points, produces in this conductor a current of one ampere, the conductor not being a source of any emf.

For insulators having high resistance, much bigger units kilo-ohm or kΩ (10^3 ohm) and mega-ohm or MΩ (10^6 ohm) are used. In case of very small resistances smaller units like milli-ohm (10^{-3} ohm) or micro-ohm (10^{-6} ohm) are employed.

1.5. OHM'S LAW

As the rate of flow of water through a pipe is directly proportional to the effective pressure (*i.e.* difference of pressure at two ends) and inversely proportional to the frictional resistance, similarly the current flowing through a conductor is directly proportional to the potential difference across the ends of the conductor and inversely proportional to the conductor resistance. This relation was discovered by Georg Simon Ohm and so it is known as Ohm's law.

If I is the current flowing through a conductor of resistance R across which a potential difference V is applied then according to Ohm's law

$$I \propto V \quad \text{and} \quad I \propto \frac{1}{R} \quad \text{or} \quad I \propto \frac{V}{R} \quad \text{or} \quad I = \frac{V}{R} \qquad ...(1.2)$$

where V is in volts, R is in ohms and I is in amperes.

Ohm's law may be defined as follows:

Physical state i.e. temperature etc. remaining the same, the current flowing through a conductor is directly proportional to the potential difference applied across its ends.

Or

The ratio of potential difference applied across a conductor and current flowing through it remains constant provided physical state i.e. temperature etc. of the conductor remains unchanged.

$$i.e. \quad \frac{V}{I} = \text{Constant} = R \qquad ...(1.3)$$

where R is known as the resistance of the conductor.

Ohm's law may be alternatively expressed as

$$V = IR \qquad ...(1.4)$$

Equations (1.2), (1.3) and (1.4) give Ohm's law in three forms with which the student should be familiar.

Ohm's law cannot be applied to circuits consisting of electronic tubes or transistors because such elements are not bilateral *i.e.* they behave in different way when the direction of flow of current is reversed as in case of a diode. Ohm's law also cannot be applied to circuits consisting of nonlinear elements such as powdered carbon, thyrite, electric arc etc. For example, for silicon carbide, the relationship between applied voltage (or potential difference) V and current flowing I is given as $V = K I^m$ where K and m are constants and m is less than unity.

1.6. LAWS OF RESISTANCE

The resistance of a wire depends upon its length, area of x-section, type of material, purity and hardness of material of which it is made of and the operating temperature.

Resistance of a wire is

(*a*) directly proportional to its length, l *i.e.* $R \propto l$

(*b*) inversely proportional to its area of x-section, a *i.e.* $R \propto 1/a$

Combining above two facts we have $R \propto l/a$

$$\text{or } R = \rho \frac{l}{a} \qquad ...(1.5)$$

where ρ (rho) is a constant depending upon the nature of the material and is known as the *specific resistance or resistivity* of the material of the wire.

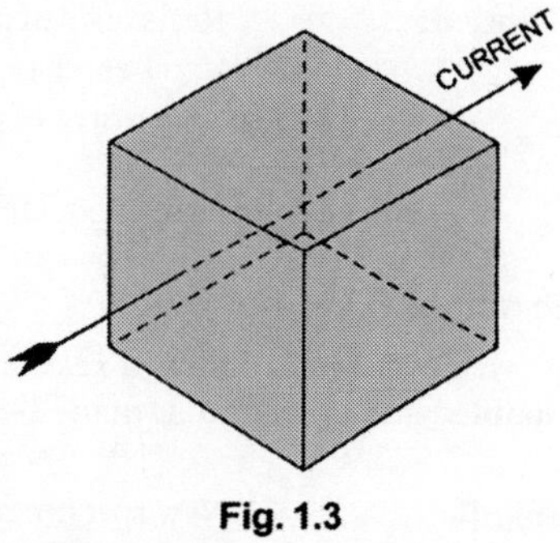

Fig. 1.3

To determine the nature of the constant ρ let us imagine a conductor of unit length and unit cross-sectional area, for example a cube whose edges are each of length one unit, and let the current flow into the cube at right angles to one face and out at the other face. Then putting $l = 1$ and $a = 1$ in Eq. (1.5) we have $R = \rho$. *Hence resistance of a material of unit length having unit cross-sectional area is defined as the resistivity or specific resistance of the material.*

Specific resistance or resistivity of a material is also defined as the resistance between opposite faces of a unit cube of that material.

Resistivity is measured in ohm-metres (Ω-*m*) or ohms per metre cube in mks (or SI) system and ohm-cm (Ω-cm) or ohm per cm cube in cgs system.

$$1\ \Omega\text{-m} = 100\ \Omega\text{-cm}$$

1.7. CONDUCTANCE AND CONDUCTIVITY

The reciprocal of resistance *i.e.* 1/R is called the *conductance* and is denoted by English letter G. It is defined as the inducement offered by the conductor to the flow of current and is measured in siemens (S). Earlier, the unit of conductance was mho (℧).

$$1 \text{ siemen} = 1 \text{ mho}$$

From Eq. (1.5)

$$G = \frac{1}{R} = \frac{1}{\rho \frac{l}{a}} = \frac{1}{\rho} \cdot \frac{a}{l} = \sigma \frac{a}{l} \qquad \text{...(1.6)}$$

where $\sigma = \frac{1}{\rho}$ and is known as *specific conductance* or *conductivity* of the material. Hence *conductivity* is the reciprocal of the resistivity and is defined as the conductance between the two opposite faces of a unit cube. The unit of conductivity is siemens/metre (S/m).

Example 1.1

A coil consists of 2,000 turns of copper wire having a cross-sectional area of 0.8 mm². The mean length per turn is 80 cm, and the resistivity of copper is 0.02 μ Ω-m. Find the resistance of the coil.

Solution:

Length of the coil, l = Number of turns × mean length per turn = 2,000 × 0.8 = 1,600 m

Cross-sectional area of wire, $a = 0.8 \text{ mm}^2 = 0.8 \times 10^{-6} \text{ m}^2$

Resistivity of copper, $\rho = 0.02\ \mu\Omega\text{-m} = 2 \times 10^{-2} \times 10^{-6}\ \Omega\text{-m} = 2 \times 10^{-8}\ \Omega\text{-m}$

Resistance of the coil, $R = \rho\frac{l}{a} = 2 \times 10^{-8} \times \frac{1{,}600}{0.8 \times 10^{-6}} = 40\ \Omega$ **Ans.**

Example 1.2

The resistance of a conductor 1 mm² in cross-section and 20 m long is 0.346 Ω. Determine the specific resistance of the conductor material. [*Bangalore Univ. Elec. Circuits-I, 1991*]

Solution:

Resistance of conductor, $R = 0.346\ \Omega$

Length of conductor, $l = 20$ m

Cross-sectional area of conductor, $a = 1 \text{ mm}^2 = 1 \times 10^{-6} \text{ m}^2$

Specific resistance of conductor material, $\rho = \frac{R \times a}{l} = \frac{0.346 \times 1 \times 10^{-6}}{20} = 1.73 \times 10^{-8}\ \Omega\text{-m}$ **Ans.**

Example 1.3

A wire of length 1 m has a resistance of 2 Ω. Obtain the resistance if specific resistance is doubled, diameter is doubled and the length is made three times of the first.

[*Pb. Technical Univ. Basic Electrical and Electronics Engineering Second Semester 2004-05*]

Solution:

New specific resistance, $\rho_2 = 2\rho_1$

New length of wire, $l_2 = 3\,l_1$

New cross-sectional area, $a_2 = \frac{\pi}{4}\left(d_2^2\right) = \frac{\pi}{4}(2d_1)^2 = 4\,a_1$

New resistance, $R_2 = \frac{\rho_2 l_2}{a_2} = \frac{2\rho_1 \times 3l_1}{4a_1} = 1.5\frac{\rho_1 l_1}{a_1} = 1.5\,R_1 = 1.5 \times 2 = 3\ \Omega$ **Ans.**

Example 1.4

Determine the resistance of a copper tube having external diameter 8 cm; thickness 5 mm and length 5 metres. It is given that specific resistance of copper is 1.70 μ Ω-cm.

Solution :

Resistivity of copper, $\rho = 1.7 \times 10^{-8}$ Ω-m

Length of copper tube, l = 5 metres

External diameter of tube, D = 8 cm = 0.08 metre

Internal diameter of tube, d = D – 2 × thickness of tube = 0.08 – 2 × 0.005 = 0.07 m

$$\text{Cross-sectional area of tube, } a = \frac{\pi}{4}(D^2 - d^2) = \frac{\pi}{4}(0.08^2 - 0.07^2) = 0.001178 \text{ m}^2$$

$$\text{Resistance of copper tube, } R = \rho\frac{l}{a} = \frac{1.7\times10^{-8}\times5}{0.001178} = 72.12\ \mu\Omega \quad \textbf{Ans.}$$

Example 1.5

A rectangular metal strip has the following dimensions:

x = 10 cm, y = 0.5 cm, z = 0.2 cm

Determine the ratio of resistances R_x, R_y and R_z between the respective pairs of opposite faces.

Solution: Let the resistivity of the material of rectangular strip be ρ ohms per cm cube.

Resistance between the opposite faces ADHE and BCGF

$$R_x = \frac{\rho \times \text{length of AB}}{\text{Area of face ADHE}}$$

$$= \frac{\rho\times10}{0.5\times0.2} = 100\rho \qquad ...(i)$$

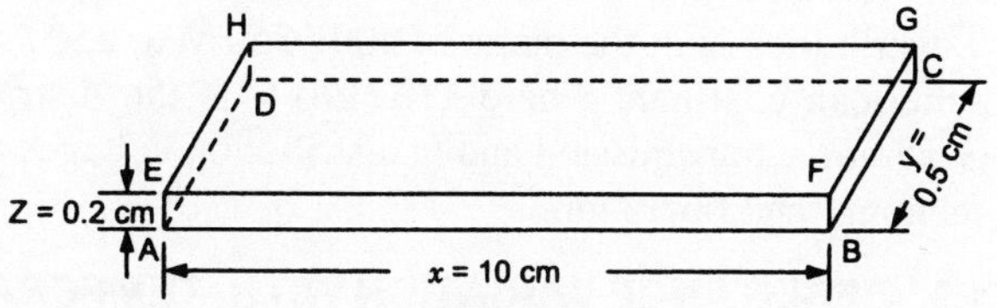

Fig. 1.4

Resistance between the opposite faces ABFE and CGHD

$$R_y = \frac{\rho\times\text{length of BC}}{\text{Area of face ABFE}} = \frac{\rho\times0.5}{10\times0.2} = 0.25\rho \qquad ...(ii)$$

Resistance between the opposite faces BCDA and EFGH,

$$R_z = \frac{\rho\times\text{length of AE}}{\text{Area of face BCDA}} = \frac{\rho\times0.2}{10\times0.5} = 0.04\rho \qquad ...(iii)$$

Ratio of resistances R_x, R_y and R_z is given as

$$R_x : R_y : R_z :: 100\ \rho : 0.25\ \rho : 0.04\rho$$

or $R_x : R_y : R_z$ :: 10,000 : 25 : 4 **Ans.**

1.8. TYPES OF SUPPLY

There are two types of supply viz. dc and ac.

The voltage or current available from batteries or solar cells is *direct* in the sense that the polarity remains the same. Such sources are called direct current sources. A plot of output of such sources with respect to time is a straight line parallel to time axis as illustrated in Fig. 1.5. This shows that the voltage / current output of a dc source is constant with respect to time, unless the chemicals in the battery are exhausted or the light incident on solar cell varies.

Power obtainable from batteries or solar cells is very small as compared to the total power we need. The dc power in bulk can be generated by using dc generators. Standard dc voltages are 230 and 460 V.

A current (or voltage) is called *alternating* if it periodically changes its direction and magnitude whereas direct currents are steady and in one direction. In more restricted sense, alternating current is a periodically varying current, the average value of which, over a period, is zero.

At present a large percentage of electrical energy (nearly all) being used for domestic, commercial and industrial purposes is generated as alternating current because of technical and economical reasons (refer to Art. 5.1).

For industrial applications of direct current like electrolytic processes, welding processes and variable speed motor drives, the present trend is to generate ac and then convert it into dc by converters. For conversion of ac into dc the dc generator as a part of motor (ac)-generator (dc) set has to compete the SCR rectifiers and other various other power controlled devices which usually are cheaper in cost, compact in size, relatively noise-free in operation and require minimum maintenance, though they suffer from the disadvantages of having poor power factor, harmonic generation, poor braking etc.

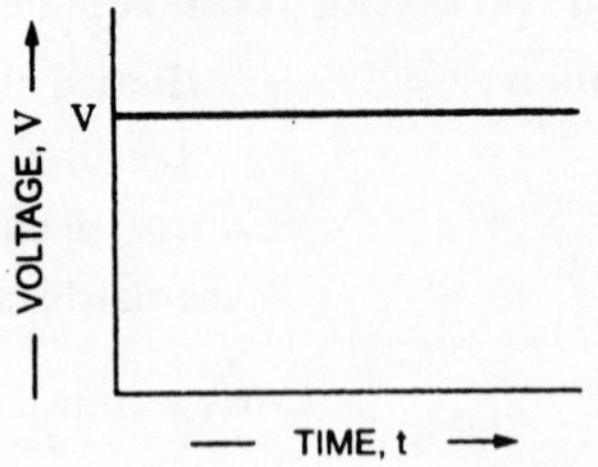

(*a*) *Output Characteristic of Battery*

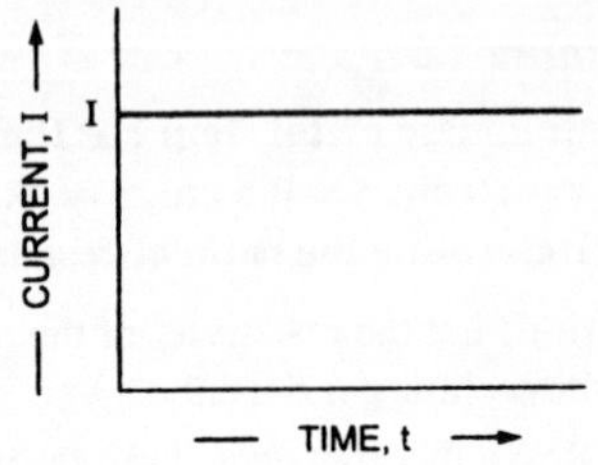

(*b*) *Output Characteristic of Solar Cell*

Fig. 1.5

In our country the standard voltage for low tension (*lt*) distribution is 240 / 415 V (240 V between phase and neutral wires and 415 V from line-to-line). This is a reasonably safe voltage. Higher the voltage used, lower is the installation cost *i.e.* economy is achieved but increase of voltage means more hazard to human life and property. Hence, a compromise is made. All over the world the standard voltage used for *lt* distribution is in the range of 200 / 350 V to 250 / 440 V except in American continent where 110 /190 V is the distribution voltage. Generation, transmission and *ht* distribution voltages are governed by economy and operation.

1.9. RESISTANCE VARIATION WITH TEMPERATURE

The resistance of all pure metallic conductors increases with the increase in temperature but the resistance of the insulators and non-metallic materials generally decreases with the increase in temperature.

If the resistance of any pure metal is plotted on a temperature base, it is found that over the range of temperature from 0 to 100 °C the graph is practically a straight line, as illustrated in Fig. 1.6. If this straight line is extended, it cuts the temperature axis at some temperature, $-t_0$ °C, known as *inferred zero resistance temperature*. This does not mean that the resistance of the metal is actually zero at that temperature, but $-t_0$ °C is the temperature at which the resistance would be zero if the rate of decrease between 100 and 0 °C were maintained constant at all temperatures. From the similarity of the triangles in Fig. 1.6.

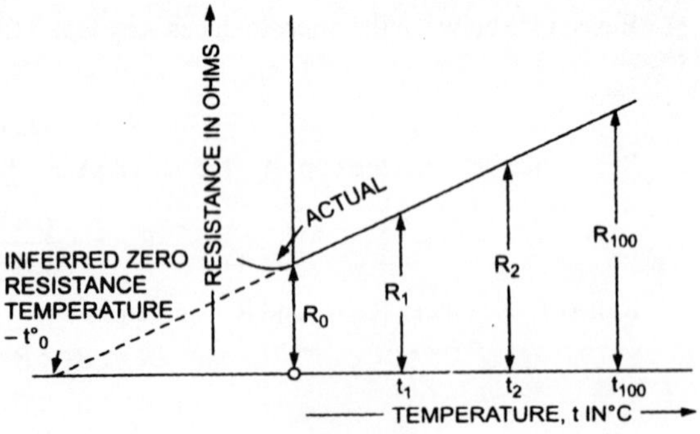

Fig. 1.6

$$\frac{R_2}{R_1} = \frac{t_0 + t_2}{t_0 + t_1} \qquad ...(1.7)$$

where R_1 and R_2 are the resistances at temperatures t_1 °C and t_2 °C respectively. Thus, if the resistance R_1 for any temperature t_1 °C is known, then resistance for any other temperature t_2 can be computed from above equation provided that t_0 for that particular material is known. Inferred zero resistance temperatures, t_0 for various materials are given in Table 1.1.

Table 1.1 Resistivities, Temperature Coefficients and Inferred Zero Resistance Temperatures

Material	*Resistivity in Ω-m at 20 °C (× 10^{-8})*	*Temperature Coefficient of Resistance in ohms per ohm per °C rise in temperature at 20 °C*	*Inferred Zero Resistance Temperature in °C*
Aluminium, annealed	2.82	0.0039	– 236
Aluminium, hard drawn	2.92	0.0038	– 243
Brass	6 – 8	0.002	– 480
Carbon	3,000 – 7,000	– 0.0005	—
Constantan or Eureka (60% copper and 40% nickel)	49	0.000008	– 125,000
Copper, annealed	1.724	0.00393	– 234.5
Copper, hard drawn	1.79	0.00382	– 242
German Silver (80% Cu, 12% Ni, 4% Zn)	33	0.0004	– 2,480
Gold	2.44	0.0034	– 274
Iron	12	0.005	– 180
Lead	22	0.0039	– 236
Manganin (84% Cu, 12% Mn, 4% Ni)	42 to 74	0.00003	– 33,000
Mercury	96	0.00089	– 1,100
Molybdenum	5.7	0.0033	– 280
Nichrome (60% Cu, 25% Fe,15 % Cr)	100	0.00044	– 2.250
Nickel	7.8	0.006	– 147
Phosphor, Bronze	5–10	0.0035	– 266
Platinum	10	0.003	– 310
Platinum–iridium	24.6	0.0012	– 814
Silver (99.98% pure)	1.6	0.0038	– 243
Tungsten	5.51	0.0045	– 200
Zinc	6.3	0.004	– 230

Note: It *has been found that the resistivity of the metallic conductors is significantly affected by variations in temperature but it is also affected by other factors like tension and pressure to some extent. For most metals except lithium and calcium resistivity decreases with the increase in pressure. However, resistivity increases with increase in tension.*

The variation of resistance with temperature is often utilized in determining temperature variations. For example, in testing of an electric machine, the resistance of the coil is measured both before and after the test run, and the increase in resistance is a measure of the rise in temperature. For computation of temperature rise Eq. (1.7) may be transposed to the following form

$$t_2 - t_1 = \frac{R_2 - R_1}{R_1}(t_0 + t_1) \qquad ...(1.8)$$

A temperature of 20 °C has been adopted as the *standard reference temperature* for measurement of resistance, and the hand books give the resistance of the various materials at that temperature. Consequently, when a designer is computing the resistance of any conductor from its dimensions, the initial temperature t, at which the resistance is known, is generally 20 °C.

1.10. TEMPERATURE COEFFICIENT OF RESISTANCE

Let a metallic conductor having a resistance of R_0 at 0 °C be heated to t °C and let its resistance at this temperature be R_t. From Eq. (1.7)

$$\frac{R_t}{R_0} = \frac{t_0 + t}{t_0 + 0}$$

$$\text{or } R_t = R_0 + \frac{1}{t_0} R_0 t$$

$$\text{or change in resistance, } \Delta R = R_t - R_0 = \frac{1}{t_0} R_0 t = \alpha_0 R_0 t \qquad ...(1.9)$$

where $\alpha_0 = \frac{1}{t_0}$ and is called the temperature coefficient of resistance of the material at 0 °C

From Eq. (1.9) it may be concluded that change in resistance due to change in temperature

(*a*) varies directly as its initial resistance,

(*b*) varies directly as rise in temperature, and

(*c*) depends on the nature of the material of the conductor.

The Eq. (1.9) may be rewritten as

$$\alpha_0 = \frac{\Delta R}{R_0 t} \qquad ...(1.10)$$

So the temperature coefficient of resistance may be defined as the ratio of increase in resistance per degree rise of temperature to the original resistance.

If R_0 is the resistance of any conductor at 0 °C and α_0 is the temperature coefficient of resistance at 0°, then resistance at t° C is given as

$$R_t = \text{Original resistance + increase in resistance} = R_0 + R_0 \alpha_0 t = R_0 (1 + \alpha_0 t) \qquad ...(1.11)$$

The above expression holds good for both increase as well as decrease in temperature.

It is to be noted that

(*i*) temperature coefficient of resistance for all pure metallic conductors is positive *i.e.* the resistance of all pure metallic conductors increase with the increase in temperature, that of non-metallic materials such as of carbon is negative i.e. the resistance of non-metallic materials such as of carbon decreases with the increase in temperature. The temperature coefficient of resistance of alloys like constantan and manganin is negligible.

(*ii*) temperature coefficient of resistance is not constant but depends on the initial temperature on which the increment in resistance is based. When the increment is based on the resistance measured at 0 °C, then the temperature coefficient of resistance has the value of α_0. At any other temperature t °C, value of temperature coefficient of resistance is α_t and so on. For any material the temperature coefficient of resistance at 0 °C *i.e.* α_0 has the maximum value.

The temperature coefficient of resistance at any temperature t_1 is given as

$$\alpha_1 = \frac{1}{\frac{1}{\alpha_0} + t_1} \qquad ...(1.12)$$

The temperature coefficient of resistance at temperature t_2 in terms of temperature coefficient of resistance at temperature t_1 is given as

$$\alpha_2 = \frac{1}{\frac{1}{\alpha_1} + (t_2 - t_1)} \qquad ...(1.13)$$

If R_1 is the resistance of any conductor at t_1 °C and α_1 is the temperature coefficient of resistance at t_1 °C, then resistance of the conductor at t_2 °C is given

$$R_2 = R_1 [1 + \alpha_1 (t_2 - t_1)] \qquad ...(1.14)$$

(*iii*) As the resistance of the material changes with the change in temperature so it is obvious that resistivity of the material depends on temperature.

Knowledge of temperature coefficient of resistance is utilized in determining the temperature rise of electrical machines, as already described in Art. 1.9.

Values of resistivity, temperature coefficient of resistance at 20 °C and inferred zero resistance temperature for various materials are given in Table 1.1.

Example 1.6

Show that if α_0 be the resistance temperature coefficient of a conductor at 0 °C, then resistance temperature coefficient at t_1 °C is given as

$$\alpha_1 = \frac{1}{\frac{1}{\alpha_0} + t_1}$$

Solution : Let R_0, R_1 and R_2 be the resistances of a conductor at 0°C, t_1 °C and t_2 °C temperatures respectively, then from Eq. (1.11)

Resistance at t_1 °C, $R_1 = R_0\ (1 + \alpha_0 t_1)$ ...(*i*)

Resistance at t_2 °C, $R_2 = R_0\ (1 + \alpha_0 t_2)$ ...(*ii*)

Also $R_2 = R_1\ [1 + \alpha_1 (t_2 - t_1)]$ ... Refer to Eq. (1.14)

or $\frac{R_2}{R_1} = 1 + \alpha_1 (t_2 - t_1)$ ...(*iii*)

Dividing Eq. (*ii*) by Eq. (*i*), we get

$$\frac{R_2}{R_1} = \frac{1+\alpha_0 t_2}{1+\alpha_0 t_1} = \frac{1+\alpha_0 t_2 + \alpha_0 t_1 - \alpha_0 t_1}{1+\alpha_0 t_1}$$

$$= \frac{1+\alpha_0 t_1}{1+\alpha_0 t_1} + \frac{\alpha_0 t_2 - \alpha_0 t_1}{1+\alpha_0 t_1} = 1 + \frac{\alpha_0 (t_2 - t_1)}{1+\alpha_0 t_1} = 1 + \frac{\alpha_0}{1+\alpha_0 t_1}(t_2 - t_1) \quad ...(iv)$$

Comparing Eqs. (*iii*) and (*iv*) we get

$$1 + \alpha_1\ (t_2 - t_1) = 1 + \frac{\alpha_0}{1+\alpha_0 t_1}(t_2 - t_1)$$

or $\alpha_1 = \frac{\alpha_0}{1+\alpha_0 t_1} = \frac{1}{\frac{1}{\alpha_0} + t_1}$ (Proved)

Example 1.7

Show that if α_1 is the resistance temperature coefficient of a conductor at t_1 °C then resistance temperature coefficient at t_2 °C is given by

$$\frac{1}{\frac{1}{\alpha_0} + (t_2 - t_1)}$$

Solution: Let R_1, R_2 and R_3 be the resistances of a conductor at temperatures t_1, t_2 and t_3°C respectively.

Then $R_2 = R_1\ [1 + \alpha_1\ (t_2 - t_1)]$ ...(*i*)

and $R_3 = R_1\ [1 + \alpha_1 (t_3 - t_1)]$ ...(*ii*)

Also $R_3 = R_2\ [1 + \alpha_2 (t_3 - t_2)]$

or $\frac{R_3}{R_2} = 1 + \alpha_2\ (t_3 - t_2)$ ...(*iii*)

Dividing Eq. (*ii*) by Eq. (*i*) we get

$$\frac{R_3}{R_2} = \frac{1+\alpha_1 (t_3 - t_1)}{1+\alpha_1 (t_2 - t_1)} = \frac{1+\alpha_1 (t_2 - t_1) + \alpha_1 (t_3 - t_1) - \alpha_1 (t_2 - t_1)}{1+\alpha_1 (t_2 - t_1)} = 1 + \frac{\alpha_1 (t_3 - t_2)}{1+\alpha_1 (t_2 - t_1)} \quad ...(iv)$$

Comparing Eqs. (*iii*) and (*iv*) we get

$$1 + \alpha_2 \ (t_3 - t_2) = 1 + \frac{\alpha_1 (t_3 - t_2)}{1 + \alpha_1 (t_2 - t_1)}$$

$$\text{or} \quad \alpha_2 = \frac{\alpha_1}{1 + \alpha_1 (t_2 - t_1)} = \frac{1}{\frac{1}{\alpha_1} + (t_2 - t_1)} \qquad \text{(Proved)}$$

Example 1.8

A coil of relay is made of copper wire. At a temperature of 20 °C, the resistance of the coil is 400 Ω. Calculate the resistance of the coil at temperature of 80 °C. The temperature coefficient of copper is 0.0038 Ω/Ω/°C at 0 °C.

Solution : Temperature coefficient of resistance at 0 °C, $\alpha_0 = 0.0038 \ \Omega / \Omega / °C$

$$\text{Temperature coefficient of resistance at 20 °C, } \alpha_{20} = \frac{1}{\frac{1}{\alpha_0} + t} = \frac{1}{\frac{1}{0.0038} + 20} = 0.003532$$

Resistance of wire at 20 °C, $R_1 = 400 \ \Omega$

Resistance of wire at 80 °C, $R_2 = R_1 [1 + \alpha_{20} (t_2 - t_1)]$

$= 400 [1 + 0.003532 (80 - 20)] = 484.8 \ \Omega$ **Ans.**

Example 1.9

A platinum coil has a resistance of 3.146 Ω at 40 °C and 3.767 Ω at 100 °C. Find the resistance at 0 °C and the temperature coefficient of resistance at 40 °C. [*Allahabad Univ. Elec. Science II, 1993*]

Solution: Let the temperature coefficient of resistance of platinum at 0 °C be α_0 and resistance of platinum coil at 0 °C be R_0, then

$$\text{Resistance at 40 °C, } R_1 = R_0 \ (1 + 40 \ \alpha_0) \qquad ...(i)$$

$$\text{Resistance at 100 °C, } R_2 = R_0 \ (i + 100 \ \alpha_0) \qquad ...(ii)$$

Dividing Eq. (*ii*) by Eq. (*i*) we have

$$\frac{R_2}{R_1} = \frac{1 + 100 \ \alpha_0}{1 + 40 \ \alpha_0}$$

$$\text{or} \quad \frac{3.767}{3.146} = \frac{1 + 100 \ \alpha_0}{1 + 40 \ \alpha_0}$$

$$\text{or} \quad \alpha_0 = 0.00379$$

$$\text{Temperature coefficient of resistance at 40 °C, } \alpha_{40} = \frac{1}{\frac{1}{\alpha_0} + 40} = \frac{1}{\frac{1}{0.00379} + 40} = 0.00329 \ \Omega/\Omega/ °C \quad \textbf{Ans.}$$

Substituting $R_1 = 3.146$ and $\alpha_0 = 0.00379$ in Eq. (*i*) we have

$$3.146 = R_0 \ (1 + 40 \times 0.00379)$$

$$\text{or} \quad R_0 = \frac{3.146}{1 + 40 \times 0.00379} = 2.732 \ \Omega \quad \textbf{Ans.}$$

Example 1.10

A coil has resistance of 18 Ω when its mean temperature is 20 °C and of 20 Ω when its mean temperature is 50 °C. Find its mean temperature rise when its resistance is 21 Ω and the surrounding temperature is 15 °C.

[*Allahabad Univ. Electrical Technology, 1992*]

Solution: Let R_0 be the resistance of the coil at 0 °C and α_0 be its temperature coefficient of resistance at 0 °C

$$\text{Resistance at 20 °C, } 18 = R_0 \ (1 + 20 \ \alpha_0) \qquad ...(i)$$

$$\text{Resistance at 50 °C, } 20 = R_0 \ (1 + 50 \ \alpha_0) \qquad ...(ii)$$

Dividing Eq. (*ii*) by Eq. (*i*) we have

$$\frac{20}{18} = \frac{1 + 50 \ \alpha_0}{1 + 20 \ \alpha_0}$$

$$\text{or } \alpha_0 = \frac{1}{250} = 0.004$$

If t °C is the temperature of coil when its resistance is 21 Ω, then

$$21 = R_0 (1 + 0.004\,t) \qquad ...(iii)$$

Dividing Eq. (*iii*) by Eq. (*ii*) we have

$$\frac{21}{20} = \frac{1 + 0.004t}{1 + 50 \times 0.004}$$

$$\text{or } t = 65\ °C$$

$$\text{Temperature rise} = t - \text{surrounding temperature} = 65 - 15 = 50\ °C \quad \textbf{Ans.}$$

Example 1.11

A semicircular ring of copper has inner radius of 8 cm, radial thickness 4 cm and axial thickness 6 cm. Calculate the resistance of the ring at 50 °C between its two end faces. Specific resistance of copper at 20 °C = 1.724 × 10^{-6} Ω-cm. Resistance temperature coefficient of copper at 0 °C = 0.0043/ °C.

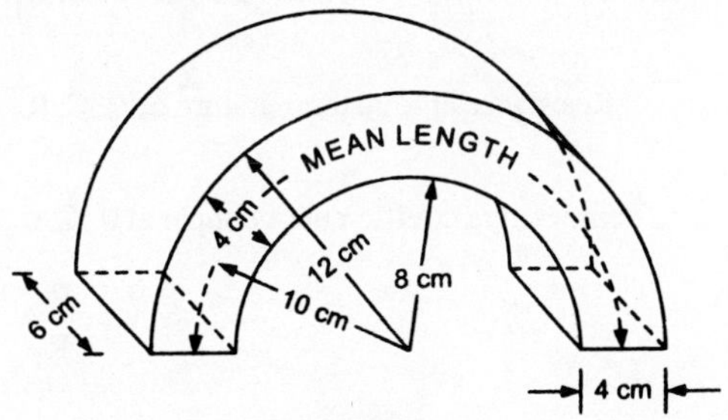

Fig. 1.7

Solution : The semi-circular ring is shown in Fig. 1.7.

$$\text{Mean radius of ring} = \frac{8+12}{2} = 10 \text{ cm}$$

$$\text{Mean length between end faces} = \pi \times 10 = 31.4 \text{ cm} = 0.314 \text{ m}$$

$$\text{Cross-section of the ring, } a = 4 \times 6 = 24 \text{ cm}^2 = 0.0024 \text{ m}^2$$

$$\text{Temperature coefficient of resistance at 0 °C, } \alpha_0 = 0.0043 \text{ per °C}$$

$$\text{Temperature coefficient of resistance at 20 °C, } \alpha_{20} = \frac{1}{\frac{1}{\alpha_0} + t_1} = \frac{1}{\frac{1}{0.0043} + 20} = 0.00396/\ °C$$

$$\text{Specific resistance at 20 °C, } \rho_{20} = 1.724 \times 10^{-6}\ \Omega\text{-cm} = 1.724 \times 10^{-8}\ \Omega\text{-m}$$

$$\text{Specific resistance at 50 °C, } \rho_{50} = \rho_{20}\,[1 + \alpha_{20}(t_2 - t_1)] = 1.724 \times 10^{-8}\ [1 + 0.00396\,(50 - 20)]$$

$$= 1.93 \times 10^{-8}\ \Omega\text{-m}$$

$$\text{Resistance of ring at 50 °C, } R_{50} = \rho_{50}\frac{l}{a} = \frac{1.93 \times 10^{-8} \times 0.314}{0.0024} = 2.525\ \mu\Omega \quad \textbf{Ans.}$$

Example 1.12

A copper conductor has its specific resistance of 1.6 × 10^{-6} Ω-cm at 0 °C and a resistance temperature coefficient of $\frac{1}{254.5}$ per °C at 20 °C. Find (*i*) the specific resistance and (*ii*) the resistance temperature coefficient at 60 °C.

[*Pune Univ. F.Y. Engg. Nov. 1988*]

Solution: Temperature coefficient of resistance at 20 °C, $\alpha_{20} = \frac{1}{254.5}$ Ω/Ω/°C

$$\text{Temperature coefficient of resistance at 60 °C, } \alpha_{60} = \frac{1}{\frac{1}{\alpha_{20}} + (t_2 - t_1)} = \frac{1}{254.5 + (60 - 20)} = \frac{1}{294.5}\ \Omega/\Omega/°C \quad \textbf{Ans.}$$

If α_0 is the temperature coefficient of resistance at 0 °C, then temperature coefficient of resistance at 20 °C

$$\alpha_{20} = \frac{1}{\frac{1}{\alpha_0} + 20}$$

$$\text{or } \alpha_0 = \frac{1}{\frac{1}{\alpha_{20}} - 20} = \frac{1}{254.5 - 20} = \frac{1}{234.5}\ \Omega/\Omega/°C$$

Specific resistance at 60 °C, $\rho_{60} = \rho_o (1 + \alpha_o t) = 1.6 \times 10^{-6} \left(1 + \frac{60}{234.5}\right) = 2.0094 \times 10^{-6}$ Ω-cm **Ans.**

Note : *From the above example it is obvious that the specific resistance of a given material does not remain constant, but changes with the change in temperature.*

Example 1.13

A potential difference of 200 V is applied to a copper field coil at a temperature of 15 °C and the current is 10 A. What will be the mean temperature of the coil when the current has fallen to 8 A ? Applied voltage being the same. Given $\alpha = \frac{1}{234.5}$ per °C at 0 °C.

Solution:

Resistance at 15 °C, $R_1 = \frac{V}{I_1} = \frac{200}{10} = 20\ \Omega$

Resistance at final temperature of t_2 °C, $R_2 = \frac{V}{I_2} = \frac{200}{8} = 25\ \Omega$

Temperature coefficient of copper at 0 °C, $\alpha_0 = \frac{1}{234.5}$

Since $\frac{R_2}{R_1} = \frac{1 + \alpha_0 t_2}{1 + \alpha_0 t_1}$,

$$\therefore \quad \frac{25}{20} = \frac{1 + \frac{t_2}{234.5}}{1 + \frac{15}{234.5}}$$

or $t_2 = 77.38$ °C **Ans.**

Example 1.14

The filament of a 60 W, 230 V lamp has a normal working temperature of 2,000 °C. Take the temperature coefficient of the matside to be 0.005. Find the approximate current which flows at instant of switching on the supply to the cold lamp. [*Pb. Technical Univ; Basic Electrical and Electronics Engineering First Semester 2005-06*]

Solution: Resistance of 60 W metal filament, $R_{2000} = \frac{V^2}{W} = \frac{(230)^2}{60} = 881.67\ \Omega$

Let the resistance of 60 W metal filament at room temperature (say 20 °C) be R_2. Temperature coefficient of resistance (say at room temperature of 20 °C) be

$$\alpha_{20} = 0.005$$

So $R_{2000} = R_{20} [1 + \alpha_{20} (t_2 - t_1)]$

or $R_{20} = \frac{R_{2000}}{1 + \alpha_{20}(t_2 - t_1)} = \frac{881.67}{1 + 0.005\,(2{,}000 - 20)} = \frac{881.67}{10.9} = 80.887\ \Omega$

Current flowing at the instant of switching on the bulb

$$I_{20} = \frac{V}{R_{20}} = \frac{230}{80.887} = 2.843 \text{ A} \quad \textbf{Ans.}$$

Example 1.15

Two coils connected in series have resistances of 600 Ω and 300 Ω and temperature coefficients of 0.1% and 0.4% per °C at 20 °C respectively. Find the resistance of combination at a temperature of 50 °C. What is the effective temperature coefficient of the combination at 50 °C ? [*Pb. Univ. Applied Electricity 1972 Supp.*]

Solution: Resistance of 600 Ω resistance at 50 °C $= 600\,[1 + (50 - 20) \times 0.001] = 618\ \Omega$

Resistance of 300 Ω resistance at 50 °C $= 300\,[1 + (50 - 20) \times 0.004] = 336\ \Omega$

Resistance of combination at 50 °C, $R_{50} = 618 + 336 = 954\ \Omega$ **Ans.**

Resistance of combination at 20 °C, $R_{20} = 600 + 300 = 900\ \Omega$

Resistance temperature coefficient of combination at 20 °C,

$$\alpha_{20} = \frac{\frac{R_{50}}{R_{20}} - 1}{t_2 - t_1} = \frac{\frac{954}{900} - 1}{50 - 20} = \frac{0.06}{30} = 0.002$$

$\because R_{50} = R_{20} [1 + \alpha_{20} (t_2 - t_1)]$

Effective temperature coefficient of combination at 50 °C

$$\alpha_{50} = \frac{1}{\frac{1}{\alpha_{20}} + (t_2 - t_1)} = \frac{1}{\frac{1}{0.002} + (50 - 20)} = \frac{1}{530} \text{ per °C } \textbf{Ans.}$$

1.11. SERIES CIRCUITS

When the resistors are connected end to end, so that they form only one path for the flow of current, then resistors are said to be connected in series and such circuits are known as *series circuits.*

Let resistors R_1, R_2 and R_3 be connected in series, as shown in Fig. 1.8, and the potential difference of V volts be applied between extreme ends A and D to cause flow of current of I amperes through all the resistors R_1, R_2 and R_3.

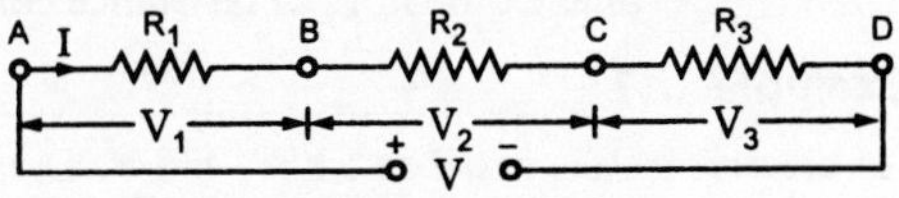

Fig. 1.8 *Series Circuit*

Now according to Ohm's law

Voltage drop across resistor R_1, $V_1 = I R_1$

Voltage drop across resistor R_2, $V_2 = I R_2$

Voltage drop across resistor R_3, $V_3 = I R_3$

Voltage drop across whole circuit, V = Voltage drop across resistor R_1 + voltage drop across resistor R_2 + voltage drop across resistor R_3

$$i.e.\ V = I R_1 + I R_2 + I R_3 = I (R_1 + R_2 + R_3)$$

$$\text{or} \quad \frac{V}{I} = R_1 + R_2 + R_3 \qquad \text{...(1.15)}$$

And according to Ohm's law $\frac{V}{I}$ gives the whole circuit resistance, say R

$\therefore$ Effective resistance of the series circuit, $R = R_1 + R_2 + R_3$ ...(1.16)

Thus, when a number of resistors are connected in series, the equivalent resistance is given by the arithmetic sum of their individual resistances.

$$i.e.\ R = R_1 + R_2 + R_3 + \dots R_n \qquad \text{...(1.17)}$$

From the above discussions for a series circuit we conclude that

1. same current flows through all parts of the circuit,
2. applied voltage is equal to the sum of voltage drops across the different parts of the circuit,
3. different resistors have their individual voltage drops,
4. voltage drop across individual resistor is directly proportional to its resistance, current being the same in each resistor,
5. voltage drops are additive,
6. resistances are additive,
7. powers are additive.

Series circuits are common in electrical equipment. The tube filaments in small radios are usually in series. Current controlling devices are wired in series with the controlled equipment. Fuses are in series with the equipment they protect. A thermostat switch is in series with the heating element in an electric iron. Automatic house-heating equipment has a thermostat, electromagnet coils, and safety cut-outs in series with a voltage source. Rheostats are placed in series with the coils in large motors for motor current control.

Example 1.16

Four coils having resistances of 3, 5, 10 and 12 ohms are connected in series across 120 V. Determine (*i*) equivalent resistance of the circuit (*ii*) current flowing through the circuit and (*iii*) voltage drop across individual coils.

Solution: The circuit diagram is shown in Fig. 1.9

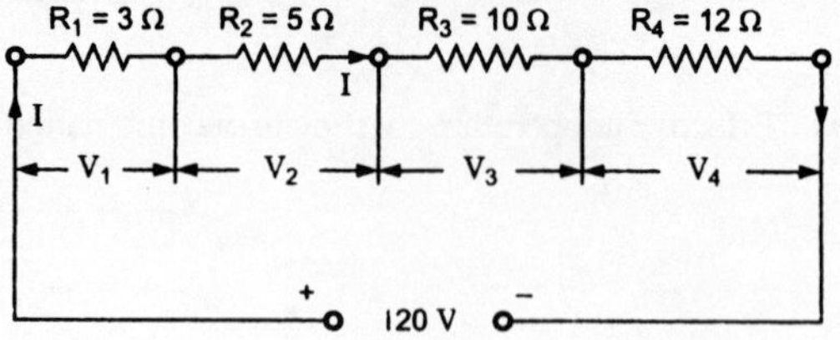

Fig. 1.9

(*i*) Equivalent resistance of the circuit,

$$R = R_1 + R_2 + R_3 + R_4$$
$$= 3 + 5 + 10 + 12 = 30\ \Omega \text{ Ans.}$$

PD applied across the circuit, V = 120 volts

(*ii*) Current flowing through the circuit,

$$I = \frac{V}{R} = \frac{120}{30} = 4 \text{ A Ans.}$$

(*iii*) Voltage drop in 3 Ω resistance coil, $V_1 = I R_1 = 4 \times 3 = 12$ V **Ans.**

Voltage drop in 5 Ω resistance coil, $V_2 = I R_2 = 4 \times 5 = 20$ V **Ans.**

Voltage drop in 10 Ω resistance coil, $V_3 = I R_3 = 4 \times 10 = 40$ V **Ans.**

Voltage drop in 12 Ω resistance coil, $V_4 = I R_4 = 4 \times 12 = 48$ V **Ans.**

Example 1.17

An electric iron is rated as 1 kW, 250 V. Calculate the current taken by it if it is connected to 220 V supply.

Solution: Power rating of electric iron, P = 1 kW = 1,000 watts

Voltage rating, V = 250 V

$$\text{Current rating, } I = \frac{P}{V} = \frac{1,000}{250} = 4 \text{ A}$$

$$\text{Resistance of electric iron, } R = \frac{V}{I} = \frac{250}{4} = 62.5\ \Omega$$

Resistance of electric iron can also be determined from the relation

$$R = \frac{V^2}{P} = \frac{(250)^2}{1,000} = 62.5\ \Omega$$

Current drawn by electric iron when connected across 220 V supply

$$I' = \frac{V'}{R} = \frac{220}{62.5} = 3.52 \text{ A Ans.}$$

Example 1.18

Three resistors are connected in series across a 12 V battery. The first resistor has the value of 1 ohm, second has a voltage drop of 4 V and third has a power dissipation of 12 W. Calculate the value of each resistance and circuit current.

Solution: Let the three resistors be of R_1 (= 1 Ω), R_2 and R_3 ohms, current flowing through the three resistors R_1, R_2 and R_3 be of I amperes and voltage drops across resistors R_1, R_2 and R_3 be of V_1, V_2 and V_3 volts respectively. The circuit is shown in Fig. 1.10.

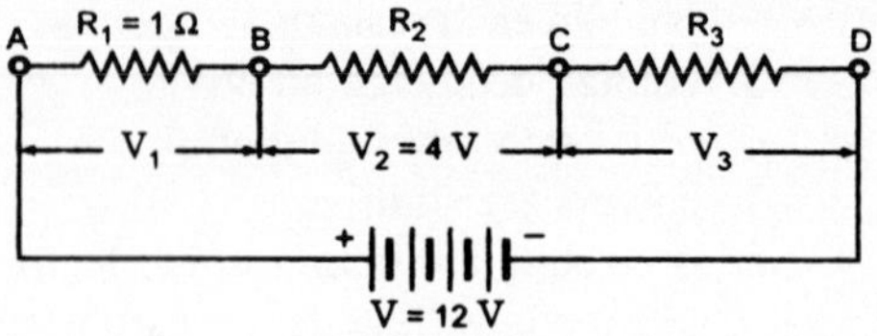

Fig. 1.10

$$\text{Now } V_1 = I R_1 = I \text{ volts}$$
$$V_2 = I R_2 = 4 \text{ volts}$$
$$V_3 = \frac{\text{Power dissipation}}{I} = \frac{12}{I}$$
$$\text{Since } V = V_1 + V_2 + V_3$$
$$\text{So } 12 = I + 4 + \frac{12}{I}$$
$$\text{or } I^2 - 8I + 12 = 0$$
$$\text{or } (I - 6)(I - 2) = 0$$
$$\text{or } I = 6 \text{ or } 2 \text{ A Ans.}$$

When I = 6 A;
$$R_2 = \frac{V_2}{I} = \frac{4}{6} = \frac{2}{3}\ \Omega$$
$$\text{and } R_3 = \frac{P}{I^2} = \frac{12}{6^2} = \frac{1}{3}\ \Omega$$
i.e. $R_1 = 1\ \Omega;\ R_2 = \frac{2}{3}\ \Omega$ and $R_3 = \frac{1}{3}\ \Omega$ **Ans.**

When I = 2 A;
$$R_2 = \frac{V_2}{I} = \frac{4}{2} = 2\ \Omega$$
$$\text{and } R_3 = \frac{P}{I^2} = \frac{12}{2^2} = 3\ \Omega$$
i.e. $R_1 = 1\ \Omega;\ R_2 = 2\ \Omega$ and $R_3 = 3\ \Omega$ **Ans.**

Example 1.19

A 100 V, 60 watt bulb is to be operated from a 220 V supply. What is the resistance to be connected in series with the bulb to glow normally ?

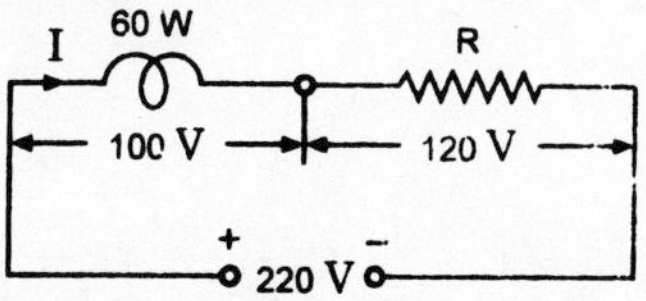

Fig. 1.11

Solution: Rated power of lamp, P = 60 W

Rated voltage of lamp, V = 100 V

Current drawn by the lamp, when operated on rated voltage, *i.e.*
$$\text{Rated current, } I = \frac{P}{V} = \frac{60}{100} = 0.6\ A$$

Lamp will operate normally on 220 V also if the current flowing through the lamp remains the rated current *i.e.* 0.6 A.

Let the resistance connected in series with the lamp to make it glow normally on 220 V be of R ohms, as shown in Fig. 1.11.

Now since the resistance R is in series with the lamp, the same current will flow through the resistance R, as in the lamp *i.e.* 0.6 A and voltage drop across series resistance R will be equal to supply voltage less voltage drop across the lamp (*i.e.* rated voltage of the lamp)

or Voltage drop across the series resistance, IR = Supply voltage – rated voltage of the lamp = 220 – 100 = 120 V
$$\text{or } R = \frac{120}{I} = \frac{120}{0.6} = 200\ \Omega \ \textbf{Ans.}$$

1.12. PARALLEL CIRCUITS

When a number of resistors are connected in such a way that one end of each of them is joined to a common point and the other ends being joined to another common point, as shown in Fig. 1.12, then resistors are said to be connected in parallel and such circuits are known as *parallel circuits.* In these circuits current is divided into as many paths as the number of resistances.

Let the resistors R_1, R_2 and R_3 be connected in parallel, as shown in Fig. 1.12, and the potential difference of V volts be applied across the circuit.

Since potential difference across each resistor is same and equal to potential difference applied to the circuit *i.e.* V

∴ According to Ohm's law

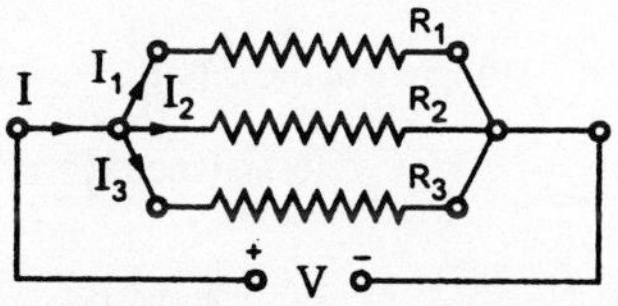

Fig. 1.12 *Parallel Circuit*

$$\text{Current in resistor } R_1,\ I_1 = \frac{V}{R_1} \quad ....(1.18)$$
$$\text{Current in resistor } R_2,\ I_2 = \frac{V}{R_2} \quad ....(1.19)$$
$$\text{Current is resistor } R_3,\ I_3 = \frac{V}{R_3} \quad ...(1.20)$$

Adding Eqs. (1.18), (1.19) and (1.20) we have
$$I_1 + I_2 + I_3 = \frac{V}{R_1} + \frac{V}{R_2} + \frac{V}{R_3} = V\left(\frac{1}{R_1} + \frac{1}{R_2} + \frac{1}{R_3}\right) \quad ...(1.21)$$

and since $I_1 + I_2 + I_3 = I$, the total current flowing through the circuit

$$\text{so } I = I_1 + I_2 + I_3 = V\left(\frac{1}{R_1}+\frac{1}{R_2}+\frac{1}{R_3}\right)$$

$$\text{or } \frac{I}{V} = \frac{1}{R_1}+\frac{1}{R_2}+\frac{1}{R_3}$$

and since $\frac{I}{V} = \frac{1}{R}$ where R is the equivalent resistance of the whole circuit.

$$\therefore \frac{1}{R} = \frac{1}{R_1}+\frac{1}{R_2}+\frac{1}{R_3} \quad ...(1.22)$$

Thus, when a number of resistors are connected in parallel, the reciprocal of the equivalent resistance is given by the arithmetic sum of the reciprocals of their individual resistances.

In general if n resistors of resistances $R_1, R_2, R_3, ... R_n$ are connected in parallel, then equivalent resistance R of the circuit is given by the equation

$$\frac{1}{R} = \frac{1}{R_1}+\frac{1}{R_2}+\frac{1}{R_3}+ ... + \frac{1}{R_n} \quad ...(1.23)$$

$$\text{Also } G = G_1 + G_2 + G_3 + ... G_n \quad ...(1.24)$$

where $G = \frac{1}{R}$, $G_1 = \frac{1}{R_1}$, $G_2 = \frac{1}{R_2}$, $G_3 = \frac{1}{R_3}$ and so on.

From the above discussions for a parallel circuit we conclude that

1. same voltage acts across all branches of the circuit,
2. different resistors (or branches) have their individual currents,
3. total circuit current is equal to the sum of individual currents through the various resistors (or branches),
4. branch currents are additive,
5. conductances are additive,
6. powers are additive,
7. the reciprocal of the equivalent or combined resistance is equal to the sum of the reciprocals of the resistances of the individual branches.

Parallel circuits are very common in use. Various lamps and appliances in a house are connected in parallel, so that each one can be operated independently. A series circuit is an "all or none" circuit, in which either every thing operates or nothing operates. For individual control, devices are wired in parallel.

Example 1.20

The equivalent resistance of four resistors joined in parallel is 20 Ω. The currents flowing through them are 0.6, 0.3, 0.2 and 0.1 ampere. Find the value of each resistor.

Solution: Equivalent resistance of the whole circuit, R = 20 Ω

Total current flowing through the circuit,

$$I = I_1 + I_2 + I_3 + I_4$$
$$= 0.6 + 0.3 + 0.2 + 0.1 = 1.2 \text{ A}$$

PD across the circuit, $V = IR = 1.2 \times 20 = 24$ V

Resistance $R_1 = \frac{V}{I_1} = \frac{24}{0.6} = 40\ \Omega$ **Ans.**

Resistance $R_2 = \frac{V}{I_2} = \frac{24}{0.3} = 80\ \Omega$ **Ans.**

Resistance $R_3 = \frac{V}{I_3} = \frac{24}{0.2} = 120\ \Omega$ **Ans.**

Resistance $R_4 = \frac{V}{I_4} = \frac{24}{0.1} = 240\ \Omega$ **Ans.**

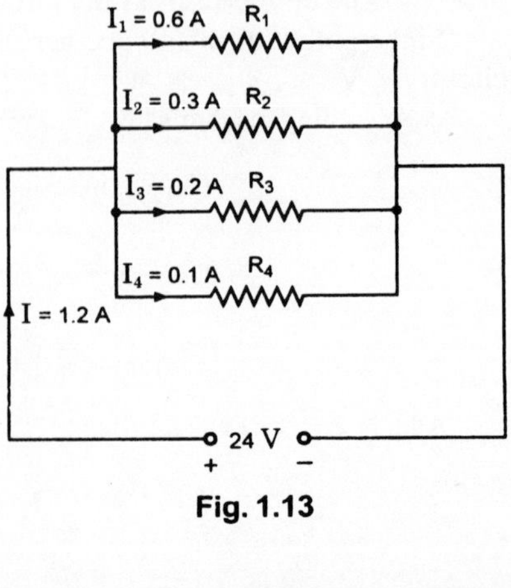

Fig. 1.13

Example 1.21

Two bulbs of 250 W, 230 V each, are connected in parallel across a 200 V supply. Calculate the total power drawn from the supply.

Solution:

Wattage rating of bulbs, $P = 250$ watts

Voltage rating of bulbs, $V = 230$ V

$$\text{Resistance of each bulb, } R = \frac{V^2}{P} = \frac{(230)^2}{250} = 211.6\ \Omega$$

$$\text{Total resistance of parallel combination, } R' = \frac{R \times R}{R + R} = \frac{R}{2} = \frac{211.6}{2} = 105.8\ \Omega$$

$$\text{Total power drawn from the supply} = \frac{(V')^2}{R'} = \frac{(200)^2}{105.8} = 378.07 \text{ W} \quad \textbf{Ans.}$$

1.13. CURRENT DISTRIBUTION IN PARALLEL CIRCUITS

Let two resistors of resistances R_1 and R_2 be connected in parallel across a pd of V volts. According to Ohm's law

$$\text{Current flowing through resistor } R_1,\ I_1 = \frac{V}{R_1} \qquad ...(1.25)$$

$$\text{Current flowing through resistor } R_2,\ I_2 = \frac{V}{R_2} \qquad ...(1.26)$$

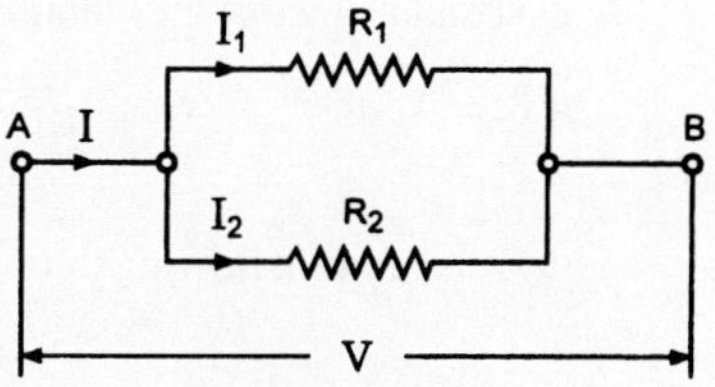

Fig. 1.14

Dividing Eq. (1.25) by Eq. (1.26) we have

$$\frac{I_1}{I_2} = \frac{V/R_1}{V/R_2} = \frac{R_2}{R_1} \qquad ...(1.27)$$

Hence current flowing through each resistor, when connected in parallel, is inversely proportional to their respective resistances.

Since conductance is reciprocal of resistance and if G_1 and G_2 are the respective conductances of resistors R_1 and R_2 then

$$\frac{1}{G_1} = R_1 \text{ and } \frac{1}{G_2} = R_2$$

Substituting the above values of R_1 and R_2 in Eq. (1.27) we have

$$\frac{I_1}{I_2} = \frac{1/G_2}{1/G_1} = \frac{G_1}{G_2} \qquad ...(1.28)$$

Adding 1 on both sides of the above equation, we have

$$\frac{I_1}{I_2} + 1 = \frac{G_1}{G_2} + 1$$

$$\text{or} \quad \frac{I_1 + I_2}{I_2} = \frac{G_1 + G_2}{G_2}$$

$$\text{or} \quad I_2 = \frac{I_1 + I_2}{G_1 + G_2} \times G_2$$

$$\text{Since } I_1 + I_2 = I \text{ and } G_1 + G_2 = \frac{1}{R_1} + \frac{1}{R_2} = \frac{1}{R} = G$$

$$\therefore \quad I_2 = I \times \frac{G_2}{G} \qquad ...(1.29)$$

$$\text{and} \quad I_1 = I \times \frac{G_1}{G} \qquad ...(1.30)$$

Hence, current in any branch of a parallel circuit is directly proportional to its respective conductance and is equal to the total current flowing through the circuit multiplied by the ratio of the conductance of the branch to that of the circuit.

The same relation holds good for parallel circuit consisting of more than two resistors and is very useful for its solution.

Let us consider a circuit consisting of resistances R_1, R_2, R_3 and R_4 ohms respectively connected in parallel across a potential difference of V volts, as shown in Fig. 1.15.

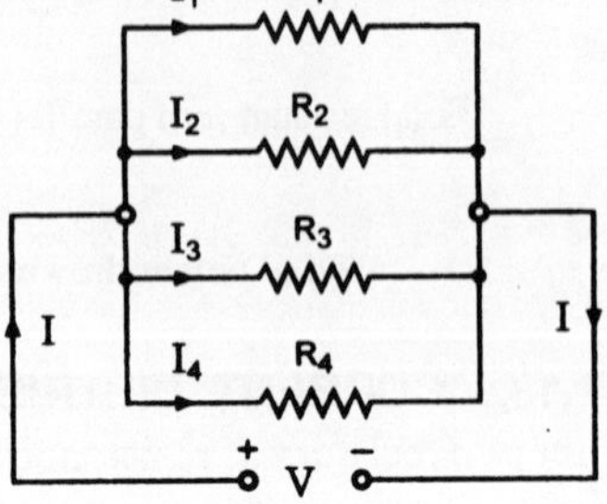

Fig. 1.15

$$\text{Conductance of first resistor, } G_1 = \frac{1}{R_1}$$

$$\text{Conductance of second resistor, } G_2 = \frac{1}{R_2}$$

$$\text{Conductance of third resistor, } G_3 = \frac{1}{R_3}$$

$$\text{Conductance of fourth resistor, } G_4 = \frac{1}{R_4}$$

Total conductance of the whole circuit is given by

$$G = \frac{1}{R} = \frac{1}{R_1} + \frac{1}{R_2} + \frac{1}{R_3} + \frac{1}{R_4} = G_1 + G_2 + G_3 + G_4$$

$$\text{Total current flowing through the circuit, } I = \frac{V}{R} = V \times G$$

$$\text{or } V = \frac{I}{G} \qquad \text{...(1.31)}$$

$$\text{Current flowing through resistance } R_1, I_1 = \frac{V}{R_1} = \frac{I}{G} \times G_1 = \frac{G_1}{G} I$$

$$\text{Similarly current flowing through resistance } R_2, I_2 = V \times G_2 = I \times \frac{G_2}{G}$$

$$\text{Current flowing through resistance } R_3, I_3 = V \times G_3 = I \times \frac{G_3}{G}$$

$$\text{and current flowing through resistance } R_4, I_4 = V \times G_4 = I \times \frac{G_4}{G}.$$

Example 1.22

Two resistances, one of 30 Ω and another of unknown value are connected in parallel, the total power dissipated in the circuit is 450 watts when the applied voltage is 90 volts. Find the value of unknown resistance.

Solution : Power dissipated in the circuit, P = 450 watts

Voltage applied across the circuit, V = 90 volts

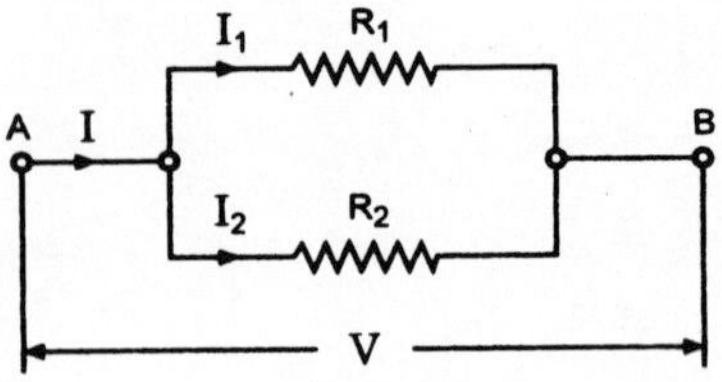

Fig. 1.16

$$\text{Power dissipated in 30 Ω resistance, } P_1 = \frac{V^2}{R_1} = \frac{(90)^2}{30} = 270 \text{ watts}$$

$$\text{Power dissipated in the unknown resistance, } P_2 = P - P_1 = 450 - 270 = 180 \text{ watts}$$

$$\text{The value of unknown resistance, } R_2 = \frac{V^2}{P_2} = \frac{(90)^2}{180} = 45\ \Omega \quad \textbf{Ans.}$$

Example 1.23

Four wires of same material, the same cross-sectional area and the same length when placed in parallel give a total resistance of 0.25 Ω. If the same four wires are connected in series what will be the effective resistance ?

Solution: Since the length, area of cross-section and material of each wire are the same, therefore, resistance of each wire will be same, let it be R ohms

When four wires, each of resistance R ohms are connected in parallel, the equivalent resistance will be

$$\frac{1}{1/R + 1/R + 1/R + 1/R} = \frac{R}{4}\ \Omega$$

Since the total resistance of the circuit consisting of four similar wires is given as 0.25

$$\text{so} \quad \frac{R}{4} = 0.25$$

$$\text{or} \quad R = 4 \times 0.25 = 1\ \Omega$$

When these wires are connected in series, their equivalent resistance will become equal to R + R + R + R *i.e.* 4 R or 4 × 1 = 4 Ω **Ans.**

Example 1.24

A lead wire and an iron wire are connected in parallel. Their specific resistances are in the ratio 49:24. The former carries 80 percent more current than the latter and the latter is 47 percent longer than the former. Determine the ratio of their cross-sectional areas. [*Nagpur Univ. Elec. Engineering, 1993*]

Solution: Let suffix 1 represent lead and suffix 2 represent iron

$$\text{It is given than} \quad \frac{I_2}{I_1} = \frac{1}{1.8}; \quad \frac{l_2}{l_1} = 1.47; \quad \frac{\rho_2}{\rho_1} = \frac{24}{49}$$

Since in parallel combination current in individual resistor is proportional to its conductance

$$\text{so} \quad \frac{G_2}{G_1} = \frac{I_2}{I_1}$$

$$\text{or} \quad \frac{a_2}{\rho_2\, l_2} \cdot \frac{\rho_1\, l_1}{a_1} = \frac{1}{1.8}$$

$$\text{or} \quad \frac{a_2}{a_1} = \frac{1}{1.8} \times \frac{\rho_2}{\rho_1} \times \frac{l_2}{l_1} = \frac{1}{1.8} \times \frac{24}{49} \times 1.47 = 0.4 \ \textbf{Ans.}$$

Example 1.25

Two coils connected in series have a resistance of 18 Ω and when connected in parallel have a resistance of 4 Ω. Find the resistance of each coil.

Solution: Let the resistances of the coils be R_1 and R_2 ohms

Equivalent resistance when connected in series

$$R_1 + R_2 = 18\ \Omega \quad ...(i)$$

Equivalent resistance when connected in parallel

$$\frac{1}{\frac{1}{R_1} + \frac{1}{R_2}} = \frac{R_1 \times R_2}{R_1 + R_2} = 4\ \Omega \quad ...(ii)$$

Multiplying Eqs. (*i*) and (*ii*) we get

$$R_1 R_2 = 72 \quad ...(iii)$$

$$\text{and} \quad R_1 - R_2 = \sqrt{(R_1 + R_2)^2 - 4R_1R_2} = \sqrt{18^2 - 4 \times 72} = \pm 6\ \Omega \quad ...(iv)$$

Adding Eqs. (*i*) and (*iv*) we get

$$\left.\begin{aligned} 2\,R_1 &= 24 \text{ or } 12\ \Omega \\ \text{or } R_1 &= 12 \text{ or } 6\ \Omega \\ \text{and } R_2 &= 6 \text{ or } 12\ \Omega \end{aligned}\right\} \textbf{Ans.}$$

Example 1.26

A 50 Ω and a 100 Ω resistors are each rated at 100 watts. What is the maximum voltage that may be applied without causing overheating of either resistor when they are connected in (*i*) series; (*ii*) in parallel.

Solution: (*i*) Rating current of 50 Ω resistor, $I_1 = \sqrt{\frac{P_1}{R_1}} = \sqrt{\frac{100}{50}} = \sqrt{2}\ A$

$$\text{Rating current of 100 } \Omega \text{ resistor, } I_2 = \sqrt{\frac{P_2}{R_2}} = \sqrt{\frac{100}{100}} = 1 \text{ A}$$

When the resistors are connected in series current flowing through them will be same, so the maximum current that can be allowed to flow through them will be rating current of the resistor having lower rating current *i.e.* 1 A

$$\text{Combined resistance, } R = R_1 + R_2 = 50 + 100 = 150 \ \Omega$$

$$\text{Maximum applied voltage} = I R = 1 \times 150 = 150 \text{ volts } \textbf{Ans.}$$

(*ii*) When the resistors of 50 Ω and 100 Ω are connected in parallel across supply of voltage say V volts, the power dissipated in 50 Ω resistor will be more than that dissipated in 100 ohm resistor since power dissipated in resistors connected in parallel is inversely proportional to the value of their resistances,

$$\therefore \quad \frac{V^2}{R_1} = P$$

$$\text{or } V = \sqrt{P R_1} = \sqrt{100 \times 50} = 70.71 \text{ volts } \textbf{Ans.}$$

1.14. SERIES-PARALLEL CIRCUITS

So far, only simple series and simple parallel circuits have been considered. Practical electric circuits very often consist of combinations of series and parallel resistances. Such circuits may be solved by the proper application of Ohm's law and the rules for series and parallel circuits to the various parts of the complex circuit. There is no definite procedure to be followed in solving complex circuits, the solution depends on the known facts concerning the circuit and the quantities which one desires to find. One simple rule may usually be followed, however—reduce the parallel branches to an equivalent series branch and then solve the circuit as a simple series circuit.

For example consider a series-parallel circuit shown in Fig. 1.17 for solution.

First of all equivalent resistances of all parallel branches are determined separately *e.g.* of branches AB and CD by the law of parallel circuits discussed in Art. 1.12.

$$\text{Equivalent resistance of parallel branch AB, } R_{AB} = \frac{1}{\frac{1}{R_1} + \frac{1}{R_2}} = \frac{R_1 R_2}{R_1 + R_2}$$

$$\text{and equivalent resistance of parallel branch CD, } R_{CD} = \frac{1}{\frac{1}{R_4} + \frac{1}{R_5} + \frac{1}{R_6}} = \frac{R_4 R_5 R_6}{R_5 R_6 + R_4 R_6 + R_4 R_5}$$

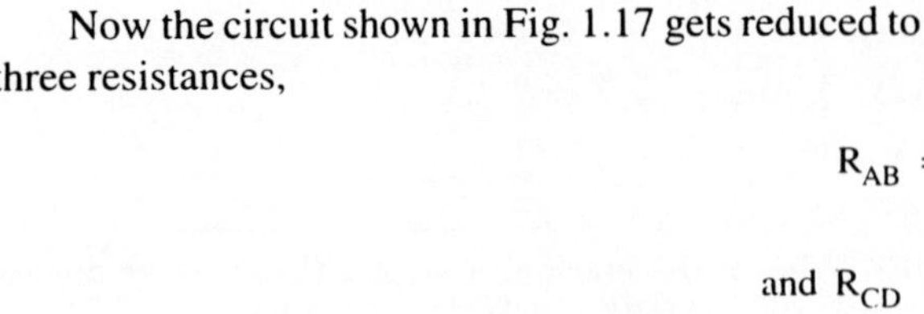

Fig. 1.17

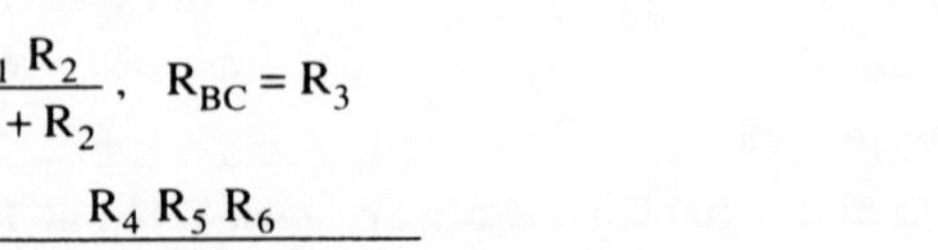

Fig. 1.18

Now the circuit shown in Fig. 1.17 gets reduced to a simple series circuit shown in Fig. 1.18 consisting of three resistances,

$$R_{AB} = \frac{R_1 R_2}{R_1 + R_2}, \quad R_{BC} = R_3$$

$$\text{and } R_{CD} = \frac{R_4 R_5 R_6}{R_5 R_6 + R_4 R_6 + R_4 R_5}$$

$$\text{Total resistance of circuit, } R_T = R_{AB} + R_{BC} + R_{CD}$$

Now circuit current may be determined from the relation

$$I = \frac{V}{R_T}$$

After knowing I, potential differences across branches AB, BC and CD are determined from the relations

$$\text{PD across branch AB, } V_{AB} = I.R_{AB} = \frac{V}{R_T} R_{AB}$$

$$\text{PD across branch BC, } V_{BC} = I.R_{BC} = \frac{V}{R_T} R_{BC}$$

$$\text{and PD across branch CD, } V_{CD} = IR_{CD} = \frac{V}{R_T} R_{CD}.$$

After determination of potential difference across each parallel branch, the currents in the various resistances are determined from the relations

$$\text{Current in resistance } R_1 = I_1 = \frac{V_{AB}}{R_1}$$

$$\text{Current in resistance } R_2 = I_2 = \frac{V_{AB}}{R_2}$$

$$\text{Current in resistance } R_3 = I_3 = I$$

$$\text{Current in resistance } R_4 = I_4 = \frac{V_{CD}}{R_4}$$

$$\text{Current in resistance } R_5 = I_5 = \frac{V_{CD}}{R_5}$$

$$\text{Current in resistance } R_6 = I_6 = \frac{V_{CD}}{R_6}$$

Thus equivalent resistance of the whole circuit, voltage drop across each branch and currents in the various resistors may be determined.

Example 1.27

Two resistances of 20 Ω and 30 Ω respectively are connected in parallel. These two parallel resistances are further connected in series with a resistance of 15 Ω. If the current through the 15 Ω resistance is 3 A find (*a*) the currents through the 20 Ω and 30 Ω resistances respectively (*b*) the voltage across the whole circuit (*c*) the total power consumed.

Solution: Equivalent resistance of branch AB, $R_{AB} = \dfrac{1}{\dfrac{1}{20}+\dfrac{1}{30}} = 12\ \Omega$

Effective resistance of the circuit, $R_{eff} = R_{AB} + R_{BC} = 12 + 15 = 27\ \Omega$

Circuit current, I = Current through 15 Ω resistance = 3 A

(*b*) Voltage across the whole circuit, $V = I R_{eff} = 3 \times 27 = 81$ V **Ans.**

(*c*) Total power consumed, $P = VI = 81 \times 3 = 243$ watts **Ans.**

(*a*) Voltage drop across branch AB, $V_{AB} = I R_{AB} = 3 \times 12 = 36$ V

Current through 20 Ω resistance, $I_1 = \dfrac{V_{AB}}{R_1} = \dfrac{36}{20} = 1.8$ A **Ans.**

Current through 30 Ω resistance, $I_2 = \dfrac{V_{AB}}{R_2} = \dfrac{36}{30} = 1.2$ A **Ans.**

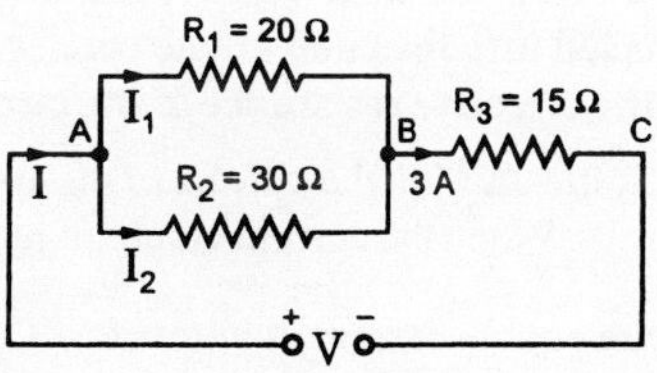

Fig. 1.19

Example 1.28

For the circuit shown in Fig. 1.20, using the method of series-parallel combination, find V_1 and I_2.

Solution : Total resistance of the circuit,

R_{eq} = 25 + 2.5 + equivalent resistance of parallel combination of resistors of 10, 50 and 20 Ω

$$= 25 + 2.5 + \frac{1}{\frac{1}{10}+\frac{1}{50}+\frac{1}{20}}$$

$$= 25 + 2.5 + \frac{100}{10+2+5} = \frac{1{,}135}{34}\ \Omega$$

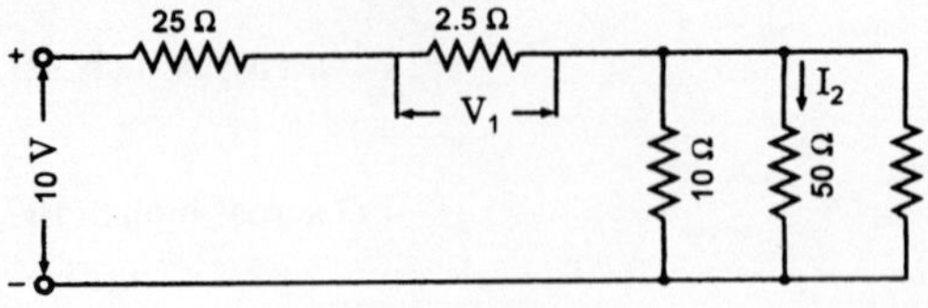

Fig. 1.20

Applied voltage, V = 10 V

$$\text{Current drawn from the supply, } I = \frac{V}{R_{eq}} = \frac{10}{1{,}135/34} = \frac{10\times 34}{1{,}135} = \frac{68}{227}\ \text{A}$$

$$\text{Voltage drop across resistor of 2.5 Ω, } V_1 = I\times 2.5 = \frac{68}{227}\times 2.5 = \frac{170}{227}\ \text{V}$$ **Ans.**

$$\text{Voltage drop across parallel combination, } V_2 = \frac{68}{227}\times\frac{100}{17} = \frac{400}{227}\ \text{V}$$

$$\text{Current through 50 Ω resistor, } I_2 = \frac{V_2}{50} = \frac{400}{227\times 50} = \frac{8}{227}\ \text{A}$$ **Ans.**

Example 1.29

A resistance R is connected in series with a parallel combination of two resistances 12 Ω and 8 Ω. Calculate R if the power dissipated in the circuit is 70 W when the applied voltage is 20 V across the circuit.

Solution: Equivalent resistance of the parallel circuit AB,

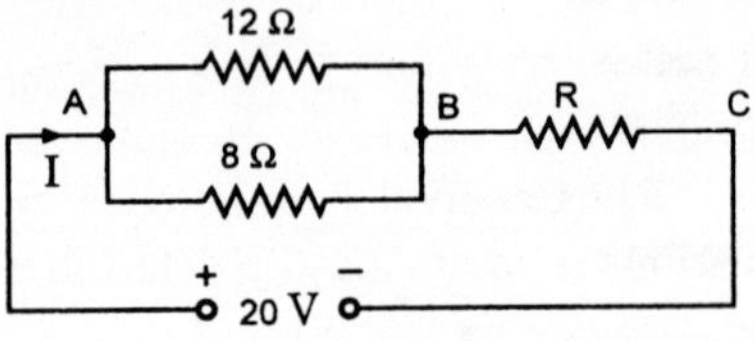

Fig. 1.21

$$R_{AB} = \frac{12\times 8}{12+8} = 4.8\ \Omega$$

Total resistance of the circuit = (R + 4.8) ohms

Total power dissipated in the circuit, P = 70 W

Voltage applied across the circuit V = 20 V

$$\text{Current through the circuit, } I = \frac{P}{V} = \frac{70}{20} = 3.5\ \text{A}$$

$$\text{Total resistance of the circuit, } (R + 4.8) = \frac{V}{I} = \frac{20}{3.5} = 5.714\ \Omega$$

or R = 5.714 – 4.8 = 0.914 Ω **Ans.**

Example 1.30

Twelve identical wires of resistance 6 Ω each are arranged to form the edges of cube. A current of 40 milli-amperes is led into the cube at one corner and out at opposite corner. Calculate the pd developed between these corners and the effective resistance of the circuit.

Solution : Let the current of 40 mA be led into the cube at corner A and out at the opposite corner E.

When the current comes at junction A, it divides itself into three paths, namely AB, AC and AH. Since the resistances are equal, current in wire AB = Current in wire AC = Current in wire AH = $\frac{40}{3}$ mA.

When the current reaches next junction points B, C and H currents are divided again along two paths at each junction points *i.e.* BD and BG at B, CD and CF at C and HG and HF at H. So, each section will carry half of the current of $\frac{40}{3}$ mA *i.e.* $\frac{20}{3}$ mA.

At next junction points F, D and G currents coming from two sides add up and begin to flow in FE, DE and GE respectively. Each section carries currents of $\frac{40}{3}$ mA. All these three conductors meet at junction point E and current from all paths come and total current of 40 mA gets out at junction point E.

One of the parallel paths is along AB, BD and DE. Taking this path and adding voltage drops in each conductor of the path, we shall get potential drop across A and E.

PD across A and E

= Voltage drop in AB + voltage drop in BD + voltage drop in DE

$$= \frac{0.04}{3} \times 6 + \frac{0.02}{3} \times 6 + \frac{0.04}{3} \times 6 = 0.2 \text{ V} \quad \textbf{Ans.}$$

Current between A and E = 40 mA = 0.04 A

Equivalent resistance between A and E

$$= \frac{\text{PD across A and E}}{\text{Current between A and E}} = \frac{0.2}{0.04} = 5\ \Omega \quad \textbf{Ans.}$$

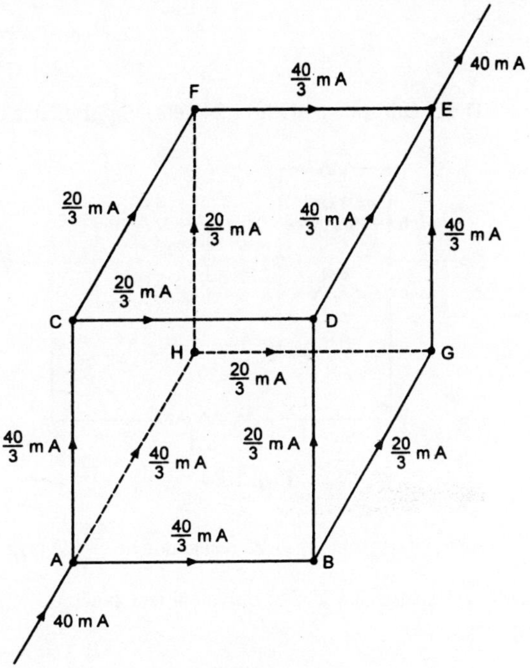

Fig. 1.22

Example 1.31

A battery of unknown emf is connected across resistances, as shown in Fig. 1.23. The voltage drop across the 8 Ω resistor is 20 V. What will be the current reading in the ammeter ? What is the emf of the battery ?

[*Bangladesh Univ. Basic Electrical. Engg. 1990*]

Solution : Voltage drop across 8 Ω resistor = 20 V

$\therefore$ Current through 8 Ω resistor, $I = \frac{20}{8} = 2.5$ A

The current is divided into two parts at point A, I_1 flowing through resistance of 11 Ω and I_2 flowing through resistances of 15 and 13 ohms in series.

$\therefore$ Current flowing along path ABC,

$$I_2 = I \times \frac{R_{AC}}{R_{AC} + R_{ABC}} = 2.5 \times \frac{11}{11+28} = 0.705 \text{ A}$$

Hence ammeter will read 0.705 A **Ans.**

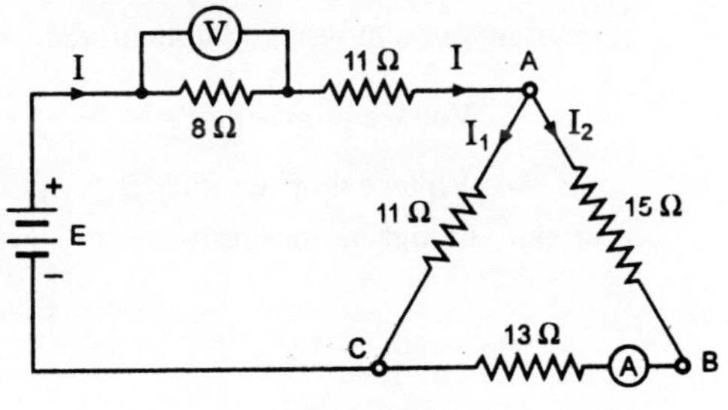

Fig. 1.23

$$\text{Total circuit resistance, } R = 8 + 11 + \frac{11 \times 28}{11 + 28} = 26.9\ \Omega$$

$$\text{EMF of the battery, } E = IR = 2.5 \times 26.9 = 67.25 \text{ V} \quad \textbf{Ans.}$$

Example 1.32

If 20 V be applied across AB shown in Fig. 1.24, calculate the total current, the power dissipated in each resistor and the value of the series resistance to halve the total current. [*Allahabad Univ. Elec. Science II, 1992*]

Solution : As seen from Fig. 1.24

$$R_{AM} = \frac{1}{\frac{1}{2} + \frac{1}{4} + \frac{1}{6} + \frac{1}{8}} = \frac{24}{12+6+4+3} = \frac{24}{25}\ \Omega$$

$$R_{MB} = \frac{1}{\frac{1}{3}+\frac{1}{6}} = \frac{6}{2+1} = 2\ \Omega$$

The circuit shown in Fig. 1.24 gets simplified to a circuit shown in Fig. 1.25.

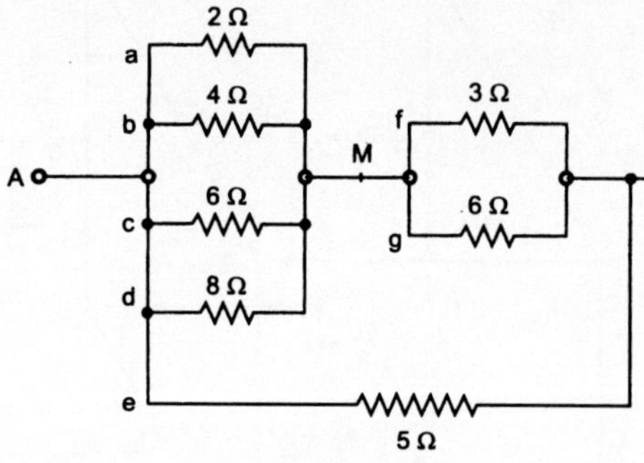

Fig. 1.24

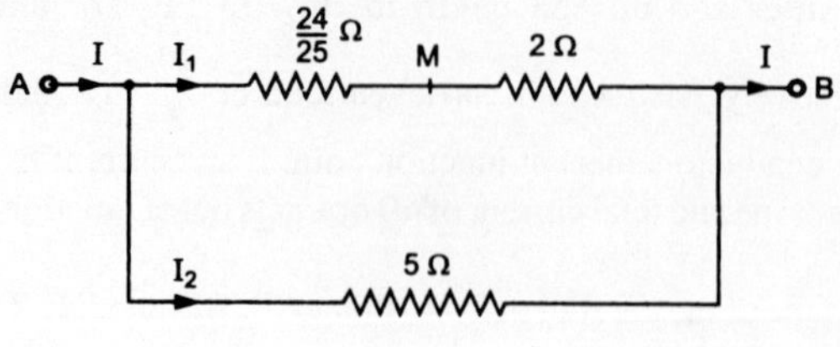

Fig. 1.25

As obvious from Fig. 1.25 resistance of $\frac{24}{25}$ Ω is in series with a resistance of 2 Ω and whole combination is in parallel with a resistance of 5 Ω. So equivalent resistance,

$$R_{eq} = \frac{1}{\frac{1}{5}+\frac{1}{\frac{24}{25}+2}} = \frac{1}{\frac{1}{5}+\frac{25}{74}} = \frac{370}{74+125} = \frac{370}{199}\ \Omega$$

$$\text{Voltage applied across AB, } V_{AB} = 20\text{ V}$$

$$\text{Total current drawn by the circuit, } I = \frac{V_{AB}}{R_{eq}} = \frac{20}{370/199} = \frac{20\times 199}{370} = 10.76\text{ A } \textbf{Ans.}$$

$$\text{Current through 5 } \Omega \text{ resistance, } I_2 = \frac{V_{AB}}{5} = \frac{20}{5} = 4\text{ A}$$

$$\text{Current through series combination, } I_1 = I - I_2 = 10.76 - 4 = 6.76\text{ A}$$

$$\text{Voltage drop across AM, } V_{AM} = I_1 \times R_{AM} = 6.76 \times \frac{24}{25} = 6.48\text{ V}$$

$$\text{Voltage drop across MB, } V_{MB} = I_1 \times R_{MB} = 6.76 \times 2 = 13.52\text{ V}$$

So currents through various resistors are given below

$$I_a = \frac{V_{AM}}{R_a} = \frac{6.48}{2} = 3.24\text{ A};\ I_b = \frac{6.48}{4} = 1.62\text{ A};$$

$$I_c = \frac{6.48}{6} = 1.08\text{ A};\ I_d = \frac{6.48}{8} = 0.81\text{ A};$$

$$I_e = I_2 = 4\text{ A};\ I_f = \frac{V_{MB}}{3} = \frac{13.52}{3} = 4.51\text{ A}$$

$$\text{and } I_g = \frac{13.52}{6} = 2.25\text{ A}$$

Power dissipated through various resistors is given below

$$P_a = I_a^2 R_a = (3.24)^2 \times 2 = 21\text{ W } \textbf{Ans.}$$

$$P_b = (1.62)^2 \times 4 = 10.5\text{ W } \textbf{Ans.}$$

$$P_c = (1.08)^2 \times 6 = 7\text{ W } \textbf{Ans.}$$

$$P_d = (0.81)^2 \times 8 = 5.25\text{ W } \textbf{Ans.}$$

$$P_e = (4)^2 \times 5 = 80 \text{ W} \quad \textbf{Ans.}$$
$$P_f = (4.51)^2 \times 3 = 61 \text{ W} \quad \textbf{Ans.}$$
$$P_g = (2.25)^2 \times 6 = 30.5 \text{ W} \quad \textbf{Ans.}$$

The series resistance required to halve the total current

$$= R_{eq} = \frac{370}{199}\ \Omega \quad \textbf{Ans.}$$

Example 1.33

What is the value of the unknown resistor R in Fig. 1.26 if the voltage drop across 500 Ω resistor is 2.5 volts ? All resistances are in ohms. *[Indore Univ. Elec. Technology April 1990]*

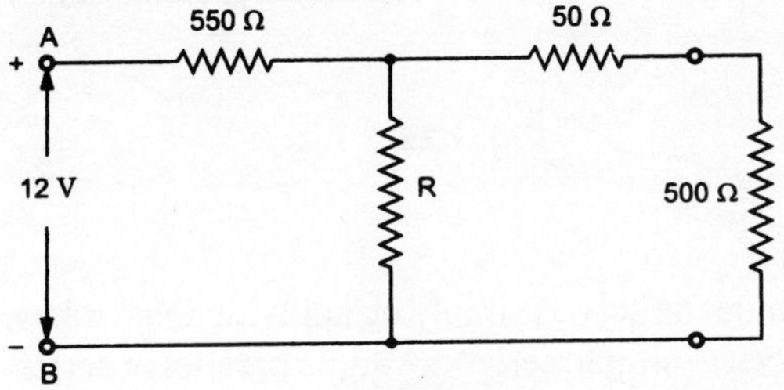

Fig. 1.26

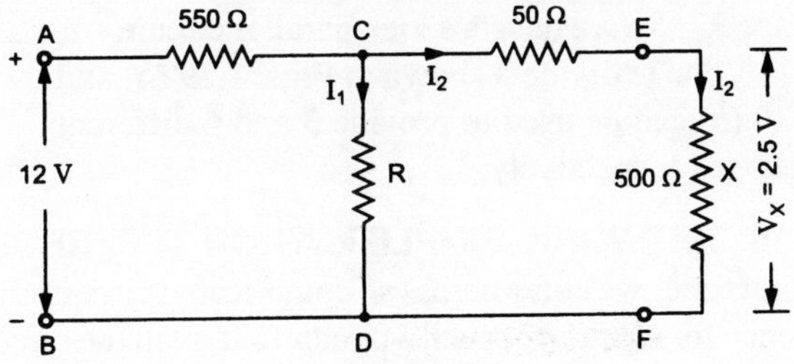

Fig. 1.27

Solution: Voltage drop across 500 Ω resistor, $V_X = 2.5$ V

$R_X = 500\ \Omega$

Current through 500 Ω resistor, $I_2 = \dfrac{V_X}{R_X} = \dfrac{2.5}{500} = 5$ mA

Current through 50 Ω resistor $= I_2 = 5$ mA ∵ 50 Ω resistor and 500 Ω resistor are in series

Voltage drop in 50 Ω resistor $= 50\ I_2 = 50 \times 5 \times 10^{-3} = 0.25$ V

Voltage drop across resistor R (or branch CD) = Voltage drop across CE + voltage drop across EF $= 0.25 + 2.5 = 2.75$ V

Voltage drop across 550 Ω resistor, $V_{AC} = 12 - V_{CD} = 12 - 2.75 = 9.25$ V

Circuit current, $I = \dfrac{V_{AC}}{550} = \dfrac{9.25}{550} = 0.0168$ A

Current through resistor R, $I_1 = I - I_2 = 0.0168 - 0.005 = 0.0118$ A

Unknown resistance, $R = \dfrac{V_{CD}}{I_1} = \dfrac{2.75}{0.0118} = 233\ \Omega$ **Ans.**

1.15. VOLTAGE DIVIDER

The voltage (or potential) divider is a high resistance connected across the supply mains and is employed to provide a variable voltage from a constant voltage supply. The variable voltage is available across a fixed point A and variable point C (refer to Fig. 1.28). Coils of resistance wire are wound uniformly on an insulating tube and a sliding contact is mounted on a conducting bar fastened about 1 cm above the tube. With terminal C open, that is, not connected to any thing, the voltage V_{AC} can be changed in small increments from zero to V_{AB}. The resistance per mm of wire between A and B is constant throughout its length and, therefore, the resistance between points A and C (R_{AC}) is the same fraction of the total resistance (R_{AB}) as the distance from A to C is a fraction of the distance from A to B. Expressing the distances as length of wires between the points, we have

$$\frac{V_{AC}}{V_{AB}} = \frac{I\,R_{AC}}{I\,R_{AB}} = \frac{R_{AC}}{R_{AB}} = \frac{l_{AC}}{l_{AB}} \qquad ...(1.32)$$

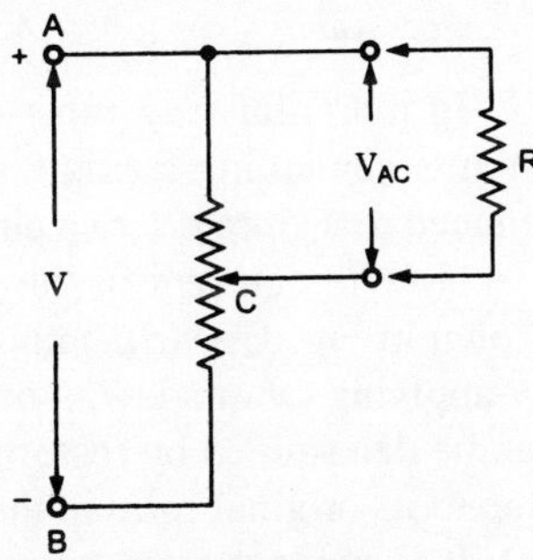

Fig. 1.28

Remember that the above relation holds good only when nothing is connected between the slider and either end of the potential divider. If a load, represented by the resistance R_L, is supplied with current and is connected between A and C, the current in the two sections AC and CB will no longer be equal and section CB will carry more current than that carried by section AC. The current in section CB will be the sum of the current in upper section AC and the current in load resistance R_L. So we have a series-parallel circuit.

Voltage dividers shown in Figs. 1.29 (*a*) and 1.29 (*b*) can be used to provide 3 and 6 different voltages respectively.

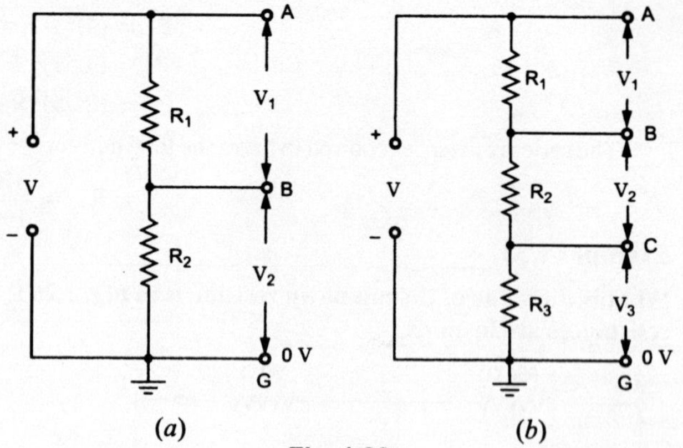

Fig. 1.29

1.16. NETWORK SIMPLIFICATION (OR REDUCTION)

Sometimes we come across so complicated circuits that they cannot be solved simply by applying Ohm's law. Hence for solution of such a circuit first of all the circuit is reduced to simple series or simple parallel or series-parallel circuit and then solved by applying Ohm's law.

Network reduction or simplification is a process by which currents and voltages acting in a circuit composed of resistors having a series, parallel or series-parallel combinations can be determined.

It may be seen from Fig. 1.30 (*a*) that resistances R_3, R_4 and R_5 are connected in series and the same current I_3 flows through them because the circuit does not branch off at points C and D. By virtue of this fact, the three resistances R_3, R_4 and R_5 can be combined into a single equivalent resistance given by the expression

$$R' = R_3 + R_4 + R_5$$

Thus a simpler equivalent circuit shown in Fig. 1.30 (*b*) is obtained. From circuit diagram shown in Fig. 1.30 (*b*) it can be seen that resistance R_2 is connected in parallel with the series combination of resistances R_3, R_4 and R_5 *i.e.* equivalent resistance R'. Resistances R_2 and R' can be combined into a single equivalent resistance given by the expression

$$R'' = \frac{R_2 R'}{R_2 + R'} = \frac{R_2(R_3 + R_4 + R_5)}{R_2 + (R_3 + R_4 + R_5)}$$

Thus the circuit is further simplified [Fig. 1.30 (*c*)]. Resistances R_1, R'' and R_6 in the circuit of Fig. 1.30 (*c*) are seen to be connected in series. Replacing all the three resistances by a single equivalent resistance connected across the battery (supply source), a simple series circuit [Fig. 1.30 (*d*)] is obtained where in

$$R''' = R_1 + R_6 + R''$$

In a similar way most of the series-parallel circuits containing a single supply source can be reduced to a simple series circuit.

Now the current in the simple series circuit, shown in Fig. 1.30 (*d*), can be readily determined by applying Ohm's law, Now the branch currents can be determined by restoring the circuit step by step to its original form in the reverse order.

It is evident from circuit diagram shown in Fig. 1.30 (*c*)

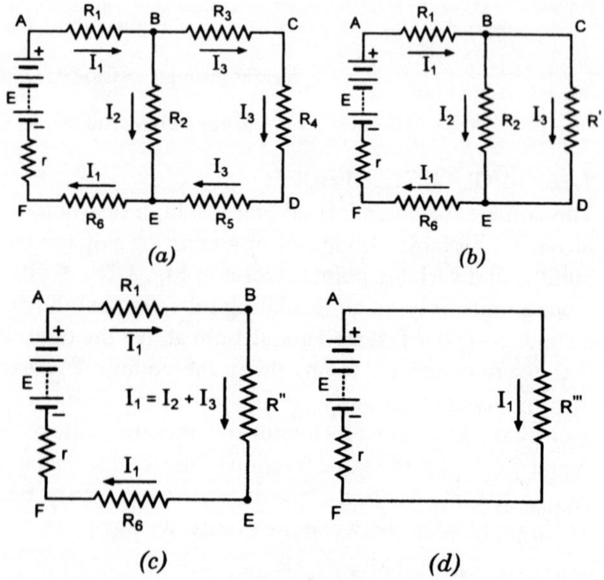

Fig. 1.30

$$I_1 = I_2 + I_3$$

Further more, the voltage across points B and E will be $I_1 R''$. Knowing the voltage across points B and E, currents I_2 and I_3 can be easily determined from the expressions

$$I_2 = \frac{I_1 R''}{R_2} \text{ and } I_3 = \frac{I_1 R''}{R'} = \frac{I_1 R''}{R_3 + R_4 + R_5}$$

The details of the process can further be understood by the following examples.

Example 1.34

What is the equivalent resistance of the ladder network shown in Fig. 1.31 (*i*) with the 75 Ω load resistor connected as shown and (*ii*) with the load resistor disconnected ?

Solution : (i) *With the 75 Ω load resistor connected as shown in Fig. 1.31.*

The resistances in series and parallel, starting from the farthest point are combined with the aid of a series of diagrams shown in Fig. 1.32.

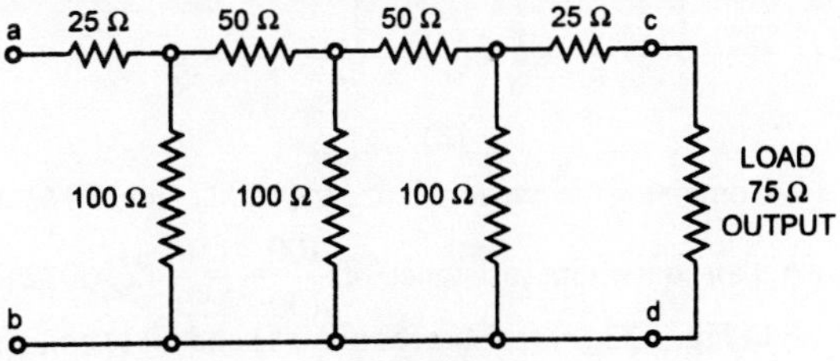

Fig. 1.31

Resistances of 25 ohms and 75 ohms between k and *c* and *c* and *d* are in series giving a resistance of 100 Ω connected in parallel with a resistance of 100 Ω connected between k and *l* [Fig. 1.32 (*b*)]

Combining the two resistances of 100 Ω each connected in parallel we get an equivalent resistance of $\frac{100 \times 100}{100 + 100} = 50\ \Omega$ between k and *l* [refer Fig. 1.32 (*c*)].

(i) *With Load Resistor Connected*

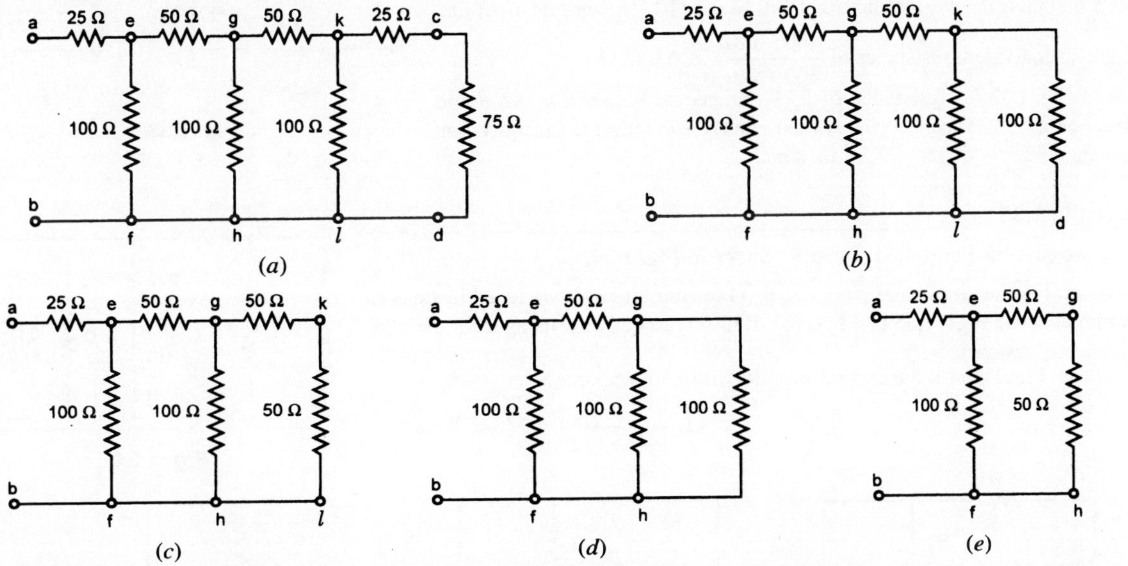

In Fig. 1.32 (*c*) resistances of 50 Ω and 50 Ω connected between *g* and k and k and *l* respectively are in series giving a resistance of 100 Ω connected in parallel with resistance of 100 Ω connected between *g* and *h* [Fig. 1.32 (*d*)].

Combining two resistances of 100 Ω each connected in parallel we get equivalent resistance of 50 Ω connected between *g* and *h*.

In Fig. 1.32 (*e*) resistances of 50 Ω and 50 Ω connected between *e* and *g* and *g* and *h* are in series giving a resistance of 100 Ω connected in parallel with resistance of 100 Ω connected between *e* and *f* [Fig. 1.32 (*f*)].

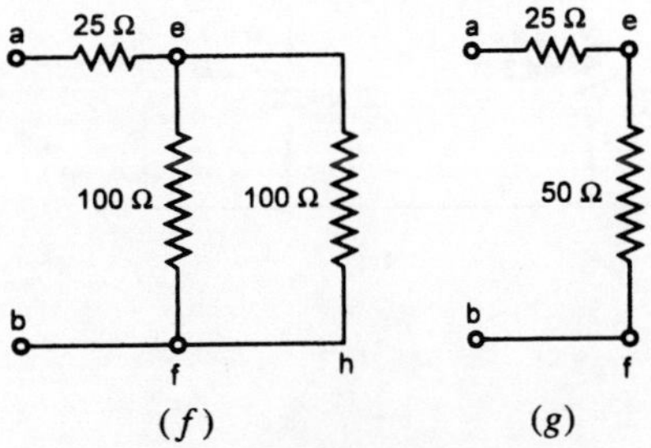

Fig. 1.32

Combining two resistances of 100 Ω each connected in parallel we get equivalent resistance of 50 Ω connected between *e* and *f* [Fig. 1.32 (*g*)].

In Fig. 1.32 (*g*) resistances of 25 Ω and 50 Ω connected between *a* and *e* and *e* and *f* are in series giving an equivalent resistance of 25 + 50 *i.e.* 75 Ω **Ans.**

(ii) *With Load Resistor Disconnected*

The circuit is shown in Fig. 1.33 (*a*). Resistances of 50 Ω and *100 Ω connected between g and* k *and* k and *l* respectively are in series giving a resistance of 150 Ω connected in parallel with a resistance of 100 Ω connected between *g* and *h* [Fig. 1.33 (*b*)].

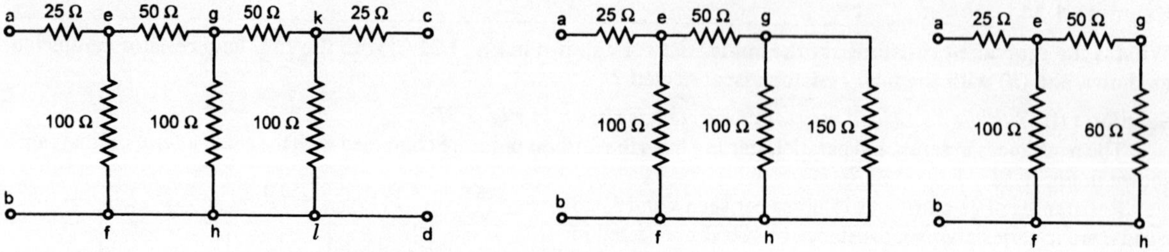

(*a*) (*b*) (*c*)

Combining the two resistances of 100 Ω and 150 Ω connected in parallel we get an equivalent resistance of $\dfrac{100 \times 150}{100 + 150}$ *i.e.* 60 Ω [Fig. 1.33 (*c*)].

In Fig. 1.33 (*c*) resistances of 50 Ω and 60 Ω connected between *e* and *g* and *g* and *h* respectively are in series giving a resistance of 110 Ω connected in parallel with a resistance of 100 Ω between *e* and *f* [Fig. 1.33 (*d*)].

Combining the two resistances of 100 Ω and 110 Ω connected in parallel we get an equivalent resistance of $\dfrac{100 \times 110}{100 + 110}$ *i.e.* 52.38 Ω

In Fig. 1.33 (*e*) resistance of 25 Ω connected between *a* and *e* and resistance of 52.38 Ω connected between *e* and *f* are in series, so the equivalent resistance = 25 + 52.38 = 77.38 Ω **Ans.**

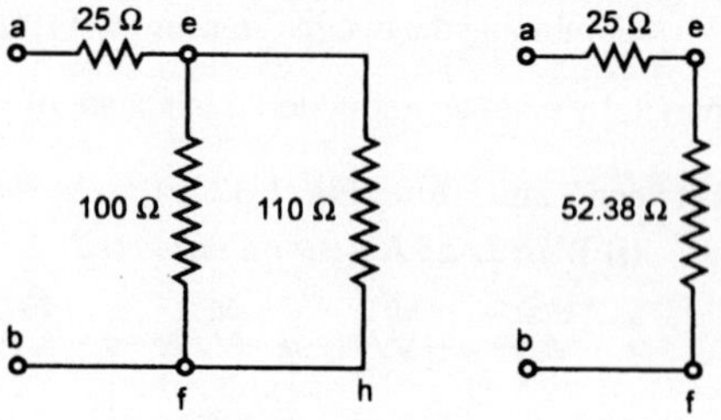

(*d*) (*e*)

Fig. 1.33

Example 1.35

Find the current I supplied by 50 V source in Fig. 1.34.

Solution: For the circuit shown in Fig. 1.34 the circuit can be reduced in steps to circuits shown in Figs. 1.35 (*a*), 1.35 (*b*), 1.35 (*c*) respectively, using series parallel reduction of resistances.

From Fig. 1.35 (*c*) the current supplied from 50 V source,

$$I = \frac{50}{2 + 1.66} = 13.66 \text{ A } \textbf{Ans.}$$

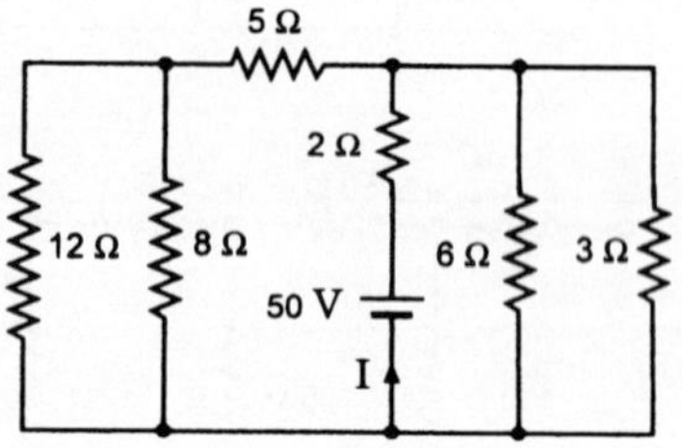

Fig. 1.34

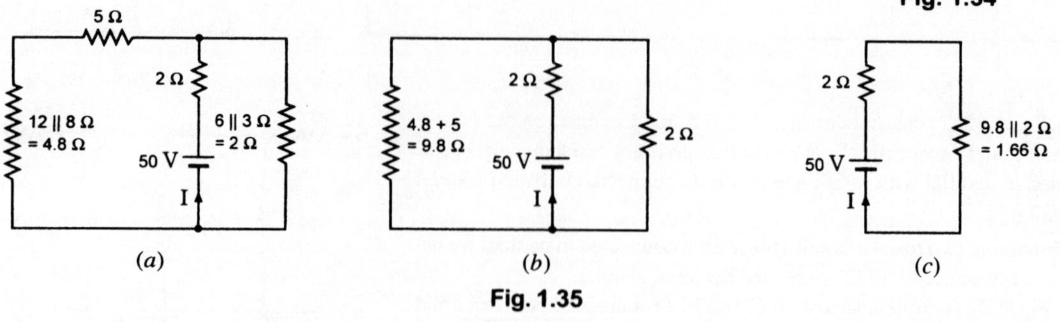

(*a*) (*b*) (*c*)

Fig. 1.35

Example 1.36

What is the difference of potential between X and Y in the network shown in Fig. 1.36 (*a*).

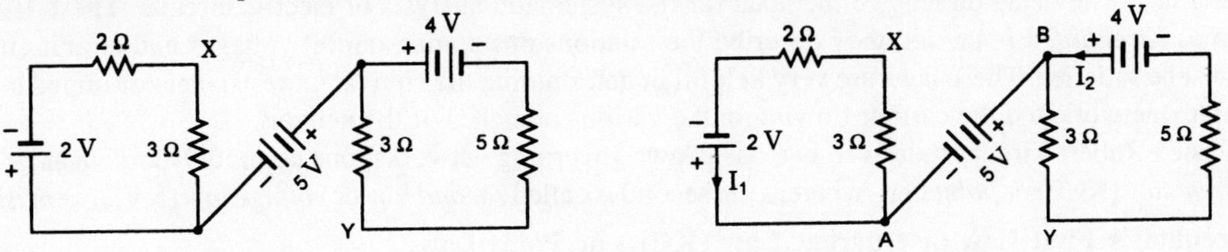

(*a*) (*b*)

Fig. 1.36

Solution: Current flowing through 2 Ω and 3 Ω resistors, $I_1 = \frac{2}{2+3} = 0.4$ A

Current flowing through 3 Ω and 5 Ω resistors, $I_2 = \frac{4}{3+5} = 0.5$ A

Voltage rise from X to Y, $V_{xy} = V_{XA} + V_{AB} - V_{BY}$ .... Refer to Fig. 1.36 (*b*)

$= 3\,I_1 + 5 - 3\,I_2 = 3 \times 0.4 + 5 - 3 \times 0.5 = 4.7$ V **Ans.**

Example 1.37

Determine the current supplied by the source in the circuit shown in Fig. 1.37.

[*R.G.P.V. Basic Electrical Engineering Jan.-Feb. 2006*]

Solution: The given circuit can be solved simply by simplification, as illustrated in Figs. 1.38 (*a*), (*b*), (*c*), (*d*), (*e*) and (*f*).

Thus effective resistance of the circuit, $R = \frac{58}{55}\,\Omega$.

Current supplied by the source, $I = \frac{V}{R} = \frac{30}{58/55} = \frac{30 \times 55}{58} = 28.45$ A **Ans.**

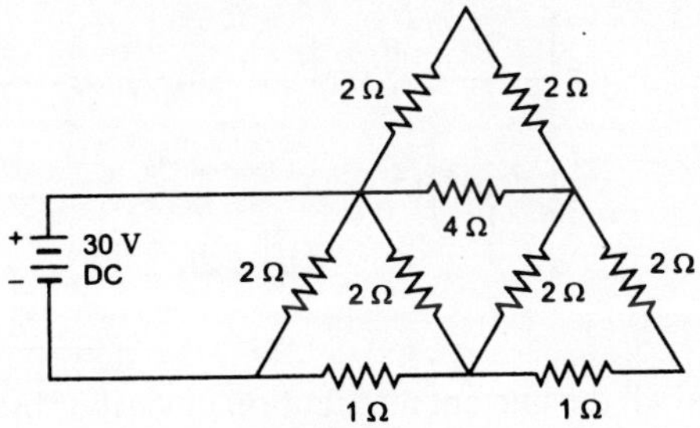

Fig. 1.37

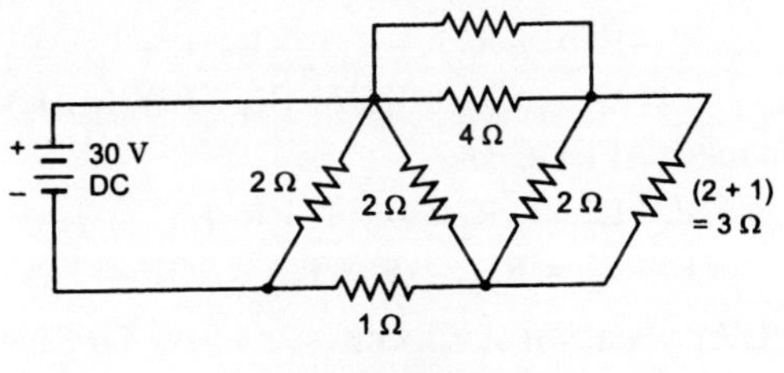

Fig. 1.38 (*a*)

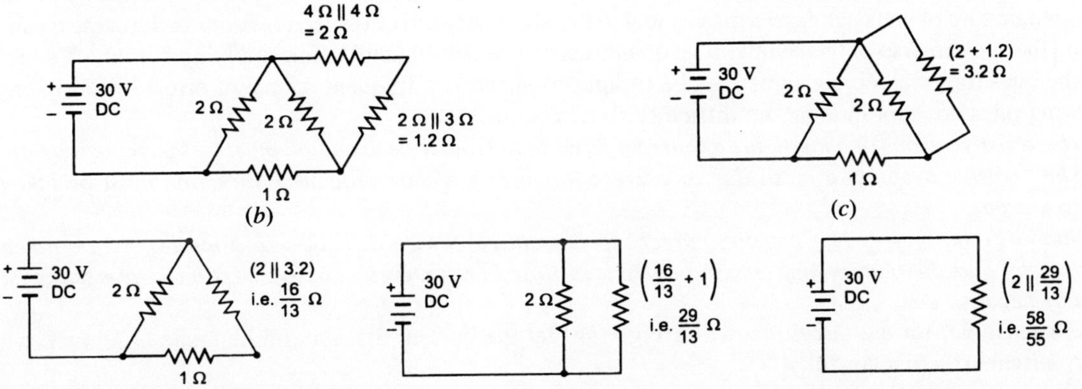

(*b*) (*c*) (*d*) (*e*) (*f*)

Fig. 1.38

1.17. KIRCHHOFF'S LAWS

The basic laws, that electric circuits follow rationally from the nature of the electrical quantities, have already been defined. They lead directly to methods for the systematic analysis of electric circuits. These laws are known as *Kirchhoff's laws*, and they describe the relationships among circuit voltages and circuit currents that must be satisfied. These laws are very helpful in determining the equivalent resistance (or impedance) of a complex network and the current flowing in the various branches of the network.

Gustov Robert Kirchhoff derived two basic laws governing networks, one commonly known as *first law* or *current law* (KCL) or *point law,* whereas the second is called *second law* or *voltage law* (KVL) or *mesh law.*

1. Kirchhoff's First Law or Current Law (KCL) or Point Law. According to this law in any network of wires carrying currents, the algebraic sum of all currents meeting at a point (or junction) is zero or the sum of incoming currents towards any point is equal to the sum of outgoing currents away from that point. If I_1, I_2, I_3, I_4, I_5, and I_6 are the currents meeting at junction O, flowing in the directions of arrowheads marked on them (Fig. 1.39), taking incoming currents as positive and outgoing currents as negative, according to Kirchhoff's first law (KCL)

$$I_1 - I_2 - I_3 + I_4 + I_5 - I_6 = 0$$

$$\text{or} \quad I_1 + I_4 + I_5 = I_2 + I_3 + I_6$$

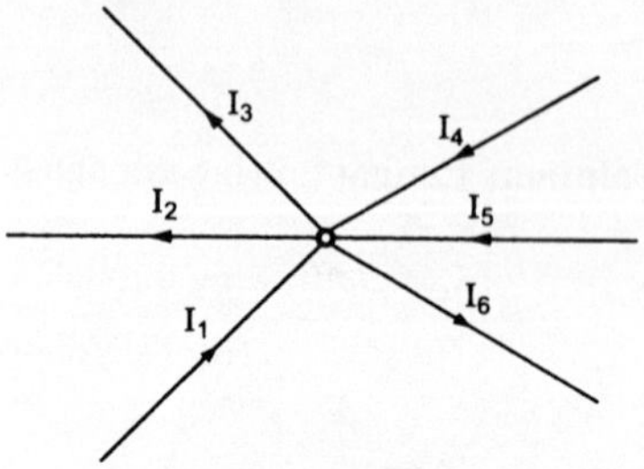

Fig. 1.39

2. Kirchhoff's Second Law or Voltage Law (KVL) or Mesh Law. According to this law in any closed circuit or mesh the algebraic sum of emfs acting in that circuit or mesh is equal to the algebraic sum of the products of the currents and resistances of each part of the circuit.

If the circuit shown in Fig. 1.40 is considered, then according to Kirchhoff's second law (or KVL)

In mesh AFCBA,

$$E_1 = R_1 (I_1 + I_2) + R_2 (I_1 + I_2) + R_5 I_1$$

$$\text{or } E_1 = (R_1 + R_2)(I_1 + I_2) + R_5 I_1$$

In mesh FEDCF,

$$-E_2 = -R_3 I_2 - R_4 I_2 - R_2 (I_1 + I_2) - R_1 (I_1 + I_2)$$

$$\text{or } E_2 = R_3 I_2 + R_4 I_2 + (R_1 + R_2)(I_1 + I_2)$$

and in mesh AFEDCBA,

$$E_1 - E_2 = -R_3 I_2 - R_4 I_2 + R_5 I_1$$

$$\text{or } E_1 - E_2 = R_5 I_1 - (R_3 + R_4) I_2$$

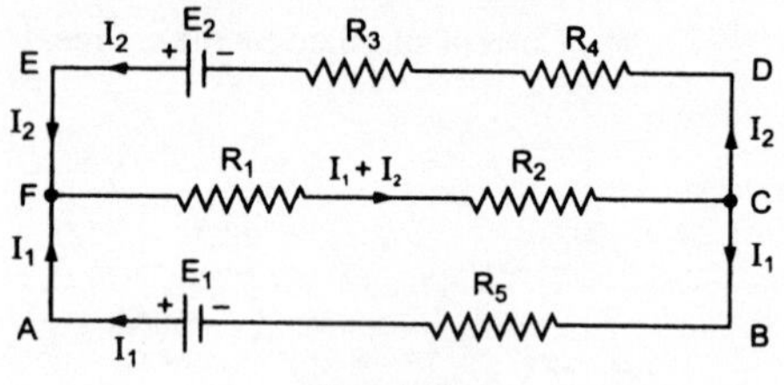

Fig. 1.40

1.17.1. Application of Kirchhoff's Laws To Circuits. First of all the current distribution in various branches of the circuit is made with directions of their flow complying with first law of Kirchhoff. Then Kirchhoff's second law is applied to each mesh (one by one) separately and algebraic equations are obtained by equating the algebraic sum of emfs acting in a mesh equal to the algebraic sum of respective drops in the same mesh. By solving the equations so obtained unknown quantities can be determined. While applying Kirchhoff's second law, the question of algebraic signs may be troublesome and is a frequent source of error. If, however, the following rules are kept in mind, no difficulty should occur.

The resistive drops in a mesh due to current flowing in clockwise direction must be taken positive drops.

The resistive drops in a mesh due to current flowing in counterclockwise direction must be taken as negative drops.

Similarly the battery emf causing current to flow in clockwise direction in a mesh must be taken as positive emf and the battery emf causing current to flow in counterclockwise direction in a mesh must be taken as negative emf.

For example, for the circuit shown in Fig. 1.40 let the current distribution be made as shown, which satisfy Kirchhoff's first law fully.

Taking first, mesh AFCBA for the application of Kirchhoff's second law, we see that there is only one emf acting in the mesh (E_1) and since it tries to send current in clockwise direction so E_1 be taken as positive, similarly all the resistive drops *i.e.* $R_1 (I_1 + I_2)$, $R_2 (I_1 + I_2)$ and $R_5 I_1$ are clockwise, so these must be taken as positive.

∴ According to Kirchhoff's second law in mesh AFCBA

$$E_1 = R_1 (I_1 + I_2) + R_2 (I_1 + I_2) + R_5 I_1$$

In mesh FEDCF, there is only one emf acting in the mesh (E_2) and since it tries to send current in counterclockwise direction through the mesh under consideration, it may be taken as negative. Since all of the resistive drops $R_3 I_2$, $R_4 I_2$, $R_2 (I_1 + I_2)$ and $R_1 (I_1 + I_2)$ are counterclockwise, these may be taken as negative.

Hence according to Kirchhoff's second law in mesh FEDCF we get,

$$-E_2 = -R_3 I_2 - R_4 I_2 - R_2 (I_1 + I_2) - R_1 (I_1 + I_2)$$

$$\text{or } E_2 = (R_3 + R_4) I_2 + (R_1 + R_2)(I_1 + I_2)$$

In mesh AFEDCBA, emf E_1 tries to cause current in clockwise direction so be taken as positive and emf E_2 tries to cause current in counterclockwise direction so be taken as negative. Similarly resistive drop $R_5 I_1$ being clockwise be taken as positive and resistive drops $R_3 I_2$ and $R_4 I_2$ being counterclockwise be taken as negative.

$$\text{Hence } E_1 - E_2 = -R_3 I_2 - R_4 I_2 + R_5 I_1$$

$$= R_5 I_1 - (R_3 + R_4) I_2$$

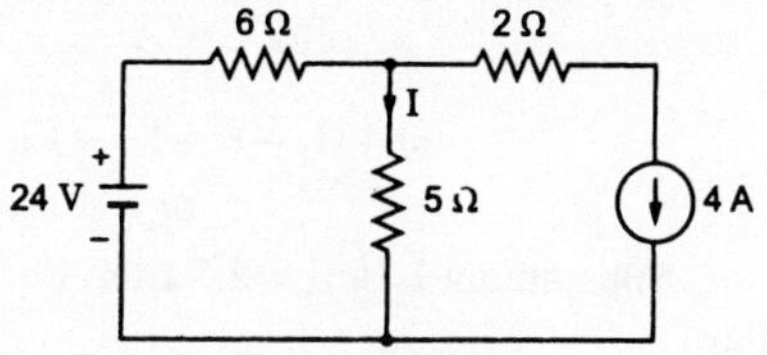

Fig. 1.41

Example 1.38

Find the current I in the circuit given in Fig. 1.41.

[*U.P. Technical Univ. Electrical Engineering Second Semester 2006-07*]

Solution: Applying Kirchhoff's current law at junction B we have

Current supplied by 24 V battery, $I_1 = (I + 4)$ amperes

Applying Kirchhoff's voltage law to mesh ABEFA we have

$$6 I_1 + 5 I = 24$$

$$\text{or } 6(I + 4) + 5 I = 24$$

$$\text{or } 11 I + 24 = 24$$

$$\text{or } I = 0 \text{ **Ans.**}$$

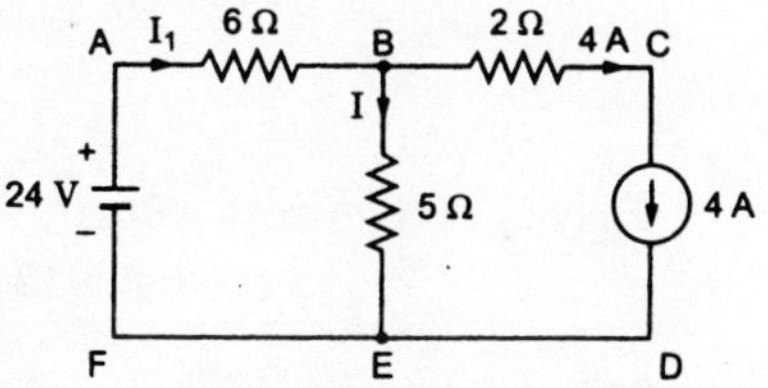

Fig. 1.42

Example 1.39

Figure 1.43 shows two batteries connected in parallel each represented by an emf along with its internal resistance. A load resistance of 6 Ω is connected across the ends of the batteries. Calculate the current through each battery and the load.

[*U.P. Technical Univ. Electrical Engineering July 2002*]

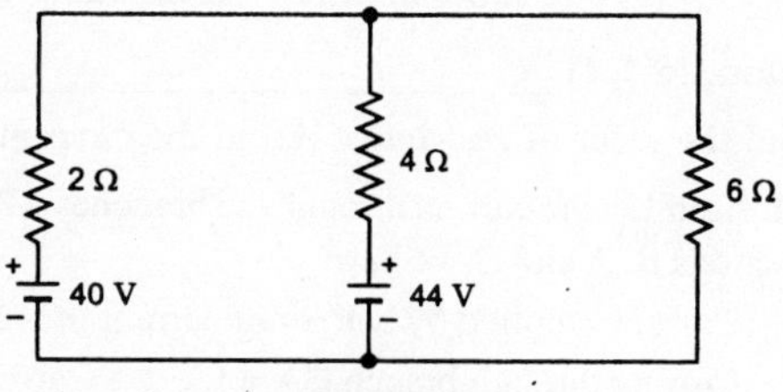

Fig. 1.43

Solution: Let the current distribution in the network be as shown in Fig. 1.44.

Applying KVL to meshes ABEFA and BCDEB we have

$$2 I_1 - 4 I_2 = 40 - 44$$

$$\text{or } 2 I_2 - I_1 = 2 \quad ...(i)$$

$$\text{and } 4 I_2 + 6(I_1 + I_2) = 44$$

$$\text{or } 5 I_2 + 3 I_1 = 22 \quad ...(ii)$$

Solving Eqs. (*i*) and (*ii*) we have

$$I_1 = \frac{34}{11} \text{ A **Ans.**}$$

$$I_2 = \frac{28}{11} \text{ A **Ans.**}$$

$$\text{Current through load, } I_L = I_1 + I_2 = \frac{34}{11} + \frac{28}{11} = \frac{62}{11} \text{ A **Ans.**}$$

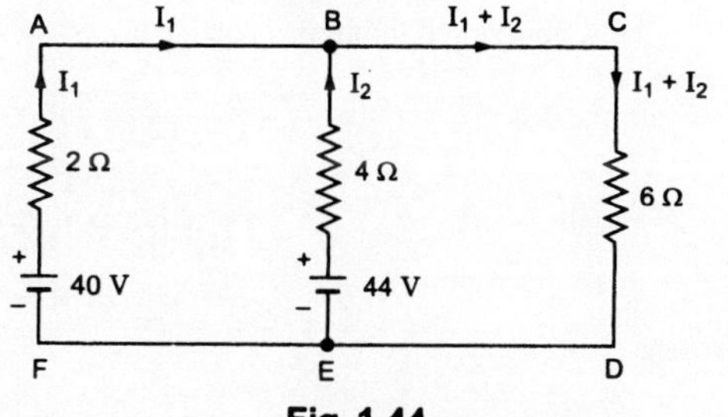

Fig. 1.44

Example 1.40

In Fig. 1.45 the potential of point A = – 30 V. Using Kirchhoff's laws find (*a*) value of V and (*b*) power dissipated by 5 Ω resistance.

Solution: The potential of point A is – 30 V and potential of point G is zero, being grounded. So the pd across 12 Ω resistor connected between terminals A and H is 30 V and current flowing through 12 Ω resistor is $\frac{30}{12}$ *i.e.* 2.5 A from terminal H to A.

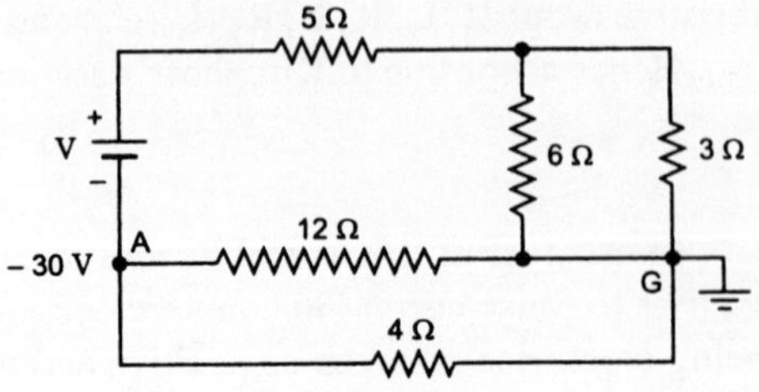

Fig. 1.45

Let the current distribution in the other branches of the circuit be, as shown in Fig. 1.46. Applying Kirchhoff's current law (KCL) to node H we have

$$I_2 + I_3 = 2.5 \quad ...(i)$$

Applying Kirchhoff's voltage law to meshes CHGDC and AFEGHA we have

$$6\,I_2 - 3\,(I_1 - I_2) = 0$$

$$\text{or } I_2 = \frac{I_1}{3} \quad ...(ii)$$

$$\text{and } 4\,(I_1 - I_2 - I_3) - 12 \times 2.5 = 0$$

$$\text{or } I_1 - (I_2 + I_3) = 7.5$$

Substituting $I_2 + I_3 = 2.5$ from Eq. (*i*) in the above equation we have

$$I_1 - 2.5 = 7.5$$

$$\text{or } I_1 = 10\text{ A}$$

$$\text{and } I_2 = \frac{I_1}{3} = \frac{10}{3}\text{ A}$$

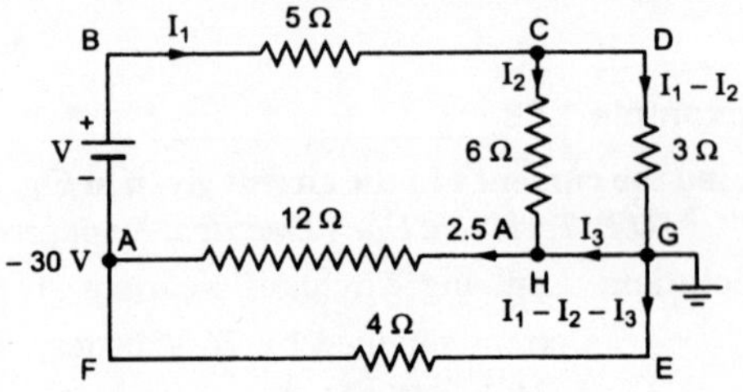

Fig. 1.46

Applying KVL to mesh ABCHA we have

$$5\,I_1 + 6\,I_2 + 12 \times 2.5 = V$$

(*a*) $$\text{or } V = 5\,I_1 + 6\,I_2 + 30$$

$$= 5 \times 10 + 6 \times \frac{10}{3} + 30 = 100\text{ V } \textbf{Ans.}$$

(*b*) Power dissipated in 5 Ω resistor $= I_1^2 \times 5 = 10^2 \times 5 = 500$ W **Ans.**

Example 1.41

Find the value of resistance R and the current I_R in Fig. 1.47 when the branch AD carries no current.

Solution: Let the current through the branches AB and DB of the circuit be I_1 and I_2 amperes respectively. Applying KCL to junctions B, A and D we have

Current supplied by battery or current in branch BC = $I_1 + I_2$

Current through branch CA = I_1 ∵ Current through branch AD is zero

Current through branch CD = I_2

Now applying Kirchhoff's voltage law to meshes ABDA, CADC and BCDB we have

$$3\,I_1 - 12\,I_2 + 0 = 0$$

$$\text{or } I_1 = 4\,I_2 \quad ...(i)$$

$$2\,I_1 + 0 - R\,I_2 = 0$$

$$\text{or } R = \frac{2\,I_1}{I_2} = 2 \times 4 = 8\ \Omega \textbf{ Ans.}$$

$$\because \text{ from Eq. } (i)\ \frac{I_1}{I_2} = 4$$

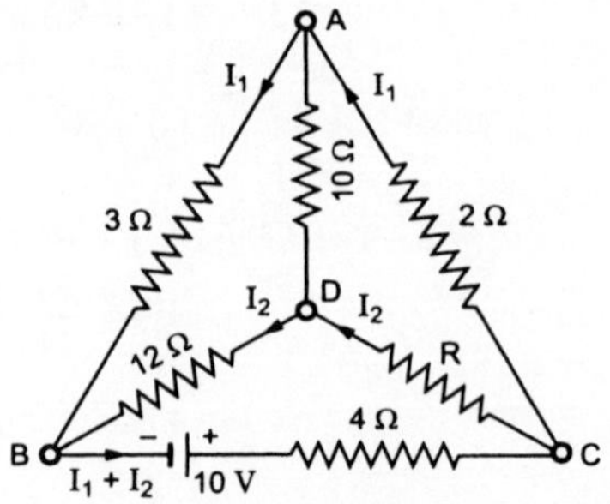

Fig. 1.47

and $4(I_1 + I_2) + R I_2 + 12 I_2 = 10$

Substituting $I_1 = 4 I_2$ from Eq. (*i*) and $R = 8\ \Omega$, as determined above in the above equation we have

$$4(4 I_2 + I_2) + 8 I_2 + 12 I_2 = 10$$

$$\text{or } I_2 = \frac{10}{40} = 0.25 \text{ A } \textbf{Ans.}$$

Example 1.42

Applying Kirchhoff's current law, determine current I_S in the electric circuit of Fig. 1.48. Take $V_0 = 16$ V.

[U.P. Technical Univ. Elec. Engineering, February 2001]

Solution : Let the current flowing through the various branches of the circuit, be as shown in the Fig. 1.48.

Now applying Kirchhoff's current law to nodes A and B we have

$$I_1 = I_2 + I_S \quad ...(i)$$

$$\text{and } I_2 + I_3 = \frac{V_1}{4} \quad ...(ii)$$

Also voltage of node B = $V_0 = 16$ V

and, therefore, $4 I_2 + V_1 = 16$ V ...(*iii*)

$$I_1 = \frac{V_1}{6} \quad ...(iv)$$

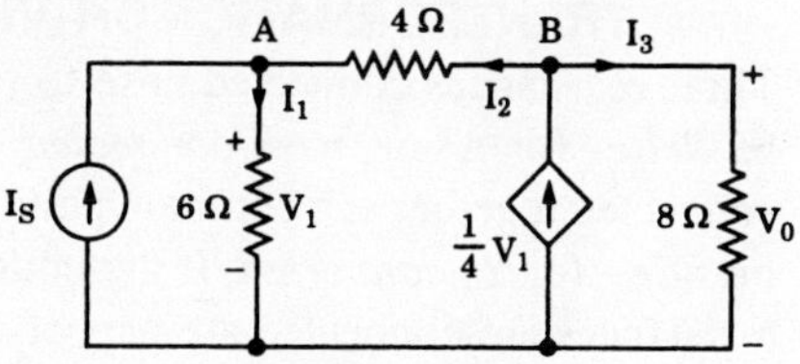

Fig. 1.48

Solving Eqs. (*i*), (*ii*), (*iii*) and (*iv*) we have

$V_1 = 12$ V; $I_1 = 2$ A; $I_2 = 1$ A and $I_s = I_1 - I_2 = 2 - 1 = 1$ A

So $I_s = 1$ A **Ans.**

Example 1.43

Find the total power delivered to the circuit by two sources in Fig. 1.49.

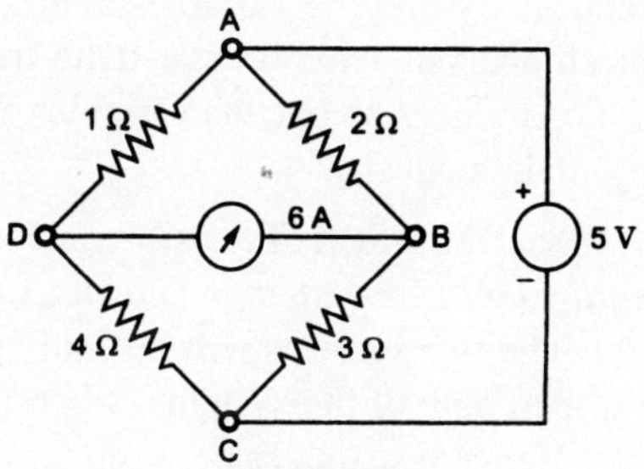

Fig. 1.49

Solution: The given circuit is redrawn with assumed distribution of current in Fig. 1.50. Applying KVL to meshes AEFCBA and ADCFEA we have

$$2 I_2 + (I_2 + 6) \times 3 = 5$$

$$\text{or } I_2 = \frac{-13}{5} = -2.6 \text{ A}$$

$$\text{and } I_1 + 4(I_1 - 6) = 5$$

$$\text{or } I_1 = \frac{29}{5} = 5.8 \text{ A}$$

Total power delivered by the two sources

= Total power absorbed by the circuit

$= I_2^2 \times 2 + (I_2 + 6)^2 \times 3 + (I_1 - 6)^2 \times 4 + I_1^2 \times 1$

$= (-2.6)^2 \times 2 + (-2.6 + 6)^2 \times 3 + (5.8 - 6)^2 \times 4 + (5.8)^2 \times 1$

= 82 W **Ans.**

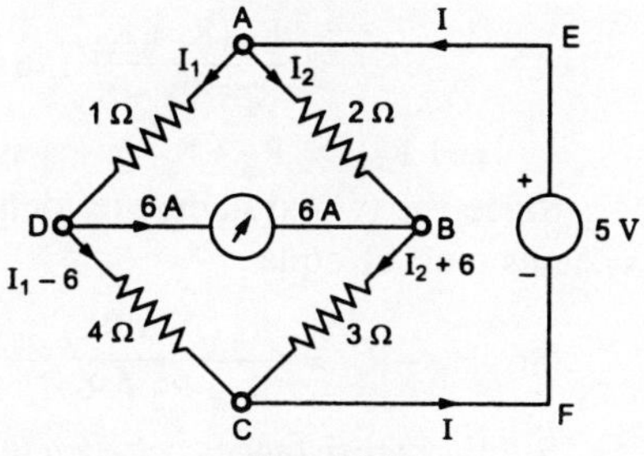

Fig. 1.50

Example 1.44

Calculate the current through the galvanometer in the following bridge (Fig. 1.51).

[U.P. Technical Univ. Electrical Engineering January 2003]

Solution: Assume current distribution in the bridge network as shown in Fig. 1.52.

Applying Kirchhoff's second law to meshes ABDA, BCDB and ABCA we have

$$I_1 + 4 I_3 - 2 I_2 = 0 \quad ...(i)$$

$$2(I_1 - I_3) - 3(I_2 + I_3) - 4 I_3 = 0$$

$$\text{or } 2 I_1 - 3 I_2 - 9 I_3 = 0 \quad ...(ii)$$

$$\text{and} \quad I_1 + 2(I_1 - I_3) = 2$$
$$\text{or} \quad 3I_1 - 2I_3 = 2 \qquad ...(iii)$$

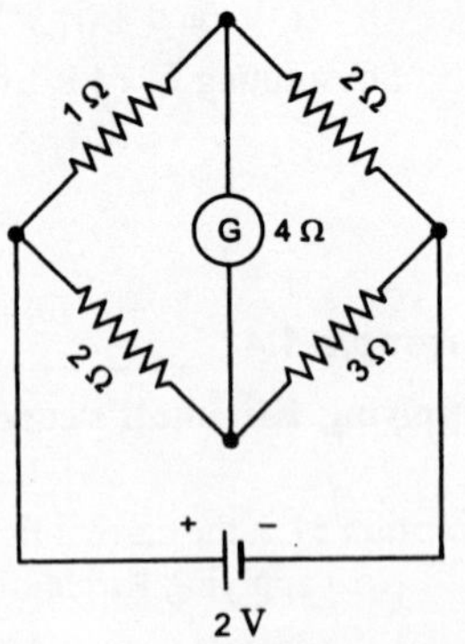

Fig. 1.51

Solving Eqs. (*i*), (*ii*) and (*iii*) we get

$$I_3 = \frac{1}{44} \text{ A} ; I_2 = \frac{17}{44} \text{ A and } I_1 = \frac{30}{44} \text{ A}$$

So current through galvanometer $= I_3 = \frac{1}{44}$ A **Ans.**

1.18. NETWORK REDUCTION BY DELTA-STAR TRANSFORMATION OR VICE VERSA

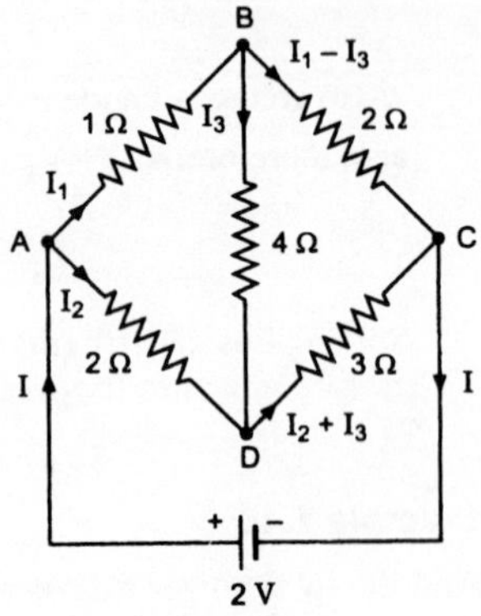

Fig. 1.52

Three resistances connected nose-to-tail as shown in Fig. 1.53 (*a*), are said to be *delta-* (or Δ) or *mesh-connected* (they form a mesh). Three resistances connected together at a common point O, as shown in Fig. 1.53 (*b*), are said to *be star* – (or Y) *connected*. If the nodes (A, B and C) to which the two sets of resistances are connected are part of a larger network, it is possible to assign values to the two sets of resistances so that they have exactly the same effect on the network. If, therefore, delta-connected resistances are part of a network it is possible to substitute them by the star-connected ones and vice versa. The obvious advantages are that a delta-star transformation eliminates a mesh which reduces by one the variables and equations necessary to solve a network by mesh analysis whereas star-delta transformation eliminates a node (node O) which reduces by one the variables and equations necessary to solve a network by node analysis.

1.18.1. Delta-Star Transformation. *The replacement of delta or mesh by equivalent star system is known as delta-star transformation.*

The two systems will be equivalent or identical if the resistances measured between any pair of lines is same in both of the systems, when the third line is open.

Hence resistances between terminals B and C,

$$R_{BC} = R_3 \parallel (R_1 + R_2)$$
$$= \frac{R_3(R_1 + R_2)}{R_1 + R_2 + R_3} \quad \text{in delta system}$$

and $R_{BC} = R_B + R_C$ in star system

Since the two systems are identical, resistances measured between terminals B and C in both of the systems must be equal.

$$\text{So} \quad R_B + R_C = \frac{R_3(R_1 + R_2)}{R_1 + R_2 + R_3} \qquad ...(1.33)$$

Similarly resistances between terminals C and A being equal in the two systems

$$R_C + R_A = \frac{R_1(R_2 + R_3)}{R_1 + R_2 + R_3} \qquad ...(1.34)$$

And resistance between terminals A and B

$$R_A + R_B = \frac{R_2(R_1 + R_3)}{R_1 + R_2 + R_3} \qquad ...(1.35)$$

Adding Eqs. (1.33), (1.34) and (1.35) we have

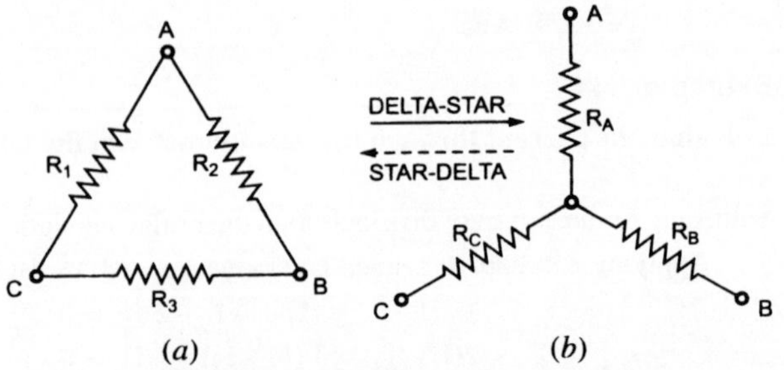

Fig. 1.53 *Delta-Star Transformation*

$$2(R_A + R_B + R_C) = \frac{2(R_1 R_3 + R_2 R_3 + R_1 R_2)}{R_1 + R_2 + R_3}$$

$$\text{or } R_A + R_B + R_C = \frac{R_1 R_3 + R_2 R_3 + R_1 R_2}{R_1 + R_2 + R_3} \qquad ...(1.36)$$

Subtracting Eqs. (1.33), (1.34) and (1.35) from Eq. (1.36) we have respectively

$$R_A = \frac{R_1 R_2}{R_1 + R_2 + R_3} \qquad ...(1.37)$$

$$R_B = \frac{R_2 R_3}{R_1 + R_2 + R_3} \qquad ...(1.38)$$

$$R_C = \frac{R_3 R_1}{R_1 + R_2 + R_3} \qquad ...(1.39)$$

These relationships may be expressed as follows:

The equivalent star resistance connected to a given terminal is equal to the product of the two delta resistances connected to the same terminal divided by the sum of the delta-connected resistances.

If the three delta-connected resistances have the same value R_D, the three resistances in the equivalent star for identical systems will be

$$R_S = \frac{R_D R_D}{R_D + R_D + R_D} = \frac{R_D}{3} \qquad ...(1.40)$$

1.18.2. Star-Delta Transformation. Multiplying Eqs. (1.37) and (1.38), (1.38) and (1.39) and (1.39) and (1.37) and then adding them we get

$$R_A R_B + R_B R_C + R_C R_A = \frac{R_1 R_2^2 R_3 + R_1 R_3^2 R_2 + R_1^2 R_2 R_3}{(R_1 + R_2 + R_3)^2}$$

$$= \frac{R_1 R_2 R_3 (R_2 + R_3 + R_1)}{(R_1 + R_2 + R_3)^2} = \frac{R_1 R_2 R_3}{R_1 + R_2 + R_3} \qquad ...(1.41)$$

Dividing Eq. (1.41) by Eqs. (1.37), (1.38) and (1.39) separately we have

$$R_3 = \frac{R_A R_B + R_B R_C + R_C R_A}{R_A} = R_B + R_C + \frac{R_B R_C}{R_A} \qquad ...(1.42)$$

$$R_1 = \frac{R_A R_B + R_B R_C + R_C R_A}{R_B} = R_A + R_C + \frac{R_A R_C}{R_B} \qquad ...(1.43)$$

$$\text{and } R_2 = \frac{R_A R_B + R_B R_C + R_C R_A}{R_C} = R_A + R_B + \frac{R_A R_B}{R_C} \qquad ...(1.44)$$

The above relationship may be expressed as below:

The equivalent delta resistance between two terminals is the sum of the two star resistances connected to those terminals plus the product of the same divided by the third star resistance.

If the three star-connected resistances have the same value, say R_S, the three resistances of the equivalent delta for identical systems will be

$$R_D = R_S + R_S + \frac{R_S R_S}{R_S} = 3 R_S \qquad ...(1.45)$$

The advantage of delta-star transformation may be shown by reference to network of Fig. 1.54. Figure 1.54 (*a*) illustrates the network before conversion, where the dotted lines are drawn around the delta connection to be transformed into a star. Fig. 1.54 (*b*) illustrates the same network after transformation. The currents in the transformed form [Fig. 1.54 (*b*)] are much simpler to determine.

The advantage of star-delta transformation can be illustrated by reference to network of Fig. 1.55. Figure 1.55 (*a*) illustrates the network prior to transformation, with the dotted lines around the star to be transformed. Fig. 1.55 (*b*) illustrates the same network after transformation. The original network is now reduced to a simple series parallel connection of resistances.

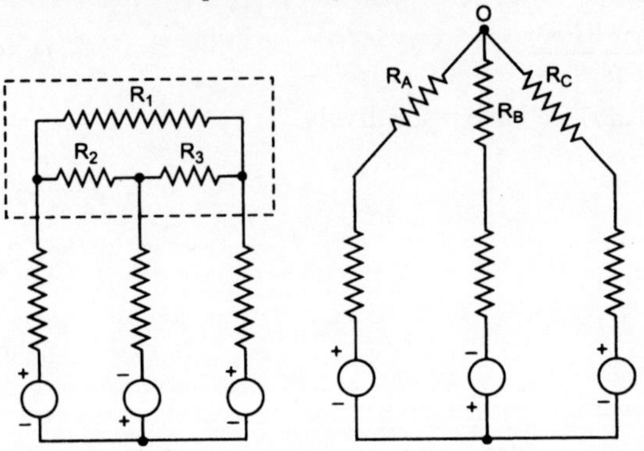

(*a*) (*b*)

Fig. 1.54

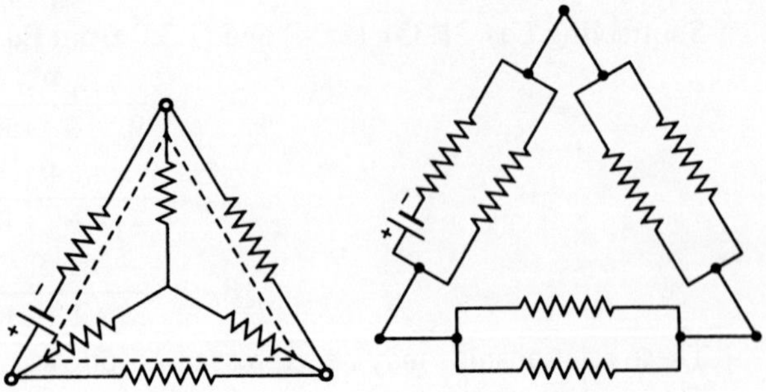

(*a*) (*b*)

Fig. 1.55

Example 1.45

Three resistances *r*, 2 *r*, and 3 *r* are connected in delta. Determine the resistances for an equivalent star connection.

[*U.P. Technical Univ. Electrical Engineering January 2003*]

Solution: Resistances for the equivalent star-connection (Fig. 1.56) are worked-out as below:

$$R_A = \frac{R_1 R_2}{R_1 + R_2 + R_3} = \frac{r \times 2r}{r + 2r + 3r} = \frac{r}{3} \quad \textbf{Ans.}$$

$$R_B = \frac{R_2 R_3}{R_1 + R_2 + R_3} = \frac{2r \times 3r}{r + 2r + 3r} = r \quad \textbf{Ans.}$$

$$R_C = \frac{R_3 R_1}{R_1 + R_2 + R_3} = \frac{3r \times r}{r + 2r + 3r} = \frac{r}{2} \quad \textbf{Ans.}$$

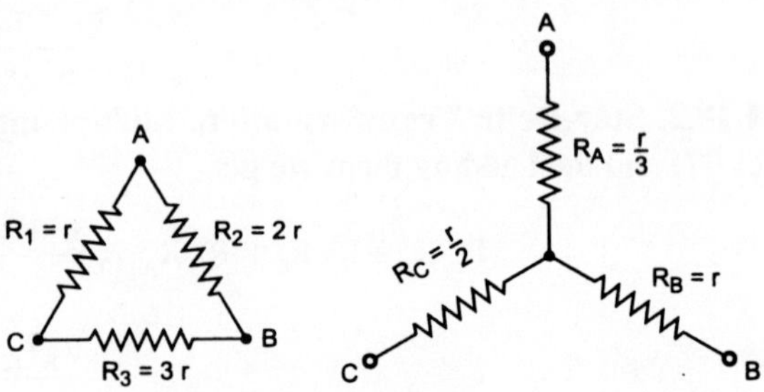

(*a*) *Delta-Connected Resistors* (*b*) *Equivalent Star-Connected Resistors*

Fig. 1.56

Example 1.46

Find the resistance between terminals XY of the bridge circuit shown in Fig. 1.57 (a) by using delta-star transformations. [*U.P. Technical Univ. Electrical Engineering July 2002*]

Solution : As illustrated in Fig. 1.57 (*b*), delta ACB has been reduced to its equivalent star

$$R_A = \frac{R_{AC} R_{AB}}{R_{AC} + R_{AB} + R_{CB}} = \frac{4 \times 6}{4 + 6 + 2} = 2\ \Omega$$

$$R_B = \frac{R_{AB} R_{BC}}{R_{AC} + R_{AB} + R_{CB}} = \frac{4 \times 2}{4 + 6 + 2} = \frac{2}{3}\ \Omega$$

$$R_C = \frac{R_{CA} R_{BC}}{R_{CA} + R_{AB} + R_{CB}} = \frac{6 \times 2}{4 + 6 + 2} = 1\ \Omega$$

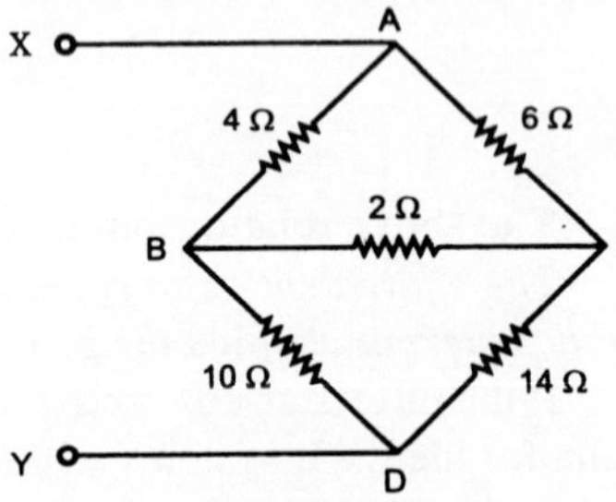

Fig. 1.57 (*a*)

Hence the given circuit is reduced to the circuit shown in Fig. 1.57 (*c*). As seen, there are two parallel paths between terminals N and D, one of resistance 1 + 14 = 15 Ω and the other is of resistance $\frac{2}{3}$ + 10 *i.e.* $\frac{32}{3}$ Ω.

$$\text{Their combined resistance} = \frac{15 \times \frac{32}{3}}{15 + \frac{32}{3}} = \frac{480}{77}\ \Omega$$

$$\text{Total resistance of network between terminals X and Y} = 2 + \frac{480}{77} = 8.234\ \Omega \quad \textbf{Ans.}$$

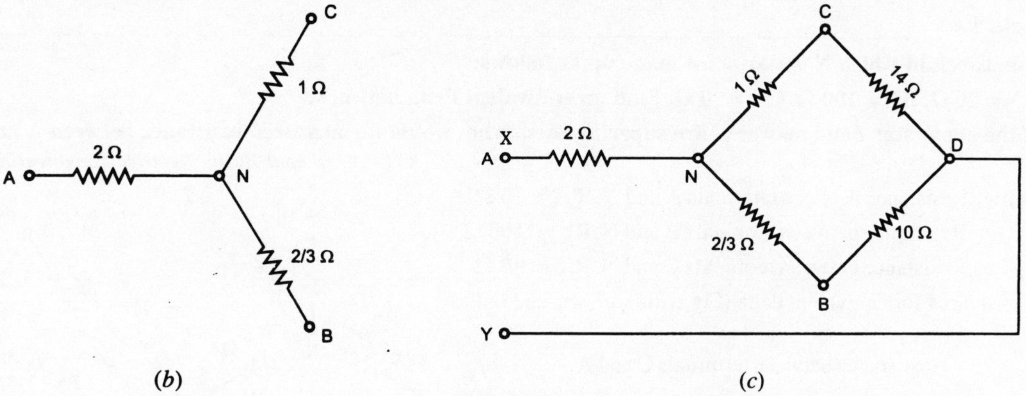

Fig. 1.57

Example 1.47

Find the resistance at the A–B terminals in the electric circuit of Fig. 1.58 using Δ–Y transformation.

[U.P. Technical Univ. Elec. Engineering February 2001]

Solution: As shown in Fig. 1.59, delta ACD has been reduced to its equivalent star

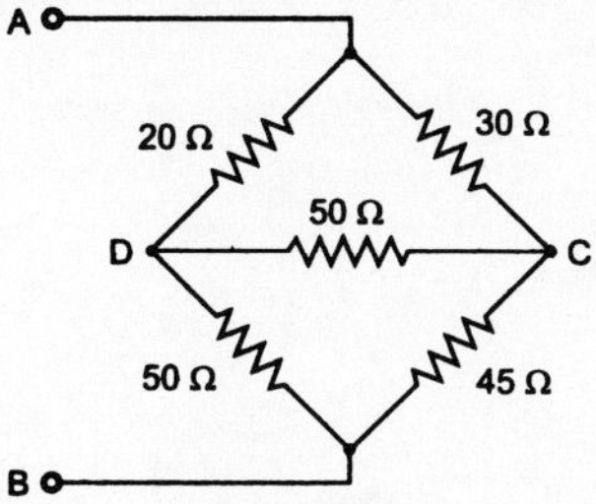

Fig. 1.58

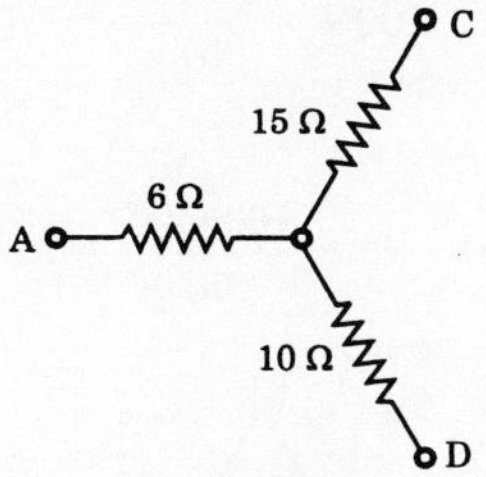

Fig. 1.59

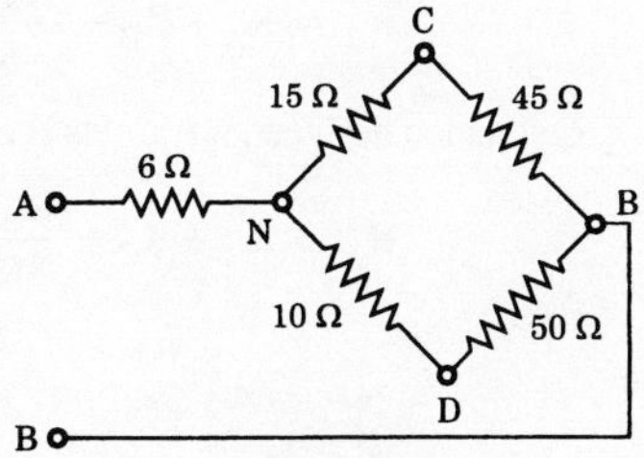

Fig. 1.60

$$R_A = \frac{R_{AC}\,R_{AD}}{R_{AC}+R_{CD}+R_{DA}} = \frac{30\times 20}{30+50+20} = 6\ \Omega$$

$$\text{Similarly } R_C = \frac{30\times 50}{100} = 15\ \Omega \quad \text{and } R_D = \frac{20\times 50}{100} = 10\ \Omega$$

Hence the given circuit is reduced to the circuit shown in Fig. 1.60. As seen, there are two parallel paths between points N and B, one is of resistance 15 + 45 = 60 Ω and the other is also of 10 + 50 = 60 Ω.

$$\text{Their combined resistance} = \frac{1}{\frac{1}{60}+\frac{1}{60}} = 30\ \Omega$$

So total resistance of the network between terminals A and B

$$= 30 + 6 = 36\ \Omega \text{ Ans.}$$

Example 1.48

Find the current drawn I from the source voltage of 100 V in the circuit shown in Fig. 1.61.

[U.P. Technical Univ. Elec. Engineering First Semester 2004-05]

Solution: From solution of above Example 1.47, the Wheatstone bridge is equivalent to a resistor of 36 Ω

Hence total resistance of the circuit, R = 10 + 36 = 46 Ω

$$\text{Current drawn, } I = \frac{V}{R} = \frac{100}{46} = 2.174 \text{ A } \textbf{Ans.}$$

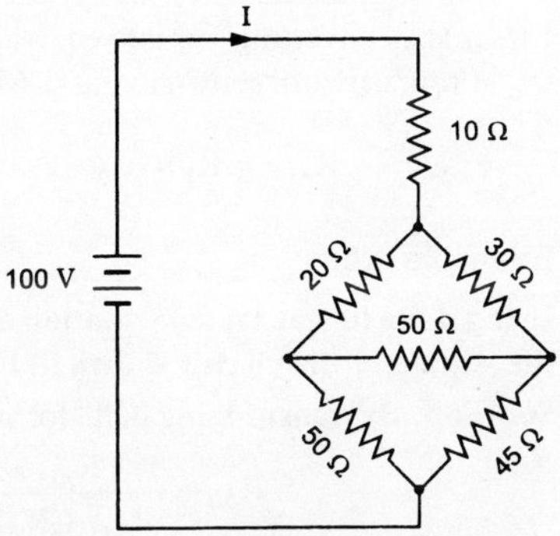

Fig. 1.61

Example 1.49

A star-network in which N is star point made up as follows:

AN = 70 Ω, BN = 100 Ω, CN = 90 Ω. Find an equivalent delta network.

If the above star-delta networks are superimposed, what would be measured resistance between A and C ?

[*R.G.P.V. Bhopal Basic Electrical Engineering 2001*]

Solution: Resistance between terminals A and N, $R_A = 70\ \Omega$

Resistance between terminals B and N, $R_B = 100\ \Omega$

Resistance between terminals C and N, $R_C = 90\ \Omega$

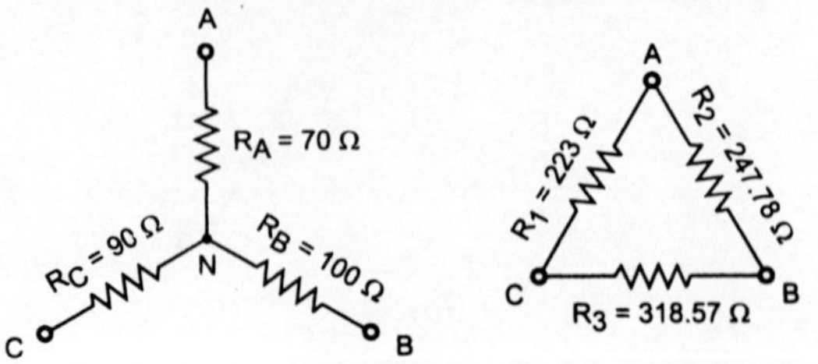

Fig. 1.62

Resistances for equivalent delta (Fig. 1.62) are worked out as below:

Resistance between terminals C and A,

$$R_1 = R_A + R_C + \frac{R_A R_C}{R_B}$$

$$= 70 + 90 + \frac{70 \times 90}{100} = 223\ \Omega \quad \textbf{Ans.}$$

Resistance between terminals A and B,

$$R_2 = R_A + R_B + \frac{R_A R_B}{R_C} = 70 + 100 + \frac{70 \times 100}{90} = 247.78\ \Omega \quad \textbf{Ans.}$$

Resistance between terminals B and C,

$$R_3 = R_B + R_C + \frac{R_B R_C}{R_A} = 100 + 90 + \frac{100 \times 90}{70} = 318.57\ \Omega \quad \textbf{Ans.}$$

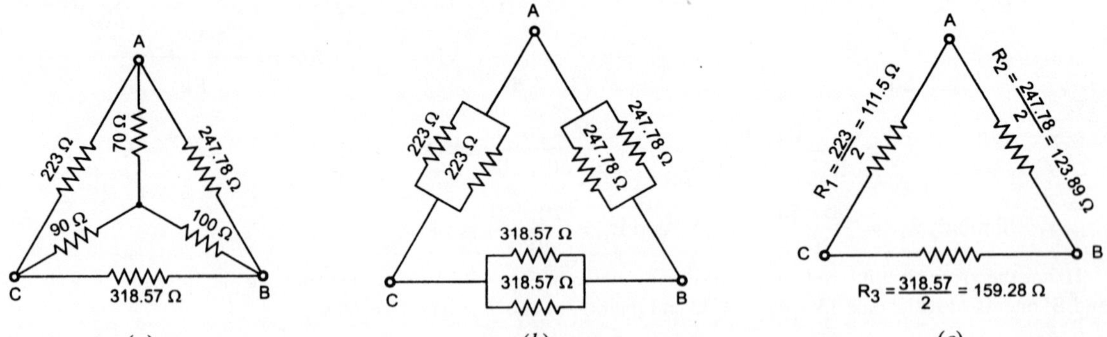

Fig. 1.63

Now, when star and delta networks are superimposed, the equivalent network becomes as shown in Figs. 1.63 (*b*) and 1.63 (*c*) by converting star into equivalent delta.

From network shown in Fig. 1.63 (*c*) resistance measured between terminals A and C,

$$R_{AC} = R_1 \,||\, (R_2 + R_3) = 111.5 \,||\, (123.89 + 159.285) = 111.5 \,||\, 283.175 = \frac{111.5 \times 283.175}{111.5 + 283.175} = 80\ \Omega \quad \textbf{Ans.}$$

Example 1.50

Using delta to star transformation determine the resistance between terminals *a-b* and the total power drawn from the supply in the circuit shown in Fig. 1.64. [*U.P. Technical Univ. Electrical Engineering First Semester 2006-07*]

Solution: By transforming delta formed by resistors of 8 Ω, 7 Ω and 3 Ω into star we have

$$R_A = \frac{R_{AC} \times R_{AB}}{R_{AC} + R_{AB} + R_{BC}} = \frac{3 \times 8}{3 + 8 + 7} = \frac{4}{3}\ \Omega$$

Similarly $R_B = \frac{8 \times 7}{18} = \frac{28}{9}\ \Omega$

and $R_C = \frac{3 \times 7}{18} = \frac{7}{6}\ \Omega$

Now the given circuit is reduced to that as shown in Fig. 1.66.

Resistance between terminals *a* and *b*,

$$R = \frac{4}{3} + \left(\frac{7}{6} + 4\right) \parallel \left(\frac{28}{9} + 10\right)$$

$$= \frac{4}{3} + \frac{31}{7} \parallel \frac{118}{9} = 1.3333 + 3.7062 = 5.04\ \Omega \textbf{ Ans.}$$

Total power drawn, $P = \frac{V^2}{R} = \frac{10^2}{5.04} = 19.84$ watts **Ans.**

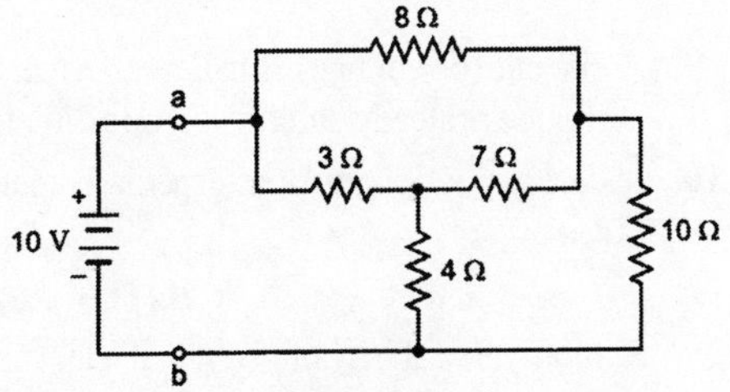

Fig. 1.64

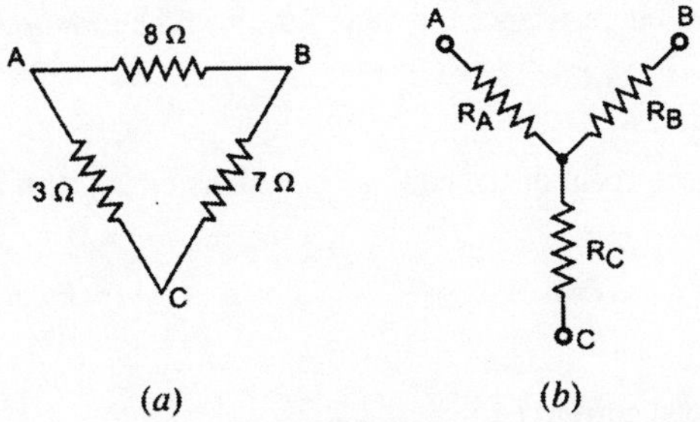

Fig. 1.65

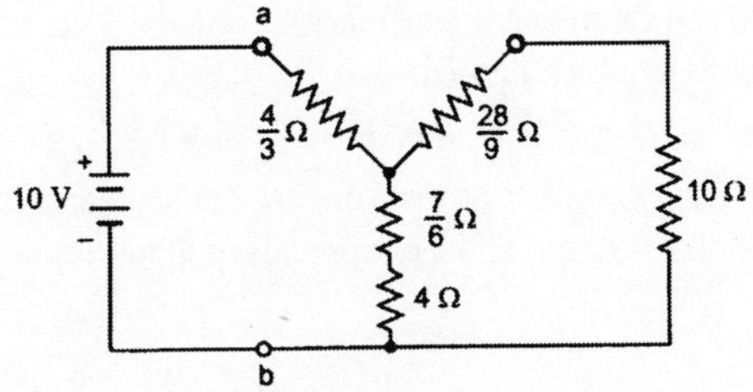

Fig. 1.66

HIGHLIGHTS

1. All matters are made of electrons, protons and neutrons.
2. The controlled movement of electrons (or drift) through a substance is called the electric current. Current is expressed in amperes.
3. Electromotive force (emf) or potential difference is that which causes the current to flow in a closed circuit. The unit of potential difference in SI system is volt.
4. Resistance may be defined as that property of substance which opposes (or restricts) the flow of an electric current (or electrons) through it. The unit of resistance in SI system is ohm (Ω).
5. Ohm's law may be given as

   $$\frac{V}{I} = \text{Constant} = R \qquad \text{for no change in physical state}$$

6. Laws of resistance are summed up as

   $$R = \rho \frac{l}{a}$$

   where ρ is the specific resistance or resistivity of the material of wire, which may be defined as the resistance between opposite faces of a unit cube of that material.
   Resistivity is measured in ohm-metres (Ω-m) or ohms per metre cube.
7. Conductance G is reciprocal of resistance *i.e.* $G = \frac{1}{R}$. The unit of conductance is siemens (S).
8. Conductivity σ is reciprocal of resistivity ρ *i.e.* $\sigma = \frac{1}{\rho}$. The unit of conductivity is siemens per metre.

9. A current (or voltage) is called alternating if it periodically changes its direction and magnitude whereas direct currents are steady and flow in one direction only.
10. In case of good conductors, especially metals, the variations in resistance with temperature (within normal range) are given by

$$R_t = R_0 (1 + \alpha_0 t)$$

Similarly, resistivity of given material varies as

$$\rho_t = \rho_0 (1 + \alpha_0 t)$$

11. Temperature coefficient of resistance is not constant but depends upon the initial temperature on which the increment in resistance is based. The values of temperature coefficient for different temperatures can be had from the relations

$$\alpha_1 = \frac{1}{\frac{1}{\alpha_0} + t_1} \text{ and } \alpha_2 = \frac{1}{\frac{1}{\alpha_1} + (t_2 - t_1)}$$

12. When a number of resistors are connected in series, the equivalent resistance is given by the arithmetic sum of their individual resistances

$$i.e. \; R = R_1 + R_2 + R_2 + ... + R_n$$

13. When a number of resistors are connected in parallel, the reciprocal of the equivalent resistance is given by the arithmetic sum of the reciprocals of their individual resistances

$$i.e. \; \frac{1}{R} = \frac{1}{R_1} + \frac{1}{R_2} + \frac{1}{R_3} + ... + \frac{1}{R_n}$$

14. When two resistances R_1 and R_2 are in parallel and carry a total current I, then

$$\text{current through resistance } R_1, I_1 = \frac{R_2}{R_1 + R_2} I$$

$$\text{and current through resistance } R_2, I_2 = \frac{R_1}{R_1 + R_2} I$$

15. When n resistors, each of R ohms, are connected in parallel, then total resistance is given by $R_T = \frac{R}{n}$ ohms.
16. Current in any branch of a parallel circuit is directly proportional to its conductance and is equal to the total current flowing through the circuit multiplied by the ratio of the conductance of the branch to that of the circuit.
17. The potential divider is a high resistance connected across the supply mains and is used to provide a variable voltage from a constant supply voltage.
18. According to *Kirchhoff's first law* (*or current law*), the algebraic sum of currents in two or more conductors at a point (junction) is always zero

$$i.e. \; \Sigma I = 0$$

While applying above law, incoming currents are taken as positive and outgoing currents as negative.

According to *Kirchhoff's second law* (*or voltage law*), the algebraic sum of emfs acting in any closed circuit or mesh is equal to the algebraic sum of the products of currents and resistances of each part of that closed circuit or mesh *i.e.*,

$$\Sigma IR = \Sigma \text{ emf}$$

19. *In delta-star transformation*, the equivalent star resistance connected to a given terminal is equal to the product of the two delta resistances connected to the same terminal divided by the sum of the delta connected resistances.
20. *In star-delta transformation*, the equivalent delta resistance between two terminals is the sum of the two star resistances connected to those terminals plus the product of the same divided by the third star resistance.

EXERCISES

1. What is Ohm's law ? State its limitations.
[Pb. Technical Univ. Basic Electrical and Electronics Engineering Dec. 2005]
2. State Ohm's law and given its significance.
[Pb. Technical Univ. Basic Electrical and Electronics Engineering May 2007]
3. State laws of resistance and hence define specific resistance.
4. Discuss the temperature dependence of resistance.
[Pb. Technical Univ. Basic Electrical and Electronics Engineering Dec. 2005]
5. Derive an expression for resistance of a conductor material and explain the effect of temperature on it.
[Pb. Technical Univ. Basic Electrical and Electronics Engineering May 2008]
6. Define resistance temperature coefficient. How does its value change with temperature ?
7. Define temperature coefficient of resistance. Write down the names of the materials having negative temperature coefficient.
8. Compute equivalent resistance of three resistances R_1, R_2 and R_3 connected in (*i*) series (*ii*) parallel.
9. State the two fundamental laws of circuit analysis. **[Pb. Technical Univ. Electrical Engineering May 2002]**
10. State and explain Kirchhoff's laws.
[Pb. Technical Univ. Basic Electrical and Electronics Engineering Dec. 2006–2007]
11. Derive the necessary equations for converting a delta network into an equivalent star network.
[Pb. Technical Univ. Basic Electrical and Electronics Engineering May 2006]
12. What is the use of delta-star and star-delta transformation in the circuits ? Explain the method of transforming a star network into a delta network.
13. What is the use of delta-star and star-delta transformation in the circuit ? Explain the method of transforming a star network into delta network. **[M.D. Univ. Rohtak Electrical Technology May 2002]**
14. How a star network is converted into a delta network ? Explain with example.
[U.P. Technical Univ. Electrical Engineering Second Semester 2006-07]
15. Obtain equivalent Y from Δ in Y-Δ transformation.
[U.P. Technical Univ. Electrical Engineering First Semester 2007-08]

SHORT ANSWER TYPE QUESTIONS WITH ANSWERS

Q. 1. What is electric current ?

Ans. The controlled movement of electrons (or drift) through a substance is called the electric current.

Q. 2. What is difference between emf and potential difference ?

Ans. EMF (electromotive force) is the force that causes a current to flow in an electric circuit while the potential difference between any two points in an electric circuit is that difference in their electrical state which tends to cause flow of electricity from one point to the other.

Q. 3. What is meant by electrical resistance ?

Ans. Resistance may be defined as that property of a substance which opposes (or restricts) the flow of electric current (or electrons) through it.

Q. 4. What is Ohm's law ? **[P.T.U. B.E.E.E. May 2006]**

Ans. Physical state *i.e.* temperature rise etc. remaining the same the current flowing through a conductor is directly proportional to the potential difference applied across its ends.

Q. 5. What are the limitations of Ohm's law ?

Ans. Ohm's law cannot be applied to the circuits consisting of (*i*) electronic tubes or transistors (because such elements are not bilateral) and (*ii*) nonlinear elements such as powdered iron, electric arc etc.

Q. 6. How does the resistance of a homogeneous material having constant length vary with the changing cross-sectional area ?

Ans. Resistance of a homogeneous material having constant length is inversely proportional to its area of x-section *i.e.* $R \alpha \frac{1}{a}$.

Q. 7. What is meant by specific resistance of material ?

Ans. The specific resistance, also called the resistivity, is equal to the resistance of the material of unit length having unit cross-sectional area.

Q. 8. The length of a conductor is doubled and its area of x-section is also doubled, then the resistance will

(*i*) increase four times (*ii*) remain unchanged

(*iii*) decrease to four times (*iv*) change at random. **[P.T.U. B.E.E.E. May 2006]**

Ans. Resistance remain unchanged because $R_2 = \rho\frac{l_2}{a_2} = \rho \times \frac{2l_1}{2a_1} = \rho\frac{l_1}{a_1} = R.$

Q. 9. How does the resistance of metals and semiconductors varies with temperature ?

Ans. The resistance of metals increases with the increase in temperature while the resistance of semiconductors decreases with the increase in temperature.

Q. 10. What is the effect of temperature on the resistance of the conductor ? **[P.T.U. B.E.E.E. Dec. 2007]**

Ans. Pleae refer to Art. 1.9.

Q. 11. What is the effect of temperature on resistivity ? **[P.T.U. B.E.E.E. May 2007]**

Ans. Please refer to Art. 1.9.

Q. 12. State factors affecting the resistivity of conductors.

Ans. Resistivity of a material depends mainly on 3 factors viz. temperature, alloying and mechanical stressing.

The resistivity of all pure metallic conductors increases with the increase in temperature.

Alloying is another factor which affects the resistivity of a material. By adding traces of some impurities to a metal, its resistivity increases.

The resistivity of a material also changes under the influence of mechanical treatment. Although mechanical stressing increases the resistivity, annealing restores the electrical conductivity by establishing regularity in crystal structure.

Q. 13. Give reasons why are all equipments connected in parallel to the supply.

[P.T.U. B.E.E. May 2006]

Ans. All equipments are connected in parallel so that each one have same voltage across it and can be operated independently.

Q. 14. Give the relationship between resistances connected in delta and equivalent star systems ?

Ans. The equivalent star resistance connected to a given terminal is equal to the product of the two delta resistances connected to the same terminal divided by the sum of the delta connected resistances.

Q. 15. Give the relationship between resistances connected in star and equivalent delta systems ?

Ans. The equivalent delta resistance between two terminals is the sum of the two star resistances connected to those terminals plus the product of the same divided by the third star resistance.

PROBLEMS

1. A gas-filled lamp has a resistance of 18 Ω when not burning. When it is burning with a voltage of 120 V, its resistance is 200 Ω. Calculate the current (*i*) at the instant of switching on and (*ii*) normal current when burning.
[Ans. (*i*) 6.667 A (*ii*) 0.6 A**]**
2. Calculate the length of copper wire, 1.5 mm in diameter, to have a resistance of 0.3 ohm, given that resistivity of copper is 0.017 microohm-m.
[Ans. 31.185 m**]**
3. Find the ratio of sectional area of copper wire to that of aluminium wire to have the same resistance and length if their resistivities are 1.72 μΩ-cm and 2.8 μΩ-cm respectively.
[Ans. 1 : 0.6143**]**

4. A conductor has a cross-section of 10 cm^2 and specific resistance of 7.5 μΩ-cm at 0 °C. What will be its resistance in ohms per km when the temperature is 40 °C. Take temperature coefficient of the material 0.005 per °C. [**Ans.** 0.09 Ω]

5. The shunt field coil of a generator has a resistance of 200 Ω at 20 °C. Find its resistance at 80° C, if the temperature coefficient of resistance of the material of wire at 20 °C is 0.004. [**Ans.** 248 Ω]

6. It is found that the resistance of a coil of wire increases from 40 Ω at 15 °C to 50 Ω at 60 °C. Calculate the resistance temperature coefficient of the material and the resistance of the coil at 0 °C.

[**Ans.** $\frac{1}{165}$ per °C; 36.667Ω]

7. The resistance of the field coils with copper conductors of a dynamo is 120 Ω at 25 °C. After working for 6 hours on full load, the resistance of the coil increases to 140 Ω. Calculate the mean temperature rise of the field coil. Take the temperature coefficient of the conductor material as 0.0042 at 0 °C.
[Bangalore Univ. Elements of Elec. Engineering 1991] [**Ans.** 43.8 °C]

8. Four resistors A, B, C and D are connected in parallel. If A is of 56 Ω; B is of 50 Ω; C is of 40 Ω and takes a current of 2.5 A and the total current drawn from the supply is 9.0 A, Determine the resistance of D in ohms. [**Ans.** 36.84 Ω]

9. Two resistors of equal values when placed in parallel gives an effective resistance equal to R. What will be the effective resistance if the two resistors are connected in series. [**Ans.** 4 R]

10. A resistor of value 1.6 Ω is connected in series with two resistors of value 4 Ω and 6 Ω which are connected in parallel. Find the total resistance of the series-parallel combination. [**Ans.** 4 Ω]

11. Two resistances of 4 Ω and 6 Ω in parallel are in series with another resistor of 12 Ω. If the current flowing in 12-Ω resistor is 2 A, determine (*i*) the current flowing through 4 Ω and 6 Ω resistors, (*ii*) voltage across the whole circuit and (*iii*) power drawn from the supply. [**Ans.** (*i*) 1.2 A and 0.8 A; (*ii*) 28.8 V; and (*iii*) 57.6 watts]

12. In the circuit shown in Fig. 1, find the voltage across and the current in each element.
[**Ans.** Current through 2 Ω and 4 Ω resistors = 2 A and voltage across these resistors are 4 V and 8 V respectively. Current through 4 Ω and 8 Ω resistors = 1 A and voltage across them are 4 V and 8 V respectively]

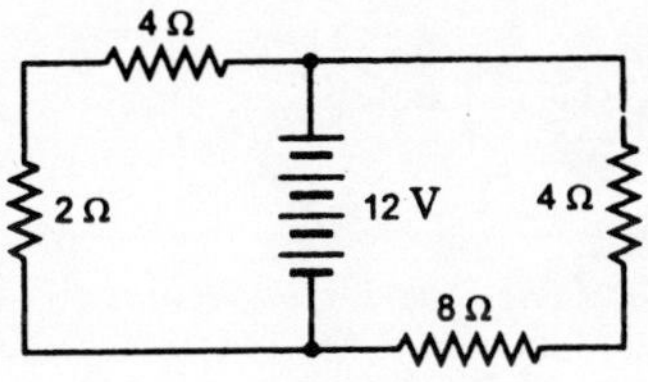

Fig. 1

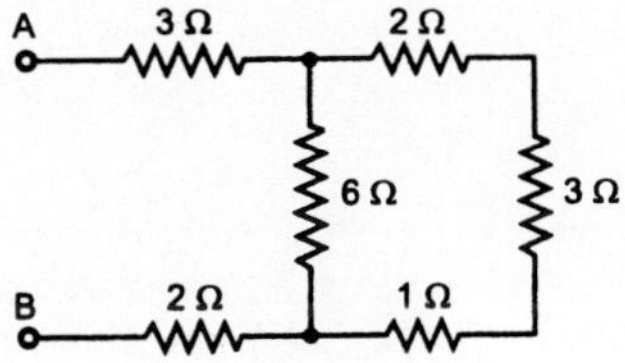

Fig. 2

13. How much current a battery of 8 volts will give when it is connected across AB in the circuit diagram given below (Fig. 2) [**Ans.** 1 A]

14. What is the equivalent resistance of the network shown in Fig. 3. [**Ans** 600 Ω]

15. In the given circuit (Fig. 4) find the current in each branch and direction of current.

[**Ans.** $\frac{96}{41}$ A, $\frac{60}{41}$ A, $\frac{36}{41}$ A]

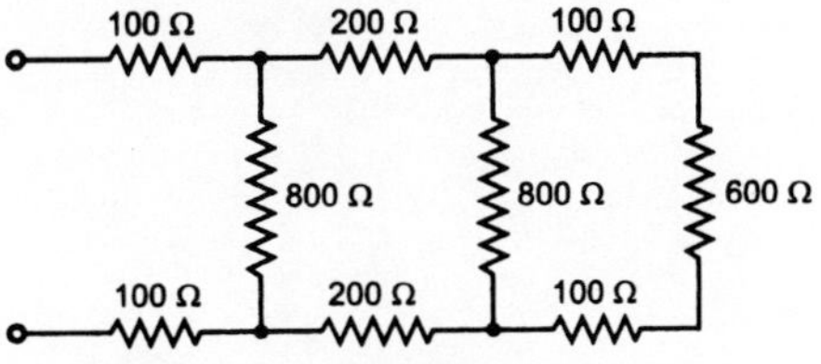

Fig. 3

16. Calculate the current in the battery, the current in each branch and pd across AB of the network shown in Fig. 5.

[**Ans.** $\frac{95}{73}$ A, $\frac{70}{73}$ A, $\frac{25}{73}$ A, $\frac{10}{73}$ A, $\frac{10}{73}$ V]

17. Determine the branch currents in the network of Fig. 6 when the value of each branch resistance is one ohm.
[Allahabad Univ. Electrical Technology 1992]
[**Ans.** 6.25 A, 6.25 A, 0, 1.25 A, 1.25 A, 7.5 A]

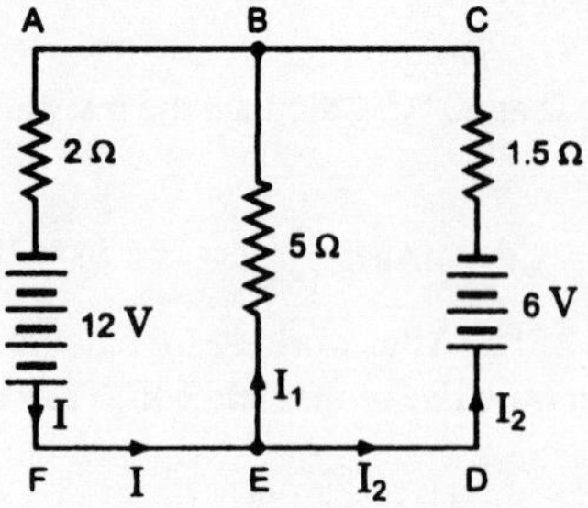

Fig. 4

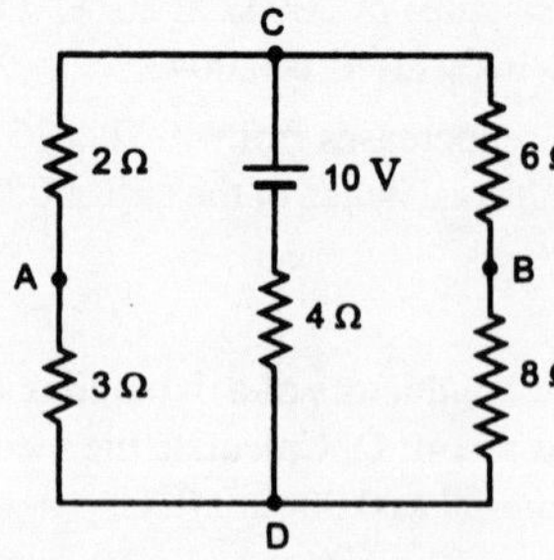

Fig. 5

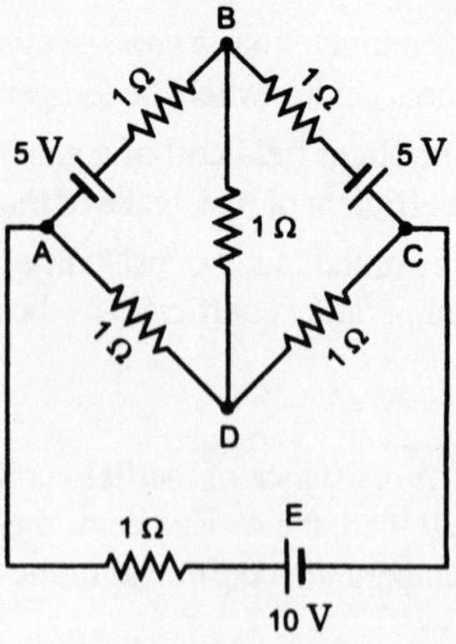

Fig. 6

18. A Wheatstone bridge consists of AB = 400 Ω, BC = 220 Ω, CD = 225 Ω and DA = 100 Ω. A 2 V cell is connected between A and C and a galvanometer of 200 Ω resistance between B and D. Estimate the current through the galvanometer by applying Kirchhoff's laws. **[Pb. Univ. Electrical Engineering-I Nov. 1983]**

[Ans. 2/703 A from D to B**]**

19. Three resistances of 20 Ω each are connected in star. Find the equivalent delta resistance. If a source of emf of 120 V is connected across any two terminals of the equivalent delta-connected resistance find the current supplied by the source. **[Calcutta Univ. Elec. Engineering 1980]**

[Ans. 60 Ω; 3 A**]**

20. Three resistances of 20, 25 and 30 ohms are connected in delta. Calculate the corresponding resistances in an equivalent star connection. **[Ans.** 8 Ω, 6.67 Ω and 10 Ω**]**

❑❑❑

UNIT-II

Electrostatics and Capacitance

CONTENTS IN THIS UNIT

Chapter 2

Electrostatics and Capacitance

2.1. ELECTROSTATICS

When two charged bodies are separated by some insulating medium, the charges (*i.e.* electrons) do not move but remain *static* on the body. This is known as *static electricity*. In other words, static electricity may be defined as *electricity at rest*. The term *static electricity* properly applies to an isolated charge of electricity produced by friction. The presence of static electricity manifests itself by attraction or repulsion.

Electrostatics is the study of static electricity, that is electricity at rest. Since a quantity of electricity is known as an electric charge, electrostatics means the study of the effects of stationary electric charges. Two kinds of electric charges exist, viz. positive and negative charges. When a body acquires an excess of positive charge it becomes positively charged and if it acquires a surplus of negative charge it becomes negatively charged. Bodies can be electrified by friction, induction or conduction.

From experiments it was revealed that by rubbing two dissimilar materials together, such as a glass rod and silk or an ebonite rod with fur, the rods become charged with opposite kinds of electricity. The experiments show that the two electrified glass rods repel each other and that the two electrified ebonite rods also repel each other but an electrified glass rod attracts a charged ebonite rod. These simple experiments indicate that the glass and ebonite rods acquire different kinds of electricity when they are electrified by friction. The kind of charge on the glass rod is regarded as positive whilst that on ebonite rod as negative.

When a glass rod is rubbed with silk, the work done in overcoming friction detaches electrons from some of the glass molecules, and these electrons transfer to the silk. One removal of silk, the glass becomes positively charged since it is then deficient in electrons. The silk meanwhile has acquired the electrons and must therefore carry an amount of negative charge equal to that amount of positive charge carried by the glass rod. Similarly the ebonite rod acquires electrons from the fur to become negatively charged.

As already mentioned in chapter 1, the SI unit of electric charge is the *coulomb* which is defined as the quantity of electricity carried in one second by an electric current of one ampere.

2.1.1. Electrostatic Induction or Electrification By Induction. If a positively charged conducting body A be brought near a perfectly insulated conducting body B, which has initially no charge, the free electrons of the conducting body will be caused to drift in a direction opposite to that of the field. As a result, there will be an excess of electrons at the end near the + vely charged conducting body, giving this end a negative charge, while the opposite end will lack electrons and its charge will be positive. The separation of charges on a conducting body due to the action of the charged body is known as *electrostatic induction* and the charges thus induced are known as *induced charges*. The negative and positive charges induced on conducting body B are of equal magnitude.

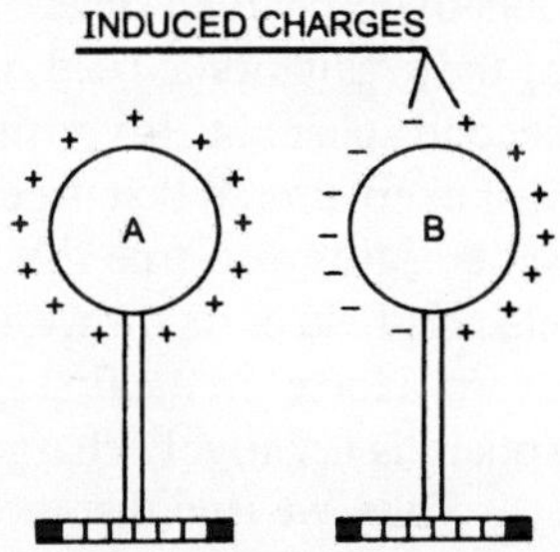

Fig. 2.1 *Electrostatic Induction*

The negative induced charge on conducting body B and positive inducing charge on conducting body A are said to be *bound charges*, because each of them remains held by each other and cannot go away. The positive charge on conducting body B is said to be free charge, as it can be conducted away from the body.

If the conducting body B having induced charge is connected to earth by a connector, as illustrated in Fig. 2.2, the positive charge of conducting body B will go to earth, leaving behind the bound negative charge on the conducting body B. The negative charge will also go to earth if conducting body A is removed.

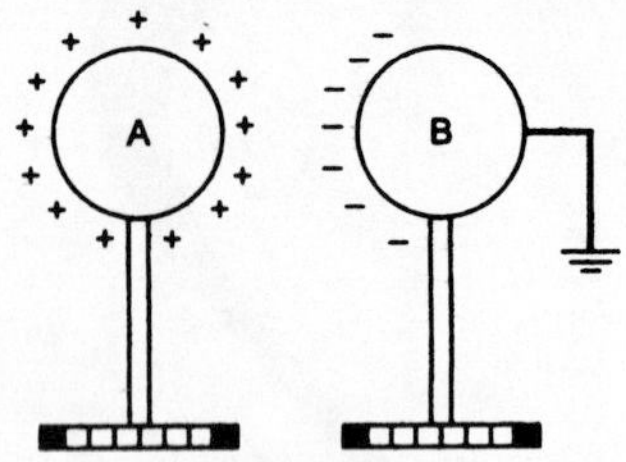

Fig. 2.2

2.2. ELECTROSTATIC FIELD

This is a well known fact that an object held in the hand falls to the ground when it is released. This is because of a *gravitational field* existing above the earth's surface, which exerts a force on the object causing it to move towards the earth. The force is exerted on the object while it is in our hand and we are required to exert an equal force opposite in the direction to the 'force of gravity' to prevent the object from moving downward.

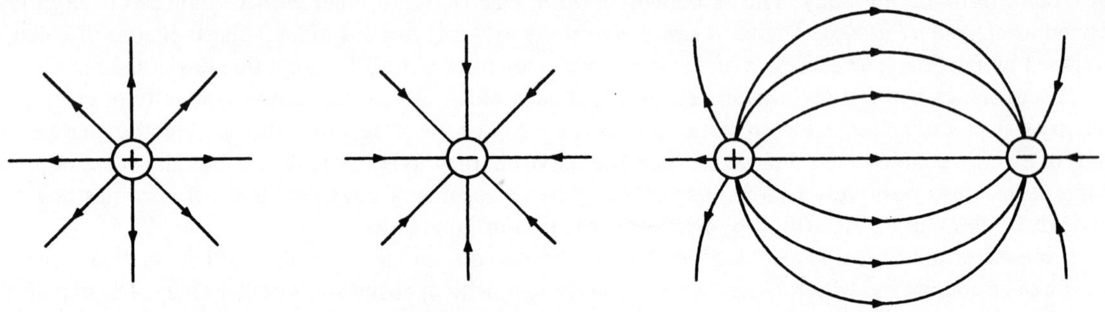

(a) Lines of Force Emanated From a Positive Charge *(b) Lines of Force Ending on a Negative Charge* *(c) Lines of Force When Two Unlike Charges Are Placed Near Each Other*

Now imagine a small sphere charged with positive electricity and suspended by an insulating thread. It produces an electric field in space all around it. If we bring a second sphere suspended by an insulating thread charged with negative electricity near it, the second sphere will be acted on by a force that urges it to move toward the positively charged sphere. We say the electric field caused by the positively charged sphere exerts a force on the negatively charged sphere in a manner similar to the action of the gravitational field on the object held in hand. If the second sphere is also positively charged, the electric field will exert a force that urges it to move away from the first sphere. It is also true that the second sphere produces an electric field of its own, and it is correct to say that its field exerts a force on the first sphere. This force will urge the first sphere to move towards second sphere if second sphere is negatively charged and away from the second sphere if second sphere is positively charged.

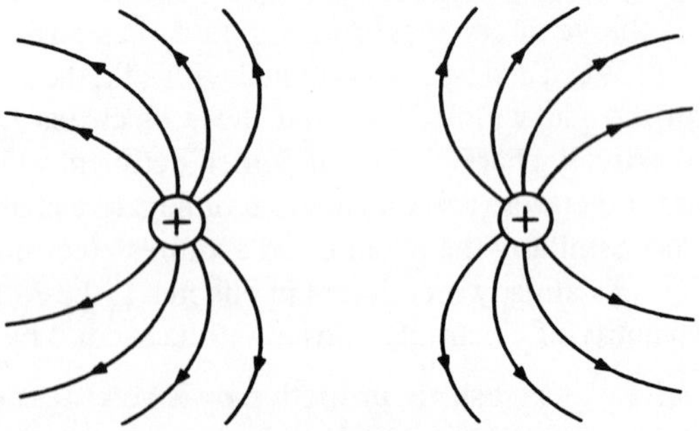

(d) Lines of Force When Two Like Charges Are Placed Near Each Other

Fig. 2.3

Thus, we find that the space around a charge is always under stress, and experiences a force on another charge when placed there. The *region or space in which the stress exists is called the electric (or dielectric or electrostatic) field.* The stress in the space is represented by the lines of force, which are emanated from positive charge and end on negative charge. The lines of force of same direction repel each other but lines of force of opposite directions attract each other. The direction of the lines of force at any point is the direction of movement of a unit positive charge placed at that point if made to move freely. They are always normal to the surface of the body from where they originate or terminate.

2.3. COULOMB'S LAW

We have seen that when two charged bodies are placed near each other, a mechanical force is experienced on them.

The law governing the force, that is experienced when two charged bodies are placed near each other, is called the Coulomb's law, which is quite similar to Newton's gravitational law.

According to this law *the force between the two charges (i) is directly proportional to the product of the charges (ii) inversely proportional to the square of the distance between them and (iii) depends on the nature of the medium surrounding the charges.*

Mathematically force F between two point charges of strength Q_1 and Q_2 placed distance d apart is given by

$$F \propto \frac{Q_1Q_2}{d^2}$$

Fig. 2.4

$$\text{or } F = K\frac{Q_1Q_2}{d^2}$$

where K is any constant. The value of this constant depends upon the surrounding medium and the system of units employed.

In SI system the force is measured in newtons, distance in metres, charge in coulombs and the force experienced between the two charges is given by an equation

$$F = \frac{Q_1Q_2}{4\pi\varepsilon_0\varepsilon_r d^2} \qquad ...(2.1)$$

where ε_0 is the permittivity of evacuated space or air and $= 8.854 \times 10^{-12}$ F/m and ε_r is the relative permittivity of the surrounding medium with respect to evacuated space or air.

The force is a vector *i.e.*, it has magnitude as well as direction. Coulomb also furnished the additional fact that the force acts along the line joining the charges and is repulsive (or outward) if the charges are of the same polarity (or nature) and attractive (or inward) if they are of opposite polarity (or nature).

Substituting $Q_1 = Q_2 = Q$ (say), $d = 1$ m, $\varepsilon_r = 1$ and $F = 9 \times 10^9$ newtons in Eq. (2.1), we get

$$9 \times 10^9 = \frac{Q^2}{4\pi \times 8.854 \times 10^{-12} \times 1 \times (1)^2}$$

$$\text{or } Q = \pm 1 \text{ coulomb}$$

Hence one coulomb of charge may be defined as that charge which when placed in air at distance of one metre from an equal and similar charge, is repelled with a force of 9×10^9 newtons.

Equation (2.1) may be written as

$$F_{medium} = \frac{Q_1Q_2}{4\pi\varepsilon_0\varepsilon_r d^2} \text{ newtons}$$

The force between the two charges in vacuum will be

$$F_{vacuum} = \frac{Q_1Q_2}{4\pi\varepsilon_0 d^2} \text{ newtons} \qquad \text{since } \varepsilon_r \text{ for vacuum (or air)} = 1$$

$$\text{So } \frac{F_{vacuum}}{F_{medium}} = \varepsilon_r \qquad ...(2.2)$$

i.e. relative permittivity of a medium is the ratio of forces experienced between two charges placed same distance apart in vacuum and medium.

Example 2.1

State and explain Coulomb's law.
Compute the force between two charges of 1 coulomb each placed 50 cm apart in vacuum.

Solution:

Charge $Q_1 = Q_2 = 1C$

Distance between charges, $d = 50 \text{ cm} = 0.5 \text{ m}$

Relative permittivity of medium (vacuum), $\varepsilon_r = 1$

$$\text{Force between two charges, } F = \frac{Q_1Q_2}{4\pi\varepsilon_0\,\varepsilon_r\,d^2} = \frac{1\times1}{4\pi\times8.854\times10^{-12}\times1\times(0.5)^2}$$

$= 36{,}000 \times 10^6$ newtons **Ans.**

Example 2.2

Find the force of interaction between two charges spaced 10 cm apart in a vacuum. The charges are 4×10^{-8} and 6×10^{-5} coulomb respectively. If the same charges are separated by the same distance in kerosine ($\varepsilon_r = 2$), what is the corresponding force of interaction ?

Solution:

Charge $Q_1 = 4 \times 10^{-8}$ C

Charge $Q_2 = 6 \times 10^{-5}$ C

Distance between charges, $d = 10 \text{ cm} = 0.1 \text{ m}$

Relative permittivity of vacuum, $\varepsilon_r = 1$

$$\text{Force between two charges, } F = \frac{Q_1Q_1}{4\pi\varepsilon_0\,\varepsilon_r\,d^2} = \frac{4\times10^{-8}\times6\times10^{-5}}{4\pi\times8.854\times10^{-12}\times1\times(0.1)^2} = 2.157 \text{ newtons } \textbf{Ans.}$$

When charges are placed in kerosine, the value of ε_r will be 2 and force between charges will be F/2 *i.e.* 2.157/2 = 1.0785 N **Ans.**

2.4. FIELD STRENGTH OR FIELD INTENSITY OR ELECTRIC INTENSITY

The electric field strength or intensity is a vector quantity, and as such it is defined at any point in a field by magnitude and direction.

The force experienced on a unit positive charge placed at a distance of *d* metres from a charge of Q coulombs is obtained by substituting $Q_1 = Q$ and $Q_2 = 1$ in Eq. (2.1). Thus,

$$F = \frac{Q}{4\pi\varepsilon_0\,\varepsilon_r\,d^2}$$

This describes a vector field and is called the *electric field strength or electric intensity*. Thus, *electric field strength (or electric intensity)* is defined as the vector force on a unit charge. This is represented by **E** and the magnitude of electric intensity is given as

$$E = \frac{Q}{4\pi\varepsilon_0\,\varepsilon_r d^2} = \frac{9\times10^9\,Q}{\varepsilon_r\,d^2} \text{ newtons/coulomb} \quad \text{...(2.3)}$$

$$\because \quad \frac{1}{4\pi\varepsilon_0} = \frac{1}{4\pi\times8.854\times10^{-12}} = 9\times10^9$$

A plot of vector **E** about a single concentrated charge Q is shown in Fig. 2.5. Figure 2.5 (*a*) shows the plot for positive charge and Fig. 2.5 (*b*) for a negative. The respective lengths of vectors represent the magnitude to a scale.

If a field is due to several charges (Q_1, Q_2, Q_3, . . .), the electric intensity will be the vector sum of electric intensities due to each charge taken separately *i.e.*

$$\mathbf{E} = \mathbf{E_1} + \mathbf{E_2} + \mathbf{E_3} + \ldots \quad \text{...(2.4)}$$

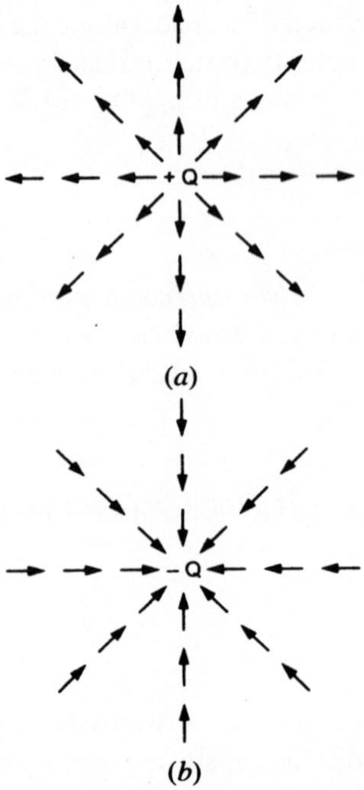

Fig. 2.5

2.5. ELECTRIC FLUX

Actually lines of force emanated from a charge are infinite but it was suggested by Faraday that electric field should be assumed to be composed of very small closed areas. From each point of this small area a line of force, known as *tube* of *force,* is emerged and total number of such tubes of force radiated out by a charge of Q coulombs is Q and is known as *electric flux.*

$$\text{Electric flux } (\psi) = Q \text{ coulombs}$$

$$\therefore \text{ Flux emanated } (\psi) = \text{Charge Q and is measured in coulombs.}$$

2.6. ELECTRIC FIELD DUE TO A POINT CHARGE

Let there be a point charge of Q coulombs placed at the centre of an imaginary sphere of radius *r* metres. The electric intensity at any point on the surface is given by

$$E = \frac{Q}{4\pi\varepsilon_0\varepsilon_r r^2}$$

The direction of this electric field intensity is normal to the surface of the sphere at that point.

Though the number of lines of force crossing the sphere at that point are infinite but practically the tubes of force crossing per unit cross-section area is $\varepsilon_0\ \varepsilon_r$ times the field intensity at that point or tubes of force per unit area

$$= \varepsilon_0\,\varepsilon_r E = \varepsilon_0\,\varepsilon_r \frac{Q}{4\pi\varepsilon_0\,\varepsilon_r r^2} = \frac{Q}{4\pi r^2}$$

Surface area of sphere $= 4\pi r^2$

Therefore, total number of tubes of force emanated from charge Q,

$$\psi = \frac{Q}{4\pi r^2} \times 4\pi r^2 = Q \text{ coulombs} \qquad ...(2.5)$$

Hence the flux emanating from any charge is equal to its strength in coulombs.

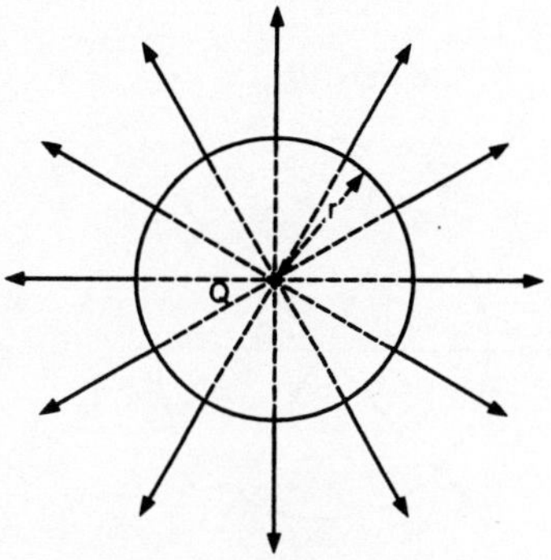

Fig. 2.6

2.7. ELECTRIC FLUX DENSITY

Electric flux density is defined as the number of tubes of force per unit area emanated normally from the surface. Electric flux density is represented by D and is measured in coulombs/m^2.

Let there be a point charge of Q coulombs placed at the centre of an imaginary sphere of radius *r* metres. The number of lines of force per unit area of the surface of the sphere,

$$D = \frac{Q}{4\pi r^2}$$

Thus, electric flux density D is measured in coulombs per square metre. The electric flux density D varies directly with the charge Q and inversely with the square of distance from the charge. It is also independent of the relative permittivity ε_r of the dielectric medium. On the other hand, the electric intensity E depends on the relative permittivity ε_r.

If the surrounding medium has a relative permittivity ε_r then the electric field intensity normal to the surface of the sphere is $\frac{Q}{4\pi\varepsilon_0\varepsilon_r r^2}$ where ε_0 is the permittivity of evacuated space or air.

$$\text{Hence } D = \varepsilon_0\,\varepsilon_r\,E \qquad ...(2.6)$$

$$\text{or } E = \frac{D}{\varepsilon_0\varepsilon_r} \text{ newtons/coulomb}$$

2.8. GAUSS'S THEOREM

If a point charge of +Q coulombs is considered to be placed at the centre of a sphere of radius r metres, as shown in Fig. 2.7 (*a*) the electric flux emanated from the charge Q is Q coulombs and is normal to the surface of the sphere.

Now, if a point charge of + Q coulombs is considered placed at any point (other than centre 0) as shown in Fig. 2.7 (*b*) the electric flux or lines of force emanated from this charge is Q but not normal to the surface. This flux can, however, be resolved along the normal to the surface known as cos θ components and another along the perpendicular to the normal to the surface known as sin θ components. If all the sin θ components of the flux all over the surface are added, their resultant is zero while the sum of all cos θ components of lines of force all over the surface is equal to Q.

Thus the total flux traversing a surface completely surrounding the charge Q coulombs is Q. This is known as Gauss's theorem.

If there are number of charges $+ Q_1, + Q_2, - Q_3, +Q_4, - Q_5$ etc. as shown in Fig. 2.7 (*c*), then according to Gauss's theorem the total flux crossing the surface is given as

$$\psi = + Q_1 + Q_2 - Q_3 + Q_4 - Q_5$$

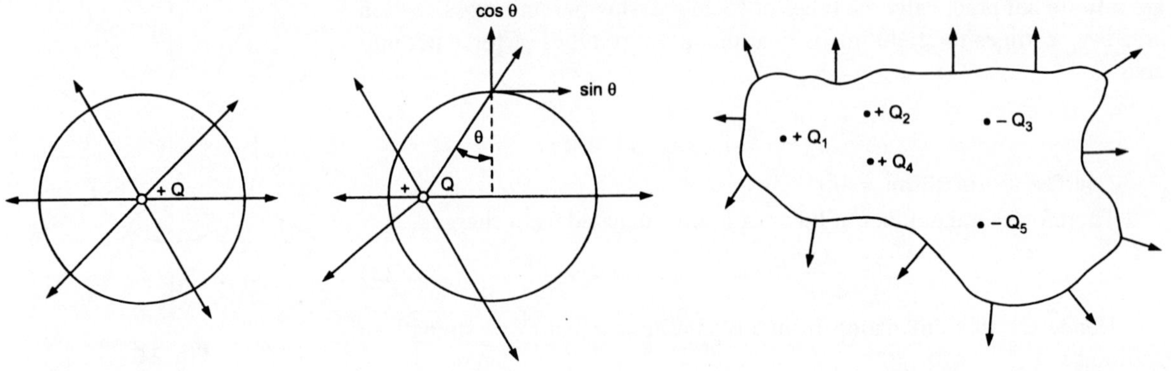

(*a*) (*b*) (*c*)

Fig. 2.7

2.9. COULOMB'S THEOREM

Consider a conducting surface of very small surface area δs having charge density of σ coulombs per unit of surface area.

Total charge on small surface $\delta s = \sigma\, \delta s$ coulombs

According to Gauss's theorem the total flux, ψ radiating from the charge of $\sigma\, \delta s$ coulombs = $\sigma\, \delta s$ coulombs. Since in conducting surface no flux can exist inside and whole of the flux comes outward normally,

$$\text{Flux density at the surface} = \frac{\psi}{S} = \frac{\sigma \delta s}{\delta s} = \sigma$$

$$\text{Field intensity at any point close to the surface, E} = \frac{D}{\varepsilon_0 \varepsilon_r} = \frac{\sigma}{\varepsilon_0 \varepsilon_r} \text{ N/C}$$

Hence according to Coulomb's theorem, the electric intensity at a surface of a conducting body, having a charge density of σ coulombs per unit of surface area is $\frac{\sigma}{\varepsilon_0 \varepsilon_r}$ where $\varepsilon_0\, \varepsilon_r = \varepsilon$ the absolute permittivity of the medium surrounding the conductor.

2.10. POTENTIAL

When the mass is moved above the ground level, work is done against the force of gravity and is equal to the product of force of gravity and vertical distance moved. This work done is added in the potential energy of the mass. Similarly if the mass is moved downward, work is done by the force of gravity and the potential energy of the mass is reduced by this amount of work done. Hence the potential energy or potential of the mass depends upon the position of the mass with respect to earth.

As already discussed, an electric charge causes an electric field around it which theoretically extends up to infinity. If any charge is introduced in this electric field, it will be attracted or repelled according to the nature of the charge and when it will be moved, the work will be done against or by the force acting on the charge due to electric field. Hence potential in an electric field is exactly the same as potential in the gravitational field.

In general potential at any point in an electric field is defined as the work done in moving a unit positive charge from infinity to that point.

2.11. POTENTIAL DIFFERENCE

The potential difference between two points in an electric field is defined as the work done in moving a unit positive charge from the point of lower potential to that of higher potential.

If work done in bringing a positive charge of one coulomb from one point to another point in an electric field is one joule then the potential difference between the two points is said to be unity (volt).

2.12. POTENTIAL GRADIENT

Practically the electric field intensity is not uniform, but varies from point to point. Let us consider that the electric field intensity between two points separated by a very small distance *ds* is constant and let it be **E**. The work done in moving unit +ve charge from one point to the other up the gradient of potential is **E ds** and, therefore, potential difference dV is given as

$$dV = -\mathbf{E}\,ds \quad \text{(minus sign is used because of work is being done against the force due to electric field)}$$

$$\text{or } \mathbf{E} = -\frac{dV}{ds} = -g, \text{ potential gradient}$$

Hence electric field intensity at any point is equal to the negative potential gradient at that point.

2.13. POTENTIAL AT A POINT

Let us consider a positive charge of one coulomb at a distance of s metres from the charge of Q coulombs placed in air. The force acting on this unit + ve charge is given by

$$E = \frac{Q}{4\pi\varepsilon_0 s^2}$$

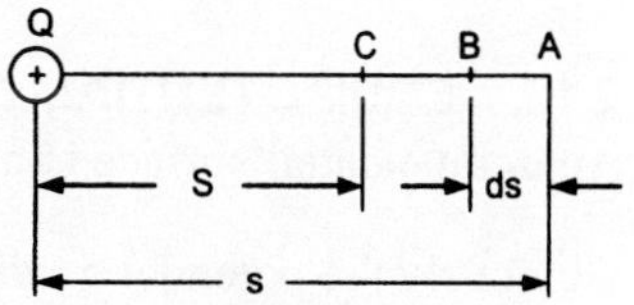

Fig. 2.8

Work done in moving this positive charge of one coulomb towards the charge of Q coulombs through a distance *ds*

$$dw = E(-ds) = \frac{Q}{4\pi\varepsilon_0 s^2} \times (-ds)$$

The negative sign is taken because ds is considered along the negative direction of *s*.

Total work done in bringing the + ve charge of one coulomb from infinity to any point, S metres from the charge of Q coulombs is given by

$$W = \int_{\infty}^{S} -E\,ds = -\int_{\infty}^{S} \frac{Q}{4\pi\varepsilon_0 s^2}\,ds = \frac{Q}{4\pi\varepsilon_0 S}$$

By definition, potential at any point,

V = The work done in bringing a + ve charge of one coulomb from infinity to that point

$$= \frac{Q}{4\pi\varepsilon_0 S} = 9 \times 10^9 \frac{Q}{S} \text{ volts in air}$$

$$\text{and } V = \frac{Q}{4\pi\varepsilon_0 \varepsilon_r S} = 9 \times 10^9 \frac{Q}{\varepsilon_r S} \text{ volts in medium of relative permittivity } \varepsilon_r.$$

Potential at a Point due to Number of Charges. As already proved, the potential at any point at a distance of S metres from the charge of Q coulombs is given by

$$V = 9 \times 10^9 \frac{Q}{\varepsilon_r S}$$

Similarly the potential at any point due to number of charges Q_1, Q_2, Q_3, Q_4 etc. placed at distances S_1, S_2, S_3, S_4 etc. respectively from it is given as

$$V = \frac{9 \times 10^9}{\varepsilon_r}\left[\frac{Q_1}{S_1} + \frac{Q_2}{S_2} + \frac{Q_3}{S_3} + \frac{Q_4}{S_4} + \ldots\right]$$

2.14. POTENTIAL AT A CHARGED SPHERE

To determine the potential at any point outside the sphere, the charge should be assumed to be concentrated at the centre of it.

Therefore, the potential at a point, S metres from the centre of the sphere having the charge Q coulombs

$$= \frac{Q}{4\pi \varepsilon_0 S} = 9 \times 10^9 \frac{Q}{S} \text{ volts}$$

To determine the potential at the surface of sphere of radius *r* metres substitute $S = r$ in above equation for potential,

∴ Potential at the surface of a sphere of radius *r* metres and consisting of charge Q,

$$V = \frac{Q}{4\pi \varepsilon_0 r} = 9 \times 10^9 \frac{Q}{r} \text{ volts}$$

2.15. POTENTIAL AND ELECTRIC FIELD INTENSITY INSIDE THE SPHERE

Experimentally it has been found that when a charge is given to the sphere or to any body, it resides only on its outer surface or *within a conducting body charge, flux and field intensity is zero and all the points within the surface are at the same potential as the point on the surface.*

∴ Potential at any point within the surface of the sphere

$$= \text{Potential at the surface of the sphere} = \frac{9 \times 10^9 Q}{r} \text{ volts.}$$

2.16. EQUIPOTENTIAL SURFACE

An equipotential surface is a surface in an electric field such that all points on it are at the same potential.

Obviously the potential gradient $\frac{dv}{ds}$, therefore, electric intensity E along equipotential surface being equal to $\frac{dv}{ds}$, is zero. *Hence the surface in an electric field at right angle to the lines of force or flux everywhere is equipotential surface.*

For example, in the case of a spherical charged body, any number of points situated at the same distance from its centre will have the same potential. Thus, different spherical surfaces around a charged sphere will be equipotential surfaces.

2.17. CAPACITANCE

Capacitance is the property of a capacitor to store electric charge required to raise its potential through unity.

When a charge of Q coulombs is given to a conductor and if its potential rises from 0 to V volts then the capacitance of the conductor will be given by $C = \frac{Q}{V}$.

In SI system the capacitance is measured in farads.

If one coulomb of charge is required to raise the potential of the conductor through one volt, then the conductor is said to be of one farad capacity.

By definition one farad is equal to one coulomb per volt.

Farad, the unit of capacitance is very large, so microfarad or micro-microfarad is usually used.

$$1\ \mu F = 10^{-6}\ F \text{ and } 1\ \mu\mu F = 10^{-12}\ F$$

2.18. CAPACITORS

Two metal plates, separated by an insulating material (or dielectric), constitute a capacitor or condenser (the former is the preferred term), namely an arrangement which has the capacity to store electricity as an excess of electrons on one plate and deficiency on the other. The time of storage of electricity is very short in many cases—even as short as one nanosecond (10^{-9} s) or less, while in other cases it may be in milliseconds, seconds or minutes. Some capacitors employed in power circuits have their amounts of charge varying slightly, but they remain in a charged condition many hours at a time.

There are various types of capacitors, such as spherical, cylindrical, parallel-plate capacitors but in all cases capacitor consists essentially of two insulated conductors, separated by an insulator or dielectric and its working depends on the action of induction.

The capacitance of a capacitor depends solely on the configuration, size and spacing of the plates and the type of dielectric used (permittivity of the dielectric used).

2.19. RELATIVE PERMITTIVITY OR SPECIFIC INDUCTIVE CAPACITY OR DIELECTRIC CONSTANT

Relative permittivity, also known as *specific inductive capacity or dielectric constant,* is that property of a dielectric which governs the capacity (or capacitance) of a system (say, a capacitor) using this dielectric. The capacitance of a capacitor depends on the surface area of plates (larger the area, greater is the capacitance), the separation between the plates or the thickness of the dielectric (thicker the dielectric, smaller is the capacitance) and also on the nature of the material of the dielectric. It is the material of the dielectric that is described by its *relative permittivity.*

From experiments it is found that if the air is replaced by a dielectric of relative permittivity ε_r in the capacitor, the value of the capacitance is greatly increased.

Relative permittivity is the ratio of the capacitance of a capacitor having a certain dielectric to the capacitance of the same capacitor with vacuum (or air) as medium and is represented by the symbol ε_r. Materials having high values of relative permittivity are preferred for capacitors where economy of space is important.

Values of relative permittivities and dielectric strengths of some of the most important insulating materials (dielectrics) are given below in Table 2.1

TABLE 2.1

Dielectric	*Relative Permittivity*	*Dielectric Strength in kV/mm*
Air	1	3
Bakelite	4.5 – 5.5	10 – 25
Ebonite	2.8	30 – 100
Fibre	2.5	4
Glass	5 – 8	10 – 50
Mica	5 – 9	60 – 230
Oil	2 – 5	5 – 20
Paper	2 – 4	30 – 40
Paraffin	3 – 5	32
Porcelain	6 – 7.5	1.5 – 10
Rubber	2 – 3.5	10 – 35
Slate	6 – 7.5	0.2 – 2.5

2.20. DIELECTRIC STRENGTH AND BREAKDOWN POTENTIAL

When very high electric field is applied across a dielectric material, a considerable number of electrons may get excited to energies within the conduction band. These electrons are accelerated rapidly by the high electric field in the dielectric and attain high kinetic energies. Some of the kinetic energy is transferred by collisions to the valence electrons which are thereby excited to the conduction band. If a large number of electrons initiate this process, it multiplies itself and an avalanche of electrons is loosened in the conduction band. The current through the dielectric increases rapidly and causes localized melting, burning or vapourization of material leading to irreversible degradation and perhaps even permanent failure of the dielectric material. This results in high electrical conductivity and the complete loss of the charge storing capacity of the dielectric. This phenomenon is known as *electrical breakdown of dielectric*. For the initiation of the process conduction of electrons is required and which may originate in a number of ways *i.e.* impurity atoms can donate electrons to the conduction band, interconnecting pores in the dielectric provide direct breakdown as a result of electrical gas discharge.

Electric strength (or dielectric strength) is the property of a material which enables it to withstand electric stress without damage. It is usually expressed in terms of the minimum *electric stress* (potential difference per unit distance) that will cause failure or breakdown of the dielectric under certain conditions of service. The dielectric strength is given as $\frac{V}{t}$ kV/m (or kV/mm) where V is the breakdown potential and t is the thickness of the dielectric (or insulating material). The value of the dielectric strength of a dielectric decreases with the increase in thickness. Dielectric strength is usually adversely affected by moisture, contamination, elevated temperature, ageing and mechanical stress. Dielectric strengths of some of the most important dielectrics are given in Table 2.1.

Example 2.3

It is found experimentally that an electric field of 3,000 volts per mm in air causes electrical breakdown of air. Determine the greatest charge that can be placed on the sphere of one metre diameter. What is the potential of the sphere for that charge ?

Solution: Let the greatest charge, that can be placed on the sphere, be of Q coulombs

$$\text{Breakdown voltage, } g = 3{,}000 \text{ V/mm} = 3 \times 10^6 \text{ V/m}$$

$$\text{Since electric intensity, } E = \text{Potential gradient, } g$$

$$\frac{Q}{4\pi\varepsilon_0\varepsilon_r r^2} = g$$

$$\text{or } Q = 4\pi\,\varepsilon_0\,\varepsilon_r r^2 \times g = 4\pi \times 8.854 \times 10^{-12} \times 1 \times (0.5)^2 \times 3 \times 10^6 = 83.45\ \mu\text{C} \textbf{ Ans.}$$

Potential of the sphere for the charge of 83.45×10^{-6} C

$$= \frac{83.45 \times 10^{-6}}{4\pi \times 8.854 \times 10^{-12} \times 1 \times 0.5} = 1{,}500 \text{ kV} \textbf{ Ans.}$$

2.21. CAPACITANCE OF AN ISOLATED SPHERICAL CONDUCTOR

Consider a perfectly insulated spherical conductor of radius r metres placed in a medium of relative permittivity ε_r and at an infinite distance from all conductors.

If a charge of Q coulombs of electricity is given to this conductor, the potential of surface will be given by

$$V = \frac{Q}{4\pi\varepsilon_0\varepsilon_r r}$$

$$\text{Capacitance, } C = \frac{Q}{V} = \frac{Q}{\frac{Q}{4\pi\varepsilon_0\varepsilon_r r}} = 4\pi r\varepsilon_0\varepsilon_r = \frac{\varepsilon_r r}{9 \times 10^9} \text{ Farads}$$

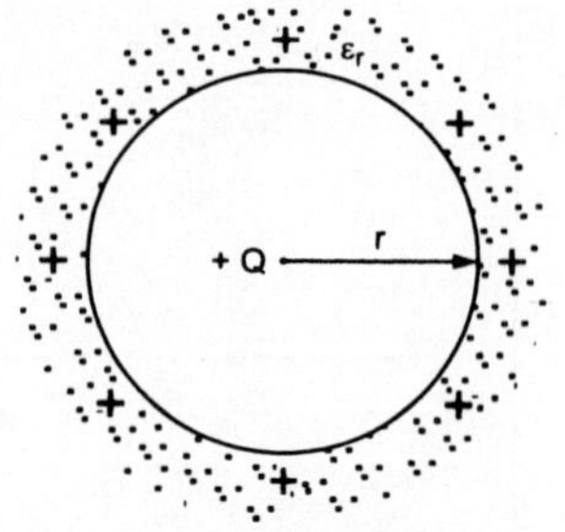

Fig. 2.9

When the conductor is placed in air, $\varepsilon_r = 1$, the capacitance of the spherical conductor is given as

$$C = \frac{r}{9 \times 10^9} \text{ Farads} \quad \text{...(2.7)}$$

2.22. CAPACITANCE OF CONCENTRIC SPHERICAL CAPACITOR

(a) When the outer sphere is earthed. Consider a spherical capacitor consisting of two concentric spheres A and B of radii a and b metres respectively separated by a dielectric of relative permittivity ε_r outer sphere being earthed.

If a + ve charge of Q coulombs is given to the inner sphere, a charge of – Q coulombs will be induced on the inner surface of the outer sphere B and a charge of + Q coulombs induced on the outer surface of sphere B will go to earth.

The electric intensity at any point P in the dielectric, r metres from the centre, due to charge on the inner sphere is $\frac{Q}{4\pi\varepsilon_0\varepsilon_r r^2}$ newtons. The electric intensity at point P due to charge on the outer sphere is zero since point P is within the sphere.

∴ PD between spheres A and B = Work done in carrying a + ve charge of one coulomb from sphere A to B

$$= \int_a^b \frac{Q\,dr}{4\pi\,\varepsilon_0\,\varepsilon_r r^2} = \frac{-Q}{4\pi\,\varepsilon_0\varepsilon_r}\left[\frac{1}{r}\right]_a^b = \frac{Q}{4\pi\,\varepsilon_0\varepsilon_r}\left(\frac{1}{a} - \frac{1}{b}\right) = \frac{Q\,(b-a)}{4\pi\,\varepsilon_0\varepsilon_r ab}$$

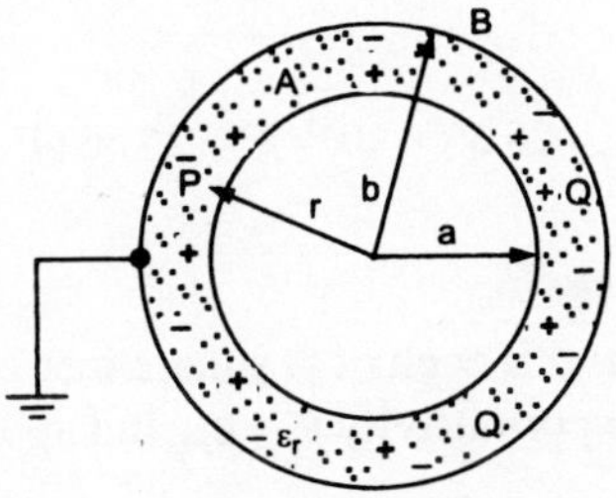

Fig. 2.10 *Concentric Spherical Capacitor*

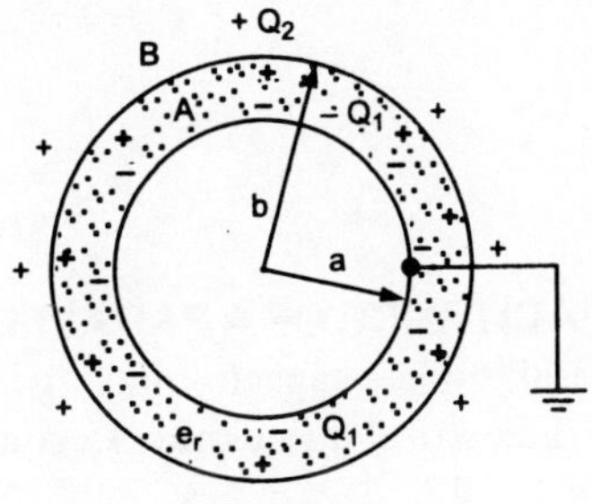

Fig. 2.11

$$\text{Capacitance of the condenser, } C = \frac{Q}{V} = \frac{Q}{\dfrac{Q\,(b-a)}{4\pi\,\varepsilon_0\varepsilon_r ab}} = 4\pi\,\varepsilon_0\varepsilon_r \frac{ab}{b-a} \text{ farads}$$

Capacitance of spherical condenser with air as medium,

$$C = 4\pi\,\varepsilon_0 \frac{ab}{b-a} = \frac{1}{9 \times 10^9} \times \frac{ab}{b-a} \text{ farads} \quad \text{...(2.8)}$$

(b) When the inner sphere is earthed. Consider a spherical capacitor consisting of two concentric spheres of radii a and b metres separated by a dielectric of relative permittivity ε_r, inner sphere being earthed.

If a positive charge of Q coulombs is given to the outer sphere B, it will distribute itself over both its inner and outer surfaces.

Let the charges of Q_1 and Q_2 coulombs be at the inner and outer surfaces respectively of sphere B where $Q = Q_1 + Q_2$.

The charge + Q_1 on the inner surface of outer sphere B will induce a charge of – Q_1 coulombs on the outer surface of sphere A, and + Q_1 coulombs on the inner surface of sphere A, which will go to earth.

Now there are two capacitors connected in parallel.

(*i*) One capacitor consists of outer surface of sphere B and earth having capacitance $C_1 = 4\pi\varepsilon_o\varepsilon_r b$ farads and

(*ii*) Second capacitor consisting of inner surface of outer sphere B and the outer surface of inner sphere A having capacitance

$$C_2 = 4\pi\,\varepsilon_o\varepsilon_r\,\frac{ab}{b-a}\ \text{farads}$$

$$\text{Total capacitance, } C = C_1 + C_2 = 4\pi\,\varepsilon_o\varepsilon_r b + 4\pi\varepsilon_o\varepsilon_r\frac{ab}{b-a} = 4\pi\,\varepsilon_o\varepsilon_r\,\frac{b^2}{b-a}\ \text{farads} \quad ...(2.9)$$

Example 2.4

There are two concentric conducting spherical surfaces of radii *a* and *b* (*a* < *b*). The inner spherical surface carries a charge Q and the outer surface is grounded. What is the potential of the inner spherical surface ?

Solution: Potential of the inner spherical surface = Potential between inner and outer spherical surfaces

$$= \frac{Q}{4\pi\varepsilon_0\,\varepsilon_r}\left(\frac{1}{a}-\frac{1}{b}\right)$$

Example 2.5

Calculate the value of capacitance of a spherical capacitor consisting of two concentric spheres of radii 60 mm and 80 mm with air as dielectric medium between the two. Derive the formula used.

Solution: Capacitance of spherical condenser with air as medium

$$C = 4\pi\varepsilon_0\frac{ab}{b-a} = \frac{1}{9\times10^9}\times\frac{ab}{b-a}\ \text{Farads} \quad \text{... Refer to Eq. (2.8)}$$

$$= \frac{1}{9\times10^9}\times\frac{60\times10^{-3}\times80\times10^{-3}}{80\times10^{-3}-60\times10^{-3}} = 26.66\times10^{-12}\ \text{F} = 26.66\ \text{pF} \quad \textbf{Ans.}$$

2.23. CAPACITANCE OF A PARALLEL-PLATE CAPACITOR

Consider a parallel-plate capacitor having two conducting plates A and B each of A square metres in area and separated by dielectrics of thickness t_1, t_2 and t_3 metres and relative permittivity ε_{r_1}, ε_{r_2} and ε_{r_3} respectively, as shown in Fig. 2.12.

If a charge of + Q coulombs is given to plate A then a charge of – Q coulombs will be induced on inner surface of plate B,

$$\text{Surface charge density on plate A, } \sigma = \frac{Q}{A}\ \text{coulombs/m}^2$$

According to Coulomb's theorem, electric field strength at a point close to the surface.

$$E = \frac{\sigma}{\varepsilon_0\varepsilon_r} = \frac{1}{\varepsilon_0\varepsilon_r}\cdot\frac{Q}{A}\ \text{V/m}$$

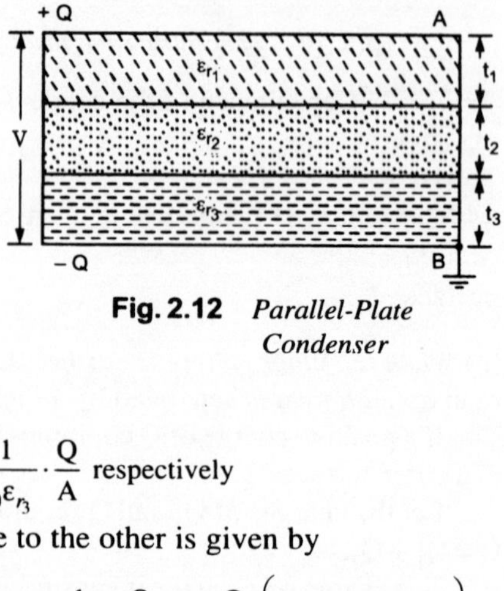

Fig. 2.12 *Parallel-Plate Condenser*

Therefore, electric field intensity in dielectrics of relative permittivity ε_{r_1}, ε_{r_2} and ε_{r_3} are

$$E_1 = \frac{1}{\varepsilon_0\varepsilon_{r_1}}\cdot\frac{Q}{A},\ E_2 = \frac{1}{\varepsilon_0\,\varepsilon_{r_2}}\cdot\frac{Q}{A}\ \text{and}\ E_3 = \frac{1}{\varepsilon_0\varepsilon_{r_3}}\cdot\frac{Q}{A}\ \text{respectively}$$

Total work done in moving a unit positive charge from one plate to the other is given by

$$W = E_1t_1 + E_2t_2 + E_3t_3 = \frac{1}{\varepsilon_0\varepsilon_{r_1}}\cdot\frac{Q}{A}t_1 + \frac{1}{\varepsilon_0\varepsilon_{r_2}}\cdot\frac{Q}{A}t_2 + \frac{1}{\varepsilon_0\varepsilon_{r_3}}\cdot\frac{Q}{A}t_3 = \frac{Q}{\varepsilon_0 A}\left(\frac{t_1}{\varepsilon_{r_1}}+\frac{t_2}{\varepsilon_{r_2}}+\frac{t_3}{\varepsilon_{r_3}}\right)$$

By definition pd between the plates A and B

= Work done in moving a unit positive charge from one plate to another

$$\text{or } V = \frac{Q}{\varepsilon_0 A}\left[\frac{t_1}{\varepsilon_{r1}} + \frac{t_2}{\varepsilon_{r2}} + \frac{t_3}{\varepsilon_{r3}}\right]$$

$$\text{or } \frac{Q}{V} = \frac{\varepsilon_0 A}{\left[\frac{t_1}{\varepsilon_{r1}} + \frac{t_2}{\varepsilon_{r2}} + \frac{t_3}{\varepsilon_{r3}}\right]}$$

and since capacitance of a condenser, $C = \frac{Q}{V}$

$$\therefore \text{ Capacitance of parallel-plate condenser, } C = \frac{Q}{V} = \frac{\varepsilon_0 A}{\left[\frac{t_1}{\varepsilon_{r1}} + \frac{t_2}{\varepsilon_{r2}} + \frac{t_3}{\varepsilon_{r3}}\right]} \text{ farads} \quad ...(2.10)$$

Case (a) When the plates are d metres apart by a dielectric of relative permittivity ε_r.

Such a capacitor is shown in Fig. 2.13.

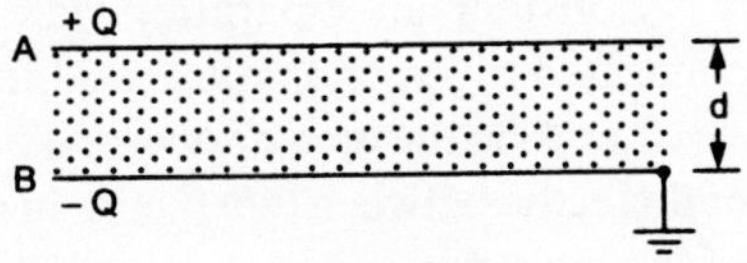

Fig. 2.13

Substituting $\varepsilon_{r1} = \varepsilon_{r2} = \varepsilon_{r3} = \varepsilon_r$

and $t_1 + t_2 + t_3 = d$ in Eq. (2.10) we get

$$\text{Capacitance, } C = \frac{\varepsilon_0 A}{\frac{t_1}{\varepsilon_r} + \frac{t_2}{\varepsilon_r} + \frac{t_3}{\varepsilon_r}} = \frac{\varepsilon_0 \varepsilon_r A}{t_1 + t_2 + t_3} = \frac{\varepsilon_0 \varepsilon_r A}{d} \text{ farads} \quad ...(2.11)$$

Case (b) When the plates are d metres apart and medium is air.

Substituting $\varepsilon_r = 1$ in Eq. (2.11) we get

$$C = \frac{\varepsilon_0 A}{d} \text{ farads} \quad ...(2.12)$$

Case (c) When a plate of some substance having thickness t metres and of relative permittivity ε_r *is introduced between plates of an air parallel-plate-capacitor.*

Such a capacitor is shown in Fig. 2.14. A and B are two conducting plates each of area A square metres placed parallel to one another and distance *d* metres apart, a slab of thickness *t* and relative permittivity ε_r is being introduced between these plates.

If a charge of + Q coulombs is given to plate A then a charge of – Q coulombs will be induced on the inner surface of plate B.

Surface charge density on plate A,

$$\sigma = \frac{Q}{A} \text{ coulombs/m}^2$$

According to Coulomb's theorem, electric field strength at a point close to the surface,

$$E = \frac{\sigma}{\varepsilon_0 \varepsilon_r} = \frac{1}{\varepsilon_0 \varepsilon_r}\frac{Q}{A} \text{ V/m}$$

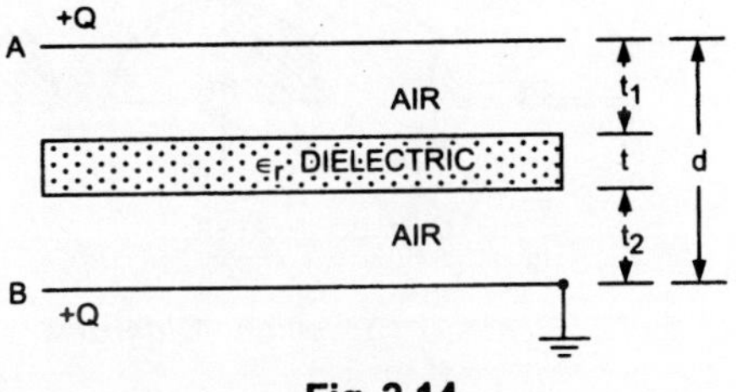

Fig. 2.14

Therefore, electric field intensities in air and slab are $\frac{1}{\varepsilon_0}\frac{Q}{A}$ and $\frac{1}{\varepsilon_0 \varepsilon_r}\frac{Q}{A}$ respectively.

PD between plates A and B, V = Total work done in moving a unit + ve charge from one plate to the other.

$$= \frac{1}{\varepsilon_0}\frac{Q}{A} \times t_1 + \frac{1}{\varepsilon_0 \varepsilon_r}\frac{Q}{A} t + \frac{1}{\varepsilon_0}\frac{Q}{A} t_2 = \frac{Q}{\varepsilon_0 A}\left(t_1 + t_2 + \frac{t}{\varepsilon_r}\right)$$

Since total distance $= t_1 + t_2 + t_3 = d$

$\therefore \quad t_1 + t_2 = d - t$

$\therefore$ PD between plates A and B, $V = \frac{Q}{\varepsilon_0 A}\left(d - t + \frac{t}{\varepsilon_r}\right)$

By definition capacitance, $C = \frac{Q}{V} = \frac{Q}{\frac{Q}{\varepsilon_0 A}\left(d - t + \frac{t}{\varepsilon_r}\right)} = \frac{\varepsilon_0 A}{d - t\left(1 - \frac{1}{\varepsilon_r}\right)}$ ...(2.13)

2.23.1. Multiplate Capacitor. The capacitance of a capacitor is usually increased by connecting foils into two groups fitting into each other and separated by a dielectric, as shown diagrammatically in Fig. 2.15. Such an arrangement is called the *multiplate capacitor.*

Consider an n-parallel plate capacitor, alternate plates being connected together as illustrated in Fig. 2.15.

Let A be the area of one side of each plate in m^2, t be the thickness of dielectric between two plates, in metres and ε_r be relative permittivity of the dielectric.

From Fig. 2.15 showing 9-plate capacitor, it is obvious that 5 plates are connected to A and 4 plates to B; each side of the four plates connected to B is in contact with the dielectric, whereas only one side of each of the outer plates is in contact with it. As a consequence, the useful surface area of each set of plates is 8 A square metres. For n plates, obviously the useful area of each set will be $(n - 1)$ A square metres.

$\therefore$ Capacitance, $C = \frac{\varepsilon_0 \varepsilon_r (n-1) A}{t}$ farads ...(2.14)

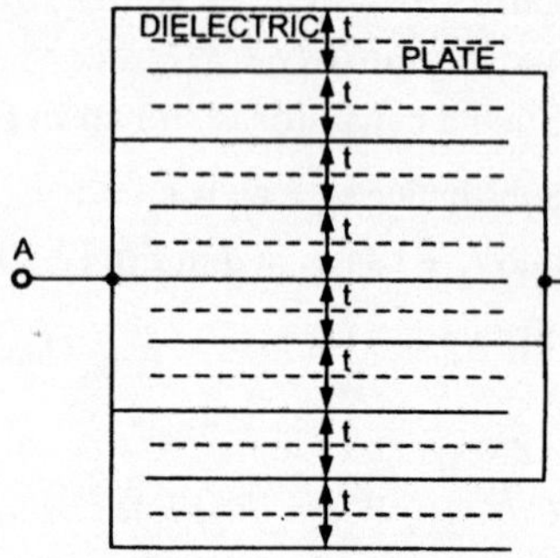

Fig. 2.15 *n-parallel-plate Capacitor*

2.23.2. Variable Capacitor. An elementary variable capacitor consists of two sets of copper or aluminium plates (which may have the shape of half discs). Each set is mounted on a common shaft, one set being fixed and the other, which interleaves with the former, being movable. The capacitance of the capacitor can be easily varied by varying the degree of interleaving, which is possible by rotation of movable plate shaft. Such variable-capacitance air capacitors, shown in Fig. 2.16, are widely used in radio work.

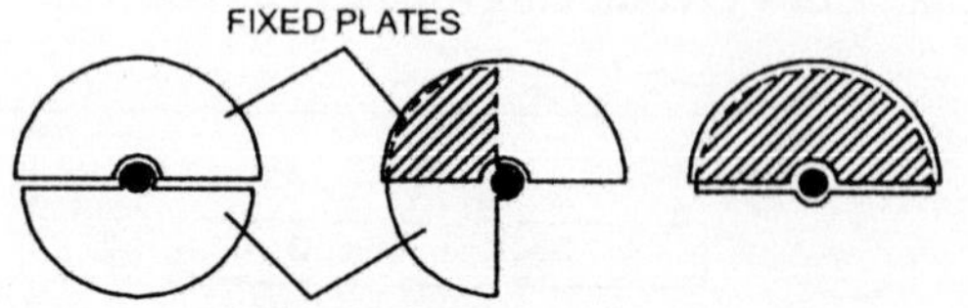

(a) Schematic Diagrams

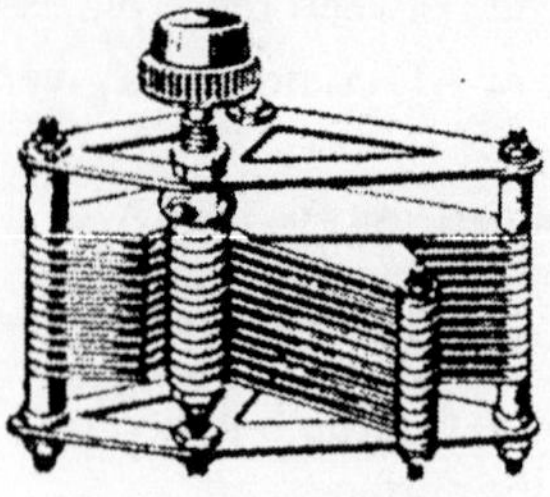

(b) General View

Fig. 2.16 *Variable Capacitor*

Example 2.6

Determine the capacitance of a parallel-plate capacitor composed of tin foil sheet 25 centimetre square separated by a glass dielectric 0.5 cm thick with relative permittivity 6.

Solution: Surface area of plates, $A = 0.25 \times 0.25 = 0.0625$ m^2

Dielectric thickness, t = 0.5 cm = 0.005 m

$$\text{Capacitance of the capacitor, } C = \frac{\varepsilon_0\varepsilon_r A}{t} = \frac{8.854\times10^{-12}\times6\times0.0625}{0.005} = 664.05\ \mu\text{F } \textbf{Ans.}$$

Example 2.7

An 1 μF parallel-plate capacitor with air as the dielectric is charged to a potential of 200 V. The distance between the two parallel plates is 1 cm. Calculate

(*i*) the electric stress within the dielectric,

(*ii*) the electric stress on the plate surface,

(*iii*) the charges on the plates.

Calculate the values (*i*), (*ii*) and (*iii*) in above if the air dielectric is replaced by paper of relative permittivity 4.

Solution: Capacitance of capacitor, C = 1×10^{-6} F

Dielectric (air) relative permittivity, ε_r = 1

Dielectric thickness, t = 1 cm = 0.01 m

(*i*) and (*ii*) Electric stress within the dielectric = Electric stress on the plate surface,

$$E = \frac{V}{t} = \frac{200}{0.01} = 20{,}000\ \text{V/m or } 20\ \text{kV/m } \textbf{Ans.}$$

(*iii*) Charge on the plate, Q = CV = $1 \times 10^{-6} \times 200$ = 200 μC **Ans.**

When air is replaced by paper of relative permittivity 4, the electric stress will remain unchanged but the charge on plate will change due to change in the capacitance of the parallel plate capacitor.

New value of capacitance, C ' = ε_r C = $4 \times 1 \times 10^{-6} = 4 \times 10^{-6}$ F

Charge on the plate, Q′ = C′ V = $4 \times 10^{-6} \times 200$ = 800 μC **Ans.**

Example 2.8

The capacitance of a capacitor formed by two parallel metal plates each 200 cm² in area separated by a dielectric 4 mm thick is 0.0004 μF. A pd of 20 kV is applied. Calculate(*i*) the total charge on the plates (*ii*) potential gradient in the dielectric (*iii*) the relative permittivity of the dielectric and (*iv*) the flux density.

Solution: Capacitance of the capacitor, C = 0.0004 μF = 4×10^{-10} F

Surface area of plates, A = 200 cm² = 0.02 m²

Dielectric thickness, t = 4 mm = 0.004 m

(*i*) Total charge on plates, Q = C V = $4 \times 10^{-10} \times 20{,}000$ = 8 μC **Ans.**

$$\text{(ii) Potential gradient in dielectric, } g = \frac{V}{t} = \frac{20{,}000}{0.004} = 5\times10^{6}\ \text{V/m } \textbf{Ans.}$$

$$\text{(iii) Dielectric relative permittivity, } \varepsilon_r = \frac{C\times t}{\varepsilon_0 A} = \frac{4\times10^{-10}\times0.004}{8.854\times10^{-12}\times0.02} = 9.035\ \textbf{Ans.}$$

$$\text{(iv) Flux density, } D = \frac{Q}{A} = \frac{8\times10^{-6}}{0.02} = 4\times10^{-4}\ \text{C/m}^2\ \textbf{Ans.}$$

Example 2.9

A parallel-plate capacitor has plates of area of 2 m² spaced by 3 slabs of relative perimittivities 2, 3, 6 and thickness 0.4 mm, 0.6 mm and 1.2 mm respectively. Calculate the capacitance of the capacitor.

Solution: From Eq. (2.10) capacitance of a composite capacitor,

$$C = \frac{\varepsilon_0 A}{\dfrac{t_1}{\varepsilon_{r_1}}+\dfrac{t_2}{\varepsilon_{r_2}}+\dfrac{t_3}{\varepsilon_{r_3}}} = \frac{8.854\times10^{-12}\times2}{\dfrac{0.4\times10^{-3}}{2}+\dfrac{0.6\times10^{-3}}{3}+\dfrac{1.2\times10^{-3}}{6}} = 2.9513\times10^{-8}\ \text{F } \textbf{Ans.}$$

Example 2.10

A capacitor is composed of two plates separated by a sheet of insulating material 3 mm thick and of $\varepsilon_r = 4$. The distance between the plates is increased to allow the insertion of a second sheet 5 mm thick and of relative permittivity ε_r. If the capacitance of the capacitor so formed is one-half of the original capacitance, find the value of ε_r.

Solution: *Before insertion of second sheet*

Distance between the plates, $d = 3 \text{ mm} = 0.003 \text{ m}$

Relative permittivity of the medium, $\varepsilon_{r_1} = 4$

$$\text{Capacitance of the capacitor, } C_1 = \frac{\varepsilon_0 \varepsilon_{r_1} A}{d} = \frac{4\varepsilon_0 A}{0.003} = \frac{4,000}{3} \varepsilon_0 A \quad ...(i)$$

After insertion of second sheet

Thickness of first sheet, $t_1 = 0.003$ m

Thickness of second sheet, $t_2 = 0.005$ m

Relative permittivity of first sheet, $\varepsilon_{r_1} = 4$

Relative permittivity of second sheet $= \varepsilon_{r_2}$ (say)

$$\text{Capacitance of the capacitor, } C_2 = \frac{\varepsilon_0 A}{\frac{t_1}{\varepsilon_{r_1}} + \frac{t_2}{\varepsilon_{r_2}}} = \frac{\varepsilon_0}{\frac{0.003}{4} + \frac{0.005}{\varepsilon_{r_2}}} \quad ...(ii)$$

$$\because C_2 = \frac{1}{2} C_1$$

$$\therefore \frac{\varepsilon_0 A}{\frac{0.003}{4} + \frac{0.005}{\varepsilon_{r_2}}} = \frac{1}{2} \times \frac{4,000}{3} \varepsilon_0 A$$

$$\text{or } 0.00075 + \frac{0.005}{\varepsilon_{r_2}} = 0.0015$$

$$\text{or } \varepsilon_{r_2} = \frac{0.005}{0.0015 - 0.00075} = 6.67 \textbf{ Ans.}$$

Example 2.11

Determine the dielectric flux in microcoulombs between two parallel plates each 0.35 metre square with an air gap of 1.5 mm between them, the pd being 3,000 V. A sheet of insulating material 1 mm thick is inserted between the plates. the relative permittivity of the insulating material being 6. Find out the potential gradient in the insulating material and also in air if the voltage across the plate is raised to 7,500 volts.

Solution: Surface area of plates, $A = (0.35)^2 = 0.1225 \text{ m}^2$

Thickness of dielectric (air), $t = 1.5 \text{ mm} = 0.0015$ m

Relative permittivity of dielectric (air), $\varepsilon_r = 1$

$$\text{Capacitance, } C = \frac{\varepsilon_0 \varepsilon_r A}{t} = \frac{8.854 \times 10^{-12} \times 1 \times 0.1225}{0.0015} = 723 \times 10^{-12} \text{ F}$$

Dielectric flux, $\psi = Q = CV = 723 \times 10^{-12} \times 3,000 = 2.17\ \mu\text{C}$ **Ans.**

Let the potential gradient in air be g then potential gradient in insulating material of relative permittivity 6 will be $g/6$

and applied voltage $V = g_1 t_1 + g_2 t_2$

$$\text{or } 7,500 = (0.0015 - 0.001)\, g + \frac{0.001\, g}{6}$$

$$\text{or } g = \frac{7,500}{0.0005 + 0.000167} = \frac{7,500}{0.000667} = 11.25 \times 10^6 \text{ V/m}$$

Potential gradient in air, $g_1 = g = 11.25 \times 10^6$ V/m **Ans.**

Potential gradient in insulating material, $g_2 = \frac{g}{6} = \frac{11.25 \times 10^6}{6} = 1.875 \times 10^6$ V/m **Ans.**

Example 2.12

Two perfect insulated capacitors are connected in series. One is an air capacitor with a plate area of 0.01 m², the plate being 1 mm apart, the other has a plate area of 0.001 m², the plates separated by a solid dielectric of 0.1 mm thickness with a dielectric constant of 5. Determine the voltage across the combination if the potential gradient in the air capacitor is 200 V/mm.

Solution: Capacitance of air capacitor, $C_1 = \frac{\varepsilon_0 \times \varepsilon_{r_1} \times A_1}{t_1} = \frac{8.854 \times 10^{-12} \times 1 \times 0.01}{0.001} = 88.54 \times 10^{-12}$ F

Capacitance of dielectric capacitor, $C_2 = \frac{\varepsilon_0 \times \varepsilon_{r_2} \times A_2}{t_2} = \frac{8.854 \times 10^{-12} \times 5 \times 0.001}{0.0001} = 442.7 \times 10^{-12}$ F

Voltage across air capacitor, $V_1 = g_1 t_1 = 200 \times 1 = 200$ V

Charge on air capacitor, $Q_1 = C_1 V_1 = 88.54 \times 10^{-12} \times 200 = 177.08 \times 10^{-10}$ C

Voltage across dielectric capacitor, $V_2 = \frac{Q_2}{C_2} = \frac{Q_1}{C_2} = \frac{177.08 \times 10^{-10}}{442.7 \times 10^{-12}} = 40$ V

(Since in series combination all the capacitors have the same charge)

Voltage across the combination, $V = V_1 + V_2 = 200 + 40 = 240$ V **Ans.**

Example 2.13

A parallel-plate capacitor has two plates of 1,500 cm² separated by 5 mm with air as dielectric. If a layer of dielectric 2 mm thick and relative permittivity 3 is introduced between the plates, what should be the new separation between the plates so that the capacitance remains the same.

Solution: Capacitance of air capacitor, $C_1 = \frac{\varepsilon_0 \varepsilon_r \times A}{t_1} = \frac{8.854 \times 10^{-12} \times 1 \times 1,500 \times 10^{-4}}{5 \times 10^{-3}} = 2.6562 \times 10^{-10}$ F

Let the new separation be d metres. Capacitance of a composite capacitor is given as

$$C = \frac{\varepsilon_0 \times A}{d - t\left(1 - \frac{1}{\varepsilon_r}\right)}$$ ... Refer to Eq. (2.13)

$$\therefore\ 2.6562 \times 10^{-10} = \frac{8.854 \times 10^{-12} \times 1,500 \times 10^{-4}}{d - 0.002\left(1 - \frac{1}{3}\right)}$$

or $d = \frac{8.854 \times 10^{-12} \times 1,500 \times 10^{-4}}{2.6562 \times 10^{-10}} + 0.00133 = 0.00633$ m or 6.33 mm **Ans.**

Example 2.14

A multiplate capacitor consists of 12 plates, with alternative plates connected to each other to form two terminals, across which a voltage of 1,000 V is applied. The distance between adjacent plates is 4 mm and area of each plate is 100 cm². The dielectric used has a relative permittivity of 4. Calculate the energy stored in this capacitor.

Solution: Number of plates, $n = 12$

Area of each plate, $A = 100\ \text{cm}^2 = 0.01\ \text{m}^2$

Distance between adjacent plates, $t = 4$ mm $= 0.004$ m

Capacitance of capacitor, $C = \frac{(n-1)\varepsilon_0 \varepsilon_r A}{t} = \frac{(12-1) \times 8.854 \times 10^{-12} \times 4 \times 0.01}{0.004} = 973.94\ \mu\mu$F

Energy stored $= \frac{1}{2} CV^2 = \frac{1}{2} \times 973.94 \times 10^{-12} \times (1,000)^2 = 487 \times 10^{-6}$ J **Ans.**

2.24. CAPACITANCE OF A CYLINDRICAL CAPACITOR

The cable can be considered to be two coaxial cylinders of inner diameter d and outer diameter D. In actual cable, d represents the diameter of core and D represents the inner diameter of lead sheath which is at earth potential. Let the relative permittivity of dielectric in between the core and sheath be ε_r.

Let the charge per metre length of the cable on the outer surface of the core be + Q coulombs and on the inner surface of the lead sheath be – Q coulombs. For all practical purposes, the charge of + Q coulombs/m on the surface of the core can be assumed to be located along its axis. The metal sheath is earthed.

Surface area of the coaxial cylinder of radius x metres and length one metre is $2\pi x$ metres^2.

Therefore, electric field intensity at a point x metres from the centre of the inner cylinder,

$$E_x = \frac{Q}{2\pi\varepsilon_0\varepsilon_r x} = \frac{Q}{2\pi\,\varepsilon_0\varepsilon_r}\cdot\frac{1}{x}\ \text{V/m}$$

Potential difference between the capacitor plates, (between core and sheath),

$$V = \int_{d/2}^{D/2} E_x\,dx = \int_{d/2}^{D/2} \frac{Q}{2\pi\,\varepsilon_0\,\varepsilon_r}\cdot\frac{1}{x}\,dx = \frac{Q}{2\pi\,\varepsilon_0\,\varepsilon_r}\log_e\frac{D}{d}\ \text{volts}$$

Capacitance of cable,

$$C = \frac{Q}{V} = \frac{Q}{\dfrac{Q}{2\pi\,\varepsilon_0\varepsilon_r}\log_e\dfrac{D}{d}}\ \text{F/m}$$

$$= \frac{2\pi\,\varepsilon_0\varepsilon_r}{\log_e\dfrac{D}{d}}$$

$$= \frac{2\pi\times 8.854\times 10^{-12}\,\varepsilon_r}{2.303\log_{10}\dfrac{D}{d}}$$

$$= \frac{\varepsilon_r\,10^{-9}}{41.4\log_{10}\dfrac{D}{d}}\ \text{F/m or}\ \frac{0.024\,\varepsilon_r}{\log_{10}\dfrac{D}{d}}\ \mu\text{F/km} \qquad \text{...(2.15)}$$

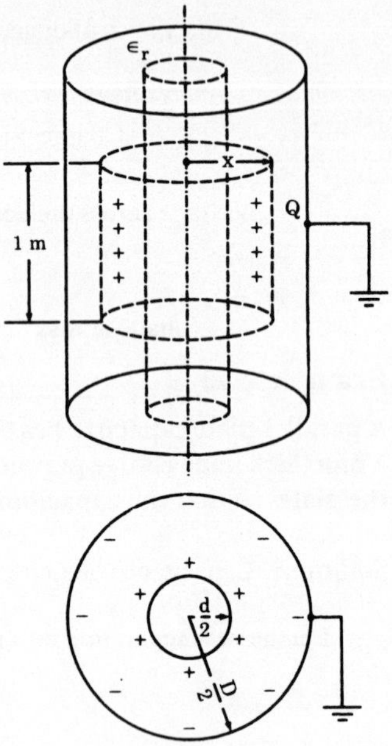

Fig. 2.17

In case of capacitor having compound dielectric, the expression for capacitance becomes

$$C = \frac{2\pi\varepsilon_0}{\sum \log_e\dfrac{D}{d}\Big/\varepsilon_r}\ \text{F/m}$$

2.25. ENERGY STORED IN A CAPACITOR

When a capacitor is in the process of being charged, small amounts of charge are continually added until the total charge on the capacitor becomes Q coulombs.

Consider a capacitor of capacitance C farads being charged from a source of V volts.

Let at any instant of charging the charge attained by it and potential difference across its plates be q coulombs and v volts respectively.

Now by definition of capacitance, $C = \dfrac{q}{v}$

or $v = \dfrac{q}{C}$

Now if the charge of the capacitor is raised by small amount dq coulombs, the work equivalent to vdq is done which appears as stored energy in the condenser.

Therefore, energy stored during the interval of raising the charge by small amount dq,

$$dw = v\,dq = \frac{q}{C}\,dq$$

Work done in giving the charge of Q coulombs to uncharged condenser

$$W = \int_0^Q \frac{q}{C}\,dq = \frac{1}{C}\left[\frac{q^2}{2}\right]_0^Q = \frac{Q^2}{2C} \text{ joules}$$

$$\text{Also } W = \frac{Q}{2}\cdot\frac{Q}{C} = \frac{1}{2}QV = \frac{1}{2}CV^2$$

$$\therefore \text{ Energy stored in a capacitor} = W = \frac{Q}{2}\cdot\frac{Q}{C} = \frac{1}{2}QV = \frac{1}{2}CV^2 \qquad ...(2.16)$$

2.25.1. Energy Stored Per Cubic Metre of Volume of Dielectric In a Capacitor. Consider a capacitor having a dielectric of thickness t metre, and plates of area A square metres. The energy stored per cubic metre in the capacitor

$$= \frac{\text{Total energy stored}}{\text{Volume of the dielectric}} = \frac{1}{2}CV^2 \div At \text{ joules} = \frac{\frac{1}{2}\varepsilon A}{t}V^2 \div At \qquad \because C = \frac{\varepsilon A}{t}$$

$$= \frac{1}{2}\varepsilon\left(\frac{V}{t}\right)^2 = \frac{1}{2}\varepsilon\,\mathbf{E}^2 = \frac{1}{2}\mathbf{DE} = \frac{1}{2}\mathbf{D}^2/\varepsilon \text{ joules} \qquad ...(2.17)$$

$$\because \mathbf{E} = \frac{V}{t} = \mathbf{D}/\varepsilon$$

Example 2.15

A spherical conductor of radius 10 cm is charged up to 150 V potential. Calculate the amount of energy stored up with the conductor.

Solution: Capacitance of spherical conductor, $C = \frac{r}{9\times10^9} = \frac{0.1}{9\times10^9} = 11.1 \times 10^{-12}$ F

Energy stored in the conductor $= \frac{1}{2}CV^2 = \frac{1}{2} \times 11.1 \times 10^{-12} \times (150)^2 = 1.25 \times 10^{-7}$ J **Ans.**

Example 2.16

A condenser is connected across 220 V dc supply for 0.3 second, the current remains steady at 0.25 A. Determine its charge and capacity.

Solution: Charge on the condenser, $Q = It = 0.25 \times 0.3 = 0.075$ C **Ans.**

Capacity of the condenser, $C = \frac{Q}{V} = \frac{0.075}{220} = 340.9\ \mu$F **Ans.**

Example 2.17

A capacitor of capacity one microfarad is charged to 10 kV and then discharged through a wire. Find the heat produced in the wire in calories.

Solution: Voltage across the capacitor, V = 10 kV = 10,000 V

Heat produced in the wire = Energy discharged by the capacitor

$= \frac{1}{2}CV^2 = \frac{1}{2} \times 1 \times 10^{-6} \times (10{,}000)^2 = 50$ J or $\frac{50}{4.2} = 11.9$ calories **Ans.**

Example 2.18

A 2 μF capacitor is charged to 600 V. If the capacitor is then discharged to 450 V, determine how much energy has been removed from the capacitor. If the discharge occurs uniformly in 3 minutes, calculate power.

Solution: Voltage to which the capacitor has been charged, $V_1 = 600$ V

Voltage to which the capacitor has been discharged, $V_2 = 450$ V

Energy stored in the capacitor, before discharge, $W_1 = \frac{1}{2}CV_1^2 = \frac{1}{2} \times 2 \times 10^{-6} \times (600)^2 = 0.36$ J

Energy stored in the capacitor, after discharge, $W_2 = \frac{1}{2}CV_2^2 = \frac{1}{2} \times 2 \times 10^{-6} \times (450)^2 = 0.2025$ J

Energy removed from the capacitor $= W_1 - W_2 = 0.36 - 0.2025 = 0.1575$ J

Duration of discharge, t = 3 minutes = 3 × 60 = 180 seconds

$$\text{Power, P} = \frac{\text{Energy removed in J or W-}s}{t \text{ in seconds}}$$

$$= \frac{0.1575}{180} = 875 \times 10^{-6} \text{ W or } 875 \ \mu\text{W}$$ **Ans.**

2.26. CAPACITORS IN SERIES

When the negative plate of one capacitor is connected to the positive plate of another capacitor, as shown in Fig. 2.18, the capacitors are said to be connected in series.

If several capacitors connected in series are put across a potential difference of V volts, as shown, then the electrons will flow from plate G to plate F and so on up to plate A from where they will pass to the positive terminal of the source of supply. Thus all the capacitors will get the same charge, let it be Q coulombs.

Let the potential differences across the capacitors of capacitances C_1, C_2 and C_3 be V_1, V_2, V_3 respectively.

Now potential difference across the capacitor of capacitance C_1, $V_1 = \frac{Q}{C_1}$

Similarly PD across the capacitor of capacitance C_2, $V_2 = \frac{Q}{C_2}$

and PD across the capacitor of capacitance C_3, $V_3 = \frac{Q}{C_3}$

PD applied across the system, $V = V_1 + V_2 + V_3$

$$\text{or } V = \frac{Q}{C_1} + \frac{Q}{C_2} + \frac{Q}{C_3}$$

$$\text{or } \frac{V}{Q} = \frac{1}{C_1} + \frac{1}{C_2} + \frac{1}{C_3}$$

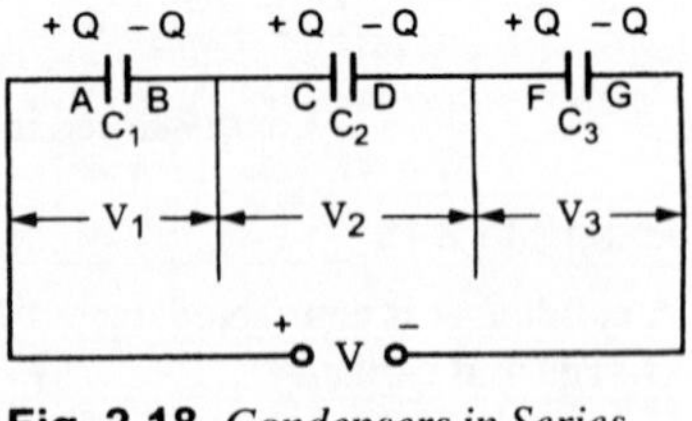

Fig. 2.18 *Condensers in Series*

If C is the equivalent capacitance of the system then by definition

$$C = \frac{\text{Charge on the system}}{\text{PD across the system}} = \frac{Q}{V}$$

$$\text{or } \frac{1}{C} = \frac{V}{Q} = \frac{1}{C_1} + \frac{1}{C_2} + \frac{1}{C_3} \quad ...(2.18)$$

Thus, when a number of capacitors are connected in series the reciprocal of the equivalent capacitance of the system is given by the arithmetic sum of the reciprocals of their individual capacitances.

Example 2.19

A 10 μF, a 20 μF and a 40 μF capacitor are connected in series to a 399 V source of emf.

(*i*) What is the equivalent capacitance ?

(*ii*) What is the magnitude of charge across each capacitor ?

(*iii*) What is the potential difference across each capacitor ?

Solution: Equivalent capacitance of series combination of 10, 20 and 40 μF capacitors,

$$C = \frac{1}{\frac{1}{C_1} + \frac{1}{C_2} + \frac{1}{C_3}} = \frac{1}{\frac{1}{10} + \frac{1}{20} + \frac{1}{40}} = \frac{40}{7} = 5.71 \ \mu\text{F}$$ **Ans.**

P D across the combination, V = 399 V

$$\text{Charge on each capacitor, } Q = CV = \frac{40}{7} \times 10^{-6} \times 399 = 2{,}280\ \mu\text{ C } \textbf{Ans.}$$

$$\text{P D across 10 } \mu\text{F capacitor, } V_1 = \frac{Q}{C_1} = \frac{2{,}280 \times 10^{-6}}{10 \times 10^{-6}} = 228 \text{ V } \textbf{Ans.}$$

$$\text{P D across 20 } \mu\text{F capacitor, } V_2 = \frac{Q}{C_2} = \frac{2{,}280 \times 10^{-6}}{20 \times 10^{-6}} = 114 \text{ V } \textbf{Ans.}$$

$$\text{P D across 40 } \mu\text{F capacitor, } V_3 = \frac{Q}{C_3} = \frac{2{,}280 \times 10^{-6}}{40 \times 10^{-6}} = 57 \text{ V } \textbf{Ans.}$$

Example 2.20

Three capacitors of 100 μF, 150 μF and 200 μF are connected in series, across a voltage of 500 V. Calculate the charge on each capacitor and energy stored in each.

Solution: Equivalent capacitance of series combination of 100, 150 and 200 μF capacitors,

$$C = \frac{1}{\frac{1}{C_1} + \frac{1}{C_2} + \frac{1}{C_3}} = \frac{1}{\frac{1}{100} + \frac{1}{150} + \frac{1}{200}} = \frac{600}{13} = 46.154\ \mu\text{F } \textbf{Ans.}$$

P D across the combination, V = 500 V

$$\text{Charge on each capacitor, } Q = CV = 46.154 \times 10^{-6} \times 500 = 0.023 \text{ C } \textbf{Ans.}$$

$$\text{Energy stored in 100 } \mu\text{F capacitor, } W_1 = \frac{Q^2}{2C_1} = \frac{(0.023)^2}{2 \times 100 \times 10^{-6}} = 2.645 \text{ J } \textbf{Ans.}$$

$$\text{Energy stored in 150 } \mu\text{F capacitor, } W_2 = \frac{Q^2}{2C_2} = \frac{(0.023)^2}{2 \times 150 \times 10^{-6}} = 1.76 \text{ J } \textbf{Ans.}$$

$$\text{Energy stored in 200 } \mu\text{F capacitor, } W_3 = \frac{Q^2}{2C_3} = \frac{(0.023)^2}{2 \times 200 \times 10^{-6}} = 1.323 \text{ J } \textbf{Ans.}$$

2.27. CAPACITORS IN PARALLEL

When one plate of each capacitor is connected to the + ve terminal of the supply mains and other plate of each capacitor to the – ve terminal of the supply mains, as shown in Fig. 2.19 the *capacitors are said to be connected in parallel.*

Let such combination consisting of three capacitors of capacitances C_1, C_2, C_3 respectively be connected across a potential difference of V volts.

Potential difference across capacitor = PD applied across the combination

Now charge on the capacitor of capacitance, C_1, $Q_1 = C_1 V$

Similarly charge on the capacitor of capacitance C_2, $Q_2 = C_2 V$

and charge on the capacitor of capacitance C_3, $Q_3 = C_3 V$

If C is the equivalent capacitance of the system then total charge drawn,

$$Q = CV = Q_1 + Q_2 + Q_3$$

$$\text{or } CV = C_1V + C_2V + C_3V$$

$$\text{or } C = C_1 + C_2 + C_3 \quad ...(2.19)$$

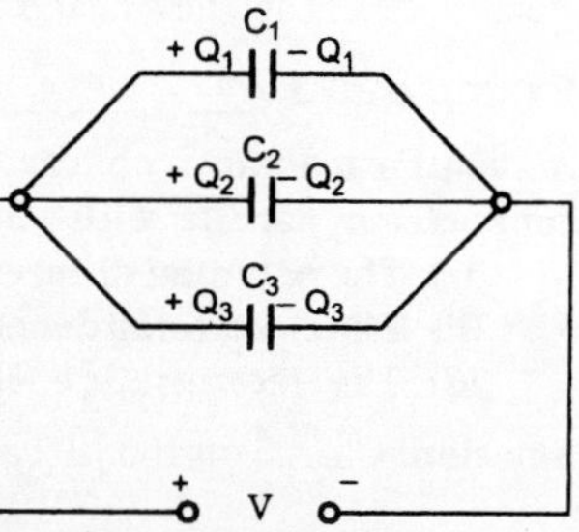

Fig. 2.19 *Condensers in Parallel*

Thus, when a number of capacitors are connected in parallel the capacitance of the system is given by the arithmetic sum of their individual capacitances.

Example 2.21

Two capacitors have capacitances of 6 μF and 10 μF respectively.

(i) Find the total capacitance when they are connected (a) in parallel (b) in series.

(ii) When the above two capacitors are connected in series across a 200 V supply, find the potential difference across each capacitor and the charge on each capacitor.

Solution: Equivalent capacitance,

(a) When connected in parallel, $C_P = C_1 + C_2 = 6 + 10 = 16\ \mu F$ **Ans.**

(b) When connected in series, $C_S = \dfrac{1}{\dfrac{1}{C_1}+\dfrac{1}{C_2}} = \dfrac{1}{\dfrac{1}{6}+\dfrac{1}{10}} = 3.75\ \mu F$ **Ans.**

(ii) Charge on each capacitor, when connected in series

$$Q = C_S \times V = 3.75 \times 10^{-6} \times 200 = 0.00075 \text{ C} \textbf{ Ans.}$$

PD across 6 μF capacitor, $V_1 = \dfrac{Q}{C_1} = \dfrac{0.00075}{6\times10^{-6}} = 125$ V **Ans.**

PD across 10 μF capacitor, $V_2 = \dfrac{Q}{C_2} = \dfrac{0.00075}{10\times10^{-6}} = 75$ V **Ans.**

Example 2.22

Two capacitors of capacitances 8 μF and 2 μF are connected in series across a 100 V dc supply. Now if the supply voltage is removed and the capacitors are then connected in parallel, what will be the final charge on each capacitor ?

Solution: Equivalent capacitance when capacitors are connected in series,

$$C_S = \frac{1}{\dfrac{1}{C_1}+\dfrac{1}{C_2}} = \frac{C_1 C_2}{C_1+C_2} = \frac{8\times2}{8+2} = 1.6\ \mu F$$

Charge on each capacitor, $Q = C_s V = 1.6 \times 10^{-6} \times 100 = 1.6 \times 10^{-4}$ C

Total charge on both capacitors $= 2Q = 2 \times 1.6 \times 10^{-4} = 3.2 \times 10^{-4}$ C

When joined in parallel a redistribution of charge takes place and the potential difference across both capacitors becomes same say, V volts

Assuming no loss of charge

Total charge $= 3.2 \times 10^{-4}$ C

Total capacitance $C_P = C_1 + C_2 = 8 + 2 = 10\ \mu F$

PD across each capacitor, $V = \dfrac{\text{Total charge}}{\text{Total capacitance}} = \dfrac{3.2\times10^{-4}}{10\times10^{-6}} = 32$ V

Hence new charge on 8 μF capacitor $= 8 \times 10^{-6} \times 32 = 2.56 \times 10^{-4}$ C **Ans.**

and new charge on 2 μF capacitor $= 2 \times 10^{-6} \times 32 = 0.64 \times 10^{-4}$ C **Ans.**

Example 2.23

A 100 μF capacitor is charged from a 200 V dc supply. After it is fully charged, it is disconnected and immediately connected in parallel with a 50 μ F capacitor. Calculate the following:

(i) The potential difference across the combination.

(ii) The electrostatic energy before the capacitors are connected in parallel.

(iii) The electrostatic energy after the capacitors are connected in parallel.

Solution: Charge on 100 μF capacitor when fully charged to 200 V,

$$Q = CV = 100 \times 10^{-6} \times 200 = 0.02 \text{ C}$$

When the capacitors are connected in parallel, the equivalent capacitance, C′ becomes (100 + 50), *i.e.* 150 μF and the charge is divided between the two capacitors (*i*) Potential difference across the combination, $V' = \frac{Q}{C'} = \frac{0.02}{150 \times 10^{-6}} = 133.333$ V **Ans.**

(*ii*) The electrostatic energy before the capacitors are connected in parallel

= Electrostatic energy stored in 100 μF capacitor when charged to 200 V

$$= \frac{1}{2} C (V)^2 = \frac{1}{2} \times 100 \times 10^{-6} \times (200)^2 = 2 \text{ joules } \textbf{Ans.}$$

(*iii*) The electrostatic energy after the capacitors are connected in parallel

$$= \frac{1}{2} C' (V)^2 = \frac{1}{2} \times 150 \times 10^{-6} \times (133.333)^2 = 1.333 \text{ J } \textbf{Ans.}$$

Example 2.24

The total capacitance of two capacitors is 0.03 μF when joined in series and 0.16 μF when connected in parallel. Find the capacitance of each capacitor.

Solution: Let the capacitances of the two capacitors be C_1 and C_2

When connected in series equivalent capacitance,

$$C_S = \frac{1}{\frac{1}{C_1} + \frac{1}{C_2}} = \frac{C_1 C_2}{C_1 + C_2} = 0.03 \ \mu F \qquad ...(i)$$

When connected in parallel, equivalent capacitance,

$$C_P = C_1 + C_2 = 0.16 \ \mu F \qquad ...(ii)$$

Multiplying Eqs. (*i*) and (*ii*) we get

$$C_1 C_2 = 0.03 \times 0.16 = 0.0048 \times 10^{-12} \qquad ...(iii)$$

$$\text{Now } C_1 - C_2 = \sqrt{(C_1 + C_2)^2 - 4 C_1 C_2} = \sqrt{(0.16 \times 10^{-6})^2 - 4 \times 0.0048 \times 10^{-12}}$$

$$= \pm 0.08 \ \mu F \qquad ...(iv)$$

Adding Eqs. (*ii*) and (*iv*) we have

C_1 = 0.12 μF or 0.04 μF **Ans.**

Substituting value of C_1 in Eq. (*ii*) we have

C_2 = 0.04 μF or 0.12 μF **Ans.**

Example 2.25

Calculate the equivalent capacitance in the circuit of Fig. 2.20, across the points AB.

Solution: Equivalent capacitance of the parallel combination of capacitors of 1, 3 and 2 μF,

$$C_P = 1 + 3 + 2 = 6 \ \mu F$$

Equivalent capacitance of the circuit shown in Fig. 2.20.

= Equivalent capacitance of series combination of 6, 3 and 2 μF capacitors

$$= \frac{1}{\frac{1}{6} + \frac{1}{3} + \frac{1}{2}} = 1 \ \mu F \ \textbf{Ans.}$$

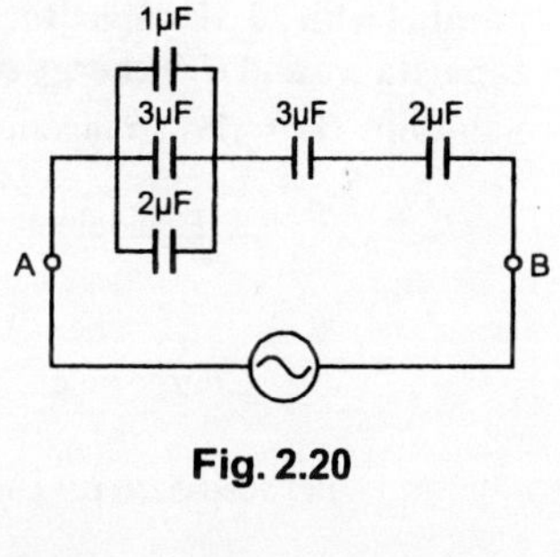

Fig. 2.20

Example 2.26

Two capacitors A and B are connected in series across 100 V supply and it is observed that the pd across them are 60 V and 40 V respectively. A capacitor of 2 μF capacitance is now connected in parallel with A and pd across B rises to 90 V. Determine the capacitances of A and B.

Solution: Let the capacitances of the capacitors A and B be C_A and C_B respectively in microfarads and potential differences across them be V_A and V_B respectively.

When the capacitors A and B are connected in series, charge on both of them will be the same.

$$i.e.\ C_A V_A = C_B V_B = Q$$

$$\text{or } 60\, C_A = 40\, C_B \qquad \because V_A = 60 \text{ V and } V_B = 40 \text{ V}$$

$$\text{or } C_B = 1.5\, C_A \qquad ...(i)$$

When a capacitor of 2 μF is connected in parallel with capacitor A, the equivalent capacitance of capacitor A and capacitor of 2 μF is

$$C'_A = C_A + 2$$

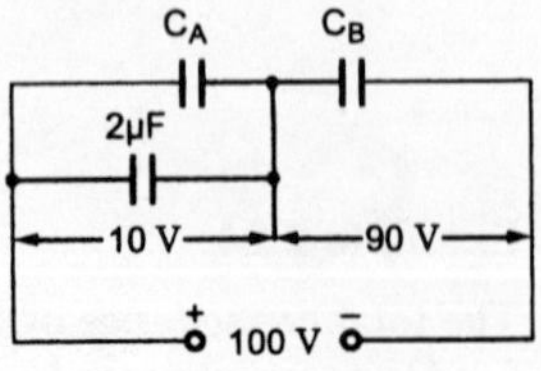

Fig. 2.21

Capacitor C_B remains as it was

Voltage across capacitor B, $V'_B = 90$ V

Voltage across parallel combination of capacitor A and capacitor of 2 μF,

$$V'_A = 100 - 90 = 10 \text{ V}$$

New charges on C'_A and C_B will be equal

$$i.e.\ V'_A C'_A = V'_B C_B$$

$$\text{or } 10\,(C_A + 2) = 90\, C_B$$

$$\text{or } C_A + 2 = 9\, C_B = 9 \times 1.5\, C_A = 13.5\, C_A \qquad \because \text{From Eq. } (i)\ C_B = 1.5\, C_A$$

$$\text{or } C_A = \frac{2}{12.5} = 0.16\ \mu\text{F} \textbf{ Ans.}$$

$$\text{and } C_B = 1.5\, C_A = 1.5 \times 0.16 = 0.24\ \mu\text{F} \textbf{ Ans.}$$

Example 2.27

Given some capacitors of 0.1 μF capable of withstanding voltage up to 15 V each, calculate the number of capacitors needed if it is desired to obtain a capacitance of 0.1 μF for use in circuit involving 60 V.

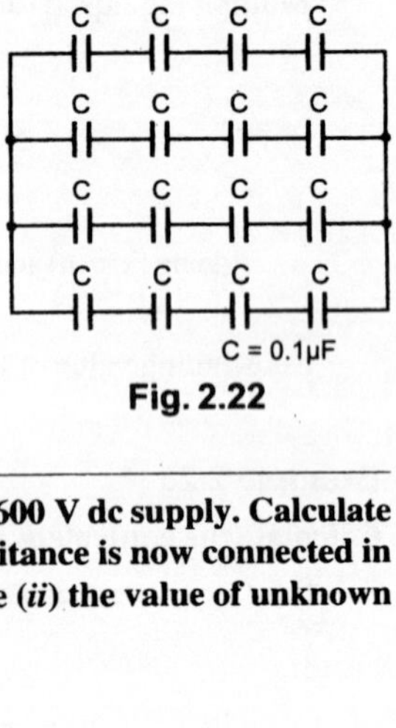

Fig. 2.22

Solution: It is given that each capacitor can withstand maximum voltage of 15 V, therefore, the number of capacitors required to be connected in series to withstand a voltage of 60 V $= \frac{60}{15} = 4$

The equivalent capacitance of the above combination $= \frac{0.1}{4} = 0.025$ μF.

Since the total capacitance required is 0.1 μF, therefore, such four combinations will be required to be connected in parallel. The arrangement is shown in Fig. 2.22.

Total number of capacitors required = 4 × 4 = 16 **Ans.**

Example 2.28

Two capacitors having capacitances of 20 μF and 30 μF are connected in series across a 600 V dc supply. Calculate (*i*) the potential difference across each capacitor. If a third capacitor of unknown capacitance is now connected in parallel with 20 μF capacitor such that the pd across 30 μF capacitor is 400 V. Calculate (*ii*) the value of unknown capacitance and (*iii*) energy stored in the third capacitor.

Solution: The series arrangement of capacitors is shown in Fig. 2.23 (*a*)

$$\text{Equivalent capacitance, } C = \frac{1}{\frac{1}{C_1} + \frac{1}{C_2}} = \frac{1}{\frac{1}{20} + \frac{1}{30}} = 12\ \mu\text{F}$$

$$\text{Charge on each capacitor, } Q = CV = 12 \times 10^{-6} \times 600 = 7{,}200 \times 10^{-6} \text{ C}$$

$$\text{PD across 20 μF capacitor, } V_1 = \frac{Q}{C_1} = \frac{7{,}200 \times 10^{-6}}{20 \times 10^{-6}} = 360 \text{ V} \textbf{ Ans.}$$

$$\text{PD across 30 μF capacitor, } V_2 = \frac{Q}{C_2} = \frac{7{,}200 \times 10^{-6}}{30 \times 10^{-6}} = 240 \text{ V} \textbf{ Ans.}$$

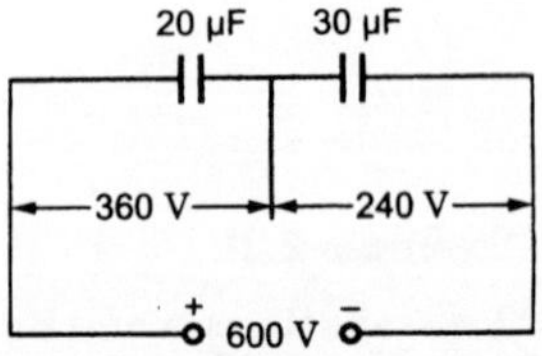

Fig. 2.23 (*a*)

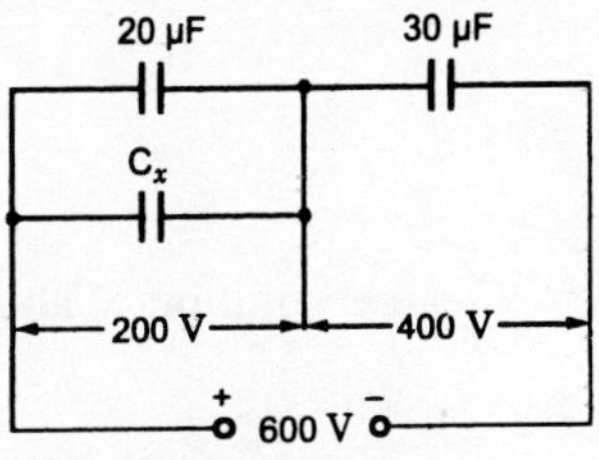

Fig. 2.23 (*b*)

Let the unknown capacitor connected across 20 μF capacitor be of C_x microfarads. The connections are shown in Fig. 2.23 (*b*).

Voltage across 30 μF capacitor, $V'_2 = 400$ V

Charge on 30 μF capacitor, $Q' = C_2 V'_2 = 30 \times 10^{-6} \times 400 = 0.012$ C

Also $Q' = (C_1 + C_x)\, V$

or $0.012 = (20 + C_x) \times 10^{-6} \times (600 - 400)$

or $C_x = \dfrac{0.012}{200 \times 10^{-6}} - 20 = 40$ μF **Ans.**

Energy stored in third capacitor $= \dfrac{1}{2} C_x (V'_1)^2 = \dfrac{1}{2} \times 40 \times 10^{-6} \times (200)^2 = 0.8$ J **Ans.**

Example 2.29

Determine the capacitance of the condenser combination between terminals of Fig. 2.24. The capacitances shown are in microfarads.

Solution: Referring to Fig. 2.24 the branches AB, BC, CD and DA form the four arms of a Wheatstone bridge.

Since $\dfrac{C_{AB}}{C_{BC}} = \dfrac{15}{30}$ and $\dfrac{C_{AD}}{C_{DC}} = \dfrac{10}{20}$

Therefore, the bridge is balanced and no current would flow through BD. The branch BD, therefore, can be removed. The remaining network will become, as shown in Fig. 2.25.

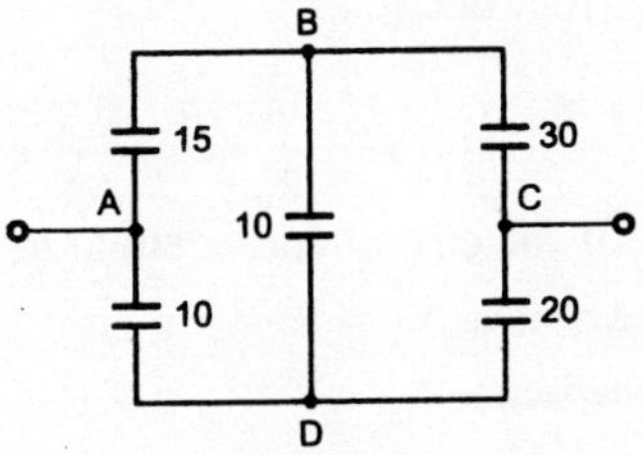

Fig. 2.24

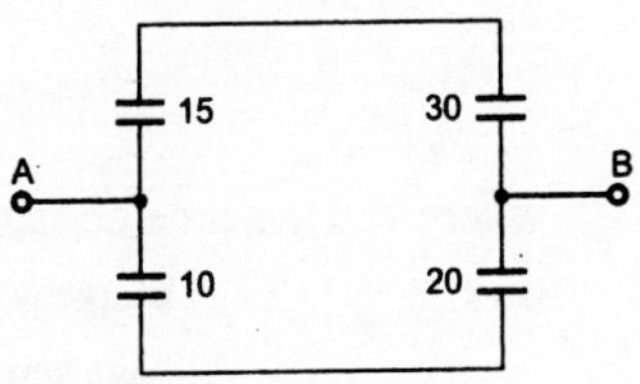

Fig. 2.25

The capacitance between terminals A and B (Fig. 2.25) is given as

$$C = \frac{1}{\frac{1}{15} + \frac{1}{30}} + \frac{1}{\frac{1}{10} + \frac{1}{20}} = 10 + 6.667 = 16.667 \text{ μF } \textbf{Ans.}$$

2.28. CHARGING AND DISCHARGING OF A CAPACITOR

Consider a circuit consisting of a resistor of resistance R ohms and a capacitor of a capacitance C farads connected in series with a switch S and a battery of V volts, as shown in Fig. 2.26.

2.28.1. Charging of a Capacitor. Before putting on the switch S in position '1', the potential difference across the capacitor was zero but at the instant, the switch S is put in position '1', the pd across the capacitor plates starts rising and it will continue rising until eventually it becomes equal to battery voltage V. Charging current is maximum at the instant of putting the switch S in position '1', then gradually diminishes and finally ceases when pd across the capacitor plates becomes equal and opposite to that of battery.

Let at any instant pd across the capacitor be v_C, charge acquired by the capacitor by q coulombs and charging current be i amperes

Voltage drop across resistor, $v_R = Ri$

Voltage drop across capacitor, $v_C = \dfrac{q}{C}$

Applied voltage, $V = v_R + v_C = Ri + \dfrac{q}{C}$

or $Ri + \dfrac{q}{C} = V$

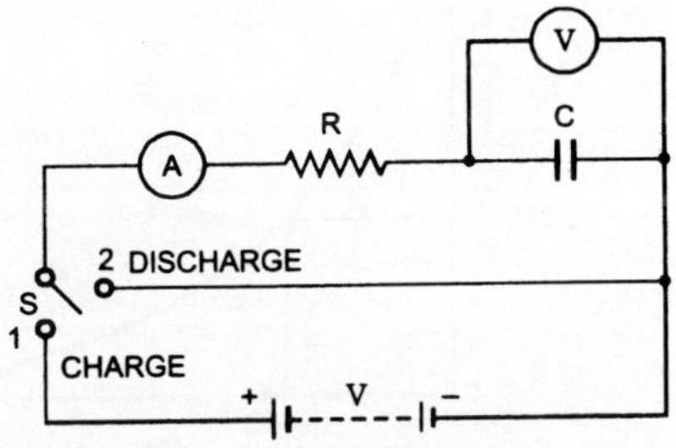

Fig. 2.26

Differentiating above expression w.r.t. 't' we have

$$R\frac{di}{dt} + \frac{1}{C}\frac{dq}{dt} = 0$$

$$\text{or} \quad R\frac{di}{dt} + \frac{1}{C}i = 0 \qquad \because \frac{dq}{dt} = i$$

$$\text{or} \quad \frac{di}{dt} + \frac{1}{RC}i = 0$$

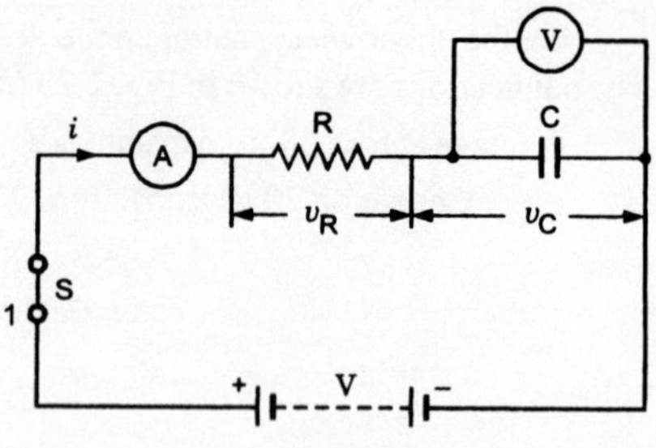

Fig. 2.27

General solution of above differential equation is

$$i = A\, e^{\frac{-t}{CR}}$$

where A is a constant and whose value can be determined from initial conditions.

At the instant of putting the switch S on, *i.e.* when $t = 0$, $i = \frac{V}{R}$ because the capacitor behaves as a short-circuit at this instant.

$$\text{So } i = A\, e^0 \quad \text{or} \quad A = \frac{V}{R}$$

and the expression for the current becomes

$$i = \frac{V}{R}\, e^{\frac{-t}{CR}} = \frac{V}{R}\, e^{-t/\tau} \qquad ...(2.20)$$

where τ is the time constant of the circuit and is equal to RC

$$\text{Voltage across resistor, } v_R = Ri = V\, e^{-t/\tau} \qquad ...(2.21)$$

$$\text{Voltage across capacitor, } v_C = V - v_R = V - V\, e^{-t/\tau} = V\,(1 - e^{-t/\tau}) \qquad ...(2.22)$$

Voltage across capacitor can also be given as

$$v_C = \frac{1}{C}\int_0^t i\, dt = \frac{1}{C}\int \frac{V}{R}\, e^{-t/CR}\, dt = \frac{1}{C}\cdot\frac{V}{R}\left[e^{-t/CR}\right]_0^t \times (-RC)$$

$$= -V e^{-t/CR}\Big|_0^t = -V\,(e^{-t/CR} - 1) = V\,(1 - e^{-t/\tau})$$

At $t = 0^+$ $\qquad i = \frac{V}{R} = I_{max}$; $v_R = V$ and $v_C = 0$

At $t = \infty$ $\qquad i = 0$; $v_R = 0$; and $v_C = V$

At $t = \tau = CR$ $\qquad i = \frac{V}{R}\, e^{-1} = 0.368\, \frac{V}{R}$

$$v_R = V\, e^{-1} = 0.368\, V \text{ and } v_C = 0.632\, V$$

The curve for i is shown in Fig. 2.28 while for v_R and v_C are shown in Fig. 2.29.

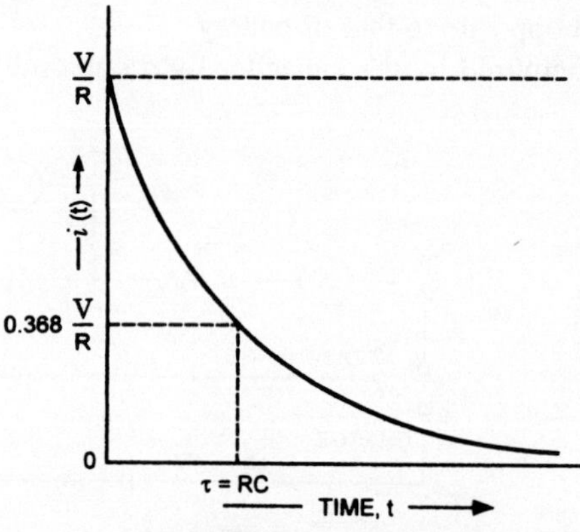

Fig. 2.28 *For i (t)*

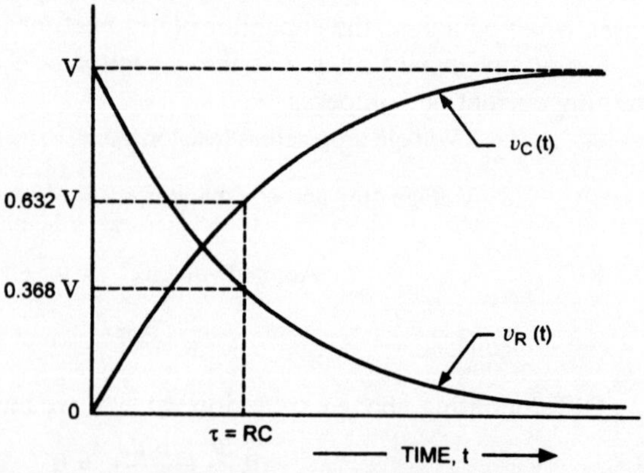

Fig. 2.29 *For v_R (t) and v_C (t)*

Substituting $v_C = \frac{q}{C}$ and $V = \frac{Q}{C}$ in Eq. (2.22) we have

$$\frac{q}{C} = \frac{Q}{C}(1 - e^{-t/\tau})$$

$$\text{or} \quad q = Q(1 - e^{-t/\tau}) \qquad ...(2.23)$$

i.e. charge increases following exponential law like growth of *pd* across the capacitor plates in which steady value is reached after infinite time.

Initial rate of rise of voltage across the capacitor C can be determined by differentiating Eq. (2.22) w.r.t. '*t*' for $t = 0$

$$\frac{dv_C}{dt} = \frac{V}{\tau} e^{-0} = \frac{V}{CR} \text{ volts/second} \qquad ...(2.24)$$

If initial rate of rise of voltage were maintained, then time required to attain voltage V would have been $\frac{V}{V/CR} = CR$. This time duration is called *time constant of the circuit* and is represented by symbol τ.

2.28.2. Discharging of a Capacitor. Now if the switch S is removed from position 1 and put in position '2' as shown in Fig. 2.30, the capacitor will get discharged through resistance R *i.e.* pd across the capacitor, v_C will decrease, current will flow in opposite direction to that of during charging and will decrease till it ceases when the capacitor is fully discharged.

At the instant of putting the switch S in position '2' after removing from position '1' when the capacitor has been fully charged

i.e. when $t = 0$ $v_C = V$ and $i = \frac{-V}{R}$

and $v_R + v_C$ = Total voltage across the circuit

or $iR + \frac{q}{C} = 0$

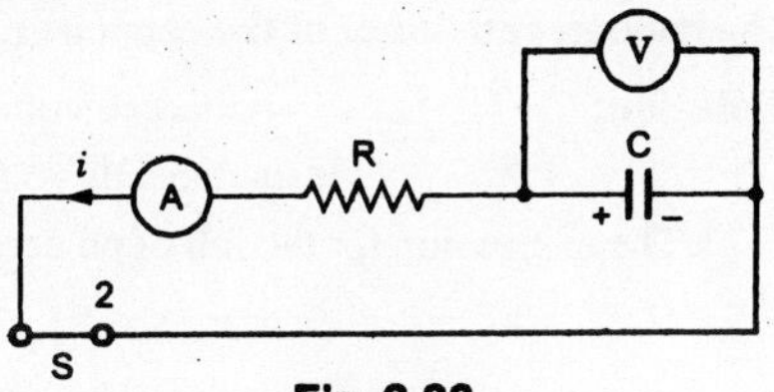

Fig. 2.30

Differentiating above equation w.r.t. time '*t*' we have

$$R\frac{di}{dt} + \frac{1}{C}\frac{dq}{dt} = 0$$

$$\text{or} \quad \frac{di}{dt} + \frac{1}{RC} i = 0$$

General solution of above differential equation is

$$i = A e^{-t/CR}$$

$$\text{When } t = 0, i = \frac{-V}{R}$$

$$\text{So } i = \frac{-V}{R} = A e^0 \quad \text{or} \quad A = \frac{-V}{R}$$

and the expression for current becomes

$$i = \frac{-V}{R} e^{-t/CR} = \frac{-V}{R} e^{-t/\tau} \qquad ...(2.25)$$

Voltage across capacitor, $v_C = -v_R = -iR = V e^{-t/\tau}$ ...(2.26)

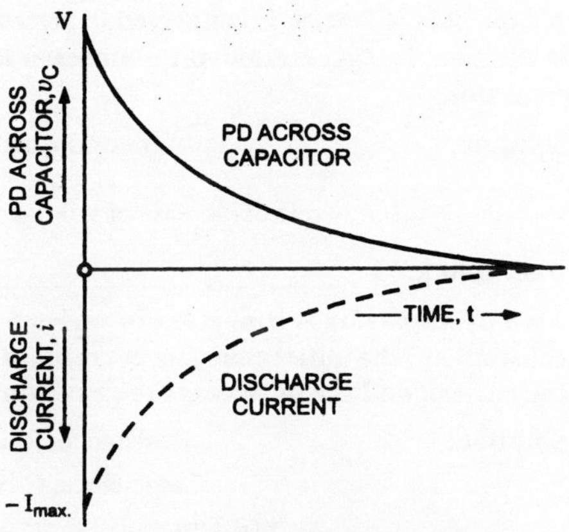

Fig. 2.31

Substituting $v_C = \frac{q}{C}$ and $V = \frac{Q}{C}$ in above Eq. (2.26) we have

$$\frac{q}{C} = \frac{Q}{C} e^{-t/\tau}$$

or $q = Qe^{-t/\tau}$ ...(2.27)

The fall of potential across capacitor and discharge current are shown in Fig. 2.31.

Example 2.30

A condenser of 8 μF capacitance is connected to a dc source through a resistance of 1 MΩ. Calculate the time taken for the condenser to receive 95% of its final charge. Sketch the charge time curve.

Solution: Capacitance of condenser,

$C = 8 \times 10^{-6}$ F

Resistance, $R = 1 \times 10^6\ \Omega$

Time constant, $\tau = CR = 8 \times 10^{-6} \times 1 \times 10^6 = 8$ seconds

Instantaneous charge on the capacitor is given by

$q = Q(1 - e^{-t/\tau}) = Q(1 - e^{-t/8})$

when $q = 0.95$ Q

$0.95\ Q = Q(1 - e^{-t/8})$

or $1 - e^{-t/8} = 0.95$

or $t = -8 \log_e 0.05 = 24$ seconds **Ans.**

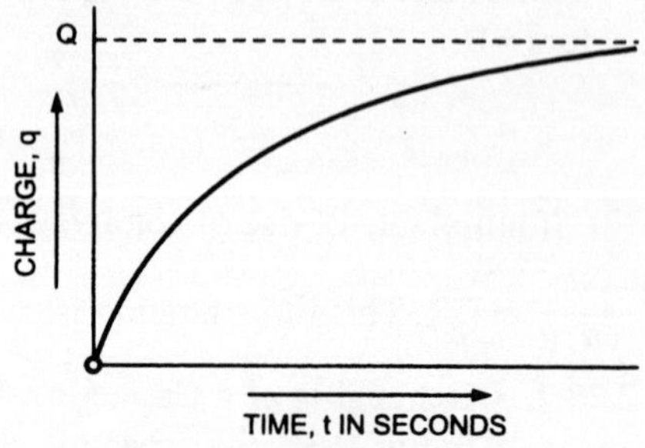

Fig. 2.32

Example 2.31

A capacitor of capacitance 12 μF is allowed to discharge through its own leakage resistance and a fall of pd from 120 V to 100 V is recorded in a time interval of 300 seconds by an electrostatic voltmeter connected in parallel. Calculate the leakage resistance of the capacitor.

Solution: Charged voltage, $V = 120$ V

Instantaneous voltage, $v = 100$ volts when $t = 300$ seconds

The expression for the fall of pd across a capacitor under discharge is

$$v = Ve^{-t/\tau}$$

$$\therefore\ 100 = 120\, e^{-300/\tau}$$

$$\text{or } \tau = \frac{300}{\log_e 1.2} = \frac{300}{0.1823} = 1{,}645 \text{ seconds}$$

and leakage resistance of capacitor, $R = \dfrac{\tau}{C} = \dfrac{1{,}645}{12 \times 10^{-6}} = 137$ MΩ **Ans.** $\because \tau = CR$

Example 2.32

A capacitor of 200 μF is connected in series with a resistance of 8,000 Ω. Estimate the time constant of the circuit. If the combination is suddenly connected to a 240 V dc supply, find the initial rate of rise of potential across the capacitor.

Solution: Time constant of the circuit, $\tau = CR = 200 \times 10^{-6} \times 8{,}000 = 1.6$ s **Ans.**

Initial rate of increase of voltage across capacitor $= \dfrac{V}{\tau}$ volts/second $= \dfrac{240}{1.6} = 150$ volts/second **Ans.**

Example 2.33

An 8 μF capacitor is connected in series with a 0.5 MΩ resistor across a 200 V dc supply. Calculate (*i*) the time constant, (*ii*) the initial charging current (*iii*) the time taken for the pd across the capacitor to grow to 160 V and (*iv*) the current and the pd across the capacitor in 4 seconds after it is connected to the supply.

Solution: Resistance, $R = 0.5\ M\Omega = 5 \times 10^5\ \Omega$

Capacitance, $C = 8\ \mu F = 8 \times 10^{-6}$ F

(*i*) Time constant, $\tau = RC = 5 \times 10^5 \times 8 \times 10^{-6} = 4.0$ seconds **Ans.**

(*ii*) Initial charging current $= I_{max} = \frac{V}{R} = \frac{200}{5 \times 10^5} = 0.4$ mA **Ans.**

(*iii*) Let the time taken be *t* seconds for the pd across the capacitor to grow 160 V

$$v_C = V(1 - e^{-t/\tau})$$

$$\text{or } 160 = 200(1 - e^{-t/4}) \quad \text{... Refer to Eq. (2.22)}$$

$$\text{or } t = -4 \log_e 0.2 = 6.438 \text{ seconds } \textbf{Ans.}$$

(*iv*) Current in 4 seconds after it is connected to supply

$$i = I_{max} e^{-t/\tau} \quad \text{... Refer to Eq. (2.20)}$$

$$= \frac{V}{R} e^{-t/\tau} = 0.4 \times 10^{-3} \times e^{-4/4} = 0.147 \text{ mA } \textbf{Ans.}$$

PD across the capacitor 4 seconds after it is connected to supply

$$v_C = V[1 - e^{-t/\tau}] \quad \text{... Refer to Eq. (2.22)}$$

$$= 200(1 - e^{-4/4}) = 126.42 \text{ V } \textbf{Ans.}$$

Example 2.34

A cable 10 km long and of capacitance 2.5 μF discharges through its insulation resistance of 50 MΩ. By what percentage the voltage would have fallen 1, 2 and 5 minutes respectively after disconnection from its bus-bars.

[*Pb. Univ Applied Electricity 1973 Supp.*]

Solution:

$$\text{Capacitance of cable, } C = 2.5 \times 10^{-6} \text{ F}$$

$$\text{Insulation resistance of cable, } R = 50 \times 10^6 \ \Omega$$

$$\text{Time constant, } \tau = CR = 2.5 \times 10^{-6} \times 50 \times 10^6 = 125 \text{ seconds}$$

During discharge the instantaneous voltage is given by the equation

$$v = Ve^{-t/\tau} = Ve^{-t/125}$$

When $t = 1$ minute *i.e.* 60 seconds, $v_1 = Ve^{-60/125} = 0.618$ V

When $t = 2$ minutes *i.e.* 120 seconds, $v_2 = Ve^{-120/125} = 0.383$ V

When $t = 5$ minutes *i.e.* 300 seconds, $v_3 = Ve^{-300/125} = 0.09$ V

When $t = 1$ minute, percentage fall in voltage across capacitor

$$= \frac{V - 0.618 \text{ V}}{V} \times 100 = 38.2\% \textbf{ Ans.}$$

When $t = 2$ minutes, percentage fall in voltage across capacitor $= \frac{V - 0.383 \text{ V}}{V} \times 100 = 61.7\%$ **Ans.**

When $t = 5$ minutes, percentage fall in voltage across capacitor $= \frac{V - 0.09 \text{ V}}{V} \times 100 = 91\%$ **Ans.**

HIGHLIGHTS

1. Static electricity may be defined as electricity at rest. The term "static electricity" properly applies to an isolated charge of electricity produced by friction.
2. Coulomb's law of electrostatic force between two point charges is

$$F = \frac{Q_1Q_2}{4\pi\varepsilon_0\varepsilon_r d^2} \quad \text{where } \varepsilon_0 = 8.854 \times 10^{-12} \text{ F/m}$$

3. Electric field intensity at any point distant *d* metres from a point charge of Q coulomb is

$$E = \frac{Q}{4\pi\varepsilon_0\varepsilon_r d^2} = \frac{9 \times 10^9 \times Q}{\varepsilon_r d^2} \text{ newtons/coulomb}$$

4. Electric flux is measured in coulombs and is equal to the charge

i.e. $\psi = Q$ coulombs

5. The electric flux density D is measured in C/m^2 and is given by

$$D = \frac{\psi}{A} = \varepsilon_0 \varepsilon_r E \ C/m^2$$

Also, $D = \sigma \ C/m^2$ where σ is charge density

6. According to Gauss's theorem

The total flux crossing the surface enclosing charges $Q_1, Q_2, ..., Q_n$ is given by

$$\psi = \sum_{i=1}^{n} Q_i$$

7. Electric field intensity at any point is equal to the negative potential gradient at that point *i.e.*

$$E = -\frac{dV}{ds} = -g, \text{ the potential gradient}$$

8. The potential at any point distant d metres from a point charge of Q coulombs is given by

$$V = \frac{Q}{4\pi\varepsilon_0\varepsilon_r d} = 9 \times 10^9 \frac{Q}{\varepsilon_r d} \text{ volts}$$

9. The surface in an electric field at right angle to the lines of force or flux everywhere is equipotential surface.
10. The electric intensity E, potential gradient g and dielectric strength are all numerically equal.
11. Capacitance of a capacitor is given by the ratio of its charge to the pd across its plates *i.e.* C = Q/V farads. By definition one farad is equal to one coulomb per volt (*i.e.* 1 F = 1 C/V).
12. The ratio of electric flux density in a vacuum (or free space) to the electric field intensity is termed the permittivity of the free space, represented by ε_0 and its value is 8.854×10^{-12} F/m.
13. Relative permittivity ε_r is ratio of capacitance of a capacitor having a certain dielectric to the capacitance of the same capacitor with vacuum (or air) as medium.
14. Capacitance of an isolated spherical conductor is given by $C = 4\pi\varepsilon_0\varepsilon_r r = \frac{\varepsilon_r r}{9 \times 10^9}$ F
15. Capacitance of concentric spherical capacitor with outer sphere grounded is given by

$$C = 4\pi\varepsilon_0\varepsilon_r \frac{ab}{b-a} \text{ farads} = \frac{\varepsilon_r}{9 \times 10^9} \frac{ab}{a-b} \text{ farads}$$

16. Capacitance of concentric spherical capacitor with inner sphere grounded is given by

$$C = 4\pi\varepsilon_0\varepsilon_r \frac{b^2}{b-a} \text{ farads}$$

17. Capacitance of a parallel plate capacitor is given by

$$(i)\ C = \frac{\varepsilon_0 A}{\frac{t}{\varepsilon_{r_1}} + \frac{t_2}{\varepsilon_{r_2}} + \frac{t_3}{\varepsilon_{r_3}}} \text{ farads}$$

when capacitor plates are separated by different dielectrics of different thickness,

$$(ii)\ C = \frac{\varepsilon_0 \varepsilon_r A}{d} \text{ farads,}$$

when the capacitor plates are d metres apart by dielectric of relative permittivity ε_r

$$(iii)\ C = \frac{\varepsilon_0 A}{d} \text{ farads} \quad \text{when medium is air or vacuum.}$$

18. Capacitance of a multiple parallel-plate capacitor having n plates is given by

$$C = \frac{(n-1)\,\varepsilon_0\varepsilon_r A}{d} \text{ farads}$$

19. Capacitance of a cylindrical capacitor is given by

$$C = \frac{0.024\,\varepsilon_r}{\log_{10} \frac{D}{d}} \ \mu\text{F/km}$$

20. Energy stored in a capacitor is given by

$$W = \frac{Q^2}{2C} = \frac{1}{2} Q V = \frac{1}{2} C V^2$$

21. Energy stored per cubic metre of volume of dielectric in a capacitor is given by

$$W = \frac{1}{2} \varepsilon E^2 = \frac{1}{2} D E = \frac{1}{2} \frac{D^2}{\varepsilon} \text{ joules}$$

22. Capacitance of combination of capacitors connected in series is given by

$$\frac{1}{C} = \frac{1}{C_1} + \frac{1}{C_2} + \frac{1}{C_3} + ...$$

23. Capacitance of combination of capacitors connected in parallel is given by

$$C = C_1 + C_2 + C_3 + ...$$

24. During charging of a capacitor
(*i*) Voltage across the capacitor rises according to equation $v_C = V(1 - e^{-t/\tau})$
(*ii*) Charge increases according to equation $q = Q(1 - e^{-t/\tau})$
(*iii*) Charging current diminishes according to equation $i = I_{max}\, e^{-t/\tau}$

25. During discharging of a capacitor
(*i*) Voltage across the capacitor falls according to equation $v = V\, e^{-t/\tau}$
(*ii*) Charge decreases according to equation, $q = Q\, e^{-t/\tau}$
(*iii*) Current decreases according to equation $i = -I_{max}\, e^{-t/\tau}$

where τ is the time constant of the R-C circuit which is defined as the time during which capacitor voltage rises to 63.2% of its final steady value.

EXERCISES

1. What is static electricity ? Explain the phenomenon of electrostatic induction.
2. State Coulomb's law for force between two point charges. Hence define unit charge in SI system.
3. Define electric potential, field intensity and flux density and give the interrelationship between them.
4. Define and explain the terms (*i*) electric field (*ii*) electric field intensity (*iii*) electric flux (*iv*) electric flux density (*v*) electric potential (*vi*) potential gradient and (*vii*) equipotential surface.
5. Define (*a*) field strength and (*b*) potential at a point in electric field and state the units in which they are measured.
6. Define and explain the terms: capacitance, dielectric strength, dielectric constant and farad.
7. The intervening medium between two concentric metallic spheres is filled with a dielectric medium of perimittivity ε. Derive an expression for the capacitance of the spherical capacitor. Find also the electrostatic energy stored by this capacitor.
8. A parallel plate capacitor has plate area A and plate separation '*d*'. The space between plates is filled up to a thickness x ($< d$) with a dielectric constant k. Calculate the capacitance of system.
9. Derive an expression for the capacitance of an n-parallel plate capacitor.
10. Derive an expression of capacitance for unit length of coaxial cylinders in terms of radii of the inner and outer cylinders and permittivity of the medium.
11. Derive an expression for the capacitance of a single core cable.
12. Find an expression for the energy stored in a condenser of capacitance C farads charged to a pd of V volts.
13. Derive an expression for the equivalent capacitance of a group of capacitors when they are connected (*i*) in parallel (*ii*) in series.
14. A condenser of capacitance C_1 at potential V_1 is connected to another condenser of capacitance C_2 at potential V_2. Show that the loss of energy becomes equal to $\frac{C_1 C_2}{C_1 + C_2}\left[\frac{V_1 - V_2}{\sqrt{2}}\right]^2$.

15. Explain the charging and discharging phenomenon of a capacitor through a resistor.

16. A capacitor 'C' in series with a resistor 'R' is connected to a dc source of E volts. Find an expression for the charge q as a function of time 't' ?

17. A condenser is charged through a large non-reactive resistance by a battery of constant voltage V. Derive an expression for the instantaneous voltage on the condenser. Draw its volt-time curve.

SHORT ANSWER TYPE QUESTIONS WITH ANSWERS

Q. 1. What is static electricity ?

Ans. Static electricity may be defined as electricity at rest. The term "static electricity" properly applies to an isolated charge of electricity produced by friction.

Q. 2. What is Coulomb's law relating force between two charges placed near each other ?

Ans. Mathematically force F between two point charges of strength Q_1 and Q_2, placed distance d apart is given by

$$F = \frac{Q_1 Q_2}{4\pi\varepsilon_0\varepsilon_r d^2}$$

Q. 3. On what factors does the force acting between two point charges placed near each other depend ?

Ans. The force acting between two point charges placed near each other depends upon (*i*) the strength of the two charges (*ii*) distance between the charges and (*iii*) nature of medium surrounding the charges.

Q. 4. Is electric field a vector quantity ?

Ans. Yes, electric field intensity is a vector quantity.

Q. 5. What is relation between electric flux density and electric field intensity ?

Ans. Electric flux density, $D = \varepsilon_o \varepsilon_r E$.

Q. 6. What is Gauss's theorem ?

Ans. According to Gauss's theorem the total electric flux ψ from a closed surface is equal to the total charge enclosed by the surface.

Q. 7. What is Coulomb's theorem ?

Ans. According to Coulomb's theorem, the electric intensity at a surface of a conducting body, having a charge density of σ coulomb's per unit of surface area is $\frac{\sigma}{\varepsilon_0\varepsilon_r}$ where $\varepsilon_0\,\varepsilon_r = \varepsilon$, the absolute permittivity of the medium surrounding the conductor.

Q. 8. What is relationship between electric field intensity and potential gradient ?

Ans. Electric field intensity at any point is equal to the negative potential gradient at that point *i.e.* $E = -g$.

Q. 9. What is permittivity of free space ?

Ans. Permittivity of free space, $\varepsilon_0 = 8.854 \times 10^{-12}$ F/m.

Q. 10. What is equipotential surface ?

Ans. The surface in an electric field at right angle to the lines of force or flux everywhere is equipotential surface.

Q. 11. How much work will be done in transferring a charge of Q coulombs over an equipotential surface ?

Ans. Zero.

Q. 12. What is the capacitance and what is its unit ?

Ans. Capacitance is the property of a capacitor to store electric charge required to raise its potential through unity and is given by

$$C = \frac{Q}{V} \quad \text{and its unit is farad}$$

Q. 13. If air is replaced by a dielectric permittivity ε_r in a condenser of capacity C with air as medium, what will be effect on its capacity ?

Ans. Capacity of the condenser will increase and become ε_r C.

Q. 14. What is a meant by dielectric strength of a medium ?

Ans. The dielectric strength is the property of a medium which enables it to withstand electric stress without damage. It is usually expressed in terms of the minimum electric stress (potential difference per unit distance) that will cause failure or breakdown of the medium under certain conditions of service.

Q. 15. There are two concentric conducting spherical surfaces of radii a and b ($a < b$). The inner spherical surface carries a charge Q and the outer surface is grounded. What is the potential of the inner spherical surface ?

Ans. $V = \frac{Q}{4\pi\varepsilon_0 \varepsilon_r}\left(\frac{1}{a} - \frac{1}{b}\right)$ Volts.

PROBLEMS

1. Determine the force between two charges each of one coulomb when they are spaced 0.5 metre distance in air. [**Ans.** 36×10^9 newtons]
2. Calculate the capacitance of a condenser which has plates of 0.2 m^2 in area separated by a distance of 2.5×10^{-5} metre, the dielectric having a permittivity of 3.5. [**Ans.** 0.248 μF]
3. A parallel-plate capacitor has two circular plates each having area of 120 cm^2, separated by 1.5 mm of dielectric. Its capacitance is 350 μF. If a potential difference of 11,000 volts is applied across the capacitor, what will be the flux density and the dielectric constant of the dielectric. [Take absolute permittivity = 8.854×10^{-12} F/m] [**Ans.** 320.8×10^{-6} C/m^2 : 4.94]
4. A parallel-plate capacitor has plates of area 2 m^2 spaced by 3 slabs of different dielectric materials. The relative perimittivities are 2, 3, and 6 and the thickness are 0.4, 0.6 and 1.2 mm respectively. Calculate the combined capacitance and the electric stress in each material when the applied voltage is 1,000 volts. [**Ans.** 0.0295×10^{-6} F; 833 kV/m, 555 kV/m, and 278 kV/m]
5. A capacitor consists of two metal plates, each 40 cm × 40 cm spaced 6 mm apart. The space between the metal plates is filled with a glass plate 5 mm thick and a layer of paper 1 mm thick. The dielectric constants of glass and paper are 8 and 2 respectively. Calculate (*i*) the capacitance (*ii*) the potential gradient in each dielectric in kV/mm due to a p d of 10 kV between the metal plates. [**Ans.** 0.00126 μF: 1.11 kV/mm, 4.44 kV/mm]
6. A 5 mm thick slab of insulating material is inserted between the plates of a parallel plate capacitor. It was found necessary to increase the separation between the plates by 2.6 mm to maintain the value of its capacitance to original value. Determine the dielectric constant of the material of the slab. [**Ans.** 2.08]
7. In a capacitor with two plates separated by an insulator 3 mm thick and of relative permittivity of 4, the distance between the plates is increased to allow the insertion of a second insulator 5 mm thick and relative permittivity ε. If the capacitance so formed is one-third of the original capacitance, find ε. [**Ans.** 3.33]
8. A capacitor is composed of two plates separated by two dielectric materials of thickness of 4 mm and 5 mm having perimittivities of 4 and 5 respectively. This capacitor is connected across 10 kV supply. Find the potential gradient along the dielectric materials. [**Ans.** 1.25×10^6 V/m; 1×10^6 V/m]
9. A parallel-plate condenser is made up of 31 metal strips, each of 10 cm radius separated by sheets of mica of dielectric constant 6 and each of thickness 0.02 mm. Find its capacity. [**Ans.** 2.5 μF]
10. A capacitor consists of 10 similar square aluminium plates each 10 cm × 10 cm mounted parallel to each other. Alternate plates are electrically connected to give two terminals. What is the capacitance between the terminals when distance between any two plates is 1 cm and the dielectric is air ? [**Ans.** 79.686×10^{-12} F]
11. Determine the capacitance of a 5 km long cable having an inner core of diameter 10 mm and impregnated paper insulation of thickness 7 mm and relative permittivity 4. [**Ans.** 1.26 μF]
12. A capacitor of 5 μF is charged to a potential difference of 8,000 V. Determine the energy stored in (*i*) joules (*ii*) watt-hours. [**Ans.** 160 J, 0.0445 watt-hours]
13. Two condensers of capacity 2 μF and 4 μF respectively are connected in series. A potential difference of 900 V is applied between the extreme terminals. Find the pd across each condenser. [**Ans.** 600 V, 300 V]
14. Consider that three capacitors having capacitances of 5, 10 and 12 microfarad are connected in series across 600 V dc mains. Determine (*i*) equivalent capacitance of the combination (*ii*) charge on each capacitor (*iii*) potential across each capacitor (*iv*) stored energy in each of the capacitor. [**Ans.** (*i*) 2.6087 μF (*ii*) 0.001565 C (*iii*) 313 V, 156.5 V, 130.5 V, (*iv*) 0.245 J, 0.122 J, 0.102 J]
15. Three capacitors of capacitances 15 μF, 20 μF and 30 μF are connected in series. A potential difference of 400 volts is applied across the combination. Find (*i*) the equivalent capacitance (*ii*) voltage distribution across each capacitor (*iii*) energy stored. [**Ans.** (*i*) 6.667 μF (*ii*) 177.78 V, 133.33 V, 88.89 V, (*iii*) 0.533 J]

16. Two capacitors of 4 μF and 6 μF are connected in series across apd of 100 V. Determine the charge on the capacitor and total energy stored in them. **[Ans.** 240 μC; 0.012 J]

17. Two capacitors when connected in parallel across a 250 V supply have charges of 2,000 and 2,500 microcoulombs. Find the energy stored in each capacitor. **[Ans.** 0.25 J; 0.3125 J]

18. Three capacitors A, B and C are charged as follows :
A = 10 μF, 100 V B = 15 μF, 150 V and C = 25 μF, 200 V
They are connected in parallel with terminals of like polarities together. Find the voltage across the combination. **[Ans.** 165 V]

19. The total capacitance of two capacitors is 0.03 μF when joined in series and 0.16 μF when connected in parallel. Calculate the capacitance of each capacitor. **[Ans.** 0.12 μF and 0.04 μF]

20. Find the equivalent capacitance between the points A and B of the network shown in Fig. 1.

$$\left[\textbf{Ans.}\ \frac{512}{219}\ \mu F\right]$$

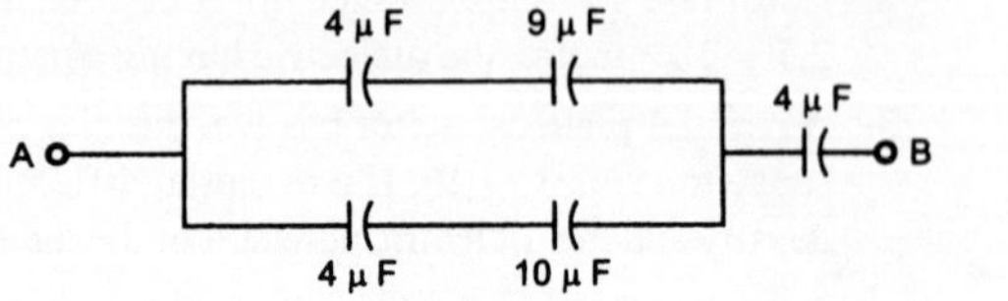

Fig. 1

21. Three capacitors A, B and C, have capacitances 10, 50 and 25 μF, respectively. Calculate (*a*) when connected in parallel (*b*) when connected in series across 250 V supply (*i*) total capacitance (*ii*) charge on each capacitor (*iii*) voltage across each capacitor, in both cases.

[Ans. (*a*) (*i*) 85 μF (*ii*) 0.0025 C, 0.0125 C, 0.00625 C, (*iii*) 250 V (*b*) (*i*) 6.25 μF (*ii*) 0.0015625 C (*iii*) 156.25 V, 31.25 V and 62.5 V]

22. A capacitor of 4 μF capacitance is charged to a pd of 400 V and then connected in parallel with an uncharged capacitor of 2 μF capacitance. Calculate the pd across the parallel capacitors. **[Ans.** 266.67 V]

23. A 9 μF capacitor is connected in series with a parallel combination of two capacitors of values of 4 μF and 2 μF respectively (*i*) Determine the capacitance of the combination (*ii*) If a pd of 20 V is maintained across the combination, determine the charge on the 9 μF and the energy stored in the 4 μF capacitor.

[Ans. (*i*) 3.6 μF (*ii*) 72 μC (*iii*) 288 × 10^{-6} J]

24. Two capacitors A and B are placed (*i*) in series, (*ii*) in parallel, C_A = 100 μF, C_B = 50 μF. Find the maximum stored energy in the circuit when a 240 V, 50 Hz supply is applied. **[Ans.** (*i*) 1.92 J (*ii*) 8.63 J]

25. How should three capacitors of capacitances 7 μF, 4 μF and 12 μF be connected to give a capacitance of 10 μF ?

[Ans. Capacitance of 7 μF in parallel with the capacitances of 4 μF and 12 μF in series]

26. A 3 μF capacitor is charged to a potential difference of 200 V and then connected in parallel with another 2 μF uncharged capacitor. Determine the potential difference across the parallel combination and the energy stored in each capacitor before and after the connection. **[Ans.** 120 V, 0.06 J, 0.00216 J, 0.00144 J]

27. A capacitor comprises of two equal metallic plates 15 cm in diameter, placed parallel to another at a distance of 7.5 mm apart. It is connected to dc voltage of 500 V, disconnected and then immersed in oil of dielectric constant 4. Determine energy stored in the capacitor before and after immersion. Account for the difference.

[Ans. 26 × 10^{-7} J; 6.5 × 10^{-7} J]

28. Two square metal plates 50 cm × 50 cm with a separation of 1 cm constitute a capacitor. The metal plates are separated by a slab of 0.8 cm thick, remaining distance of 0.2 cm being filled with air. Find the capacitance of the system. If the voltage of 600 V is applied to the capacitor, find the energy stored in it. Relative permittivity of sulphur is 4.

[Ans. 553.4 × 10^{-12} F; 99.6 μJ]

29. A capacitor of 3 μF capacitance is connected through a resistance of 1 MΩ to a constant voltage. Find how long will it take before the capacitor receives 99% of its final charge. **[Ans.** 13.8 seconds]

30. A capacitor of 0.1 μF is charged from a 200 V battery through a series resistance of 1,000 Ω. Determine

(*i*) the time for the capacitor to receive 63.2% of its final charge,
(*ii*) the charge received in this time,
(*iii*) the initial rate of charging and
(*iv*) the rate of charging when the charge is 63.2% of its final charge.

[Ans. (*i*) 10^{-4} second (*ii*) 12.64 μC (*iii*) 2 × 10^6 V/s (*iv*) 0.736 × 10^6 V/s]

❑❑❑

UNIT-III
Electromagnetics

CONTENTS IN THIS UNIT

Electromagnetics: Magnetic circuits; Comparison between electric and magnetic circuits; Series/parallel magnetic circuit calculations; Magnetic hysteresis; Hysteresis and eddy current loss; Magnetic materials; Electromagnetic induction; Statically and dynamically induced emfs in brief. Fleming's right hand rule; Left hand rule; Coefficients of self and mutual induction; Coefficient of coupling; Series/parallel combinations of inductances; Rise and decay of current in inductive circuits; Force experienced by a current carrying conductor placed in magnetic field.

Chapter

3 Electromagnetics

3.1. INTRODUCTION

Practically all electric power machinery (*e.g.* transformers, generators, motors) depend for their operation upon the magnetism produced by the magnetic circuits. The closed path followed by magnetic flux is called a *magnetic circuit* (Fig. 3.1) just as the closed path followed by current is called an electric circuit. There is a lot of similarity between the magnetic and electric circuits, and will be shown later on during discussion.

A magnetic circuit consists of a structure composed for the most part of high permeability magnetic material. The presence of high permeability material causes the magnetic flux to be confined to the paths defined by the structure, much as currents are confined to the conductors of an electric circuit. A simple example of a magnetic circuit is shown in Fig. 3.1. The core is assumed to be composed of magnetic material whose permeability is much greater than that of the surrounding air. The core is of uniform cross-section and is excited by a winding having N turns and carrying a current of I amperes. This winding develops a magnetic field in the core, as illustrated in the figure. The magnetic field can be visualized in terms of flux lines, which form closed loops interlinking with the winding. According to basic law of magnetic field, called the *Ampere's circuital law* (sometimes referred to as *Ampere's work law*) the line integral of **H** around a closed path is equal to the net current enclosed by that path *i.e.*

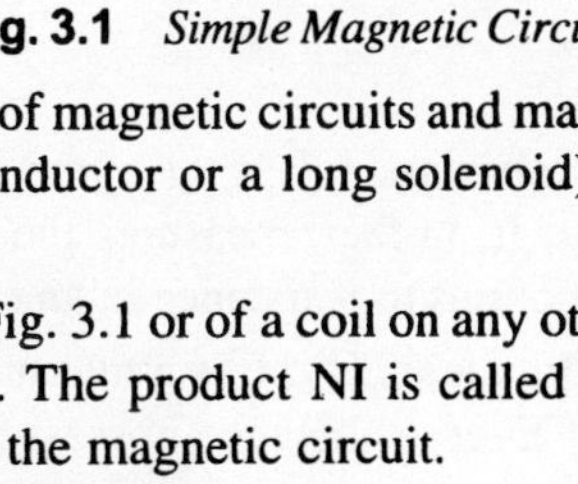

Fig. 3.1 *Simple Magnetic Circuit*

$$\oint \mathbf{H}\, dl = \Sigma \mathrm{I} \qquad ...(3.1)$$

The above law is very comprehensive and provides a basis for calculation of magnetic circuits and makes it possible in some cases (such as in the cases of a long current carrying conductor or a long solenoid) to determine readily the strength of the magnetic field.

3.1.1. Magnetic Circuit Concepts. The flux producing ability of the coil in Fig. 3.1 or of a coil on any other magnetic circuit is proportional to the number of turns N and the current I. The product NI is called the *magnetomotive force* (mmf) and determines the amount of flux developed in the magnetic circuit.

MMF of the magnetic circuit is the magnetic potential difference that tends to force flux around the magnetic circuit and is analogous to the electromotive force (emf) in an electric circuit.

$$\text{MMF} = \text{NI ampere-turns} \qquad ...(3.2)$$

The resulting flux ϕ in the magnetic circuit, besides being dependent on the mmf, is also a function of the opposition of the iron to carrying flux. This opposition is called the *reluctance* S of the magnetic circuit.

As in the case with resistance in the electric circuit, reluctance S of the magnetic circuit is directly proportional to length l, inversely proportional to cross-sectional area a, and dependent on the nature of material of the magnetic circuit. The reluctance of the magnetic circuit is given as

$$S = \frac{l}{\mu a} \text{ ampere-turns/weber} \quad ...(3.3)$$

when the flux ϕ is constant over the length and uniform over the area. The quantity μ expresses the property of the magnetic material called its *permeability*. Permeability is a measure of the receptiveness of the material of having magnetic flux developed in it. For free space, the permeability μ_0 equals $4\pi \times 10^{-7}$ H/m in the SI system. The relative permeability of magnetic material μ_r may range up to thousands.

Total flux developed in the circuit is given as

$$\phi = \frac{\text{MMF}}{S} \text{ webers} \quad ...(3.4)$$

The above Eq. (3.4) is sometimes referred to as *Ohm's law* for the magnetic circuit. It serves to emphasize the mathematical analogy between the magnetic circuit and the electric circuit. Analogous quantities in the two circuits are listed in Art 3.1.4.

Magnetic circuits differ from electric circuits in one important respect. The reluctance of the magnetic circuit containing iron or ferromagnetic material depends upon the flux carried by it. With the increase in flux, a large change in mmf is required to develop the same change in flux.

3.1.2. Terminology

1. Magnetic Field. Magnetic field is the region around a magnet within which the influence of the magnet is felt. The space around a lode stone, around a compass needle, around the earth and around a permanent magnet are examples of magnetic field.

2. Magnetic Flux. A magnetic field over a region in space is represented graphically by magnetic lines of flux (or simply lines of force). The total lines of magnetic flux in any particular magnetic field, passing through a surface, is called *magnetic flux*. It is symbolised by ϕ and is measured in webers.

3. Flux Density. It is defined as the magnetic flux passing per unit area through any material through a plane at right angles to the direction of magnetic flux. It is measured in Tesla (or Wb/m^2).

4. MMF of the magnetic circuit is defined as the magnetic potential that drives or tends to drive flux around the magnetic circuit and is analogous to the emf in an electric circuit. It is measured in ampere-turns (AT) or simply in amperes.

5. Magnetic Field Intensity. MMF per unit length (along the path of magnetic flux) is called the magnetic field intensity (H) and is given as

$$H = \frac{\text{MMF}}{\text{Length}} = \frac{NI}{l} \text{ AT/m}$$

6. Reluctance. It is the name given to that property of material which opposes the creation of magnetic flux in it. It, in fact, measures the resistance offered to the passage of magnetic flux through a material and is analogous to resistance in an electric circuit. It is measured in ampere-turns / Wb or amperes / Wb.

7. Permeability. It is a measure of receptiveness of the material of having magnetic flux developed in it.

Every substance possesses a certain power of conducting magnetic flux. For example, iron is a better conductor for magnetic flux than air (or vacuum). Permeability of a material (μ) is thus its conducting power for magnetic flux. It is equal to the ratio of flux density B and the magnetic field strength H i.e. $\mu = \frac{B}{H}$.

8. Relative Permeability. Relative permeability of a material is defined as the ratio of the number of lines of magnetic flux per unit area in the given medium to the number of lines of magnetic flux per unit area, when medium is replaced by vacuum.

Relative permeability of free space is unity and of magnetic material may range up to thousands.

3.1.3. Determination of Ampere-Turns. In any magnetic circuit, flux created is given as

$$\phi = \frac{\text{MMF}}{\text{Reluctance}} = \frac{\text{AT}}{l/\mu a} = \frac{\text{AT}}{l/\mu_0\,\mu_r\,a}$$

$$\text{or AT required} = \frac{\phi}{a} \times \frac{1}{\mu_0\,\mu_r} \times l = \frac{\text{B}}{\mu_0\,\mu_r} \times l = \text{H} \times l \text{ ampere-turns} \qquad ...(3.5)$$

Hence for determination of AT for a magnetic circuit

(*i*) First find field strength H in each part of the magnetic circuit, (*ii*) find the length of various parts of magnetic circuit, (*iii*) find the number of ampere-turns required for the various parts of magnetic circuit from the relation AT = Hl where l is the length of the part in metres and lastly, (*iv*) find total number of ampere-turns for the whole series magnetic circuit by adding ampere-turns determined for various paths in magnetic circuit.

This will be more clear to the students from examples given in this chapter.

As both permeability and reluctance may change from one operating condition to another. Hence direct numerical application of the reluctance concept and Eqs. (3.4) and (3.5) is rare. For quantitative analysis, graphical methods are generally used because they can easily be adapted to the nonlinearities involved (Art 3.8).

3.1.4. Comparison Between Magnetic and Electric Circuits

	Magnetic Circuit	**Electric Circuit**
	SIMILARITIES	
Circuit	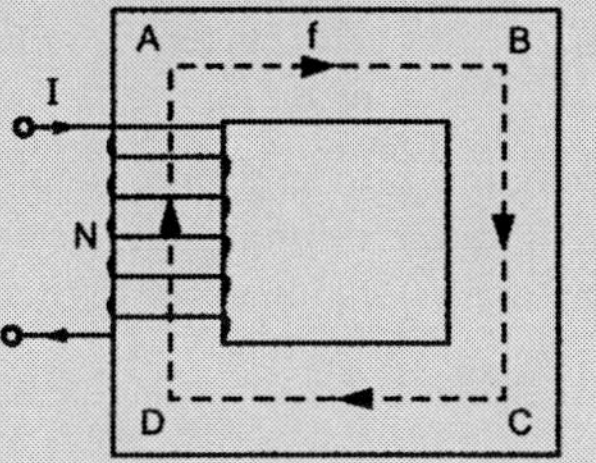	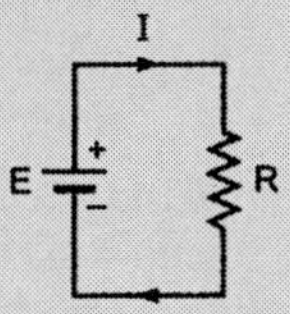
1. Definition	The closed path followed by magnetic flux is called a magnetic circuit	The closed path followed by an electric current is called an *electric circuit*
2. Driving force	**MMF** is required to establish flux ϕ in the magnetic circuit and is measured in ampere-turns (AT) or amperes	**EMF** is required to cause flow of current in an electric circuit and is measured in volts
3. Response	Flux, $\phi = \dfrac{\text{Driving force}}{\text{Magnetic reluctance}} = \dfrac{\text{AT}}{\text{S}}$ webers	Current, $\text{I} = \dfrac{\text{Driving force}}{\text{Electric resistance}} = \dfrac{\text{E}}{\text{R}}$ amperes
4. Impedance	Reluctance, $\text{S} = \dfrac{l}{\mu_0\mu_r a} = \dfrac{l}{\mu a}$ AT/Wb	Resistance, $\text{R} = \rho\,\dfrac{l}{a}$ ohms
(a) For series circuits	$\text{S} = \text{S}_1 + \text{S}_2 + \text{S}_3 + ...$	$\text{R} = \text{R}_1 + \text{R}_2 + \text{R}_3 + ...$
(b) For parallel circuits	$\text{S} = \dfrac{1}{\dfrac{1}{\text{S}_1} + \dfrac{1}{\text{S}_2} + \dfrac{1}{\text{S}_3} + ...}$	$\text{R} = \dfrac{1}{\dfrac{1}{\text{R}_1} + \dfrac{1}{\text{R}_2} + \dfrac{1}{\text{R}_3} + ...}$
5. Admittance	Permeance $= \dfrac{1}{\text{Reluctance}}$ in Wb/AT	Conductance $= \dfrac{1}{\text{Resistance}}$ in siemens

6. Proportionality constant	Reluctivity $= \dfrac{1}{\text{Permeability}} = \dfrac{1}{\mu}$ in m/H	Resistivity, $\rho = \dfrac{1}{\text{Conductivity}}$ in ohm-metre
7. Density	Flux density, $B = \mu \times$ magnetic field intensity $= \mu H$ is tesla or Wb/m^2	Current density, $J = \dfrac{\text{Electric field intensity}}{\rho}$ in A/m^2
8. Field intensity	Magnetic field intensity, $H = \dfrac{MMF}{l} = \dfrac{NI}{l}$ in AT/m	Electric field intensity $= \dfrac{E}{l}$ in volts/metre
9. Drop	MMF drop $= \phi \times$ reluctance $= \phi S$	Voltage drop $= I \times$ Resistance $= IR$

DISSIMILARITIES	
1. Flux does not actually flow in a magnetic circuit.	Current does flow in an electric circuit.
2. Permeability does not largely vary from material to material and there is hardly any material which can act as an insulator to the magnetic flux.	Conductivity varies largely from material to material so much so that some materials are insulators to electric current and some are very good conductors.
3. For a particular temperature the permeability depends up on the flux density (or total flux).	For a particular temperature, conductivity (or resistivity) is constant and independent of current strength.
4. Flux can pass through air.	Current would not flow through the air until an arc is struck.
5. Residual flux persists after removal of mmf.	The current is reduced to zero after removal of source of emf.
6. There is no waste of energy due to reluctance in the magnetic circuit and, therefore, energy is required only to create the magnetic flux but not to maintain it.	In an electric circuit, resistance causes heat to be generated resulting in waste of energy and, therefore, energy is required as long as the current flows.

3.1.5. Magnetic Circuits With Air-Gaps. Energy-conversion devices which incorporate a moving element have necessarily air gaps in their magnetic circuits. Air-gaps are also provided in the magnetic circuits to avoid saturation. A magnetic circuit with an air gap is shown in Fig. 3.2. An air-gap is nothing else but a volume of air between two magnetic surfaces. The length of air gap l_g equals the distance between the two magnetic surfaces. The area of x-section of any one of the surfaces gives the air-gap area, a_g. When the air-gap length l_g is much smaller than the dimensions of the adjacent core faces, the magnetic flux ϕ is constrained essentially to reside in the core and the air gap and is continuous through out the magnetic circuit. Thus, the configuration shown in Fig. 3.2 can be analysed as a magnetic circuit with two series components, a magnetic or iron core of permeability μ and mean length l_i and an air-gap of permeability μ_0 and length l_g. Since the permeability of air is constant, the air-gap is a linear part of the magnetic circuit and the flux density in the air-gap is proportional to the mmf across the air-gap. The necessary mmf is calculated separately for the air-gap and the iron portions and then added to determine the total mmf.

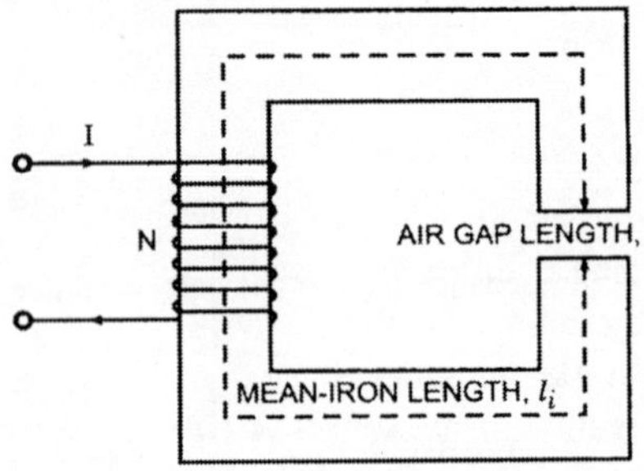

Fig. 3.2 *Magnetic Circuit With Air Gap*

3.1.6. Composite Magnetic Circuits. Consider a circular ring made from different materials of lengths l_1, l_2 and l_3, cross-sectional areas a_1, a_2 and a_3 and relative permeability μ_{r1}, μ_{r2} and μ_{r3} respectively with a cut of length l_g known as air-gap. The total reluctance is the arithmetic sum of individual reluctances as they are joined in series.

∴ Total reluctance,

$$S = \frac{l_1}{\mu_0 \mu_{r_1} a_1} + \frac{l_2}{\mu_0 \mu_{r_2} a_2} + \frac{l_3}{\mu_0 \mu_{r_3} a_3} + \frac{l_g}{\mu_0 a_g}$$

$$\text{Total mmf} = \phi \times S$$

$$= \phi \left[\frac{l_1}{\mu_0 \mu_{r_1} a_1} + \frac{l_2}{\mu_0 \mu_{r_2} a_2} + \frac{l_3}{\mu_0 \mu_{r_3} a_3} + \frac{l_g}{\mu_0 a_g} \right]$$

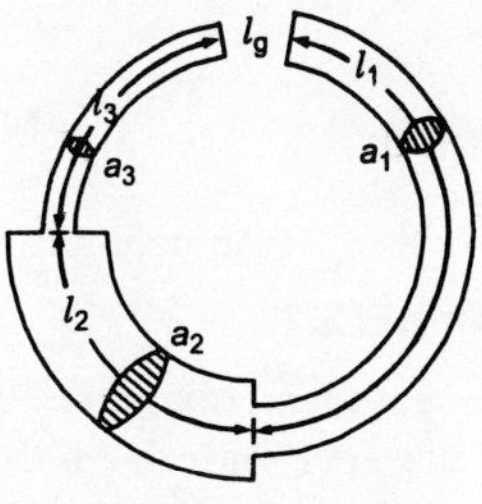

Fig. 3.3 *Composite Magnetic Circuit*

or Total ampere-turns required

$$= H_1 l_1 + H_2 l_2 + H_3 l_3 + H_g l_g \quad ...(3.6)$$

= Sum of ampere-turns required for individual parts of the magnetic circuit.

3.1.7. Parallel Magnetic Circuits. In series circuits, all parts of the magnetic circuit carry same flux and total ampere-turns required to create a given flux is the arithmetic sum of the ampere-turns required for individual parts of the circuit.

But if the various paths of the magnetic circuit are in parallel, as shown in Fig. 3.4 the ampere-turns required for the combination is equal to the ampere-turns required to create the given flux in one path.

For example for the circuit shown in Fig. 3.4 paths ABCD and AFED are in parallel, so ampere-turns required to create flux ϕ_1 in path ABCD is equal to ampere-turns required to create flux ϕ_2 in path AFED and also equal to the ampere-turns required for both of the paths.

Hence total ampere-turns required for magnetic circuit shown in Fig. 3.4.

= AT for path DA + AT for path ABCD

= AT for path DA + AT for path AFED

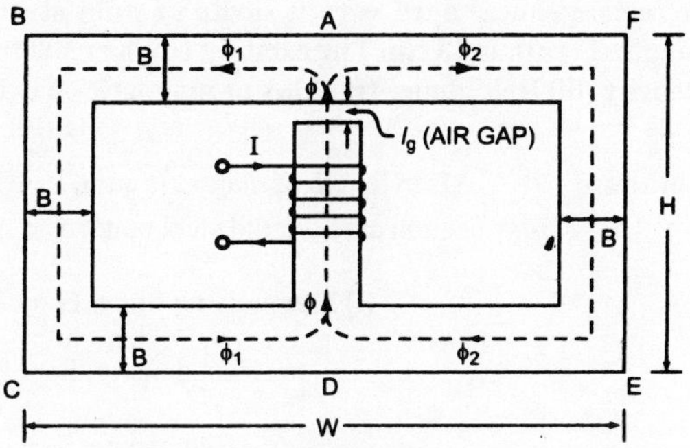

Fig. 3.4

3.1.8. Magnetic Leakage and Fringing. Leakage flux is the flux, which follows a leaking path, as shown in Fig. 3.5. Flux in the air gap is known as *useful flux* which is utilized for various useful purposes. For the purpose of calculations the iron is supposed to carry whole of the flux throughout its entire length.

The ratio of total flux (flux in the iron path) to the useful flux (flux in the air) is known as *leakage factor*.

$$\textit{i.e.}\ \text{Leakage factor} = \frac{\text{Total flux}}{\text{Useful flux}}$$

$$\text{or Leakage factor} = \frac{\text{Flux in the iron path}}{\text{Flux in the air gap}}$$

It is also seen from Fig. 3.5 that the useful flux passing across the gap tends to bulge outwards, thereby increasing the effective area of gap and reducing the flux density in the gap. This effect is referred to as *fringing*; and the longer the air gap, the greater is the fringing.

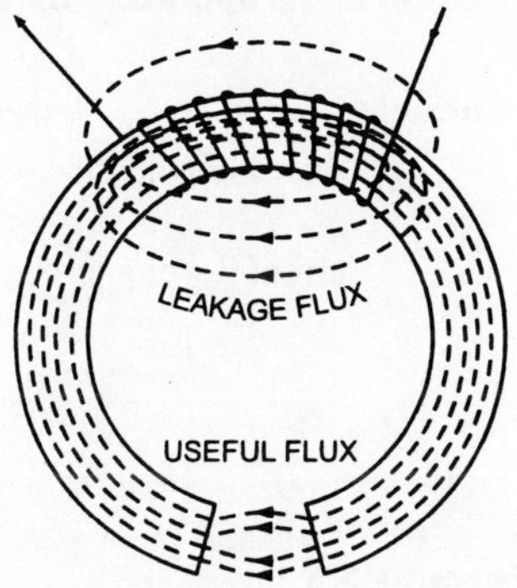

Fig. 3.5

In electrical machines (such as in generators and motors), magnetic leakage is undesirable as it causes increase in weight (not decrease in their power efficiency) and cost of manufacture. Though magnetic leakage cannot be avoided completely but can be reduced to the minimum by placing the magnetizing or exciting coils as close as possible to the air gap or to the points in the magnetic circuit where the flux is to be utilized for useful purposes.

The value of leakage factor for modern electric machinery is about 1.2.

Example 3.1

An electromagnet has an gap of 4 mm and flux density in the gap is 1.3 Wb/m². Determine the ampere-turns for the gap.

[U.P. Technical Univ. Electrical Engineering Second Semester 2006–07]

Solution:

$$\text{Magnetising force, } H_g = \frac{B}{\mu_0 \mu_r} = \frac{1.3}{4\pi \times 10^{-7} \times 1} = 1.035 \times 10^6 \text{ AT/m} \qquad \because \mu_r \text{ for air} = 1$$

$$\text{Ampere-turns required for the gap} = H_g\, l_g = 1.035 \times 10^6 \times 4 \times 10^{-3} = 4{,}140 \textbf{ Ans.}$$

Example 3.2

A ring of mean diameter 30 cm is wound with 200 turns of copper wire carrying a current of 2 A. The x-section of the magnetic material of the ring is 12 cm² and its relative permeability is 1,000. Determine the flux through it.

Solution:

$$\text{Length of iron ring, } l = \pi D = \pi \times 0.3 = 0.942 \text{ m}$$

$$\text{Area of x-section of iron path, } a = 12 \text{ cm}^2 = 0.0012 \text{ m}^2$$

$$\text{Relative permeability, } \mu_r = 1{,}000$$

$$\mu_0 = 4\pi \times 10^{-7} \text{ H/m}$$

$$\text{Flux through iron path, } \phi = \frac{NI}{l/\mu_r \mu_0 a} = \frac{200 \times 2}{0.942/1{,}000 \times 4\pi \times 10^{-7} \times 0.0012} = 0.64 \text{ mWb } \textbf{Ans.}$$

Example 3.3

A rectangular shaped core is made of mild steel plate 15 mm × 20 mm cross-section. The mean length of the magnetic path is 18 cm. The exciting coil has 300 turns and current 0.7 A. Calculate: (i) Magnetising force (ii) Flux density (iii) Reluctance (iv) Flux of magnetic circuit. Assume relative permeability of mild steel as 940.

[R.G.P.V. Basic Electrical Engineering Jan./Feb; 2006]

Solution:

$$\text{Mean length of magnetic path, } l = 18 \text{ cm} = 0.18 \text{ m}$$

$$\text{Cross-section area of mild steel plate, } a = 15 \text{ mm} \times 20 \text{ mm} = 300 \text{ mm}^2 = 3 \times 10^{-4} \text{ m}^2$$

$$(i) \text{ Magnetising force, } H = \frac{NI}{l} = \frac{300 \times 0.7}{0.18} = 1{,}166.6667 \text{ AT/m } \textbf{Ans.}$$

$$(ii) \text{ Flux density, } B = \mu_0 \mu_r H = 4\pi \times 10^{-7} \times 940 \times 1{,}166.6667 = 1.378 \text{ T } \textbf{Ans.}$$

$$(iii) \text{ Reluctance, } S = \frac{l}{\mu_0 \mu_r a} = \frac{0.18}{4\pi \times 10^{-7} \times 940 \times 3 \times 10^{-4}} = 50.794 \times 10^4 \text{ AT/Wb } \textbf{Ans.}$$

$$(iv) \text{ Flux, } \phi = B \times a = 1.378 \times 3 \times 10^{-4} = 4.134 \times 10^{-4} \text{ Wb } \textbf{Ans.}$$

Example 3.4

A coil of insulated wire of 500 turns and of resistance 4 Ω is closely wound on iron ring. The ring has a mean diameter of 0.25 m and a uniform cross-sectional area of 700 mm². Calculate the total flux in the ring when a dc supply of 6 V is applied to the ends of the winding. Assume a relative permeability of 550.

[U.P. Technical Univ. Electrical Engineering July 2002]

Solution:

$$\text{Length of iron ring, } l = \pi D = \pi \times 0.25 = 0.25\,\pi \text{ m}$$

$$\text{Area of cross-section of iron path, } a = 700 \text{ mm}^2 = 0.7 \times 10^{-3} \text{ m}^2$$

$$\text{Current flowing through the winding, } I = \frac{\text{Voltage applied across coil}}{\text{Resistance of coil}} = \frac{6}{4} = 1.5 \text{ A}$$

$$\text{Total flux in the ring, } \phi = \frac{NI \times \mu_r \mu_0 a}{l} = \frac{500 \times 1.5 \times 550 \times 4\pi \times 10^{-7} \times 0.7 \times 10^{-3}}{0.25\,\pi}$$

$$= 0.462 \text{ mWb } \textbf{Ans.}$$

Example 3.5

A wrought iron bar 30 cm long and 2 cm in diameter is bent into a circular shape as given in Fig. 3.6. It is then wound with 500 turns of wire. Calculate the current required to produce a flux of 0.5 mWb in magnetic circuit with an air gap of 1 mm; μ_r (iron) = 4,000 (assume constant).

[U.P. Technical Univ. Electrical Engineering Second Semester 2004-05]

Solution:

Flux to be produced, $\phi = 0.5 \text{ mWb} = 0.5 \times 10^{-3}$ Wb

Area of cross-section of core, $a = \pi \times 10^{-4}$ m^2

$$\text{Flux density required, B} = \frac{\phi}{a} = \frac{0.5\times10^{-3}}{\pi\times10^{-4}} = \frac{5}{\pi} = 1.5915 \text{ T}$$

Airgap length, $l_g = 1$ mm $= 0.001$ m

Core length, $l_i = 30$ cm $= 0.3$ m

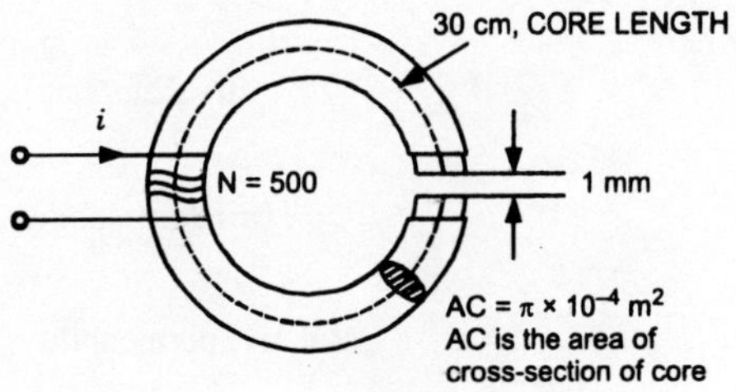

Fig. 3.6

$$\text{Total ampere-turns required, AT} = \frac{B}{\mu_0}l_g + \frac{B}{\mu_0\mu_r}l_i = \frac{1.5915}{4\pi\times10^{-7}}\times0.001 + \frac{1.5915}{4\pi\times10^{-7}\times4{,}000}\times0.3$$

$$= 1{,}266 + 95 = 1{,}361$$

$$\text{Current required, I} = \frac{AT}{N} = \frac{1{,}361}{500} = 2.72 \text{ A } \textbf{Ans.}$$

Example 3.6

A circular ring 20 cm in diameter has an air-gap 1 mm wide cut in it. The area of cross-section of the ring is 3.6 cm². Calculate the value of the direct current needed in a coil of 1,000 turns uniformly wound round the ring to create a flux of 0.5 mWb in the air-gap. Neglect fringing and assume relative permeability for the iron as 650.

[U.P. Technical Univ. Electrical Engineering First Semester 2006-07]

Solution:

Flux to be produced, $\phi = 0.5 \text{ mWb} = 0.5 \times 10^{-3}$ Wb

Area of x-section, $a = 3.6 \text{ cm}^2 = 3.6 \times 10^{-4}$ m^2

$$\text{Flux density required, B} = \frac{\phi}{a} = \frac{0.5\times10^{-3}}{3.6\times10^{-4}} = 1.389 \text{ T}$$

Air gap length, $l_g = 1 \text{ mm} = 1 \times 10^{-3}$ m

Length of iron path, $l_i = \pi d - l_g = \pi \times 0.20 - 1 \times 10^{-3} = 0.62732$ m

$$\text{Total ampere-turns required, AT} = \frac{B}{\mu_0}l_g + \frac{B}{\mu_0\mu_r}l_i = \frac{1.389}{4\pi\times10^{-7}}\times1\times10^{-3} + \frac{1.389}{4\pi\times10^{-7}\times650}\times0.62732$$

$$= 1{,}105 + 1{,}067 = 2{,}172$$

$$\text{Current required, I} = \frac{AT}{N} = \frac{2{,}172}{1{,}000} = 2.172 \text{ A } \textbf{Ans.}$$

Example 3.7

A ring of ferromagnetic material has a rectangular cross-section. The inner diameter is 7.4 in., the outer diameter is 9 in., and the thickness is 0.8 in. There is a coil of 600 turns wound on the ring. When the coil carries a current of 2.5 A, the flux produced in the ring is 1.2 × 10⁻³ Wb. Find: (i) Magnetic field intensity (ii) Reluctance (iii) Permeability.

[U.P. Technical Univ. Electrical Engineering Second Semester 2007-08]

Solution:

Number of turns in the coil, N = 600

Current carried by coil, I = 2.5 A

Flux produced, $\phi = 1.2 \times 10^{-3}$ Wb

Mean length of magnetic flux path is that of a circle midway between the inside and outside diameter i.e.

$$l = \frac{\pi(7.4+9)}{2} = 25.45 \text{ in} = 25.45 \times 0.0254 = 0.64635 \text{ m}$$

Cross-sectional area of the toroidal core, $a = (9 - 7.4) \times 0.8 = 1.28 \text{ in}^2 = 8.26 \times 10^{-4}$ m^2

$$(i)\ \text{Magnetic flux density, B} = \frac{\phi}{a} = \frac{1.2\times10^{-3}}{8.26\times10^{-4}} = 14.5\ \text{Wb/m}^2\ \textbf{Ans.}$$

$$\text{(ii) Reluctance, S} = \frac{\text{MMF}}{\phi} = \frac{\text{NI}}{\phi} = \frac{600\times2.5}{1.2\times10^{-3}} = 1{,}250\times10^3\ \text{AT/Wb}\ \textbf{Ans.}$$

$$\text{(iii) Permeability, }\mu = \frac{l}{Sa} = \frac{0.64635}{1{,}250\times10^3\times8.26\times10^{-4}} = 0.000623\ \text{H/m}\ \textbf{Ans.}$$

$$\text{Relative permeability, }\mu_r = \frac{\mu}{\mu_0} = \frac{0.000623}{4\pi\times10^{-7}} = 496\ \textbf{Ans.}$$

Example 3.8

A cast steel electromagnet has an air gap length of 3 mm and an iron path of length 40 cm. Find the number of ampere-turns necessary to produce a flux density of 0.7 Wb/m² in the gap. Neglect leakage and fringing. Assume ampere-turns required for air gap to be 70% of the total ampere-turns. [*R.G.P.V. Basic Electrical Engineering 2001*]

Solution:

$$\text{Air-gap length, } l_g = 3\ \text{mm} = 3\times10^{-3}\ \text{m}$$

$$\text{Flux density required in the air-gap, B}_g = 0.7\ \text{Wb/m}^2$$

$$\text{Magnetising force, H}_g = \frac{\text{B}}{\mu_0\mu_r} = \frac{0.7}{4\pi\times10^{-7}\times1} = 5.57\times10^5\ \text{AT/m} \qquad \because \mu_r \text{ for air} = 1$$

$$\text{Ampere-turns required for the air-gap, AT}_g = \text{H}_g\, l_g = 5.57\times10^5\times3\times10^{-3} = 1{,}671$$

Since ampere-turns required for the air-gap are 70% of the total ampere-turns

$$i.e.\ \text{AT}_g = 0.7\,\text{AT (total ampere-turns required)}$$

$$\text{So total ampere-turns required, AT} = \frac{\text{AT}_g}{0.7} = \frac{1{,}671}{0.7} = 2{,}387\ \textbf{Ans.}$$

Example 3.9

A steel ring of 25 cm diameter and of circular section 3 cm in diameter has an air gap of 1.5 mm length. It is wound uniformly with 750 turns of wire carrying a current of 2.1 A. Calculate (i) magnetomotive force (ii) flux density in air gap (iii) magnetic flux (iv) relative permeability of steel ring. Assume that iron path takes about 35% of the total magnetomotive force. [*R.G.P.V. Basic Electrical Engineering February 2005*]

Solution:

$$\text{Area of x-section, } a = \frac{\pi}{4}(d)^2 = \frac{\pi}{4}(3\times10^{-2})^2 = 0.0007\ \text{m}^2$$

$$(i)\ \text{Magnetomotive force, AT} = \text{N}\times\text{I} = 750\times2.1 = 1{,}575\ \textbf{Ans.}$$

$$\text{Ampere-turns provided for iron path, AT}_i = 35\%\text{ of total AT} = 0.35\times1{,}575 = 551$$

$$\text{Ampere-turns provided for air-gap, AT}_g = \text{AT} - \text{AT}_i = 1{,}575 - 551 = 1{,}024$$

$$(ii)\ \text{Flux density in the air-gap, B}_g = \frac{\text{AT}_g\times\mu_0}{l_g} = \frac{1{,}024\times4\pi\times10^{-7}}{1.5\times10^{-3}} = 0.858\ \text{T}\ \textbf{Ans.} \qquad \because \text{AT}_g = \frac{\text{B}_g}{\mu_0} l_g$$

$$(iii)\ \text{Magnetic flux, }\phi = \text{B}_g\times a = 0.858\times0.0007 = 0.6\ \text{mWb}\ \textbf{Ans.}$$

Assuming negligible leakage we have

$$\text{Flux density in iron path, B}_i = \text{B}_g = 0.858\ \text{T}$$

$$\text{Length of iron path, } l_i = \pi\times\text{D} - l_g = \pi\times0.25 - 0.0015 = 0.784\ \text{m}$$

$$(iv)\ \text{Relative permeability of steel, }\mu_r = \frac{\text{B}_i l_i}{\mu_0\,\text{AT}_i} = \frac{0.858\times0.784}{4\pi\times10^{-7}\times551} = 970\ \textbf{Ans.} \qquad \because \text{AT}_i = \frac{\text{B}_i}{\mu_0\mu_r} l_i$$

Example 3.10

A steel ring of circular section of 1 cm in radius and having a mean circumference of 94.3 cm and air-gap of 1 mm long. It is uniformly wound with an exciting coil consisting of 600 turns and excited with a current of 2.5 amp. Neglecting magnetic leakage, calculate:

(i) MMF (ii) magnetic flux (iii) reluctance (iv) flux density (v) relative permeability of steel.

Assume that steel part takes about 40% of total amp-turns. [*R.G.P.V. Basic Electrical Engineering, Jan./Feb; 2006*]

Solution:

$$\text{Area of x-section, } a = \pi \times r^2 = \pi \times (0.01)^2 = 0.000314 \text{ m}^2$$

$$(i)\ \text{MMF, AT} = \text{N} \times \text{I} = 600 \times 2.5 = 1{,}500 \quad \textbf{Ans.}$$

$$\text{Ampere-turns provided for steel path, AT}_i = 40\% \text{ of total AT} = 0.4 \times 1{,}500 = 600$$

$$\text{AT provided for air gap, AT}_g = \text{AT} - \text{AT}_i = 1{,}500 - 600 = 900$$

(*iv*) Flux density, B = Flux density in the air gap neglecting magnetic leakage

$$= \text{B}_g = \frac{\text{AT}_g \times \mu_0}{l_g} = \frac{900 \times 4\pi \times 10^{-7}}{1 \times 10^{-3}} = 1.131 \text{ T} \quad \textbf{Ans.}$$

$$(ii)\ \text{Magnetic flux, } \phi = \text{B} \times a = 1.131 \times 0.000314 = 0.355 \text{ mWb} \quad \textbf{Ans.}$$

$$(iii)\ \text{Reluctance of magnetic circuit, S} = \frac{\text{NI}}{\phi} = \frac{1{,}500}{0.355 \times 10^{-3}} = 4.225 \times 10^6 \text{ AT/Wb} \quad \textbf{Ans.}$$

$$(v)\ \text{Relative permeability of steel, } \mu_r = \frac{\text{B}_i l_i}{\mu_0 \text{AT}_i} = \frac{1.131 \times 0.943}{4\pi \times 10^{-7} \times 600} = 1{,}415 \quad \textbf{Ans.}$$

Example 3.11

A cast steel ring has *a* circular cross-section of 3 cm in diameter and a mean circumference of 80 cm. A 1 mm air-gap is cut in the ring which is wound with a coil of 600 turns. Estimate the current required to establish a flux of 0.75 mWb in the air-gap. Neglect leakage.

Magnetization data:

H (AT/M)	200	400	600	800	1,000	1,200	1,400	1,600	1,800
B (T)	0.1	0.32	0.6	0.9	1.08	1.18	1.27	1.32	1.36

[U.P. Technical Univ. Electromechanical Energy Conversion-I, 2004-2005]

Solution:

$$\text{Area of x-section, } a = \frac{\pi}{4}(d)^2 = \frac{\pi}{4}(0.03)^2 = 0.0007 \text{ m}^2$$

$$\text{Flux density in the air-gap, B}_g = \frac{\Phi_g}{a} = \frac{0.75 \times 10^{-3}}{0.0007} = 1.07 \text{ T}$$

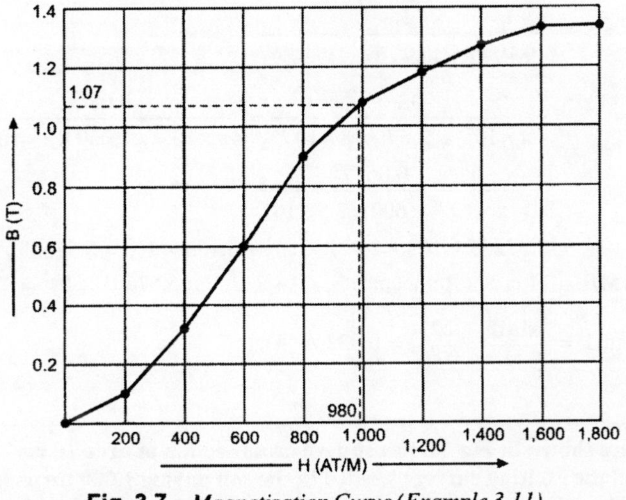

Fig. 3.7 *Magnetization Curve (Example 3.11)*

$$\text{Ampere-turns required for air-gap, AT}_g = \frac{\text{B}_g}{\mu_0} \times l_g = \frac{1.07}{4\pi \times 10^{-7}} \times 1 \times 10^{-3} = 850$$

Flux density in iron path, $B_i = B_g = 1.07$ T ∵ leakage in negligible.

From Fig. 3.7, magnetising force, H corresponding to flux density B of 1.07 T is 980 AT/m

AT required for iron-path, $AT_i = H \times l_i = 980 \times (80 - 0.1) \times 10^{-2} = 783$

Total ampere-turns required, $AT = AT_g + AT_i = 850 + 783 = 1{,}633$

$$\text{Current required, } I = \frac{AT}{N} = \frac{1{,}633}{600} = 2.72 \text{ A } \textbf{Ans.}$$

Example 3.12

What are the similarities between electrical circuits and magnetic circuits ?

An iron ring of mean length 50 cm and relative permeability 300 has an air gap of 1 mm. If the ring is provided with a winding of 200 turns and a current of 1 A is allowed to flow through, find the flux density across the airgap.

[*U.P. Technical Univ. Elec. Engineering June 2001*]

Solution : Let the flux density through the iron as well as air gap be B Wb/m^2

Total ampere-turns provided, $AT = N \times I = 200 \times 1 = 200$

Total ampere-turns required = AT required for air gap + AT required for iron ring

$$\text{or Total ampere-turns provided} = \frac{B}{\mu_0} l_g + \frac{B}{\mu_0 \mu_r} l_i$$

$$\text{or } 200 = \frac{B}{\mu_0}\left(l_g + \frac{l_i}{\mu_r}\right) = \frac{B}{\mu_0}\left(0.001 + \frac{0.5}{300}\right)$$

$$\text{or Flux density, } B = \frac{200 \times \mu_0}{0.001 + \frac{0.5}{300}} = \frac{200 \times 4\pi \times 10^{-7}}{0.001 + \frac{0.5}{300}} = 0.0942 \text{ T } \textbf{Ans.}$$

Example 3.13

An iron ring is made up of three parts:

$l_1 = 10$ cm, $A_1 = 8\ cm^2$, $l_2 = 8$ cm, $A_2 = 3\ cm^2$, $l_3 = 6$ cm, $A_3 = 2.5\ cm^2$.

It is wound with a coil of 250 turns. Calculate current required to produce flux of 0.4 mWb.

$\mu_1 = 2{,}670$, $\mu_2 = 1{,}050$, $\mu_3 = 600$. [*U.P. Technical Univ. Electrical Engineering Second Semester 2007-08*]

Solution:

$$\text{Total reluctance, } S = \frac{l_1}{\mu_0 \mu_{r_1} a_1} + \frac{l_2}{\mu_0 \mu_{r_2} a_2} + \frac{l_3}{\mu_0 \mu_{r_3} a_3}$$

$$= \frac{0.1}{4\pi \times 10^{-7} \times 2{,}670 \times 8 \times 10^{-4}} + \frac{0.08}{4\pi \times 10^{-7} \times 1{,}050 \times 3 \times 10^{-4}}$$

$$+ \frac{0.06}{4\pi \times 10^{-7} \times 600 \times 2.5 \times 10^{-4}}$$

$$= 0.37255 \times 10^5 + 2.021 \times 10^5 + 3.183 \times 10^5 = 5.57655 \times 10^5 \text{ AT/Wb}$$

$$\text{MMF} = \text{Flux } \phi \times \text{Reluctance } S = 0.4 \times 10^{-3} \times 5.57655 \times 10^5 = 223 \text{ ampere-turns}$$

$$\text{Current required, } I = \frac{\text{MMF}}{N} = \frac{223}{250} = 0.892 \text{ A } \textbf{Ans.}$$

Example 3.14

A rectangular magnetic core shown in Fig 3.8 has square cross-section of area 16 cm^2. An air-gap of 2 mm is cut across one of its limbs. Find the exciting current needed in the coil having 1,000 turns wound on the core to create an air-gap flux of 4mWb. The relative permeability of the core is 2,000.

[*U.P. Technical Univ. Electrical Engineering February 2002*]

Solution: Flux to be created, ϕ = 4 mWb

$= 0.004$ Wb

Area of x-section, $a = 16 \text{ cm}^2$

$= 16 \times 10^{-4} \text{ m}^2$

Fluse density required, $B = \dfrac{\phi}{a}$

$$= \frac{4 \times 10^{-3}}{16 \times 10^{-4}}$$

$= 2.5$ T

Length of air-gap $l_g = 2$ mm

$= 2 \times 10^{-3}$ m

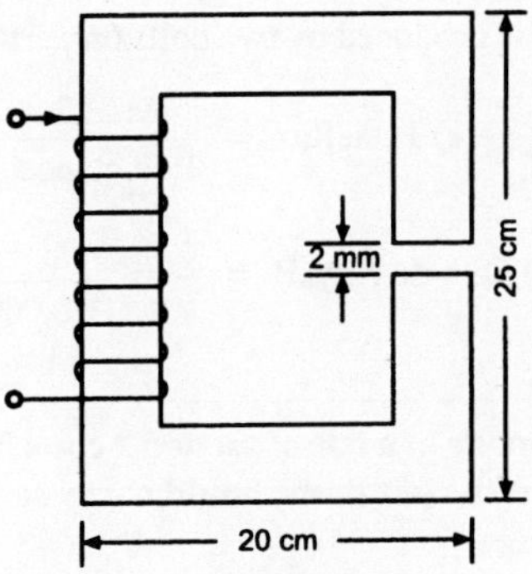

Fig. 3.8

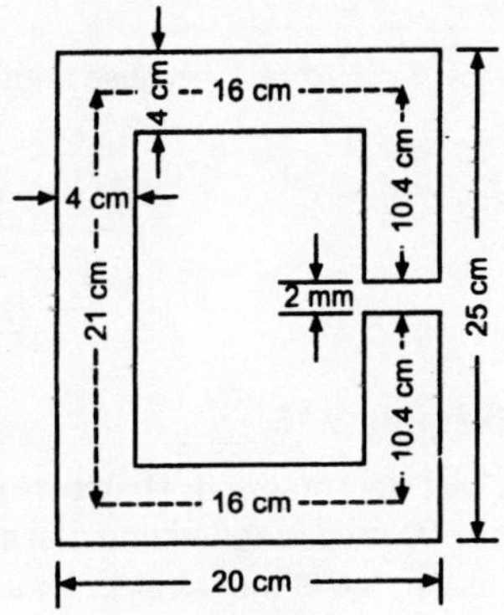

Fig. 3.9

Each side of cross-section $= \sqrt{16} = 4$ cm

$$\text{Length of iron-path, } l_i = \left(25 - 2\times\frac{4}{2} + 20 - 2\times\frac{4}{2}\right)\times 2 - 0.2 = 73.8 \text{ cm} = 0.738 \text{ m}$$

$$\text{Total ampere-turns required, AT} = \frac{B}{\mu_0} l_g + \frac{B}{\mu_0 \mu_r} l_i = \frac{2.5\times 2\times 10^{-3}}{4\pi\times 10^{-7}} + \frac{2.5\times 0.738}{2{,}000\times 4\pi\times 10^{-7}} = 3{,}979 + 734 = 4{,}713$$

Number of turns on the coil = 1,000

$$\text{Exciting current required, I} = \frac{AT}{N} = \frac{4{,}713}{1{,}000} = 4.713 \text{ A}$$ **Ans.**

Example 3.15

The magnetic circuit frame shown in Fig. 3.10 is built of iron of square cross-section of 3 cm. Each air gap is 2 mm wide and each of the coil is wound with 1,000 turns. The relative permeability of part A and B may be taken as 1,000 and 1,200 respectively. Calculate (a) reluctance of part A (b) reluctance of two air gaps (c) total reluctance of complete magnetic path (d) total mmf (e) total flux and (e) flux density.

Solution: Cross-sectional area,

$a = (0.03)^2 = 0.0009 \text{ m}^2$

Mean length flux path in part A,

$l_A = 20 - (1.5 + 1.5) = 17$ cm or 0.17 m

Mean length flux path in part B,

$l_B = (20 - 1.5 - 1.5) + (10 - 1.5)$

$+ (10 - 1.5)$

$= 34$ cm or 0.34 m

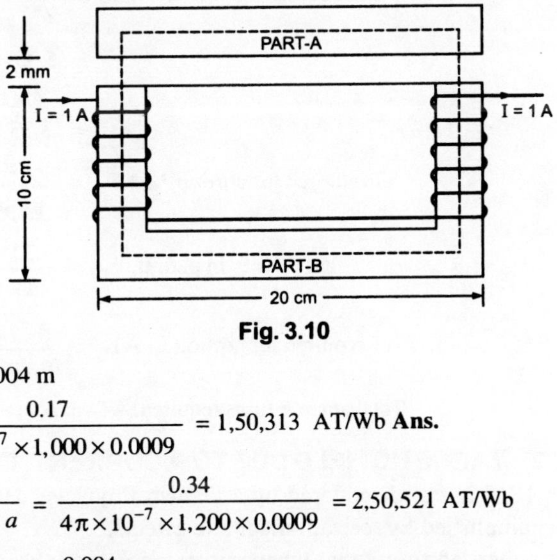

Fig. 3.10

Relative permeability of part A, $\mu_{rA} = 1{,}000$

Relative permeability of part B, $\mu_{rB} = 1{,}200$

Length of two air-gaps, $l_g = 2 \times 2 = 4 \text{ mm} = 0.004$ m

$$(a) \text{ Reluctance of part A, } S_A = \frac{0.17}{4\pi\times 10^{-7}\times 1{,}000\times 0.0009} = 1{,}50{,}313 \text{ AT/Wb}$$ **Ans.**

$$\text{Reluctance of part B, } S_B = \frac{l_B}{\mu_0 \mu_{rB} \times a} = \frac{0.34}{4\pi\times 10^{-7}\times 1{,}200\times 0.0009} = 2{,}50{,}521 \text{ AT/Wb}$$

$$(b) \text{ Reluctance of two air-gaps, } S_g = \frac{l_g}{\mu_0 a} = \frac{0.004}{4\pi\times 10^{-7}\times 0.0009} = 35{,}36{,}776 \text{ AT/Wb}$$ **Ans.**

$$(c) \text{ Total reluctance, } S = S_A + S_B + S_g = 1{,}50{,}313 + 2{,}50{,}521 + 35{,}36{,}776$$

$$= 39{,}37{,}610 \text{ AT/Wb}$$ **Ans.**

(*d*) Total mmf $= 2 \times 1{,}000 \times 1 = 2{,}000$ AT **Ans.**

because mmf is produced by two coils on part B each having 1,000 turns and carrying a current of 1 A

$$(e)\ \text{Total flux} = \frac{\text{MMF}}{\text{Reluctance}} = \frac{2{,}000}{39{,}37{,}610} = 5.08 \times 10^{-4}\ \text{Wb.}$$

$$(f)\ \text{Flux density, B} = \frac{\phi}{a} = \frac{5.08\times10^{-4}}{0.0009} = 0.564\ \text{Wb/m}^2\ \text{(or Tesla)}\ \textbf{Ans.}$$

Example 3.16

A cast steel magnetic structure made of a bar of section 2 cm × 2 cm is shown in Fig. 3.11. Determine the current that the 500 turn magnetizing coil on the left limb should carry so that a flux of 2 m Wb is produced in the right limb. Take $\mu_r = 600$ and neglect leakage.

Solution: Flux created by magnetizing coil, say ϕ, is divided at junction point into two paths depending upon the reluctances of portions B and C respectively.

$$\text{Reluctance of portion B, } S_B = \frac{l_B}{\mu_0\,\mu_r \times a} = \frac{0.15}{\mu_0\,\mu_r\,a}$$

$$\text{Reluctance of portion C, } S_C = \frac{l_C}{\mu_0\,\mu_r\,a} = \frac{0.25}{\mu_0\,\mu_r\,a}$$

$$\text{Flux in portion C, } \phi_C = 2\text{mWb} = 0.002\ \text{Wb (given)}$$

$$\text{Flux in portion B, } \phi_B = \phi_C \times \frac{S_C}{S_B} = 2 \times 10^{-3} \times \frac{0.25/\mu_0\mu_r a}{0.15/\mu_0\mu_r a} = 3.33 \times 10^{-3}\ \text{Wb}$$

$$\text{Total flux in portion A, } \phi = \phi_B + \phi_C$$

$$= 3.33 \times 10^{-3} + 2 \times 10^{-3}$$

$$= 5.33 \times 10^{-3}\ \text{Wb}$$

$$\text{Flux density in portion A, } B_A = \frac{\phi}{a}$$

$$= \frac{5.33\times10^{-3}}{4\times10^{-4}}$$

$$= 13.33\ \text{Wb/m}^2$$

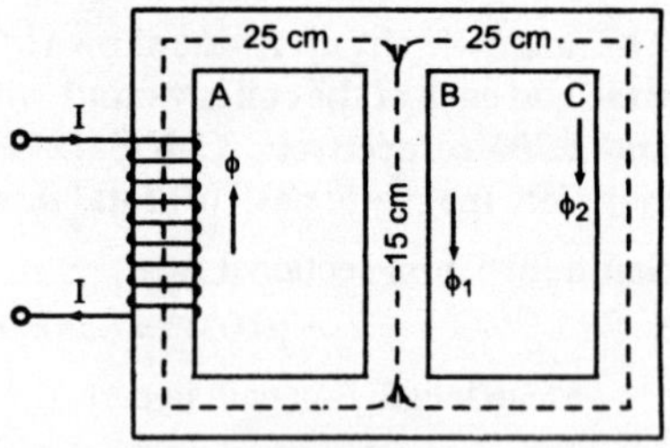

Fig. 3.11

$$\text{AT required for portion A, } AT_A = \frac{B_A \times l_A}{\mu_0\,\mu_r} = \frac{13.33\times0.25}{4\pi\times10^{-7}\times600} = 4{,}416$$

$$\text{Flux density in path B, } B_B = \frac{3.33\times10^{-3}}{4\times10^{-4}} = 8.33\ \text{Wb/m}^2$$

$$\text{AT required for portion B, } AT_B = \frac{B_B\,l_B}{\mu_0\,\mu_r} = \frac{8.33\times0.15}{4\pi\times10^{-7}\times600} = 1{,}658$$

$$\text{Total ampere-turns required, AT} = AT_A + AT_B = 4{,}416 + 1{,}658 = 6{,}074\ \textbf{Ans.}$$

3.2. MAGNETIC FIELD DUE TO A CURRENT CARRYING CONDUCTOR

In 1819 it was discovered by a Danish Physicist, Hans Christian Oersted that an electric current is always accomplished by certain magnetic effects.

Oersted found that when current is passed through a conductor placed above the magnetic needle, the needle turns in a certain direction, as shown in Fig. 3.12. He also found that when the direction of flow of current is reversed the magnetic needle also deflects in opposite direction.

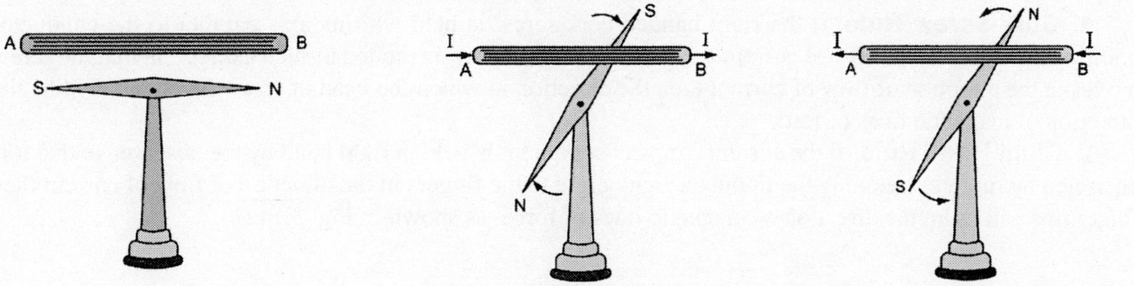

(a) No Current Flow (b) Current Flowing From A to B (c) Current Flowing From B to A

Fig. 3.12

Further investigation showed that the field around the current carrying conductor consists of lines of force, which encircles the conductor. It can be proved experimentally by passing a current carrying conductor AB in the card board and plotting the field with the help of magnetic needle on it, as shown in Figs. 3.13 (*a*) and 3.13 (*b*).

From Figs. 3.13 (*a*) and 3.13 (*b*) it is observed that when the current is passed through conductor in upward direction, the direction of lines of force is counterclockwise direction (observed from the top of the conductor) and when the current is passed through the conductor in downward direction, the direction of lines of force is clockwise (observed from the top of the conductor).

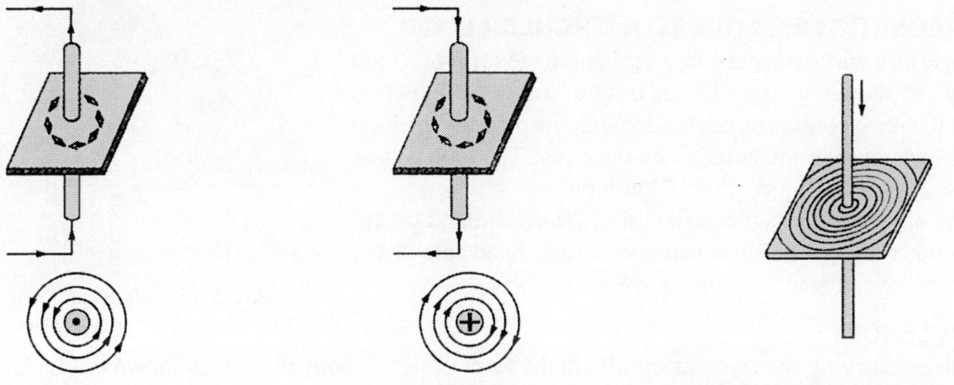

(*a*) (*b*) (*c*) *Magnetic Field Around A Current Carrying Conductor*

Fig. 3.13

The properties of the lines of magnetic induction around a current carrying conductor are summarized as below:

(*i*) Lines of magnetic induction are circles, symmetrical about, and concentric with, the axis of the conductor.

(*ii*) The spacing between the lines of induction decreases as we move closer to the conductor [Fig. 3.13 (*c*)]

(*iii*) The direction of lines of magnetic induction depends on the direction of flow of current through the conductor.

(*iv*) Magnetic induction or flux density depends upon the strength (or magnitude) of the current flowing through the conductor.

3.3. DETERMINATION OF DIRECTION OF MAGNETIC FIELD AROUND A CURRENT CARRYING CONDUCTOR

The direction of lines of force (magnetic field) around a straight current carrying conductor may be determined by any of the following rules:

1. Cork Screw Rule. If the right handed cork screw is held with its axis parallel to the conductor pointing the direction of flow of current and the head of the screw is rotated in such a direction that the screw moves in the direction of flow of current then the direction in which the head of screw is rotated, will be the direction of magnetic lines of force.

2. Right Hand Rule. If the current carrying conductor is held in right hand by the observer so that it is encircled by fingers stretching the thumb at right angle to the fingers in the direction of flow of current then finger tips will point the direction of magnetic lines of force, as shown in Fig. 3.14 (*b*).

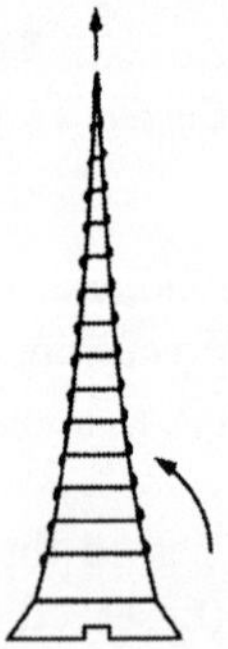

(*a*) *Cork Screw Rule*

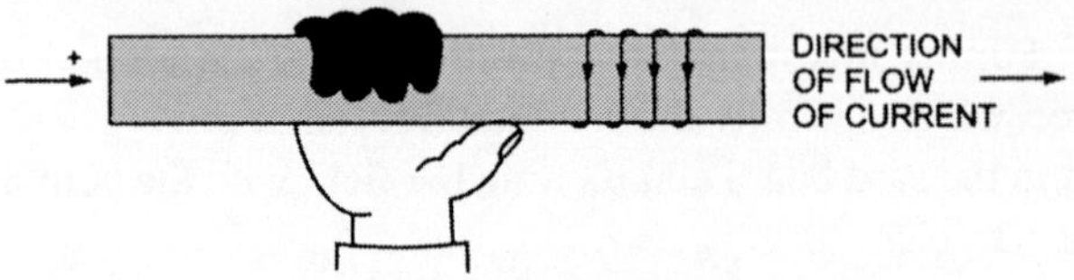

(*b*) *Right Hand Rule*

Fig. 3.14

3.4. MAGNETIC FIELD DUE TO A CIRCULAR LOOP

If a single turn wire carrying current is bent in the form of a loop (or ring), as shown in Fig. 3.15, the lines of magnetic induction around it will be concentric circles, leaving the plane of the loop (or ring) on one side and entering on the other. The loop acts as the true magnet having north and south poles.

The direction of magnetic field may be determined by applying either of the two rules namely (*i*) right hand rule or (*ii*) cork screw rule, already explained in Art. 3.3.

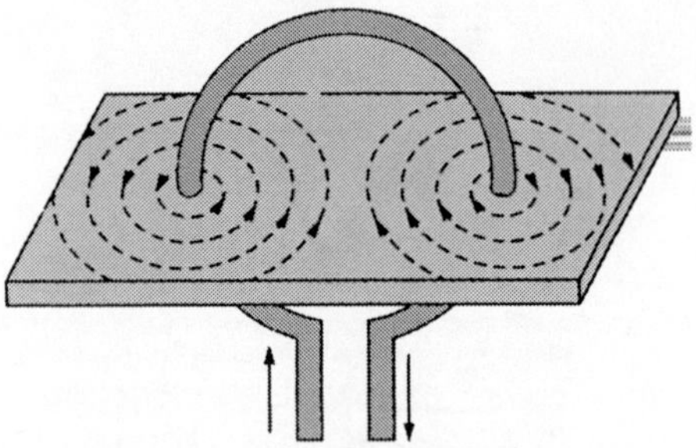

Fig. 3.15 *Magnetic Field Due To a Circular Loop*

3.5. SOLENOID

The current carrying wire wound spirally in the form of helix about an axis, as shown in Fig. 3.16, is known as *solenoid* or *coil.* Magnetic field produced due to current carrying solenoid is fairly uniform over a small region in the middle of the coil. It acts just like a bar magnet having north and south poles.

There are several methods used to determine the polarity of the solenoid.

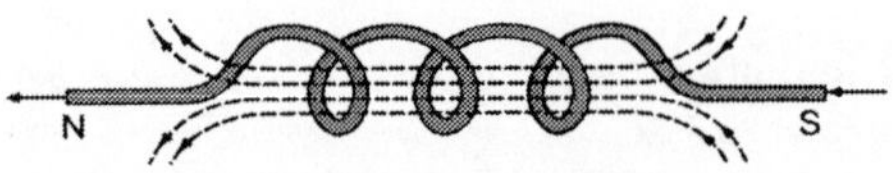

Fig. 3.16 *Solenoid*

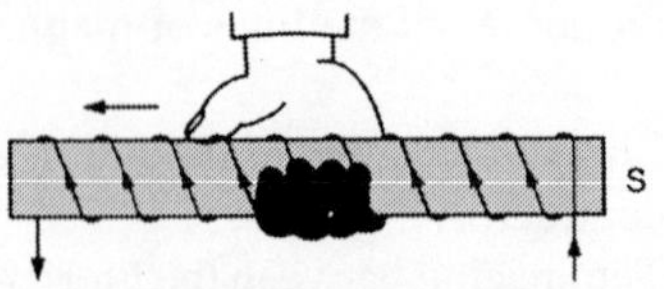

Fig. 3.17 *Helix Rule*

1. By Use of Compass Needle. If one of the poles (say north pole) of the compass needle be brought into close proximity to one of the poles of the current carrying solenoid of unknown polarity the action of the compass needle will immediately classify the pole as north or south depending upon whether the needle is repelled or attracted.

2. Helix Rule. If the helix is held in right hand in such a manner that the finger tips point in the direction of flow of current and thumb is outstretched longitudinally along the coil, it will point towards north pole (Fig. 3.17).

3.6. FORCE ON A CURRENT CARRYING CONDUCTOR LYING IN THE MAGNETIC FIELD

In Fig. 3.18 (*a*) a uniform magnetic field between the two opposite poles is shown.

In Fig. 3.18 (*b*) the cross-section of a conductor carrying current in inward direction placed between the two magnets, the field being temporarily removed, is shown. By applying the right-hand thumb rule, the direction of the field around the conductor is found to be clockwise.

If the current carrying conductor shown in Fig. 3.18 (*b*) is placed in the magnetic field shown in Fig. 3.18 (*a*), the resultant magnetic field would be similar to that shown in Fig. 3.18 (*c*).

The lines of force above the conductor are strengthened, since they are in same direction but the lines of force below the conductor are weakened because the two fields below the conductor are opposite in directions and hence tends to destroy each other. Magnetic lines like rubber bands have a tendency to strengthen out and, therefore, a force is experienced on the conductor in the downward direction, as shown in Fig. 3.18 (*c*).

If the direction of current is reversed in the conductor, as shown in Fig. 3.18 (*d*), the direction of force experienced is reversed. In this case the lines of force above the conductor are weakened while those below the conductor are strengthened.

Hence it is observed that when a current carrying conductor is placed at right angle to the direction of magnetic field, a mechanical force is experienced on the conductor in a direction perpendicular to both the direction of magnetic field and flow of current.

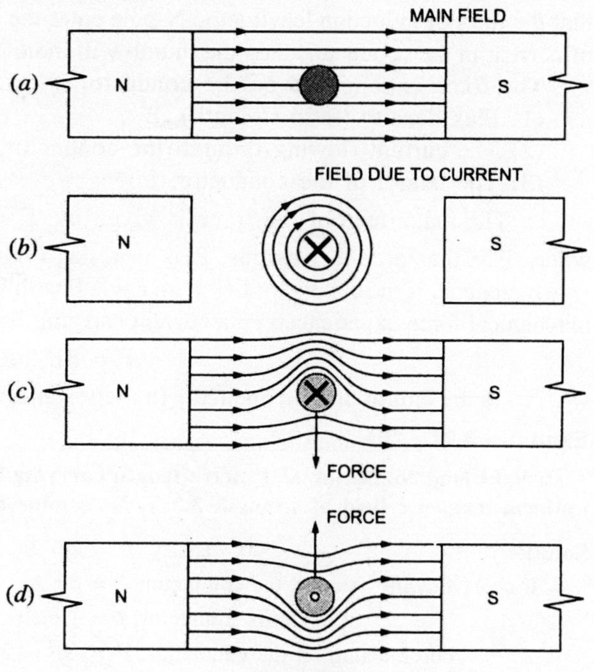

Fig. 3.18

The direction of this force can be determined by applying *Fleming's Left Hand Rule* which states that if the thumb, forefinger and middle finger of the left hand are stretched in such a way that they are at right angles to each other mutually and forefinger points towards the direction of the magnetic field, middle finger towards the direction of the flow of current then thumb will point the direction of force acting on the conductor [Fig. 3.19 (*a*)].

If the current in the conductor is reversed, keeping the direction of magnetic field unchanged the direction of force will reverse.

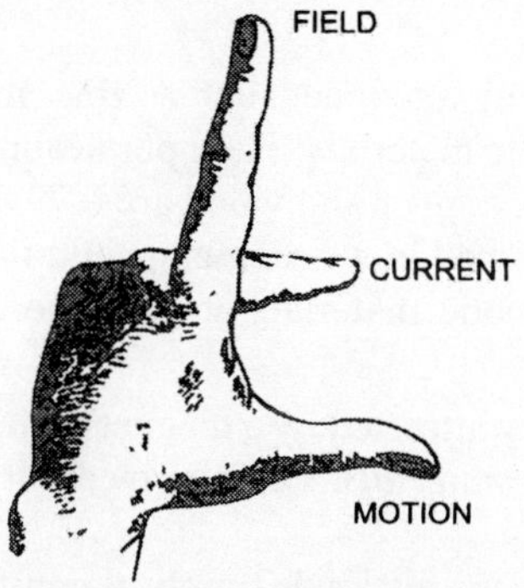

(*a*) *Fleming's Left Hand Rule*

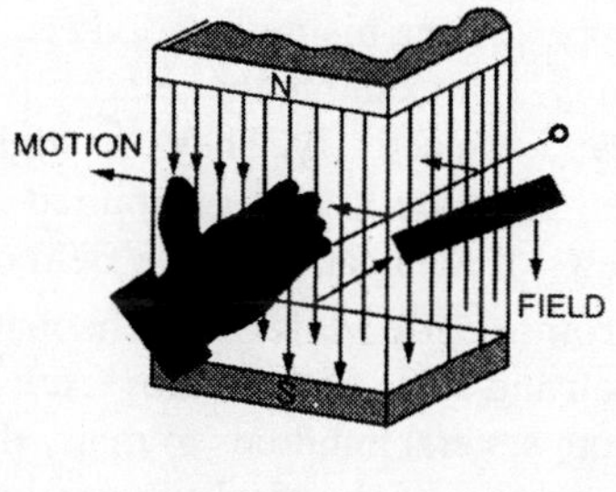

(*b*)

Fig. 3.19

Similarly, if the direction of the magnetic field is reversed, keeping the direction of flow of current in the conductor unchanged, the direction of force will reverse.

It should be noted that no force is exerted on a conductor when it lies parallel to the magnetic field.

Figure 3.19 (*b*) shows another method of finding the direction of force acting on a current carrying conductor placed in a magnetic field. It is known as *Flat left hand rule*. If the palm of the left hand is so placed that the lines of induction leaving the N-pole enter the palm and the four fingers are in the direction of flow of current in the conductor, then the thumb will indicate the direction of force acting on the conductor.

The force experienced on the conductor is directly proportional to

(1) Flux density (field strength), B

(2) The current flowing through the conductor, I and

(3) The length of the conductor, l

$\therefore$ The magnitude of the force is given by, $F = BIl$ (3.7)

where F is the force in newtons, B is in teslas (Wb/m^2), I is in amperes and l is in metres.

In general, if the conductor lies at an angle θ with a magnetic field of flux density B webers/metre^2, then mechanical force experienced on a current carrying conductor is given by

$$F = BIl \sin \theta \text{ newtons} \tag{3.8}$$

where l is the length of the conductor in metres and I is the current carried by the conductor in amperes.

Example 3.17

A straight long conductor of 1 metre length carrying a current of 60 amperes is placed at right angle to a uniform magnetic field of strength 2.5 T. Determine the mechanical force acting on the conductor.

Solution: Flux density, B = 2.5 T

Current flowing through the conductor, I = 60 A

Length of conductor, l = 1 metre

Force acting on the conductor, $F = BIl = 2.5 \times 60 \times 1 = 150$ N **Ans.**

3.7. CLASSIFICATION OF MAGNETIC MATERIALS

Although all materials have magnetic properties of some kind being either *diamagnetic*, *paramagnetic* or *ferromagnetic*, the term "magnetic material" is customarily applied only to substances which exhibit ferromagnetism.

1. Paramagnetic Materials. The materials, which are not strongly attracted by a magnet, such as aluminium, tin, platinum, magnesium, manganese etc. are known as *paramagnetic materials.* Their relative permeability is small but positive. For example, the relative permeabilties for aluminium, air and platinum are 1.00000065, 1.0000031 and 1.00036 respectively. Such materials are slightly magnetized when placed in a strong magnetic field and act in the direction of the magnetic field. These materials have little application in the field of electrical engineering.

2. Diamagnetic Materials. The materials which are repelled by a magnet such as zinc, mercury, lead, sulphur, copper, silver, bismuth, wood etc., are known as diamagnetic materials. Their permeability is slightly less than unity. For example the relative permeabilities of bismuth, copper and wood are 0.99983, 0.999995 and 0.9999995 respectively. They are slightly magnetized when placed in a strong magnetic field and act in the direction opposite to that of applied magnetic field. Diamagnetic materials are unimportant from the point of views of application in the field of electrical engineering.

3. Ferromagnetic Materials. The materials, which are strongly attracted by a magnet such as iron, steel, nickel, cobalt and some of their alloys, are known as *ferromagnetic materials.* Their permeability is very high (varying from several hundreds to many thousands).

For the purpose of electrical engineering it will suffice to classify the materials simply as non-ferromagnetic and ferro-magnetic materials. The former includes materials of relative permeabilities practically equal to unity while the latter have relative permeabilities many times greater than unity. Paramagnetic and diamagnetic materials fall in the former (*i.e.* non-ferromagnetic) class.

Ferromagnetic materials are of two types: (*a*) those easily magnetized, called the *soft magnetic materials* and (*b*) those retaining their magnetism with great tenacity, designated as *hard magnetic materials.*

(*a*) Soft Ferromagnetic Materials. They have high relative permeability, low coercive force, easily magnetized and demagnetized and have extremely small hysteresis. Soft ferromagnetic materials are iron and its alloys with nickel, cobalt, tungsten and aluminium. Ease of magnetization and demagnetization makes them highly suitable for applications involving changing magnetic flux as in electromagnets, electric motors, generators, transformers, inductors, telephone receivers, relays etc. They are also useful for magnetic screening. Their properties may be greatly enhanced through careful manufacture, and by heating and slow annealing so as to achieve a high degree of crystal purity. Large magnetic moment at room temperature makes soft ferromagnetic materials extremely useful for magnetic circuits but ferromagnetics are very good conductors and suffer energy loss from eddy currents produced within them. There is additional energy loss due to the fact that magnetization does not proceed smoothly but in minute jumps. This loss is called *magnetic residual loss* and it depends purely on the frequency of the changing flux density and not on its magnitude.

(*b*) **Hard Ferromagnetic Materials.** They have relatively low permeability, and very high coercive force. These are difficult to magnetize and demagnetize. Typical hard ferromagnetic materials include cobalt steel and various ferromagnetic alloys of nickel, aluminium and cobalt. They retain high percentage of their magnetization and have relatively high hysteresis loss. They are highly suited for use as permanent magnets in loud speakers etc. Sometimes they are also called the *permanent magnetic materials.* During manufacturing, a high degree of dislocation is introduced. Hard ferromagnetic materials are frequently heated to high temperature and then quenched in suitable liquid to introduce strains.

4. Ferrites. This is a special group of ferromagnetic materials that occupy an intermediate position between ferromagnetic and nonmagnetic materials. They consist of extremely fine particles of a ferromagnetic material possessing high permeability, and are held together with a binding resin. The magnetization produced in the ferrites is large enough to be of commercial value but their magnetic saturations are not as high as those of ferromagnetic materials. As in the case of ferromagnetics, ferrites may be soft or hard ferrites.

(*a*) **Soft Ferrites.** Ceramic magnets, also called ferromagnetic ceramics and ferrites, are made of an iron oxide, Fe_2O_3, with one or more divalent oxides such as NiO, MnO or ZnO. These magnets have a square hysteresis loop and high resistance to demagnetization and are valued for magnets for computing machines where a high resistance is desired. The great advantage of ferrites is their high resistivity. Commercial ferrites can have resistivities as high as 10^9 ohm-cm. Eddy currents resulting from alternating fields are therefore, reduced to a minimum, and the range of application of these magnetic materials is extended to high frequencies, even to microwaves. Ferrites are carefully made by mixing powdered oxides, compacting and sintering at high temperatures. High-frequency transformers in television and frequency modulated receivers are almost always made with ferrite cores. Ferrites, with large magnetostrictive effects, are sometimes used in electro-mechanical transducers. In high-frequency applications, magnetostriction in ferrites can lead to undesirable noise.

(*b*) **Hard Ferrites.** These are ceramic permanent magnetic materials. The most important family of hard ferrites has the basic composition $MO.Fe_2O_3$, where *M* is a barium (Ba) ion or a strontium (Sr) ion. These materials have a hexagonal structure and low in cost and density. Hard ferrites are used in generators, relays and motors. Electronic applications include magnets for loud speakers, telephone ringers and receivers. They are also used in holding devices for door closers, seals, latches and in several toy designs.

There is another category of magnetic materials, known as *super-paramagnetic,* made from powdered iron or other magnetic particles. These materials are used in transformers for electronics and cores for inductors. Permalloy (molybdenum-nickel-iron powder) is the best known example of this important category of magnetic materials.

3.8. MAGNETISATION OR B-H CURVES

The curve drawn giving relationship between induction density B and magnetising force H is known as *B-H curve* or *magnetising curve.* Fig. 3.20 shows the general shape of B-H curve of a magnetic material. In general it has four distinct regions *oa*, *ab*, *bc* and the region beyond *c*. During the region *oa* the increase in flux density is very small, in region *ab* the flux density B increases almost linearly with the magnetising force H, in region *bc* the increase in flux density is again small and in region beyond point *c*, the flux density B is almost constant. The flat part of the B-H curve corresponds to *magnetic saturation* of the material. The nonlinearity of the curve indicates that the relative permeability $\mu_r \left(\frac{B}{\mu_0 H}\right)$ of a magnetic material is not* constant but depends very largely upon the flux density B.

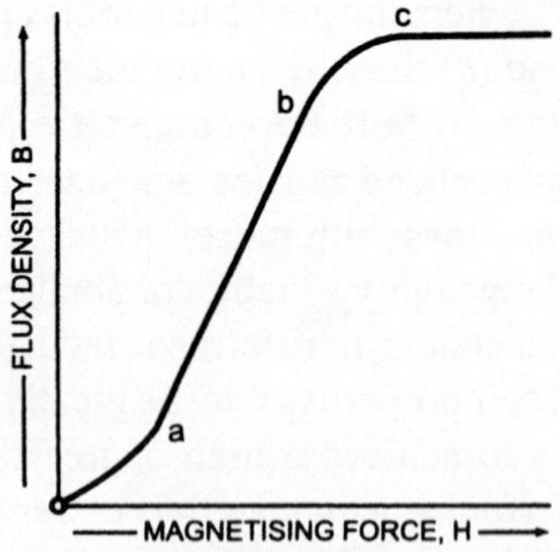

Fig. 3.20

The B-H curve depends on the material only and not upon the dimensions of a specific piece. Magnetising or B-H curves for silicon alloy (electrical sheet) steel, cast steel and cast iron etc., are shown in Fig. 3.21. In making magnetic calculations, values of H and μ_r corresponding to working flux density B are to be taken. The use of B-H curves permits the calculations of magnetic circuits with a fair degree of accuracy.

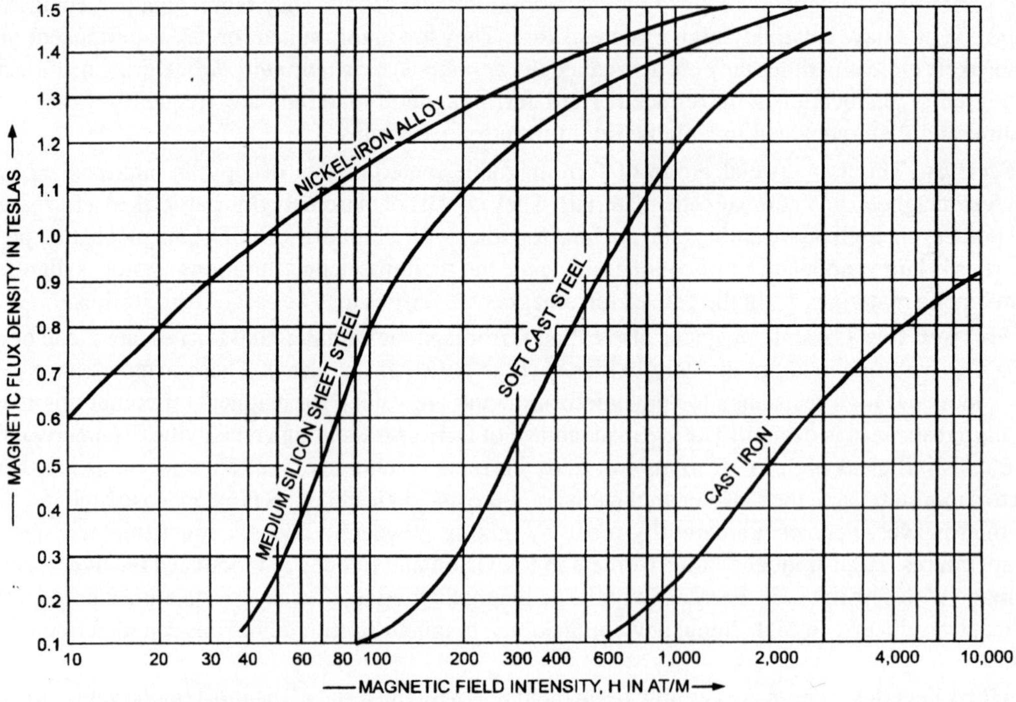

Fig. 3.21 *Magnetization Curves of Typical Ferromagnetic Materials*

3.9. MAGNETIC HYSTERESIS

If a magnetic substance is magnetised in a strong magnetic field, it retains a considerable portion of magnetism after the magnetic force has been withdrawn. The phenomenon of lagging of magnetisation or induction flux density behind the magnetising force is known as *magnetic hysteresis*.

Let a core of specimen of iron be wound with a number of turns of a wire and current be passed through the solenoid. A magnetic field of intensity H proportional to the current flowing through the solenoid is produced. Let magnetising force H is increased from zero to a certain maximum value and then gradually

* Relative permeability μ_r for all non-magnetic materials (*e.g.* brass, air etc.) is constant; being equal to 1. Thus for non-magnetic materials $B = \mu_0 H$. Since μ_0 is constant, B α H or B-H curve for a non-magnetic material is a straight line passing through origin.

reduced to zero. If the values of flux density B in the core corresponding to various values of magnetising force H are determined and B-H curves are drawn for increasing and decreasing values of magnetising force H then it will be observed that B-H curve obtained for decreasing values of H lies above that obtained for increasing values of H.

While decreasing the magnetising force H, when H is brought to zero the induction density B, is represented by OC and is called as *residual magnetism.* The power of retaining the residual magnetism is called the *retentivity* of the material.

Now if the direction of flow of current is reversed, the magnetising force H is reversed. Let the current be increased in the negative direction until the induction density B becomes zero. At this instant *i.e.* when B = 0, the demagnetising force H = OD, which is required to neutralize the residual magnetism, and is known as *coercive force.* If the demagnetising force H is further increased to the previous maximum value and again gradually decreased to zero, reversed and further increased in original or positive direction to the maximum value, a closed loop ACDEFGA is obtained which is usually known as *hysteresis loop* or *magnetic cycle.*

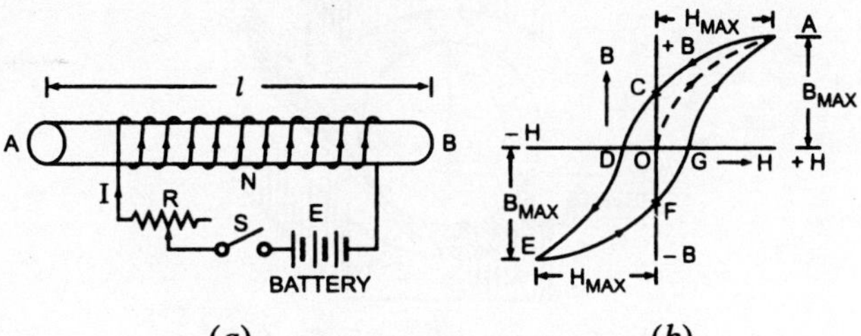

(*a*) (*b*)

Fig. 3.22

It is to be noted from the hysteresis loop that B lags behind H. The two never attain zero value simultaneously.

Hysteresis is especially pronounced in materials of high residual magnetism, such as hard steel. In most cases, hysteresis is a liability as it causes dissipation of heat, waste of energy, and humming due to change in polarity and rotation of element magnets in the material.

Hysteresis loops for hard steel, wrought iron and cast steel and for alloyed sheet steel are shown in Fig. 3.23.

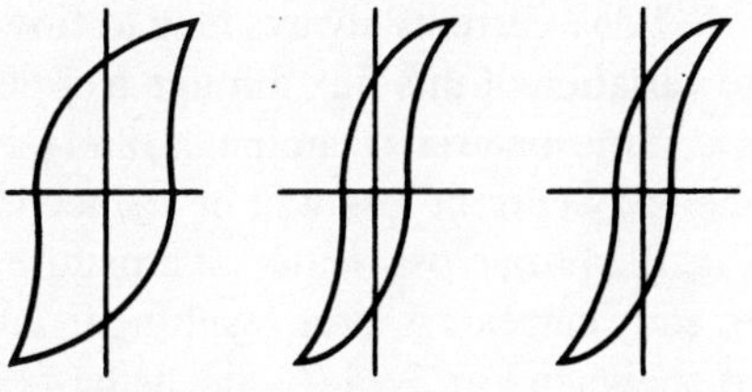

(*a*) (*b*) (*c*)

Fig. 3.23 *Hysteresis Loops*

Loop (*a*) is for hard steel. Due to its high retentivity power and large coercive force, this material is well suited for permanent magnets. Since the area of hysteresis loop for hard steel is large, hard steel is not suitable for rapid reversals of magnetisation. Certain alloys of steel, aluminium and nickel known as *alnico alloys* are extremely suitable for permanent magnets.

Loop (*b*) is for wrought iron and cast steel which rises steeply. Hence these materials have high magnetic permeability and good retentivity, therefore, these materials are suitable for cores of electromagnets.

Loop (*c*) is for iron, low carbon steel, silicon alloys, permalloy or Mumetal sheets. Since the permeability of these materials is very high and hysteresis losses are very low, these materials are most suitable for transformer cores and armatures, which are subjected to rapid reversals of magnetisation. Silicon alloys and permalloy (78.5% Ni; 21% iron with small quantities of copper, molybdenum, chromium, cobalt and manganese etc.) are better for use as compared to iron and low carbon steel.

3.10. EDDY CURRENTS

We have seen that whenever the flux, linking with a closed electric circuit changes, an emf is induced in the circuit and a current flows, the value of which depends on the emf around the circuit and the resistance of the circuit. It is not necessary that the circuit be a wire and that the flux passes entirely through it. If a solid block of metal is traversed by a varying flux, metallic circuits in the block itself, which are linked by the flux, will carry current. If the magnetic circuit is made up of iron and if the flux in the circuit is variable, currents will be induced by induction in the iron circuit itself. All such currents are known as *eddy currents.*

Eddy currents result in a loss of power, with consequent heating of the material. The magnitude of power loss due to eddy currents is often a matter of considerable importance in electrical engineering. Electrical machinery usually involves varying fluxes. In dc as well as ac machines, we have an armature built of iron, which is either revolved in a permanent magnetic field or through which a revolving magnetic field passes. In ac engineering the flux linking the core of transformer also varies. In either case there will be losses in the iron due to hysteresis as well as due to eddy currents.

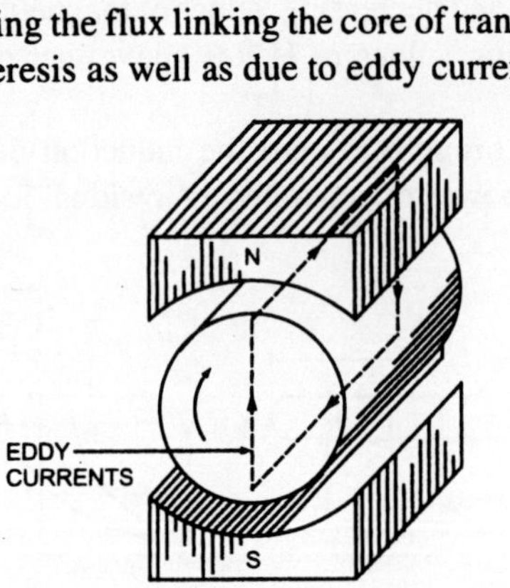

(*a*) *Eddy Currents Induced in The Solid Armature of an Electric Machine*

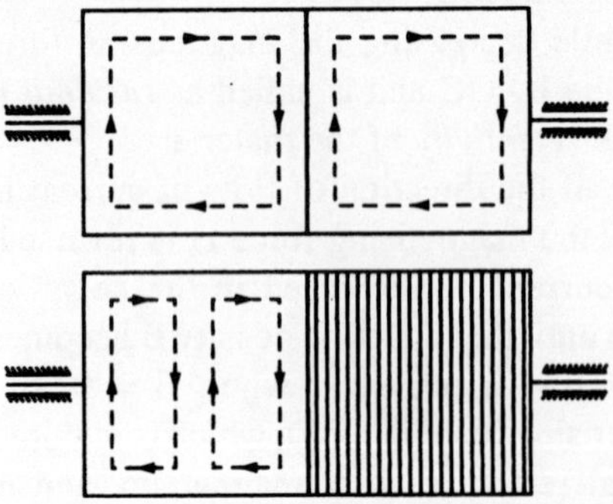

(*b*) *Effect of a Split of Magnetic Path on Eddy Currents (Direction of Eddy Currents is Shown By Arrows)*

Fig. 3.24

Eddy currents always tend to flow in planes perpendicular to the magnetic flux, as they are induced due to variation of this flux through a circuit. If the material of the magnetic circuit is subdivided into thin sheets *i.e.*, if the material is laminated, and these sheets are placed parallel to the flux and insulated from one another, the eddy current loss will be reduced. When the core body is one continuous solid iron piece, as shown in Fig. 3.24 (*a*), cross-section of armature is large, therefore, resistance is very small and consequently magnitude of eddy currents is large resulting in large power loss. If the same core is splitted up into thin insulated sheets, as shown in Fig. 3.24 (*b*), the path of eddy currents is splitted up into several paths of high resistance because cross-section of such paths is very less. Hence magnitude of eddy currents and eddy current loss is considerably reduced. That is why armature of electrical machines (generators and motors) are built up of sheet-steel laminations (usually between 0.45 mm and 0.50 mm thick) assembled on a shaft so that the armature is subdivided into insulated sheets which are parallel to the working flux of the machine. The core of transformer is also built up of thin laminations due to the same reasons.

3.11. HYSTERESIS AND EDDY CURRENT LOSSES

When magnetic circuits are subjected to time-varying flux densities, there are two causes of power loss in the form of heat in the iron core. These losses are significant in determining the heating, rating, and efficiency of rotating electrical machines, transformers, and ac operated devices.

The first loss is associated with the phenomenon of hysteresis, discussed in Art 3.9, and is an expression of the fact that when ferromagnetic material is involved, not all the energy of the magnetic field is returned to the circuit when the mmf is removed. It is known as *hysteresis loss.* When the flux varies from + B, to – B, [Fig. 3.22 (*b*)] at the frequency *f*, the hysteresis loss per unit volume of material may be shown to be proportional to the area of the hysteresis loop and to the number of loops traversed per second.

3.11.1. Determination of Hysteresis Loss. Consider a ring of specimen circumference *l* metres, cross-sectional area *a* metre^2 and having N turns of an insulated wire. Let the current flowing through the coil be of I amperes.

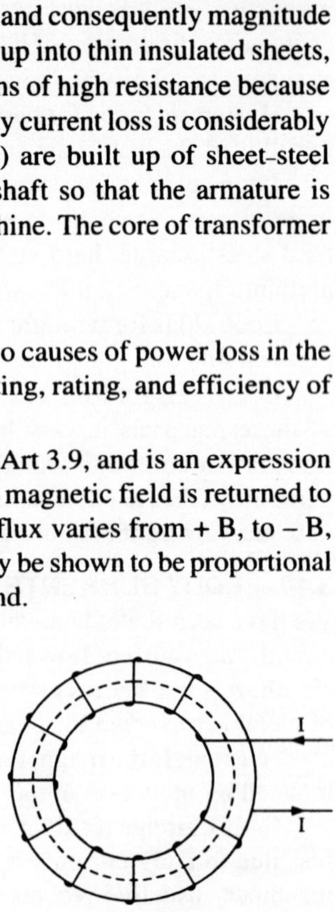

Fig. 3.25

$$\text{Magnetizing force, } H = \frac{NI}{l} \quad \text{or} \quad I = \frac{Hl}{N}$$

Let the flux density at this instant be B.

Total flux through the ring, $\phi = Ba$ webers.

When the current flowing through the solenoid alters, the flux produced in the iron ring also alters, so the emf (e') is induced, whose value is given by

$$e' = -N\frac{d\phi}{dt} = -N\frac{d(Ba)}{dt} = -Na\frac{dB}{dt}$$

According to Lenz's law this induced emf will oppose the flow of current, therefore, in order to maintain the current I in the coil, the source of supply must have an equal and opposite emf.

Hence applied emf, $e = -e' = Na\frac{dB}{dt}$

Energy consumed in short time dt, during which flux density has changed

$$= e\,I\,dt = Na\frac{dB}{dt} \times I \times dt$$

$$= Na\frac{dB}{dt} \times \frac{Hl}{N} \times dt = alH\,dB \text{ joules} \qquad \because \; I = \frac{Hl}{N}$$

Thus total work done or energy consumed during one complete cycle of magnetization,

$$W = al\int_o^{B_{max}} H\,dB$$

Now al is the volume of the ring and HdB is the area of elementary strip of B-H curve shown in Fig. 3.26 and $\int H\,dB$ is the total area enclosed by hysteresis loop.

∴ Energy consumed per cycle

= Volume of ring × area of the hysteresis loop ...(3.9)

This energy expended in taking specimen through a magnetic cycle is wasted and since it appears as heat, it is termed as hysteresis loss.

For determination of hysteresis loss, the area of the loop is measured in units of H and B i.e. in AT/m and Wb/m^2. To measure the area of the loop in units of H and B, the area of the loop is measured in square metres and multiplied by the scales of H and B.

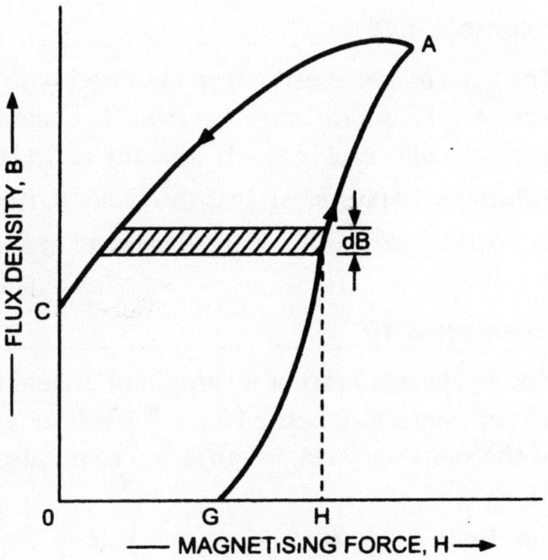

Fig. 3.26

For example if one metre represents x AT/m on H-axis and y Wb/m^2 on B-axis, then hysteresis loss,

$$W_h = \text{Area of hysteresis loop in } m^2 \times x \times y \text{ joules/m}^3\text{/cycle} \qquad ...(3.10)$$

Hysteresis loss can also be determined by *Steinmetz method*, according to which hysteresis loss per cubic metre per cycle of magnetization of a magnetic material depends upon (*i*) the maximum value of the flux density and (*ii*) the magnetic quality of the material.

$$\therefore \quad \text{Hysteresis loss } \alpha\ (B_{max})^{1.6} \text{ J/m}^3\text{/cycle} = \eta\,(B_{max})^{1.6} \text{ J/m}^3\text{/cycle}$$

$$= \eta\,(B_{max})^{1.6} f\,v \text{ J/s or watts} \qquad ...(3.11)$$

where η is a constant for a given specimen and a given range of flux density and is known as Steinmetz hysteresis coefficient, f is the frequency of reversals of magnetization and v is the volume of magnetic material in m^3.

3.11.2. Eddy Current Losses. The second loss arises from the fact that the core itself is composed of conducting material, so that the voltage induced in it by the varying flux produces circulating currents in the iron. These are called *eddy currents* and are accompanied by i^2r loss in the core called the *eddy current loss*. Since the eddy currents depend upon the rate of change of flux as well as the resistance of the path, it is reasonable to expect this loss to vary as the square of both the maximum flux density and the frequency. Eddy current loss is given by the expression

$$P_e = K_e f^2 (B_{max})^2 t^2 \text{ v watts or joules/second} \qquad ...(3.12)$$

where K_e is the *eddy current coefficient* and depends upon the type of core material, f is the number of complete magnetization cycles per second, B_{max} is the maximum flux density in teslas (Wb/m^2), t is the thickness of laminations in metres and v is the volume of core material in m^3.

To increase the core resistance and thereby minimize eddy currents, magnetic cores subjected to alternating fluxes are assembled from thin sheets with an insulating layer (surface oxide or varnish) between successive laminations. Eddy currents also have a magnetic effect, tending to make the flux density at the centre lower than at the surface. This effect, known as *screening effect*, is negligible in properly laminated cores at power frequencies but may be significant at higher frequencies.

Hysteresis and eddy current losses, taken together, are known as *core or iron loss*. Core loss is present in dc machines as well as ac machines, for the rotor iron of a rotating dc machine contains flux which varies cyclically in both magnitude and direction. Many machines are operated at constant voltage and constant frequency or speed and, as a result, have substantially constant core losses regardless of load being supplied. The operating voltage and frequency of a power transformer, for example, are normally constant, by virtue of eq. (3.12), the maximum flux and flux density are, therefore, constant, and the core loss remains the same whether the transformer is loaded or not.

Example 3.18

The area of hysteresis curve obtained with a certain specimen of iron was 9.3 cm². The coordinates were such that 1 cm = 1,000 AT/m and 1 cm = 0.2 T. Calculate (*i*) the hysteresis loss per cubic metre per cycle and (*ii*) the hysteresis loss per cubic metre at a frequency of 50 Hz.

Solution: (*i*) Hysteresis loss/m³/cycle = Area of hysteresis loss in B-H units = 9.3 × 1,000 × 0.2 = 1,860 J/m³/cycle **Ans.**

(*ii*) Hysteresis loss per cubic metre at a frequency of 50 Hz

= Hysteresis loss/m³/cycle × frequency = 1,860 × 50 = 93,000 watts or 93 kW **Ans.**

Example 3.19

The hysteresis loop of a sample of sheet steel subjected to a maximum flux density of 1.3 Wb/m² has an area of 93 cm², the scales being 1 cm = 0.1 Wb/m² and 1 cm = 50 AT/m. Calculate the hysteresis loss in watts when 1,500 cm³ of the same material is subjected to an alternating flux of 1.3 Wb/m² peak value of a frequency of 65 Hz.

[*Allahabad Univ. Electro-mechanics, 1992*]

Solution: Hysteresis loss/m³/cycle = Area of hysteresis loss in B-H units = 93 × 0.1 × 50 = 465 J/m³/cycle

Number of reversals per second, f = 65

Volume of the material, v = 1,500 cm³ = 0.0015 m³

Hysteresis loss, p_h = Hysteresis loss /m³/cycle × f × v = 465 × 65 × 0.0015 = 45.34 W **Ans.**

Example 3.20

Calculate the hourly loss of energy in kWh in a specimen of iron, the hysteresis loop of which is equivalent in area to 250 J/m³, frequency of 50 Hz, specific gravity of iron 7.5; weight of specimen 10 kg.

[*Nagpur Univ. Electrical Engineering Materials 1991, M.S. Univ. of Baroda Electro-Technics I, 1976*]

Solution:

$$\text{Volume of iron} = \frac{\text{Mass}}{\text{Density}} = \frac{10 \times 1{,}000}{7.5} = \frac{10^5}{75}\,\text{cm}^3 = \frac{1}{750}\,\text{m}^3$$

Number of cycles per hour = 50 × 60 × 60 = 180,000

Energy loss per cubic metre of iron volume per cycle

= Area of hysteresis loop = 250 J

Total energy loss per hour = Area of hysteresis loop × volume of iron × number of cycles per hour

$$= 250 \times \frac{1}{750} \times 180{,}000 = 60{,}000\text{ J} = \frac{60{,}000}{36 \times 10^5}\text{ kWh} = 0.0166\text{ kWh}$$ **Ans.**

Example 3.21

Calculate the loss of energy caused by hysteresis in one hour in 50 kg of iron if the peak lux density reached is 1.3 T (Wb/m²) and the frequency is 25 Hz. Assume Steinmetz coefficient as 628 J/m³ and density of iron as 7.8 × 10³ kg/m³. What will be the area of B-H curve of this specimen if 1 cm = 12.5 AT/m and 1 cm = 0.1 T (Wb/m²)?

[*Madras Univ. Electrical Engineering Nov. 1970; 1987*]

Solution: Hysteresis loss/second $= \eta\,(B_{max})^{1.6} f v = 628 \times (1.3)^{1.6} \times 25 \times \dfrac{50}{7.8 \times 10^3} = 153$ J/s

Hysteresis loss per hour $= 153 \times 60 \times 60 = 550{,}800$ J **Ans.**

Hysteresis loss per m^3 per cycle as per Steinmetz law

$$= \eta\,(B_{max})^{1.6} = 628 \times 1.3^{1.6}$$

Also hysteresis loss per m^3 per cycle $= xy \times$ area of B-H loop

$$\therefore \text{Area of B-H loop} = \frac{\text{Hysteresis loss/m}^3\text{/cycle}}{xy} = \frac{628 \times 1.3^{1.6}}{12.5 \times 0.1} = 764 \text{ cm}^2 \textbf{ Ans.}$$

Example 3.22

The eddy current loss in a magnetic material subjected to maximum flux density of 1.2 Wb/m², alternating sinusoidally at a frequency of 50 Hz is 1,600 watts. Determine the eddy current loss when the maximum flux density is 1.5 W/b m² and the frequency is 60 Hz.

Solution: Eddy current loss, $P_{e_1} = 1{,}600$ watts

Maximum flux density, $B_{max\,1} = 1.2$ T

Frequency, $f_1 = 50$ Hz

Maximum flux density, $B_{max\,2} = 1.5$ T

Frequency, $f_2 = 60$ Hz

Since eddy current loss, $P_e \propto B_{max}^2 \times f^2$

$$\therefore\ P_{e_2} = P_{e_1} \times \left(\frac{B_{max2}}{B_{max1}}\right)^2 \times \left(\frac{f_2}{f_1}\right)^2 = 1{,}600 \times \left(\frac{1.5}{1.2}\right)^2 \times \left(\frac{60}{50}\right)^2 = 3{,}600 \text{ watts } \textbf{Ans.}$$

HIGHLIGHTS

1. The closed path followed by magnetic flux is called a *magnetic circuit*. There is a lot of similarity between the magnetic and electric circuits, as mentioned in Art 3.1.4.
2. **MMF** of the magnetic circuit is a magnetic potential that tends to force flux around the magnetic circuit and is analogous to the emf in an electric circuit.

   $$\text{MMF} = \text{NI ampere-turns}$$
3. **Reluctance** is the name given to that property of material which opposes the creation of magnetic flux in it. It is measured in ampere-turns/Wb or amperes/Wb. It may be given as

   $$S = \frac{l}{\mu_o \mu_r a} \text{ AT/Wb}$$

   where l is the length of magnetic path, a is x-sectional area, μ_o is the permeability of the evacuated space or air and is equal to $4\pi \times 10^{-7}$ H/m, and μ_r is the relative permeability of the medium with respect to evacuated space or air.
4. Flux produced ϕ, mmf and reluctance are related as follows

   $$\phi = \frac{\text{MMF}}{\text{Reluctance}} = \frac{\text{NI}}{l} \times \mu_o \mu_r a$$
5. Permeance of a magnetic circuit is the reciprocal of reluctance and is given as

   $$\text{Permeance} = \frac{1}{\text{Reluctance}} = \frac{\mu_o \mu_r a}{l}$$
6. For a series magnetic circuit

   Total ampere-turns required $= H_1 l_1 + H_2 l_2 + H_3 l_3 + \ldots$

   where H_1 is the magnetizing force required to create flux through part l_1 of the magnetic circuit and so on.

7. For parallel magnetic circuit,

AT required = AT required for either parallel magnetic path.

8. Leakage factor is the ratio of total flux (flux in the iron path) and useful flux (flux in the air-gap).
9. The current carrying wire wound spirally in the form of helix is known as solenoid or coil.
10. When a current carrying conductor is placed at right angle to the direction of magnetic field, a mechanical force is experienced on the conductor in a direction perpendicular to both the direction of magnetic field and flow of current. The direction of this force can be determined by Fleming's left hand rule.
11. According to Fleming's left hand rule if the thumb, forefinger and middle finger of the left hand are stretched in such a way that they are at right angles to each other mutually and forefinger points towards the direction of the magnetic field, middle finger towards the direction of the flow of current then thumb will point the direction of force acting on the conductor (Fig. 1).

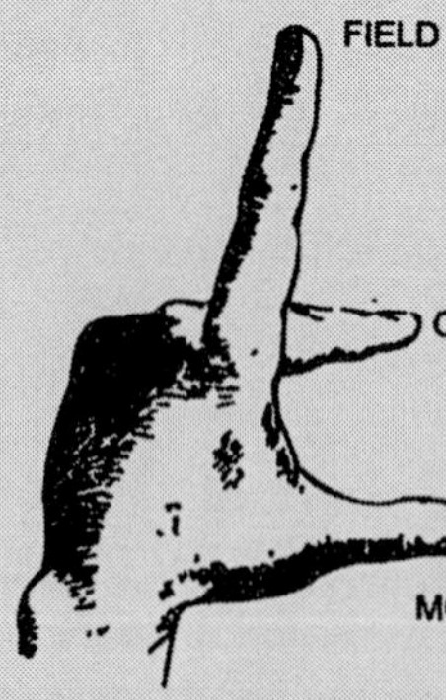

Fig. 1 *Fleming's Left Hand Rule*

12. Although all materials have magnetic properties of some kind being either *diamagnetic*, *paramagnetic* or *ferromagnetic*, the term "magnetic material" is customarily applied only to substances which exhibit ferromagnetism. Ferromagnetic materials are of two types: (*a*) those easily magnetized, called the soft magnetic materials and (*b*) those retaining their magnetism with great tenacity, designated as hard magnetic materials.
13. The curve drawn giving relationship between induction density B and magnetizing force H is known as B-H curve or magnetizing curve.
14. The phenomenon of lagging of magnetization or induction flux density B behind the magnetizing force H is known as magnetic hysteresis.
15. The energy loss associated with the phenomenon of hysteresis is called the hysteresis loss and is given as

$$\text{Hysteresis loss, } P_n = \eta(B_{max})^{1.6} f \, v \text{ joules/second or watts}$$

where η = Steinmetz hysteresis coefficient, v is the volume of core in m^3, B_{max} is the peak magnitude of flux density and *f* is the frequency of magnetic cycles per second.

Hysteresis loss is also given as

$$P_h = xy \times (\text{area of loop in mm}^2) \times f \, v \text{ joules}$$

where x and y = Scale factors of H and B

16. If the magnetic circuit is made up of iron and if the flux in the circuit is variable, currents, known as *eddy currents*, will be induced in the iron circuit itself.

Power loss due to eddy currents is given as

$$P_e = K_e(B_{max})^2 f^2 t^2 \, v \text{ watts}$$

where K_e is the eddy current coefficient, *f* is the number of complete magnetization cycles per second, *t* is the thickness of laminations, v is the volume of material and B_{max} is the maximum value of flux density.

EXERCISES

1. Discuss analogy between electric and magnetic circuits. On what factors does the magnetic reluctance depend ? Is it constant at all flux densities ?
2. Explain the following terms used in magnetic circuit:
   Reluctance, MMF, Flux density and Magnetic field strength.
   [R.G.P.V. Bhopal Basic Electrical Engineering Jan./Feb; 2006]
3. Bring out the analogy between electric and magnetic circuits. Point out major difference, if any between them.
   [Pb Technical Univ. Basic Electrical and Electronics Engineering Second Semester 2004-05]
4. Explain magnetic and electric circuits. Give analogy between them.
   [U.P. Technical Univ. Electrical Engineering Second Semester 2006–07]

5. Compare electric and magnetic circuits with respect to their similarities and dissimilarities.
[Nagpur Univ. Summer 2004]

6. Derive an expression for the electrodynamic force on a conductor carrying current and lying in the magnetic field.

7. State and explain Fleming's left hand rule.

8. Describe the classification of magnetic materials. Give the examples for each class.
[U.P.S.C. I.E.S. Electrical Engineering I, 2004]

9. What are the ferrites and ferromagnetic materials ? Give their properties and applications.
[U.P.S.C. I.E.S. Electrical Engineering I, 1995]

10. Distinguish between soft and hard magnetic materials and classify them suitably. Indicate the properties sought in each case and suggest their applications. **[U.P.S.C. I.E.S. Electrical Engineering I, 1992]**

11. Explain B-H curve of a magnetic material. **[R.G.P.V. Bhopal Basic Electrical Engineering June 2007]**

12. How is B-H curve of a ferromagnetic material different from that of a non-magnetic material ? Name all the salient regions of B-H curve of magnetic material. **[U.P. Technical Univ. Electrical Engineering First Semester 2003-04]**

13. What do you understand by magnetic hysteresis ? Differentiate between hard and soft magnetic materials.
[U.P.S.C. I.E.S. Electrical Engineering-I, 2005]

14. Explain the terms 'hysteresis', 'retentivity' and 'coercive force' in the case of ferromagnetics. Show that the area of the hysteresis loop is proportional to the work done in carrying a magnetic substance through a cycle of magnetization.

15. Explain the effect of hysteresis with suitable diagrams. Also explain the terms residual magnetism, coercive force.
[R.G.P.V. Bhopal Basic Electrical Engineering February 2005]

16. Explain hysteresis and hysteresis loss. Write down the factors affecting them. What is Steinmetz law and what for it is used ?

17. What do you understand by hysteresis and eddy current losses ? State the factors on which they depend. How can they be reduced ? **[Pb. Univ. Elements of Electrical Engineering April 1975]**

18. Explain fringing of magnetic flux, magnetic leakage, saturation of ferromagnetic materials, B-H curves, hysteresis and eddy current losses.
[R.G.P.V. Bhopal Basic Electrical Engineering June 2003]

19. What are hysteresis and eddy current losses ? What are the undesirable effects of eddy current and hysteresis loss ? How can they be minimised ? Mention some applications of eddy currents.
[R.G.P.V. Bhopal Basic Electrical Engineering June 2002]

20. Explain what are hysteresis and eddy current losses. Discuss how these losses are reduced in transformer cores ?
[R.G.P.V. Bhopal Basic Electrical Engineering June 2004]

SHORT ANSWER TYPE QUESTIONS WITH ANSWERS

Q. 1. On what factors does the magnetic reluctance depend ?

Ans. The reluctance of the magnetic circuit is directly proportional to length l, inversely proportional to cross-sectional area and dependent on the nature of material of the magnetic circuit.

Q. 2. Is magnetic reluctance constant for all flux densities ?

Ans. No.

Q. 3. What is meant by leakage and fringing ? **[R.G.P.V. Bhopal December 2001]**

Ans. Leakage flux is the flux, which follows a leaking path, as shown in Fig. 3.5.

Useful flux passing across the gap tends to bulge outwards, as shown in Fig. 3.5., thereby increasing the effective area of gap and reducing the flux density in the gap. This effect is referred to as fringing; and the longer the air gap, the greater is the fringing.

Q. 4. What is meant by leakage factor ?

Ans. The ratio of total flux (flux in the iron path) to the useful flux (flux in the air gap) is known as leakage factor.

Q. 5. How will you find the direction of magnetic field around a current carrying conductor ?

Ans. By Cork Screw rule or right hand rule.

Q. 6. How will you determine the polarity of a solenoid ?

Ans. By using compass needle or by using helix rule.

Q. 7. On what factors does the force experienced by a current carrying conductor placed in a magnetic field depend ?

Ans. The force experienced on a current carrying conductor placed in a magnetic field is directly proportional to (*i*) flux density (field strength), B (*ii*) the magnitude of current flowing through the conductor (*iii*) the length of the conductor and (*iv*) sine of the angle at which the conductor lies with the magnetic field *i.e.* $F = BIl \sin \theta$ newtons.

Q. 8. What is Fleming's left hand rule ?

Ans. Fleming's left hand rule is used to determine the direction of mechanical force experienced by a current carrying conductor placed in a magnetic field.

Fleming's left hand rule states that if the thumb, forefinger and middle finger of the left hand are stretched in such a way that they are at right angles to each other mutually and forefinger points towards the direction of the magnetic field, middle finger towards the direction of flow of current then thumb will point in the direction of force acting on the conductor [Fig. 3.19(*a*)].

Q. 9. Give the relative permeabilities of paramagnetic materials, diamagnetic materials and ferromagnetic materials.

Ans.

Type of Materials	*Relative Permeability*
Paramagnetic	Small but positive. Slightly highermaterials than unity; say 1.00036 for platinum.
Diamagnetic	Slightly less than unity; say 0.999995 materials for copper.
Ferromagnetic	Very high (varying from several materials hundreds to thousands)

Q. 10. What are ferrites ?

Ans. Ferrites is a special group of ferromagnetic materials, that occupy an intermediate position between ferromagnetic and non-magnetic materials.

Q. 11. What is magnetization curve ?

Ans. The curve drawn between induction density B and the magnetizing force H is known as B-H or magnetization curve.

Q. 12. What information is obtained from the loop? **[U.P. Technical Univ. First Semester 2005-06]**

Ans. The area enclosed by the hysteresis loop gives the following information regarding hysteresis loss occurring in the sample under testing. Hysteresis loss/m^3/cycle = Area of hysteresis loss in B-H scales.

Q. 13. Draw hysteresis loops for (*i*) permanent magnets (*ii*) transformer and (*iii*) Ferrite.

[U.P. Technical Univ. First Semester 2005-06]

Ans. Hysteresis loops for permanent magnets, transformer core and ferrite are shown below in Figs. 1(*a*), (*b*), (*c*) respectively.

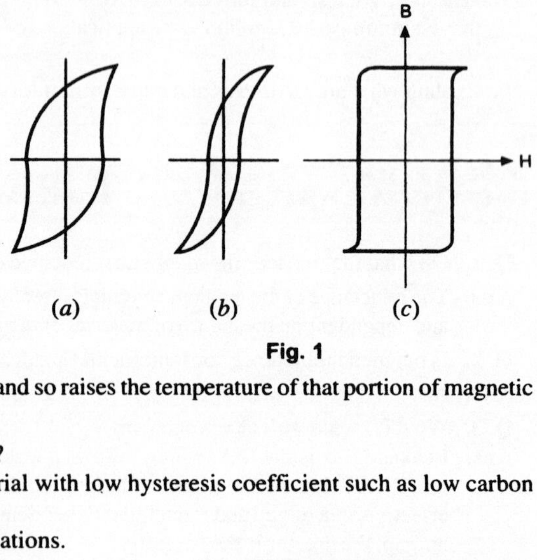

Fig. 1

Q. 14. What is hysteresis loss ? On what factors does it depend ?

Ans. The energy expended in taking a specimen through a magnetic cycle is known as hysteresis loss.

Hysteresis loss per cubic metre per cycle of magnetization of a magnetic material depends upon (*i*) the maximum value of flux density and (*ii*) the magnetic quality of the material.

Q. 15. What are the undesirable effects of eddy current and hysteresis loss ? **[R.G.P.V. Bhopal June 2002]**

Ans. Eddy current and hysteresis loss appears in the form of heat and so raises the temperature of that portion of magnetic circuit, in which they occur.

Q. 16. How can eddy current and hysteresis losses be minimised ?

Ans. Hysteresis loss can be minimised by choosing a core material with low hysteresis coefficient such as low carbon steel, silicon alloys, alloy steel.

Eddy current loss can be minimised by using thin laminations.

Q. 17. Mention some applications of eddy currents. **[R.G.P.V. Bhopal June 2002]**

Ans. Use of eddy currents is made in (*i*) eddy current heating (*ii*) eddy current braking and (*iii*) to develop damping torque in measuring instruments.

PROBLEMS

1. A magnetic circuit has a uniform cross-sectional area of 5 cm^2 and a length of 25 cm. A coil of 120 turns is wound uniformly over the magnetic circuit. When the current in the coil is 1.5 A, the total flux is 0.3 mWb; when the current in the coil is 5 A, the total flux is 0.6 mWb. For each value of current, calculate (*i*) the magnetising force (*ii*) the relative permeability of iron 1,500. **[Ans.** 2,546 AT]

2. A magnetic circuit consists of an iron ring of mean circumference 80 cm with cross-sectional area 12 cm^2 throughout. A current of 2 A in the magnetising coil of 200 turns produces a total flux of 1.2 mWb in the iron. Calculate (*i*) the flux density in the iron (*ii*) the absolute and relative permeability of iron (*iii*) the reluctance of the circuit. **[Ans.** (*i*) 1.0 T (*ii*) 0.002; 1,592 (*iii*) 3.334×10^6 AT/Wb]

3. Calculate the relative permeability of an iron when the exciting current taken by the 600 turn coil is 1.2 A, and the total flux produced is 1 mWb. The mean circumference of the ring is 0.5 m and the area of cross-section is 10 cm^2. **[Ans.** 552]

4. A cast iron ring with a mean diameter of 10 cm and a cross-sectional area of 3 cm^2 has a radial gap of 0.15 cm. Calculate the current required through the winding of 1,000 turns wound uniformly over the ring to produce a flux of 2×10^{-4} Wb in the air gap. Take relative permeability of cast iron as 250. **[Ans.** 1.462 A]

5. A cast steel ring has a circular x-section 3 cm diameter and a mean circumference of 80 cm. The ring is uniformly wound with a coil of 600 turns.
   (*i*) Estimate the current required to produce a flux of 0.5 mWb in the ring.
   (*ii*) If a saw cut of 2 mm wide is made in the ring, find approximately the flux produced by the current found in (*i*).
   Take ampere-turns per metre at the above flux density 670. **[Ans.** (*i*) 0.893A (*ii*) 0.1614 m Wb]

6. A steel ring of 25 cm mean diameter and of circular section 3 cm in diameter has an air gap of 1.5 mm length. It is wound uniformly with 700 turns of wire carrying a current of 2A. Calculate (*i*) magnetomotive force. (*ii*) flux density (*iii*) magnetic flux (*iv*) reluctance and (*v*) relative permeability of steel ring.
   Neglect magnetic leakage and assume that iron path takes about 35 percent of the total magnetomotive force. **[Ans.** (*i*) 1,400 AT (*ii*) 0.762 T (*iii*) 0.539 m Wb (*iv*) 2.6×10^6 AT/Wb (*v*) 968]

7. An iron ring of mean diameter 22 cm and cross-section 10 cm^2 has an air gap 1 mm wide. The ring is wound uniformly with 200 turns of wire. The permeability of ring material is 1,000. A flux of 0.16 m Wb is required in the gap. What current should be passed through the wire ? **[Pb. Univ. Elec. Engineering-1, April 1982]** **[Ans.** 1.075 A]

8. A flux density of 1 Wb/m^2 is required in a 2 mm air gap of an electromagnet with an iron path of 100 cm. Calculate the ampere–turns required for the same. Assume μ for iron = 1,900 and fringing around the air gap is neglected. **[Pb. Univ. Elec. Circuits & Systems Nov.; 1983]** **[Ans.** 2,110]

9. An iron ring has cross–section 3 cm^2 and a mean diameter of 25 cm. An air gap of 0.4 mm has been made by a saw cut across the section. The ring is wound with 200 turns through which a current of 2 A is passed. If the total flux is 21×10^{-5} webers, find μ for iron assuming no leakage. **[Pb. Univ. Elec. Circuits & Systems April 1983]** **[Ans.** 2,470]

10. An iron ring has a mean circumferential length of 60 cm with an air gap of 1 mm and a uniform winding of 300 turns. When a current of 1 A flows through the coil, find the flux density. The relative permeability of iron is 300. Assume $\mu = 4\pi \times 10^{-7}$ H/m. **[Delhi Univ. Electrical Engg. I, 1970]** **[Ans.** 0.1256 T]

11. An iron ring has a x-section of 4 cm^2. An air-gap of 0.5 mm has been made by a cut across its section. The effective iron path is 75 cm. The ring is wound with a coil of 500 turns through which a current of 1.5 A is passed. If the total flux in the air gap is 0.3 m Wb, find the relative permeability of iron. **[Ans.** 991.2]

12. A magnetic ring has a mean circumference of 1.5 m and is of 0.01 m^2 in cross-section and is wound with 175 turns. A saw cut of 4 mm width is made in the ring. Calculate the magnetising current required to produce a flux of 0.8 m Wb in the air-gap. Assume permeability of iron as 400 and leakage factor of 1.25. **[Pb. Univ. Applied Electricity Nov; 1974]** **[Ans.** 3.16 A]

13. An iron ring of mean circumference 80 cm is made from round iron of area 8 cm^2. It has a saw cut of 2 mm wide and is wound with 500 turns. Find the current required to produce a flux of 0.8 m Wb across the air-gap. Assume a relative permeability for iron as 625 and leakage factor of 1.25. **[Ans.** 5.73 A]

14. The magnetic circuit shown in Fig. 1 is built-up of iron of square cross-section, 2 cm wide. Each air gap is 2 mm wide, μ for part A may be taken as 900 and for B as 1250. Each of the exciting coils is wound with 1200 turns and the exciting coil is carrying current 0.8 ampere. Find:

(*i*) Reluctance of A
(*ii*) Reluctance of B
(*iii*) Reluctance of two air gaps
(*iv*) Total reluctance
(*v*) Total flux
(*vi*) Flux density (neglecting leakage). **[R.G.P.V. Bhopal Basic Electrical Engineering June, 2007]**

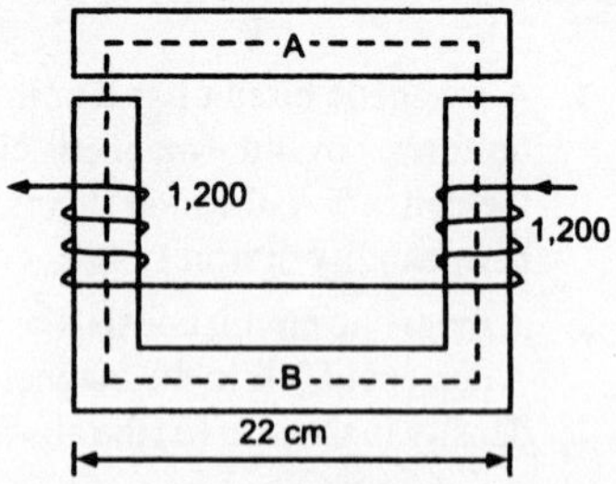

Fig. 1

[Ans. (*i*) 442,100 AT/Wb (*ii*) 636,620 AT/Wb (*iii*) 7,957,750 AT/Wb (*iv*) 9,036,470 AT/Wb (*v*) 0.2125 mWb (*vi*) 0.53 Wb/m^2 or Tesla]

15. Calculate the hourly loss of energy in a specimen of iron, the hysteresis loop of which is equivalent in area to 250 J/m^3. Frequency 25 Hz, density of iron 7.5, weight 10 kg. **[Pb. Univ. Elements of Elec. Engg. April 1983]**
[Ans. 30,000 J]

16. The hysteresis loop of a sample of sheet steel subjected to a maximum flux density of 1.3 T has an area of 93 cm^2, the scales being 1 cm = 0.1 T and 1 cm = 50 AT/m. Calculate the hysteresis loss in watts when 1,500 cm^3 of the same material is subjected to an alternating flux density of 1.3 T peak value at a frequency of 65 Hz.
[M.S. Univ. of Baroda 1970]
[Ans. 45.34 watts]

17. The hysteresis loop for a specimen of iron weighing 12 kg is equivalent in area to 300 J/m^3. Find the loss of energy per hour at 25 Hz. Take density of iron as equal to 7.5 g/cc. **[Gujarat Univ. October 1972]**
[Ans. 43,200 J]

18. The hysteresis loop of certain magnetic material is drawn at a maximum flux density of 1.0 Wb/m^2 with the following scales.

$$1 \text{ cm} = 250 \text{ AT/cm and } 1 \text{ cm} = 0.1 \text{ Wb/m}^2$$

The area of the loop is found to be 50 cm^2. If the density of the material is 7.8 gm/cm^3, calculate the loss in watts per kg due to hysteresis, when the material is subjected to a maximum alternating flux density of 1 Wb/m^2 at 50 Hz.
[Pb. Univ. Applied Electricity 1972]
[Ans. 800 watts]

19. The area of the hysteresis loop obtained with a certain specimen of iron was 9.3 cm^2. The coordinates were such that 1 cm = 1,000 AT/m and 1 cm = 0.2 Wb/m^2. Calculate (*a*) the hysteresis loss per m^3 per cycle and (*b*) the hysteresis loss per m^3 at a frequency of 50 Hz if the maximum flux density were 1.5 Wb/m^2 (*c*) the hysteresis loss per m^3 for a maximum flux density of 1.2 Wb/m^2 and a frequency of 30 Hz, assuming the loss to be proportional to $B_{max}^{1.8}$.
[Allahabad Univ. Electrical Technology 1991]
[Ans. (*a*) 1,860 J/m^3/cycle (*b*) 93,000 W/m^3 (*c*) 37.36 kW]

20. The flux in a magnetic core is alternating sinusoidally at a frequency of 600 Hz. The maximum flux density is 2 Tesla and the eddy current loss is 15 watts. Find the eddy current loss in the core if the frequency is raised to 800 Hz and the maximum flux density is reduced to 1.5 Tesla.
[U.P. Technical Univ; Electromechanical Energy Conversion-I, 2004-05]
[Ans. 15 watts]

❑❑❑

Chapter 4

Electromagnetic Induction

4.1. PRODUCTION OF EMF

An electromotive force (emf) may be produced by (*i*) chemical action (*ii*) heating of thermojunction or (*iii*) electromagnetic action.

Here we will be discussing last method of producing emf (*i.e.* electromagnetic induction) in detail.

It has been shown in Art. 3.6 that when a current carrying conductor is suitably placed in a magnetic field a force is exerted on the conductor in a direction perpendicular to both the direction of flow of current and that of magnetic field. Faraday discovered that reverse of the above is also true *i.e.* an emf can be produced by moving a conductor in a magnetic field. This will be clear from the following experiment.

When two ends of a coil are connected to a galvanometer G and a bar magnet N S is brought nearer to the coil, deflection in the galvanometer is observed. The deflection of galvanometer indicates the flow of current through the coil and this deflection will persist so long as the magnet is moved. If the magnet movement is stopped, the deflection will reduce to zero. Now if the magnet is moved away from the coil, galvanometer will again show deflection but in opposite direction.

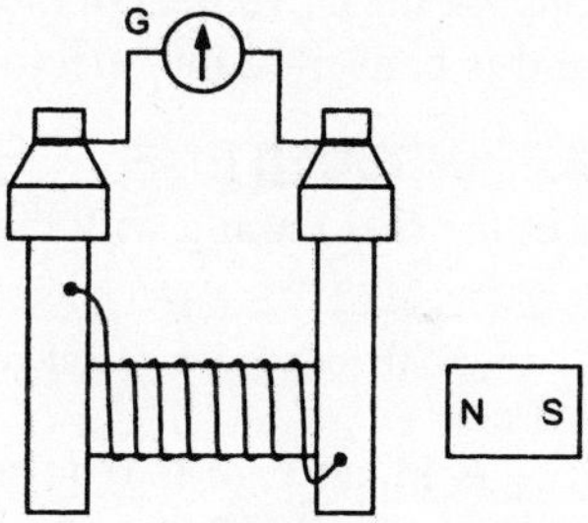

Fig. 4.1

When the magnet is moved towards the coil, linking flux with the coil increases, therefore, an emf is induced in the coil, which causes flow of current through the coil and galvanometer shows deflection. Similarly when the magnet is moved away from the coil, the linking flux decreases, therefore, an emf is induced but in direction opposite to previous one. But when the movement of the magnet is stopped, though flux links with the coil, but there is no change in linking flux, therefore, no emf is induced in the coil.

The same results will be obtained if bar magnet is kept stationary and coil is moved towards or away from the bar magnet.

Hence, for the production of an emf, the change in the magnitude of the linking flux is essential.

4.2. FARADAY'S LAWS OF ELECTROMAGNETIC INDUCTION

Faraday's First Law. *This law states that when the flux linking with the coil or circuit changes an emf is induced in it or whenever the magnetic flux is cut by the conductor an emf is induced in the conductor.*

Faraday's Second Law. *This law states that the magnitude of emf induced is directly proportional to the rate of change of flux linkage or to the product of number of turns and rate of change of flux linking the coil.*

i.e. induced emf, $e \propto N.\frac{d\phi}{dt}$ where N $\frac{d\phi}{dt}$ is the product of number of turns and rate of change of linking flux and is termed as *rate of change of flux linkage*.

4.3. LENZ'S LAW

This law states that the direction of induced emf is such that the current produced by it sets up a magnetic field opposing the motion or change producing it.

The above statement will be clarified from the following experiment.

When the bar magnet is moved towards the coil, as shown in Fig. 4.2, an emf is induced in the coil, which causes an electric current to flow through the coil in the direction shown in Fig. 4.2. Due to flow of this electric current through the coil magnetic field is produced, north pole adjacent to north pole of the bar magnet and south pole on the other end, which can be determined from helix rule explained in Art. 3.5. Now since similar poles repel each other, the movement of the bar magnet towards the coil is opposed by the field produced. Mechanical energy is expended in moving the bar magnet towards the coil against this force of repulsion and this expended energy is converted into electrical energy.

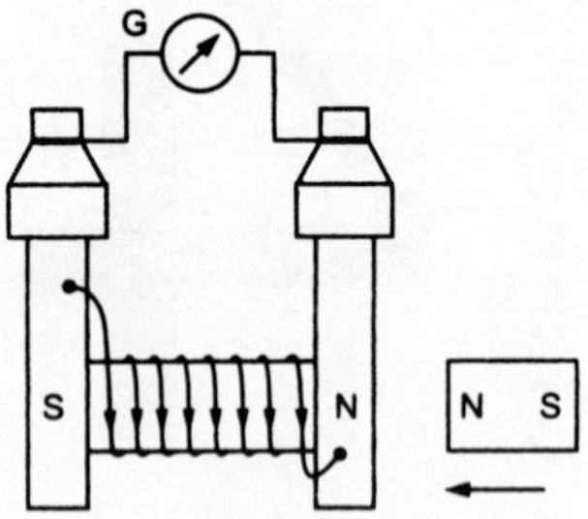

Fig. 4.2

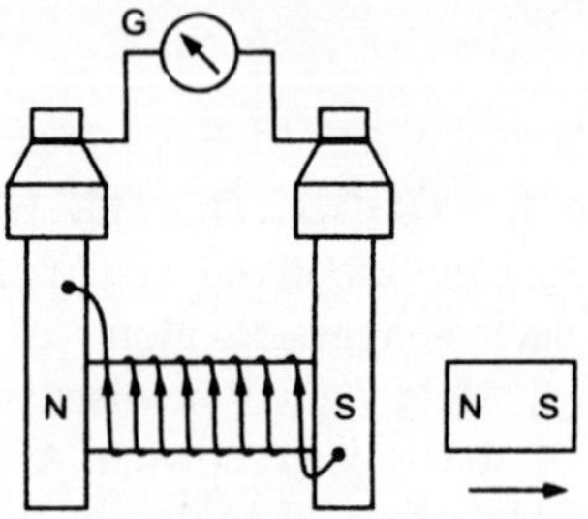

Fig. 4.3

The same results will be observed when the bar magnet is moved away from the coil.

When the bar magnet is moved away from the coil, an emf is induced in the coil, current flows through it in the direction opposite to that in which it was flowing when the magnet was moved towards the coil, so magnetic field will be produced of opposite polarity *i.e.* south pole adjacent to north pole of the bar magnet. Therefore, force of attraction will be exerted between the coil and bar magnet, which will oppose the movement of the bar magnet away from the coil. Mechanical energy is expended in this case also and is converted into electrical energy.

4.4. INDUCED EMF

Let the flux linking with the coil of turns N be changed by an amount $d\phi$ in short time dt.

$$\text{EMF induced, } e = \text{Rate of change of flux linkage} = \text{Number of turns} \times \text{rate of change of flux} = N\frac{d\phi}{dt}$$

A minus sign is required to be placed before the right hand side quantity of above expression just to indicate the phenomenon explained by Lenz's law, therefore, expression for induced emf may be written as

$$e = -N\frac{d\phi}{dt} \text{ volts} \quad \text{...(4.1)}$$

Example 4.1

A coil has 400 turns. Find the induced voltage in it if the flux changes from 0.2 mWb to 1 mWb in 0.2 second.

Solution: Change of flux in 0.2 s, $d\phi = 0.001 - 0.0002 = 0.0008$ Wb

$$\text{Rate of change of flux, } \frac{d\phi}{dt} = \frac{0.0008}{0.2} = 0.004 \text{ Wb/s}$$

$$\text{Induced voltage, } e = N\frac{d\phi}{dt} = 400 \times 0.004 = 1.6 \text{ V } \textbf{Ans.}$$

Example 4.2

A coil of resistance 100 Ω is placed in a magnetic field of 1 mWb. The coil has 100 turns and a galvanometer of 400 Ω resistance is connected in series with it. Find the average emf and the current if the coil is moved from given field to a field of 0.2 mWb in 0.1 second.

Solution: Number of turns on the coil, N = 100

Change of flux in 0.1 s, $d\phi = 0.001 - 0.0002 = 0.0008$ Wb

$$\text{Rate of change of flux, } \frac{d\phi}{dt} = \frac{0.0008}{0.1} = 0.008 \text{ Wb/s}$$

$$\text{Induced emf, } e = N\frac{d\phi}{dt} = 100 \times 0.008 = 0.8 \text{ V } \textbf{Ans.}$$

$$\text{Current through the coil, I} = \frac{e}{\text{Coil resistance + galvanometer resistance}}$$

$$= \frac{e}{100 + 400} = \frac{0.8}{500} = 0.0016 \text{ A or } 1.6 \text{ mA } \textbf{Ans.}$$

4.5. DYNAMICALLY INDUCED EMF

We have learnt that when the flux linking with the coil or circuit changes, an emf is induced in the coil or circuit.

EMF can be induced by changing the flux linking in two ways :

(*i*) By increasing or decreasing the magnitude of the current producing the linking flux. In this case there is no motion of the conductor or of coil relative to the field and, therefore, emf induced in this way is known as *statically induced emf.*

(*ii*) By moving a conductor in a uniform magnetic field and emf produced in this way is known as *dynamically induced emf.*

Consider a conductor of length l metres placed in a uniform magnetic field of density B Wb/m^2, as shown in Fig. 4.4 (*a*).

Let this conductor be moved with velocity v m/s in the direction of the field, as shown in Fig. 4.4 (*b*). In this case no flux is cut by the conductor, therefore, no emf is induced in it.

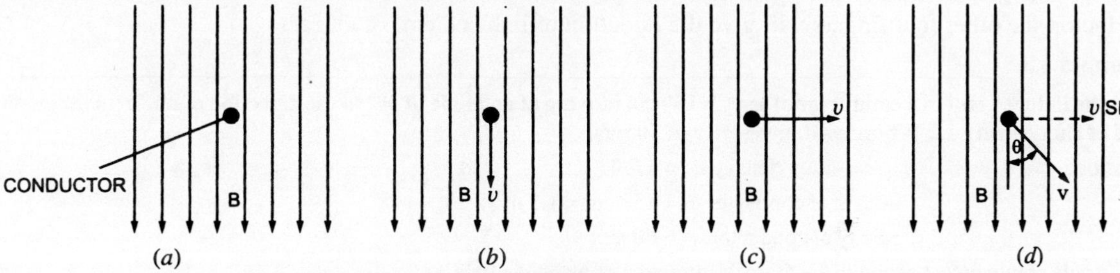

Fig. 4.4

Now if this conductor is moved with velocity v m/s in a direction perpendicular to its own length and perpendicular to the direction of the magnetic field, as shown in Fig. 4.4 (*c*) flux is cut by the conductor, therefore, an emf is induced in the conductor.

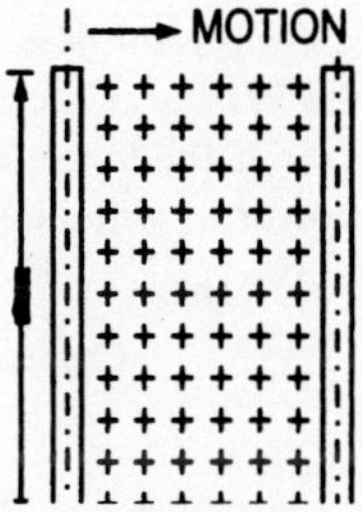

Fig. 4.5

$$\text{Area swept per second by the conductor} = l \times v \text{ m}^2\text{/s}$$

$$\text{Flux cut per second} = \text{Flux density} \times \text{area swept per second} = Blv$$

$$\text{Rate of change of flux, } \frac{d\phi}{dt} = \text{Flux cut per second} = Blv \text{ Wb/s}$$

$$\text{Induced emf, } e = \frac{d\phi}{dt} = Blv \text{ volts}$$

If the conductor is moved with velocity v metres per second in a direction perpendicular to its own length and at an angle θ to the direction of magnetic field, as shown in Fig. 4.4 (*d*).

The magnitude of emf induced is proportional to the component of the velocity in a direction perpendicular to the direction of the magnetic field and induced emf is given by

$$e = Blv \sin\theta \text{ volts} \quad \text{...(4.2)}$$

The direction of this induced emf is given by Fleming's right hand rule.

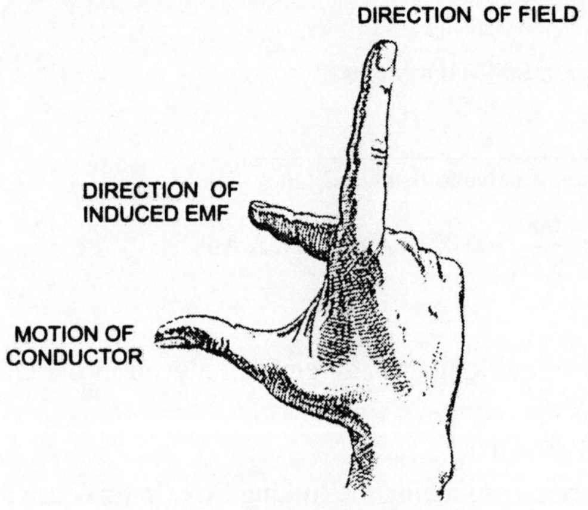

(*a*) *Fleming's Right Hand Rule*

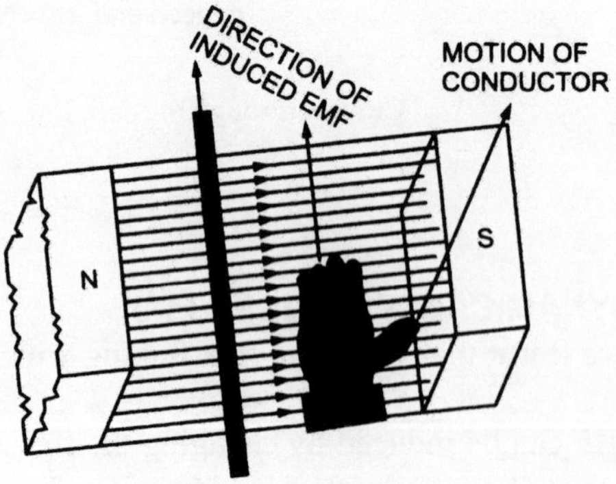

(*b*) *Right Hand Flat Palm Rule*

Fig. 4.6 *Determination of Direction of Induced EMF*

If the thumb, forefinger and middle finger of right hand are held mutually perpendicular to each other, forefinger pointing into the direction of the field and thumb in the direction of motion of conductor then the middle finger will point in the direction of the induced emf as shown in Fig. 4.6 (*a*).

Fig. 4.6 (*b*) illustrates another way of determination of induced emf, known as *right hand flat palm* rule. This law states that if right hand is so placed in a magnetic field along the conductor that the magnetic lines of force emerging from the north pole enter the palm and the thumb points in the direction of motion of conductor, the other four fingers will give the direction of induced emf or current.

Example 4.3

Find the induced emf in a conductor of length 150 cm moving at an angle of 30° to the direction of uniform magnetic field of flux density 1.2 Wb/m² with a velocity of 60 m/s.

Solution: Flux density $B = 1.2$ T

Length of conductor, $l = 150$ cm $= 1.5$ m

Velocity of conductor, $v = 60$ m/s

Angle of movement of conductor from the direction of magnetic field,

$$\theta = 30°$$

EMF induced, $e = Blv \sin\theta = 1.2 \times 1.5 \times 60 \times \sin 30° = 108 \times 0.5 = 54$ V **Ans.**

Example 4.4

A conductor of active length 30 cm carries a current of 100 A and lies at right angles to a magnetic field of strength 0.4 Wb/m². Calculate the force in newtons exerted on it.

If the force causes the conductor to move at a velocity of 10 m/s, calculate (*i*) the emf induced in it and (*ii*) the power in watts developed by it.

Solution: Length of conductor, $l = 30$ cm $= 0.3$ m

Current flowing through conductor, $I = 100$ A

Field strength, $B = 0.4$ Wb/m^2

Force exerted on the conductor, $F = BIl = 0.4 \times 100 \times 0.3 = 12$ N **Ans.**

Conductor velocity, $v = 10$ m/s

Induced emf, $e = Blv = 0.4 \times 0.3 \times 10 = 1.2$ V **Ans.**

Power developed, $P = F \times v = 12 \times 10 = 120$ watts **Ans.**

Example 4.5

A square coil of 10 cm side and with 100 turns is rotated at a uniform speed of 500 rpm about an axis at right angle to a uniform field of 0.5 T. Calculate the instantaneous value of induced emf when the plane of coil is (*i*) at right angle to the plane of the field (*ii*) at 30° to the plane of the field and (*iii*) in the plane of the field.

Solution: As there are two active sides in the coil

$\therefore$ emf induced in the coil is given by $e = 2Blv \sin\theta$ volts

where θ is the angle of the coil movement with the direction of the field.

(*a*) *When plane of the coil is at right angle to the plane of the field.*

The coil moves at an angle of 0° with the direction of the field *i.e.* moves just parallel to the lines of force or $\theta = 0$ or $\sin\theta = 0°$

$\therefore$ emf induced, $e = 0$ **Ans.**

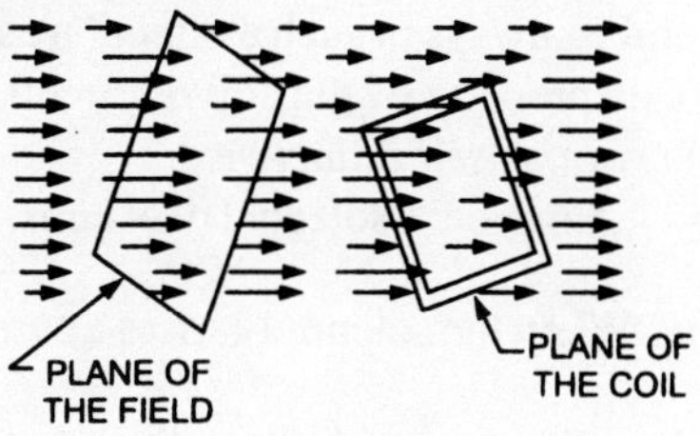

Fig. 4.7

(*b*) *When the plane of the coil makes an angle of 30° with the plane of the field*

The coil moves at an angle of (90° – 30°) with the direction of the field *i.e.* $\theta = 60°$ and $\sin 60° = \frac{\sqrt{3}}{2}$

Flux density, $B = 0.5$ T

Length of coil, $l = 10$ cm $= 0.1$ m

Number of turns $N = 100$

$$\text{Velocity, } v = \omega r \text{ m/s} = \frac{2\pi rpm}{60} r \text{ m/s}$$

$$= \frac{2\pi \times 500}{60} \times 0.05$$

$$= 2.62 \text{ m/s}$$

EMF induced, $e = 2\,Blv\,N \sin\theta$

$$= 2 \times 0.5 \times 0.1 \times 2.62 \times 100 \times \frac{\sqrt{3}}{2}$$

$$= 22.65 \text{ volts } \textbf{Ans.}$$

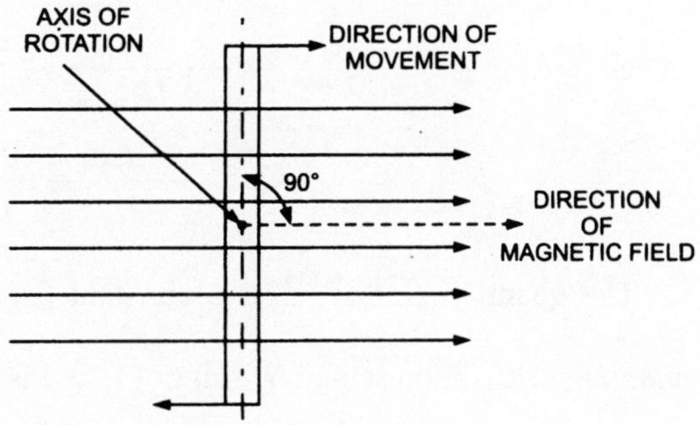

(*a*)

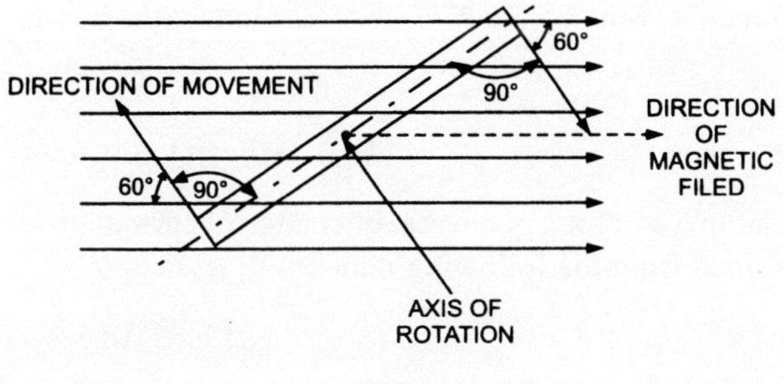

(*b*)

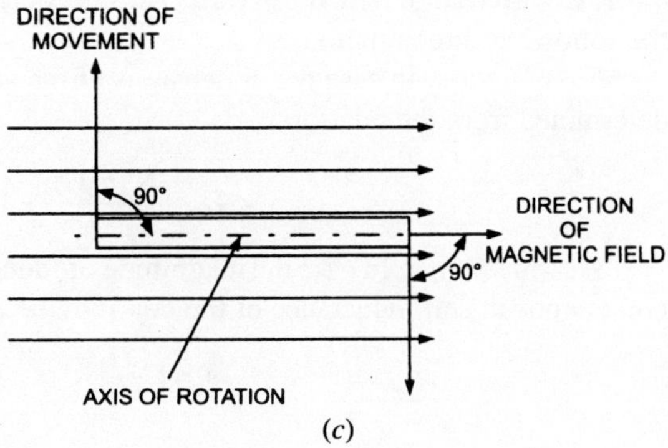

(*c*)

Fig. 4.8

(*c*) *When the plane of the coil is in the plane of the field*

In this position, the coil moves at right angle to the field direction

i.e. $\theta = 90°$ and $\sin 90° = 1$

$\therefore$ EMF induced, $e = 2\,B\,l\,v\,N \sin\theta = 2 \times 0.5 \times 0.1 \times 2.62 \times 100 \times 1 = 26.2$ V **Ans.**

4.6. STATICALLY INDUCED EMF

Statically induced emf may be (*a*) self induced emf or (*b*) mutually induced emf.

4.6.1. Self Induced EMF. When the current flowing through the coil is changed, the flux linking with its own winding changes and due to the change in linking flux with the coil, an emf, known as *self induced emf, is induced.*

Since according to Lenz's law, an induced emf acts to oppose the change that produces it, a self induced emf is always in such a direction as to oppose the change of current in the coil or circuit in which it is induced. This property of the coil or circuit due to which it opposes any change of the current in the coil or circuit, is known as *self inductance*.

Consider a solenoid of N turns, length l metres, area of x-section a square metres and of relative permeability μ_r. When the solenoid carries a current of i amperes, a magnetic field of flux $\dfrac{Ni}{l/\mu_0 \mu_r a}$ webers is set up around the solenoid and links with it.

If the current flowing through the solenoid is changed, the flux produced by it will change and, therefore, an emf will be induced.

$$\text{Self induced emf, } e = -N\frac{d\phi}{dt} = -N\frac{d}{dt}\left[\frac{Ni}{l/\mu_r \mu_0 a}\right]$$

$$= -N\frac{N}{l/\mu_r \mu_0 a}\frac{di}{dt}$$

$$= -\frac{N^2 \mu_r \mu_0 a}{l} \cdot \frac{di}{dt} \qquad ...(4.3)$$

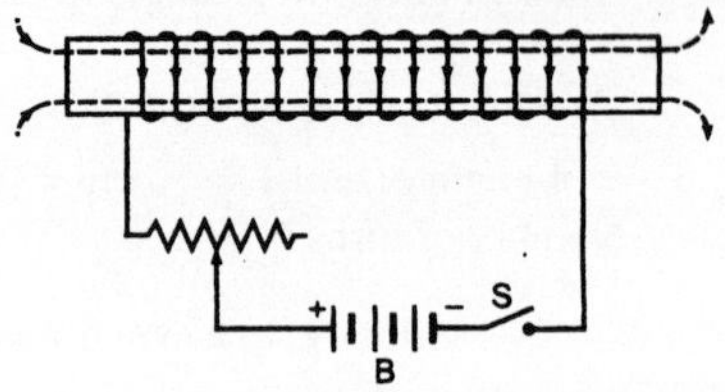

Fig. 4.9

The quantity $\dfrac{N^2 \mu_r \mu_0 a}{l}$ is a constant for any given coil or circuit and is called *coefficient of self inductance.* It is represented by symbol L and is measured in henries.

$$\text{Hence self induced emf, } e = -L\frac{di}{dt} \qquad \text{where } L = \frac{N^2 \mu_r \mu_0 a}{l} \text{ henrys}$$

4.6.2. Coefficient of Self Induction. The coefficient of self induction (L) can be determined from any one of the following three relations.

First Method. In case the dimensions of the solenoid are given, the coefficient of self induction may be determined from the relation

$$L = \frac{N^2 a \mu_r \mu_0}{l} \text{ henrys} \qquad ...(4.4) \qquad \text{[Refer to Art. 4.6.1]}$$

Second Method. In case the magnitude of induced emf in a coil for a given rate of change of current in the coil is known, self inductance of the coil may be determined from the following relation.

$$e = L\frac{di}{dt} \qquad \text{...[Refer Art. 4.6.1]}$$

$$\text{or } L = \frac{e}{di/dt} \qquad ...(4.5)$$

Third Method. In case the number of turns of the coil and flux produced per ampere of current in the coil is known, the self inductance of the coil may be determined from the following relation

$$L = \frac{N\phi}{i} \qquad ...(4.6)$$

The above relation can be derived as follows:

Magnetic flux produced in a coil of N turns, length l metres, area of x-section a metres^2 and relative permeability μ_r when carrying a current of i amperes is given by

$$\phi = \frac{Ni}{l/\mu_r \mu_0 a} = \frac{N}{l/\mu_r \mu_0 a} i$$

and self inductance of the coil, $L = \frac{N^2 a \mu_r \mu_0}{l} = \frac{N}{i} \frac{N a \mu_r \mu_0}{l} i = \frac{N\phi}{i}$

From the above relation it is obvious that the self inductance of a coil or circuit is equal to weber-turns per ampere in the coil or circuit.

In the above relation if $N\phi = 1$ Wb-turn and $i = 1$ A then L = 1 H.

Hence a coil is said to have a self inductance of one henry if a current of 1 A, when flowing through it, produces flux linkage of 1 wb-turn in it.

4.6.3. Mutually Induced EMF. Consider two coils A and B placed closed together so that the flux created by one coil completely links with the other coil. Let coil A have a battery and switch S and coil B be connected to the galvanometer G.

When switch S is opened, no current flows through coil A, so no flux is created in coil A, *i.e.* no flux links with coil B, therefore, no emf is induced across coil B, the fact is indicated by galvanometer zero deflection. Now when the switch S is closed current in coil A starts rising from zero value to a finite value, the flux is produced during this period and increases with the increase in current of coil A, therefore, flux linking with the coil B increases and an emf, known as *mutually induced emf,* is produced in coil B, the fact is indicated by galvanometer deflection. As soon as the current in coil A reaches its finite value, the flux produced or flux linking with coil B becomes constant, so no emf is induced in coil B, and galvanometer pointer returns back to zero position. Now if the switch S is opened, current will start decreasing, resulting in decrease in flux linking with coil B, an emf will be again induced but in direction opposite to previous one, this fact will be shown by the galvanometer deflection in opposite direction.

Hence whenever the current in coil A changes, the flux linking with coil B changes and an emf, known as mutually induced emf, is induced in coil B.

Consider coil A of turns N_1 wound on a core of length l metres, area of cross-section a square metres and relative permeability μ_r. When the current of i_1 amperes flows through it, a flux of $\frac{N_1 i_1}{l/\mu_0 \mu_r a}$ is set up around the coil. Let whole of the flux produced due to flow of current in coil A be linked with the coil B having N_2 turns and placed near by coil A.

Mutually induced emf,

$$e_m = -\text{Rate of change of flux linkage of coil B}$$

$$= -N_2 \times \text{rate of change of flux in coil A}$$

$$= -N_2 \frac{d}{dt}\left[\frac{N_1 i_1}{l/\mu_0 \mu_r a}\right]$$

$$= -\frac{N_1 N_2 a \mu_0 \mu_r}{l} \cdot \frac{di_1}{dt} \quad ...(4.7)$$

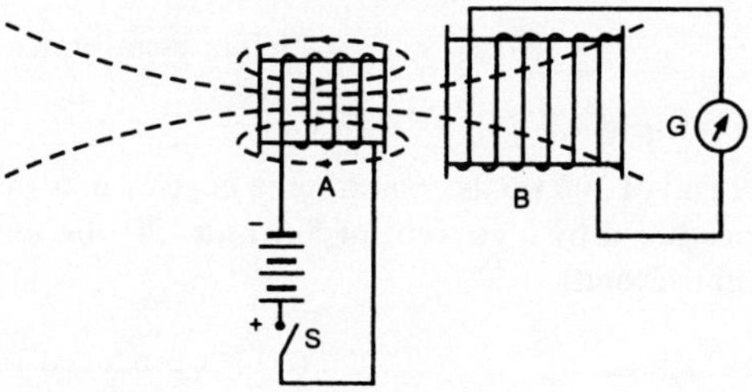

Fig. 4.10

The quantity $\frac{N_1 N_2 a \mu_0 \mu_r}{l}$ is called the *coefficient of mutual induction* of coil B with respect to coil A. It is represented by symbol M and is measured in henrys.

Hence mutually inducd emf, $e_m = -M \frac{di_1}{dt}$ where $M = \frac{N_1 N_2 a \mu_0 \mu_r}{l}$ henrys

4.6.4. Coefficient of Mutual Induction. Mutual inductance may be defined as the ability of one coil or circuit to induce an emf in a nearby coil by induction when the current flowing in the first coil is changed. The action is also reciprocal i.e. the change in current flowing through second coil will also induce an emf in the first coil. The ability of reciprocal induction is measured in terms of the coefficient of mutual induction M.

The coefficient of mutual induction (M) can be determined from any one of the following three relations.

First Method. In case the dimensions of the coils are given, the coefficient of mutual induction may be determined from the relation

$$L = \frac{N_1 N_2 a \mu_0 \mu_r}{l} \text{ henrys} \quad \text{...(4.8) [Refer to Art. 4.6.3]}$$

Second Method. In case the magnitude of induced emf in the second coil for a given rate of change of current in the first coil is known, mutual inductance between the coil may be determined from the following relation

$$e_m = M\frac{di_1}{dt} \quad \text{[Refer to Art. 4.6.3]}$$

$$\text{or } M = \frac{e_m}{di_1/dt} \quad \text{...(4.9)}$$

Third Method. In case the number of turns of the coil and flux linking with this coil per ampere of current in another coil is known, the mutual inductance of the coil may be determined from the following relation

$$M = N_2 \frac{\phi_2}{i_1} \text{ Henry} \quad \text{...(4.10)}$$

Example 4.6

A solenoid has 1,200 turns and carries a current of 2 A. The iron core has a length of 0.4 m and cross-section of 80 cm^2, the relative permeability is 1,000. Calculate the self induced emf in the solenoid, if the current is switched off in 0.01 second.

Solution: Area of x-section of iron core, $a = 80 \text{ cm}^2 = 0.008 \text{ m}^2$

$$\text{Self inductance of coil, } L = \frac{N^2 a \mu_r \mu_0}{l} = \frac{(1{,}200)^2 \times 0.008 \times 1{,}000 \times 4\pi \times 10^{-7}}{0.4} = 36.191 \text{ H}$$

$$\text{Rate of change of current, } \frac{di}{dt} = \frac{2-0}{0.01} = 200 \text{ A/s}$$

$$\text{Self induced emf, } e = L\frac{di}{dt} = 36.191 \times 200 = 7{,}238 \text{ V } \textbf{Ans.}$$

Example 4.7

A coil of 300 turns, wound on a core of non-magnetic material, has an inductance of 10 mH. Calculate (*i*) the flux produced by a current of 5 A and (*ii*) the average value of emf induced when a current of 5 A is reversed in 8 milliseconds.

Solution:

$$(i) \text{ Flux produced, } \phi = \frac{iL}{N} = \frac{5 \times 0.01}{300} = 0.1667 \text{ mWb } \textbf{Ans.} \qquad \because L = \frac{N\phi}{i}$$

$$\text{Rate of change of current, } \frac{di}{dt} = \frac{5-(-5)}{0.008} = 1{,}250 \text{ A/s}$$

$$\text{Induced emf, } e = L\frac{di}{dt} = 0.01 \times 1{,}250 = 12.5 \text{ V } \textbf{Ans.}$$

Example 4.8

A coil of 1,000 turns is wound on a laminated core of steel having a cross-section of 5 cm^2. The core has an air gap of 2 mm cut at right angle. What value of current is required to have an air gap flux density of 0.5 T ? Permeability of steel may be taken as infinity. Determine the coil inductance.

[*U.P. Technical Univ. Electrical Engineering First Semester 2003-04*]

Solution: Total ampere-turns required, $AT = \frac{B}{\mu_0 \mu_r} l_i + \frac{B}{\mu_0} l_g = 0 + \frac{0.5}{4\pi \times 10^{-7}} \times 2 \times 10^{-3} = 796$ $\quad \because \mu_r = \infty$

Current required, $i = \frac{AT}{N} = \frac{796}{1,000} = 0.796$ A **Ans.**

Inductance of coil, $L = \frac{N\phi}{i} = \frac{1,000 \times 0.5 \times 5 \times 10^{-4}}{0.796} = 0.314$ H **Ans.**

Example 4.9

A toroid has a core of square x-section 2,500 mm² in area and a mean diameter of 250 mm. The core material has a relative permeability 1,000. Calculate the number of turns wound on the core to obtain an inductance of 1 H.

Solution: Length of core, $l = \pi \times D = \pi \times 0.25 = 0.25\,\pi$ metres

Area of x-section of core, $a = 2,500\ mm^2 = 0.0025\ m^2$

Relative permeability of core material, $\mu_r = 1,000$

$\mu_0 = 4\pi \times 10^{-7}$ H/m

Self inductance of coil, L = 1 H

$$\because L = \frac{N^2 a \mu_0 \mu_r}{l},$$

Number of turns on the coil, $N = \sqrt{\frac{L \times l}{\mu_0 \mu_r a}} = \sqrt{\frac{1 \times 0.25\pi}{4\pi \times 10^{-7} \times 1,000 \times 0.0025}} = 500$ **Ans.**

Example 4.10

The resistance and inductance of a coil are 3 Ω and 0.1 m H respectively. What potential difference exists at the terminals of the solenoid at the instant when the current is 1 A, but increasing at the rate of 10,000 A per second ?

Solution: Resistance of the coil, R = 3 Ω

Inductance of the coil, $L = 0.1\ mH = 1 \times 10^{-4}$ H

Current flowing through the coil, I = 1.0 A

Rate of increase of current, $\frac{di}{dt} = 10,000$ A/s

Potential difference across the solenoid, V = Voltage drop in resistance + induced voltage in the coil

$$= iR + L\frac{di}{dt} = 1 \times 3 + 1 \times 10^{-4} \times 10,000 = 4 \text{ V } \textbf{Ans.}$$

Example 4.11

An iron ring of 0.15 metre diameter and 0.001 m² in cross–section with a saw cut 2 mm wide is wound with 300 turns of wire. The gap flux density is 1 T. The relative permeability of the iron is 500. Determine the exciting current and inductance. Ignore leakage and fringing.

Solution: Flux density, B = 1 T

Relative permeability for iron, $\mu_r = 500$

Length of iron path $= \pi \times 0.15 - 0.002 = 0.469$ m

Total ampere-turns required, $AT = H_g l_g + H_i l_i = \frac{B}{\mu_0} l_g + \frac{B}{\mu_0 \mu_r} l_i$

$$= \frac{1}{4\pi \times 10^{-7}} \times 0.002 + \frac{1}{4\pi \times 10^{-7} \times 500} \times 0.469$$

$$= 1,591 + 747 = 2,338$$

$$\text{Exciting current, } i = \frac{AT}{N} = \frac{2{,}338}{300} = 7.76 \text{ A } \textbf{Ans.}$$

$$\text{Flux, } \phi = B \times a = 1 \times 0.001 = 0.001 \text{ Wb}$$

$$\text{Inductance, } L = \frac{N\phi}{i} = \frac{300 \times 0.001}{7.76} = 38.66 \text{ mH } \textbf{Ans.}$$

Example 4.12

A circuit has 1,000 turns enclosing a magnet circuit 20 cm² in cross-section. With 4 A, the flux density is 1.0 Wb/m² and with 9 A it is 1.4 Wb/m². Find the mean value of the inductance between these current limits and the induced emf if the current falls from 9 A to 4 A in 0.05 second.

Solution: Flux corresponding to the current of 4 A, $\phi_1 = 1 \times 20 \times 10^{-4} = 0.002$ Wb

$$\text{Self inductance corresponding to the current of 4 A, } L_1 = \frac{N\phi_1}{i_1} = \frac{1{,}000 \times 0.002}{4} = 0.5 \text{ H}$$

Flux corresponding to the current of 9 A, $\phi_2 = 1.4 \times 20 \times 10^{-4} = 0.0028$ Wb

$$\text{Self inductance corresponding to the current of 9 A, } L_2 = \frac{N\phi_2}{i_2} = \frac{1{,}000 \times 0.0028}{9} = 0.31 \text{ H}$$

$$\text{Mean value of self inductance, } L = \frac{L_1 + L_2}{2} = \frac{0.5 + 0.31}{2} = 0.405 \text{ H } \textbf{Ans.}$$

$$\text{Rate of change of current, } \frac{di}{dt} = \frac{9-4}{0.05} = 100 \text{ A/s}$$

$$\text{EMF induced, } e = L\frac{di}{dt} = 0.405 \times 100 = 40.5 \text{ V } \textbf{Ans.}$$

Example 4.13

Find the inductance of a coil in which a current of 0.2 A increasing at the rate of 0.4 A per second represents a power flow of 0.4 watt. [*D.E.I Agra Basic Electrical Engineering, May 1989*]

Solution:

Power flow, $p = 0.4$ watt

Current, $i = 0.2$ A

$$\text{Induced emf, } e = \frac{p}{i} = \frac{0.4}{0.2} = 2 \text{ V} \qquad \because p = ei$$

$$\text{Rate of increase of current, } \frac{di}{dt} = 0.4 \text{ A/s}$$

$$\text{Inductance of coil, } L = \frac{e}{di/dt} = \frac{2}{0.4} = 5 \text{ H } \textbf{Ans.}$$

Example 4.14

Two coils having 100 and 150 turns respectively are wound side by side on a closed iron circuit of section 125 cm² and mean length 200 cm. If the permeability of iron is 2,000, calculate (a) self inductance of each coil (b) mutual inductance between them (c) the emf induced in the second coil if current in first coil changes from 0 to 5 A in 0.02*s*.

Solution: Area of x-section of iron circuit, $a = 125 \text{ cm}^2 = 0.0125 \text{ m}^2$

Length of iron circuit, $l = 200 \text{ cm} = 2.0$ m

$$(a) \text{ Self inductance of first coil, } L_1 = \frac{N_1^2 \mu_0 \mu_r a}{l} = \frac{(100)^2 \times 4\pi \times 10^{-7} \times 2{,}000 \times 0.0125}{2.0} = 157.1 \text{ m H } \textbf{Ans.}$$

$$\text{Self inductance of second coil, } L_2 = \frac{N_2^2 \mu_0 \mu_r a}{l} = \frac{(150)^2 \times 4\pi \times 10^{-7} \times 2{,}000 \times 0.0125}{2.0} = 353.4 \text{ mH } \textbf{Ans.}$$

(*b*) Mutual inductance between coil, $M = \frac{N_1 N_2 \mu_0 \mu_r a}{l} = \frac{100 \times 150 \times 4\pi \times 10^{-7} \times 2{,}000 \times 0.0125}{2.0}$

$= 235.6$ mH **Ans.**

Rate of change of current in first coil, $\frac{di_1}{dt} = \frac{5-0}{0.02} = 250$ A/s

EMF induced in second coil, $e_2 = M\frac{di_1}{dt} = 235.6 \times 10^{-3} \times 250 = 58.9$ V **Ans.**

Example 4.15

Two coils having 100 and 50 turns respectively are wound on a core with $\mu = 4{,}000\ \mu_0$. Effective core length = 60 cm and core area =9 cm^2. Find the mutual inductance between the coils.

[*U.P. Technical Univ. Electrical Engineering Second Semester 2004-05*]

Solution: Number of turns, $N_1 = 100$ and $N_2 = 50$

Core area, $a = 9\ \text{cm}^2 = 9 \times 10^{-4}\ \text{m}^2$

Core length, $l = 60$ cm $= 0.6$ m

$\mu = 4{,}000\ \mu_0 = 4{,}000 \times 4\pi \times 10^{-7} = 5.0265 \times 10^{-3}$ H/m

Mutual inductance between the coils, $M = \frac{N_1 N_2 a \mu}{l} = \frac{100 \times 50 \times 9 \times 10^{-4} \times 5.0265 \times 10^{-3}}{0.6} = 37.7$ mH **Ans.**

Example 4.16

Two coils have a mutual inductance of 0.3 H. If the current in one coil is varied from 5 A to 2 A in 0.4 s, calculate (*i*) the average emf induced in the second coil (*ii*) the change of flux linked with the second coil assuming that it is wound with 200 turns.

Solution: Mutual inductance, $M = 0.3$ H

Rate of change of current in one coil, $\frac{di_1}{dt} = \frac{5-2}{0.4} = 7.5$ A/s

(*i*) EMF induced in second coil, $e_2 = M\frac{di_1}{dt} = 0.3 \times 7.5 = 2.25$ V **Ans.**

(*ii*) Since mutually induced emf, $e_2 = N_2 \frac{d\phi_2}{dt}$

∴ Change of flux linked with the second coil, $\frac{d\phi_2}{dt} = \frac{e_2}{N_2} = \frac{2.25}{200} = 0.01125$ Wb/s **Ans.**

Example 4.17

The self-inductance of a coil of 500 turns is 0.25 H. If 60% of the flux is linked with a second coil of 10,000 turns, calculate

(*i*) The mutual inductance of the two coils.

(*ii*) EMF induced in the second coil when current in the first coil changes at the rate of 100 A/s.

Solution: Self-inductance of the first coil, $L_1 = 0.25$ H

Number of turns on first coil, $N_1 = 500$

Flux per ampere of current in first coil, $\frac{\phi_1}{i_1} = \frac{L_1}{N_1} = \frac{0.25}{500} = 0.5$ mWb $\quad \because L_1 = N_1 \frac{\phi_1}{i_1}$

Flux linked with second coil per ampere of current in first coil

$$\frac{\phi_2}{i_1} = \frac{0.6\,\phi_1}{i_1} = 0.6 \times 0.5 = 0.3 \text{ mWb}$$

(*i*) The mutual inductance of the two coils, $M = \frac{N_2 \phi_2}{i_1} = 10{,}000 \times 0.3 \times 10^{-3} = 3$ H **Ans.**

(*ii*) EMF induced in the second coil, $e_2 = M\frac{di_1}{dt} = 3 \times 100 = 300$ V **Ans.**

Example 4.18

Two identical 1,000 turn coils X and Y lie in parallel planes such that 60% of the flux produced by one coil links the other. A current of 5 A in X produces in it a flux of 0.05 mWb. If the current in X changes from 6 A to – 6 A in 0.01*s*; what is the emf induced in Y ? Calculate the self inductance of each coil and their mutual inductance.

Solution: Flux produced per ampere of current in coil X, $\frac{\phi_1}{i_1} = \frac{0.05 \times 10^{-3}}{5} = 1 \times 10^{-5}$ wb

Number of turns on coil X, $N_1 = 1{,}000$

Number of turns on coil Y, $N_2 = 1{,}000$

Self inductance of coil X (or Y), $L = N_1 \frac{\phi_1}{i_1} = 1{,}000 \times 10^{-5} = 0.01$ H **Ans.**

Flux linking in coil Y per ampere of current in coil X, $\frac{\phi_2}{i_1} = \frac{0.6 \times 0.05 \times 10^{-3}}{5} = 0.6 \times 10^{-5}$ Wb

Mutual inductance between the coils, $M = N_2 \frac{\phi_2}{i_1} = 1{,}000 \times 0.6 \times 10^{-5} = 0.006$ H **Ans.**

Rate of change of current in coil X, $\frac{di_1}{dt} = \frac{6-(-6)}{0.01} = 1{,}200$ A/s

EMF induced in coil Y, $e_2 = M \frac{di_1}{dt} = 0.006 \times 1{,}200 = 7.2$ V **Ans.**

4.7. MUTUAL COUPLING IN A SIMPLE MAGNETIC CIRCUIT

The self inductance of a circuit is intimately associated with the magnetic field linking the circuit. The self inductance emf may be thought of as the emf induced in the circuit by a magnetic field produced by the circuit current.

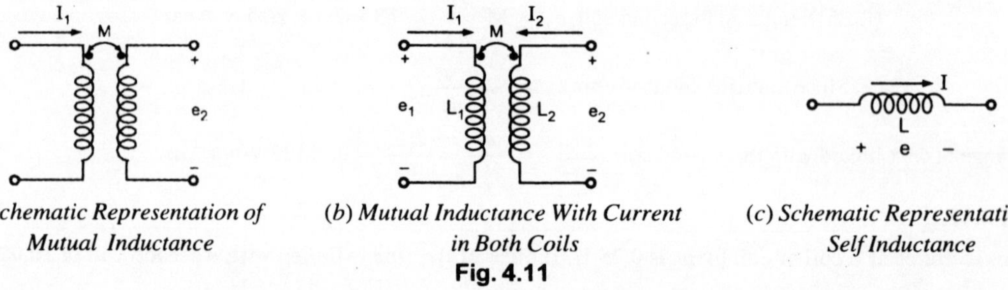

(*a*) *Schematic Representation of Mutual Inductance* (*b*) *Mutual Inductance With Current in Both Coils* (*c*) *Schematic Representation of Self Inductance*

Fig. 4.11

Since a magnetic field exists in the region around the current that develops it, there is also may a possibility that an emf be induced in the other circuits linked by the field. Two circuits linked by the same magnetic field are said to be *coupled* to each other. The circuit element used to represent *magnetic coupling* is shown in Fig. 4.11 (*a*) and is called *mutual inductance.* It is represented by symbol M and is measured in henrys, already stated in Art 4.6.3. The volt-ampere relationship is one which gives the induced emf in one circuit by a current in another and is given as

$$e_2 = M\frac{di_1}{dt}$$

A similar equation can, of course, be written giving an emf e_1 induced by a current i_2. The two dots, called *polarity markings,* in Fig. 4.11 (*a*) are used for indicating the direction of the magnetic coupling between the two coils : note from the figure and from equation given above that by matching the dots, the directions of currents and voltage drops are made to correspond with those of Fig. 4.11 (*b*) and Eq. (4.8) for self inductance.

If currents are present in both coupled circuits, emfs of self inductance and mutual inductance are induced in each circuit. The self induced emf, have the directions shown in Fig. 9.36 (*c*) the mutually induced emfs, follow the pattern of Fig. 4.11 (*a*). Fig. 4.11 (*b*) shows two such elements. Each coil is characterized by its own self inductance; the combination has a mutual inductance indicating the coupling between the coils. The volt-ampere relationships for this case are given below :

$$e_1 = L_1 \frac{di_1}{dt} + M\frac{di_2}{dt} \qquad ...(4.11)$$

$$\text{and } e_2 = L_2 \frac{di_2}{dt} + M\frac{di_1}{dt} \qquad ...(4.12)$$

Coupling between two closed circuits permits the transfer of energy between the circuits through the medium of the mutual magnetic field. This phenomenon is the basis on which transformers operate.

4.8. COEFFICIENT OF COUPLING

When two coils are placed near each other, all the flux produced by one coil does not link with the other coil, as assumed in Art. 4.6.3, only a certain portion (say, K) of flux produced by one coil links with other coil, K being less than unity. K is called the *coefficient of coupling.*

Flux created in coil A due to current of i_1 amperes,

$$\phi_1 = \frac{N_1 i_1}{l/\mu_0 \mu_r a} \text{ Wb}$$

$$\text{Flux linking with coil B, } \phi_2 = K\phi_1 = \frac{K N_1 i_1}{l/\mu_0 \mu_r a} \text{ Wb}$$

$$\text{Coefficient of mutual inductance, } M = \frac{N_2 \phi_2}{i_1} = \frac{K N_1 N_2}{l/\mu_0 \mu_r a} \qquad ...(i)$$

$$\text{Coefficient of self inductance of coil A, } L_1 = \frac{N_1 \phi_1}{i_1} = \frac{N_1^2}{l/\mu_0 \mu_r a} \qquad ...(ii)$$

$$\text{Coefficient of self inductance of coil B, } L_2 = \frac{N_2 \phi_2}{i_2} = \frac{N_2^2}{l/\mu_0 \mu_r a} \qquad ...(iii)$$

Multiplying Eqs. (*ii*) and (*iii*) and taking square root of both sides we get

$$\sqrt{L_1 L_2} = \frac{N_1 N_2}{l/\mu_0 \mu_r a} \qquad ...(iv)$$

Comparing Eqs. (*i*) and (*iv*) we get

$$M = K\sqrt{L_1 L_2}$$

$$\text{or } K = \frac{M}{\sqrt{L_1 L_2}} \qquad ...(4.13)$$

When the coils are tightly coupled *i.e.* when the flux due to one coil links with the other coil completely, the coefficient of coupling, K is unity and the coefficient of mutual inductance M is given as

$$M = \sqrt{L_1 L_2} \qquad ...(4.14)$$

When the flux due to one coil does not link with the other coil at all, the value of coefficient of coupling, K is zero.

Example 4.19

Two coils A of 1,200 turns and B of 800 turns lie near each other so that 60 percent of the flux produced in one links the other. It is found that a current of 5 A in A produces a flux of 0.25 mWb while the same current in B produces a flux of 0.15 mWb. Determine the mutual inductance and coefficient of coupling between the coils.

[*Kanpur Univ. Electrical Science, 1978*]

Solution:

Number of turns on coil A, $N_1 = 1{,}200$

Number of turns on coil B, $N_2 = 800$

$$\text{Flux per ampere of current in coil A, } \frac{\phi_1}{i_1} = \frac{0.25\times10^{-3}}{5} = 0.05 \times 10^{-3} \text{ Wb}$$

$$\text{Self inductance of coil A, } L_1 = N_1 \frac{\phi_1}{i_1} = 1{,}200 \times 0.05 \times 10^{-3} = 0.06 \text{ H}$$

$$\text{Flux per ampere of current in coil B, } \frac{\phi_2}{i_2} = \frac{0.15\times10^{-3}}{5} = 0.03 \times 10^{-3} \text{ Wb}$$

$$\text{Self inductance of coil B, } L_2 = N_2 \frac{\phi_2}{i_2} = 800 \times 0.03 \times 10^{-3} = 0.024 \text{ H}$$

$$\text{Flux linking coil B per ampere of current in coil A, } \frac{\phi_2}{i_1} = \frac{0.25\times10^{-3}}{5} \times 0.6 = 0.03 \times 10^{-3} \text{ Wb}$$

$$\text{Mutual inductance between the coils, } M = N_2 \frac{\phi_2}{i_1} = 800 \times 0.03 \times 10^{-3} = 0.024 \text{ H } \textbf{Ans.}$$

$$\text{Coefficient of coupling, } K = \frac{M}{\sqrt{L_1 L_2}} = \frac{0.024}{\sqrt{0.06\times0.024}} = 0.6325 \textbf{ Ans.}$$

Example 4.20

Define coefficient of coupling and calculate it for two coils having self inductances of 60 mH and 80 mH. The mutual inductance between them is 40 mH

Solution:

Self inductance, $L_1 = 60$ mH $= 0.06$ H

Self inductance, $L_2 = 80$ m H $= 0.08$ H

Mutual inductance, $M = 40$ mH $= 0.04$ H

$$\text{Coefficient of coupling, } K = \frac{M}{\sqrt{L_1 L_2}} = \frac{0.04}{\sqrt{0.06\times0.08}} = 0.577 \textbf{ Ans.}$$

4.9. INDUCTORS IN SERIES

Figure 4.12 (*a*) represents two inductors (inductance coils) in series with the axis of one coil perpendicular to the axis of the other. In this case there is no mutual inductance *i.e.* $M = 0$. If the coils are connected is series, represented as L_1 and L_2 in Fig. 4.12 (*b*), so that when the current enters the dot end of coil L_1 and leaves, it must enter L_2 at its dotted end, the fluxes of two coils will add. Such a series connection is known as *series aiding*. If the connections to inductor L_2 are reversed so that the current must enter its undotted end, their fluxes will oppose each other. This series connection is known as *series opposing*. It has been assumed that the axes of the inductance coils are on the same straight line.

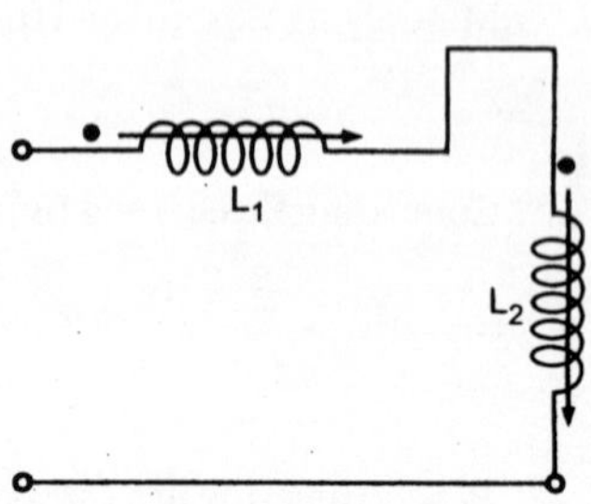

(*a*)

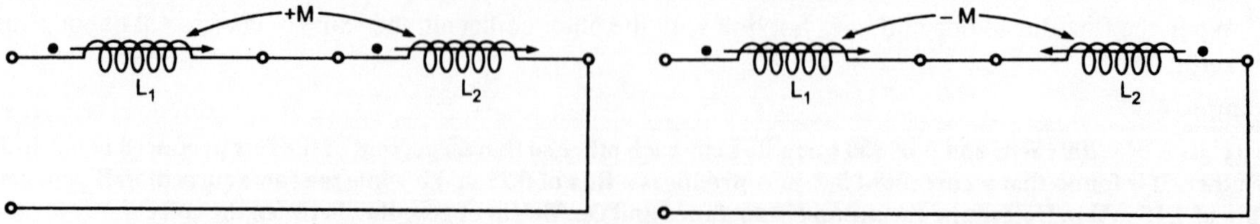

(*b*) *Series Aiding Connections* (*c*) *Series Opposing Connections*

Fig. 4.12

Series-aiding. For the series aiding connections, the total emf induced in each of the coils L_1 and L_2 is due to coil's self inductance and the emf induced by the other coil. So

$$e_1 = L_1 \frac{di}{dt} + M\frac{di}{dt} = (L_1 + M)\frac{di}{dt}$$

$$\text{Similarly } e_2 = (L_2 + M)\frac{di}{dt}$$

So the total induced emf in the circuit is given as

$$e = e_1 + e_2 = (L_1 + L_2 + 2\,M)\frac{di}{dt}$$

$$\text{or } \frac{e}{di/dt} = L_1 + L_2 + 2\,M = \text{Total inductance of the circuit, L}$$

$$\text{or } L = L_1 + L_2 + 2\,M \quad \text{...(4.15)}$$

Series Opposing. For the series opposing connections the mutually induced emf opposes the self induced emf. So

$$e_1 = L_1 \frac{di}{dt} - M\frac{di}{dt} = (L_1 - M)\frac{di}{dt}$$

$$\text{and } e_2 = L_2 \frac{di}{dt} - M\frac{di}{dt} = (L_2 - M)\frac{di}{dt}$$

$$e = e_1 + e_2 = (L_1 + L_2 - 2\,M)\frac{di}{dt}$$

$$\text{or } \frac{e}{di/dt} = (L_1 + L_2 - 2\,M)$$

$$\text{or } L_1 = (L_1 + L_2 - 2\,M) \quad \text{... (4.16)}$$

Example 4.21

A primary coil having an inductance of 100 μH is connected in series with a secondary coil of 240 μH and the total inductance of the combination is measured to be 146 μH. Determine the coefficient of coupling.

Solution: Self inductance of first coil, $L_1 = 100 \times 10^{-6}$ H

Self inductance of second coil, $L_2 = 240 \times 10^{-6}$ H

Total inductance of combination, $L = 146 \times 10^{-6}$ H

Since the total inductance of the combination is less than the arithmetic sum of the self inductances of coils, the coils have been connected in such a way that fluxes are opposing each other. Hence equivalent inductance of combination is given by the relation

$$L = L_1 + L_2 - 2M$$

$$\text{or Mutual inductance between coils, } M = \frac{L_1 + L_2 - L}{2} = \frac{100 \times 10^{-6} + 240\times10^{-6} - 146\times10^{-6}}{2} = 97\times10^{-6}\text{H}$$

$$\text{Coefficient of coupling } K = \frac{M}{\sqrt{L_1 L_2}} = \frac{97\times10^{-6}}{\sqrt{100\times10^{-6}\times240\times10^{-6}}} = 0.626 \textbf{ Ans.}$$

Example 4.22

The total inductance of two coils, A and B when connected in series is 0.5 H or 0.2 H, depending on the relative directions of the current in coils. Coil A, when isolated from coil B has a self inductance of 0.2 H. Calculate (*a*) the mutual inductance between the two coils (*b*) the self inductance of coil B (*c*) the coupling factor between the coils and (*d*) the possible values of the induced emf in coil A when the current is decreasing at 1,000 A/s in the series circuit.

[Hyderabad Univ Electrical Technology 1992]

Solution: Equivalent inductance of combination when the coils are connected in additive series,

$$L_1 + L_2 + 2M = 0.5 \text{ H} \quad ...(i)$$

Equivalent inductance of combination when the coils are connected in subtractive series

$$L_1 + L_2 - 2M = 0.2 \text{ H} \quad ...(ii)$$

Subtracting Eq. (*ii*) from Eq. (*i*) we have

$$(a) \quad 4M = 0.3 \text{ H}$$

$$\text{or } M = 0.075 \text{ H } \textbf{Ans.}$$

Adding Eqs. (*i*) and (*ii*) we have

$$2(L_1 + L_2) = 0.7 \text{ H}$$

$$(b) \text{ or } L_2 = \frac{0.7}{2} - L_1 = 0.35 - 0.2 = 0.15 \text{ H } \textbf{Ans.}$$

$$(c) \text{ Coefficient of coupling, } K = \frac{M}{\sqrt{L_1 L_2}} = \frac{0.075}{\sqrt{0.2 \times 0.15}} = 0.433 \text{ or } 43.3\% \textbf{ Ans.}$$

$$(d) \text{ Induced emf in coil A, } e_1 = L_1 \frac{di}{dt} \pm M \frac{di}{dt} = (L_1 \pm M) \frac{di}{dt}$$

$$= (0.2 \pm 0.075) \times 1{,}000 = 275 \text{ V or } 125 \text{ V } \textbf{Ans.}$$

Example 4.23

Two identical coils with terminals T_1 T_2 and T_3 T_4 respectively are placed side by side. The inductances measured under different sets of connections are as follows ;

When T_2 is connected to T_3 and inductance measured between T_1 and T_4 is 4 H.
When T_2 is connected to T_4 and inductance measured between T_1 and T_3 is 0.8 H.
Determine the self inductance of each coil and the coefficient of coupling.

[*M.S. Univ. Baroda Electro-technics-I, 1975*]

Solution: Self inductance of coil with terminals T_1 T_2 = Self inductance of coil B with terminals T_3 T_4 = L (say)

∵ the coils are identical

Let the mutual inductance between the coils be M

Equivalent inductance between T_1 and T_4 (T_2 connected to T_3),

$$L + L + 2M = 4 \text{ H}$$

$$\text{or } L + M = 2\text{H} \quad ...(i)$$

Equivalent inductance between T_1 and T_3 (T_2 connected to T_4)

$$L + L - 2M = 0.8 \text{ H}$$

$$\text{or } L - M = 0.4 \text{ H} \quad ...(ii)$$

Solving Eqs. (*i*) and (*ii*) we have

$$\text{Self inductance of each coil, } L = \frac{2 + 0.4}{2} = 1.2 \text{ H } \textbf{Ans.}$$

$$\text{Mutual inductance between coil, } M = \frac{2 - 0.4}{0.2} = 0.8 \text{ H}$$

$$\text{Coefficient of coupling, } K = \frac{M}{\sqrt{L_1 L_2}} = \frac{0.8}{\sqrt{1.2 \times 1.2}} = 0.667 \text{ or } 66.7\% \textbf{ Ans.}$$

4.10. INDUCTORS IN PARALLEL

Let us consider two coils of inductances L_1 and L_2 connected in parallel, as shown in Fig. 4.13.

The supply current divides into two components i_1 and i_2 flowing through two coils.

$$i.e.\ i = i_1 + i_2$$

$$\text{or} \quad \frac{di}{dt} = \frac{di_1}{dt} + \frac{di_2}{dt}$$

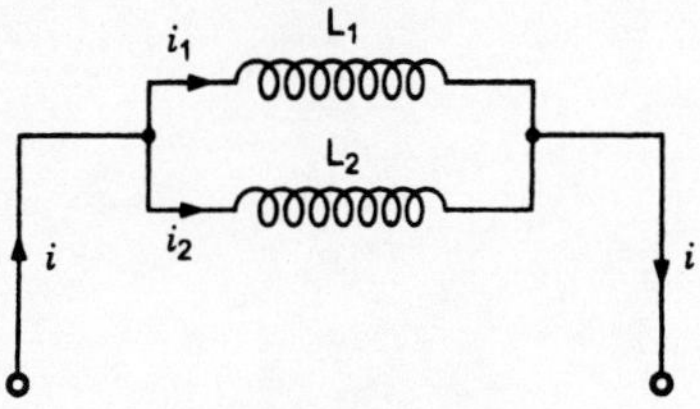

Fig. 4.13

$$\text{Self induced emf in coil A, } e_{LA} = -L_1 \frac{di_1}{dt}$$

Mutually induced emf in coil A due to change of current in coil B,

$$e_{MA} = -M \frac{di_2}{dt}$$ where M is the mutual coefficient of inductance

$$\text{Resultant emf induced in coil A} = e_{LA} + e_{MA} = -\left(L_1 \frac{di_1}{dt} + M \frac{di_2}{dt}\right)$$

$$\text{Similarly resultant emf induced in coil B} = -\left(L_2 \frac{di_2}{dt} + M \frac{di_1}{dt}\right)$$

As both coils are connected in parallel, therefore, resultant emf induced in both of the coils must be equal

$$\therefore \quad L_1 \frac{di_1}{dt} + M \frac{di_2}{dt} = L_2 \frac{di_2}{dt} + M \frac{di_1}{dt}$$

$$\text{or} \quad (L_1 - M) \frac{di_1}{dt} = (L_2 - M) \frac{di_2}{dt}$$

$$\text{or} \quad \frac{di_1}{dt} = \left(\frac{(L_2 - M)}{(L_1 - M)}\right) \frac{di_2}{dt} \qquad ...(i)$$

$$\text{or} \quad \frac{di_1}{dt} + \frac{di_2}{dt} = \left(\frac{(L_2 - M)}{(L_1 - M)} + 1\right) \frac{di_2}{dt}$$

$$\text{or} \quad \frac{di}{dt} = \left(\frac{L_1 + L_2 - 2M}{L_1 - M}\right) \frac{di_2}{dt} \qquad ...(ii)$$

If L is the equivalent inductance of the combination then induced emf

$$= -L \frac{di}{dt}$$

Since induced emf in parallel combination = Induced emf in either of the coils,

$$\therefore L \frac{di}{dt} = L_1 \frac{di_1}{dt} + M \frac{di_2}{dt}$$

$$\text{or } L = \frac{L_1 \dfrac{di_1}{dt} + M \dfrac{di_2}{dt}}{\dfrac{di}{dt}}$$

Substituting $\frac{di_1}{dt} = \left[\frac{L_2 - M}{L_1 - M}\right] \frac{di_2}{dt}$ from Eq. (*i*) and $\frac{di}{dt} = \left[\frac{L_1 + L_2 - 2M}{L_1 - M}\right] \frac{di_2}{dt}$ from Eq. (*ii*) we get

$$L = \frac{L_1 \left[\dfrac{L_2 - M}{L_1 - M}\right] \dfrac{di_2}{dt} + M \dfrac{di_2}{dt}}{\left[\dfrac{L_1 + L_2 - 2M}{L_1 - M}\right] \dfrac{di_2}{dt}}$$

$$\text{or } L = \frac{L_1L_2 - L_1M + L_1M - M^2}{L_1 + L_2 - 2M}$$

$$\text{or } L = \frac{L_1L_2 - M^2}{L_1 + L_2 - 2M} \quad \text{when mutual flux helps the individual flux} \quad ...(4.17)$$

$$\text{or } L = \frac{L_1L_2 - M^2}{L_1 + L_2 + 2M}$$

when mutual flux opposes the individual flux. ...(4.18)

Example 4.24

Two coils of self inductances 150 mH and 250 mH and of mutual inductance 120 mH are connected in parallel. Determine the equivalent inductance of the combination if (*i*) mutual flux helps the individual flux (*ii*) mutual flux opposes the individual flux.

Solution: Self inductance of coil A, L_1 = 150 mH = 0.15 H

Self inductance of coil B, L_2 = 250 mH = 0. 25 H

Mutual inductance, M = 120 mH = 0.12 H

(*i*) Equivalent inductance of the combination when mutual flux helps the individual flux

$$L = \frac{L_1L_2 - M^2}{L_1 + L_2 - 2M} = \frac{0.15 \times 0.25 - (0.12)^2}{0.15 + 0.25 - 2 \times 0.12} = 0.144 \text{ H } \textbf{Ans.}$$

(*ii*) Equivalent inductance of the combination when mutual flux opposes the individual flux.

$$L = \frac{L_1L_2 - M^2}{L_1 + L_2 + 2M} = \frac{0.15 \times 0.25 - (0.12)^2}{0.15 + 0.25 + 2 \times 0.12} = 0.036 \text{ H } \textbf{Ans.}$$

4.11. EFFECTS OF SELF INDUCTION IN A DC CIRCUIT

In case of non-inductive coil, when a pd of V volts is applied across it, the current immediately attains the value of V/R amperes, where R is the resistance of the coil, because magnetic field is not set up due to flow of current in such a coil, so there is no self induced emf which may oppose the applied voltage V.

But in case of an inductive coil, when potential difference of V volts is applied across it, current starts rising, but due to rise in current flowing through the coil, self induced emf is produced in the coil, which opposes the applied voltage and increase in current. Theoretically, the current takes infinite time to reach its maximum value given by

$$I_{max} = \frac{V}{R}$$

In case of non-inductive coil applied voltage is to overcome the resistive drop, IR only but in case of inductive coil applied voltage is to overcome the back emf set up by self induction of coil in addition to resistive drop IR.

4.11.1. Rise of Current. Consider an inductive coil of resistance R ohms and inductance L henrys connected across a source of supply of V volts.

Let at any instant current flowing through the coil be of *i* amperes.

Resistive drop = i R volts

$$\text{Back emf, } e' = -L\frac{di}{dt}$$

Applied voltage is to overcome resistive drop as well as back emf. Hence while current is increasing, we have

$$V = iR + L\frac{di}{dt}$$

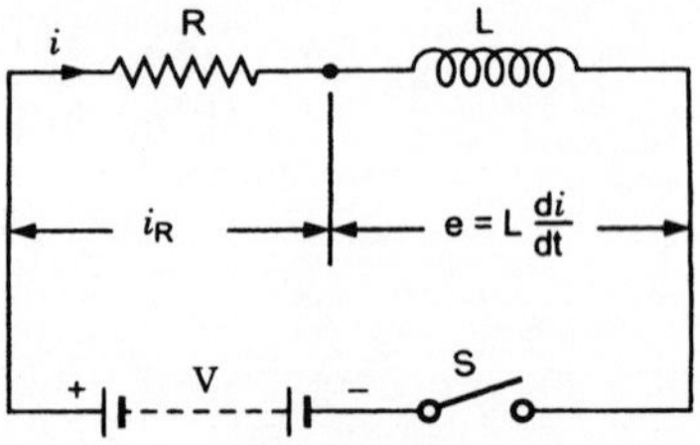

Fig. 4.14

$$\text{or } V - iR = L\frac{di}{dt}$$

$$\text{or } \frac{di}{V - iR} = \frac{dt}{L}$$

$$\text{or } \frac{-R\,di}{V - iR} = \frac{-R}{L}dt$$

Integrating both sides we get

$$\log_e (V - iR) = \frac{-R}{L}t + K$$

where K is a constant of integration

From initial conditions, when $t = 0$, $i = 0$

$$\therefore K = \log_e V$$

$$\text{Hence } \log_e (V - iR) = \frac{-R}{L}t + \log_e V$$

$$\text{or } \log_e \frac{V - iR}{V} = \frac{-R}{L}t$$

$$\text{or } \frac{V - iR}{V} = e^{\frac{-R}{L}t}$$

$$\text{or } V - iR = V.e^{\frac{-R}{L}t}$$

$$\text{or } iR = V\left(1 - e^{\frac{-R}{L}t}\right)$$

$$\text{or } i = \frac{V}{R}\left(1 - e^{\frac{-R}{L}t}\right) = I_{max}(1 - e^{-t/\tau}) \quad ...(4.19)$$

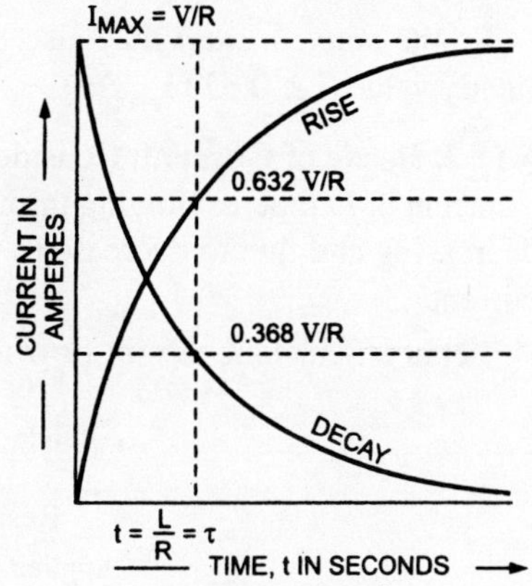

Fig. 4.15

where $I_{max} = \frac{V}{R}$, the final steady-state value of current and $\tau = \frac{L}{R}$, the time constant

(The time constant is also represented by English letter "T")

The equation $i = I_{max}(1 - e^{-t/\tau})$ is an exponential expression and is often referred to as *Helmholtz expression.*

Immediately after the switch is closed, the rate of rise of current *i.e.,* initial rate of rise of current can be determined by differentiating Eq. (4.19) w.r.t. for $t = 0$. Hence we get

$$\frac{di}{dt} = \frac{V}{R\tau}e^{-t/\tau} = \frac{V}{R\times\frac{L}{R}}e^0 = \frac{V}{L} \qquad \because \tau = \frac{L}{R} \text{ and } t = 0$$

If this rate of rise of current were maintained then time required to attain the maximum value of current $I_{max} = V/R$ would have been $\frac{V/R}{V/L} = \frac{L}{R}$. This time is called the *time constant* of the circuit.

Hence the *time constant,* τ may be defined as the *time required for the current to rise to its final steady-state value if it continued rising at its initial rate* (*i.e.* V/L)

If we substitute $t = \tau = \frac{L}{R}$ in Eq. (4.19), we get

$$i = I_{max}(1 - e^{-1}) = 0.632\, I_{max}$$

So the *time constant* may also be defined as the time required for the current to reach 0.632 of its final steady value (*i.e.* 0.632 I_{max}).

4.11.2. Decay of Current. Consider an inductive coil of resistance R ohms, inductance L henrys and having a current of I amperes flowing through it. If the source of supply is removed from its circuit, the current starts decreasing and due to decrease in current, self induced emf will be set up, which will oppose the decay of current.

Let at any instant current flowing through the coil be of *i* amperes

$$\text{Self induced emf, } e' = -L\frac{di}{dt}$$

$$\text{Resistive drop} = iR$$

$$\text{Applied voltage, V} = 0$$

$$\text{Since applied voltage, V} = \text{Resistive drop + back emf}$$

$$\therefore\ 0 = iR + L\frac{di}{dt}$$

$$\text{or}\quad \frac{di}{i} = \frac{-R}{L}dt$$

Integrating both sides we get

$$\log_e i = \frac{-R}{L}t + K' \qquad \text{where } K' \text{ is a constant of integration}$$

From initial conditions *i.e.* when $t = 0$, $i = I_{max}$

$$\therefore\ K' = \log_e I_{max}$$

$$\text{Hence } \log_e i = \frac{-R}{L}t + \log_e I_{max}$$

$$\text{or}\quad \log_e \frac{i}{I_{max}} = \frac{-R}{L}t$$

$$\text{or}\quad i = I_{max}\, e^{\frac{-R}{L}t} = I_{max}\, e^{-t/\tau} \qquad \text{...(4.20) where } \tau = \frac{L}{R}\text{, the time constant}$$

The rise and decay of current are shown in Fig. 4.15.

Example 4.25

If a coil has resistance of 10 ohms and inductance 1 H, estimate the value of current after 0.1 second of switching on 100 V dc supply.

Solution:

$$\text{Time constant of coil, } \tau = \frac{L}{R} = \frac{1}{10} = 0.1 \text{ second}$$

$$\text{Current after 0.1 second of switching, } i = \frac{V}{R}(1 - e^{-t/\tau}) = \frac{100}{10}(1 - e^{0.1/0.1}) = 6.32 \text{ A } \textbf{Ans.}$$

Example 4.26

A choke with an inductance of 20 H and a resistance of 16 Ω is connected through a switch to a 24 V source of emf.

(*i*) What is the initial current when the switch is closed ?

(*ii*) What is the initial rate of change of current when the switch is closed ?

(*iii*) What is the final steady-state current ?

(*iv*) What is the current at the instant of 1.25 s after switching ?

Solution:

(*i*) Initial current when the switch is closed *i.e.* $t = 0$

$$i = I_{max}(1 - e^{-t/\tau}) = I_{max}(1 - e^0) = 0 \textbf{ Ans.}$$

(*ii*) Rate of change of current when the switch is closed *i.e.* when $t = 0$

$$\frac{di}{dt} = \frac{V}{L} = \frac{24}{20} = 1.2 \text{ A/s } \textbf{Ans.}$$

(*iii*) Final steady-state current, $I_{max} = \frac{V}{R} = \frac{24}{16} = 1.5$ A **Ans.**

Time constant, $\tau = \frac{L}{R} = \frac{20}{16} = 1.25$ seconds

(*iv*) Current after 1.25 seconds of switching, $i = \frac{V}{R}(1 - e^{-t/\tau}) = 1.5(1 - e^{-1.25/1.25}) = 1.5 \times 0.632 = 0.948$ A **Ans.**

Example 4.27

A coil has a time constant of 1 second and an inductance of 8 H. If the coil is connected to a 100 volt source, determine:

(*i*) the rate of rise of current at the instant of switching (*ii*) the steady value of the current and (*iii*) the time taken for the current to reach 60% of the steady value of current. [*M.S. Univ. Baroda Electro-Technics I, 1975*]

Solution: Time constant of circuit, $\tau = \frac{L}{R} = 1$ second

Resistance of circuit, $R = \frac{L}{\tau} = \frac{8}{1} = 8\ \Omega$

(*i*) The rate of rise of current at the instant of switching,

$$\frac{di}{dt} = \frac{V}{L} = \frac{100}{8} = 12.5 \text{ A/s } \textbf{Ans.}$$

(*ii*) Steady value of current, $I_{max} = \frac{V}{R} = \frac{100}{8} = 12.5$ A **Ans.**

Let the time taken for the current to reach 60% of the steady value of current *i.e.* 0.6 × 12.5 or 7.5 A be *t* seconds

$$\therefore\ i = I_{max}(1 - e^{-t/\tau})$$

$$\therefore\ 7.5 = 12.5(1 - e^{-t/1})$$

$$\text{or } t = -\log_e 0.4 = 0.916 \text{ second } \textbf{Ans.}$$

Example 4.28

A coil of resistance 20 Ω and inductance 0.8 H is connected to a 200 V dc supply. Find (*i*) rate of change of current at the instant of closing switch (*ii*) final steady value of current (*iii*) time constant of the circuit (*iv*) time taken for the current to rise to half its final value, and (*v*) energy stored in the field when steady-state is reached.

Solution: (*i*) Rate of change of current at the instant of switching,

$$\frac{di}{dt} = \frac{V}{L} = \frac{200}{0.8} = 250 \text{ A/s } \textbf{Ans.}$$

(*ii*) Final steady value of current, $I_{max} = \frac{V}{R} = \frac{200}{20} = 10$ A **Ans.**

(*iii*) Time constant of circuit, $\tau = \frac{L}{R} = \frac{0.8}{20} = 0.04$ second **Ans.**

(*iv*) Equation for instantaneous current while current is rising may be written as

$$i = I_{max}(1 - e^{-t/\tau})$$

When $i = \frac{1}{2} I_{max}$;

$$\frac{1}{2} I_{max} = I_{max}(1 - e^{-t/0.04})$$

$$\text{or } e^{-25t} = 0.5$$

$$\text{or } t = -0.04 \log_e 0.5 = 0.028 \text{ second } \textbf{Ans.}$$

(v) Energy stored in the field when steady-state is reached

$$= \frac{1}{2} LI^2_{max} = \frac{1}{2} \times 0.8 \times (10)^2 = 40 \text{ J } \textbf{Ans.}$$

Example 4.29

The time constant of a certain inductive coil was found to be 0.0025 second. With a resistance of 80 Ω added in series, a new time constant of 0.0005 second was obtained. Determine the inductance and resistance of the coil. Determine the steady value of the current and the time taken by the current to attain 63.2% of the steady value of the current when the coil alone is connected across a constant voltage of 200 volts.

Solution: Time constant in the first case, $\tau_1 = \frac{L}{R} = 0.0025$ s ...(i)

Time constant when a resistance of 80 Ω is added, $\tau_2 = \frac{L}{R+80} = 0.0005$ s ..(ii)

Dividing Eq. (i) by Eq. (ii) we get

$$\frac{R+80}{R} = \frac{0.0025}{0.0005} = 5$$

$$\text{or } R = 20\ \Omega \textbf{ Ans.}$$

$$\text{and } L = R\tau_1 = 20 \times 0.0025 = 0.05 \text{ H } \textbf{Ans.}$$

Steady value of current, $I_{max} = \frac{V}{R} = \frac{200}{20} = 10$ A **Ans.**

Instantaneous current is given by equation $i = \frac{V}{R}(1 - e^{t/0.0025})$

$$\text{or } 0.632\,I = I(1 - e^{-t/0.0025})$$

$$\text{or } e^{-t/0.0025} = 0.368$$

$$\text{or } t = -0.0025 \log_e 0.368 = 0.0025 \text{ second } \textbf{Ans.}$$

Example 4.30

A circuit of resistance R ohms and inductance L henries has direct voltage of 230 V applied to it. 0.3 second after switching on, the current in circuit was found to be 5 A. After the current has reached its final steady value, the circuit was suddenly short-circuited. The current was again found to be 5 A at 0.3 second after short-circuiting the coil. Find the value of R and L. *[Bombay Univ. Basic Electricity 1976]*

Solution: The instantaneous current is given by equations

$$i = I_{max}(1 - e^{-t/\tau}) \text{ while growing}$$

$$\text{and } i = I_{max}\, e^{-t/\tau} \text{ while decaying}$$

$$\therefore\ 5 = I_{max}(I - e^{-0.3/\tau}) \quad \text{...(}i\text{) while growing}$$

$$\text{and } 5 = I_{max}\, e^{-0.3/\tau} \quad \text{...(}ii\text{) while decaying}$$

Comparing Eqs. (i) and (ii) we have

$$1 - e^{-0.3/\tau} = e^{-0.3/\tau}$$

$$\text{or } e^{-0.3/\tau} = 0.5$$

$$\text{or } \tau = \frac{-0.3}{\log_e 0.5} = \frac{-0.3}{0.6931} = 0.4328 \text{ second}$$

Substituting $\tau = 0.4328$ in Eq. (ii) we have

$$5 = I_{max}\, e^{-0.3/0.4328}$$

$$\text{or } I_{max} = \frac{5}{3^{-0.3/0.4328}} = \frac{5}{0.5} = 10 \text{ A}$$

Resistance of circuit, $R = \frac{V}{I_{max}} = \frac{230}{10} = 23\ \Omega$ **Ans.**

Inductance of circuit, $L = \tau R = 0.4328 \times 23 = 9.9544$ H **Ans.**

HIGHLIGHTS

1. When the flux linking the circuit changes, an emf is induced, and its magnitude is given as

$$e = N\frac{d\phi}{dt} = \frac{N(\phi_2 - \phi_1)}{dt}$$

2. The magnitude of dynamically induced emf in a conductor moving at right angles to a uniform magnetic field is given as

$$e = Blv \text{ volts}$$

If the conductor moves at an angle θ to the magnetic field, then

$$e = Blv \sin\theta \text{ volts}$$

The direction of dynamically induced emf is given by Fleming's right hand rule. According to Fleming's right hand rule:

If the thumb, forefinger and middle finger of right hand are held mutually perpendicular to each other, forefinger pointing into the direction of the field and thumb in the direction of motion of conductor then the middle finger will point in the direction of the induced emf as shown in Fig 1.

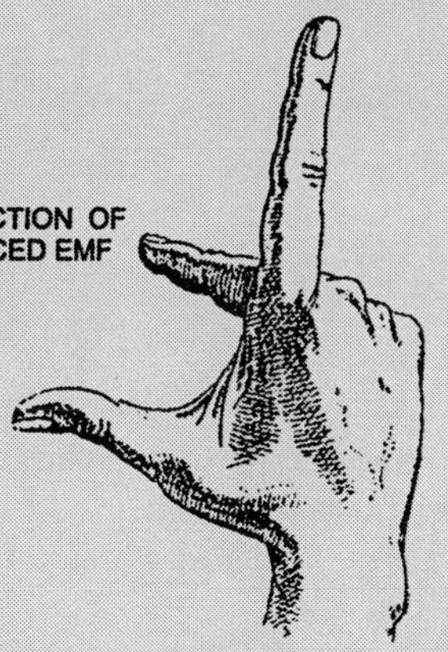

Fig. 1

3. When the current flowing through the coil is changed, the flux linking with its own winding changes and due to the change in linking flux with the coil, an emf, known as *self induced emf, is induced.* The magnitude of self-induced emf, e in a coil of N turns is given by

$$e = N\frac{d\phi}{dt} = L\frac{di}{dt}$$

where L is known as *self inductance* of the coil and is given as

$$L = \frac{e}{di/dt} = \frac{N\phi}{i} = \frac{N^2\mu_o\mu_r a}{l} = \frac{N^2}{\text{Reluctance, S}}$$

4. Whenever the current in coil A (placed near another coil B) changes, the flux linking with coil B changes and an emf, known as mutually induced emf, is induced in coil B.

The magnitude of mutually induced emf is given as

$$e = M\frac{di_1}{dt} \text{ volts}$$

where $\frac{di_1}{dt}$ is the rate of change of current in coil A and M is mutual inductance of the two coils in henry and is given as

$$M = \frac{e_m}{di_1/dt} = \frac{N_2\phi_2}{i_1} = \frac{N_1N_2 a\mu_o\mu_r}{l} = \frac{N_1N_2}{\text{Reluctance, S}}$$

5. The coefficient of coupling between two coils is given as

$$K = \frac{M}{\sqrt{L_1L_2}}$$

where L_1 and L_2 are the individual inductances of the coils and M is the mutual inductance between the coils.

6. When two coils of inductance L_1 and L_2 are connected in series and M is the mutual inductance between them, the total inductance of the circuit is given as

$$L = L_1 + L_2 + 2M \quad \text{in series aiding}$$

and

$$L = L_1 + L_2 - 2M \quad \text{in series opposing}$$

7. In an inductive circuit, the current builds up according to the relation

$$i = I_{max}(1 - e^{-t/\tau})$$

where $\tau = \dfrac{L}{R}$, the time constant of the circuit

Similarly, the current decays according to the relation

$$i = I_{max} \cdot e^{-t/\tau}$$

8. Time constant, τ may be defined as the time required for the current to rist to its final steady-state value, if it is continued rising at its initial rate of V/L.

EXERCISES

1. State Faraday's laws of electromagnetic induction. **[GATE 1994]**
2. State Faraday's laws of electromagnetic induction. Distinguish between statically induced emf and dynamically induced emf with examples. **[V.T.U. Balgaun Karnataka Univ. February 2002]**
3. State and explain Faraday's laws of electromagnetic induction. Lenz's law. Fleming's right hand rule and Fleming's left hand rule. **[V.T.U. Belgaum Karnataka Univ. Summer 2003]**
4. Explain what is meant by self inductance and mutual inductance. Define the units in which each is mearsured.
5. Define "self inductance" and " mutual inductance". Obtain an expression for the mutual inductance of two coaxial solenoids of wires closely wound one upon the other. **[Kanpur Univ. Second B. Tech. Electrical Science 1979]**
6. Explain the concept of mutual inductance. Define coefficient of coupling and derive the expression between self inductances of two coils, mutual inductance between them and the coefficient of coupling.
7. Define mutual inductance (M) and show that

$$K = \frac{M}{\sqrt{L_1 L_2}}, K \leq 1$$

where L_1 and L_2 are inductances of coil 1 and coil 2 respectively and K is coupling coefficient.

[Pb Technical Univ. Basic Electrical and Electronics Engineering First Semester 2005-06]

8. Two coils of self inductances L_1 and L_2 are placed side by side so that the mutual inductance between them is M. If they are connected in series addition derive for the expression for the net inductance of the coils.
9. Distinguish between (*i*) Statically and dynamically induced emfs. (*ii*) Electric and magnetic circuits.
10. Derive an expression for the voltage across resistance when R-L is suddenly connected to a dc voltage. Indicate the physical significance of 'time constant' for such a circuit. **[Pb. Univ. Elements of Elec. Engineering 1985]**
11. Derive the equation for an R-L circuit which gives the rise and decay of current as function of time. What is meant by the time constant of a circuit ? **[Pb. Univ. Applied Electricity 1973]**
12. Define the time constant of a series R-L circuit. If a constant dc voltage is suddenly applied across such a circuit, find an expression for the current flow. **[Pb. Univ. Elements of Electrical Engineering Nov. 1975]**

SHORT ANSWER TYPE QUESTIONS WITH ANSWERS

Q. 1. What is indicated by the negative sign before N $\dfrac{d\phi}{dt}$ in expression $e = -N\dfrac{d\phi}{dt}$?

Ans. The negative sign before N $\dfrac{d\phi}{dt}$ in expression $e = -N\dfrac{d\phi}{dt}$ indicates the phenomenon explained by Lenz's law.

Q. 2. What is Lenz's law ?

Ans. The direction of emf induced in a conductor or coil is governed by Lenz's law which states that the direction of induced emf or voltage is such that the current produced by it sets up a magnetic field opposing the motion or change producing it.

Q. 3. What is the difference between a statically induced emf and a dynamically induced emf ? **[P.T.U. January 2000]**

Ans. When flux linking with any coil or circuit changes, an emf is induced in the coil or circuit, called the *electromagnetically induced emf.*

EMF induced in a coil or circuit by increasing or decreasing magnitude of the linking flux (no relative motion between coil or circuit and magnetic field) is known as statically *induced emf.*

The emf induced in a conductor or coil due to relative motion between it and the magnetic field, that produces flux, is known as dynamically induced emf.

Q. 4. What is the application of Fleming's right hand rule ?

Ans. Fleming's right hand rule is used in determination of direction of emf induced in a conductor moving in a uniform magnetic field.

Q. 5. What is Fleming's right hand rule ?

Ans. According to Fleming's right hand rule if the thumb, forefinger and middle finger of the right hand are held mutually perpendicular to each other, forefinger pointing into the direction of the field and thumb in the direction of motion of conductor then middle finger will point in the direction of the induced emf, as shown in Fig. 4.6 (*a*).

Q. 6. A conductor of effective length $l(m)$ moves in the direction of a uniform magnetic field B (Tesla) at a constant speed v (ms^{-1}). What is the magnitude of the emf induced in the conductor ?

Ans. No., as no flux is cut by the conductor.

Q. 7. What is meant by coefficient of coupling between the coils ?

Ans. When the two coils are placed near each other, all the flux produced by one coil does not link with the other coil, only a certain portion (say, K) of flux produced by one coil link with the other coil, K being less than unity. K is called the coefficient of coupling.

Q. 8. What is the effect of self-induction in a dc circuit ?

Ans. The self induction opposes the change of current in a dc circuit.

Q. 9. What is meant by the 'time constant' of a series R-L circuit ?

Ans. The time constant τ of a series R-L circuit is defined as the time required for the current to rise to its final steady-state value if it continued rising at its initial rate of V/L amperes per second.

The time constant may also be defined as the time required for the current to reach 0.632 of its final steady-state value (*i.e.* $0.632\ I_{max}$).

Q. 10. If a constant dc voltage is suddenly applied across an R-L series circuit, what is the initial current ?

Ans. Zero.

PROBLEMS

1. A 0.5 m long wire moves at right angles to its length at 40 m/s in uniform magnetic field of 1 Wb/m^2. Calculate the emf induced in the conductor when the direction of motion is (*i*) perpendicular to field, (*ii*) inclined at 30° to the direction of field. **[Ans.** (*i*) 20 V (*ii*) 10 V**]**
2. A square coil of 10 cm side and with 100 turns is rotated at a uniform speed of 500 rpm about an axis at right angle to a uniform field of 0.5 Wb/m^2. Calculate the instantaneous value of induced emf when the plane of the coil is (*i*) at right angle to the plane of the field (*ii*) in the plane of the field (*iii*) at 45° with the field direction. **[Ans.** (*i*) zero. (*ii*) 26.2 V (*iii*) 18.53 V**]**
3. A coil of 150 turns is linked with a flux of 0.001 Wb when carrying a current of 10 A. If the current is uniformly reversed in 0.1 s, calculate the induced emf. **[Pb.Univ. Elements of Elec. Engineering Nov; 1975]** **[Ans.** 3 V **]**
4. Determine the approximate resistance and inductance of an iron-cored solenoid of mean diameter 1 cm, 1 metre long, wound with 1,000 turns of copper wire 0.5 mm in diameter. What potential difference exists at the terminals of the solenoid when the current is 2 amperes, but increasing at the rate of 25 A/s ? Relative permeability of iron = 400 and specific resistance of copper $1.725 \times 10^{-8}\ \Omega$ - m. **[Ans.** 2.76 Ω; 39.48 mH; 6.5 V**]**
5. The resistance and inductance of a coil are 3 Ω and 0.1 mH respectively. What potential difference exists at the terminals of the solenoid at the instant when the current is 1 A but is increasing at the rate of 10,000 A per second? **[Ans.** 4 V**]**

6. A coil of 300 turns, wound on a core of non-magnetic material has an inductance of 10 mH. Calculate (*i*) the flux produced by a current of 5 A and (*ii*) the average value of emf induced when a current of 5 A is reversed in 8 milli-seconds. **[Ans.** (*i*)0.1667mWb (*ii*)12.5 V]

7. A circuit has 1,000 turns enclosing a magnetic circuit 20 cm^2 in section. With 4 A, the flux density is 1.0 Wb/m^2 and with 9 A it is 1.4 Wb/m^2. Find the mean value of the inductance between these current limits and the induced emf if the current falls from 9 A to 4 A in 0.05 s. **[Osmania Univ. Elec. Engg. –I Dec. 1971]**
[Ans. 0.16 H H,16 V]

8. A steel ring has the following particulars:
Mean diameter of the ring = 30 cm ; flux density of 1 Wb/m^2 is produced by 4,000 AT/m. The cross-section of the ring is 10 cm^2, number of turns wound = 600. If a gap of 1cm is cut in the ring and the flux density of 1 Wb/m^2 is maintained in the air gap, determine the inductance. **[Gujarat Univ. Elec. Circuits & Fields-I, October 1972]**
[Ans. 30.8 mH]

9. Two coils having 30 and 600 turns respectively are wound side by side on a closed iron circuit of area of cross-section 100 cm^2 and mean length 200 cm. Estimate the mutual inductance between the coils if the relative permeability of the iron is 2,000. If a current of zero ampere grows to 20 A in a time of 0.02 second in the first coil, find the emf induced in the second coil. **[J.N.T. Univ. Warangal Electrical Engineering I,1974]**
[Ans. 226 mH; 226 V]

10. Two identical coils A and B each having 1,000 turns lie in parallel planes such that 60% of the lines of force produced by one coil link the other. A current of 5 amperes in coil A produces in it a flux of 0.05 mWb. If the current in A changes from + 6 A to – 6 A in 0.01 second, what will be the magnitude of the emf induced in B ? Calculate the self and mutual inductances of each coil. **[A.P.S. Univ.Rewa Electrical Engineering 1976]**
[Ans. 0.01 H; 0.006 H: 7.2 V]

11. Two coils A and B of 600 and 100 turns respectively are wound uniformly around a wooden ring of mean circumference 80 cm. The x-sectional area of the ring is 4 cm^2. Calculate (*a*) self inductance of each coil (*b*) mutual inductance between the coils (*c*) emf induced in the coil B when a current of 2 A in coil A is reversed in 0.01 second.
[Ans. (*a*) 226.2 μ H ; 6.283 μ H; (b) 37.7 μ H; (c) 15.08 mV]

12. Two coils having 50 and 500 turns respectively are wound side by side on a closed iron circuit of section 50 cm^2 and mean length 120 cm. Estimate the mutual inductance between the coils if the permeability of iron is 1,000. Also find the self-inductance of each coil and the coefficient of coupling. **[Pb. Univ. Elec. Engg-1 April 1982]**
[Ans. 0.13 H; 0.013 H; 1.3 H; 1.0]

13. The combined inductance of two coils connected in series is 1.2 H and 0.2 H, depending on the relative directions of the currents in the coils. If one of the coils when isolated has a self inductance of 0.4 H, determine (*a*) self inductance of the other coil (*b*) the mutual inductance and (*c*) the coefficient of couplings. **[Ans.** (*a*) 0.3 H (*b*) 0.25 H (c) 0.722]

14. The combined inductance of two coils connected in series is 0.6 H and 0.1 H depending on the relative directions of currents in the coils. If one of the coils, when isolated has a self inductance of 0.2 H, calculate (*a*) mutual inductance and (*b*) coupling coefficient. **[Ans.** (*a*) 0.125 H; (*b*) 0.7217]

15. Two coils with terminals T_1, T_2 and T_3, T_4 respectively are placed side by side. When measured separately, the inductance of first coil is 1,200 mH and that of the second is 800 mH. When T_2 is joined to T_3, the inductance between T_1 and T_4 is 2,500 mH. What is the mutual inductance between the two coils ? Also determine the inductance between T_1 and T_3 when T_2 is joined to T_4. **[Nagpur Univ. Electrical Circuits 1991]**
[Ans. 250 mH; 1,500 mH]

16. The field winding of a dc machine takes a steady current of 10 A when connected to a dc source of 230 V. On switching the voltage to the field it is observed that it takes 0.3 second to reach 5 A. Find the value of the inductance of the field. Find also the time for the current to fall to 30% of final steady value if the field terminals are short-circuited.
[Ans. 9.96 H; 0.52 second]

17. A coil of inductance 10 H and resistance 2 Ω is supplied at 20 V. Find (*a*) time constant of the circuit (*b*) the average emf induced when the circuit is opened if the current falls to zero in 0.5 second (*c*) how much time would it take for the current to reach a value of 2 A after switching off the supply. **[Pb. Univ. Applied Electricity 1973]**
[Ans. (*i*) 5 seconds (*ii*) 200 V (*iii*) 8 seconds]

18. A coil of resistance 25 Ω and inductance 0.5 H is switched on to a dc voltage of 220 V. Calculate (*i*) initial rate of rise of current, (*ii*) time constant, and (*iii*) steady-state current. **[Pb. Univ. Engg.-1, April 1981]**
[Ans. (*i*) 440 ampere/second, (*ii*) 0.02 second, (*iii*) 8.8 A]

❑❑❑

UNIT-IV

Single Phase AC Circuits

Contents in This Unit

Single Phase AC Circuits: Generation of alternating voltages and currents, their equations; Definitions, RMS and average values; Vector representation of alternating quantities; Addition and subtraction of vectors; Complex algebra; Phasor relations between voltage and current in each of resistance, inductance and capacitance; AC series and parallel circuits; Power and power factor; Methods of circuit solution (analytically and vectorially); Resonance in series and parallel circuits.

Chapter 5

Fundamentals of Alternating Currents

5.1. INTRODUCTION

A current (or voltage) is called *alternating* if it reverses periodically in direction, and its magnitude undergoes a definite cycle of changes in definite intervals of time. Each cycle of alternating current (or voltage) consists of two half cycles, during one of which the current (or voltage) acts in one direction; while during the other in opposite direction. In more restricted sense, alternating current is a periodically varying current, the average value of which, over a period, is zero. The direct current always flows in one direction, and its magnitude remains unaltered. In order to produce an alternating current through an electric circuit, a source capable of reversing the emf periodically (ac generator) is required while for generating dc in an electric circuit, a source capable of developing a constant emf is required such as a battery or dc generator. The graphical representations of alternating current and direct current are given in Figs. 5.1(*a*) and (*b*) respectively.

At present a large percentage of the electrical energy (nearly all) being used for domestic and commercial purposes is generated as alternating current. In fact, almost the whole of the vast amount of electrical energy used throughout the world for every imaginable purpose is generated by alternating current generators. This is not due to any superiority of alternating current over direct current in the sphere of applicability to industrial and domestic use. In fact there are certain types of works for which alternating current is unsuitable and, therefore, direct current is absolutely necessary such as for electroplating, charging of storage batteries, refining of copper, refining of aluminium, electrotyping, production of industrial gases by electrolysis, municipal traction etc. In some power applications, the ac motor is unsatisfactory such as for metal rolling mills, paper making machines, high speed gearless elevators, automatic machine tools and high-speed printing presses. Direct current required for these applications is nowadays derived from an ac supply by the use of suitable convertors or rectifiers. For lighting and heating dc and ac are equally useful. The reasons for generation of electrical energy in the form of alternating current are given below:

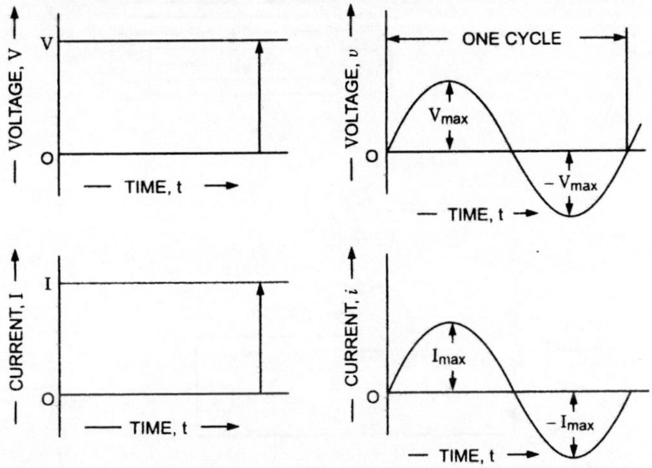

(*a*) *DC Voltage and Current* (*b*) *AC Voltage and Current*

Fig. 5.1

1. AC generators have no commutator and can, therefore, be built in very large units to run at high speeds producing high voltages (as high as 11,000 volts), so that the construction and operating cost per kW is low, whereas dc generator capacities and voltages are limited to comparatively low values.
2. Alternating current can be generated at comparatively high voltages and can be raised and lowered readily by a static machine called the *transformer* which makes the transmission and distribution of electrical energy economical. In direct current use of transformers is not possible.
3. AC induction motor is cheaper in initial cost and in maintenance since it has got no commutator and is more efficient than dc motor for constant speed work, so it is desirable to generate power as alternating current.
4. The high transmission efficiency in ac makes the generation of electrical energy economical by generating it in large quantities in a single station and distributing over a large territory.
5. The switchgear (*e.g.* switches, circuit breakers etc.) for ac system is simpler than that required in a dc system.
6. The maintenance cost of ac equipment is less.

5.2. GENERATION OF ALTERNATING EMF

We know that an alternating emf can be generated either by rotating a coil within a stationary magnetic field, as illustrated in Fig. 5.2 (*a*) or by rotating a magnetic field within a stationary coil, as illustrated in Fig. 5.2 (*b*). The emf generated, in either case, will be of sinusoidal waveform. The magnitude of emf generated in the coil depends upon the number of turns on the coil, the strength of magnetic field and the speed at which the coil or magnetic field rotates. The former method is employed in case of small ac generators while the later one is employed for large sized ac generators.

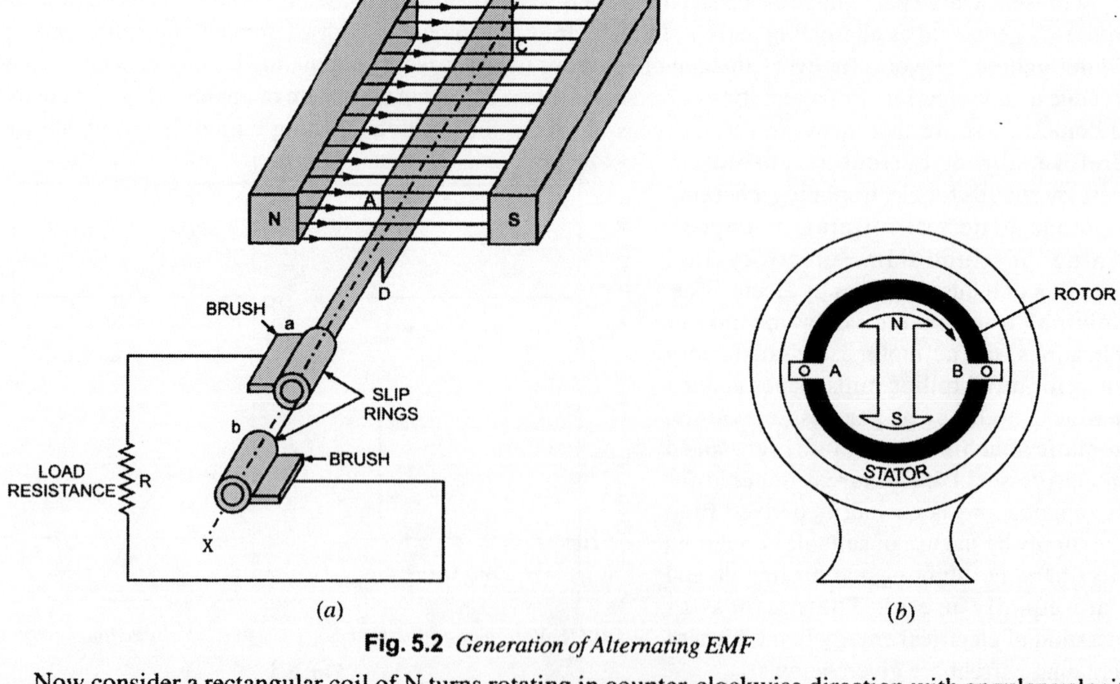

(*a*) (*b*)

Fig. 5.2 *Generation of Alternating EMF*

Now consider a rectangular coil of N turns rotating in counter-clockwise direction with angular velocity of ω radians per second in a uniform magnetic field, as illustrated in Fig. 5.3.

Let the time be measured from the instant of coincidence of the plane of the coil with the X-axis. At this instant maximum flux, Φ_{max} links with the coil. Let the coil assume the position, as shown in Fig. 5.3, after moving in counter-clockwise direction for t seconds.

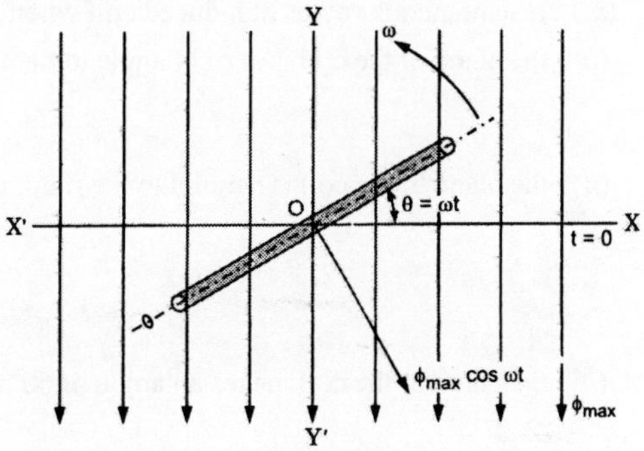

Fig. 5.3

The angle θ through which the coil has rotated in t seconds = ωt

In this position, the component of flux along perpendicular to the plane of coil = $\Phi_{max} \cos \omega t$.

Hence, flux linkages of the coil at this instant

= Number of turns on coil × linking flux

i.e; instantaneous flux linkages = $N\,\Phi_{max} \cos \omega t$

Since emf induced in a coil is equal to the rate of change of flux linkages with minus sign,

EMF induced at any instant,

$$e = -\frac{d}{dt}[N\,\Phi_{max} \cos \omega t]$$

$$= \Phi_{max}\,N\,\frac{d}{dt}[-\cos \omega t] = \Phi_{max}\,N\,\omega \sin \omega t \quad ...(5.1)$$

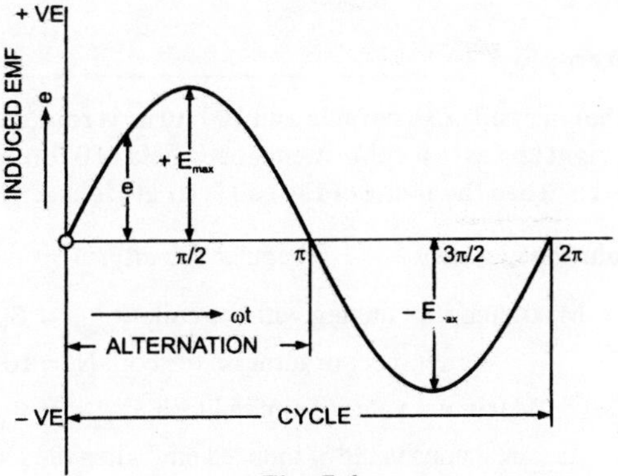

Fig. 5.4

when $\omega t = 0$, $\sin \omega t = 0$, therefore, induced emf is zero,

when $\omega t = \frac{\pi}{2}$, $\sin \frac{\pi}{2} = 1$, therefore, induced emf is maximum,

which is denoted by E_{max} and is equal to $\Phi_{max}\,N\,\omega$

Substituting $\Phi_{max} N\omega = E_{max}$ in Eq. (*i*) we have

Instantaneous emf, $e = E_{max} \sin \omega t$ ...(5.2)

So the emf induced varies as the sine function of the time angle ωt, and if emf induced is plotted against time, a curve of sine wave shape is obtained as illustrated in Fig. 5.4. Such an emf is called the *sinusoidal* emf. The sine curve is completed when the coil rotates through an angle of 2π radians. The induced emf e will have maximum value, represented by E_{max}, when the coil has turned through $\frac{\pi}{2}$ radians (or 90°) in counter-clockwise direction from the reference axis (*i.e.* OX axis).

Example 5.1

A coil having 200 turns and area of x-section 250 cm² is rotated about its axis at right angle to a uniform magnetic field of strength 0.5 T at a speed of 1,200 rpm. Determine (*i*) maximum value of emf induced (*ii*) equation for instantaneous induced emf and (*iii*) instantaneous values of induced emf when (*a*) the plane of the coil is at right angle to the field. (*b*) the plane of the coil is parallel to the field and (*c*) the plane of the coil is at an angle of 60° to the field.

Solution: Angular velocity, $\omega = \frac{2\pi \times RPM}{60} = \frac{2\pi \times 1{,}200}{60} = 40\,\pi$ radians/s

Maximum flux linking with the coil, $\Phi_{max} = B_{max} \times$ area of coil $= 0.5 \times 250 \times 10^{-4} = 0.0125$ Wb

(*i*) Maximum value of induced emf, $E_{max} = \Phi_{max} N\omega = 0.0125 \times 200 \times 40\,\pi = 314$ V **Ans.**

(*ii*) Equation for instantaneous induced emf $= E_{max} \sin \omega t = 314 \sin 40\,\pi t$ **Ans.**

(*iii*) Instantaneous values of induced emf when

(*a*) the plane of the coil is at right angle to the field *i.e.* when $\theta = \omega t = 0$

$$\therefore \quad e_0 = 314 \sin 0^\circ = 0 \textbf{ Ans.}$$

(*b*) the plane of the coil is parallel to the field, *i.e.* when $\theta = \omega t = \frac{\pi}{2}$ radians

$$\therefore \quad e' = 314 \sin \frac{\pi}{2} = 314 \text{ V } \textbf{Ans.}$$

(*c*) the plane of the coil makes an angle of 60° to the field *i.e.* when $\theta = \frac{\pi}{2} - \frac{\pi}{3} = \frac{\pi}{6}$ radians

$$\therefore \quad e'' = 314 \sin \frac{\pi}{6} = 314 \times \frac{1}{2} = 157 \text{ V } \textbf{Ans.}$$

Example 5.2

A square coil of 10 cm side and 100 turns is rotated at a uniform speed of 1,000 revolutions per minute, about an axis at right angles to a uniform magnetic field of 0.5 Wb/m². Calculate the instantaneous value of the induced electromotive force, when the plane of the coil is (*i*) at right angle to the field (*ii*) in the plane of the field.

Solution: Angular velocity, $\omega = \frac{2\pi \times \text{RPM}}{60} = \frac{2\pi \times 1{,}000}{60} = 33.33\,\pi$ radians/second

Maximum flux linking with the coil, $\Phi_{max} = B_{max} \times \text{area of coil} = 0.5 \times (0.1)^2 = 0.005$ Wb

Number of turns on the coil, N = 100

Maximum value of emf induced, $E_{max} = \Phi_{max} N\omega = 0.005 \times 100 \times 33.33\,\pi = 52.36$ V

Instantaneous value of induced emf when the plane of the coil is,

(*i*) at right angle to field *i.e.* when $\theta = \omega t = 0$

$$e_0 = 52.36 \sin 0^\circ = 0 \textbf{ Ans.}$$

(*ii*) in the plane of the field *i.e.* when $\theta = \omega t = \frac{\pi}{2}$ radians

$$e' = 52.36 \sin \frac{\pi}{2} = 52.36 \text{ V } \textbf{Ans.}$$

5.3. SINUSOIDAL QUANTITIES (EMF, VOLTAGE OR CURRENT)

It is not an accident that the bulk of electric power generated in electric power stations throughout the world and distributed to the consumers appears in the form of sinusoidal variations of voltage and current. There are many technical and economical advantages associated with the use of sinusoidal voltages and currents. For example, it will be learned that the use of sinusoidal voltages applied to appropriately designed coils results in a revolving magnetic field which has the capacity to do work. As a matter of fact it is this principle which underlies the operation of almost all the electric motors found in home appliances and about 90% of all electric motors found in commercial and industrial applications. Although other waveforms can be used in such devices, none leads to an operation which is as efficient and economical as that achieved through the use of sinusoidal quantities.

The other advantages of using sinusoidal voltages and currents are:

1. The waveform from generation to utilization remains the same if a sinusoidal waveform is generated.
2. Electromagnetic torque developed in three phase machines (generators and motors) with balanced three-phase currents is uniform (constant), and therefore, there are no oscillations in developed torque and absence of noise in operation.
3. Non-sinusoidal voltages which contain harmonic frequencies, according to Fourier analysis, are harmful to the system on account of
   (*i*) increased losses in generators, motors, transformers, and transmission and distribution systems,

(*ii*) more interference (noise) to nearby communication circuits,

(*iii*) resonance may result in overvoltages or overcurrents at many pockets on the way from generating station to consumer's premises which may damage the equipment and increase losses.

(*iv*) increased current through power factor improvement capacitors.

In practical electrical engineering it is assumed that the alternating voltages and currents are sinusoidal, though they may slightly deviate from it. The advantage of this assumption is that calculations become simple. *It may be noted that alternating voltage and current mean sinusoidal voltage and current unless stated otherwise.*

Alternating emf following sine law (*i.e.* sinusoidal emf) is illustrated in Fig. 5.4 and is expressed in the form

$$e = E_{max} \sin \omega t \qquad ...(5.3)$$

where e is the instantaneous value of alternating emf (or voltage), E_{max} is the maximum value of the alternating emf (or voltage) and ω is angular velocity of the coil.

The rotating coil moves through an angle of 2π radians in one cycle, so angular velocity $\omega = 2\pi f$ where f is the number of cycles completed per second.

Substituting $\omega = 2\pi f$ in Eq. (5.3) we have

$$e = E_{max} \sin 2\pi f t \qquad ...(5.4)$$

If the alternating emf (or voltage) given by Eq. (5.3) is applied across a load, alternating current flows through the circuit which would also vary sinusoidally *i.e.* following a sine law. The expression for alternating current is given as

$$i = I_{max} \sin \omega t = I_{max} \sin 2\pi f t \qquad ...(5.5)$$

provided the load is pure resistive.*

5.4. TERMINOLOGY

An *alternating quantity* (voltage or current) is one which changes continuously in magnitude and alternates in direction at regular intervals of time. It rises from zero to maximum positive value, falls to zero, increases to a maximum in the reverse direction and falls back to zero again, as illustrated in Fig. 5.4. The emf (or voltage) and current repeat the procedure.

The important ac terms are defined below:

5.4.1. Waveform. The shape of the curve of the voltage or current when plotted against time as abcissa (base) is called the *waveform*. The waveform of an alternating voltage varying sinusoidally is shown in Fig. 5.4. The waveform of the induced emf in an alternator differs slightly from that of sine wave but for calculation purposes it is treated as such.

5.4.2. Instantaneous Value. The value of alternating quantity (emf, voltage or current) at any particular instant is called the *instantaneous value* and is designated by a small italic letter (*e* for emf, *v* for voltage and *i* for current). The instantaneous values of an alternating quantity can be determined either from the curve or from an equation of the alternating quantity. For example, the instantaneous values of emf represented by the curve shown in Fig. 5.4 at 0, $\frac{\pi}{2}$, π and $\frac{3\pi}{2}$ are zero, $+E_{max}$, zero and $-E_{max}$ respectively.

5.4.3. Alternation and Cycle. When a periodic wave, such as sinusoidal wave, goes through one complete set of positive or negative values, it completes one *alternation* and when it goes through one complete set of positive and negative values it is said to have completed one *cycle.*

Alternation and cycle can also be defined in terms of angular measure. One alternation corresponds to 180°(or π radians) while one cycle corresponds to 360° (or 2π radians).

* The load may be resistive, inductive or capacitive. It will be shown in chapter 6 that if the load is inductive or capacitive the current equation differs in time angle.

5.4.4. Time Period and Frequency. The time taken in seconds by an alternating quantity to complete one cycle is known as *time period* or *periodic time* and is denoted by T.

The number of cycles completed per second by an alternating quantity is known as *frequency* and is denoted by *f*. In SI system the frequency is expressed in hertz (pronounced as hurts). One hertz (or Hz) is equal to one cycle per second.

The number of cycles completed per second = f

$$\text{Time period, } T = \text{Time taken in completing one cycle} = 1/f$$

$$\text{or } f = 1/T \qquad \text{...(5.6)}$$

The commercial ac power is generated at frequency of 50 Hz or 60 Hz.* The reasons of suitability of frequency of this range are:

1. The output of the equipment increases with the increase in frequency. For a given output, smaller size machines are required as compared to those for lower frequency output. Because of high power-weight ratio, relative cost of the equipment is also reduced.
2. Lower regulation, lower skin effect resulting in lower ohmic losses, lower magnetic and dielectric losses resulting in higher efficiency, lower corona loss and higher power transmission line capacity as compared to those at higher frequency.

5.4.5. Angular Velocity and Frequency. A glance at Fig. 5.4 indicates that each cycle spans 2π radians. Hence, if this quantity is divided by the time period, *angular velocity* of the sine function is obtained. It is denoted by ω and is expressed in radians per second.

$$\therefore \quad \text{Angular velocity, } \omega = \frac{\text{Angle turned}}{\text{Time taken}} = \frac{2\pi}{T} = \frac{2\pi}{1/f} = 2\pi f \text{ radians per second} \qquad \text{...(5.7)}$$

5.4.6. Electrical Time Degrees and Mechanical Degrees. It is seen that the coil must revolve past a pair of poles in order to carry the generated emf through one complete cycle. In circuit work, one complete cycle of voltage or current is designated as 360 electrical degrees or 2π electrical radians. To correlate with this designation the arc through which a coil of dynamo must rotate in order to generate one cycle of emf is called 360 electrical degrees. In a 2-pole machine one complete revolution of coil produces one cycle of emf. But in a multipolar machine, such as four, six or eight pole machine, the emf completes one cycle or 360 electrical degrees or 2π electrical radians as soon as the coil passes a pair of poles and a mechanical degree will be equal to as many electrical degrees as there are pairs of poles in the dynamo structure. In a multipolar machine, the number of cycles completed per second by generated emf,

$$f = \text{Pair of poles} \times \text{number of revolutions made per second} = \frac{P}{2} \times \frac{N}{60} = \frac{PN}{120} \qquad \text{...(5.8)}$$

where N is the speed of rotation of the coil in rpm.

5.4.7. Amplitude. The maximum value, positive or negative, which an alternating quantity attains during one cycle is called the *amplitude* of the alternating quantity. The amplitude of an alternating emf (or voltage) and current is designated by E_{max} (or V_{max}) and I_{max} respectively.

Example 5.3

What is the time period of the wave produced by a 6-pole alternator which is driven at 1,000 rpm ?

[Pb Technical Univ. Basic Electrical Engineering, June 2003]

Solution: Number of poles on alternator, P = 6

Speed of alternator, N = 1,000 rpm

$$\text{Number of cycles completed per second by generated emf, } f = \frac{PN}{120} = \frac{6 \times 1{,}000}{120} = 50$$

$$\text{Time period, } T = \frac{1}{f} = \frac{1}{50} = 0.02 \text{ second} \quad \textbf{Ans.}$$

* In India the standard frequency for power supply is 50 Hz.

5.5. DETERMINATION OF MAXIMUM VALUE AND FREQUENCY FROM EMF OR CURRENT EQUATIONS

From Eq. (5.4)

$$e = E_{max} \sin 2\pi f t$$

From above expression we observe that

(*i*) *the maximum value of an alternating emf is given by the coefficient of the sine of the time angle.*

(*ii*) *the frequency is given by coefficient of time t divided by 2π*

$$i.e.\ f = \frac{\text{Coefficient of time, } t}{2\pi} \quad ...(5.9)$$

Similarly we can also find the maximum value and frequency of the current from the equation of instantaneous values of current.

Example 5.4

An ac voltage of 50 Hz frequency has a peak value of 220 V. Write down the expression for the instantaneous value of this voltage. [*Pb Technical Univ. Electrical Engineering January 2000*]

Solution: Supply frequency, f = 50 Hz

Peak value of ac voltage, V_{max} = 220 V

Expression for instantaneous value of ac voltage (assumed sinusoidal) with θ as zero when time is zero, is given as

$$v = V_{max} \sin \omega t = V_{max} \sin 2\pi ft = 220 \sin 2\pi \times 50\, t = 220 \sin 314t \quad \textbf{Ans.}$$

Example 5.5

An alternating current of frequency 50 Hz has a maximum value of 100 A. Calculate (*a*) its value $\frac{1}{600}$ second after the instant the current is zero and its value decreasing there afterwards (*b*) how many seconds after the instant the current is zero (increasing there afterwards) will the current attain the value of 86.6 A ?

[*Allahabad Univ. Elec. Technology 1991*]

Solution: The current wave form is shown in Fig. 5.5.

The equation of the alternating current (assumed sinusoidal) with respect to the origin O is

$$i = 100 \sin 100\, \pi\, t \quad ...(i)$$

(*a*) Since the current is measured from the instant the current is zero and is decreasing there afterwards (*i.e.* from point A in Fig. 5.5), the equation for the alternating current with respect to the point A becomes

$$i = 100 \sin (100\, \pi\, t + \pi) = -100 \sin 100\, \pi\, t \quad ...(ii)$$

Substituting $t = \frac{1}{600}$ second in above equation we get the instantaneous value of current $\frac{1}{600}$ second after the instant the current is zero and decreasing there afterward

$$\text{So } i = -100 \sin 100\, \pi \times \frac{1}{600} = -100 \sin \frac{\pi}{6} = -50 \text{ A} \quad \textbf{Ans.}$$

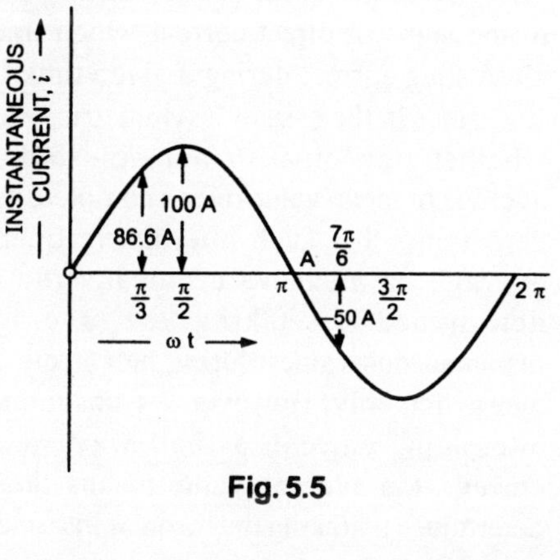

Fig. 5.5

(*b*) Let the current attain the value of 86.6 A, *t* seconds after the zero value of the current. Now substituting *i* = 86.6 A in Eq. (*i*) we get

$$86.6 = 100 \sin 100\, \pi\, t$$

$$\text{or } t = \frac{1}{100\,\pi} \sin^{-1} \frac{86.6}{100} = \frac{1}{100\,\pi} \times \frac{\pi}{3} = \frac{1}{300} \text{ second } \textbf{Ans.}$$

5.6. PLOTTING OF SINE WAVEFORM

Sine curve may be graphically drawn, as illustrated in Fig. 5.6. Draw a circle of radius equal to the maximum value of sinusoidal quantity. Divide the circumference of the circle drawn so into any number of equal parts, say 12, and draw a horizontal line AB (the base on which the sine wave is to be drawn) passing through the centre of the circle. Divide the line AB into the same number of equal parts *i.e.* 12 and number the points correspondingly. Draw perpendicular ordinates from each point. Project the points on the circle horizontally to meet the perpendicular ordinates having corresponding numbers. Draw smooth curve through these points. Curve so drawn will be of sine waveform.

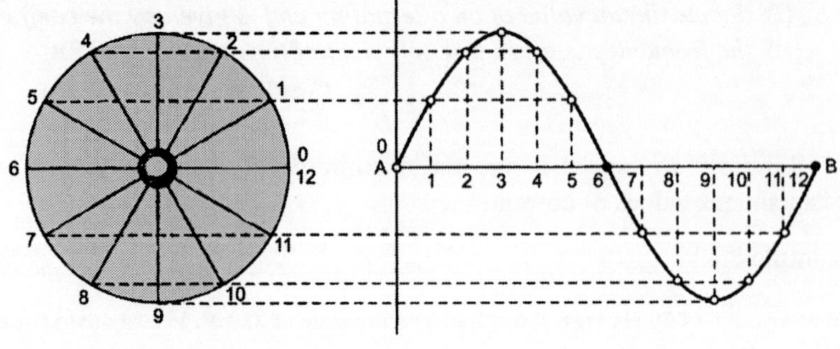

Fig. 5.6 *Simple Method of Plotting Sine Wave*

5.7. AVERAGE AND EFFECTIVE (RMS) VALUES OF ALTERNATING VOLTAGE AND CURRENT

In a dc system, the voltage and current are constant and, therefore, there is no problem in specifying their magnitude. But in case of ac system, an alternating voltage or current varies from instant to instant and so poses a problem how to specify the magnitude of an alternating voltage or current. An alternating voltage or current may possibly be expressed in terms of peak (maximum) value, average (mean) value or effective (rms) value.

In specifying an alternating voltage or current, its peak or maximum value is rarely used because it has that value only twice each cycle. Furthermore, the average or mean value cannot be used because it is positive as much as it is negative, so the average value is zero. Although the average value over half cycle might be used, it would not be as logical a choice as what we shall find *effective* (*virtual or rms*) *value* which is related to the power developed in a resistance by the alternating voltage or current.

5.7.1. Average Value of Alternating Current. The average (or mean) value of an alternating current is equal to the value of direct current which transfers across any circuit the same charge as is transferred by that alternating current during a given time.

Since in the case of a symmetrical alternating current (*i.e.* one whose two half cycles are exactly similar, whether sinusoidal or non-sinusoidal) the average or mean value over a complete cycle is zero hence for such alternating quantities average or mean value means the value determined by taking the average of instantaneous values during half cycle or one alternation only. However, for unsymmetrical alternating current, as half wave rectified current, the average value means the value determined by taking the mean of instantaneous values over the whole cycle.

The average value is determined by measuring the lengths of a number of equidistant ordinates and then taking their mean *i.e.* of $i_1, i_2, i_3 ..... i_n$ etc. which are mid-ordinates.

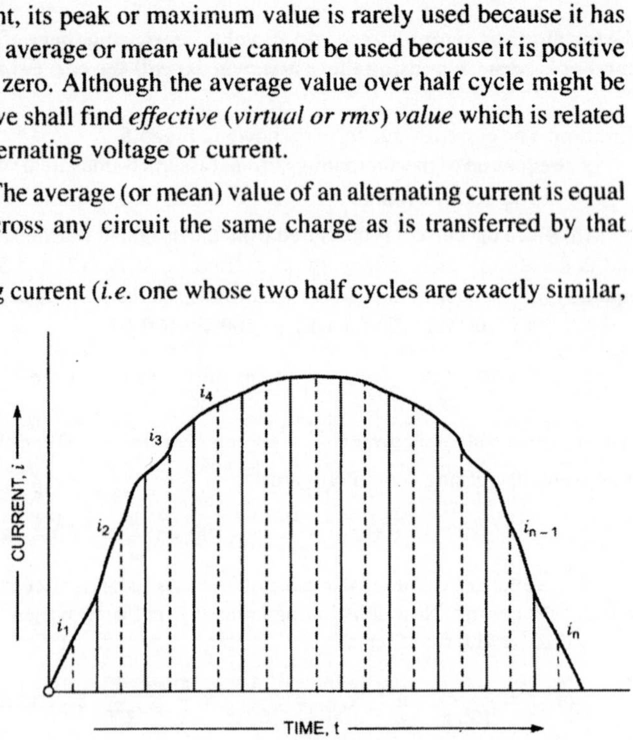

Fig. 5.7

$\because$ Average value of alternating current, $I_{av} = \dfrac{i_1 + i_2 + i_3 + i_4 \ldots + i_n}{n}$

$= \dfrac{\text{Area of one alternation (or half cycle)}}{\text{Length of base over one alternation (or half cycle)}}$

Using the integral calculus the average (or mean) value of a function $f(t)$ over a specific interval of time between t_1 and t_2 is given by

$$F_{av} = \frac{1}{t_2 - t_1} \int_{t_1}^{t_2} f(t)\, dt \qquad \ldots(5.10)$$

Any function whose cycle is repeated continuously, irrespective of its wave shape, is termed as *periodic function*, such as sinusoidal function, and its average value is given by

$$F_{av} = \frac{1}{T} \int_0^T f(t)\, dt \qquad \ldots(5.11)$$

where T is time period of periodic function.

In case of a symmetrical alternating current, whether sinusoidal or non-sinusoidal the average value is determined by taking average of one half cycle or one alternation only.

i.e., for symmetrical waveforms, $F_{av} = \dfrac{2}{T} \displaystyle\int_0^{T/2} f(t)\, dt \qquad \ldots(5.12)$

5.7.2. RMS Value or Effective Value of Alternating Current. The rms or effective value of an alternating current or voltage is given by that steady current or voltage which when flows or applied to a given resistance for a given time produces the same amount of heat as when the alternating current or voltage is flowing or applied to the same resistance for the same time.

Consider an alternating current of waveform shown in Fig. 5.8 flowing through a resistor of R ohms. Divide the base of one alternation into n equal parts and let the mid-ordinates be $i_1, i_2, i_3 \ldots i_n$ etc.

Heat produced during 1st interval $= i_1^2 R \times \dfrac{T}{n}$ joules

Heat produced during 2nd interval $= i_2^2 R \times \dfrac{T}{n}$ joules

Heat produced during 3rd interval $= i_3^2 R \times \dfrac{T}{n}$ joules

Heat produced during nth interval $= i_n^2 R \times \dfrac{T}{n}$ joules

Total heat produced in time T

$$= RT\left(\frac{i_1^2 + i_2^2 + i_3^2 + \ldots i_n^2}{n}\right) \text{ joules}$$

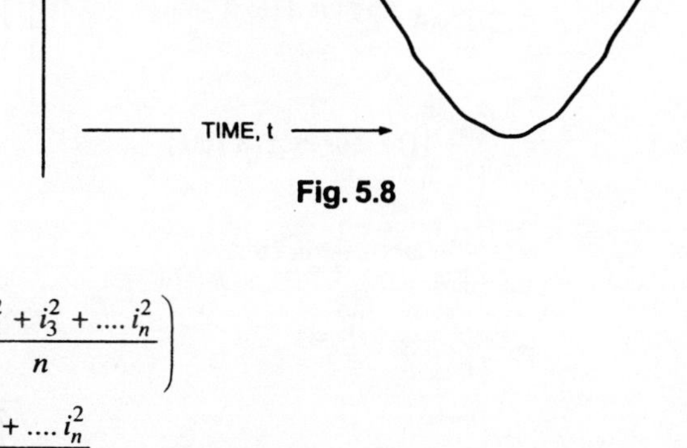

Fig. 5.8

Now if I_{eff} is the effective current, then heat produced by this current in time T = I_{eff}^2 RT joules. By definition these two expressions are equal

$$\therefore\ I_{eff}^2\, RT = RT\left(\frac{i_1^2 + i_2^2 + i_3^2 + \ldots i_n^2}{n}\right)$$

$$\text{or } I_{eff}^2 = \frac{i_1^2 + i_2^2 + i_3^2 + \ldots i_n^2}{n}$$

$$\text{or } I_{eff} = \sqrt{\frac{i_1^2 + i_2^2 + i_3^2 + \ldots i_n^2}{n}}$$

Hence the *effective* or *virtual value* of alternating current or voltage is equal to the square root of the mean of the squares of successive ordinates and that is why it is known as *root-mean-square (rms) value.*

Using the integral calculus the root mean square (rms) or effective value of an alternating quantity over a time period is given by

$$F_{rms} = \sqrt{\frac{1}{T}\int_0^T f^2(t)\,dt} \quad (5.13)$$

5.8. AVERAGE AND EFFECTIVE (RMS) VALUES OF SINUSOIDAL CURRENT AND VOLTAGE

5.8.1. Average Value For Sinusoidal Current or Voltage. The average value of a sine wave over a complete cycle is zero. Therefore, the half cycle average value is intended.

Instantaneous value of sinusoidal current is given by

$$i = I_{max} \sin \omega t$$

Considering first half cycle i.e when ωt varies from 0 to π we get,

$$I_{av} = \frac{\text{Area of first half cycle}}{\pi}$$

$$= \frac{1}{\pi}\int_0^\pi i\,d(\omega t) = \frac{1}{\pi}\int_0^\pi I_{max} \sin \omega t \; d(\omega t)$$

$$\text{or } I_{av} = \frac{I_{max}}{\pi}\left[-\cos \omega t\right]_0^\pi$$

$$= \frac{2}{\pi} I_{max} = 0.637\, I_{max} \quad \text{...(5.14)}$$

Similarly $E_{av} = 0.637\, E_{max}$

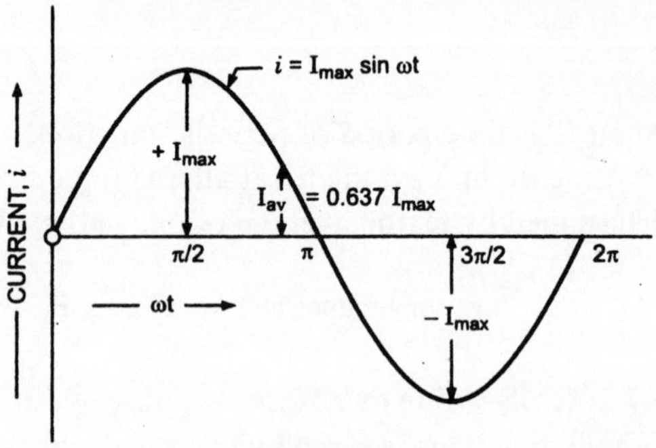

Fig. 5.9

5.8.2. Effective (RMS) Value For Sinusoidal Current or Voltage. A sinusoidal alternating current is represented by

$$i = I_{max} \sin \omega t$$

$$= \frac{\text{Area of first half cycle of } i^2}{\pi}$$

$$= \frac{1}{\pi}\int_0^\pi i^2\, d(\omega t)$$

$$= \frac{1}{\pi}\int_0^\pi I_{max}^2 \sin^2 \omega t\, d(\omega t)$$

$$= \frac{I_{max}^2}{2\pi}\int_0^\pi (1 - \cos 2\omega t)\, d(\omega t)$$

$$= \frac{I_{max}^2}{2\pi}\left[\omega t - \frac{1}{2}\sin 2\omega t\right]_0^\pi$$

$$= \frac{I_{max}^2}{2\pi} \times \pi = \frac{I_{max}^2}{2}$$

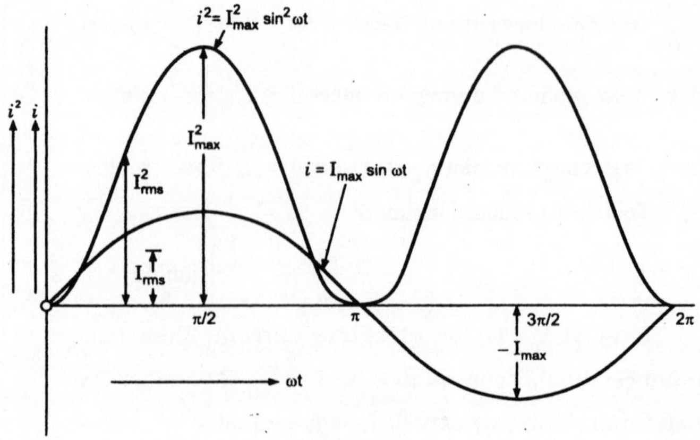

Fig. 5.10

$$\text{or } I_{rms} = \sqrt{\frac{I_{max}^2}{2}} = \frac{I_{max}}{\sqrt{2}} \quad \text{...(5.15)}$$

$$\text{Similarly, } E_{rms} = \frac{E_{max}}{\sqrt{2}}$$

Example 5.6

A sinusoidally varying alternating current of frequency 60 Hz has a maximum value of 15 amperes.

(i) Write down the equation for instantaneous value,

(ii) Find the value of current after 1/200 second,

(iii) Find the time taken to reach 10 amperes for the first time, and

(iv) Find its average value. [*U.P. Technical Univ. Electrical Engineering Second Semester 2002-03*]

Solution: Maximum value of current, $I_{max} = 15$ A

Frequency of alternating current, $f = 60$ Hz

(*i*) The equation for instantaneous value of sinusoidal alternating current with θ as zero when time is zero, is given as

$$I = I_{max} \sin 2\pi f t = 15 \sin 2\pi \times 60\, t = 15 \sin 120\,\pi\, t \textbf{ Ans.}$$

(*ii*) Substituting $t = \dfrac{1}{200}$ second in the expression

$I = 15 \sin 120\,\pi\, t$ we have value of current after $\dfrac{1}{200}$ second

$$= 15 \sin 120\,\pi \times \frac{1}{200} = 15 \sin 0.6\,\pi = 15 \times 0.951 = 14.266 \text{ A } \textbf{Ans.}$$

(*iii*) Let the instantaneous value of current be 10 A for the first time *t* seconds after the instant the current is zero and becoming positive

$$\therefore \quad 10 = 15 \sin 120\,\pi\, t$$

$$\text{or} \quad t = \frac{1}{120\pi} \sin^{-1} \frac{10}{15} = \frac{1}{120\pi} \sin^{-1} 0.6667 = 0.001936 \text{ second } \textbf{Ans.}$$

$$(iv) \text{ Average value, } I_{av} = \frac{2}{\pi} I_{max} = \frac{2}{\pi} \times 15 = 9.55 \text{ A } \textbf{Ans.}$$

Example 5.7

An ac sinusoidal current has rms value of 40 A at 50 Hz frequency. Write expression of instantaneous current and obtain its value 0.002 sec after passing through maximum positive value.

[*Pb Technical Univ. Basic Electrical & Electronics Engineering First Semester 2004-05*]

Solution: Expression for instantaneous value of an ac sinusoidal current with θ as zero when time is zero is given as

$$i = I_{max} \sin \omega t = I_{max} \sin 2\pi f t = 40\sqrt{2} \sin 2\pi \times 50t = 56.57 \sin 314t \textbf{ Ans.}$$

When time is measured from the positive maximum value, the above equation is modified into

$$v = 40\sqrt{2} \sin\left(314t + \frac{\pi}{2}\right) = 40\sqrt{2} \cos 314t = 40\sqrt{2} \cos 314 \times 0.002$$

$$= 40\sqrt{2} \cos 0.628$$

$$= 40\sqrt{2} \cos \frac{\pi}{5} = 45.77 \text{ A } \textbf{Ans.}$$

Example 5.8

An alternating current when passed through a resistance immersed in water for 5 minutes, just raised the temperature of water to boiling point. When a direct current of 4 amperes was passed through the same resistance under identical conditions it took 8 minutes to boil the water. Find the rms value of the alternating current.

Solution : Let the rms value of alternating current passed through the resistance be I_{rms} amperes.

Heat produced when an alternating current of I_{rms} amperes is passed through a resistance R immersed in water for 5 minutes

$$= (I_{rms})^2 \times R \times 5 \times 60 = 300\ I_{rms}^2\ R \text{ joules} \qquad ...(i)$$

Heat produced when a direct current of 4 A is passed through the same resistance R immersed in the water for 8 minutes

$$= (4)^2 \times R \times 8 \times 60 = 7{,}680 \text{ R joules} \qquad ...(ii)$$

Since heat produced in both of the cases is same, equating expressions (*i*) and (*ii*) we get

$$300\ I_{rms}^2\ R = 7{,}680\ R$$

$$\text{or } I_{rms} = \sqrt{\frac{7{,}680}{300}} = \sqrt{25.6} = 5.06 \text{ A} \quad \textbf{Ans.}$$

Example 5.9

An alternating current is given by $i = 20 \sin 600\, t$ amperes. Find the (*i*) frequency, (*ii*) peak value of current and (iii) the time taken from $t = 0$ for the current to reach a value of 10 A.

[*U.P. Technical Univ. Electrical Engineering First Semester 2004-05*]

Solution: (*i*) Frequency, $f = \dfrac{600}{2\pi} = 99.5$ Hz **Ans.**

(*ii*) Peak value of current, $I_{max} = 20$ A **Ans.**

(*iii*) Time taken from $t = 0$ for the current to reach a value of 10 A,

$$t = \frac{1}{600} \sin^{-1} \frac{10}{20} = \frac{\pi}{6 \times 600} s = \frac{\pi}{3{,}600} s = 0.873 \text{ ms} \quad \textbf{Ans.}$$

Example 5.10

An alternating current varying sinusoidally with a frequency of 50 Hz has an rms value of 20 A. Write down the equation for the instantaneous value and find this value (a) 0.0025 second (b) 0.0125 second after passing through a positive maximum value. At what time, measured from a positive maximum value, will the instantaneous current be 14.14 A ?

[*Allahabad Univ. Electrical Science 1992*]

Solution : The current waveform is shown in Fig. 5.11.

Peak value of current,

$$I_{max} = \sqrt{2}\, I_{rms} = \sqrt{2} \times 20 = 28.28 \text{ A}$$

The equation of alternating current with respect to origin O is

$$i = I_{max} \sin 2\pi f t = 28.28 \sin 2\,\pi \times 50\, t$$

$$= 28.28 \sin 100\,\pi\, t$$

The above equation is valid when time is measured from the instant the current is zero and increasing in positive direction. Since the time is measured from the positive maximum value (point A in Fig. 5.11), the above equation is modified to

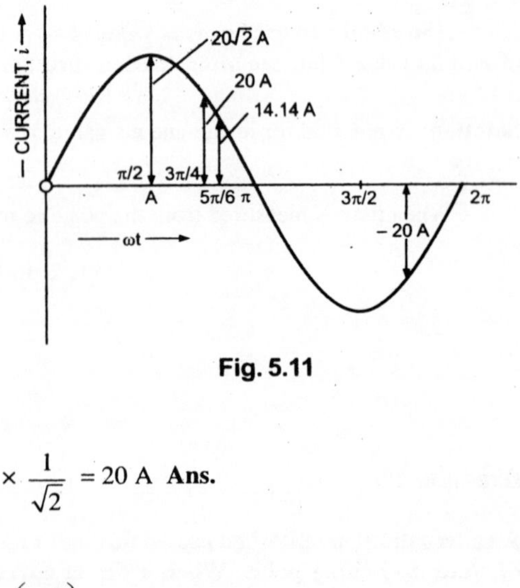

Fig. 5.11

$$i = 28.28 \sin\left(100\,\pi t + \frac{\pi}{2}\right) = 28.28 \cos 100\,\pi\, t \quad \textbf{Ans.}$$

(*a*) When $t = 0.0025$ s, the instantaneous value of current,

$$i_1 = 28.28 \cos 100\,\pi \times 0.0025 = 28.28 \cos \frac{\pi}{4} = 28.28 \times \frac{1}{\sqrt{2}} = 20 \text{ A} \quad \textbf{Ans.}$$

(*b*) When $t = 0.0125$ s, the instantaneous value of current,

$$i_2 = 28.28 \cos 100\,\pi \times 0.0125 = 28.28 \cos \frac{5\pi}{4} = 28.28 \left(\frac{-1}{\sqrt{2}}\right) = -20 \text{ A} \quad \textbf{Ans.}$$

(*c*) Substituting $i = 14.14$ A in expression $i = 28.28 \cos 100\,\pi\, t$ we have

$$14.14 = 28.28 \cos 100\,\pi\, t$$

$$\text{or } t = \frac{1}{100\,\pi} \cos^{-1} \frac{14.14}{28.28} = \frac{1}{100\,\pi} \cos^{-1} 0.5 = \frac{1}{100\,\pi} \times \frac{\pi}{3} = \frac{1}{300} \text{ second } \textbf{Ans.}$$

Example 5.11

An alternating voltage is given by $v = 141.4 \sin 314\, t$. Find (*i*) frequency (*ii*) rms value (*iii*) average value (*iv*) the instantaneous value of voltage when '*t*' is 3 ms (*v*) the time taken for the voltage to reach 100 V for the first time after passing through zero value. [*U.P. Technical Univ. Electrical Engineering First Semester 2006-07*]

Solution: Instantaneous value of alternating voltage is given by expression

$$v = 141.4 \sin 314\, t$$

(*i*) The frequency is given by the coefficient of time divided by 2 π, therefore,

$$\text{frequency, } f = \frac{314}{2\pi} = 50 \text{ Hz } \textbf{Ans.}$$

Peak value of an alternating voltage is given by the coefficient of sine of time angle, therefore,

$$V_{max} = 141.4 \text{ volts}$$

(*ii*) RMS value of given alternating voltage, $V_{rms} = \frac{V_{max}}{\sqrt{2}} = \frac{141.4}{\sqrt{2}} = 100 \text{ V}$ **Ans.**

(*iii*) Average value of given alternating voltage, $V_{av} = 0.637\, V_{max} = 0.637 \times 141.4 = 90 \text{ V}$ **Ans.**

(*iv*) Instantaneous value of voltage at 3 ms (0.003 *s*) after the instant the voltage is zero and increasing in positive direction may be obtained by substituting $t = 0.003$ second in Eq. (*i*). So we have

$$v = 141.4 \sin 314 \times 0.003 = 141.4 \sin 0.942 = 114.4 \text{ V } \textbf{Ans.}$$

(*v*) Time taken from $t = 0$ for the voltage to reach 100 V

$$t = \frac{1}{314} \sin^{-1} \frac{100}{141.4} = \frac{\pi/4}{314} \text{ second} = \frac{1}{400}\, s = 2.5 \text{ ms } \textbf{Ans.}$$

5.9. FORM FACTOR AND PEAK FACTOR

5.9.1. Form Factor. In certain cases it is convenient to have calculations at first upon the mean value of the emf over half a period, therefore, it becomes essential to have some means of connecting this mean value with the effective or rms value. The knowledge of form factor, which is defined as the ratio of effective value to the average or mean value of periodic wave is, therefore, necessary.

Mathematically form factor is given by the relation

$$\text{Form factor} = \frac{\text{Effective value}}{\text{Average value}}$$

$$\text{Form factor for sinusoidal wave, } K_f = \frac{E_{rms}}{E_{av}} = \frac{\dfrac{E_{max}}{\sqrt{2}}}{\dfrac{E_{max}}{\pi/2}} = 1.11$$

5.9.2. Peak Factor. Knowledge of peak factor of an alternating voltage is very essential in connection with determining the dielectric strength since the dielectric stress developed in any insulating material is proportional to the maximum value of the voltage applied to it.

Peak or crest or amplitude factor of a periodic wave is defined as the ratio of maximum or peak to the effective or rms value of the wave

$$i.e.\ \text{Peak factor, } K_p = \frac{\text{Maximum value}}{\text{Effective value}}$$

$$\text{Peak factor for sinusoidal wave, } K_p = \frac{E_{max}}{E_{rms}} = \frac{E_{max}}{E_{max}/\sqrt{2}} = 1.414$$

Example 5.12

The equation of an alternating current $i = 42.42 \sin 628\ t$. Determine (*i*) its maximum value (*ii*) frequency (*iii*) rms value (*iv*) average value (*v*) form factor. *[U.P. Technical Univ. Electrical Engineering Second Semester 2005-06]*

Solution: Instantaneous value of alternating current is given by the expression

$$i = 42.42 \sin 628\,t$$

(*i*) Maximum value of an alternating current is given by the coefficient of the sine of the time angle

$$i.e.\ I_{max} = 42.42 \text{ A } \textbf{Ans.}$$

(*ii*) Frequency is given by coefficient of time, *t* divided by 2π

$$\therefore \text{Frequency, } f = \frac{628}{2\pi} = 100 \text{ Hz } \textbf{Ans.}$$

$$(iii) \text{ RMS value of the alternating current, } I_{rms} = \frac{I_{max}}{\sqrt{2}} = \frac{42.42}{\sqrt{2}} = 30 \text{ A } \textbf{Ans.}$$

$$(iv) \text{ Average value of the alternating current, } I_{av} = 0.637\ I_{max} = 0.637 \times 42.42 = 27\text{A } \textbf{Ans.}$$

$$(v) \text{ Form factor } K_f = \frac{I_{rms}}{I_{av}} = \frac{30}{27} = 1.11 \textbf{ Ans.}$$

Example 5.13

An alternating voltage is $v = 100 \sin 100t$. Find (*i*) Amplitude (*ii*) Time period and frequency (*iii*) Angular velocity (*iv*) Form factor (*v*) Peak factor. *[U.P. Technical Univ. Electrical Engineering Second Semester 2007-08]*

Solution: Instantaneous value of alternating voltage is given by expression

$$v = 100 \sin 100\,t$$

(*i*) Amplitude of alternating voltage is given by the coefficient of the sine of the time angle, so

$$\text{Amplitude of given wave} = V_{max} = 100 \text{ V } \textbf{Ans.}$$

(*ii*) Frequency is given by coefficient of time, *t* divided by 2π

$$\therefore \text{Frequency, } f = \frac{100}{2\pi} = 15.9 \text{ Hz } \textbf{Ans.}$$

$$\text{Time period, T} = \frac{1}{f} = \frac{1}{15.9} = 0.063 \text{ second or 63 ms } \textbf{Ans.}$$

(*iii*) Angular velocity, $\omega = 2\pi f = 2\pi \times 15.9 = 100$ radians/second **Ans.**

$$\text{RMS value, } V_{rms} = \frac{V_{max}}{\sqrt{2}} = \frac{100}{\sqrt{2}} = 70.71 \text{ V}$$

$$\text{Average value } V_{av} = 0.637\ V_{max} = 0.637 \times 100 = 63.7 \text{ V}$$

$$(iv) \text{ Form factor, } K_f = \frac{V_{rms}}{V_{av}} = \frac{70.71}{63.7} = 1.11 \textbf{ Ans.}$$

$$(v) \text{ Peak factor, } K_p = \frac{V_{max}}{V_{rms}} = \frac{100}{70.71} = 1.4142 \textbf{ Ans.}$$

Example 5.14

Find the rms value, average value and form factor of the voltage waveform shown in Fig. 5.12.

[U.P. Technical Univ. Electrical Engineering Second Semester 2002-03]

Solution: Average value of voltage over one cycle

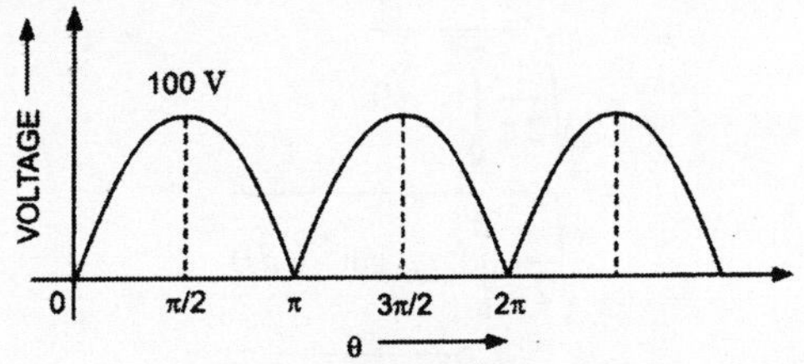

Fig. 5.12

$$= \frac{2\int_0^{\pi} V_{max} \sin\theta\, d\theta}{2\pi}$$

$$= \frac{V_{max}}{\pi}\int_0^{\pi} \sin\theta\, d\theta$$

$$= \frac{2\,V_{max}}{\pi} = \frac{2}{\pi} \times 100 = 63.66 \text{ V } \textbf{Ans.}$$

RMS value of voltage over one cycle

$$= \sqrt{\frac{2\int_0^{\pi} V_{max}^2 \sin^2 d\theta}{2\pi}} = \sqrt{\frac{V_{max}^2}{2\pi}\int_0^{\pi} 2\sin^2\theta\, d\theta} = \sqrt{\frac{V_{max}^2}{2\pi}\int_0^{\pi}(1-\cos 2\theta)\, d\theta}$$

$$= \sqrt{\frac{V_{max}^2}{2\pi}\times\pi} = \frac{V_{max}}{\sqrt{2}} = \frac{100}{\sqrt{2}} = 70.71 \text{ V } \textbf{Ans.}$$

$$\text{Form factor of the waveform shown} = \frac{\text{RMS value}}{\text{Average value}} = \frac{70.7}{63.66} = 1.11 \textbf{ Ans.}$$

Example 5.15

Calculate the average and root mean square values, the form factor and peak factor of a periodic current wave having the following values for equal time intervals over half cycle, changing suddenly from one value to the next: 0, 40, 60, 80, 100, 80, 60, 40, 0.

Solution: Since the wave is symmetrical, so considering one half cycle only. Average value

$$I_{av} = \frac{i_1 + i_2 + i_3 + i_4 + i_5 + i_6 + i_7 + i_8}{8}$$

$$= \frac{0 + 40 + 60 + 80 + 100 + 80 + 60 + 40}{8} = 57.5 \text{ A } \textbf{Ans.}$$

RMS value, I_{rms}

$$= \sqrt{\frac{i_1^2 + i_2^2 + i_3^2 + i_4^2 + i_5^2 + i_6^2 + i_7^2 + i_8^2}{8}}$$

$$= \sqrt{\frac{0 + 1{,}600 + 3{,}600 + 6{,}400 + 10{,}000 + 6{,}400 + 3{,}600 + 1{,}600}{8}}$$

= 64.42 A **Ans.**

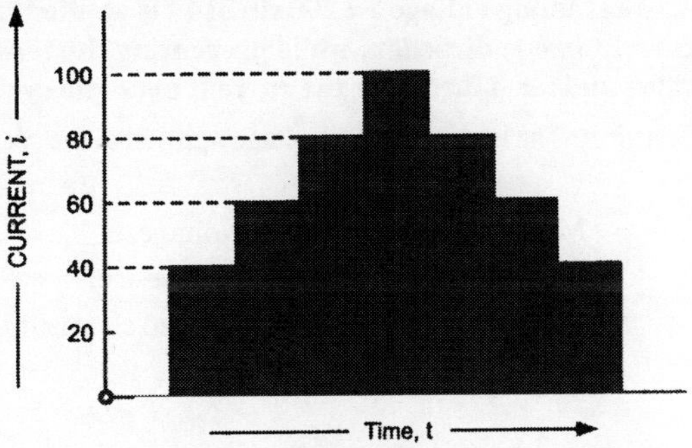

Fig. 5.13

$$\text{Peak value, } I_{max} = 100 \text{ A}$$

$$\text{Form factor} = \frac{I_{rms}}{I_{av}} = \frac{64.42}{57.5} = 1.12 \textbf{ Ans.}$$

$$\text{Peak factor} = \frac{I_{max}}{I_{rms}} = \frac{100}{64.42} = 1.552 \textbf{ Ans.}$$

5.10. RMS VALUE, AVERAGE VALUE, PEAK FACTOR AND FORM FACTOR OF HALF WAVE RECTIFIED ALTERNATING CURRENT

Half wave rectified alternating current is one whose one-half cycle has been suppressed *i.e.* one which flows for half the time during one cycle. It is illustrated in Fig. 5.14 where suppressed half cycle is shown dotted.

As mentioned earlier, for determining rms and average values of such an alternating current summation would be carried over the period for which current actually flows *i.e.* 0 to π but would be averaged for the whole cycle *i.e.* from 0 to 2π.

∴ RMS value of current,

$$I_{rms} = \sqrt{\frac{1}{2\pi}\int_0^{\pi} i^2\, d\theta}$$

$$= \sqrt{\frac{1}{2\pi}\int_0^{\pi} I_{max}^2 \sin^2\theta\, d\theta}$$

$$= \frac{I_{max}}{\sqrt{2\pi}}\sqrt{\int_0^{\pi}\sin^2\theta\, d\theta}$$

$$= \frac{I_{max}}{\sqrt{4\pi}}\sqrt{\int_0^{\pi}(1-\cos 2\theta)\, d\theta} = \frac{I_{max}}{\sqrt{4\pi}}\sqrt{\left[\theta - \frac{1}{2}\sin 2\theta\right]_0^{\pi}} = \frac{I_{max}}{\sqrt{4\pi}} \times \sqrt{\pi} = \frac{I_{max}}{2}$$

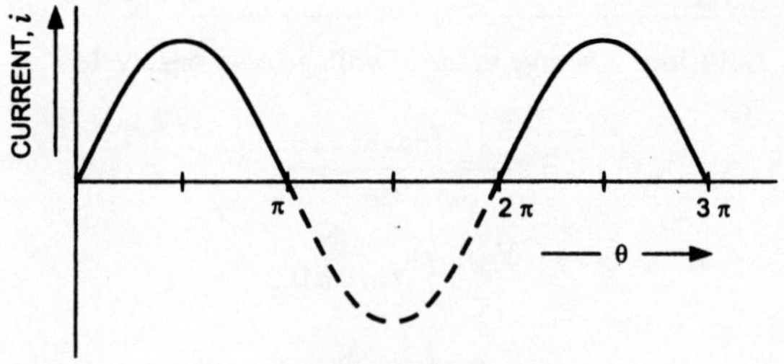

Fig. 5.14

Average value of current,

$$I_{av} = \frac{1}{2\pi}\int_0^{\pi} i\, d\theta = \frac{1}{2\pi}\int_0^{\pi} I_{max}\sin\theta\, d\theta = \frac{I_{max}}{2\pi}\left[-\cos\theta\right]_0^{\pi} = \frac{I_{max}}{\pi}$$

$$\text{Peak Factor} = \frac{I_{max}}{I_{rms}} = \frac{I_{max}}{I_{max}/2} = 2$$

$$\text{Form factor} = \frac{I_{rms}}{I_{av}} = \frac{I_{max}/2}{I_{max}/\pi} = \frac{\pi}{2} = 1.57$$

Example 5.16

An alternating voltage e = 200 sin 314 t is applied to a device which offers an ohmic resistance of 20 Ω to the flow of current in one direction, while preventing the flow of current in opposite direction. Calculate rms value, average value and form factor for the current over one cycle. *[Nagpur Univ. Elec. Engineering 1992]*

Solution: The instantaneous voltage applied to the rectifying device is given by the expression

$$e = 200 \sin 314\, t$$

Maximum value of applied voltage, E_{max} = Coefficient of the sine of time angle = 200 volts

Resistance of rectifying device, R = 20 Ω

Maximum value of half-wave rectified alternating current,

$$I_{max} = \frac{E_{max}}{R} = \frac{200}{20} = 10 \text{ amperes}$$

RMS value of half-wave rectified alternating current,

$$I_{rms} = \frac{I_{max}}{2} = \frac{10}{2} = 5 \text{ amperes } \textbf{Ans.}$$

Average value of the half-wave rectified alternating current,

$$I_{av} = \frac{I_{max}}{\pi} = \frac{10}{\pi} = 3.18 \text{ amperes } \textbf{Ans.}$$

Form factor of the half-wave rectified alternating current

$$= \frac{I_{rms}}{I_{av}} = \frac{5}{3.18} = 1.57 \textbf{ Ans.}$$

Example 5.17

A voltage $v(t) = 220\sqrt{2}\sin 100t$ is applied to the circuit shown. What is the rms value of current through the resistor R of 100 Ω ? Derive the formula used. *[Pb Technical Univ. Electrical Engineering May 2002]*

Solution: The maximum value of voltage applied to the circuit

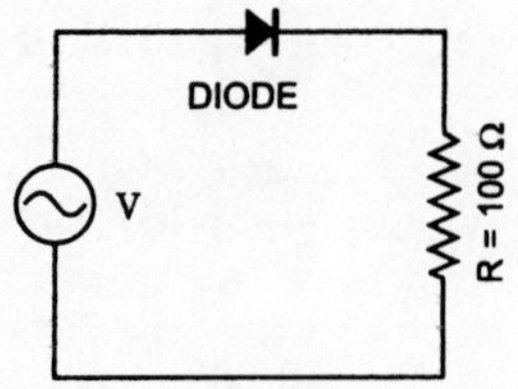

Fig. 5.15

$$V_{max} = 220\sqrt{2} \text{ V}$$

Maximum vlaue of circuit current, $I_{max} = \frac{V_{max}}{R} = \frac{220\sqrt{2}}{100} = 3.11$ A

RMS value of circuit current, $I_{rms} = \frac{I_{max}}{2} = \frac{3.11}{2} = 1.55$ A **Ans.**

Example 5.18

Define average value of alternating current. Find this value for the waveform shown in Fig. 5.16.

[Pb Technical Univ. Basic Electrical Engineering December 2002]

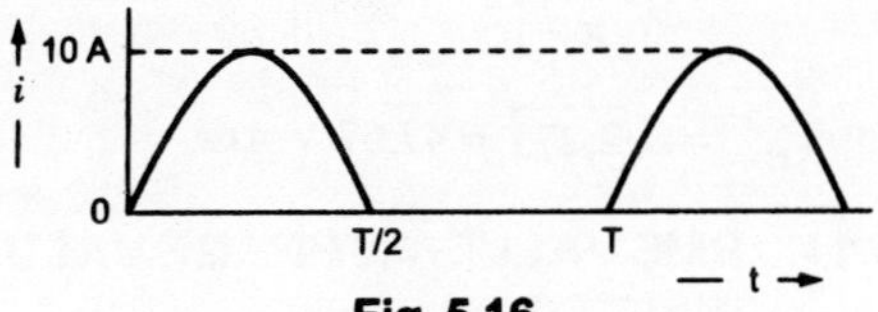

Fig. 5.16

Solution: Maximum value of alternating current, I_{max} = 10 A

Average value, $I_{av} = \frac{I_{max}}{\pi} = \frac{10}{\pi} = 3.183$ A **Ans.**

Example 5.19

Calculate the rms value of current given by $i = 10 + 5 \cos(628t + 30°)$

Solution: The waveform of the given current is shown in Fig. 5.17.

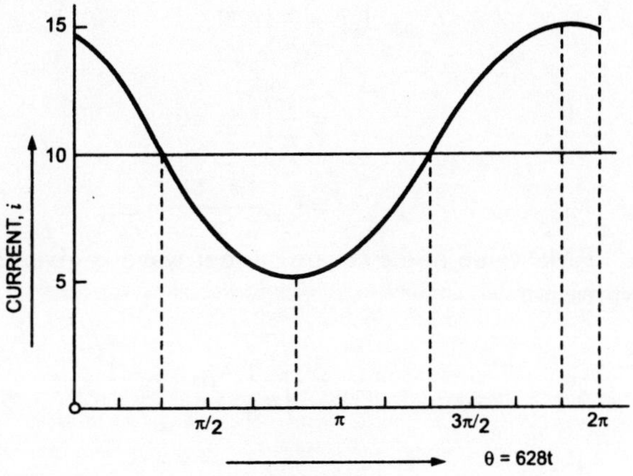

Fig. 5.17

$$i = 10 + 5\cos\left(628t + \frac{\pi}{6}\right) \quad \because 30° = \frac{\pi}{6} \text{ radians}$$

$$= 10 + 5\cos\left(\theta + \frac{\pi}{6}\right) \quad [\text{where } \theta = 628\,t]$$

$$i^2 = \left[10 + 5\cos\left(\theta + \frac{\pi}{6}\right)\right]^2$$

$$= 100 + 100\cos\left(\theta + \frac{\pi}{6}\right) + 25\cos^2\left(\theta + \frac{\pi}{6}\right)$$

$$= 100 + 100\cos\left(\theta + \frac{\pi}{6}\right) + 12.5\left[1 + \cos 2\left(\theta + \frac{\pi}{6}\right)\right]$$

$$= 112.5 + 100\cos\left(\theta + \frac{\pi}{6}\right) + 12.5\cos 2\left(\theta + \frac{\pi}{6}\right)$$

$$i_{rms}^2 = \frac{1}{2\pi}\int_0^{2\pi} i^2\, d\theta \text{ considering complete cycle} = \frac{1}{2\pi}\int_0^{2\pi}\left[112.5 + 100\cos\left(\theta + \frac{\pi}{6}\right) + 12.5\cos 2\left(\theta + \frac{\pi}{6}\right)\right] d\theta$$

$$= \frac{1}{2\pi} \times 112.5 \times 2\pi = 112.5$$

$$\text{or } I_{rms} = \sqrt{112.5} = 10.6 \text{ A } \textbf{Ans.}$$

Example 5.20

Find the average and effective values of voltage for sinusoidal waveform shown in Fig. 5.18.

[Allahabad Univ. Electrical Science-I, 1991]

Solution: The equation of the given sinusoidal waveform is $v = 100 \sin \theta$

$$V_{av} = \frac{1}{2\pi}\int_{\pi/4}^{\pi} 100 \sin\theta \, d\theta = \frac{100}{2\pi}[-\cos\theta]_{\pi/4}^{\pi} = 27.17 \text{ V } \textbf{Ans.}$$

$$V_{rms}^2 = \frac{1}{2\pi}\int_{\pi/4}^{\pi} 100^2 \sin^2\theta \, d\theta$$

$$= \frac{10,000}{4\pi}\int_{\pi/4}^{\pi} (1-\cos 2\theta)\, d\theta$$

$$= \frac{2,500}{\pi}\left[\theta - \frac{\sin 2\theta}{2}\right]_{\pi/4}^{\pi} = \frac{2,500}{\pi}\left[\pi - \frac{\pi}{4} + \frac{1}{2}\right] = 2,273$$

$$\text{or } V_{rms} = \sqrt{2,273} = 47.67 \text{ V } \textbf{Ans.}$$

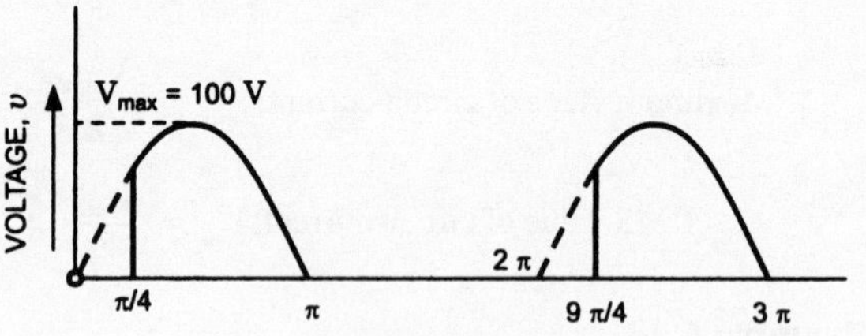

Fig. 5.18

5.11. RMS VALUE, AVERAGE VALUE, PEAK FACTOR AND FORM FACTOR OF SQUARE WAVEFORM

Square current wave is shown in Fig. 5.19. Instantaneous value of current can be expressed as

$$i = I_{max} \quad \text{for} \quad 0 < \theta < \pi$$

Average value of the square current wave is given by expression

$$I_{av} = \frac{1}{\pi}\int_0^{\pi} i \, d\theta$$

$$= \frac{1}{\pi}\int_0^{\pi} I_{max} \, d\theta$$

$$= I_{max}\frac{(\pi - 0)}{\pi} = I_{max}$$

RMS value of the square current wave is given by the expression

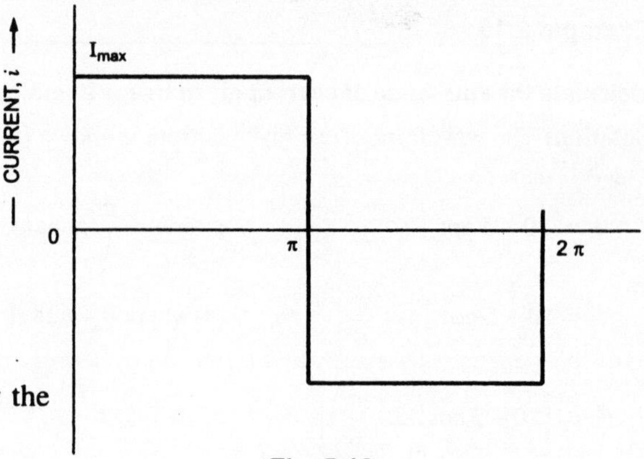

Fig. 5.19

$$I_{rms} = \sqrt{\frac{1}{\pi}\int_0^{\pi} i^2 d\theta}$$

$$= \sqrt{\frac{1}{\pi}\int_0^{\pi} I_{max}^2 \, d\theta} = \sqrt{\frac{1}{\pi} I_{max}^2 (\pi - 0)} = I_{max}$$

$$\text{Form factor, } K_f = \frac{\text{RMS value}}{\text{Average value}} = \frac{I_{max}}{I_{max}} = 1$$

$$\text{Peak factor, } K_p = \frac{\text{Peak value}}{\text{RMS value}} = \frac{I_{max}}{I_{max}} = 1$$

5.12. RMS VALUE, AVERAGE VALUE, FORM FACTOR AND PEAK FACTOR OF A TRIANGULAR WAVEFORM

Let the maximum value of the current be I_{max} amperes.

Since $i = I_{max}$ when $\theta = \pi$. Hence, expression for the instantaneous current can be written as

$$i = \qquad \text{for } 0 < \theta < \pi$$

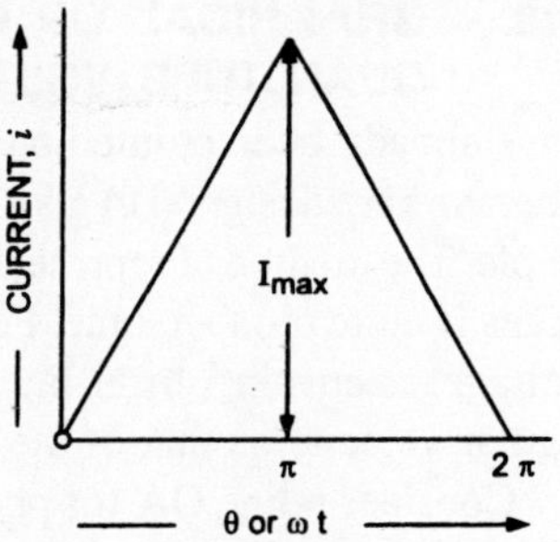

Fig. 5.20

RMS value of the current wave, $I_{rms} = \sqrt{\frac{1}{\pi}\int_0^{\pi} i^2 \, d\theta}$

$$= \sqrt{\frac{1}{\pi}\int_0^{\pi} \frac{I_{max}^2}{\pi^2}\theta^2 \, d\theta} = \frac{I_{max}}{\sqrt{3}}$$

Average value of the current wave, $I_{av} = \frac{1}{\pi}\int_0^{\pi} i \, d\theta$

$$= \frac{1}{\pi}\int_0^{\pi} \frac{I_{max}}{\pi}\theta \, d\theta = \frac{I_{max}}{2}.$$

Form factor, $K_f = \frac{I_{rms}}{I_{av}} = \frac{I_{max}/\sqrt{3}}{I_{max}/2} = \frac{2}{\sqrt{3}} = 1.155$ **Ans.**

Peak factor, $K_p = \frac{I_{max}}{I_{rms}} = \frac{I_{max}}{I_{max}/\sqrt{3}} = \sqrt{3}$ **Ans.**

Example 5.21

Calculate from first principles the rms value of a triangular voltage wave in which the voltage rises from 0 to V_{max} and completes the cycle by falling instantly back to zero. Give also the ratio of rms to mean value for the waveform (sawtooth wave). *[Pb Univ. April 1974]*

Solution: Triangular voltage wave, in which the voltage rises from 0 to V and completes the cycle by falling instantly back to zero (*i.e.* sawtooth voltage wave) is shown in Fig. 5.21.

As seen from Fig. 5.21 the instantaneous value of voltage can be expressed as follows:

$v = k\theta$ where k is the slope of the curve.

Let the maximum value of the voltage be V_{max}.

Since $v = V_{max}$ when $\theta = 2\pi$

$$\therefore k = \frac{V_{max}}{2\pi}$$

and the expression for instantaneous voltage becomes

$$v = \frac{V_{max}}{2\pi}\theta$$

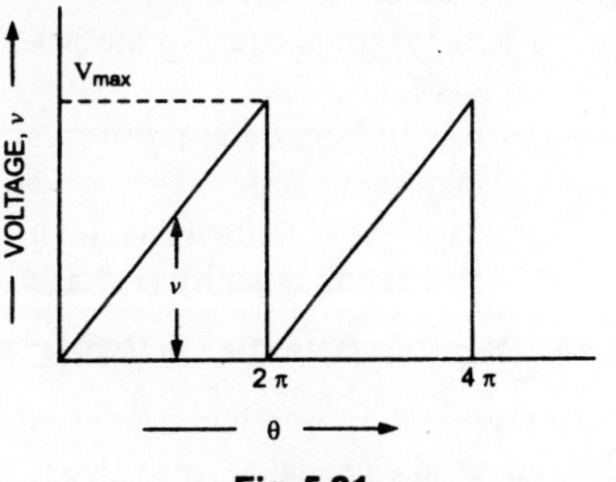

Fig. 5.21

RMS value of the voltage wave, $V_{rms} = \sqrt{\frac{1}{2\pi}\int_0^{2\pi} v^2 \, d\theta}$

$$= \sqrt{\frac{1}{2\pi}\int_0^{2\pi} \frac{V_{max}^2}{4\pi^2}\theta^2 \, d\theta} = \sqrt{\frac{V_{max}^2}{8\pi^3}\left[\frac{\theta^3}{3}\right]_0^{2\pi}} = V_{max}\sqrt{\frac{8\pi^3/3}{8\pi^3}} = \frac{V_{max}}{\sqrt{3}} \quad \textbf{Ans.}$$

Mean value of the voltage wave, $V_{av} = \frac{1}{2\pi}\int_0^{2\pi} v \, d\theta = \frac{1}{2\pi}\int_0^{2\pi} \frac{V_{max}}{2\pi}\theta \, d\theta = \frac{V_{max}}{4\pi^2}\left[\frac{\theta^2}{2}\right]_0^{2\pi} = \frac{V_{max}}{2}$

Ratio of rms value to mean value for the given waveform

$$= \frac{V_{rms}}{V_{av}} = \frac{V_{max}/\sqrt{3}}{V_{max}/2} = \frac{2}{\sqrt{3}} = 1.155 \textbf{ Ans.}$$

5.13. GRAPHICAL OR PHASOR REPRESENTATION OF ALTERNATING (SINUSOIDAL) QUANTITIES (VOLTAGES AND CURRENTS)

It has already been pointed out that it is assumed that alternating voltages and currents follow sine law and generators are designed to give emfs having sine waveform. The above said assumption makes the calculations simple. The method of representing alternating quantities by waveform or by the equations giving instantaneous values is quite cumbersome. For solution of ac problems it is advantageous to represent a sinusoidal quantity (voltage or current) by a line of definite length rotating in counter-clockwise* direction with the same angular velocity as that of the sinusoidal quantity. Such a rotating line is called the *phasor*.**

Consider a line OA (or phasor as it is called) representing to scale the maximum value of an alternating quantity, say emf *i.e.* OA = E_{max} and rotating in counter-clockwise direction at an angular velocity ω radians/second about the point O, as shown in Fig. 5.22. An arrow head is put at the outer end of the phasor, partly to indicate which end is assumed to move and partly to indicate the precise length of the phasor when two or more phasors happen to coincide.

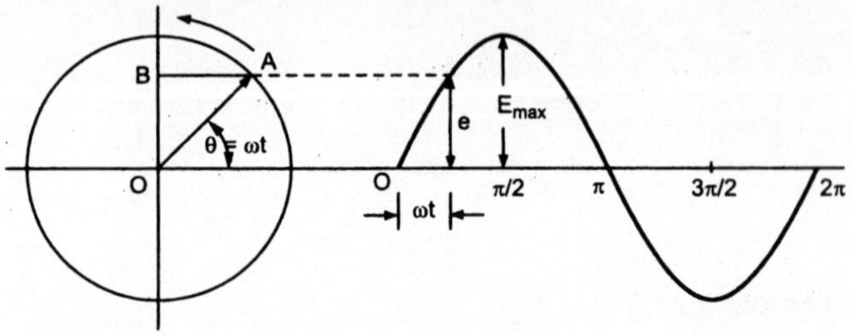

Fig. 5.22 *Phasor Representation of An Alternating Quantity*

Figure 5.22 shows OA when it has rotated through an angle θ, being equal to ω*t*, from the position occupied when the emf was passing through its zero value. The projection of OA on Y-axis,

$$OB = OA \sin\theta = E_{max} \sin \omega t = e, \text{ the value of the emf at that instant}$$

Thus, the projection of OA on the vertical axis represents to scale the instantaneous value of emf.

It will be seen that the phasor OA rotating in counter-clockwise direction will represent a sinusoidal quantity (voltage or current) if

(*i*) its length is equal to the peak or maximum value of the sinusoidal voltage or current to a suitable scale.

(*ii*) it is in horizontal position at the instant the alternating quantity (voltage or current) is zero and increasing and

(*iii*) its angular velocity is such that it completes one revolution in the same time as taken by the alternating quantity (voltage or current) to complete one cycle.

5.14. PHASOR DIAGRAM USING RMS VALUES

Since there is definite relation between maximum value and rms value ($E_{max} = \sqrt{2}\, E_{rms}$), the length of phasor OA can be taken equal to rms value if desired. But it should be noted that in such cases, the projection of the rotating phasor on the vertical axis will not give the instantaneous value of that alternating quantity.

The phasor diagram drawn in rms values of the alternating quantities helps in understanding the behaviour of the ac machines under different loading conditions.

* It is a standard convention that the phasor is rotated in counter-clockwise direction — a convention that is in harmony with the general use of polar coordinates.

** Strictly speaking, sinusoidal quantities are scalar quantities varying periodically with time and, according to the definition of vector quantities, they are not true vectors. Voltage is simply energy or work per coulomb and cannot be classified as a vector. Current is also not a vector quantity because it is merely the flow of electrons through a wire. But keeping in mind that any sinusoidal quantity at a given frequency is completely specified by its amplitude and phase angle, its similarity to a vector quantity is evident, since the amplitude may be considered as the magnitude and the phase angle as the direction of a vector. To account for the difference the term phasor has been adopted, instead of the term vector, for representing graphically the magnitude and phase of a sinusoidal current or voltage.

5.15. PHASE AND PHASE ANGLE

By phase of an alternating current is meant the fraction of the time period of that alternating current that has elapsed since the current last passed through the zero position of reference. The phase angle of any quantity means the angle the phasor representing the quantity makes with the reference line (which is taken to be at zero degrees or radians). For example the phase angle of current I_2 in Fig. 5.23 is $(-\Phi)$.

5.16. PHASE DIFFERENCE

When two alternating quantities, say, two emfs, or two currents or one voltage and one current are considered simultaneously, the frequency being the same, they may not pass through a particular point at the same instant. One may pass through its maximum value at the instant when the other passes through the value other than its maximum one. These two quantities are said to have a *phase difference*. Phase difference is always given either in degrees or in radians.

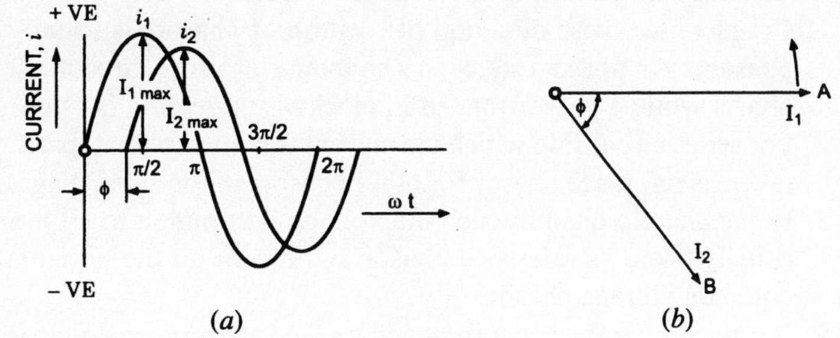

(*a*) (*b*)

Fig. 5.23

The phase difference is measured by the angular distance between the points where the two curves cross the base or reference line in the same direction.

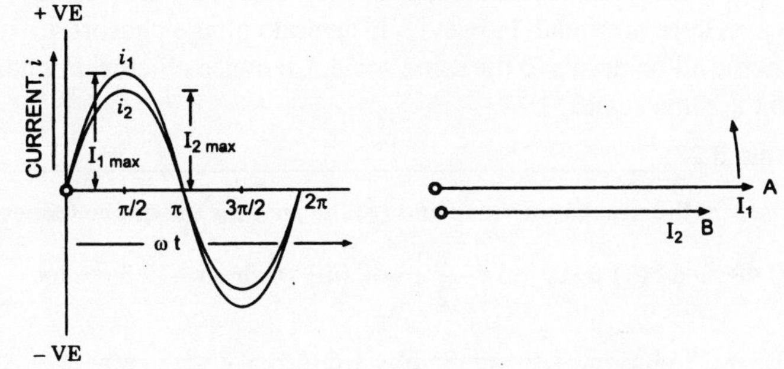

(*a*) (*b*)

Fig. 5.24

The quantity ahead in phase is said to lead the other quantity while the second quantity is said to lag behind the first one. In Fig. 5.23 (*b*) current I_1 represented by phasor OA leads the current I_2 represented by phasor OB by Φ or current I_2 lags behind the current I_1 by Φ. The leading current I_1 goes through its zero and maximum values first and the current I_2 goes through its zero and maximum values after time angle Φ. The two waves representing these two currents are shown in Fig. 5.23 (*a*). If I_1 is taken as reference phasor, the two currents can be expressed as

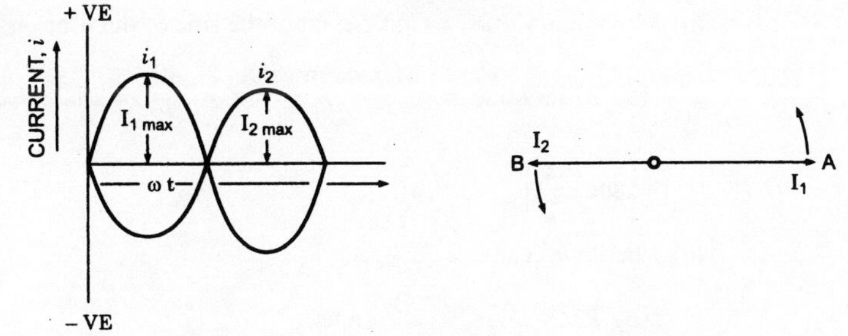

(*a*) (*b*)

Fig. 5.25

$$i_1 = I_{1\,max} \sin \omega t$$

$$\text{and } i_2 = I_2 \sin (\omega t - \Phi)$$

The two quantities are said to be in phase with each other if they pass through zero values at the same instant and rise in the same direction, as shown in Fig. 5.24. But the two quantities passing through zero values at the same instant but rising in opposite directions, as shown in Fig. 5.25 are said to be in phase opposition, *i.e.* phase difference is 180°. When the two alternating quantities have a phase difference of 90° or $\frac{\pi}{2}$ radians they are said to be in quadrature.

5.17. CONVENTIONS FOR DRAWING PHASOR DIAGRAMS

As already mentioned, the alternating quantities (voltages and currents) in practice are represented by straight lines having definite direction and length. Such lines are called the *phasors* and the diagrams in which phasors represent currents, voltages and their phase difference are known as *phasor diagrams*.

Though phasor diagrams can be drawn to represent either maximum or effective values of voltages and currents but since effective values are of much more importance, phasor diagrams are mostly drawn to represent effective values.

In order to achieve consistent and accurate results it is essential to follow certain conventions. Some of the common conventions in this regard are enlisted below:

1. Counter-clockwise direction of rotation of phasors is usually taken as positive direction of rotation of phasors *i.e.* a phasor rotated in a counter-clockwise direction from a given phasor is said to lead the given phasor while a phasor rotated in clockwise direction is said to lag the given phasor.
2. For series circuits, in which the current is common to all parts of the circuit, the current phasor is usually taken as reference phasor for other phasors in the same diagram and drawn on horizontal line.
3. In parallel circuits in which the voltage is common to all branches of the circuit, the voltage phasor is usually taken as reference phasor and drawn on the horizontal line. Other phasors are referred to the common voltage phasor.
4. It is not necessary that current and voltage phasors are drawn to the same scale; in fact it is often desirable to draw the current phasor to a larger scale than the voltage phasor when the values of currents being represented are small. However, if several voltage phasors are to be used in the same phasor diagram, they should all be drawn to the same scale. Likewise all current phasors in the same diagram should be drawn to the same scale.

Example 5.22

Calculate (*a*) the maximum value and (*b*) the root mean square values of the following quantities:

(*i*) 40 sin ωt (*ii*) B sin $\left(\omega t - \frac{\pi}{2}\right)$ and (*iii*) 10 sin ωt –17.3 cos ωt.

Draw the phasors showing the phase difference with respect to A sin $\left(\omega t - \frac{\pi}{6}\right)$.

Solution: (*i*) For 40 sin ω *t*

(*a*) Maximum value = Coefficient of the sine of the time angle = 40 **Ans.**

(*b*) RMS value $= \dfrac{\text{Maximum value}}{\sqrt{2}} = \dfrac{40}{\sqrt{2}} = 28.28$ **Ans.**

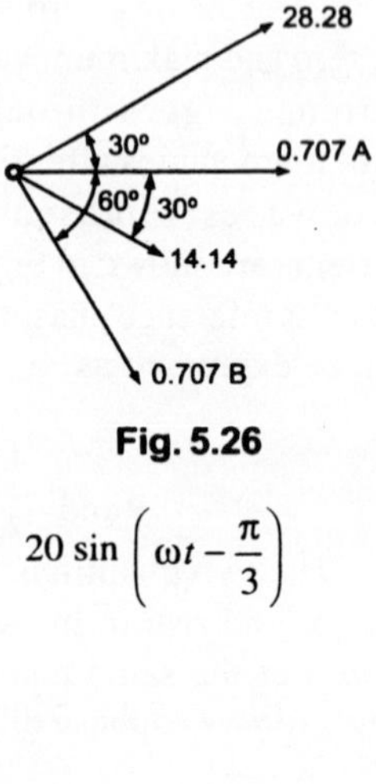

Fig. 5.26

(*ii*) For B sin $\left(\omega t - \frac{\pi}{2}\right)$

(*a*) Maximum value = B **Ans.**

(*b*) RMS value $= \dfrac{B}{\sqrt{2}}$ **Ans.**

(*iii*) For 10 sin ω*t* – 17.3 cos ω*t*, which may be written as

$$20\left(\frac{1}{2}\sin\omega t - \frac{\sqrt{3}}{2}\cos\omega t\right) \quad \text{or} \quad 20\left(\cos\frac{\pi}{3}\sin\omega t - \sin\frac{\pi}{3}\cos\omega t\right) \quad \text{or} \quad 20\sin\left(\omega t - \frac{\pi}{3}\right)$$

(*a*) Maximum value = 20

(*b*) RMS value $= \dfrac{20}{\sqrt{2}} = 14.14$ **Ans.**

Phasors showing the phase difference with respect to A sin $\left(\omega t - \frac{\pi}{6}\right)$ are shown in Fig. 5.26.

5.18. ADDITION AND SUBTRACTION OF ALTERNATING QUANTITIES

In most of the ac circuits, it is necessary to consider the combined action of several emfs or voltages acting in a series circuit and action of several currents flowing through the different branches of a parallel circuit.

Let it be required to add two currents given by the equations

$$i_1 = I_{1\,max} \sin \omega t \text{ and } i_2 = I_{2\,max} \sin (\omega t - \Phi)$$

The resultant sum may be expressed as

$$i_r = i_1 + i_2 = I_{1\,max} \sin \omega t + I_{2\,max} \sin (\omega t - \Phi)$$

but it is too awkward and gives no idea of the peak value and phase angle of the resultant current.

The currents may be added graphically by plotting their curves in the same system of coordinates and then adding the ordinates of i_1 and i_2 point by point, according to the equation $i_r = i_1 + i_2$. Evidently this method is also too cumbersome and unwieldy to be practical. This is particularly so when situation arise where more than two sinusoidal quantities are to be added.

A simpler and more direct method consists in adding the sinusoidal quantities as phasors.

Consider the phasors $I_{1\,max}$ and $I_{2\,max}$ that would generate the two curves i_1 and i_2 and let them be in a position, as shown in Fig. 5.27 at one particular instant of time. If we now add $I_{1\,max}$ and $I_{2\,max}$ by completing the parallelogram as shown, the diagonal $I_{r\,max}$ will, when rotated, generate a third sine curve. It remains to be shown that this third sine curve coincides with the waveform of i_r obtained by adding i_1 and i_2 point by point.

Now the vertical component of $I_{r\,max}$ is the sum of the vertical components of $I_{1\,max}$ and $I_{2\,max}$. Therefore, the waveform of i_r is the graph generated by rotating $I_{r\,max}$ in counter-clockwise direction.

It follows, therefore, that two or more alternating quantities may be added in the same way as forces are added, namely by constructing parallelograms or closed polygons and either measuring or calculating the lengths of the diagonals or closing sides and the magnitude of the phase angles.

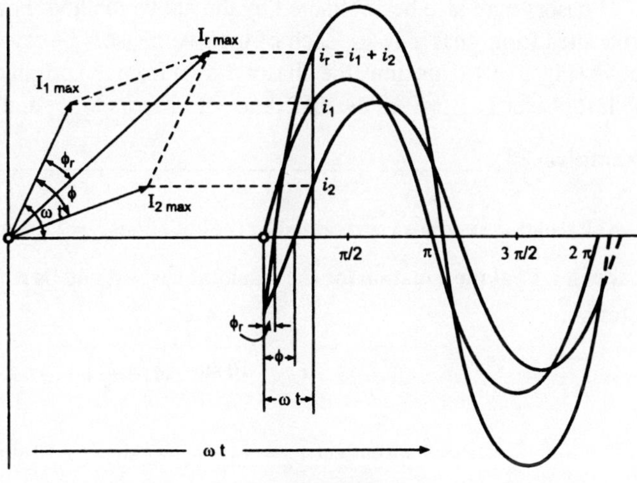

Fig. 5.27

Illustrations: The way in which the two given currents i_1 and i_2 can be added by the parallelogram rule of phasor addition is illustrated in Fig. 5.28 where the currents are shown as phasor drawn from the origin O of the system of coordinates. The resultant phasor is the diagonal of the parallelogram formed by the phasors $I_{1\,max}$ and $I_{2\,max}$.

This method is more convenient when more than two phasors are to be added, as shown in Fig. 5.29. From the end point of $I_{1\,max}$ a phasor is constructed parallel to $I_{2\,max}$ of the same magnitude and direction as the latter; then from the end point of $I_{2\,max}$ a phasor is constructed parallel to $I_{3\,max}$ and so on. Phasor $I_{r\,max}$ from the origin of the first phasor ($I_{1\,max}$) to the end point of the last phasor ($I_{5\,max}$) represents the sum of all the phasors.

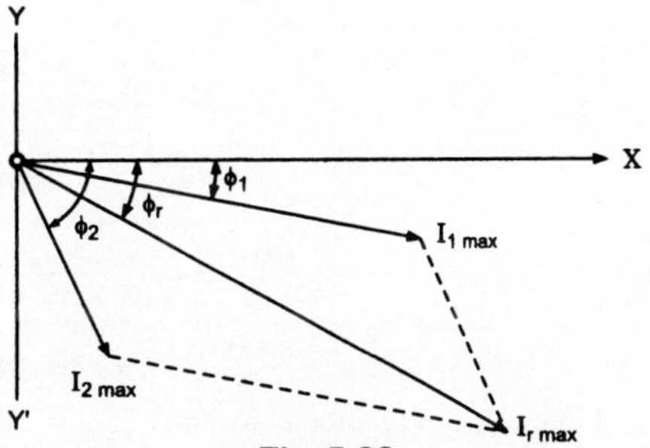

Fig. 5.28

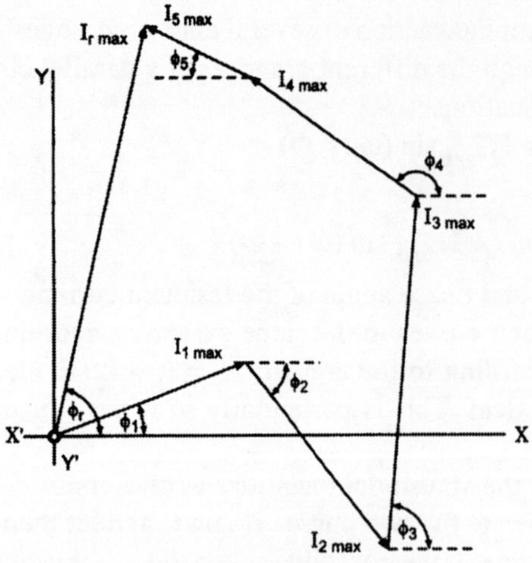

Fig. 5.29

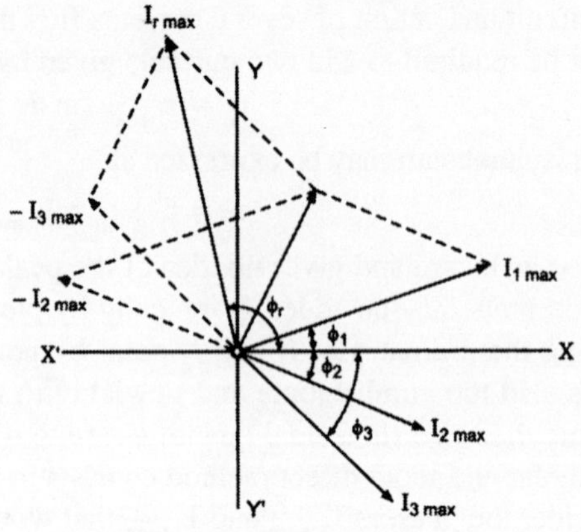

Fig. 5.30

Phasors may also be subtracted by the above method. For example if phasors $I_{2\,max}$ and $I_{3\,max}$ are to be subtracted from phasor $I_{1\,max}$ each of the two phasors be reversed in direction and then added as explained above [Fig. 5.30]. This time, the phasor drawn from the origin of the first phasor $I_{1\,max}$ to the terminal point of the last phasor $I_{3\,max}$ gives the difference of phasors $I_{2\,max}$ and $I_{3\,max}$ from phasor $I_{1\,max}$.

Example 5.23

Two alternating currents represented by the equations $i_1 = 7 \sin \omega t$ and $i_2 = 10 \sin\left(\omega t + \frac{\pi}{3}\right)$ are fed into a common conductor. Find the equation for the resultant current and its rms value.

Solution:

$$i_1 = 7 \sin \omega t$$

$$i_2 = 10 \sin\left(\omega t + \frac{\pi}{3}\right)$$

$$\text{Resultant current, } i_r = i_1 + i_2 = 7 \sin \omega t + 10 \sin\left(\omega t + \frac{\pi}{3}\right)$$

$$= 7 \sin \omega t + 10 \sin \omega t \cos\frac{\pi}{3} + 10 \cos \omega t \sin\frac{\pi}{3} = 12 \sin \omega t + 8.66 \cos \omega t$$

Multiplying and dividing RHS of the above expression by $\sqrt{(12)^2 + (8.66)^2}$ *i.e.* 14.8, we get

$$i_r = 14.8\left(\sin \omega t \times \frac{12}{14.8} + \cos \omega t \times \frac{8.66}{14.8}\right)$$

$$= 14.8 (\sin \omega t \cos \Phi_r + \cos \omega t \sin \Phi_r) \quad \text{where } \cos \Phi_r = \frac{12}{14.8} \text{ and } \sin \Phi_r = \frac{8.66}{14.8}$$

$$\text{or } i_r = 14.8 \sin (\omega t + \Phi_r) \quad \text{where } \Phi_r = \text{Tan}^{-1}\frac{8.66}{12} = 0.199\,\pi \text{ radians}$$

$$\text{or } i_r = 14.8 \sin (\omega t + 0.199\,\pi) \textbf{ Ans.}$$

$$\text{Peak value of resultant current, } I_{r\,max} = 14.8 \text{ A}$$

$$\text{RMS value of resultant current, } I_{r\,rms} = \frac{I_{r\,max}}{\sqrt{2}} = \frac{14.8}{\sqrt{2}} = 10.46 \text{ A } \textbf{Ans.}$$

***Analytical Method*:** Resolving currents $I_{1\,max}$ and $I_{2\,max}$ along X-axis and Y-axis we get

$$\text{Algebraic sum of X-components} = 7 \cos 0° + 10 \cos \frac{\pi}{3} = 7 + 5 = 12 \text{ A}$$

$$\text{Algebraic sum of Y-components} = 7 \sin 0° + 10 \sin \frac{\pi}{3} = 8.66 \text{ A}$$

$$\text{Maximum value of resultant current, } I_{r\,max} = \sqrt{(\text{X-components})^2 + (\text{Y-components})^2} = \sqrt{12^2 + (8.66)^2} = 14.8 \text{ A}$$

$$\text{and phase angle, } \Phi_r = \text{Tan}^{-1} \frac{\text{Y-components}}{\text{X-components}} = \text{Tan}^{-1} \frac{8.66}{12} = 0.199\,\pi$$

$$\therefore \text{ Resultant current, } i_r = I_{r\,max} \sin(\omega t + \Phi_r) = 14.8 \sin(\omega t + 0.199\,\pi) \textbf{ Ans.}$$

Graphical Method. Take OX as reference phasor. Draw phasor OA 3.5 cm in length along OX so as to represent current $I_{1\,max}$ of 7 amperes (to the scale 1 cm = 2 A) in magnitude as well as in direction. Then draw phasor AB = 5 cm in length making an angle of $\frac{\pi}{3}$ radians or 60° with the reference phasor so as to represent current $I_{2\,max}$ of 10 amperes in magnitude as well as in direction. The phasor OB, obtained by joining points O and B, will represent the resultant current in magnitude as well as in direction. On measurement from Fig. 5.32.

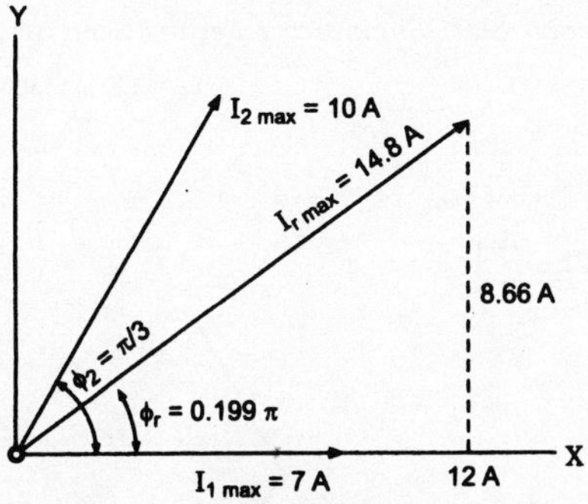

Fig. 5.31

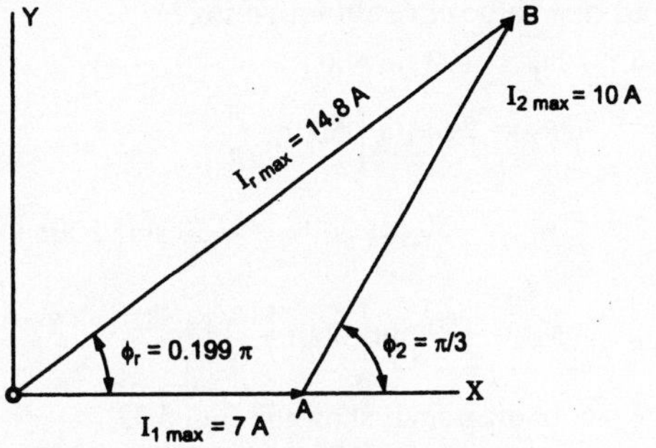

Fig. 5.32

$$\text{Maximum value of resultant current, } I_{r\,max} = \text{OB} \times 2 = 7.4 \times 2 = 14.8 \text{ A}$$

$$\text{and phase angle, } \Phi_r = 35.8° \text{ or } 0.199\,\pi \text{ radians}$$

Hence the resultant current may be expressed as

$$i_r = 14.8 \sin(\omega t + 0.199\,\pi) \textbf{ Ans.}$$

Example 5.24

Two ac voltages are represented by —

$$\boldsymbol{v_1(t) = 30 \sin(314\,t + 45°)}$$
$$\boldsymbol{v_2(t) = 60 \sin(314\,t + 60°)}$$

Calculate the resultant voltage $v(t)$ and express in the form —

$$\boldsymbol{v(t) = V_m \sin(314\,t + \phi)}$$

[U.P. Technical Univ. Electrical Engineering Second Semester 2003-04]

Solution:

$$v_1(t) = 30 \sin(314\,t + 45°)$$
$$v_2(t) = 60 \sin(314\,t + 60°)$$

Resolving voltages $V_{1\,max}$ and $V_{2\,max}$ along X-axis and Y-axis we have

$$\text{Algebraic sum of X-components} = 30 \cos 45° + 60 \cos 60° = 30 \times \frac{1}{\sqrt{2}} + 60 \times \frac{1}{2} = 51.21 \text{ V}$$

$$\text{Algebraic sum of Y-component} = 30 \sin 45° + 60 \sin 60° = 30 \times \frac{1}{\sqrt{2}} + 60 \times \frac{\sqrt{3}}{2} = 73.17 \text{ V}$$

Maximum value of resultant voltage, $V_{r\,max} = \sqrt{(\text{X-components})^2 + (\text{Y-components})^2}$

$$= \sqrt{(51.21)^2 + (73.17)^2} = 89.31 \text{ V}$$

$$\text{Phase angle, } \phi_r = \text{Tan}^{-1}\frac{\text{Y-component}}{\text{X-component}} = \text{Tan}^{-1}\frac{73.17}{51.21} = 55°$$

So expression for resultant voltage is

$$v(t) = V_{max}\sin(\omega t + \phi_r) = 89.31 \sin(314\,t + 55°) \textbf{ Ans.}$$

Example 5.25

Draw a phasor diagram showing the following voltages:

$v_1 = 100 \sin 500\,t$ $\qquad$ $v_2 = 200 \sin\left(500t + \frac{\pi}{3}\right)$

$v_3 = -50 \cos 500\,t$ $\qquad$ $v_4 = 150 \sin\left(500t - \frac{\pi}{4}\right)$

Find rms value of resultant voltage. *[U.P. Technical Univ. Elec. Engineering Second Semester 2005-06]*

Solution : $v_1 = 100 \sin 500\,t$

$$v_2 = 200 \sin\left(500t + \frac{\pi}{3}\right)$$

$$v_3 = -50\cos 500\,t = -50\sin\left(500t + \frac{\pi}{2}\right) = 50\sin\left(500t - \frac{\pi}{2}\right)$$

$$v_4 = 150 \sin\left(500t - \frac{\pi}{4}\right)$$

The phasor diagram is shown in Fig. 5.33.

Resolving voltages V_1, V_2, V_3 and V_4 along X-axis and Y-axis we have

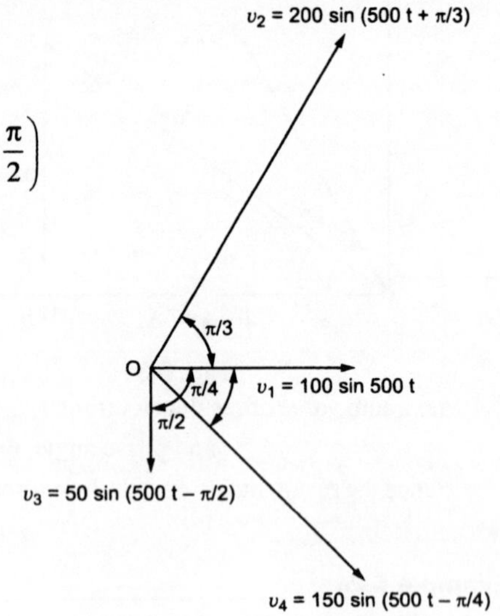

Fig. 5.33 *Phasor Diagram*

Algebraic sum of X-components

$$= 100\cos 0° + 200\cos\frac{\pi}{3} + 50\cos\left(-\frac{\pi}{2}\right) + 150\cos\left(-\frac{\pi}{4}\right)$$

$$= 100 + 200 \times 0.5 + 0 + \frac{150}{\sqrt{2}} = 306 \text{ V}$$

Algebraic sum of Y-components

$$= 100\sin 0° + 200\sin\frac{\pi}{3} + 50\sin\left(-\frac{\pi}{2}\right) + 150\sin\left(-\frac{\pi}{4}\right)$$

$$= 0 + 200 \times 0.866 - 50 - 106 = 17.2 \text{ V}$$

Maximum value of resultant voltage, $V_{r\,max} = \sqrt{(306)^2 + (17.2)^2} = 306.5$ V

RMS value of resultant voltage, $V_{r\,rms} = \frac{V_{r\,max}}{\sqrt{2}} = \frac{306.5}{\sqrt{2}} = 216.72$ V **Ans.**

Example 5.26

The instantaneous values of two alternating voltages are represented by $v_1 = 60 \sin\theta$ and $v_2 = 40\sin(\theta - \pi/3)$. Derive expressions for the instantaneous values of (*i*) the sum and (*ii*) the difference of these voltages.

[U.P. Technical Univ. Electrical Engineering July 2002]

Solution: (*a*) $v_1 = 60\sin\theta$ and $v_2 = 40\sin\left(\theta - \frac{\pi}{3}\right)$

(*i*) Maximum value of voltage, $V_{r\,max} = \sqrt{(\text{Sum of X-components})^2 + (\text{Sum of Y-components})^2}$

$$= \sqrt{\left[60\cos 0° + 40\cos\left(\frac{-\pi}{3}\right)\right]^2 + \left[60\sin 0° + 40\sin\left(\frac{-\pi}{3}\right)\right]^2}$$

$$= \sqrt{(60+20)^2 + (-34.64)^2} = \sqrt{(80)^2 + (-34.64)^2} = 87.178 \text{ V}$$

$$\text{Phase angle, } \phi = \text{Tan}^{-1} \frac{\text{Sum of Y-components}}{\text{Sum of X-components}}$$

$$= \text{Tan}^{-1} \frac{-34.64}{80} = -23.41° = -0.13\,\pi \text{ radian}$$

So expression for the instantaneous values of the sum of given voltages,

$$v_r = 87.178 \sin(\theta - 0.13\,\pi) \textbf{ Ans.}$$

(*ii*) Maximum value of voltage, $V_{rms} = \sqrt{(\text{Difference of X-components})^2 + (\text{Difference of Y-components})^2}$

$$= \sqrt{\left[60\cos 0° - 40\cos\left(\frac{-\pi}{3}\right)\right]^2 + \left[60\sin 0° - 40\sin\left(\frac{-\pi}{3}\right)\right]^2}$$

$$= \sqrt{(60-20)^2 + (34.64)^2} = \sqrt{40^2 + (34.64)^2} = 52.915 \text{ V}$$

$$\text{Phase angle, } \phi = \text{Tan}^{-1} \frac{34.64}{40} = 40.89° \text{ or } 0.227\,\pi \text{ radian}$$

So expression for the instantaneous values of the difference of given voltages,

$$v_r = 52.915 \sin(\theta + 0.227\,\pi)$$

HIGHLIGHTS

1. An *alternating quantity* (voltage or current) is one which changes continuously in magnitude and alternates in direction at regular intervals of time.

   An alternating quantity that varies sinusoidally (*i.e.* according to the sin of angle θ) is called the *sinusoidal quantity.* Sinusoidal quantities are expressed as

   $$e = E_{max} \sin\theta = E_{max} \sin \omega t = E_{max} \sin 2\pi f t$$

   $$\text{and } i = I_{max} \sin\theta = I_{max} \sin \omega t = I_{max} \sin 2\pi f t$$

2. The shape of the curve of the voltage or current when plotted against time as abscissa (base) is called the *waveform.*
3. When a periodic wave, such as sinusoidal wave, goes through one complete set of positive and negative values it is said to have completed one *cycle*. One cycle corresponds to 360° or 2 π radians.

   Alternation is one half of cycle and corresponds to 180° or π radians.
4. The time taken in seconds by an alternating quantity to complete one cycle is known as *time period* or *periodic time* (T), while the number of cycles completed per second by an alternating quantity is known as *frequency* (*f*).

   $$f = \frac{1}{T}$$

   In India the standard frequency for power supply is 50 Hz (cycles per second).
5. Angular velocity, $\omega = 2\pi f$ radians per second.
6. In a multipolar machine, the number of cycles completed per second by generated emf,

   $$f = \text{Pairs of poles} \times \text{number of revolutions made per second} = \frac{P}{2} \times \frac{N}{60} = \frac{PN}{120}$$

7. The maximum value, positive or negative, which an alternating quantity attains during one cycle is called the *amplitude* of the alternating quantity.
8. The general expression for an alternating voltage and current are given as

$$v = V_{max} \sin(\omega t \pm \phi)\,; \quad i = I_{max} \sin(\omega t \pm \phi)$$

where ϕ is the phase angle in degrees or radians.
The coefficient of the sine of the time angle gives maximum or peak value of the alternating quantity (emf, voltage or current)
Coefficient of time t divided by 2π gives the frequency of the periodic wave.
9. The *average (or mean) value* of an alternating current is expressed by that steady current (dc) which transfers across any circuit the same charge as transferred by that alternating current
Average or mean value of alternating current is given as

$$I_{av} = \frac{i_1 + i_2 + i_3 + \ldots\ldots + i_n}{n} = \frac{\text{Area of one alternation}}{\text{Length of base over one alternation}}$$

$$\text{For sinusoidal current } I_{av} = \frac{2}{\pi} I_{max} = 0.637\, E_{max}.$$

10. *Effective* or *virtual value* of alternating current or voltage is equal to the square root of the mean of the squares of successive ordinates and that is why it is known as *root-mean square* (rms) value.

$$\text{i.e. } I_{eff} \text{ or } I_{rms} = \sqrt{\frac{i_1^2 + i_2^2 + i_3^2 + \ldots + i_n^2}{n}}$$

The *rms* or *effective value* of an alternating current or voltage is given by that steady current or voltage which when flows or applied to a given resistance for a given time produces the same amount of heat as when the alternating current or voltage flowing or applied to the same resistance for the same time.

$$\text{For sinusoidal current, } I_{rms} = \frac{I_{max}}{\sqrt{2}} = 0.707\, I_{max}$$

11. The ratio of rms value to the average value of a waveform is known as *form factor.* For sinusoidal wave its value is 1.11.
12. The ratio of maximum value to the rms value of a waveform is known as peak factor. For sinusoidal wave its value is $\sqrt{2}$ or 1.414.
13. The effective and average values of different waves are given below in tabular form.

Waveform	*Sinusoidal*	*Half Wave Rectified AC*	*Square*	*Triangular*
Effective Value	$0.707\, I_{max}$	$0.5\, I_{max}$	I_{max}	$\frac{1}{\sqrt{3}} I_{max}$
Average Value	$0.637\, I_{max}$	$0.318\, I_{max}$	I_{max}	$0.5\, I_{max}$

14. By *phase* of an alternating quantity is meant the fraction of time period of that alternating quantity that has elapsed since the quantity last passed through zero position of reference.
The angular displacement between the maximum positive values of two alternating quantities having the same frequency is called the phase difference between them.
An alternating quantity that attains its positive maximum value prior to the other is called the *leading quantity.*
An alternating quantity that attains its positive maximum value after the other is called the *lagging quantity.*
15. The resultant current of two currents repersented by

$$i_1 = I_{max_1} \sin \omega t \text{ and } i_2 = I_{max_2} \sin(\omega t + \theta)$$

flowing through a common conductor is given as

$$i_r = I_{r\,max} \sin(\omega t \pm \phi)$$

where $$I_{r\,max} = \sqrt{I_{max_1}^2 + I_{max_2}^2 + 2I_{max_1} I_{max_2} \cos\theta}$$

and $$\phi = \text{Tan}^{-1} \frac{I_{max_2} \sin\theta}{I_{max_1} + I_{max_2} \cos\theta}.$$

The sign of the phase angle ϕ of the resultant current will depend upon its position w.r.t. the reference phasor.

EXERCISES

1. Explain why a large percentage of electrical energy used for commercial purposes is generated as ac.
2. With a neat sketch briefly explain how an alternating voltage is produced when a coil is rotated in a magnetic field. **[V.T.U. Belgaum Karnataka Univ., Summer 2003]**
3. Derive an expression for the instantaneous value of alternating sinusoidal emf in terms of its maximum value, angular frequency and time. **[Gujarat Univ. June/July 2003]**
4. Define and explain the following:
(*i*) Time period (*ii*) Amplitude (*iii*) Phase difference. **[Gujarat Univ. June/July 2003]**
5. Define and explain the following:
(*i*) Frequency (*ii*) Time period (*iii*) Instantaneous value (*iv*) Average value (*v*) RMS value with respect to alternating quantities (*vi*) Amplitude. **[R.G.P.V. Bhopal Basic Electrical Engineering 2002]**
6. Define the terms: Instantaneous value, Average value, RMS value, frequency and phase. **[Pb. Technical Univ. Electrical Engineering June 2000; December 2000; May 2001]**
7. What is an rms value of an ac quantity ? Obtain expression for the rms value of a sinusoidal current in terms of its maximum value. **[V.T.U. Belgaum Univ. February 2002]**
8. Derive expressions for average value and rms value of a sinusoidally varying ac voltage. **[V.T.U. Belgaum Univ. Summer 2003]**
9. Define rms value, average value and form factor of a sinusoidal alternating voltage. **[Pb. Technical Univ. Basic Electrical and Electronics Engineering First Semester 2004-05]**
10. Find average value, rms value and form factor of half wave rectified alternating current. **[U.P. Technical Univ. Electrical Engineering Second Semester 2006-07]**
11. Calculate the average value, effective value and form factor of the output voltage wave of a half wave rectifier as shown in Fig. 1. **[U.P. Technical Univ. Electrical Engineering Second Semester 2003-04]**

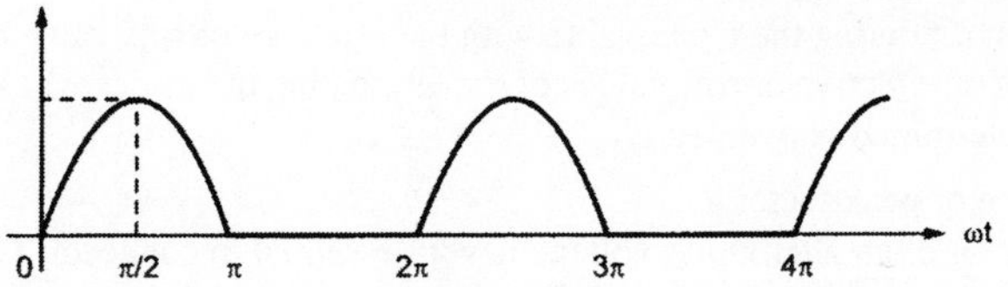

Fig. 1

12. Explain the need and method of representing alternating quantities as phasor quantity.
13. What do you understand by phasor representation of alternating quantity ? What is an alternating quantity ? When is an alternating quantity said to be lagging or leading with respect to another quantity ?
14. Explain with diagrams what do you understand by (i) in-phase (*ii*) lagging and (*iii*) leading as applied to sinusoidal ac quantities.
15. Explain the difference between ac and dc quantities. Draw the waveform and phasor diagram of two alternating quantities having a phase difference of 90°.
16. Explain the component method for the addition and subtraction of alternating quantities.

SHORT ANSWER TYPE QUESTIONS WITH ANSWERS

Q. 1. Why sinusoidal wave shape is insisted for voltages and currents while generating, transmitting and utilising ac electric power ? **[P.T.U. January 2000]**

Ans. Sinusoidal wave shape is insisted for voltages and currents while generating, transmitting and utilising ac electrical power because it has following advantages.

(*i*) Sinusoidal voltages and currents produce minimum disturbance in the electrical circuits during operation.

(*ii*) Sinusoidal voltages and currents cause less interference (noise) to nearby communication circuits (telephone lines etc.).

(*iii*) Sinusoidal voltages and currents result in low iron as well as low copper losses in transformers, and ac rotating machines for a given output. So ac machines with sinusoidal voltages and currents operate at higher efficiency.

Q. 2. What do you understand by ω ?

Ans. Each cycle of a sinusoidal wave spans 2π radians. Hence, if this quantity is divided by the time period, angular velocity of the sinusoidal wave is obtained. It is denoted by ω and is expressed in radians per second.

Q. 3. Why the rms value of an alternating current or voltage is used to denote its amplitude ?

Ans. RMS value of an alternating current or voltage is used to denote its amplitude because it is related to the power developed in a resistance by the alternating current or voltage.

Q. 4. What is rms value of an alternating current ? **[P.T.U. December 2003]**

Ans. The effective or rms value of an alternating current is given by that steady current which when flows through a given resistance for a given time produces the same amount of heat as when the alternating current is flowing through the same resistance for the same time duration.

Q. 5. Do wave shapes other than sine wave have effective value ? Explain. **[P.T.U. December 2000]**

Ans. Yes, all the wave shapes including the sine wave have effective value since in each half cycle of the wave, work is being done. Actually the effective value of an alternating wave (sinusoidal or non-sinusoidal) is defined as below:

The rms or effective value of an alternating current or voltage is given by that steady current or voltage which when flows or applied to a given resistance for a given time produces the same amount of heat as when the alternating current or voltage is flowing or applied to the same resistance for the same time.

Q. 6. Differentiate between form factor and peak factor. **[P.T.U. June 2003]**

Ans. Form factor is defined as the ratio of the effective value to the average or mean value of the periodic wave while peak factor is defined as the ratio of peak or maximum value to the effective or rms value of the periodic wave.

i.e. $K_f = \dfrac{\text{RMS value}}{\text{Average value}}$ while $K_p = \dfrac{\text{Peak value}}{\text{RMS value}}$

Q. 7. What is significance of form factor ?

Ans. Form factor is a means of relating the mean value with the effective or rms value of alternating quantity and it is useful in determination of effective or rms values of the alternating quantities whose mean or average values over half a period can be determined conveniently.

Q. 8. What is the significance of peak factor ?

Ans. Knowledge of peak factor of an alternating voltage is very essential in connection with determining the dielectric strength since the dielectric stress developed in an insulating material is proportional to the peak value of the voltage applied to it.

Q. 9. What is the significance of the phasor representation of an alternating quantity ?

Ans. The phasor representation of an alternating quantity enables us to understand its magnitude and position with respect to reference line.

Q. 10. What do you mean by phase and phase difference ?

Ans. The phase of an alternating quantity (voltage or current) at any instant is defined as the fractional part of a cycle through which the quantity has advanced while the phase difference may be defined as the angular displacement between the maximum positive values of the two phasors representing the two quantities having the same frequency.

PROBLEMS

1. For ac single phase 230 V, 50 Hz, assuming sinusoidal waveform find (*i*) maximum value (*ii*) rms value (*iii*) frequency (*iv*) instantaneous value voltage equation (*v*) voltage 0.003 second after starting from zero and rising in + ve direction.
[**Ans.** (*i*) 325.3 V (*ii*) 230 V (*iii*) 50 Hz (*iv*) 325.3 sin 314 *t* (*v*) 263.17 V]

2. $e = 300 \sin 628\, t$
Find maximum value, rms value, average value and frequency. Find the instantaneous value of voltage at $t = 0.001$ second after *e* is zero and rising in positive direction. [**Ans.** 300 V, 212.13 V, 190.98 V, 100 Hz, 176.26 V]

3. An alternating current is represented by $i = 400 \sin\left(157t + \dfrac{\pi}{6}\right)$.
Determine: (*i*) Peak value (*ii*) Average value (*iii*) RMS value (*iv*) Frequency (*v*) Time period (*vi*) Form factor (*vii*) Peak factor, and (*viii*) Phase angle of the current represented.
[**Ans.** (*i*) 400 A (*ii*) 254.65 A (*iii*) 282.84 A, (*iv*) 25 Hz (*v*) 0.04 second (*vi*) 1.11 (*vii*) 1.414 (*viii*) 30°]

4. A voltage waveform is given by the expression
$v = 200 \sin (628\, t + 60°)$. Determine:
(*i*) maximum value (*ii*) rms value (*iii*) frequency (*iv*) time period and (*v*) phase angle of the voltage.
[**Ans.** (*i*) 200 V (*ii*) 141.4 V (*iii*) 100 Hz (*iv*) 0.01s (*v*) 60°]

5. An alternating current is represented by $i = 70.7 \sin 520\, t$. Determine (*i*) the frequency; (*ii*) the current 0.0015 second after passing through zero, increasing positively, (*iii*) rms and average values.
[**Ans.** (*i*) 82.76 Hz (*ii*) 49.72 A (*iii*) 50 A; 45 A]

6. An alternating current of frequency 60 Hz has a maximum value of 120 A. Write down the equation for its instantaneous value. Find the time taken to reach 96 A for the first time. **[Kanpur Univ. 1979]**
[**Ans.** $i = 120 \sin 120\, \pi t$; 2.46 ms]

7. An alternating current of frequency 50 Hz has a maximum value of 100 A. Calculate (i) its value $\dfrac{1}{600}$ second after the instant the current is zero (ii) in how many seconds after the zero value the current attains the value of 86.6 A ? **[Kanpur Univ. 1977]**
[**Ans.** (*i*) 50 A (*ii*) $\dfrac{1}{300}$ second]

8. The ac supply at a house is 230 V, 50 Hz. Find maximum, average and rms values of voltage.
[**Ans.** 325.27 V, 207.2 V, 230 V]

9. The equation of an alternating current $i = 42.42 \sin 628\, t$. Determine (*i*) its maximum value (*ii*) frequency (*iii*) rms value (*iv*) average value (*v*) form factor. **[Pb. Univ. Elements of Electrical Engineering December 1988]**
[**Ans.** (*i*) 42.42 A (*ii*) 100 Hz (*iii*) 30 A (*iv*) 27 A (*v*) 1.11]

10. Calculate the rms value, the form factor and the peak factor of a periodic voltage having the following values for equal time intervals, changing suddenly from one value to the next:
0, 5, 10, 20, 50, 60, 50, 20, 10, 5, 0 –5, –10 V etc.
What would be the rms value of the sine wave having the same peak value?[**Ans.** 31.064 V; 1.35; 1.9315; 42.43 V]

11. Determine the rms value of a triangular wave in which the voltage rises uniformly from 0 to V volts and completes the cycle by falling instantaneously to zero. [**Ans.** $V_{max}/\sqrt{3}$]

12. A sinusoidal alternating voltage of amplitude 100 V is applied across a circuit containing a rectifying device which entirely prevents current flowing in one direction and offers a resistance of 10 Ω to the flow of current in the other direction. A hot wire ammeter is used to measure the current. Find the reading of the instrument. [**Ans.** 5 A]

13. The currents flowing in the three branches of an ac parallel circuit are :
$i_1 = 20 \sin \omega t$; $i_2 = 25 \sin\left(\omega t - \dfrac{\pi}{6}\right)$; $i_3 = 15 \sin\left(\omega t + \dfrac{\pi}{6}\right)$. Determine the total current drawn from the supply.
[**Ans.** $54.87 \sin (\omega t - 0.029\, \pi)$]

14. Three circuits in parallel take currents which can be represented by

$$i_1 = 10 \sin 314\, t;\; i_2 = 7.5 \sin\left(314t - \frac{\pi}{3}\right)$$

and

$$i_3 = 12 \sin\left(314t - \frac{\pi}{4}\right)$$

Sketch a phasor diagram to represent the three currents and their resultant. Express the resultant in the same form as the three individual currents expressed above. What is the rms value and the frequency of the resultant current ?

[Ans. 26.35 sin (314 t – 0.584); 18.63 A; 50 Hz]

15. Three voltages represented by $e_1 = 20 \sin \omega t$; $e_2 = 30 \sin\left(\omega t - \frac{\pi}{4}\right)$ and $e_3 = 40 \cos\left(\omega t + \frac{\pi}{6}\right)$ act together in a circuit.

Find an expression for the resultant voltage. Represent them by appropriate phasors.

[Madras Univ. Electrotechnics 1981; Nagpur Univ. Elec. Circuits 1991]

[Ans. 25.1 sin (ωt + 0.18 π)]

16. Three sinusoidal voltages acting in series are given by —

$$V_1 = 10 \sin 440\, t$$

$$V_2 = 10\sqrt{2}\, \sin (440\, t - 45°)$$

$$V_3 = 20 \cos 440\, t$$

Determine —

(i) an expression for the resultant voltage,

(ii) the frequency and rms. value of the resultant voltage. **[U.P. Technical Univ. Elec. Engineering February 2001]**

[Ans. v_r = 22.36 sin (440 t + 0.1476 π); 70 Hz; 15.81 V]

❑❑❑

Chapter

6 Single Phase Series Circuits

6.1. INTRODUCTION

In a dc circuit the relationship between the applied voltage V and current flowing through the circuit I is a simple one and is given by the expression I = V/R but in an a c circuit this simple relationship does not hold good. Variations in current and applied voltage set up magnetic and electrostatic effects respectively and these must be taken into account with the resistance of the circuit while determining the quantitative relations between current and applied voltage. With comparatively low-voltage, heavy-current circuits magnetic effects may be very large, but electrostatic effects are usually negligible. On the other hand with high-voltage circuits electrostatic effects may be of appreciable magnitude, and magnetic effects are also present.

In this chapter we will discuss how do the magnetic effects due to variations in current and electrostatic effects due to variations in the applied voltage affect the relationship between the applied voltage and current.

6.2. PURELY RESISTIVE CIRCUITS

A purely resistive or a non-inductive circuit is a circuit which has inductance so small that at normal frequency its reactance is negligible as compared to its resistance. Ordinary filament lamps, water resistances etc. are the examples of non-inductive resistances. If the circuit is purely non-inductive, no reactance emf *(i.e.* self-induced or back emf) is set up and whole of the applied voltage is utilised in overcoming the ohmic resistance of the circuit.

Consider an ac circuit containing a non-inductive resistance of R ohms connected across a sinusoidal voltage represented by $v = V_{max} \sin \omega t$, as shown in Fig. 6.1 (*a*).

As already said, when the current flowing through a pure resistance changes, no back emf is set up, therefore, applied voltage has to overcome the ohmic drop of i R only

$$i.e. \quad i R = v$$

$$\text{or} \quad i = \frac{v}{R} = \frac{V_{max}}{R} \sin \omega t$$

$$\text{Current will be maximum when } \omega t = \frac{\pi}{2} \text{ or } \sin \omega t = 1$$

$$\therefore \quad I_{max} = \frac{V_{max}}{R}$$

and instantaneous current may be expressed as

$$i = I_{max} \sin \omega t$$

From the expressions of instantaneous applied voltage and instantaneous current, it is evident that in a pure resistive circuit, the applied voltage and current are in phase with each other, as shown by wave and phasor diagrams in Figs. 6.1 (*b*) and 6.1 (*c*) respectively.

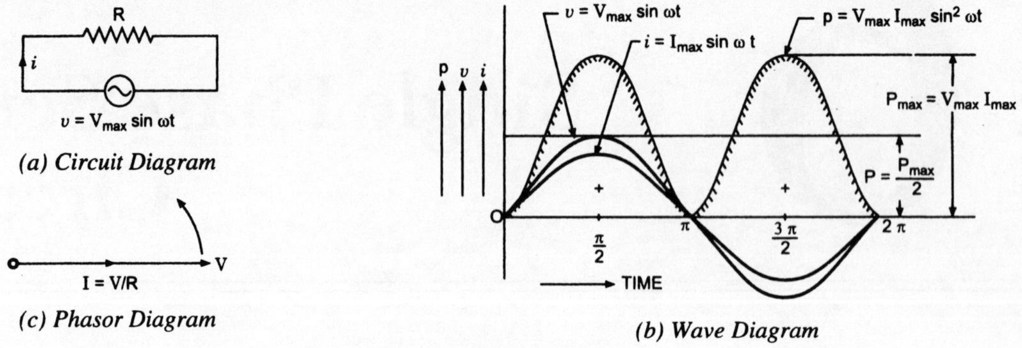

(*a*) *Circuit Diagram*

(*c*) *Phasor Diagram*

(*b*) *Wave Diagram*

Fig. 6.1 *Purely Resistive Circuit*

6.2.1. Power in Purely Resistive Circuit. The instantaneous power delivered to the circuit in question is the product of the instantaneous values of applied voltage and current.

$$i.e.\ p = v\,i = V_{max} \sin \omega t\, I_{max} \sin \omega t = V_{max}\, I_{max} \sin^2 \omega t$$

$$\text{or } p = \frac{V_{max}\, I_{max}}{2}(1 - \cos 2\,\omega\, t) \qquad \text{Since } \sin^2 \omega\, t = \frac{1 - \cos 2\,\omega t}{2}$$

$$= \frac{V_{max}\, I_{max}}{2} - \frac{V_{max}\, I_{max}}{2} \cos 2\,\omega\, t$$

$$\text{Average power, } P = \text{Average of } \frac{V_{max}\, I_{max}}{2} - \textit{average of } \frac{V_{max}\, I_{max}}{2} \cos 2\,\omega\, t$$

Since average of $\frac{V_{max}\, I_{max}}{2}$ cos 2 ω *t* over a complete cycle is zero,

$$P = \frac{V_{max}\, I_{max}}{2} = \frac{V_{max}}{\sqrt{2}} \cdot \frac{I_{max}}{\sqrt{2}} = V\, I \text{ watts}$$

where V and I are the rms values of applied voltage and current respectively.

Thus, for purely resistive circuits, the expression for power is the same as for d c circuits. From the power curve for a purely resistive circuit shown in Fig. 6.1 (*b*) it is evident that power consumed in a pure resistive circuit is not constant, it is fluctuating. However, it is always positive. This is so because the instantaneous values of voltage and current are always either both positive or negative and, therefore, the product is always positive. This means that the voltage source constantly delivers power to the circuit and the circuit consumes it.

Example 6.1

A 250 V (rms), 50 Hz voltage is applied across a circuit consisting of a pure (non-inductive) resistance of 20 Ω. Determine (*i*) the current flowing through the circuit and (*ii*) power absorbed by the circuit. Give the expressions for the voltage and current. Give the wave and phasor diagrams.

Solution: (*i*) Current flowing through the circuit, $I = \frac{V}{R} = \frac{250}{20} = 12.5$ A **Ans.**

(*ii*) Power absorbed, P = VI = 250 × 12.5 = 3,125 W or 3.125 kW **Ans.**

Peak value of applied voltage, $V_{max} = \sqrt{2}\, V_{rms} = \sqrt{2} \times 250 = 353.6$ V

Peak value of current, $I_{max} = \sqrt{2}\, I_{rms} = \sqrt{2} \times 12.5 = 17.68$ A

Expression for voltage, $v = V_{max} \sin 2\,\pi f t$

Assuming that at $t = 0$, $v = 0$ and is increasing in positive direction

$$\text{so } v = 353.6 \sin 2\pi \times 50\,t = 353.6 \sin 314\,t \textbf{ Ans.}$$

$$\text{Expression for current, } i = I_{max} \sin 2\pi f t = 17.68 \sin 314\,t \textbf{ Ans.}$$

Wave and phasor diagrams are similar to those shown in Figs. 6.1 (*b*) and 6.1 (*c*) respectively.

Example 6.2

A 100 ohm resistance is carrying a sinusoidal current given by 3 cos ω *t*.

Determine: (*i*) instantaneous power taken by resistance, (*ii*) average power.

[*U.P. Technical Univ. Electrical Engineering First Semester 2003-04*]

Solution: (*i*) Instantaneous power taken by the resistance, $p = v\,i = i\,R \times i = 100 \times 3 \cos \omega t \times 3 \cos \omega t$

$$= 900 \cos^2 \omega t = 450\,(1 + \cos 2\omega t) \textbf{ Ans.}$$

(*ii*) Average power = 450 watts **Ans.**

6.3. PURELY INDUCTIVE CIRCUITS

An inductive circuit is a coil with or without an iron core having negligible resistance. Practically pure inductance can never be had as the inductive coil has always small resistance. However, a coil of thick copper wire wound on a laminated iron core has negligible resistance and is known as a *choke coil.**

When an alternating voltage is applied to a purely inductive coil, an emf, known as self induced emf, is induced in the coil which opposes the applied voltage. Since coil has no resistance, at every instant applied voltage has to overcome this self-induced emf only.

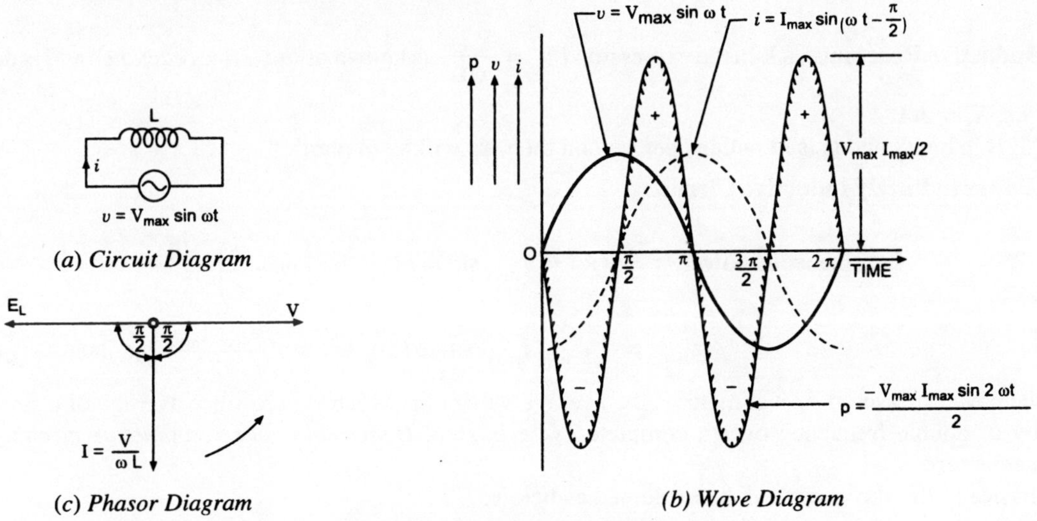

Fig. 6.2 *Purely Inductive Circuit*

$$\text{Let the applied voltage } v = V_{max} \sin \omega t$$

$$\text{and self inductance of coil} = L \text{ henry}$$

$$\text{Self induced emf in the coil, } e_L = -L\frac{di}{dt}$$

Since applied voltage at every instant is equal and opposite to the self induced emf *i.e.* $v = -e_L$

$$\therefore\ V_{max} \sin \omega t = -\left(-L\frac{di}{dt}\right)$$

* Inductors or choke coils are made of iron core because large valued flux densities can be produced in iron cores and so inductances of large values can be had. Air-cored inductors become too much bulky to give an inductance of a given value.

$$\text{or } di = \frac{V_{max}}{L} \sin \omega t \, dt$$

Integrating both sides we get $i = \frac{V_{max}}{L} \int \sin \omega t \, dt = \frac{V_{max}}{\omega L} (-\cos \omega t) + A$

where A is a constant of integration, which is found to be zero from initial conditions

$$i.e. \quad i = \frac{-V_{max}}{\omega L} \cos \omega t = \frac{V_{max}}{\omega L} \sin \left(\omega t - \frac{\pi}{2} \right)$$

Current will be maximum when $\sin \left(\omega t - \frac{\pi}{2} \right) = 1$, hence, maximum value of current,

$$I_{max} = \frac{V_{max}}{\omega L}$$

and instantaneous current may be expressed as

$$i = I_{max} \sin \left(\omega t - \frac{\pi}{2} \right)$$

From the expressions of instantaneous applied voltage and instantaneous current flowing through a purely inductive coil it is observed that the current lags behind the applied voltage by $\frac{\pi}{2}$, as shown in Fig. 6.2 (*b*) by wave diagram and in Fig 6.2 (*c*) by phasor diagram.

6.3.1. Inductive Reactance. ωL in the expression $I_{max} = \frac{V_{max}}{\omega L}$ is known as *inductive reactance* and is denoted by X_L *i.e.* $X_L = \omega L$

If L is in henry and ω is in radians per second then X_L will be in ohms.

6.3.2. Power in Purely Inductive Circuit.

$$\text{Instantaneous power, } p = v \times i = V_{max} \sin \omega t \, I_{max} \sin \left(\omega t - \frac{\pi}{2} \right)$$

$$\text{or } p = -V_{max} I_{max} \sin \omega t \cos \omega t = -\frac{V_{max} I_{max}}{2} \sin 2 \omega t$$

The power measured by wattmeter is the average value of p which is zero since average of a sinusoidal quantity of double frequency over a complete cycle is zero. *Hence, in a purely inductive circuit power absorbed is zero.*

Physically the above fact can be explained as below:

During the second quarter of a cycle the current and the magnetic flux of the coil increases and the coil draws power from the supply source to build up the magnetic field (the power drawn is positive and the energy drawn by the coil from the supply source is represented by the area between the curve p and the time axis). The energy stored in the magnetic field during build up is given as $W_{max} = \frac{1}{2} L I_{max}^2$.

In the next quarter the current decreases. The emf of self induction will, however, tend to oppose its decrease. The coil acts as a generator of electrical energy, returning the stored energy in the magnetic field to the supply source (now the power drawn by the coil is negative and the curve p lies below the time axis).

The chain of events repeats itself during the next half cycles. Thus, a proportion of power is continually exchanged between the field and the inductive circuit and the power consumed by a purely inductive coil is zero.

Example 6.3

The voltage and current through a circuit element are

$$v = 100 \sin (314\,t + 45°) \text{ volts}$$
$$i = 10 \sin (314\,t + 315°) \text{ amperes.}$$

(*i*) Identify the circuit element (*ii*) Find the value (*iii*) Obtain expression for power.

[*U.P. Technical Univ. Electrical Engineering September 2001*]

Solution: $v = 100 (\sin 314\,t + 45°)$ volts

$$i = 10 \sin (314\,t + 315°) = 10 \sin (314\,t + 315° - 360°) = 10 \sin (314\,t - 45°) \text{ amperes}$$

(*i*) From the equation for voltage and current it is observed that the circuit current *i* lags behind the applied voltage *v* by 90°. It means the circuit element is an *inductor*. **Ans.**

(*ii*) Inductive reactance of the inductor, $X_L = \frac{V_{rms}}{I_{rms}} = \frac{100/\sqrt{2}}{10/\sqrt{2}} = 10\ \Omega$ **Ans.**

and $L = \frac{X_L}{\omega} = \frac{10}{314} = 0.0318$ H or 31.8 mH **Ans.**

(*iii*) Expression for power is given as $p = \frac{-V_{max}\, I_{max}}{2} \sin 2\,\omega t$

$$= \frac{-100 \times 10}{2} \sin (2 \times 314\,t) = -500 \sin 628\,t$$ **Ans.**

6.4. PURELY CAPACITIVE CIRCUITS

When a dc voltage is impressed across thc plates of a perfect condenser, it will become charged to full voltage almost instantaneously. The charging current will flow only during the period of "build up" and will cease to flow as soon as the capacitor has attained the steady voltage of the source. This implies that for a direct current, a capacitor is a break in the circuit or an infinitely high resistance.

In Fig. 6.3 a sinusoidal voltage is applied to a capacitor. During the first quarter-cycle, the applied voltage increases to the peak value, and the capacitor is charged to that value. The current is maximum in the beginning of the cycle and becomes zero at the maximum value of the applied voltage, so there is a phase difference of 90° between the applied voltage and current. During the first quarter-cycle the current flows in the normal direction through the circuit; hence the current is positive.

In the second quarter-cycle, the voltage applied across the capacitor falls, the capacitor loses its charge, and current flows through it against the applied voltage because the capacitor discharges into the circuit. Thus, the current is negative during the second quarter-cycle and attains a maximum value when the applied voltage is zero.

The third and fourth quarter-cycles repeat the events of the first and second, respectively, with the difference that the polarity of the applied voltage is reversed, and there are corresponding current changes. In other words, an alternating current flows in the circuit because of the charging and discharging of the capacitor. As illustrated in Figs. 6.3 (*b*) and (*c*) the current begins its cycle 90 degrees ahead of the voltage, so the current in a capacitor leads the applied voltage by 90 degrees – the opposite of the inductance current-voltage relationship.

Let an alternating voltage represented by $v = V_{max} \sin \omega\, t$ be applied across a capacitor of capacitance C farads. The expression for instantaneous charge is given as

$$q = C\, V_{max} \sin \omega t$$

Since the capacitor current is equal to the rate of change of charge, the capacitor current may be obtained by differentiating the above equation

$$i = \frac{dq}{dt} = \frac{d}{dt} [C\, V_{max} \sin \omega\, t] = \omega\, C\, V_{max} \cos \omega\, t = \frac{V_{max}}{1/\omega C} \sin \left(\omega t + \frac{\pi}{2}\right)$$

Current is maximum when $t = 0$

$$\therefore \quad I_{max} = \frac{V_{max}}{1/\omega C}$$

Substituting $\frac{V_{max}}{1/\omega C} = I_{max}$ in the above equation for instantaneous current, we get

$$i = I_{max} \sin\left(\omega t + \frac{\pi}{2}\right)$$

From the equations of applied voltage and instantaneous current flowing through capacitance, it is observed that the current leads the applied voltage by $\frac{\pi}{2}$, as shown in Figs. 6.3 (*b*) and (*c*) by wave and phasor diagrams respectively.

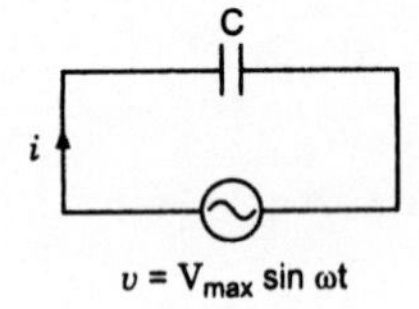

(a) Circuit Diagram

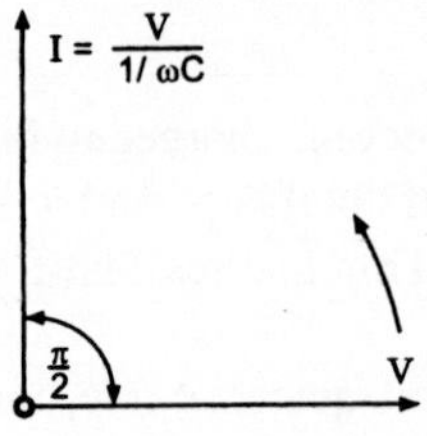

(c) Phasor Diagram

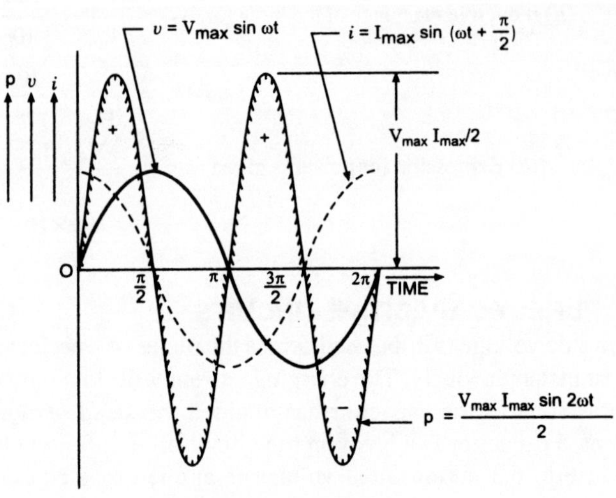

(b) Wave Diagram

Fig. 6.3 *Purely Capacitive Circuit*

6.4.1. Capacitive Reactance. $\frac{1}{\omega C}$ in the expression $I_{max} = \frac{V_{max}}{1/\omega C}$ is known as *capacitive reactance* and is denoted by X_C *i.e.* $X_C = \frac{1}{\omega C}$

If C is in farads and ω is in radians/s, then X_C will be in ohms.

6.4.2. Power in Purely Capacitive Circuit.

$$\text{Instantaneous power, } p = v\,i = V_{max} \sin \omega t.\ I_{max} \sin\left(\omega t + \frac{\pi}{2}\right) = V_{max} I_{max} \sin \omega t \cos \omega t$$

$$= \frac{V_{max} I_{max}}{2} \sin 2\omega t$$

$$\text{Average power, } P = \frac{V_{max} I_{max}}{2} \times \text{average of } \sin 2\,\omega t \text{ over a complete cycle} = 0.$$

Hence, power absorbed in a purely capacitive circuit is zero. The same is shown graphically in Fig. 6.3 (*b*). The energy taken from the supply circuit is stored in the capacitor during the first quarter-cycle and returned during the next.

The energy stored by a capacitor at maximum voltage across its plates is given by the expression

$$W_C = \frac{1}{2} C V^2_{max}$$

This can be realized when it is recalled that no heat is produced and no work is done while current is flowing through a capacitor. As a matter of fact, in commercial capacitors, there is a slight energy loss in the dielectric in addition to a minute $I^2 R$ loss due to flow of current over the plates having definite ohmic resistance. The power curve is a sine wave of double the supply frequency. Although it raises the power factor from zero to 0.002 or even a little more, but for ordinary purposes the power factor is taken to be zero. Obviously the phase angle due to dielectric and ohmic losses decreases slightly.

Example 6.4

A 50-Hz voltage of 230 V effective value is impressed on a capacitance of 26.5 μF. (*a*) Write the time equations for the voltage and the resulting current. Let the zero axis of the voltage wave be at *t* = 0. (*b*) Show the voltage and current on a time diagram. (*c*) Show the voltage and current on a phasor diagram. (*d*) Find the maximum energy stored in the capacitor. Find the relative heating effects of two current waves of equal peak value, the one sinusoidal and the other rectangular in waveform. [*Allahabad Univ. Electrical Technology 1991*]

Solution: (*a*) The time equation for the applied voltage is given by

$$v(t) = V_{max} \sin(2\pi f t + \theta) = 230\sqrt{2} \sin(2\pi \times 50t + 0°) = 325 \sin 314t \quad \textbf{Ans.}$$

$$\because V_{max} = 230\sqrt{2}\ V; f = 50 \text{ Hz and } \theta = 0°$$

$$\text{Capacitive reactance, } X_C = \frac{1}{2\pi f C} = \frac{1}{2\pi \times 50 \times 26.5 \times 10^{-6}}\ \Omega = 120\ \Omega$$

Maximum value of current flowing through the capacitor,

$$I_{max} = \frac{V_{max}}{X_C} = \frac{325}{120} = 2.71 \text{ A}$$

Phase angle $\phi = 90°$ (lead)

So the time equation for the resulting current is given by

$$i(t) = 2.71 \sin(314t + 90°) = 2.71 \cos 314t \quad \textbf{Ans.}$$

(*b*) Voltage and current on a time diagram are shown in Fig. 6.3(*b*).

(*c*) Voltage and current on a phasor diagram are shown in Fig. 6.3(*c*).

(*d*) Maximum energy stored in the capacitor,

$$W_C = \frac{1}{2} C (V_{max})^2 = \frac{1}{2} \times 26.5 \times 10^{-6} \times \left(230\sqrt{2}\right)^2 = 1.4 \text{ J} \quad \textbf{Ans.}$$

For equal peak value, say I_{max},

RMS value of current in case of rectangular waveform, $I_{rms1} = I_{max}$

RMS value of current in case of sinusoidal waveform, $I_{rms2} = \frac{I_{max}}{\sqrt{2}}$

The ratio of heating effects of two current waves is given by

$$\frac{H_{rectangular}}{H_{sinusoidal}} = \frac{(I_{rms_1})^2 \times R}{(I_{rms_2})^2 \times R} = \frac{I_{max}^2\ R}{\left(\frac{I_{max}}{\sqrt{2}}\right)^2 R} = 2 \quad \textbf{Ans.}$$

6.5. RESISTANCE—INDUCTANCE (R-L) SERIES CIRCUIT

This is the case most generally met with in practice, nearly all circuits contain both resistance and inductance.

Consider an ac circuit consisting of resistance of R ohms and inductance of L henrys connected in series, as shown in Fig. 6.4 (*a*).

Let the supply frequency be *f* and current flowing through the circuit be of I amperes (rms value).

Now voltage drop across resistance, $V_R = I R$ in phase with the current.

Voltage drop across inductance, $V_L = I X_L = I\omega L$ leading I by $\pi/2$ radians, as shown in Fig 6.4 (*c*)

The applied voltage, being equal to the phasor sum of V_R and V_L, will be given by the diagonal of the parallelogram.

$$\text{Applied voltage, } V = \sqrt{(V_R)^2 + (V_L)^2} = \sqrt{(I\,R)^2 + (I\,X_L)^2} = I\sqrt{R^2 + X_L^2} = I\,Z$$

where $X_L = \omega L = 2\,\pi f L$

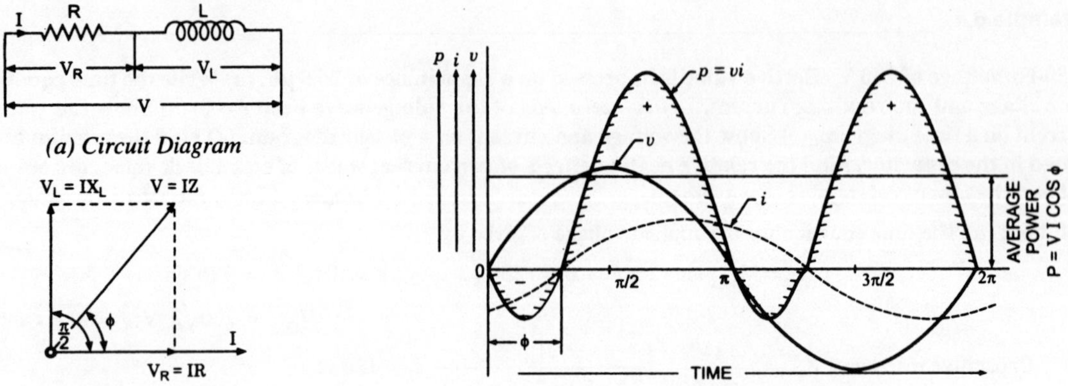

(a) *Circuit Diagram*

(c) *Phasor Diagram*

(b) *Wave Diagram*

Fig. 6.4 *Resistance and Inductance in Series*

Quantity $\sqrt{R^2 + X_L^2}$ is known as *impedance*, denoted by Z and is expressed in ohms.

From phasor diagram it is also evident that the current lags behind the applied voltage V by angle Φ, which is given by

$$\text{Tan}\,\Phi = \frac{V_L}{V_R} = \frac{I\,X_L}{I\,R} = \frac{X_L}{R}$$

$$\text{or} \quad \Phi = \text{Tan}^{-1}\frac{X_L}{R}$$

Since X_L and R are known, the value of phase angle Φ can be computed. If the applied voltage $v = V_{max} \sin \omega t$, then expression for the circuit current will be

$$i = I_{max} \sin(\omega t - \Phi)$$

where $I_{max} = \frac{V_{max}}{Z}$ and $\Phi = \text{Tan}^{-1}\frac{X_L}{R}$.

6.5.1. Impedance Triangle. If a triangle ABC is drawn so that $AB = \frac{V_R}{I} = R$, $BC = \frac{V_L}{I} = X_L$ and $AC = \frac{V}{I} = Z$, it is a triangle similar to the voltage triangle [Fig. 6.5 (*a*)]. Such a triangle is called an *impedance triangle*, which is most useful in letting one see at a glance how R, X, and Z are related to each other. The angle between Z and R sides of the impedance triangle is known as *phase angle of the circuit* and cos of this angle is known as *power factor of the circuit.*

$$\text{Power factor} = \text{Cos}\,\Phi = \frac{R}{Z}$$

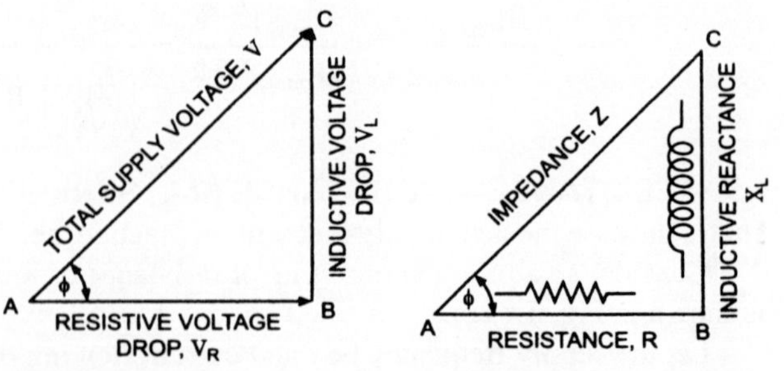

(a) *Voltage Triangle*

(b) *Impedance Triangle*

Fig. 6.5

6.5.2. Power in Resistance—Inductance (R-L) Circuit.

Instantaneous power, $p = v\,i = V_{max} \sin \omega t \times I_{max} \sin (\omega t - \Phi)$

$$= \frac{1}{2} V_{max} I_{max} [2 \sin \omega t \,.\, \sin (\omega t - \Phi)]$$

$$= \frac{1}{2} V_{max} I_{max} [\cos \Phi - \cos (2 \omega t - \Phi)]$$

$$= \frac{1}{2} V_{max} I_{max} \cos \Phi - \frac{1}{2} V_{max} I_{max} \cos (2 \omega t - \Phi)$$

Since average value of pulsating component $\frac{1}{2} V_{max} I_{max} \cos (2 \omega t - \Phi)$ over a complete cycle is zero, average power of the circuit,

$$P = \text{Constant component, } \frac{1}{2} V_{max} I_{max} \cos \Phi = \frac{V_{max}}{\sqrt{2}} . \frac{I_{max}}{\sqrt{2}} \cos \Phi = V I \cos \Phi$$

where V and I are the rms values of voltage and current and Φ is the phase angle between applied voltage V and circuit current I.

Alternatively power consumed by the circuit,

P = Power consumed by resistance   ∵ Power consumed by inductance is zero

$$= I^2 R = I (I R) = \frac{V}{Z} . I R = V I \frac{R}{Z} = V I \cos \Phi$$

Since from impedance triangle $\cos \Phi = \frac{R}{Z}$

So the power in an a c circuit is given by the product of rms values of current and voltage and cosine of the phase angle between voltage and current. Cosine of the phase angle between the voltage and current, $\cos \Phi$ is known as the *power factor* of the circuit, and is equal to $\frac{R}{Z}$ which is obvious from impedance triangle [Fig. 6.5 (*b*)]

6.6. APPARENT POWER, TRUE POWER, REACTIVE POWER AND POWER FACTOR

The product of rms values of current and voltage, VI is called the *apparent power* and is measured in volt-amperes or kilovolt amperes (kVA).

The *true power* in an a c circuit is obtained by multiplying the apparent power by the power factor and is expressed in watts or kilowatts (kW).

The product of apparent power, VI and the sine of the angle between voltage and current, sin Φ is called the *reactive power.* This is also known as *wattless power* and is expressed in reactive volt-amperes or kilovolt amperes reactive (kVAR).

i.e Apparent power, $S = VI$ volt-amperes or $\frac{VI}{1,000}$ kVA

True power, $P = VI \cos \Phi$ watts or $\frac{VI \cos \Phi}{1,000}$ kW

Reactive power, $Q = VI \sin \Phi$ VAR or $\frac{VI \sin \Phi}{1,000}$ kVAR

and $\text{kVA} = \sqrt{(\text{kW})^2 + (\text{kVAR})^2}$

The above relations can easily be followed by referring to the power diagram shown in Fig. 6.6 (*a*).

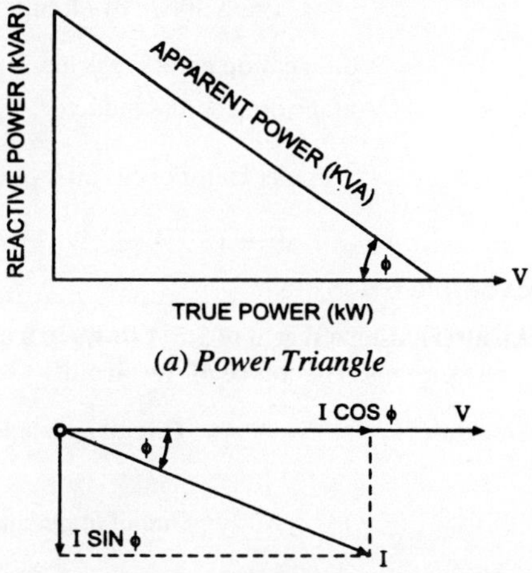

(*a*) *Power Triangle*

(*b*)

Fig. 6.6

Power factor may be defined as

(*i*) cosine of the phase angle between voltage and current or

(*ii*) the ratio of the resistance to impedance or

(*iii*) the ratio of true power to apparent power.

The power factor can never be greater than unity. The power factor is expressed either as fraction or as a percentage. It is usual practice to attach the word 'lagging' or ' leading' with the numerical value of power factor to signify whether the current lags behind or leads the voltage.

6.6.1. Active Component of Current. The current component which is in phase with circuit voltage (*i.e.* I cos Φ) and contributes to active or true power of the circuit is called the *active* (*wattful* or *in-phase*) *component* of current.

6.6.2. Reactive Component of Current. The current component which is in quadrature (or 90° out of phase) to circuit voltage (*i.e.* I sin Φ) and contributes to reactive power of the circuit, is called the *reactive* (or *wattless*) *component* of current.

6.6.3. Q-Factor Coil. Reciprocal of power factor is known as Q-factor of the coil. It is also called the quality factor or figure of merit of a coil.

$$\text{Mathematically Q-factor} = \frac{1}{\text{Power factor}} = \frac{1}{\text{Cos }\Phi} = \frac{Z}{R}$$

If R is very small in comparison to inductive reactance X_L, the

$$\text{Q-factor} = \frac{X_L}{R} = \frac{\omega L}{R}$$

$$\text{Also Q} = 2\pi \times \frac{\text{Maximum energy stored}}{\text{Energy dissipated per cycle}}$$

Example 6.5

A 4 ohm resistor is connected to a 10 mH inductor across a 100 V, 50 Hz voltage source. Find input current, voltage drops across resistor and inductor, power factor of the circuit and the real power consumed in the circuit.

[*R.G.P.V. Bhopal Basic Electrical Engineering December 2003*]

Solution:

$$\text{Circuit resistance, R} = 4\ \Omega$$

$$\text{Circuit inductance, L} = 10\text{ mH} = 10 \times 10^{-3}\text{ H}$$

$$\text{Inductive reactance of circuit, } X_L = 2\pi f L = 2\pi \times 50 \times 10 \times 10^{-3} = 3.142\ \Omega$$

$$\text{Circuit impedance, Z} = \sqrt{R^2 + X_L^2} = \sqrt{4^2 + (3.142)^2} = 5.086\ \Omega$$

$$\text{Current drawn by circuit, I} = \frac{\text{Supply voltage}}{Z} = \frac{100}{5.086} = 19.66\text{ A}\ \textbf{Ans.}$$

$$\text{Voltage drop across resistor, } V_R = IR = 19.66 \times 4 = 78.64\text{ V}\ \textbf{Ans.}$$

$$\text{Voltage drop across inductor, } V_L = I X_L = 19.66 \times 3.142 = 61.77\text{ V}\ \textbf{Ans.}$$

$$\text{Power factor of circuit, } \cos\phi = \frac{R}{Z} = \frac{4}{5.086} = 0.786\text{ (lagging)}\ \textbf{Ans.}$$

$$\text{Real power consumed, P} = VI\cos\phi = 100 \times 19.66 \times 0.786 = 1{,}546\text{ W}\ \textbf{Ans.}$$

Example 6.6

An alternating current of 1.5 A flows in a circuit when applied voltage is 300 V. The power consumed is 225 W. Find resistance and reactance of the circuit. [*U.P. Technical Univ. Electrical Engineering First Semester 2004-05*]

Solution :

$$\text{Circuit resistance, R} = \frac{\text{Power consumed}}{I^2} = \frac{225}{(1.5)^2} = 100\ \Omega\ \textbf{Ans.}$$

$$\text{Circuit impedance, Z} = \frac{V}{I} = \frac{300}{1.5} = 200\ \Omega$$

$$\text{Circuit reactance, X} = \sqrt{Z^2 - R^2} = \sqrt{(200)^2 - (100)^2} = 173.2\ \Omega\ \textbf{Ans.}$$

Example 6.7

The voltage and current through a circuit element are

$$v = 50 \sin (314\,t + 55°) \text{ volts}$$
$$i = 10 \sin (314\,t + 325°) \text{ amperes}$$

Find the value of power drawn by the element.

[*U.P. Technical Univ. Electrical Engineering Second Semester 2006-07*]

Solution: Phase angle between voltage and current = 325° – 55° = 270° *i.e.* current leads the voltage by 270° or voltage leads current by 90° *i.e.* current lags behind the voltage by 90°

$$\text{Power drawn by the element} = VI \cos \phi = \frac{V_{max}}{\sqrt{2}} \times \frac{I_{max}}{\sqrt{2}} \cos \phi = \frac{50}{\sqrt{2}} \times \frac{10}{\sqrt{2}} \times \cos 90° = 0 \textbf{ Ans.}$$

Example 6.8

A coil has resistance of 10 Ω and draws a current of 5 A when connected across 100 V, 50 Hz source. Determine the reactive power of the circuit. [*Pb. Technical Univ. Electrical Engineering May 2001*]

Solution:

$$\text{Coil resistance, } R = 10\ \Omega$$
$$\text{Coil impedance, } Z = \frac{V}{I} = \frac{100}{5} = 20\ \Omega$$
$$\text{Power factor of the circuit, } \cos \phi = \frac{R}{Z} = \frac{10}{20} = 0.5 \text{ (lagging)}$$
$$\text{Sin } \phi = \sqrt{1 - \cos^2 \phi} = \sqrt{1 - (0.5)^2} = 0.866$$
$$\text{Reactive power} = VI \sin \phi = 100 \times 5 \times 0.866 = 433 \text{ VAR } \textbf{Ans.}$$

Example 6.9

A resistance and an inductance are connected in series across a voltage $v = 283 \sin 314\,t$. The current expression is found to be $4 \sin (314t - \pi/4)$. Find the values of resistance, inductance and power factor.

[*U.P. Technical Univ. Electrical Engineering January 2003*]

Solution:

$$\text{Circuit voltage, } v = 283 \sin 314\,t \text{ volts}$$
$$\text{Circuit current, } i = 4 \sin \left(314\,t - \frac{\pi}{4}\right)$$

From the equations of circuit voltage and circuit current it is seen that current I lags behind the applied voltage V by $\frac{\pi}{4}$ radian or 45°

$$\text{So phase angle, } \phi = 45° \text{ (lagging)}$$
$$\text{Power factor of the circuit} = \text{Cos } \phi = \text{Cos } 45° = 0.707 \text{ (lagging) } \textbf{Ans.}$$
$$\text{Circuit impedance, } Z = \frac{V_{max}}{I_{max}} = \frac{283}{4} = 70.75\ \Omega$$
$$\text{Circuit resistance, } R = Z \cos \phi = 70.75 \times 0.707 = 50\ \Omega \textbf{ Ans.}$$
$$\text{Circuit reactance, } X_L = Z \sin \phi = 70.75 \times \sin 45° = 70.75 \times 0.707 = 50\ \Omega$$
$$\text{Frequency, } f = \frac{314}{2\pi} = 50 \text{ Hz}$$
$$\text{Inductance, } L = \frac{X_L}{2\pi f} = \frac{50}{2\pi \times 50} = 0.1592 \text{ H } \textbf{Ans.}$$

Example 6.10

A series circuit containing a resistance R and an inductance L is connected to a 200 V, 50 Hz source V_s. The current drawn is I amperes rms. Taking the source voltage as reference represented by $v_S = 200\sqrt{2} \sin (2\pi \times 50\,t)$, write the expressions for :

(*i*) the current through the circuit, $i(t)$
(*ii*) voltage across the resistance, $v_R(t)$

(iii) voltage across the inductance, $v_L(t)$
What is the peak value of the instantaneous power drawn?

[*R.G.P.V. Bhopal Basic Electrical Engineering December 2004*]

Solution: Source voltage, $v_s(t) = 200\sqrt{2}\sin(2\pi \times 50\, t) = 200\sqrt{2}\sin(314\, t)$

(i) Current through the circuit, $i(t) = \frac{v_s}{Z}$ lagging behind v_s by angle ϕ where $Z = \sqrt{R^2 + (\omega L)^2}$ and $\phi = \text{Tan}^{-1}\frac{\omega L}{R}$

(ii) Hence $i(t) = \frac{200\sqrt{2}}{Z}\sin(314t - \phi)$ **Ans.**

Voltage across resistance, $v_R(t) = R\, i(t)$ in phase with $i(t) = 200\sqrt{2}\frac{R}{Z}\sin(314t - \phi)$ **Ans.**

(iii) Voltage across inductance, $v_L(t) = \omega L\, i(t)$ leading $i(t)$ by $\frac{\pi}{2}$ radians $= \frac{200\sqrt{2}\,\omega L}{Z}\sin\left(314t - \phi + \frac{\pi}{2}\right)$ **Ans.**

Instantaneous power is given as $p = v_s(t)\, i(t) = 200\sqrt{2}\sin(314\, t) \times \frac{200\sqrt{2}}{Z}\sin(314t - \phi)$

$$= \frac{80{,}000}{Z}\sin(314t)\sin(314t - \phi)$$

$$= \frac{40{,}000}{Z}\cos\phi - \frac{40{,}000}{Z}\cos(628t - \phi)$$

Hence peak value of instantaneous power is $\frac{40{,}000}{Z}$ volt-amperes **Ans.**

Example 6.11

An emf, $e_0 = 141.4\sin(377\, t + 30°)$ is impressed on the impedance coil having a resistance of 4 Ω and an inductive reactance of 1.25 Ω, measured at 25 Hz. What is the equation of the current ? Sketch the waves for i, e_R, R_L and e_0.

Solution: Applied emf, $e_0 = 141.4\sin(377\, t + 30°)$

$$\text{Supply frequency}, f = \frac{\text{Coefficient of time, } t}{2\pi} = \frac{377}{2\pi} = 60 \text{ Hz}$$

Resistance of coil, R = 4 Ω

Inductive reactance of coil at supply frequency of 60 Hz,

$$X_L = \text{Coil inductive reactance at 25 Hz} \times \frac{60}{25} = 1.25 \times \frac{60}{25} = 3\ \Omega \qquad \because X_L \propto f$$

$$\text{Coil impedance, } Z = \sqrt{R^2 + X_L^2} = \sqrt{4^2 + 3^2} = 5\ \Omega$$

$$\text{Maximum value of coil current, } I_{max} = \frac{E_{0\,max}}{Z} = \frac{141.4}{5} = 28.28 \text{ A}$$

$$\text{Phase angle, } \phi = \text{Tan}^{-1}\frac{X_L}{R} = \text{Tan}^{-1}\frac{3}{4} = 36.87°$$

It means that the circuit current lags behind the impressed voltage by 36.87°. Hence the circuit current is given by the equation

$$i = I_{max}\sin(377\, t + 30° - \phi) = 28.28\sin(377\, t + 30° - 36.87°)$$

$$= 28.28\sin(377\, t - 6.87°) \textbf{ Ans.}$$

Resistive drop e_R, being in phase with circuit current i, is given by the equation

$$e_R = i\,R = 4 \times 28.28\sin(377\, t - 6.87°) = 113.12\sin(377\, t - 6.87°)$$

The inductive voltage, that leads the circuit current by 90°, is given by the equation

$$e_L = i\,X_L = 3 \times 28.28\sin(377\, t - 6.87° + 90°) = 84.84\sin(377\, t + 83.13°)$$

The waves for i, e_R, e_L and e_0 are shown in Fig. 6.7.

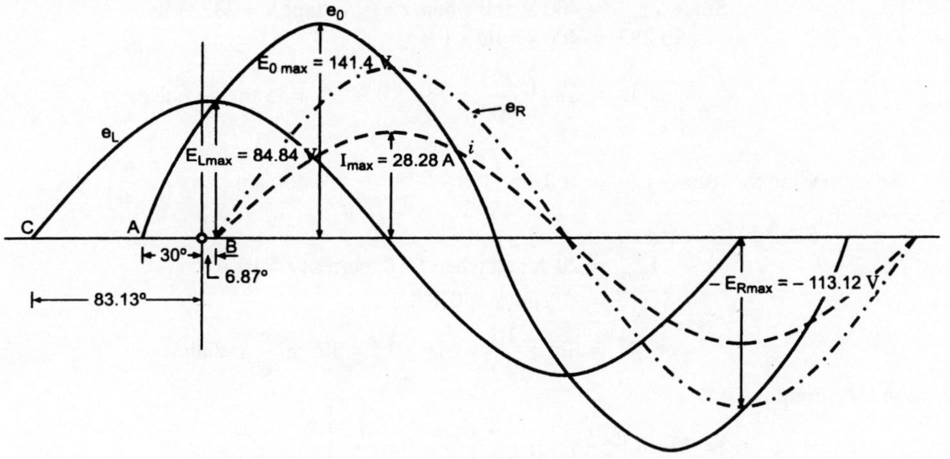

Fig. 6.7

Example 6.12

In a particular circuit a voltage of 10 V at 25 Hz produces 100 mA, while the same voltage at 75 Hz produces 60 mA. Draw the circuit diagram and insert values of the constants. At what frequency will the value of the impedance be twice that at 25 Hz ? *[R.G.P.V. Bhopal Basic Electrical Engineering June 2003]*

Solution: Since the current at high frequency decreases with the increase in frequency, the circuit is inductive.

Let the circuit parameters be resistance of R ohms and inductance L Henry.

At 25 Hz supply frequency circuit impedance, $\mathbf{Z}_1 = R + j2\pi f L = R + j\,50\pi L$

$$\text{and current } \mathbf{I}_1 = \frac{\mathbf{V}}{\mathbf{Z}_1} = \frac{10}{\sqrt{R^2 + (50\,\pi L)^2}} = 0.1 \text{ A} \qquad ...(i)$$

At 75 Hz supply frequency circuit impedance, $\mathbf{Z}_2 = R + j150\pi L$

$$\text{and current } \mathbf{I}_2 = \frac{\mathbf{V}}{\mathbf{Z}_2} = \frac{10}{\sqrt{R^2 + (150\,\pi L)^2}} = 0.06 \text{ A} \qquad ...(ii)$$

Solving Eqs. (*i*) and (*ii*) we have R = 88.2 Ω **Ans.**

and L = 0.3 H **Ans.**

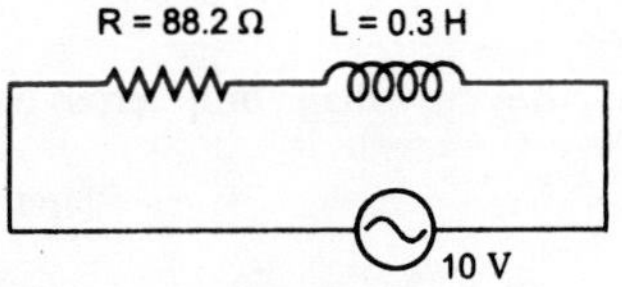

Fig. 6.8

Let the impedance at a frequency of f' be double of that at 25 Hz

$$\text{i.e. } \sqrt{R^2 + (2\pi f' L)^2} = 2\sqrt{R^2 + (2\pi \times 25 \times L)^2}$$

$$\text{or } R^2 + 4\pi^2 f'^2 L^2 = 4R^2 + 10{,}000\,\pi^2 L^2$$

$$\text{or } f' = \sqrt{\frac{3R^2 + 10{,}000\,\pi^2 L^2}{4\pi^2 L^2}}$$

$$= \sqrt{\frac{3 \times (88.2)^2 + 10{,}000\,\pi^2 \times (0.3)^2}{4\pi^2 (0.3)^2}} = 95 \text{ Hz} \quad \textbf{Ans.}$$

Example 6.13

The maximum values of the alternating voltage and current are 400 V and 20 A respectively in a circuit connected to 50 Hz supply and these quantities are sinusoidal. The instantaneous values of voltage and current are 283 V and 10 A respectively at *t* = 0 both increasing positively. (*i*) Write down the expressions for voltage and current at time *t*. (*ii*) Determine the power consumed in the circuit. *[Pune Univ. Elec. Engineering 1985]*

Solution: In general the expression for a sinusoidal voltage is

$$v = V_{max} \sin(\omega t + \Phi)$$

where Φ is the phase difference with respect to the point corresponding to $t = 0$

$$\text{Since } V_{max} = 400 \text{ V and when } t = 0, \text{ voltage v} = 283 \text{ volts}$$
$$\text{So } 283 = 400 \sin(\omega \times 0 + \Phi)$$
$$\text{or } \Phi = \text{Sin}^{-1}\frac{283}{400} = \text{Sin}^{-1} 0.7075 = 45^\circ \text{ or } \frac{\pi}{4} \text{ radian}$$

So expression for voltage is $v = 400 \sin\left(2\pi \times 50\,t + \frac{\pi}{4}\right) = 400 \sin\left(314\,t + \frac{\pi}{4}\right)$ **Ans.**

Similarly for current,

$$I_{max} = 20 \text{ A and when } t = 0, \text{ current } i = 10 \text{ A}$$
$$\text{So } 10 = 20 \sin(\omega \times 0 + \Phi')$$
$$\text{or } \Phi' = \text{Sin}^{-1}\frac{10}{20} = \text{Sin}^{-1} 0.5 = 30^\circ \text{ or } \frac{\pi}{6} \text{ radian}$$

So expression for current is

$$i = I_{max} \sin(2\pi f t + \Phi') = 20 \sin\left(2\pi \times 50t + \frac{\pi}{6}\right)$$
$$= 20 \sin\left(314t + \frac{\pi}{6}\right) \text{ **Ans.**}$$

Power consumed in the circuit,

$$P = VI\cos\Phi = \frac{V_{max}}{\sqrt{2}} \cdot \frac{I_{max}}{\sqrt{2}} \cos\left(\frac{\pi}{4} - \frac{\pi}{6}\right)$$
$$= \frac{400}{\sqrt{2}} \times \frac{20}{\sqrt{2}} \cos\frac{\pi}{12}$$
$$= 3{,}863.7 \text{ W or } 3.8637 \text{ kW}$$ **Ans.**

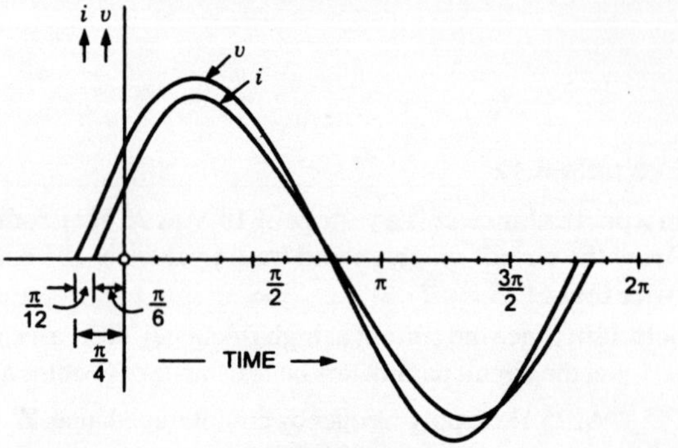

Fig. 6.9

Example 6.14

A non-inductive resistance of 10 Ω is connected in series with an inductive coil across 200 V, 50 Hz ac supply. The current drawn by the series combination is 10 A. The resistance of the coil is 2 Ω. Determine (*i*) inductance of the coil (*ii*) power factor (*iii*) voltage across the coil. [*U.P. Technical Univ. Electrical Engineering First Semester 2005-06*]

Solution: Total resistance of the circuit, R = Non-inductive resistance + resistance of coil = 10 + 2 = 12 Ω

Voltage drop across the resistance of the whole circuit,

$$V_R = IR = 10 \times 12 = 120 \text{ V}$$

Let the voltage drop across the inductance of the coil be V_L volts.

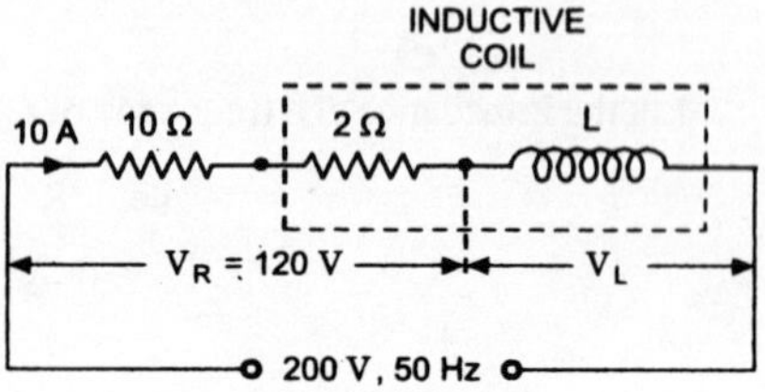

Fig. 6.10

$$\text{Now } \sqrt{V_R^2 + V_L^2} = \text{Supply voltage, } V_S$$
$$\text{or } V_L = \sqrt{V_S^2 - V_R^2}$$
$$= \sqrt{200^2 - 120^2} = 160 \text{ V}$$

$$\text{Inductive reactance of the coil, } X_L = \frac{V_L}{I} = \frac{160}{10} = 16\ \Omega$$

$$(i) \text{ Inductance of coil, } L = \frac{X_L}{2\pi f} = \frac{16}{2\pi \times 50} = 0.051 \text{ H or } 51 \text{ mH}$$ **Ans.**

$$(ii) \text{ Power factor of the coil} = \text{Cos}\left(\text{Tan}^{-1}\frac{16}{2}\right) = 0.124 \text{ (lagging)}$$ **Ans.**

$$\text{Power factor of the circuit} = \frac{R}{Z} = \frac{12}{\sqrt{12^2 + 16^2}} = 0.6 \text{ (lagging)}$$ **Ans.**

$$(iii) \text{ Voltage across the coil, } V_C = I\sqrt{2^2 + 16^2} = 10 \times \sqrt{260} = 161.245 \text{ V}$$ **Ans.**

Example 6.15

A coil is connected in series with a non-inductive resistance of 30 Ω across 240 V, 50 Hz supply. The reading of a voltmeter across the coil is 180 V and across the resistance is 130 V. Calculate (*a*) power consumed by the coil, and (*b*) *pf* of the whole circuit. *[Pb Technical Univ. Basic Electrical and Electronics Engineering Second Semester 2004-05]*

Solution: Circuit current, $I = \dfrac{\text{Voltage across non-inductive resistance}}{\text{Magnitude of non-inductive resistance}} = \dfrac{V_R}{R} = \dfrac{130}{30} = 4.333$ A

Impedance of choke coil,

$$Z = \frac{\text{Voltage across choke coil}}{\text{Circuit current, I}} = \frac{180}{4.333} = 41.54\ \Omega$$

From Δ OAB shown in Fig. 6.11 (*b*)

$$OB^2 = OA^2 + AB^2 + 2\,OA \times AB \times \cos\theta$$

$$\text{or } \cos\theta = \frac{OB^2 - OA^2 - AB^2}{2 \times OA \times AB}$$

$$= \frac{240^2 - 130^2 - 180^2}{2 \times 130 \times 180} = 0.177735$$

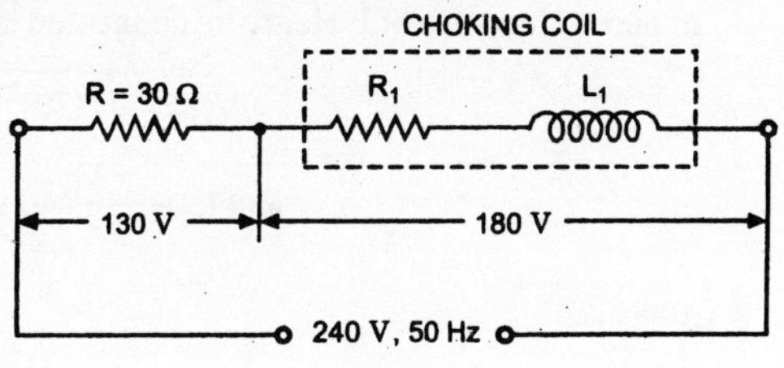

(*a*)

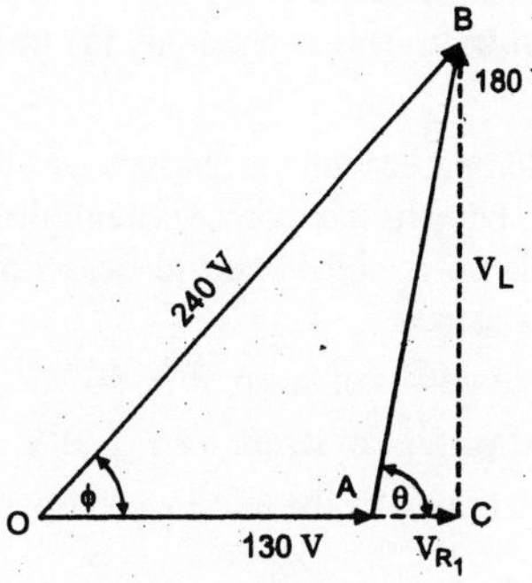

(*b*)

Fig. 6.11

Resistance of coil, $R_1 = Z\cos\theta = 41.54 \times 0.17735 = 7.367\ \Omega$

Reactance of coil $X_L = \sqrt{Z^2 - R_1^2} = \sqrt{(41.54)^2 - (7.367)^2}$

$= 40.88\ \Omega.$

(*a*) Power consumed by the coil

$= I^2R_1 = (4.333)^2 \times 7.367 = 138.34$ W **Ans.**

(b) Resistance of whole circuit

$= R + R_1 = 30 + 7.367 = 37.367\ \Omega$

Reactance of whole circuit = Reactance of coil = 40.88 Ω

Impedance of whole circuit = $\sqrt{(37.367)^2 + (40.88)^2} = 55.38\ \Omega$

Power factor of the whole circuit, $\cos\phi = \dfrac{37.367}{55.38} = 0.6747$ (lagging) **Ans.**

Example 6.16

A 500 W discharge lamp takes a current of 4 A at unity power factor. Calculate the inductance of the choke coil required to enable the lamp to work on a 250 V, 50 Hz ac mains. *[G.G.S.I.P. Delhi Univ. Electrical Science May-June 2006]*

Solution: Voltage rating of lamp, $V = \dfrac{P}{I} = \dfrac{500}{4} = 125$ V

Let the inductance of choke coil, required to enable the lamp to work on 250 V, 50 Hz mains, be L Henry.

Voltage drop across choke coil (neglecting its resistance),

$$V_L = I \times X_L = I \times 2\pi f L = 4 \times 2\pi \times 50 \times L = 400\,\pi L \text{ volts}$$

Since voltage drop across the inductance leads the current by 90° or $\dfrac{\pi}{2}$ radians while that across the lamp will be in phase with it, so

$$\sqrt{(125)^2 + (400\pi L)^2} = 250$$

$$\text{or } 400\,\pi\,L = \sqrt{(250)^2 - (125)^2} = 216.506 \text{ V } \textbf{Ans.}$$

$$L = \frac{216.506}{400\pi} = 0.1723 \text{ H } \textbf{Ans.}$$

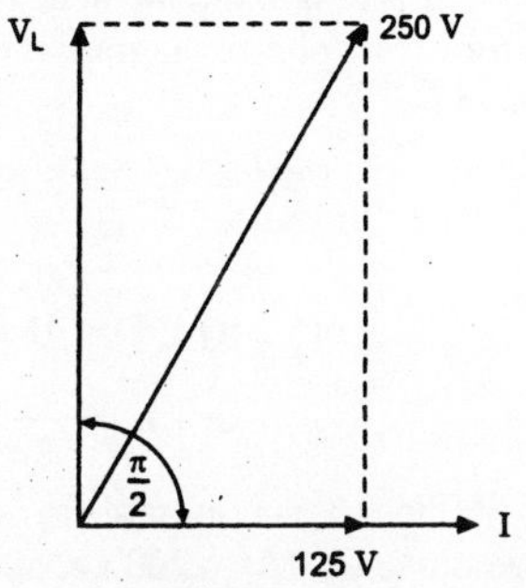

Fig. 6.12

Example 6.17

A 100 V, 80 W lamp is to be operated on 230 V, 50 Hz ac. Calculate the inductance to be connected in series with the lamp. Lamp can be taken as a pure resistance.

[U.P. Technical Univ. Electrical Engineering First Semester 2004-05]

Solution : Current drawn by the lamp when operated on rated voltage, $I = \frac{100}{80} = 1.25$ A

Let pure inductance of L Henry be connected in series with the lamp, the voltage across inductance,

$$V_L = \sqrt{(230)^2 - (100)^2} = 207.123 \text{ V}$$

$$\text{and } L = \frac{V_L}{2\pi f I} = \frac{207.123}{2\pi \times 50 \times 1.25} = 0.527 \text{ H } \textbf{Ans.}$$

Example 6.18

A 120 V, 60 W lamp is to be operated on 220 V, 50 Hz supply mains. In order that lamp should operate on correct voltage, calculate value of—

(*a*) non inductive resistance, (*b*) pure inductance. *[U.P. Technical Univ. Elec. Engineering Second Semester 2005-06]*

Solution: Compared with the resistance of the filament of a lamp its inductance is negligible. As such a lamp is considered to be a non-inductive resistance.

Rated power of lamp, P = 60 W

Rated voltage of lamp, V = 120 V

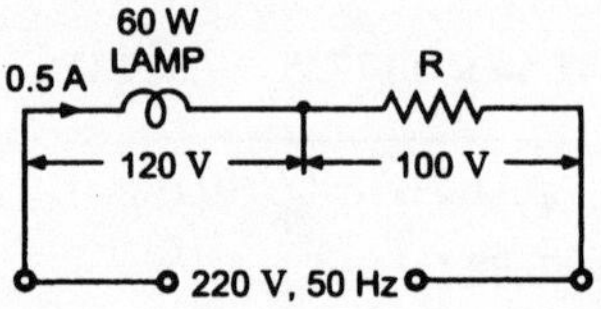

(*a*)

(*b*)

Fig. 6.13

Current drawn by the lamp when operated on rated voltage

$$I = \frac{P}{V} = \frac{60}{120} = 0.5 \text{ A}$$

(*a*) Let a non-inductive resistance of R ohms be connected in series with the lamp, as shown in Fig. 6.13 (*a*)

Since in this case voltage across the lamp and that across the resistance V_R, both are in phase, their arithmetic sum is equal to applied voltage *i.e.*

$$V_R + 120 = 220$$

$$\text{or } V_R = 220 - 120 = 100 \text{ V}$$

$$\text{Resistance, } R = \frac{V_R}{I} = \frac{100}{0.5} = 200\ \Omega \textbf{ Ans.}$$

(*b*) Let pure inductance of L henry be connected in series with the lamp, as shown in Fig. 6.14 (*a*)

Since in this case voltage drop across the inductance leads the current by 90° or $\pi/2$ radians while that across the lamp will be in phase with it, therefore, from phasor diagram shown in Fig. 6.14 (*b*) we have

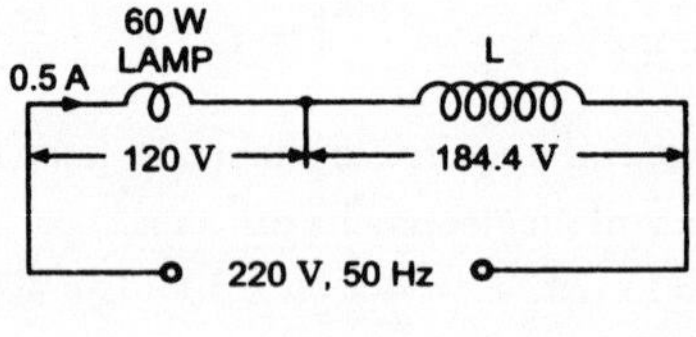

(*a*)

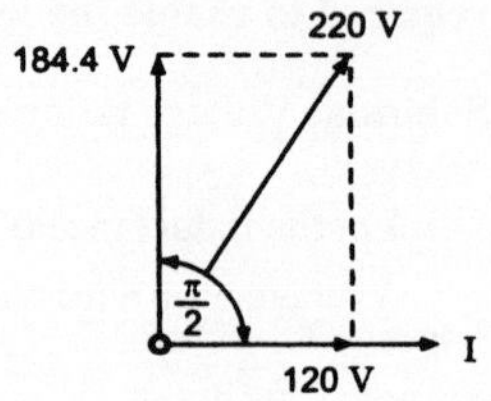

(*b*)

Fig. 6.14

$$120^2 + V_L^2 = 220^2$$

$$\text{or } V_L = \sqrt{220^2 - 120^2} = 184.4 \text{ volts}$$

$$\text{or } 2\pi f L I = 184.4 \text{ V}$$

$$\text{or } L = \frac{184.4}{2\pi \times 50 \times 0.5} = 1.174 \text{ H } \textbf{Ans.}$$

Method (*b*) is preferable to method (*a*) because in method (*b*) there is no power loss whereas in method (a) power loss comes out to be $0.5^2 \times 200$ *i.e.* 50 watts.

Example 6.19

Three coils are connected in series across a 100 V, 50 Hz supply. Values of their resistances are as follows:

$R_1 = 15\ \Omega, \quad L_1 = 0.01$ H
$R_2 = 10\ \Omega, \quad L_2 = 0.03$ H
$R_3 = 3\ \Omega, \quad L_3 = 0.06$ H

Calculate the potential drop and phase angle for each coil.

Solution: Total resistance in the circuit, $R = R_1 + R_2 + R_3 = 15 + 10 + 3 = 28\ \Omega$

Total inductance in the circuit, $L = L_1 + L_2 + L_3 = 0.01 + 0.03 + 0.06 = 0.1$ H

$$\text{Impedance of first coil, } Z_1 = \sqrt{R_1^2 + (2\pi f L_1)^2} = \sqrt{(15)^2 + (2\pi \times 50 \times 0.01)^2} = 15.325\ \Omega$$

$$\text{Impedance of second coil, } Z_2 = \sqrt{(10)^2 + (2\pi \times 50 \times 0.03)^2} = 13.742\ \Omega$$

$$\text{Impedance of third coil, } Z_3 = \sqrt{(3)^2 + (2\pi \times 50 \times 0.06)^2} = 19.087\ \Omega$$

$$\text{Impedance of whole circuit, } Z = \sqrt{R^2 + (2\pi f L)^2} = \sqrt{(28)^2 + (2\pi \times 50 \times 0.1)^2} = 42.083\ \Omega$$

$$\text{Current through the circuit, } I = \frac{V}{Z} = \frac{100}{42.083} = 2.376 \text{ A}$$

Potential drop across first coil, $V_1 = I Z_1 = 2.376 \times 15.325 = 36.417$ V **Ans.**

Potential drop across second coil,

$V_2 = I Z_2 = 2.376 \times 13.742 = 32.65$ V **Ans.**

Potential drop across third coil,

$V_3 = I Z_3 = 2.376 \times 19.087 = 45.35$ V **Ans.**

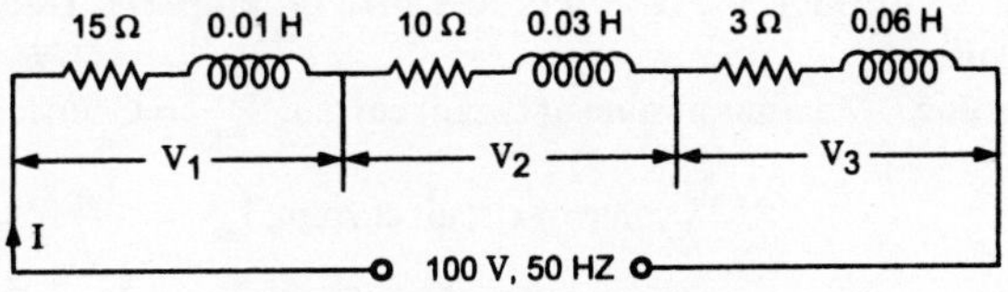

Fig. 6.15 *Circuit Diagram*

Phase angle of first coil,

$$\Phi_1 = \cos^{-1}\frac{R_1}{Z_1} = \cos^{-1}\frac{15}{15.325} = 11°51' \textbf{ Ans.}$$

Phase angle of second coil,

$$\Phi_2 = \cos^{-1}\frac{R_2}{Z_2} = \cos^{-1}\frac{10}{13.742} = 43°18' \textbf{ Ans.}$$

$$\text{Phase angle of third coil, } \Phi_3 = \cos^{-1}\frac{R_3}{Z_3} = \cos^{-1}\frac{3}{19.087} = 80°57' \textbf{ Ans.}$$

6.7. RESISTANCE — CAPACITANCE (R-C) SERIES CIRCUIT

Consider an ac circuit consisting of resistance of R ohms and capacitance of C farads connected in series, as shown in Fig. 6.16 (*a*).

Let the supply frequency be of *f* Hz and current flowing through the circuit be of I amperes (rms value).

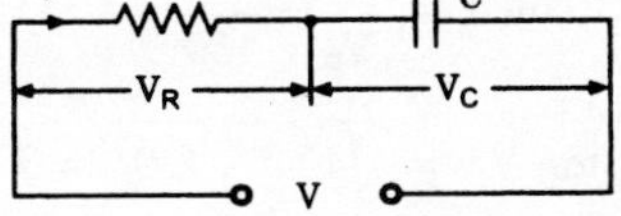

(*a*) *Circuit Diagram*

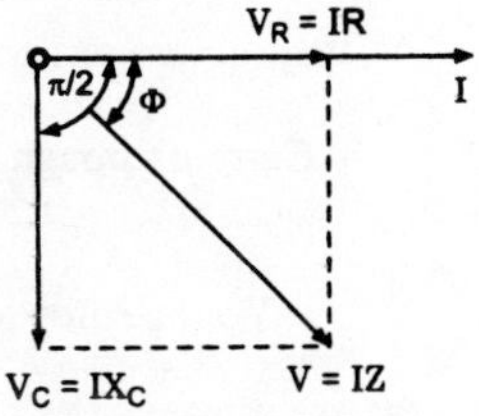

(*b*) *Phasor Diagram*

Fig. 6.16

Voltage drop across resistance,

$V_R = I R$ in phase with the current.

Voltage drop across capacitance,

$V_C = I X_C$ lagging behind I by $\frac{\pi}{2}$ radians or 90°, as shown in Fig. 6.16 (*b*).

The applied voltage, being equal to phasor sum of V_R and V_C, is given in magnitude by

$$V = \sqrt{(V_R)^2 + (V_C)^2} = \sqrt{(IR)^2 + (IX_C)^2} = I\sqrt{R^2 + X_C^2} = I Z \text{ where } Z^2 = R^2 + X_C^2$$

The applied voltage lags behind the current by an angle Φ

where $$\tan \Phi = \frac{V_C}{V_R} = \frac{I X_C}{I R} = \frac{X_C}{R} = \frac{1}{\omega RC}$$

or $$\Phi = \tan^{-1} \frac{1}{R\omega C}$$

Power factor, $\cos \Phi = \frac{R}{Z}$

If instantaneous voltage is represented by

$$v = V_{max} \sin \omega t$$

then instantaneous current will be expressed as

$$i = I_{max} \sin (\omega t + \Phi)$$

and power consumed by the circuit is given by

$$P = V I \cos \Phi$$

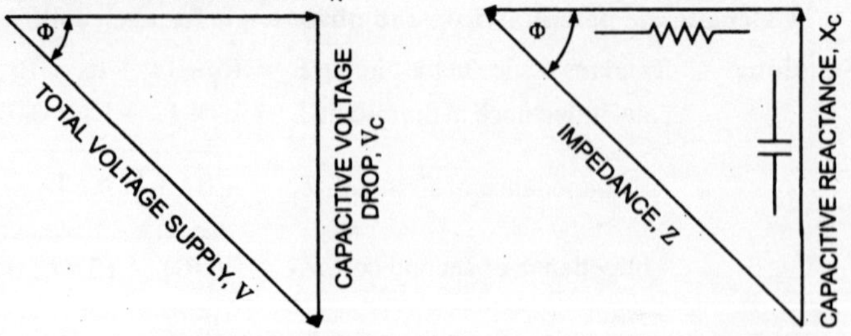

(a) Voltage Triangle *(b) Impedance Triangle*

Fig. 6.17

Voltage triangle and impedance triangle are shown in Figs. 6.17 (*a*) and 6.17 (*b*) respectively.

Example 6.20

A circuit consists of 20 Ω resistance in series with capacitance of 200 μF, connected across 50 Hz supply. The current through the circuit is 10.8 sin 314*t* amperes. Determine the voltage across each component and across the circuit. *[Pb. Technical Univ. Basic Electrical Engineering December 2002]*

Solution: Maximum value of circuit current, I_{max} = Coefficient of the sine of the time angle = 10.8 A

$$\text{RMS value of circuit current, } I_{rms} = \frac{I_{max}}{\sqrt{2}} = \frac{10.8}{\sqrt{2}} = 7.637 \text{ A}$$

$$\text{Voltage across 20 Ω resistance} = I_{rms} \times R = 7.637 \times 20 = 152.74 \text{ V} \quad \textbf{Ans.}$$

$$\text{Voltage across 200 μF capacitor} = I_{rms} X_C = \frac{7.637}{2\pi \times 50 \times 200 \times 10^{-6}} = 121.55 \text{ V} \quad \textbf{Ans.}$$

$$\text{Voltage across the circuit, } V = \sqrt{(152.74)^2 + (121.55)^2} = 195.2 \text{ V} \quad \textbf{Ans.}$$

Example 6.21

A supply of 400 V, 50 Hz is applied to a series RC circuit. Find the value of C if the power absorbed by the resistor be 500 W at 150 V. What is the energy stored in capacitor? *[R.G.P.V. Bhopal Basic Electrical Engineering June 2005]*

Solution: Supply voltage, V = 400 V

Power absorbed, P = 500 W

Voltage across resistor, V_R = 150 V

$$\text{Current through R-C series circuit, } I = \frac{P}{V_R} = \frac{500}{150} = 3.333 \text{ A}$$

$$\text{Voltage drop across capacitor, } V_C = \sqrt{400^2 - 150^2} = 370.8 \text{V}$$

$$\text{Capacitive reactance of the circuit, } X_C = \frac{V_C}{I} = \frac{370.8}{3.333} = 111.25 \, Ω$$

$$\text{Capacitance of circuit, } C = \frac{1}{2\pi f X_C} = \frac{1}{2\pi \times 50 \times 111.25} = 28.6 \, μ\text{F} \quad \textbf{Ans.} \qquad \because X_C = \frac{1}{2\pi f C}$$

$$\text{Energy stored in capacitor} = \frac{1}{2} C \times V_{C\,max}^2 = \frac{1}{2} \times 28.6 \times 10^{-6} \times (370.8\sqrt{2})^2 = 3.934 \text{ J} \quad \textbf{Ans.}$$

Example 6.22

A voltage wave $e(t) = 141.4 \sin 120t$ produces a current $i(t) = 14.14 \sin 120t + 7.07 \cos(120t + 30°)$ in a circuit. Determine (*i*) the resultant time expression of current (*ii*) the power factor and the power delivered by the source.

[*U.P. Technical Univ. Electrical Engineering February 2002*]

Solution: Current $i(t) = 14.14 \sin 120t + 7.07 \cos(120t + 30°)$

$= 14.14 \sin 120t + 7.07 \sin(120t + 30° + 90°)$

$$\text{Maximum value of current, } I_{r\,max} = \sqrt{(\text{X-component})^2 + (\text{Y-component})^2}$$

$$= \sqrt{(14.14 \cos 0° + 7.07 \cos 120°)^2 + (14.14 \sin 0° + 7.07 \sin 120°)^2}$$

$$= \sqrt{(14.14 - 3.535)^2 + (6.1228)^2} = 12.25 \text{ A}$$

Phase angle of resultant current is given as

$$\theta = \text{Tan}^{-1} \frac{\text{Y-component}}{\text{X-component}} = \text{Tan}^{-1} \frac{6.1228}{10.605} = 30°$$

(*i*) So the resultant time expression of current is

$i(t) = 12.25 \sin(120t + 30°)$ **Ans.**

Phase angle between V and I is $\phi = 30°$ (leading)

Power factor = Cos 30° = 0.866 (leading) **Ans.**

$$\text{Power delivered by source, } P = V_{rms} I_{rms} \cos\phi = \frac{V_{max}}{\sqrt{2}} \cdot \frac{I_{max}}{\sqrt{2}} \cos\phi = \frac{V_{max} I_{max}}{2} \cos\phi$$

$$= \frac{141.4 \times 12.25}{2} \cos 30° = 750 \text{ W } \textbf{Ans.}$$

Example 6.23

Voltage and current for a circuit with two elements in series are expressed as follows:

$$v(t) = 170 \sin\left(6280t + \frac{\pi}{3}\right)$$

$$i(t) = 8.5 \sin\left(6280t + \frac{\pi}{2}\right)$$

(*i*) Plot the two waveforms (*ii*) Determine the frequency in Hz. (*iii*) Determine the power factor stating its nature (*iv*) What are the values of the elements ?

[*Nagpur UNiv., April 1996*]

Solution: Peak value of applied voltage, $V_{max} = 170$ V

Peak value of circuit current, $I_{max} = 8.5$ A

(*ii*) Frequency,

$$f = \frac{\text{Coefficient of time in expression of voltage or current}}{2\pi}$$

$$= \frac{6{,}280}{2 \times 3.14} = 1{,}000 \text{ Hz } \textbf{Ans.}$$

(*iii*) Phase angle between voltage and current,

$$\phi = \left(\frac{\pi}{2} - \frac{\pi}{3}\right) = \frac{\pi}{6} \text{ or } 30°.$$

Since current leads the voltage, phase angle between V and I, $\phi = 30°$ (leading)

Power factor = Cos ϕ = Cos 30° = 0.866 (leading) **Ans.**

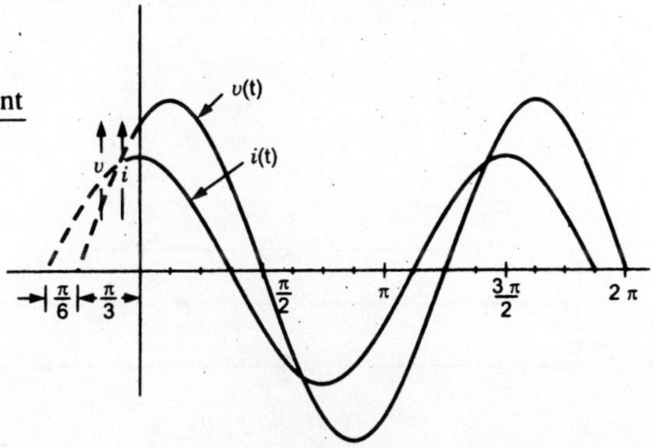

Fig. 6.18

(*i*) Voltage and current waveforms are shown in Fig. 6.18.

$$\text{Impedance, } Z = \frac{V_{rms}}{I_{rms}} = \frac{V_{max}/\sqrt{2}}{I_{max}/\sqrt{2}} = \frac{V_{max}}{I_{max}} = \frac{170}{8.5} = 20\ \Omega$$

$$\text{Resistance, } R = Z\cos\phi = 20 \times 0.866 = 17.32\ \Omega \quad \textbf{Ans.}$$

$$\text{Reactance, } X = Z\sin\phi = 20\sin 30^\circ = 10\ \Omega$$

Since the current leads the voltage, the reactive element is capacitance

$$\text{and } C = \frac{1}{2\pi f X_C} = \frac{1}{2\pi \times 1{,}000 \times 10} = 15.92\ \mu F \quad \textbf{Ans.}$$

Example 6.24

A metal-filament lamp, rated at 750 W, 100 V is to be connected in series with a capacitance across a 230 V, 50 Hz supply. Calculate the value of capacitance required. Draw phasor diagram.

[*U.P. Technical Univ. Electrical Engineering July 2002*]

Solution : Let a pure capacitance of C farads be connected in series with the lamp, as shown in Fig. 6.19 (*a*).

Since in this case voltage drop across the capacitor lags behind the current by 90° or $\frac{\pi}{2}$ radians while that across the lamp will be in phase with it, from phasor diagram shown in Fig. 6.19 (*b*), we have

$$V_C = \sqrt{(230)^2 - (100)^2} = 207\ V$$

$$I = \frac{P}{V} = \frac{750}{100} = 7.5 A$$

$$X_C = \frac{V_C}{I} = \frac{207}{7.5} = 27.6\ \Omega$$

$$\text{and } C = \frac{1}{2\pi f X_C} = \frac{1}{2\pi \times 50 \times 27.6} = 115.33\ \mu F \quad \textbf{Ans.}$$

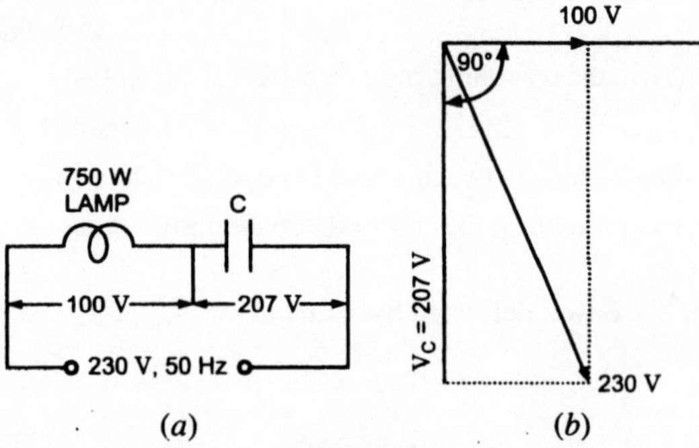

Fig. 6.19

6.8. RESISTANCE-INDUCTANCE-CAPACITANCE (R-L-C) SERIES CIRCUIT

Consider an ac circuit containing resistance of R ohms, inductance of L henries and capacitance of C farads connected in series, as shown in Fig. 6.20 (*a*).

Let the current flowing through the circuit be of I amperes and supply frequency be f Hz.

Voltage drop across resistance, $V_R = IR$ in phase with I

Voltage drop across inductance, $V_L = I\omega L$ leading I by $\pi/2$ radians or 90°

Voltage drop across capacitance, $V_C = \frac{I}{\omega C}$ or IX_C lagging behind I by $\pi/2$ radians or 90°

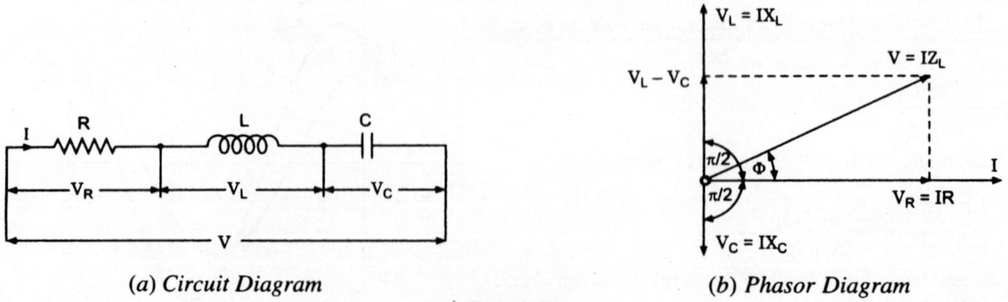

(*a*) Circuit Diagram (*b*) Phasor Diagram

Fig. 6.20

V_L and V_C are 180° out of phase with each other (or reverse in phase), therefore, when combined by parallelogram they cancel each other. The circuit can either be effectively inductive or capacitive depending upon which voltage drop (V_L or V_C) is predominant. Let us consider the case when V_L is greater than V_C.

The applied voltage *V*, being equal to the phasor sum of V_R, V_L and V_C is given in magnitude by

$$V = \sqrt{(V_R)^2 + (V_L - V_C)^2} = \sqrt{(IR)^2 + (IX_L - IX_C)^2} = I\sqrt{(R)^2 + (X_L - X_C)^2}$$

The term $\sqrt{R^2 + (X_L - X_C)^2}$ is known as *impedance* of the circuit and is represented by Z. Its unit is ohm.

Phase angle Φ between voltage and current is given by

$$\Phi = \text{Tan}^{-1}\frac{V_L - V_C}{V_R} = \text{Tan}^{-1}\frac{IX_L - IX_C}{IR} = \text{Tan}^{-1}\frac{X_L - X_C}{R} = \text{Tan}^{-1}\frac{X}{R}$$

Φ will be + ve *i.e.* applied voltage will lead the current if $X_L > X_C$ and Φ will be – ve *i.e.* applied voltage will be behind the current if $X_L < X_C$.

Power factor of the circuit is given by

$$\text{Cos}\,\Phi = \frac{R}{Z} = \frac{R}{\sqrt{R^2 + (X_L - X_C)^2}}$$

Power consumed in the circuit, $P = I^2 R$ or $V I \cos \Phi$.

6.8.1. Reactance. Inductive reactance, X_L is directly proportional to frequency being equal to ωL or $2\pi f L$ and capacitive reactance, X_C is inversely proportional to frequency being equal to $\frac{1}{\omega C}$ or $\frac{1}{2\pi f C}$.

Inductive reactance causes the current to lag behind the applied voltage, while the capacitive reactance causes the current to lead the voltage. So when inductance and capacitance are connected in series, their effects neutralize each other and their combined effect is then their difference. The combined effect of inductive reactance and capacitive reactance is called the reactance and is found by subtracting the capacitive reactance from the inductive reactance or according to equation

$$X = X_L - X_C$$

When $X_L > X_C$ *i.e.* $X_L - X_C$ is positive, the circuit is inductive and phase angle Φ is positive.

When $X_L < X_C$ *i.e.* $X_L - X_C$ is negative, the circuit is capacitive and phase angle Φ is negative.

When $X_L = X_C$ *i.e.* $X_L - X_C = 0$, the circuit is purely resistive and phase angle Φ is zero.

If the expression for applied voltage is taken as

$$v = V_{max} \sin \omega t$$

then expression for the current will be

$$i = I_{max} \sin(\omega t \pm \Phi) \qquad \text{where } I_{max} = \frac{V_{max}}{\text{Circuit impedance, Z}}$$

The value of Φ will be positive when current leads *i.e.* when $X_C > X_L$ and negative when current lags *i.e.* when $X_L > X_C$.

Example 6.25

A choke coil having a resistance of 10 ohm and inductance of 0.05 H is connected in series with a condenser of 100 μF. The whole circuit has been connected to 200 volt, 50 Hz supply. Calculate:

(*i*) impedance (*ii*) current (*iii*) power factor (*iv*) power input (*v*) apparent and reactive power of circuit.

[*R.G.P.V. Bhopal Basic Electrical Engineering Jan./Feb; 2006*]

Solution: Resistance of circuit, $R = 10\ \Omega$

Inductive reactance of circuit, $X_L = 2\pi f L = 2\pi \times 50 \times 0.05 = 15.7\ \Omega$

Capacitive reactance of circuit, $X_C = \frac{1}{2\pi fC} = \frac{1}{2\pi \times 50 \times 100 \times 10^{-6}} = 31.83\ \Omega$

(*i*) Impedance of circuit, $Z = \sqrt{R^2 + (X_L - X_C)^2} = \sqrt{10^2 + (15.7 - 31.83)^2} = 18.978\ \Omega$ **Ans.**

(*ii*) Current through the circuit, $I = \frac{\text{Supply voltage, V}}{Z} = \frac{200}{18.978} = 10.538$ A **Ans.**

Phase angle of circuit, $\phi = \text{Tan}^{-1}\frac{X_L - X_C}{R} = \text{Tan}^{-1}\frac{15.7 - 31.83}{10} = \text{Tan}^{-1}(-1.613) = -58.2^\circ$

(*iii*) Power factor of circuit, *pf* = Cos ϕ = Cos (– 58.2°) = 0.5269 (leading) **Ans.** $\because X_C > X_L$

(*iv*) Power input, P = V I cos ϕ = 200 × 10.538 × 0.5269 = 1,110 watts or 1.11 kW **Ans.**

(*v*) Apparent power, $S = \frac{VI}{1{,}000} = \frac{200 \times 10.538}{1{,}000} = 2.1076$ kVA **Ans.**

Reactive power, $Q = \sqrt{S^2 - P^2} = \sqrt{(2.1076)^2 - (1.11)^2} = 1.79$ kVAR **Ans.**

Example 6.26

A series R-L-C circuit has 100 Ω resistor, 0.318 H inductor and C, $v = 230 \times \sqrt{2} \sin \omega t$ V, $i = 2.3\sqrt{2} \sin \omega t$ A. Find (*i*) C (*ii*) V_L (*iii*) power. Take ω = 314.15 *r*/second.

[*U.P. Technical Univ. Electrical Engineering First Semester, 2007-08*]

Solution: Resistance, R = 100 Ω

Inductive reactance, $X_L = \omega L = 314.5 \times 0.318 = 100\ \Omega$ $\because \omega = 314.5$

Capacitive reactance = X_C (say)

Now impedance, $Z = \sqrt{R^2 + (X_L - X_C)^2}$

$= \frac{V_{max}}{I_{max}} = \frac{230\sqrt{2}}{2.3\sqrt{2}} = 100\ \Omega$

or $\sqrt{100^2 + (100 - X_C)^2} = 100$

or $X_C = 100\ \Omega$

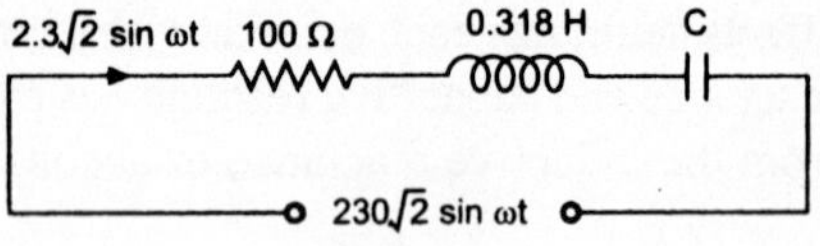

Fig. 6.21

(*i*) Capacitance of the capacitor, $C = \frac{1}{\omega X_C} = \frac{1}{314.5 \times 100} = 31.8\ \mu F$ **Ans.**

(*ii*) Voltage across inductor, $V_L = \frac{I_{max}}{\sqrt{2}} \times X_L = \frac{2.3\sqrt{2}}{\sqrt{2}} \times 100 = 230$ V **Ans.**

(*iii*) Total power consumed, P = VI watts $= \frac{230\sqrt{2}}{\sqrt{2}} \times \frac{2.3\sqrt{2}}{\sqrt{2}} = 529$ W **Ans.**

Example 6.27

A resistance R, an inductance L = 0.01 H and a capacitance C are connected in series. When a voltage $v = 400 \cos (3{,}000\, t - 10^\circ)$ V is applied to the series combination, the current flowing is $10\sqrt{2} \cos (3{,}000\, t - 55^\circ)$ A. Find R and C. [*Nagpur Univ. Electric Circuits 1992*]

Solution: $v = 400 \cos (3{,}000\, t - 10^\circ) = 400 \sin (3{,}000\, t - 10^\circ + 90^\circ) = 400 \sin (3{,}000\, t + 80^\circ)$

$i = 10\sqrt{2} \cos (3{,}000\, t - 55^\circ) = 10\sqrt{2} \sin (3{,}000\, t - 55^\circ + 90^\circ) = 10\sqrt{2} \sin (3{,}000\, t + 35^\circ)$

Phase angle, $\Phi = 80^\circ - 35^\circ = 45^\circ$ *i.e.* circuit is inductive and current lags behind voltage

$\omega = 3{,}000$

$$\text{Inductive reactance, } X_L = \omega L = 3{,}000 \times 0.01 = 30\ \Omega$$

$$\text{Circuit impedance, } Z = \frac{V_{max}}{I_{max}} = \frac{400}{10\sqrt{2}} = 20\sqrt{2} = 28.28\ \Omega$$

$$\text{Net reactance, } X = X_L - X_C = Z \sin \Phi = 28.28 \sin 45° = 28.28 \times \frac{1}{\sqrt{2}} = 20\ \Omega$$

$$\text{or Capacitive reactance } X_C = X_L - X = 30 - 20 = 10\ \Omega$$

$$\text{Circuit capacitance, } C = \frac{1}{\omega X_C} = \frac{1}{3{,}000 \times 10} = 33.33\ \mu\text{F } \textbf{Ans.}$$

$$\text{Circuit resistance, } R = Z \cos \Phi = 28.28 \cos 45° = 20\ \Omega \textbf{ Ans.}$$

Example 6.28

A coil of 0.8 pf is connected in series with 110 μF capacitor. Supply frequency is 50 Hz. The potential difference across the coil is found to be equal to that across the capacitor. Calculate the resistance and inductance of the coil. Calculate the net power factor. *[Nagpur University, November 1997]*

Solution: Let the resistance and inductance of the coil be of R ohms and L henrys. The circuit is given in Fig. 6.22.

Reactance of capacitor,

$$X_C = \frac{1}{\omega C} = \frac{1}{2\pi \times 50 \times 110 \times 10^{-6}} = 28.94\ \Omega$$

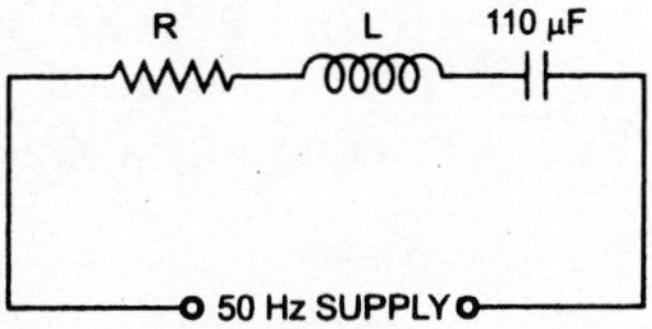

Fig. 6.22

Since voltage drop across the coil is equal to that across the capacitor which are connected in series, impedance of the coil will be equal to that of capacitor. Thus

$$\text{Coil impedance, } Z = X_C = 28.94\ \Omega$$

$$\text{Coil resistance, } R = Z \cos \phi = 28.94 \times 0.8 = 23.15\ \Omega \textbf{ Ans.}$$

$$\text{Coil reactance, } X_L = \sqrt{Z^2 - R^2} = \sqrt{(28.94)^2 - (23.15)^2} = 17.36\ \Omega$$

$$\text{Coil inductance, } L = \frac{X_L}{2\pi f} = \frac{17.36}{2\pi \times 50}\ \text{H} = 0.0553\ \text{H or } 55.3\ \text{mH } \textbf{Ans.}$$

$$\text{Total impedance of circuit, } Z_T = \sqrt{R^2 + (X_L - X_C)^2} = \sqrt{(23.15)^2 + (17.36 - 28.94)^2} = 25.885\ \Omega$$

$$\text{New power factor} = \frac{R}{Z_T} = \frac{23.15}{25.885} = 0.8943 \text{ (leading) } \textbf{Ans.}$$

Example 6.29

Find applied voltage and power loss in a circuit shown below in Fig. 6.23. The value of C is 20 μF, current I = 0.345 A. *[U.P. Technical Univ. Electrical Engineering First Semester 2007-08]*

Solution: From circuit diagram we have

$$R = \frac{\text{Voltage drop across R}}{\text{Current flowing through R}} = \frac{25\ \text{V}}{0.345} = 72.46\ \Omega$$

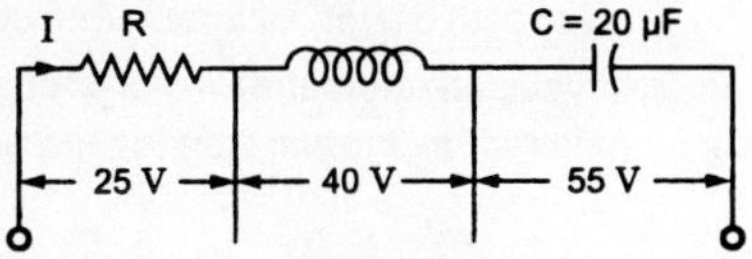

Fig. 6.23

Power loss in the circuit = Power loss in resistor

∵ power loss consumed in L and C is zero.

$$= I^2 R = (0.345)^2 \times 72.46 = 8.625\ \text{W } \textbf{Ans.}$$

$$\text{Applied voltage, } V = \sqrt{V_R^2 + (V_L - V_C)^2} = \sqrt{25^2 + (40 - 55)^2} = 29.155\ \text{V } \textbf{Ans.}$$

Example 6.30

A constant frequency sinusoidal voltage source of magnitude V_S is connected to series circuit made of a resistance of 15 Ω, a coil of winding resistance R and an inductance L and a 50 μF capacitor. The voltage across the 15 Ω

resistor is 30 V, across the coil is 50 V, across the capacitor is 40 V. The voltage across the combination of 15Ω resistor and the coil to together is 72.11 V. Determine the values of the inductance L, winding resistance R and the source voltage V. [*GATE 1999*]

Solution: Current flowing through the circuit, $I = \dfrac{\text{Voltage drop across R}}{R} = \dfrac{30}{15} = 2\ \text{A}$

Voltage drop across capacitor of 50 μF, $V_C = I X_C = \dfrac{I}{2\pi f C}$

So supply frequency, $f = \dfrac{I}{2\pi C V_C} = \dfrac{2}{2\pi \times 50 \times 10^{-6} \times 40} = 159.15\ \text{Hz}$

Voltage across the coil $= I\sqrt{(R^2 + X_L)^2} = 50$

or $R^2 + X_L^2 = \left(\dfrac{50}{I}\right)^2 = \left(\dfrac{50}{2}\right)^2 = 625$ ...(*i*)

and voltage across the combination of 15 Ω resistor and the coil,

$$I\sqrt{(15+R)^2 + X_L^2} = 72.11$$

$$\text{or } (15+R)^2 + X_L^2 = \left(\frac{72.11}{I}\right)^2$$

$$= \left(\frac{72.11}{2}\right)^2 = 1{,}300 \qquad \text{...(ii)}$$

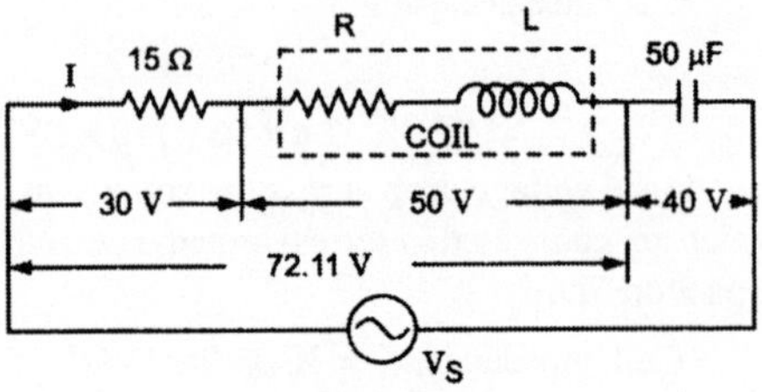

Fig. 6.24

Solving Eqs. (*i*) and (*ii*) we have

$R = 15\ \Omega$ **Ans.**

$X_L = 20\ \Omega$

Inductance of coil, $L = \dfrac{X_L}{2\pi f} = \dfrac{20}{2\pi \times 159.15} = 20\ \text{mH}$ **Ans.**

Reactance of capacitor, $X_C = \dfrac{1}{2\pi f C} = \dfrac{1}{2\pi \times 159.15 \times 50 \times 10^{-6}} = 20\ \Omega$

Since $X_L = X_C$, therefore,

Supply voltage, $V_S = IR = 2 \times (15 + 15) = 60\ \text{V}$ **Ans.**

HIGHLIGHTS

1. The path of flow of alternating current is called the ac circuit.
2. When an alternating voltage given by expression $v = V_{max} \sin \omega t$ is applied across a purely resistive circuit, the alternating current flowing through the circuit is given as

$$i = I_{max} \sin \omega t \quad \text{where } I_{max} = \frac{V_{max}}{R}$$

Power in purely resistive circuit, $P = V_{rms} \times I_{rms}$ or VI watts

Power factor of purely resistive circuit is unity.

3. When an alternating voltage given by expression $v = V_{max} \sin \omega t$ is applied across a purely inductive circuit (a coil with or without an iron core having negligible resistance), the alternating current flowing through the circuit is given as

$$i = I_{max}\left(\omega t - \frac{\pi}{2}\right)$$

where $I_{max} = \frac{V_{max}}{X_L}$; Inductive reactance X_L being equal to $2\pi f L$ or ωL ohms

and power absorbed by a pure inductor = 0

PF of a purely inductive circuit is zero.

4. In a purely capacitive circuit, if $v = V_{max} \sin \omega t$

$$i = I_{max} \sin\left(\omega t + \frac{\pi}{2}\right)$$

where $I_{max} = \frac{V_{max}}{X_C}$; the capacitive reactance X_C being equal to $\frac{1}{\omega C}$ or $\frac{1}{2\pi f C}$ ohms

Power absorbed, P = 0

PF is also zero.

5. An ac circuit can have three elements, namely R, L and C. Power is only consumed in R. The power consumed in elements L and C are zero.

The current through pure inductor lags behind the applied voltage by $\frac{\pi}{2}$ radians while the current through a pure capacitor leads the applied voltage by $\frac{\pi}{2}$ radians.

6. In an R-L series ac circuit, if $v = V_{max} \sin \omega t$

$$i = I_{max} \sin(\omega t - \phi)$$

$$I_{max} = \frac{V_{max}}{Z};\ Z = \sqrt{R^2 + \omega^2 L^2};\ \phi = \text{Tan}^{-1}\frac{\omega L}{R};$$

$$P = VI\cos\phi = I^2 R$$

Power factor, $\cos\phi = \frac{R}{Z}$ (lagging)

7. Power factor of an circuit is given by

(*i*) cosine of the phase angle between voltage and current *i.e.* $\cos\phi$

or (*ii*) the ratio of resistance to impedance *i.e.* $\frac{R}{Z}$

or (*iii*) the ratio of true power to apparent power *i.e.* $\frac{\text{True Power, P}}{\text{Apparent power, S}}$

The *apparent power* of an ac circuit is given by the product of rms values of current and voltage *i.e.*

$$S = VI \text{ volt-amperes.}$$

The *true power* in an ac circuit is obtained by multiplying the apparent power by the power factor *i.e.*

$$P = S\cos\phi = VI\cos\phi \text{ watts.}$$

The product of apparent power VI and sine of angle between voltage and current, $\sin\phi$, is called the reactive power, Q.

$$\text{and } S = \sqrt{P^2 + Q^2}$$

8. Reciprocal of power factor of a coil is known as Q-factor of the coil. It is also called the quality factor or figure of merit of a coil.

Mathematically, Q-factor $= \frac{1}{PF} = \frac{1}{\cos\phi} = \frac{Z}{R} = \frac{\omega L}{R}$ if R is very small in comparison with inductive reactance X_L.

9. In an R-C series ac circuit, if $v = V_{max} \sin \omega t$

$$i = I_{max} \sin(\omega t + \phi)$$

$$I_{max} = \frac{V_{max}}{Z};\ Z = \sqrt{R^2 + \left(\frac{1}{\omega C}\right)^2} = \sqrt{R^2 + X_C^2}$$

$$\phi = \text{Tan}^{-1}\frac{X_C}{R} = \text{Tan}^{-1}\frac{1}{\omega CR}$$

$$P = VI\cos\phi = I^2 R$$

$$\text{Power factor, } \cos\phi = \frac{R}{Z} \text{ (leading)}$$

10. In an R-L-C series ac circuit

$$\text{Impedance, } Z = \sqrt{R^2 + (X_L - X_C)^2}$$

Circuit behaves as an inductive circuit if $X_L > X_C$; as a capacitive if $X_L < X_C$ and as a purely resistive if $X_L = X_C$

$$\text{Circuit current, } I = \frac{V}{Z}$$

$$\text{Phase angle, } \phi = \text{Tan}^{-1}\frac{X_L - X_C}{R}$$

$$\text{Power consumed, } P = VI\cos\phi = I^2 R$$

$$\text{Power factor, } \cos\phi = \frac{R}{Z}$$

EXERCISES

1. Show that the instantaneous power consumed in a pure resistive circuit is not constant but it is fluctuating. **[U.P. Technical Univ. Elec. Engineering February 2001]**
2. In a purely inductive circuit, prove from first principles, that the current lags behind applied voltage by quarter of a cycle and also show that the average power demand is zero. **[U.P. Technical Univ. Electrical Engineering First Semester 2004-05]**
3. Prove that the average power consumption in a pure inductor is zero when ac voltage is applied. **[Gujarat Univ., June/July 2003]**
4. Explain that power consumed in a pure capacitor is zero. **[R.G.P.V. Bhopal Basic Electrical Engineering 2001]**
5. Deduce an expression for the average power in a single phase series R L circuit and therefrom explain the term power factor. **[V.T.U. Belgaun Karnataka Univ. February 2002]**
6. Derive expressions for instantaneous power and active power in a circuit in which the voltage and current are represented by $v = V_m \sin\theta$ and $i = I_m \sin(\theta - \phi)$ respectively. **[U.P. Technical Univ. Electrical Engineering July 2002]**
7. Differentiate among real, reactive and apparent powers. **[U.P. Technical Univ. Electrical Engineering June 2001]**
8. Explain the difference between apparent power, real power and reactive power. **[G.G.S.I.P. Delhi Univ. Electrical Science May 2005]**
9. Define : (*i*) Q factor of a coil (*ii*) Power factor (*iii*) Active, and reactive power and their significance. **[M.D. Univ. Rohtak Electrical Technology May 2002]**
10. Define the terms : (*i*) Active power (*ii*) reactive power.
    Also derive the expression for power in an ac circuit. **[R.G.P.V. Bhopal Basic Electrical Engineering 2001]**
11. Discuss critically the active, reactive and apparent power in a single phase ac circuit and their relation. **[R.G.P.V. Bhopal Basic Electrical Engineering 2001]**
12. How do you analyse series R-C circuit? Draw its phasor diagram. **[Pb. Technical Univ. Basic Electrical and Electronics Engineering First Semester 2005-06]**

13. A single phase sinusoidal ac voltage supply is fed to a series R-C circuit. Determine (*i*) the instantaneous expression of current flowing in the circuit (*ii*) tne impedance (*iii*) the power factor (*iv*) the power consumed and (*v*) the reactive volt-amperes. **[U.P. Technical Univ. Electrical Engineering First Semester 2006-07]**
14. Draw wave diagram and phasor diagram to illustrate clearly the relation between *v* and *i* in the case of (*i*) R-L series circuit (*ii*) R-C series circuit (*iii*) R-L-C series circuit. **[U.P Technical Univ. Electrical Engineering June 2001]**
15. Explain the behaviour of series R-L-C circuit with sinusoidal input. **[M.D. Univ. Rohtak Electrical Technology May 2003]**
16. Derive expressions for impedance, current and power factor for an R-L-C series circuit when applied with ac voltage. Draw also the vector diagram. **[Gujarat University June/July 2003]**

SHORT ANSWER TYPE QUESTIONS WITH ANSWERS

Q. 1. What is the phase relationship between the supply voltage and current flowing through a purely resistive circuit ?
Ans. Both supply voltage and circuit current are in phase with each other.

Q. 2. What is the phase relationship between the supply voltage and current flowing through a purely inductive circuit ?
Ans. In a purely inductive circuit, current lags behind the supply voltage by 90° or $\frac{\pi}{2}$ radians.

Q. 3. Why are the inductors usually iron cored ?
Ans. Inductors or choke coils are made of iron core because large valued flux densities can be produced in iron cores and so inductances of large value can be had. Air-cored inductors become too much bulky to provide an inductance of a required value.

Q. 4. Does an inductance draw instantaneous power as well as average power ? **[R.G.P.V. December 2002]**
Ans. An inductance draws instantaneous power but no average power. The power drawn by inductance during one quarter cycle is released in another quarter cycle, and so average power drawn by an inductance is zero.

Q. 5. What do you understand by power factor of a circuit ?
Ans. Power factor may be defined as the cosine of the phase angle between voltage and current. It may also be defined as the ratio of resistance to impedance or the ratio of true power to apparent power.

Q. 6. What do you understand by active and reactive components of current in an ac circuit ?
Ans. The current component which is in phase with circuit voltage (i.e. I cos ϕ) and contributes to active or true power of the circuit is called the active (wattfull or in-phase) component of current.

The current component which is in quadrature (or 90° out of phase) to the circuit voltage (i.e. I sin ϕ) and contributes to reactive power of the circuit is called the reactive (or wattless) component of current.

Q. 7. What are active and reactive powers ? Draw the power triangle. **[P.T.U. December 2003]**
Ans. The power which is actually consumed or utilized in an ac circuit is called the true or active power of the circuit. Power is consumed only in resistance. It is given by the product of the circuit voltage V, current I and power factor cos ϕ *i.e.* P = VI cos ϕ. It is expressed in watts.

A pure inductor and a pure capacitor do not consume any power, as in a quarter cycle whatsoever power is drawn from the supply source by these components, the same is returned to the supply source in the other quarter cycle. This power which flows back and forth (i.e. in both directions in the circuit) or reacts itself is called the reactive power. This is also known as wattless power.

The reactive power of an ac circuit is given by the product of voltage V, current I and sine of the phase angle ϕ i.e. Reactive power, Q = VI sin ϕ. It is expressed in reactive volt-amperes.

The power triangle is shown in Fig 6.6(a)

Q. 8. What is the relationship between supply voltage and circuit current in a purely capactive circuit ?
Ans. In a purely capactive circuit, current leads the applied voltage by 90° or $\pi/2$ radians.

Q. 9. What is the effect of frequency on inductive reactance ?
Ans. Inductive reactance of an inductor increases proportionately with the increase in supply frequency *i.e.* $X_L \propto f$.

Q. 10. What is the effect of frequency on capactive reactance ?

Ans. Capactive reactance of a given capacitor decreases with the increase in supply frequency. Capacitive reactance is inversely proportional to supply frequency *i.e.* $X_C \alpha \frac{1}{f}$.

Q. 11. What are the values of power factor for (*i*) purely resistive circuit (*ii*) purely inductive circuit and (*iii*) purely capacitive circuit ?

Ans. (*i*) Unity, (*ii*) Zero (lagging) (*iii*) Zero (leading).

Q. 12. What is the relationship between apparent power, true power and reactive power of an ac circuit ?

Ans. Apparent power = $\sqrt{(\text{True power})^2 + (\text{Reactive power})^2}$

Q. 13. In an R-L-C circuit on what factors depends whether the phase angle between voltage V and current I is lagging or leading ?

Ans. Phase angle in an R-L-C circuit is given as

$$\phi = \text{Tan}^{-1}\frac{X}{R} = \text{Tan}^{-1}\frac{X_L - X_C}{R}$$

i.e. the phase angle between voltage V and current I will depend upon whether the inductive reactance is greater or smaller than capacitive reactance. When X_L exceeds X_C voltage V will lead current I and when X_C exceeds X_L applied voltage V will lag behind the current I.

Q. 14. Describe the properties of (*i*) resistance (*ii*) inductance and (*iii*) capacitance used in ac circuits.

[R.G.P.V. June 2003]

Ans. The property of a substance, which opposes the flow of an electric current through it, is called the *resistance*. It is symbolised by English capital letter R and is measured in ohms (Ω).

An inductor has been defined as a physical device which is capable of storing energy by virtue of a current flowing through it.

An inductor is a circuit component which opposes the change of current flowing through it and induces a voltage when the current flowing through it varies in magnitude and/or direction.

In case of an inductor current does not change instantaneously. It offers high impedance to ac but very low impedance to dc *i.e.* it blocks ac signal but passes dc signal.

A capacitor is a physical device which is capable of storing energy by virtue of a voltage existing across it. The voltage applied across the capacitor sets up an electric field within it and the energy is stored in the electric field.

A capacitor is basically meant to store electrons (or electrical energy), and release them whenever required. *Capacitance* is a measure of a capacitor's ability to store charge and is measured in farads (F). Farad, the unit of capacitance is very large, so microfarad (μF) or micro-microfarad (μμF) is usually used

$$1\ \mu\text{F} = 10^{-6}\ \text{F and } 1\ \mu\mu\text{F, also called the pico-farad (p F),} = 10^{-12}\ \text{F.}$$

Q. 15. Explain 'resistance', 'reactance' and 'impedance'. **[R.G.P.V. December 2002]**

Ans. **Resistance** may be defined as that property of a substance which opposes (or restricts) the flow of an electric current (or electrons) through it. It is expressed in ohms.

Reactance may be inductive or capacitive.

Inductive reactance is capable of storing energy by virtue of a current flowing through it. It opposes the change of current flowing through it and induces a voltage when the current flowing through it varies in magnitude and/or direction. In inductive circuit current does not change instantaneously. It is highly resistive to ac but it does not oppose dc.

Capacitive reactance is capable of storing energy by virtue of a volage existing across it. It does not oppose the flow of ac but it is highly resistive to dc. Voltage across it cannot change instantaneously.

Inductive reactance and capacitive reactance in series or in parallel form a tuned circuit. Both of the reactances are expressed in ohms.

Impedance is a property of an ac circuit due to which it opposes the flow of current through it. It may be defined as the ratio of voltage applied across the circuit and current flowing in that circuit. It is expressed in ohms. It is given as

$$\text{Impedance} = \sqrt{(\text{Resistance})^2 + (\text{Reactance})^2}$$

PROBLEMS

1. A 50 Hz voltage of 115 V (rms) is impressed on a 50 Ω resistance. Write the time equations for the voltage and resultant current. Let the zero point of the voltage wave be at $t = 0$ **[Ans.** 162.6 sin 314t and 3.25 sin 314t]
2. A 50 Hz voltage of 230 V (rms) is impressed on an inductance of 0.2 H. Give the time equations for the voltage and the resultant current. Let the zero point of the voltage wave be at $t = 0$. Determine the maximum energy stored in the inductance.

$$\left[\textbf{Ans. } 325.3\sin 314t; 5.2\sin\left(314t - \frac{\pi}{2}\right); 2.7\text{ J}\right]$$

3. A 50 Hz voltage of 230 V rms is impressed on a capacitance of 20 μ F. Write the time equation for the voltage and the resultant current. Let the zero point of the voltage wave be at $t = 0$. Determine the energy stored in the capacitor.

$$\left[\textbf{Ans. } 325.3\sin 314t; 1.45\sin\left(314t + \frac{\pi}{2}\right); 1.06\text{ J}\right]$$

4. A 120 V ac circuit contains 10 Ω resistance and 30 Ω reactance in series. What would be the average power in the circuit? **[Pb. Univ. Elements of Electrical Engineering II December 1988]**
[Ans. 144 watts]
5. A coil has a resistance of 10 Ω and draws a current of 5 A when connected across a 100 V, 50 Hz source. Determine the power factor of the circuit. **[Pb. Technical Univ. Electrical Engineering June 2000]**
[Ans. 0.5 (lagging)]
6. A coil has a resistance of 10 Ω and draws a current of 5 A when connected across a 100 V, 50 Hz source. Determine the inductance of the coil. **[Pb. Technical Univ. Electrical Engineering December 2000]**
[Ans. 55.13 mH]
7. A 20 mH inductor is connected in series with a 10 Ω resistor across a 220 V, 50 Hz supply. Find the current and power factor of the circuit. **[Pb. Technical Univ. Electrical Engineering December 2001]**
[Ans. 18.628 A; 0.8467 (lagging)]
8. The element of 500 W electric press is designed for use on 200 V, 50 Hz supply. What value of resistance is needed to be connected in series in order that press can be operated at 230 V ac supply ?
[Pb. Technical Univ. Basic Electrical and Electronics Engineering First Semester 2004-05]
[Ans. 12 Ω]
9. In a coil, the inductance being 50 mH: the impedance is 20 ohm. If the power factor of the circuit be 0.5 lag, find the values of angular frequency and resistance of the coil. **[R.G.P.V. Bhopal Basic Electrical Engineering June 2005]**
[Ans. 346.41 radians/second; 10 Ω]
10. The voltage applied to a circuit is $v = 100 \sin(\omega t + 30°)$ and the current flowing in the circuit is $i = 15 \sin(\omega t + 60°)$. Determine the impedance, resistance, reactance, power and the power factor of the circuit.
[U.P. Technical Univ. Electrical Engineering Second Semester 2003-04]
[Ans. 6.667 Ω; 5.774 Ω; 3.333 Ω (capacitive); 649.52 W; 0.866 (leading)]
11. In an ac circuit the current is given by $i = 22 \sin(314 - \pi/6)$ A. If the voltage of the circuit is the reference quantity, determine (*i*) power factor (*ii*) rms value of current, (*iii*) frequency of current.
[U.P. Technical Univ. Electrical Engineering Jaunary 2003]
[Ans. 0.866 (lagging); 15.56 A; 50 Hz]
12. A choke coil takes a current of 2 A lagging 60° behind the applied voltage of 220 Vat 50 Hz. Calculate the inductance and resistance of coil. **[V.T. U. Belgaum Karnataka Univ. Winter 2003]**
[Ans. 0.3032 H; 55 Ω]
13. The instantaneous values of the voltage across a two element series circuit and the current flowing through it are given by $v = 100\sin\left(314t - \frac{\pi}{4}\right)$ V, $i = 20 \sin(314\,t - 90°)$A. Find the frequency and the circuit elements.
[V.T. U. Belgaum Karnataka Univ. Winter 2003]
[Ans. 50 Hz; 3.536 Ω, 11.26 mH]

14. A 120 V, 100 W lamp is to be connected to a 220 V, 50 Hz ac supply. What value of pure inductance should be connected in series in order that lamp is run on the rated voltage ?

[U.P. Technical Univ. Electrical Engineering First Semester 2003-04]

[**Ans.** 0.704 H]

15. A voltage $v(t) = 141.4 \sin(314t + 10°)$ is applied to a circuit and the steady current given by $i(t) = 14.14 \sin(314t - 20°)$ is found to flow through it. Determine (*i*) the pf of the circuit (*ii*) the power delivered to the circuit (*iii*) Draw the phasor diagram. **[Nagpur Univ., Summer 2000]**

[**Ans.** (*i*) 0.866 (lag) (*ii*) 866 W]

16. A voltage of 125 V at 60 Hz is applied across a non-inductive resistor connected in series with a condensor. The current is 2.2 A. The power loss in the resistor is 96.8 W and that in the condensor is negligible. Calculate the resistance and the capacitance. [**Ans.** 20 Ω ; 49.88 μ F]

17. A voltage wave $e(t) = 141.4 \sin 120t$ produces a current $i(t) = 14.14 \sin 120t + 7.07 \cos(120t + 30°)$ in a circuit. Determine (*i*) the resultant time expression of current (*ii*) the power factor and the power delivered by the source.

[U.P. Technical Univ. Electrical Engineering February 2002]

Ans. (*i*) $i(t) = 12.25 \sin(120t + 30°)$; (*ii*) 0.866 (leading); 750 W]

18. A voltage $v = 100 \sin 314t - 50 \cos 314t$, is applied to a circuit having R = 20 Ω in series with C = 100 μF. Obtain expression for instantaneous current, rms value of current and the power in the circuit.

[Pb. Technical Univ. Electrical Engineering 1999]

[**Ans.** $i(t) = 2.974 \sin(314t + 31.3°)$; 2.1 A; 88.32 W]

19. A tungsten filament lamp rated at 500 W, 100 V is to be connected in series with a capacitance across 220 V, 50 Hz supply. Calculate (*a*) the value of capacitor such that the voltage and power consumed by the lamp are according to the rating of the lamp (*b*) the power factor of the current drawn from the supply (*c*) draw the phasor diagram of the circuit.

[Nagpur Univ. Electrical Technology 1991]

[**Ans.** 81.2 μ F (*b*) 0.4545 (lead)]

20. A series circuit contains an inductor of 10 Ω resistance and 6 Ω reactance and a condenser of 25 Ω capacitive reactance and negligible resistance. If the voltage is 240 V, what current flows through the circuit ?

[Pb. Univ. Elements of Electrical Engineering December 1988]

[**Ans.** 11.178 A leading applied voltage by 62.24°]

21. A coil of resistance 8 Ω and inductance 0.12 henry is connected in series with a condenser of capacitance 140 microfarads across a 230 volts, 50 Hz supply. Determine:

(*i*) Impedance of the entire circuit,

(*ii*) Current flowing through the condenser,

(*iii*) Power factor of the circuit,

(*iv*) Voltage across the condenser. **[U.P. Technical Univ. Electrical Engineering Second Semester 2002-03]**

[**Ans.** (*i*) 16.97 Ω (*ii*) 13.55 A (*iii*) 0.4714 (lagging) (*iv*) 308.2 V]

22. A coil of resistance 12 Ω and inductance 0.05 H, a non-inductive resistance of 20 Ω resistance and a loss-free 40 μF capacitor are connected across a 240 V, 50 Hz sinusoidal supply. Calculate (i) the current and (ii) the power factor of the circuit. **[U.P. Technical Univ. Electrical Engineering September 2001]**

[**Ans.** (*i*) 3.36 A (*ii*) 0.448 (lagging)]

23. A series R-L-C circuit consisting of a resistance of 20 Ω, inductance 0.2 H and capacitive of 150 μ F is connected across a 230 V 50 Hz source. Calculate (i) the impedance (ii) the current (iii) the magnitude and nature of the power factor (iv) the frequency of supply to be adjusted to make power factor unity.

[U.P. Technical Univ. Electrical Engineering February 2002]

[**Ans.** (*i*) 46.167 Ω (*ii*) 4.98 A (*iii*) 0.433 (lagging) (*iv*) 29.06 Hz]

24. A resistance of 12 Ω, inductance of 0.1 H and a capacitance of 100 μF are connected in series across ac 220 V, 100 Hz supply. Calculate the current and its power factor, power consumed and phase angle mentioning whether it is leading or lagging. **[Pb. Technical Univ. Basic Elec. Engineering December 2003]**

[**Ans.** 4.54 A, 0.2478 (lagging), 247.7 W, 75.65° (lagging)]

25. A circuit having a resistance of 12 Ω, an inductance of 0.15 H and a capacitance of 100 μF in series connected across a 100 V, 50 Hz supply. Calculate the impedance, current, the phase difference between the current and supply voltage.

[V.T.U. Belgaum Karnataka Univ., Summer 2003]

[**Ans.** 19.44 Ω; 5.144 A; 51.88° (lagging)]

26. Resistor (= R), choke coil (r, L) and a capacitor of 25.2 μF are connected in series. When supplied from an ac source, it takes 0.4 A. If the voltage across the resistor is 20 V, voltage across the resistor and choke is 45 V, voltage across the choke is 35 V, and the voltage across the capacitor is 50 V. Find : (*a*) the values of r, L (*b*) applied voltage and its frequency (*c*) PF of the circuit and the active power consumed. Draw the phasor diagram. **[Nagpur Univ. April 1998]**

[Ans. (*a*) 25 Ω, 267 mH (*b*) 34.22 V, 50 Hz (*c*) 0.877 (leading), 12 W]

27. A coil of resistance 20 Ω and inductance 100 mH is connected in series with a capacitance of 40 μF across 100 V, 50 Hz ac supply. Calculate (*i*) magnitude of current (*ii*) power factor (*iii*) phase angle (*iv*) voltage across each element.

[M.D. Univ. Rohtak Electrical Technology 1st Semester (E. Scheme) New Examination 2006-07]

[Ans. (*i*) 1.92 A (*ii*) 0.3835 (leading) (*iii*) 67.45° (leading) (*iv*) 38.4 V; 60.3 V; 152.8 V]

28. A resistance of 20 Ω and inductance of 0·2 H and capacitance of 100 μF are connected in series across 220 V, 50 Hz mains. Determine the following:

(*i*) Impedance,
(*ii*) current,
(*iii*) voltage across R, L and C,
(*iv*) power in watts,
(*v*) pf and the angle of lag.

[R.G.P.V. Bhopal Basic Electrical Engineering Second Semester 2006-07]

[Ans. (*i*) 36.89 Ω (*ii*) 5.96 A, (*iii*) 119.2 V; 374.5 V; 189.7 V (*iv*) 710.5 W (*v*) 0.542 (lagging); 57.2° (lagging)]

❑❑❑

Chapter 7

Complex Algebra and Circuit Analysis

7.1. INTRODUCTION

We have seen that ac circuits cannot be solved by use of simple algebra since geometrical relations are to be taken into consideration. Though phasor diagram method is quite satisfactory for fairly simple circuits but becomes complicated when a circuit or network is made up of several branches. AC circuits can be conveniently solved by use of complex algebra. The results obtained by this method are of the same order of accuracy as those obtained by trigonometrical methods although calculations are usually much simpler.

The above system enables equations representing alternating voltages and currents and their phase relationships to be expressed in simple algebraic form. It is based upon the idea that a phasor can be resolved into two components along two axis at right angles to each other.

There are four forms or methods of representing the vector or phasor quantities viz., (*i*) symbolic form, (*ii*) trigonometrical form (*iii*) exponential form and (*iv*) polar form. All of these methods of expressing vector or phasor quantities enable those operations which are carried out graphically in a phasor diagram, to be performed analytically.

7.2. SYMBOLIC FORM

A phasor, as already mentioned above, can be resolved into two components along two axis at right angles to each other. It is preferable and convenient to resolve it along X-axis and Y-axis.

In symbolic form, a phasor is resolved into two components at right angles to each other and the component lying along Y axis is designated by $+j$ or $-j$ according to its direction.

In Fig. 7.1 phasors OA_1, OA_2, OA_3 and OA_4 are shown lying in first, second, third and fourth quadrants respectively. The phasor OA_1 lying in first quadrant may be resolved into two components a_1 along X-axis (sometimes called the axis of reals) and b_1 along Y-axis (sometimes called the axis of imaginaries or *j*-axis). The component along Y-axis should be designated by *j* and should be written as jb_1. Thus we represent the phasor OA_1

$$\boldsymbol{OA_1}^* = a_1 + j\, b_1$$

The phasor, when expressed in above form, is said to be expressed in *complex form*. In electrical engineering, the horizontal component (*i.e.* a_1) is called the *in-phase* or *active* component while the vertical component is called the *quadrature* or *reactive* component.

---

* Symbols representing phasors are printed in bold face italics, while those representing only the magnitude are printed in ordinary italics, thus:

$\boldsymbol{V}, \boldsymbol{I}, \boldsymbol{Z},$ phasors or complex numbers

$V, I, Z,$ magnitude or module

Other methods of denoting phasors are by putting bar above, dot above or below letters in ordinary italics as $\bar{V}, \bar{I}, \bar{Z}, \dot{V}, \dot{I}, \dot{Z}$ or $\underset{\cdot}{V}, \underset{\cdot}{I}, \underset{\cdot}{Z}$.

The length of line OA_1 represents, to adopted scale, the absolute value or modulus (magnitude) of the phasor and is equal to $\sqrt{a_1^2 + b_1^2}$, and its angle with the X-axis, θ_1, called the *argument* of the phasor, is given by the expression

$$\theta_1 = \text{Tan}^{-1} \frac{b_1}{a_1}.$$

Similarly phasors OA_2, OA_3 and OA_4 can be resolved into two components along axis of reals and axis of imaginaries and can be expressed as

$$OA_2 = -a_2 + j\,b_2$$
$$OA_3 = -a_3 - j\,b_3$$
$$OA_4 = a_4 - j\,b_4$$

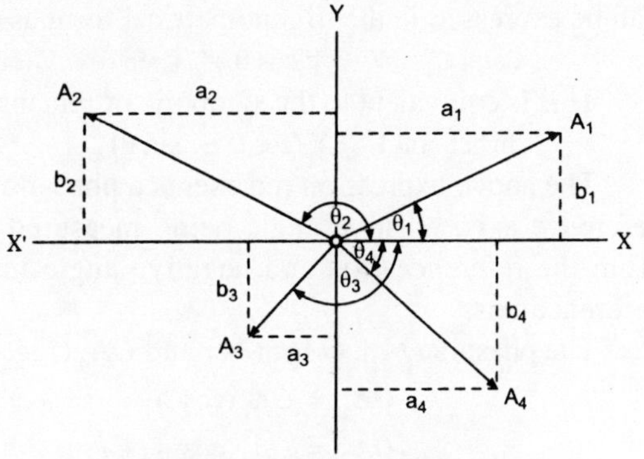

Fig. 7.1 *Resolution of Phasors*

7.2.1. Significance of Operator *j*. The letter *j* used in the above expressions is an operator, the application of which to phasor *a*, or phasor component *a*, causes its rotation through 90° in the positive (counterclockwise) direction. It is assigned the numerical value $\sqrt{-1}$. The double operation of *j* on phasor *a* rotates it in counter-clockwise direction through 180° and hence reverses its sense because $j\,j = j^2 = \left(\sqrt{-1}\right)^2 = -1$. The triple operation of *j* on phasor *a* rotates it in counter-clockwise direction through 270° and so on.

For example, if phasor V lying along X-axis is operated with *j*, the phasor V will rotate through an angle of 90° in counterclockwise direction and lie along Y-axis. If this phasor V is operated twice with operator *j* it will rotate through 180° in counterclockwise direction from X-axis and will lie along X-axis in negative direction, as illustrated in Fig. 7.2. Application of the operator *j* thrice to this phasor V will rotate it through 270° in counter-clockwise direction from X-axis and now the phasor V will lie along Y-axis in negative direction. On operating the phasor V with *j* four times, it will rotate through 360° in counterclockwise direction from the reference axis and so will come to the original position.

Hence it is seen that *j* is an operator which turns phasor *a* through an angle of 90° in counterclockwise direction, each time that it is applied and has a numerical value of $\sqrt{-1}$ so that

$$j^2 = \left(\sqrt{-1}\right)^2 = -1;$$
$$j^3 = \left(\sqrt{-1}\right)^3 = -\sqrt{-1}$$
$$\text{and } j^4 = \left(\sqrt{-1}\right)^4 = (-1)^2 = 1$$

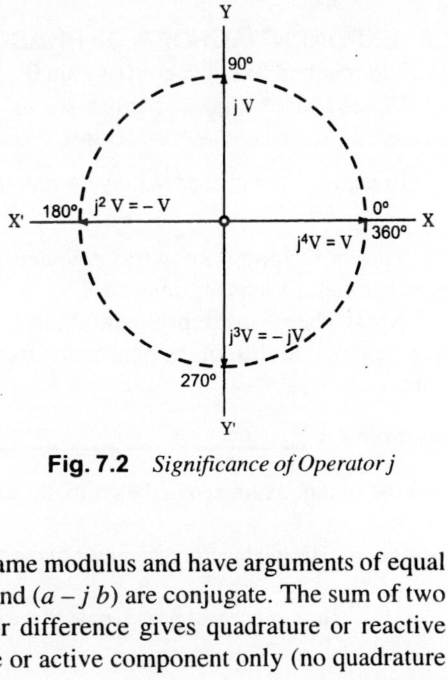

Fig. 7.2 *Significance of Operator j*

7.3. CONJUGATE COMPLEX NUMBERS

Two complex numbers are said to be conjugate when they are of same modulus and have arguments of equal magnitude and opposite sign. For example, the numbers $(a + j\,b)$ and $(a - j\,b)$ are conjugate. The sum of two conjugate numbers gives in-phase or active component and their difference gives quadrature or reactive component. The product of two conjugate numbers gives in-phase or active component only (no quadrature component) equal to the sum of the square of their magnitudes.

i.e. $(a + j\,b) + (a - j\,b) = 2a$, active component

$(a + j\,b) - (a - j\,b) = j\,2b$, reactive component

$(a + j\,b) \times (a - j\,b) = (a^2 + b^2)$, active component

7.4. TRIGONOMETRICAL FORM OF PHASOR REPRESENTATION

As may be seen from Fig. 7.3, X-component of V = $V\cos\theta$ and Y-component of $V = V\sin\theta$. Hence phasor V can be expressed in the trigonometrical form as

$$V = V\cos\theta + jV\sin\theta = V(\cos\theta + j\sin\theta)$$

This is equivalent to the symbolic or rectangular form $V = (a + jb)$ because $a = V\cos\theta$ and $b = V\sin\theta$

In general $V = V(\cos\theta \pm j\sin\theta)$

The above expression represents a phasor of numerical value V and having phase angle of $\pm\,\theta$ with the reference axis, positive angle being measured in counterclockwise direction from the reference axis and negative angle in clockwise direction from the reference axis.

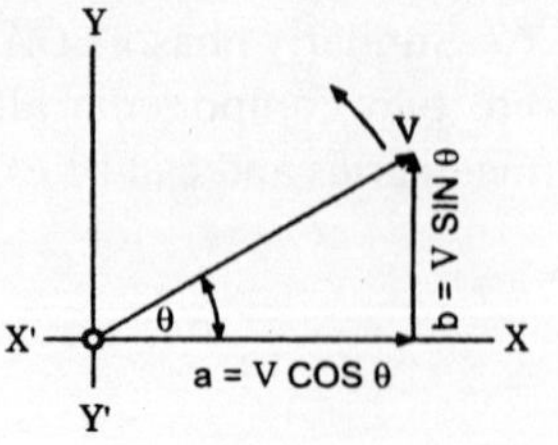

Fig. 7.3

The phasors OA_1, OA_2, OA_3 and OA_4 (Fig. 7.1) may be represented thus:

$$\boldsymbol{OA_1} = OA_1(\cos\theta_1 + j\sin\theta_1)$$
$$\boldsymbol{OA_2} = OA_2(\cos\theta_2 + j\sin\theta_2)$$
$$\boldsymbol{OA_3} = OA_3(\cos\theta_3 - j\sin\theta_3)$$
$$\boldsymbol{OA_4} = OA_4(\cos\theta_4 - j\sin\theta_4)$$

7.5. POLAR FORM OF PHASOR REPRESENTATION

The phasor V (Fig. 7.3) may be alternatively expressed as $V\angle\theta$ where V is the magnitude of the phasor and θ is the angle measured in counterclockwise direction from reference axis, called the *argument* of the phasor.

In general, the phasor V may be represented as

$$V = V\angle \pm \theta$$

The above form of phasor representation is known as *polar form.*

The phasors OA_1, OA_2, OA_3 and OA_4 (Fig. 7.1) may be represented thus $\boldsymbol{OA_1} = OA_1\angle\theta_1$; $\boldsymbol{OA_2} = OA_2\angle\theta_2$; $\boldsymbol{OA_3} = OA_3\angle -\theta_3$; $\boldsymbol{OA_4} = OA_4\angle -\theta_4$.

7.6. EXPONENTIAL FORM OF PHASOR REPRESENTATION

By Euler's equation $e^{j\theta} = \cos\theta + j\sin\theta$

Hence phasor V may be expressed as

$$V = Ve^{j\theta} \text{ where } e \text{ is the base of natural logarithm.}$$

In general, the phasor V may be represented as

$$V = Ve^{\pm j\theta}$$

The above form is known as *exponential form* of representing a phasor. It represents a phasor of modulus (or magnitude) V and argument $\pm\,\theta$.

Note: The phasor representation in either of four mathematical forms convey the same information *i.e.* magnitude (modulus) of the phasor and its phase angle (argument). So it is possible to convert one form to the other.

Example 7.1

Give the voltage expressed in the symbolic form $V = (5 + j\,12)$ volts in trigonometrical, polar and exponential forms.

Solution : The modulus or magnitude of the given voltage, $V = \sqrt{5^2 + 12^2} = 13$ V

The argument or phase angle of the given voltage, $\theta = \text{Tan}^{-1}\dfrac{12}{5}$

$= 67.38°$ or $0.374\,\pi$ radians.

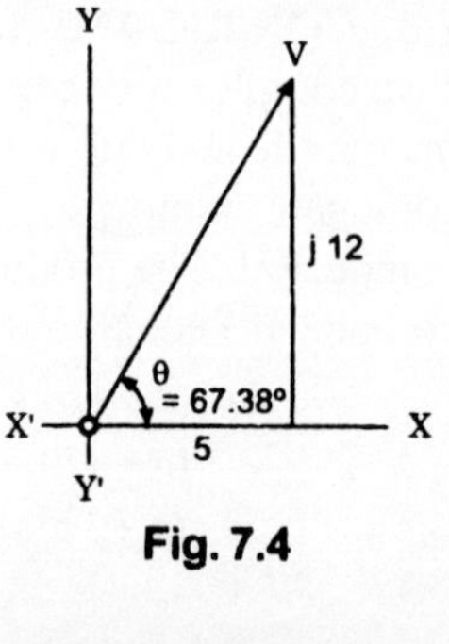

Fig. 7.4

So given voltage in

(*i*) Trigonometrical form is 13 (cos 67.38° + j sin 67.38°) volts **Ans.**

(*ii*) Polar form is $13 \angle 67.38°$ volts **Ans.**

(*iii*) Exponential form is $13\, e^{+j\,0.374\pi}$ volts **Ans.**

7.7. ADDITION AND SUBTRACTION OF PHASORS

The symbolic or rectangular form is most suitable form for addition or subtraction of phasors. If the phasors are given in polar or exponential form, they should be first converted to rectangular form and then addition or subtraction be carried out.

7.7.1. Addition of Phasors. For the addition of phasors in the rectangular form in-phase (or active) components are added together and quadrature components are added together.

Let the two voltage phasors V_1 and V_2 represented as $\mathbf{V}_1 = a_1 + j\,b_1$, and $\mathbf{V}_2 = a_2 - j\,b_2$, be added.

Then resultant voltage will be given as

$$\mathbf{V} = \mathbf{V}_1 + \mathbf{V}_2 = (a_1 + j\,b_1) + (a_2 - j\,b_2) = (a_1 + a_2) + j\,(b_1 - b_2)$$

Magnitude of resultant voltage,

$$V = \sqrt{(a_1 + a_2)^2 + (b_1 - b_2)^2}$$

Argument or phase angle of the resultant voltage,

$$\theta = \text{Tan}^{-1}\frac{b_1 - b_2}{a_1 + a_2}$$

Addition process by graphic representation is illustrated in Fig. 7.5.

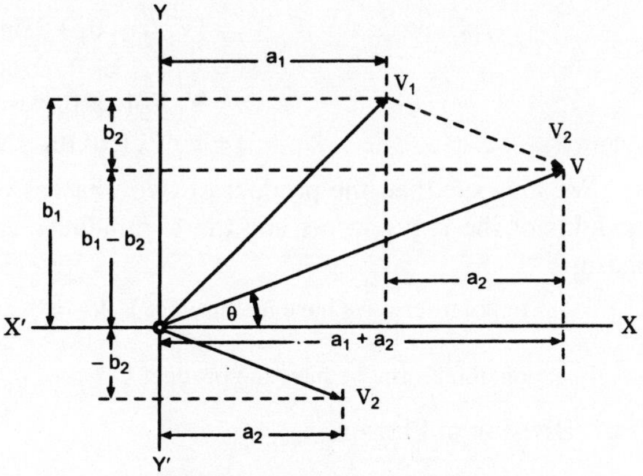

Fig. 7.5 *Addition of Phasors*

7.7.2. Subtraction of Phasors. By the rule of phasor subtraction

$$\mathbf{V} = \mathbf{V}_1 - \mathbf{V}_2 = (a_1 + j\,b_1) - (a_2 - j\,b_2) = (a_1 - a_2) + j\,(b_1 + b_2)$$

Magnitude of resultant voltage,

$$V = \sqrt{(a_1 - a_2)^2 + (b_1 + b_2)^2}$$

Argument or phase angle of resultant voltage,

$$\theta = \text{Tan}^{-1}\frac{b_1 + b_2}{a_1 - a_2}$$

Subtraction process by graphic representation is illustrated in Fig. 7.6.

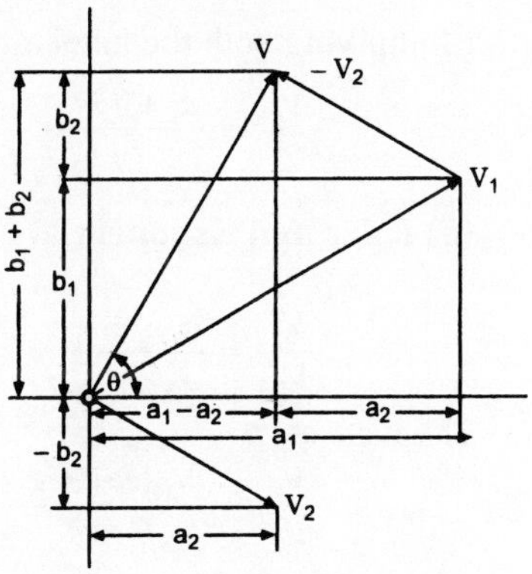

Fig. 7.6 *Subtraction of Phasors*

7.8. MULTIPLICATION AND DIVISION OF PHASORS

The $(a + j\,b)$ method of representing a phasor is particularly well suited to the process of addition and subtraction, as illustrated above, but it also lends itself readily to multiplication and division as illustrated below:

However, it is worthnoting that it is easier to multiply and divide phasors when they are expressed in polar or exponential form.

Consider two phasors given by $V_1 = (a_1 + j\,b_1)$ or $V_1 \angle \theta_1$ or $V_1 e^{j\theta_1}$

and $V_2 = (a_2 - j\,b_2)$ or $V_2 \angle -\theta_2$ or $V_2 e^{-j\theta_2}$

7.8.1. Multiplication of Phasors

(*i*) Rectangular Form.

$$V_1 \times V_2 = (a_1 + j\,b_1)(a_2 - j\,b_2) = a_1\,a_2 - j\,a_1 b_2 + j\,a_2 b_1 - j^2 b_1 b_2 = (a_1 a_2 + b_1 b_2) + j\,(a_2 b_1 - a_1 b_2)$$

$$\text{Magnitude of resultant} = \sqrt{(a_1 a_2 + b_1 b_2)^2 + (a_2 b_1 - a_1 b_2)^2}$$

$$\text{Argument, } \theta = \text{Tan}^{-1} \frac{a_2 b_1 - a_1 b_2}{a_1 a_2 + b_1 b_2}$$

***(ii)* Polar and Exponential Form.** For multiplication of phasors in polar forms, their magnitude or modulus are multiplied and arguments or phase angles are added as proved below:

$$\boldsymbol{V}_1 = V_1 \angle\theta_1 = V_1 e^{j\theta_1} = V_1 (\cos\theta_1 + j\sin\theta_1)$$

$$\boldsymbol{V}_2 = V_2 \angle -\theta_2 = V_2 e^{-j\theta_2} = V_2 (\cos\theta_2 - j\sin\theta_2)$$

$$\begin{aligned}\text{Their product} &= \boldsymbol{V}_1 \boldsymbol{V}_2 = [V_1 (\cos\theta_1 + j\sin\theta_1)]\,[V_2 (\cos\theta_2 - j\sin\theta_2)] \\ &= V_1V_2 [\cos\theta_1 \cos\theta_2 - j\cos\theta_1 \sin\theta_2 + j\sin\theta_1 \cos\theta_2 - j^2 \sin\theta_1 \sin\theta_2] \\ &= V_1 V_2 [\cos\theta_1 \cos\theta_2 + \sin\theta_1 \sin\theta_2 + j(\sin\theta_1 \cos\theta_2 - \sin\theta_2 \cos\theta_1)] \\ &= V_1 V_2 [\cos(\theta_1 - \theta_2) + j\sin(\theta_1 - \theta_2)]\end{aligned}$$

We thus see that, the product of two phasors is a phasor having modulus equal to the product of the modulus of the two phasors and the argument is equal to the algebraic sum of the arguments of the two phasors.

In polar form we have the product $\boldsymbol{V}_1\boldsymbol{V}_2 = V_1V_2 \angle\theta_1 - \theta_2$

In exponential form we have the product $\boldsymbol{V}_1\boldsymbol{V}_2 = V_1 V_2 e^{j(\theta_1 - \theta_2)}$

7.8.2. Division of Phasors.

***(i)* Rectangular Form.** $\dfrac{\boldsymbol{V}_1}{\boldsymbol{V}_2} = \dfrac{a_1 + j b_1}{a_2 - j b_2}$

Multiplying both the numerator and denominator by conjugate of $(a_2 - j b_2)$ *i.e.* by $(a_2 + j b_2)$ we have

$$\frac{\boldsymbol{V}_1}{\boldsymbol{V}_2} = \frac{a_1 + j b_1}{a_2 - j b_2} \times \frac{a_2 + j b_2}{a_2 + j b_2} = \frac{(a_1 a_2 - b_1 b_2) + j(a_1 b_2 + a_2 b_1)}{a_2^2 + b_2^2} = \frac{(a_1 a_2 - b_1 b_2)}{a_2^2 + b_2^2} + j\frac{a_1 b_2 + a_2 b_1}{a_2^2 + b_2^2}$$

***(ii)* Polar and Exponential Form.**

$$\frac{\boldsymbol{V}_1}{\boldsymbol{V}_2} = \frac{a_1 + j b_1}{a_2 - j b_2} = \frac{V_1(\cos\theta_1 + j\sin\theta_1)}{V_2(\cos\theta_2 - j\sin\theta_2)} = \frac{V_1 \angle\theta_1}{V_2 \angle -\theta_2} = \frac{V_1 e^{j\theta_1}}{V_2 e^{-j\theta_2}}$$

The quotient is

$$\begin{aligned}\frac{\boldsymbol{V}_1}{\boldsymbol{V}_2} &= \frac{V_1(\cos\theta_1 + j\sin\theta_1)}{V_2(\cos\theta_1 - j\sin\theta_2)} = \frac{V_1}{V_2}\,\frac{(\cos\theta_1 + j\sin\theta_1)}{(\cos\theta_2 - j\sin\theta_2)} \times \frac{\cos\theta_2 + j\sin\theta_2}{(\cos\theta_2 + j\sin\theta_2)} \\ &= \frac{V_1}{V_2}\left[\frac{\cos\theta_1 \cos\theta_2 - \sin\theta_1 \sin\theta_2 + j(\sin\theta_1 \cos\theta_2 + \sin\theta_2 \cos\theta_1)}{\cos^2\theta_2 + \sin^2\theta_2}\right] \\ &= \frac{V_1}{V_2}[\cos(\theta_1 + \theta_2) + j\sin(\theta_1 + \theta_2)]\end{aligned}$$

We thus see that, when we divide one phasor by another phasor, the quotient is a phasor having modulus equal to the quotient of the modulus of the two phasors and an argument equal to the algebraic difference of the arguments of the two phasors.

In polar form we have

$$\frac{\boldsymbol{V}_1}{\boldsymbol{V}_2} = \frac{V_1}{V_2} \angle\theta_1 - (-\theta_2) = \frac{V_1}{V_2} \angle\theta_1 + \theta_2$$

In exponential form we have

$$\frac{\boldsymbol{V}_1}{\boldsymbol{V}_2} = \frac{V_1}{V_2}\, e^{j(\theta_1 + \theta_2)}$$

Example 7.2

Perform the following operation and the final result may be given in the polar form :

$$(8 + j6) \times (-10 - j\,7.5)$$

[Bangalore Univ. Electrical Engineering and Electronics, 1990]

Solution: $(8 + j\,6) \times (-10 - j\,7.5) = (-80 - j\,60 - j\,60 + 45) = (-35 - j\,120)$

The resultant phasor lies in third quadrant because both the components of the phasor are negative. The magnitude and phase angle are given by

$$\text{Magnitude} = \sqrt{(-35)^2 + (-120)^2} = 125$$

$$\text{Phase angle } \theta = \text{Tan}^{-1}\frac{-120}{-35} = \angle 180° + 73.74° = \angle 253.74°$$

Hence resultant may be given as $125\ \angle 253.74°$ **Ans.**

Example 7.3

Perform the following operation and express the final result in polar form :

$$5\angle 30° + 8\angle -30°$$

[Bangalore Univ. Electrical Engineering and Electronics, 1989]

Solution:

$$5\angle 30° = 5(\cos 30° + j\sin 30°) = (4.33 + j\,2.5)$$

$$8\angle -30° = [8(\cos(-30°) + j\sin(-30°)) = (6.928 - j\,4)$$

So $5\angle 30° + 8\angle -30° = (4.33 + j\,2.5) + (6.928 - j\,4) = (11.258 - j\,1.5)$

The resultant phasor lies in fourth quadrant and its magnitude and phase angle are given by

$$\text{Magnitude} = \sqrt{(11.258)^2 + (-1.5)^2} = 11.358$$

$$\text{Phase angle } \theta = \text{Tan}^{-1}\frac{-1.5}{11.258} = \angle -7.59°$$

Hence resultant may be given as $11.358\ \angle -7.59°$ **Ans.**

Example 7.4

Calculate the following in polar form :

(*i*) **Add $(40 + j\,20)$ to $(20 + j\,120)$**
(*ii*) **Subtract $(10 + j\,30)$ from $(20 - j\,20)$**
(*iii*) **Multiply $(15 + j\,20)$ with $(20 + j\,30)$**
(*iv*) **Divide $(6 + j\,7)$ by $(5 + j\,3)$**

[Gujarat Univ. June/July 2003]

Solution: (*i*) $(40 + j\,20) + (20 + j\,120) = (60 + j\,140) = \sqrt{(60^2 + 140^2)}\ \angle \tan^{-1} 140/60 = 152.32\ \angle 66.8°$ **Ans.**

(*ii*) $(20 - j\,20) - (10 + j\,30) = (10 - j\,50) = \sqrt{10^2 + (-50)^2}\ \angle \tan^{-1} -50/10 = 50.99\ \angle -78.69°$ **Ans.**

(*iii*) $(15 + j\,20) \times (20 + j\,30) = 300 + j\,450 + j\,400 - 600 = (-300 + j\,850) = 901.4\ \angle 109.44°$ **Ans.**

Alternatively $(15 + j\,20) \times (20 + j\,30) = 25\ \angle 53.13° \times 36.056\ \angle 56.31° = 901.4\ \angle 109.44°$ **Ans.**

$$(iv)\ \frac{6 + j\,7}{5 + j\,3} = \frac{(6 + j\,7)(5 - j\,3)}{(5 + j\,3)(5 - j\,3)} = \frac{30 - j\,18 + j\,35 + 21}{34}$$

$$= \frac{51 + j\,17}{34} = (1.5 + j\,0.5) = 1.581\ \angle 18.44° \quad \textbf{Ans.}$$

$$\text{Alternatively } \frac{6 + j\,7}{5 + j\,3} = \frac{9.22\ \angle 49.40°}{5.83\ \angle 30.96°} = 1.581\ \angle 18.44° \quad \textbf{Ans.}$$

Example 7.5

Three phasors are A = 10 + j 10 ; B = 50 $\angle 120°$ and C = 8 − j 6. Evaluate $\frac{AB}{C}$ and express it in rectangular and polar forms.

Solution : Phasor $A = (10 + j10)$ have modulus of $\sqrt{10^2 + 10^2}$ *i.e.* $10\sqrt{2}$ and argument of $\tan^{-1}\frac{10}{10}$ *i.e.* 45° or $\frac{\pi}{4}$ radians.

So phasor A in polar form will be $10\sqrt{2}\ \angle 45°$.

Phasor B in polar form is $50\ \angle 120°$.

Phasor $C = (8 - j\,6)$ have modulus of $\sqrt{8^2 + (-6)^2}$ *i.e.* 10 and argument of $\tan^{-1}\frac{-6}{8}$ or − 36.87°

$$\text{So } C = 10\ \angle -36.87°$$

$$\text{Hence } \frac{AB}{C} = \frac{10\sqrt{2}\ \angle 45° \times 50\ \angle 120°}{10\ \angle -36.87°} = \frac{10\sqrt{2} \times 50}{10}\ \angle 45° + 120° - (-36.87°)$$

$$= 50\sqrt{2}\ \angle 201.87° \quad \textbf{Ans.}$$

In rectangular form $\frac{AB}{C}$ will be $= 50\sqrt{2}\cos 201.87° + j\,50\sqrt{2}\sin 201.87° = -65.622 - j\,26.34$ **Ans.**

7.9. POWERS AND ROOTS OF PHASORS

Powers and roots of phasors can be determined conveniently in polar form. If the phasor is not in polar form then it will be advisable to convert it first to polar form and then carry out algebraic operations.

7.9.1. Powers of Phasors. Let us find the fourth power of the phasor $5\ \angle 20°$, for this purpose, the phasor has to be multiplied by itself four times.

$$\text{So } \left(5\ \angle 20°\right)^4 = 5 \times 5 \times 5 \times 5\ \angle 20° + 20° + 20° + 20° = 625\ \angle 80°$$

$$\text{In general } \mathbf{A}^n = \mathrm{A}^n\ \angle n\,\theta$$

7.9.2. Roots of Phasors. Let us find the one-fourth power of the phasor $5\ \angle 20°$

$$\left(5\ \angle 20°\right)^{1/4} = (5)^{1/4}\angle 20/4 = 1.496\ \angle 5° \quad \textbf{Ans.}$$

$$\text{In general } \sqrt[n]{\mathbf{A}} = \sqrt[n]{\mathrm{A}}\ \angle \theta/n.$$

7.10. PHASOR ALGEBRA APPLIED TO SINGLE PHASE SERIES CIRCUITS

A simple series circuit having non-inductive resistance R and inductive reactance X_L is shown in Fig. 7.7 (a). Let the current phasor be taken as reference phasor and be represented as $\boldsymbol{I} = I + j\,0$ or $I\ \angle 0°$.

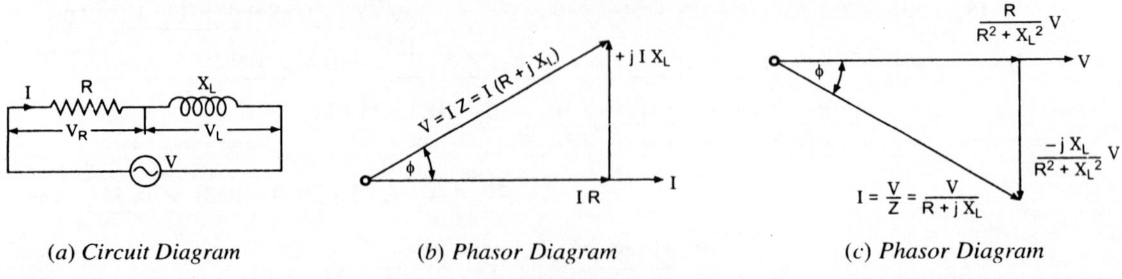

(a) *Circuit Diagram* (b) *Phasor Diagram* (c) *Phasor Diagram*

Fig. 7.7

Since voltage drop across resistance R, V_R = IR in phase with current I, it is expressed as $(IR + j\,0)$ and the voltage drop across the inductive reactance, $V_L = I\,X_L$ in quadrature with current and leading, therefore, it is expressed as $(0 + j\,IX_L)$. Hence the voltage applied across the circuit can be expressed as

$$V = V_R + V_L = (IR + j\,0) + (0 + j\,IX_L) = I\,(R + j\,X_L) = I\,Z^* = I\,Z\angle\Phi$$

Hence impedance **Z** can be written as $(R + j\,X_L)$ or $Z\angle\Phi$ where $\Phi = \text{Tan}^{-1}\dfrac{X_L}{R}$.

Thus, impedance can be written as complex quantity though it is not a complex quantity. Since impedance causes resistance and reactance voltage drops to be at right angle to each other, it is actually a complex operator.

Now let the applied voltage be taken as reference phasor and expressed as $V(1 + j\,0)$ or $V\angle 0°$ then current be determined by using impedance as phasor

$$I = \frac{V}{Z} = \frac{V + j0}{R + j\,X_L} \times \frac{R - j\,X_L}{R - j\,X_L} = \frac{VR}{R^2 + X_L^2} - j\frac{VX_L}{R^2 + X_L^2} \quad \text{using complex notation}$$

$$\text{and } I = \frac{V}{Z} = \frac{V\angle 0°}{Z\angle\Phi} = \frac{V}{Z}\angle -\Phi \qquad \text{where } \Phi = \text{Tan}^{-1}\frac{X_L}{R} \text{ using polar notation.}$$

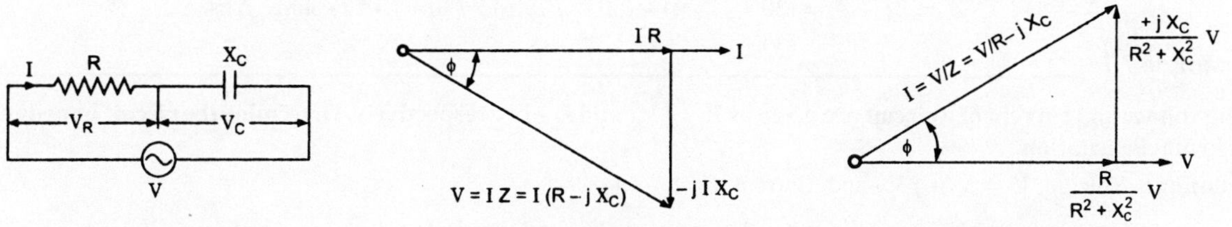

(*a*) *Circuit Diagram* (*b*) *Phasor Diagram* (*c*) *Phasor Diagram*

Fig. 7.8

In Fig. 7.8 (*a*) circuit having resistance *R* and capacitive reactance X_C connected in series are shown. Taking current *I* as reference phasor *i.e.* along *X*-axis, as shown in phasor diagram [Fig. 7.8 (*b*)]

Applied voltage,

$$V = V_R + V_C = (I\,R + j\,0) + (0 - j\,I\,X_C) = I\,(R - j\,X_C) = I\,Z = I\,Z\angle -\Phi$$

Hence impedance *Z* can be written as $(R - j\,X_C)$ or $Z\angle -\Phi$ where $\Phi = \text{Tan}^{-1}\dfrac{X_C}{R}$

Taking applied voltage as reference phasor, as shown in Fig. 7.8 (*c*), the current flowing through the circuit is given as

$$I = \frac{V}{Z} = \frac{V + j\,0}{R - j\,X_C} \times \frac{R + j\,X_C}{R + j\,X_C} = \frac{VR}{R^2 + X_C^2} + j\frac{VX_C}{R^2 + X_C^2} \quad \text{using complex notation}$$

$$\text{and } I = \frac{V}{Z} = \frac{V\angle 0°}{Z\angle -\Phi} = \frac{V}{Z}\angle\Phi \qquad \text{where } \Phi = \text{Tan}^{-1}\frac{X_C}{R} \text{ using polar notation.}$$

Thus, general expression for phasor impedance can be written as

$$Z = (R \pm j\,X) \text{ in complex notation and } Z = Z\angle \pm\Phi; \qquad \text{where } \Phi = \text{Tan}^{-1}\frac{X}{R} \text{ in polar notation.}$$

Plus sign is used for inductive circuit as inductive reactance is considered positive and minus sign is used for capacitive circuit as capacitive reactance is considered negative.

* In the expression **V** = **I Z**, **V** and **I** differ from **Z** in that they are associated with time varying quantities, whereas **Z** is a complex number independent of time and is, therefore, not a phasor in the sense that **V** and **I** are phasors. It has, however, become the practice to refer all complex numbers used in ac circuit calculations as phasors.

Example 7.6

State in symbolic notation, the impedance of each of the following circuits at a frequency of 50 Hz.

(*a*) A resistance of 20 Ω in series with an inductance of 0.1 H

(*b*) A resistance of 50 Ω in series with a capacitance of 40 μF

(*c*) Circuits (*a*) and (*b*) in series.

Solution: *For Circuit (a)*

$$\text{Resistance, } R_1 = 20\ \Omega$$

$$\text{Reactance, } X_1 = 2\pi f L = 2\pi \times 50 \times 0.1 = 31.4\ \Omega \text{ (being inductive)}$$

$$\therefore \text{Impedance, } Z_1 = (R_1 + jX_1) = (20 + j\,31.4)\ \Omega \quad \textbf{Ans.}$$

For Circuit (b)

$$\text{Resistance, } R_2 = 50\ \Omega$$

$$\text{Reactance, } X_2 = \frac{1}{2\pi f C} = \frac{1}{2\pi\times 50\times 40\times 10^{-6}} = -79.6\ \Omega \text{ (being capacitive)}$$

$$\therefore \text{Impedance, } Z_2 = (50 - j\,79.6)\ \Omega \quad \textbf{Ans.}$$

When circuits (*a*) and (*b*) are connected in series, total impedance of the circuit,

$$\mathbf{Z} = \mathbf{Z}_1 + \mathbf{Z}_2 = (20 + j\,31.4) + (50 - j\,79.6) = (70 - j\,48.2) \text{ ohms} \quad \textbf{Ans.}$$

Example 7.7

The voltage and current of a circuit are given as $V_1 + j\,V_2$ and $I_1 - j\,I_2$ respectively. Determine the circuit impedance in symbolic notation.

Solution: Voltage, $V = V_1 + j\,V_2$ and Current, $I = I_1 - j\,I_2$

$$\text{Impedance, } Z = \frac{V}{I} = \frac{V_1 + j\,V_2}{I_1 - j\,I_2} = \frac{V_1 + j\,V_2}{I_1 - j\,I_2} \times \frac{I_1 + j\,I_2}{I_1 + j\,I_2} = \frac{V_1 I_1 - V_2 I_2}{I_1^2 + I_2^2} + j\,\frac{V_2 I_1 + I_2 V_1}{I_1^2 + I_2^2} \quad \textbf{Ans.}$$

Example 7.8

A resistance of 4 Ω is connected in series with an inductive reactance of 5 Ω. Draw the impedance phasor for this circuit and express the impedance in the symbolic form, polar form and exponential form.

Solution: Circuit resistance, R = 4 Ω

Circuit reactance, X = 5 Ω (inductive)

The impedance phasor of the circuit is shown in Fig. 7.9.

$$\text{Magnitude of impedance, } Z = \sqrt{4^2 + 5^2} = 6.4\ \Omega$$

$$\text{Argument, } \theta = \text{Tan}^{-1}\frac{5}{4} = 51.34° \text{ or } 0.285\,\pi \text{ radians} \quad \textbf{Ans.}$$

$$\therefore \text{Impedance, } \mathbf{Z} = (4 + j\,5)\ \Omega \text{ in symbolic form} \quad \textbf{Ans.}$$

$$\text{or } \mathbf{Z} = 6.4\ \angle 51.34°\ \Omega \text{ in polar form} \quad \textbf{Ans.}$$

$$\text{or } \mathbf{Z} = 6.4\,e^{j\,0.285\pi} \text{ in exponential form} \quad \textbf{Ans.}$$

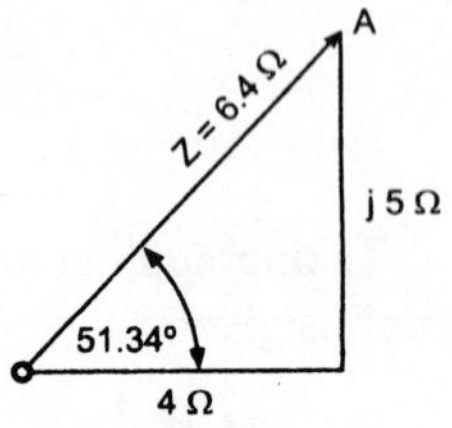

Fig. 7.9

Example 7.9

$V = 200\ \angle 36.9°$ volts ; $Z = 20\ \angle -53.1°$ ohms. Find I.

Solution : Circuit voltage, $V = 200\ \angle 36.9°$ volts

and Circuit impedance, $Z = 20\ \angle -53.1°$ ohms

$$\text{Circuit current, } I = \frac{V}{Z} = \frac{200\ \angle 36.9°}{20\ \angle -53.1°} = \frac{200}{20}\ \angle 36.9° - (-53.1°) = 10\ \angle 90° \text{ A} \quad \textbf{Ans.}$$

Example 7.10

A load having impedance of $1 + j\,1\ \Omega$ is connected to an ac voltage represented as $v = 20\sqrt{2}\cos(\omega t + 10°)$ volt.

(i) Find the current in load, expressed in the form of $i = I_m \sin(\omega t + \phi)$ A.

(ii) Find real power consumed by the load. *[U.P. Technical Univ. Elec. Engineering Second Semester 2004-05]*

Solution: Load impedance, $\mathbf{Z} = (1 + j1)\ \Omega = \sqrt{2}\angle 45°\ \Omega$

Maximum value of applied voltage, $V_{max} = 20\sqrt{2}$ V

RMS value of applied voltage, $V_{rms} = \dfrac{V_{max}}{\sqrt{2}} = \dfrac{20\sqrt{2}}{\sqrt{2}} = 20$ V

Phase angle of applied voltage, $\phi = 100°$ $\quad \because \cos(\omega t + 10°) = \text{Sin}(\omega t + 90° + 10°)$

So applied voltage, $\mathbf{V} = 20\angle 100°$ V

Load current, $\mathbf{I} = \dfrac{\mathbf{V}}{\mathbf{Z}} = \dfrac{20\angle 100°}{\sqrt{2}\angle 45°} = 10\sqrt{2}\angle 55°$ A

Peak value of load current, $I_{max} = \sqrt{2} \times 10\sqrt{2} = 20$ A

and equation for load current is given as $i = I_{max}\sin(\omega t + \phi) = 20\sin(\omega t + 55°)$ **Ans.**

Example 7.11

A 60 Hz sinusoidal voltage $v = 100\sin\omega t$ is applied to a series R-L circuit. Assuming $R = 10\ \Omega$, $L = 0.01$ H, find the steady-state current and relative phase angle. Also compute the effective magnitude and phase voltage drops across each element. *[R.G.P.V. Bhopal Basic Electrical Engineering June 2004]*

Solution: RMS value of applied voltage, $V = \dfrac{V_{peak}}{\sqrt{2}} = \dfrac{100}{\sqrt{2}} = 70.71$ V

Resistance, $R = 10\ \Omega$

Inductive reactance, $X_L = 2\pi f L = 2\pi \times 60 \times 0.01 = 3.77\ \Omega$

Impedance, $Z = \sqrt{R^2 + X_L^2} = \sqrt{10^2 + (3.77)^2} = 10.687\ \Omega$

Phase angle, $\phi = \text{Tan}^{-1}\dfrac{X_L}{R} = \text{Tan}^{-1}\dfrac{3.77}{10} = 20.656°$

Steady-state current, $\mathbf{I} = \dfrac{\mathbf{V}}{\mathbf{Z}} = \dfrac{70.71\angle 0°}{10.687\angle 20.656°}$ A Taking applied voltage as reference phasor

$= 6.62\angle -20.656°$ A

Thus, the effective value of steady-state current is 6.62 A, and the phase angle is 20.656° (lag) **Ans.**

Voltage drop across resistor, $\mathbf{V_R} = \mathbf{I}R = 6.62\angle -20.656° \times 10 = 66.2\angle -20.656°$ V **Ans.**

Voltage drop across inductance, $\mathbf{V_L} = j\,\mathbf{I}\,X_L = 6.62\angle -20.656° \times 3.77\angle 90° = 24.95\angle 69.344°$ V **Ans.**

Example 7.12

In a given R-L circuit, $R = 35\ \Omega$ and $L = 0.1$ H. Find (i) current through the circuit (ii) power factor if a 50 Hz frequency, voltage $V = 220\angle 30°$ is applied across the circuit. *[R.G.P.V. Bhopal Basic Electrical Engineering 2001]*

Solution: Resistance of R-L circuit, $R = 35\ \Omega$

Inductive reactance of R-L circuit, $X_L = 2\pi f L = 2\pi \times 50 \times 0.1 = 31.42\ \Omega$

Magnitude of impedance, $Z = \sqrt{R^2 + X_L^2} = \sqrt{35^2 + 31.42^2} = 47.03\ \Omega$

Argument, $\theta = \text{Tan}^{-1}\dfrac{X_L}{R} = \text{Tan}^{-1}\dfrac{31.42}{35} = 41.91°$

$$\text{Hence impedance, } \mathbf{Z} = (35 + j\,31.42)\ \Omega = 47.03 \angle 41.9^\circ\ \Omega$$

$$\text{Applied voltage, } \mathbf{V} = 220 \angle 30^\circ\ \text{V}$$

$$(i)\ \text{Current through the circuit, } \mathrm{I} = \frac{\mathbf{V}}{\mathbf{Z}} = \frac{220 \angle 30^\circ}{47.03 \angle 41.9^\circ} = 4.678 \angle -11.9^\circ\ \text{A}$$

Magnitude of current, I = 4.678 A **Ans.**

$$(ii)\ \text{Power factor} = \cos\theta = \cos[30^\circ - (-11.9^\circ)] = \cos 41.9^\circ = 0.7443\ \text{(lagging)}$$ **Ans.**

Example 7.13

A coil of resistance 10 Ω and inductance 0.1 H is connected in series with a condenser of capacitance 150 μF across a 200 V, 50 Hz supply. Determine the following:

***(i)* Impedance *(ii)* Current *(iii)* Power factor *(iv)* Voltage across the coil; and *(v)* Voltage across the condenser.**

Solution: Resistance of the circuit, $R = 10\ \Omega$

$$\text{Inductive reactance of the circuit, } X_L = 2\pi f L = 2\pi \times 50 \times 0.1 = 31.42\ \Omega$$

$$\text{Capacitive reactance of the circuit, } X_C = \frac{1}{2\pi fC} = \frac{1}{2\pi \times 50 \times 150 \times 10^{-6}} = 21.22\ \Omega$$

$$\text{Magnitude of impedance, } Z = \sqrt{R^2 + (X_L - X_C)^2} = \sqrt{(10^2) + (31.42 - 21.22)^2} = 14.28\ \Omega$$

$$\text{Argument, } \theta = \tan^{-1}\frac{X_L - X_C}{R} = \tan^{-1}\frac{31.42 - 21.22}{10} = 45.57^\circ$$

$$(i)\ \text{Hence impedance, } \mathbf{Z} = (10 + j\,10.2) \text{ or } 14.28 \angle 45.57^\circ\ \Omega$$ **Ans.**

$$(ii)\ \text{Current, } I = \frac{V}{Z} = \frac{200 \angle 0^\circ}{14.28 \angle 45.57^\circ} = 14 \angle -45.57^\circ\ \text{A}$$ **Ans.**

$$(iii)\ \text{Power factor} = \cos\Phi = \cos 45.57^\circ = 0.7\ \text{(lagging)}$$ **Ans.**

$$(iv)\ \text{Voltage across the coil} = I\,(R + jX_L) = 14 \angle -45.57^\circ \times (10 + j\,31.42)$$

$$= 14 \angle -45.57^\circ \times 32.97 \angle 72.34^\circ$$

$$= 14 \times 32.97 \angle -45.57^\circ + 72.34^\circ$$

$$= 461.58 \angle 26.77^\circ \text{ volts}$$ **Ans.**

$$(v)\ \text{Voltage across the capacitor} = I X_C = 14 \angle -45.57^\circ \times 21.22 \angle -90^\circ$$

$$= 297.08 \angle -135.57^\circ \text{ volts}$$ **Ans.**

Example 7.14

In a circuit, the applied voltage is found to lag the current by 30°.

***(a)* Is the power factor lagging or leading ? *(b)* What is the value of power factor *(c)* Is the circuit inductive or capacitive ?**

In the diagram of Fig. 7.10, the voltage drop across Z_1 is $(10 + j0)$ volts. Find out *(i)* the current in the circuit *(ii)* the voltage drops across Z_2 and Z_3 *(iii)* the voltage of the generator. [*Bombay Univ. Electrical Engineering-I, 1991*]

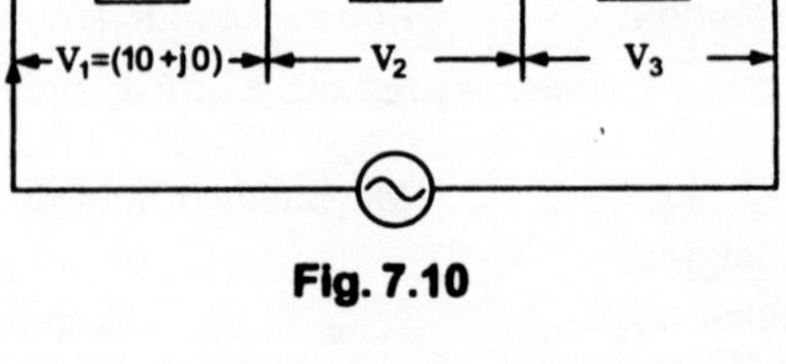

Fig. 7.10

Solution: *(a)* Since the applied voltage lags behind the current, current leads the applied voltage. It means that power factor is **leading. Ans.**

(b) Power factor = Cos ϕ = Cos 30° = 0.866 (leading) **Ans.**

(c) Since the current leads the applied voltage, the circuit is *capacitive.* **Ans.**

(*i*) Circuit current, **I** = Current through impedance $\mathbf{Z}_1$

$$= \frac{\text{Voltage drop across } \mathbf{Z}_1, \mathbf{V}_1}{\text{Impedance } \mathbf{Z}_1} = \frac{10 + j\,0}{3 + j\,4} = \frac{10 \angle 0^\circ}{5 \angle 53.15^\circ} = 2 \angle -53.13^\circ \text{ A}$$

$= (1.2 - j1.6)$ A **Ans.**

(*ii*) Voltage drop across $\mathbf{Z}_2$, $\mathbf{V}_2 = \mathbf{I} \times \mathbf{Z}_2 = 2 \angle -53.13^\circ \times (2 + j\,3.46) = 2 \angle -53.13^\circ \times 3.996 \angle 59.97^\circ$

$= 7.993 \angle 6.84^\circ = (7.936 + j\,0.952)$ V **Ans.**

Voltage drop across $\mathbf{Z}_3$, $\mathbf{V}_3 = \mathbf{I} \times \mathbf{Z}_3 = 2 \angle -53.13^\circ \times (1 - j\,7.46) = 2 \angle -53.13^\circ \times 7.527 \angle -82.365^\circ$

$= 15.05 \angle -135.495^\circ$ V

$= (-10.74 - j\,10.55)$ V **Ans.**

(*iii*) Generator voltage, $\mathbf{V} = \mathbf{V}_1 + \mathbf{V}_2 + \mathbf{V}_3 = (10 + j\,0) + (7.936 + j\,0.952) + (-10.74 - j\,10.55)$

$= (7.196 - j\,9.598) = 12 \angle -53.13^\circ$ V **Ans.**

Example 7.15

In the circuit of Fig. 7.11, applied voltage is given by (0 + *j*10) and the current is (0.8 + *j* 0.6) A. Determine the values of R and X and also indicate if X is inductive or capacitive. [*Nagpur Univ. Electrical Technology 1991*]

Solution: Applied voltage, $V = (0 + j\,10) = 10 \angle 90^\circ$ volts

Circuit current, $I = (0.8 + j\,0.6) = 1 \angle 36.9^\circ$ A

Current lags behind applied voltage by $\Phi = (90^\circ - 36.9^\circ)$ *i.e.* 53.1°

Hence the circuit shown is inductive.

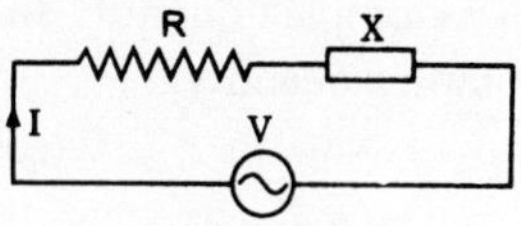

Fig. 7.11

$$\text{Circuit impedance, } Z = \frac{V}{I} = \frac{10 \angle 90^\circ}{1 \angle 36.9^\circ} = 10 \angle 53.1^\circ = (6 + j\,8)\ \Omega$$

Hence R = 6 Ω and X = 8 Ω (inductive)

7.11. POWER DETERMINATION

If the voltage and current in a circuit are expressed in complex form, power can be determined as follows:

Let the current be expressed as

$I = I_1 - j\,I_2$ and voltage as $V = V_1 + j\,V_2$.

From Fig. 7.12 it is evident that the real component of voltage, V_1 and real component of current I_1 are in phase. Since components of voltage and current may be treated as if they were acting alone, the power contributed by V_1 and I_1, $P_1 = V_1\,I_1$.

The coefficient of V_2 is $+j$ and coefficient I_2 is $-j$, these two components of voltage and current are opposite in phase and therefore, their product gives negative power.

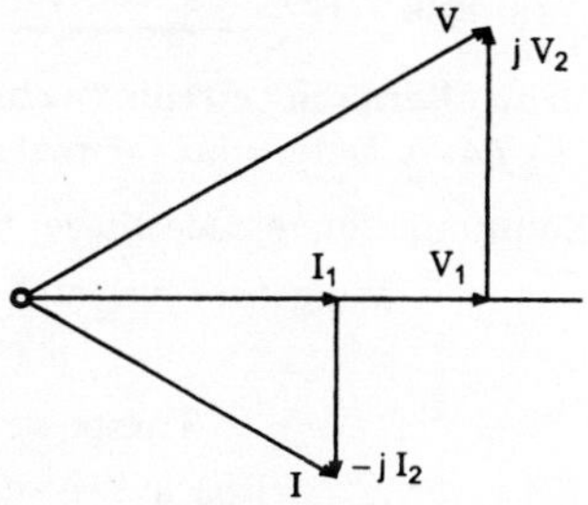

Fig. 7.12

So power $P = -V_2\,I_2$

Voltage component V_1 and current component I_2 are in quadrature so power contributed by V_1 and I_2 is zero, similarly voltage component V_2 and current component I_1 are in quadrature and, therefore, contribute no power.

∴ Total power in the circuit, $P = P_1 + P_2 = V_1\,I_1 - V_2\,I_2$

The total power is the algebraic sum of the product of the real quantities V_1 and I_1 and the product of imaginary quantities V_2 and I_2. The sign being determined in the ordinary algebraic method. The operator j must not be included in the multiplication.

Example 7.16

Determine the power consumed in a circuit carrying a current of (3 – j 4) A and connected across (80 + j 40) volt supply mains.

Solution: Supply voltage, V = (80 + j 40) volts

Circuit current, I = (3 – j 4) amperes

Power consumed, $P = V_1 I_1 + V_2 I_2 = 80 \times 3 + 40\,(-4) = 80$ watts **Ans.**

7.12. CONJUGATE METHOD OF POWER DETERMINATION

In any circuit let the applied voltage and current flowing be $V = V(\cos\alpha + j\sin\alpha)$ and $I = I(\cos\beta + j\sin\beta)$ respectively. $(\alpha - \beta)$ is the phase difference between V and I so power factor of the circuit is $\cos(\alpha - \beta)$.

Multiplying V and I we have

$$VI = V(\cos\alpha + j\sin\alpha) \times I(\cos\beta + j\sin\beta) = VI[\cos(\alpha+\beta) + j\sin(\alpha+\beta)].$$

The real part of the above expression is $VI\cos(\alpha+\beta)$ and evidently it cannot be true power of the circuit, because power factor of the circuit is $\cos(\alpha - \beta)$. Similarly $VI\sin(\alpha+\beta)$ is not correct expression of the reactive volt-amperes. Hence to determine the power by j method, it is necessary to change the sign of the imaginary part of either V or I. Thus if we change the sign of imaginary part of V, leaving I as before, we have

$$V_C = V(\cos\alpha - j\sin\alpha)$$

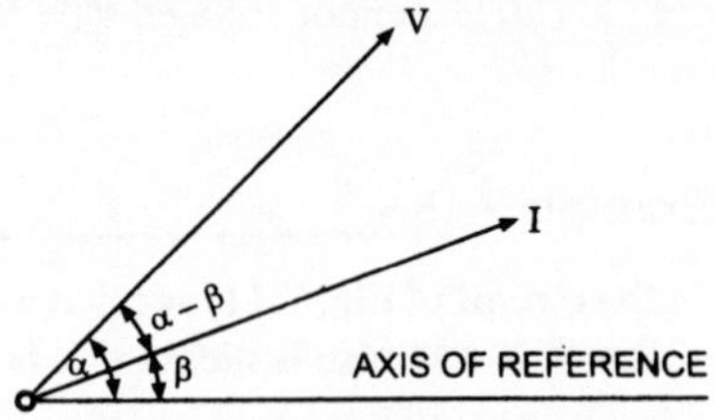

Fig. 7.13

Complex quantity so obtained is known as *conjugate* of the original complex quantity.

Now $V_C I = V(\cos\alpha - j\sin\alpha) \times I(\cos\beta + j\sin\beta) = VI[\cos(\alpha-\beta) - j\sin(\alpha-\beta)]$.

The real part of the above expression $VI\cos(\alpha-\beta)$ gives the true power and imaginary part $VI\sin(\alpha-\beta)$ gives the reactive power. It should, however, be noted that the real power obtained by the conjugate method is the same regardless of whether V or I is reversed but sign of reactive volt-amperes will depend upon it. When the conjugate of voltage is used, capacitive VARs have a positive sign and inductive VARs a negative sign. If the conjugate of the current rather than of the voltage is used the capacitive VARs have a negative sign and inductive VARs a positive sign.

The product of V and I_C is called the *complex power* where I_C is the conjugate of I.

Example 7.17

In an alternating circuit, the impressed voltage is given by V = (100 – j 50) volts, and the current in the circuit is I = (3 – j 4) A. Determine the real and reactive power in the circuit. [*Calcutta Univ. Electrical Engineering 1991*]

Solution: Impressed voltage, V = (100 – j 50)

Conjugate of current, I_C = (3 + j 4) amperes

$V I_C$ = (100 – j 50) (3 + j 4) = 500 + j 250

True power = Real part of $V I_C$ = 500 watts **Ans.**

Reactive power = j part of $V I_C$ = 250 VAR (inductive) **Ans.**

Example 7.18

The current in a circuit is given by (5 + j 10) A when the applied voltage is (225 + j 150) V. Determine (*i*) the complex expression for the impedance stating whether it is inductive or capacitive (*ii*) power (*iii*) phase angle between current and voltage. Draw the complete phasor diagram.

Solution: Circuit voltage, V = (225 + j 150) volts or $270.42 \angle 33.7°$ V

Circuit current, I = (5 + j 10) amperes or $11.18 \angle 63.4°$ A

Circuit impedance, $\mathbf{Z} = \frac{V}{I} = \frac{270.42}{11.18} \angle 33.7° - 63.4° = 24.2 \angle -29.7° \; \Omega$

or $\mathbf{Z} = 24.2 [\cos(-29.7°) + j \sin(-29.7°)] = (21 - j\,12)\;\Omega$ **Ans.**

(*i*) Since the *j* part of the impedance expression is negative hence the impedance is capacitive. **Ans.**

(*ii*) Conjugate current,

$$I_C = (5 - j\,10)$$

$$V I_C = (225 + j\,150)(5 - j\,10) = (2{,}625 - j\,1{,}500)$$

Power of the circuit,

$$P = \text{Real part of } V I_C = 2{,}625 \text{ W } \textbf{Ans.}$$

Power of the circuit can also be determined as follows :

$$P = VI \cos \Phi = 270.42 \times 11.18 \cos (63.4° - 33.7°)$$

$$= 270.42 \times 11.18 \cos 29.7° = 2{,}625 \text{ W } \textbf{Ans.}$$

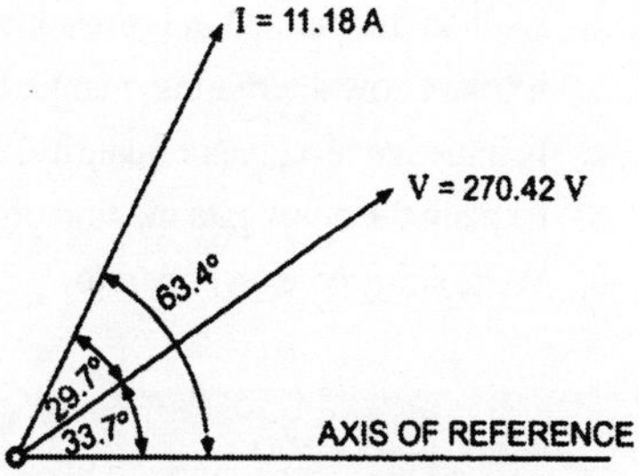

Fig. 7.14

(*iii*) Phase angle between current and voltage,

$$\Phi = \text{Argument of current} - \text{argument of voltage} = 63.4° - 33.7° = 29.7° \text{ (leading) } \textbf{Ans.}$$

Phasor diagram is shown in Fig. 7.14.

HIGHLIGHTS

1. *j* is an operator which indicates an operation of rotating the phasor through 90° in counterclockwise direction. It has numerical value of $(\sqrt{-1})$ so that

$$j^2 = (\sqrt{-1})^2 = -1; \; j^3 = (\sqrt{-1})^3 = -\sqrt{-1} \text{ and } j^4 = (\sqrt{-1})^4 = 1$$

2. Impedance can be expressed as

$\mathbf{Z} = (R \pm j X)$ in rectangular form

$\mathbf{Z} = Z \angle \pm\theta$ in polar form where $\theta = \text{Tan}^{-1} \frac{X}{R}$ and $Z = \sqrt{R^2 + X^2}$

$\mathbf{Z} = Z(\cos\theta \pm j \sin\theta)$ in trigonometrical form

and $\mathbf{Z} = Z e^{\pm j\theta}$ in exponential form

3. In symbolic method or *j*-method of solution of ac circuits, the quantities are represented in complex, triangular or polar form and then the solution is obtained.

If applied voltage is represented as

$$\mathbf{V} = V \pm j\,0 \text{ or } V \angle 0°$$

and I is a current lagging behind the voltage by angle ϕ, then it is represented as

$$\mathbf{I} = (a - j\,b) \text{ or } I(\cos\phi - j \sin\phi) \text{ or } I \angle -\phi$$

For addition and subtraction, complex form is used while for multiplication and division, polar form is used.

4. Apparent power in complex form is given as

$$S = \text{Phasor voltage} \times \text{conjugate of phasor current} = \mathbf{V}\,\mathbf{I}_C$$

EXERCISES

1. What is the use of operator j ? Explain briefly. **[G.G.S.I.P. Delhi Univ. Electrical Science May 2005]**
2. Explain the method of representing a phasor by symbolic or j notation.
3. Explain how alternating quantities may be represented by complex numbers ?
4. Is impedance a phasor quantity, if not then how it is expressed in phasor or complex form ?
5. Explain the conjugate method of determining power in ac circuits.
6. Write short note on j operator. **[R.G.P.V. Bhopal December 2002]**

SHORT ANSWER TYPE QUESTIONS WITH ANSWERS

Q. 1. What are the values of j^2 and j^4 ?

Ans. $j^2 = -1$ and $j^4 = +1$

Q. 2. What is meant by complex power ?

Ans. The product $S = V.I_C$ is called the complex power where I_C is the conjugate of I.

Q. 3. Why is impedance represented by a complex number ?

[R.G.P.V. Bhopal December 2002]

Ans. In an expression **V** = **IZ**, the impedance **Z** is not associated with time varying quantities and is not a phasor in the sense that **V** and **I** are phasors. However, it is represented as a complex number. It is represented as

$Z = R \pm jX$ +ve sign for inductive reactance and – ve sign for capacitive reactance.

Q. 4. How is complex impedance dependent on frequency ?

[R.G.P.V. Bhopal December 2002]

Ans. Complex impedance has two components viz. real or horizontal component and imaginary or vertical components. The reactance is represented by imaginary component. The inductive and capacitive reactances are given as

Inductive reactance, $X_L = \omega L = 2\pi f L$ where f is supply frequency in Hz.

Capacitive reactance, $X_C = \dfrac{1}{\omega C} = \dfrac{1}{2\pi f C}$ where f is supply frequency in Hz.

Since imaginary component of impedance depends upon the supply frequency so the impedance.

PROBLEMS

1. Give the current expressed in the polar form **I** = $10 \angle 53.1°$ amperes in symbolic, trigonometrical and exponential forms. **[Ans.** $(6 + j\,8)$ A; $10\,(\cos 53.1° + j \sin 53.1°)$ A; $10\, e^{j\,0.295\,\pi}$ A**]**
2. Divide phasor $(5 + j\,6)$ by phasor $(4 - j\,3)$. **[Ans.** $0.08 + j1.56$**]**
3. Add phasors $30 \angle 45°$ and $7.5 \angle -30°$. **[Ans.** $27.7 + j\,17.46$**]**
4. Determine $\sqrt[5]{243} \angle 120°$. **[Ans.** $3 \angle 24°$ **]**
5. Determine in symbolic, polar and exponential forms the impedance of each of the following circuits at a frequency of 50 Hz.

   (*a*) A resistance of 10 Ω in series with an inductance of 50 mH.

   (*b*) A resistance of 25 Ω in series with a capacitor of 80 μF.

(*c*) Circuits (*a*) and (*b*) in series. [**Ans.** (*a*) $(10 + j\,15.7)\ \Omega$; $18.62\ \angle 57.5^\circ\ \Omega$; $18.62\ e^{j\,0.319\,\pi}\ \Omega$

(*b*) $(25 - j\,39.8)\ \Omega$; $47\ \angle -57.87^\circ\ \Omega$; $47\ e^{-j\,0.32\,\pi}\ \Omega$

(*c*) $(35 - j\,24.1)\ \Omega$; $42.5\ \angle -34.55^\circ\ \Omega$; $42.5\ e^{-j\,0.192\,\pi}\ \Omega$]

6. Determine the current flowing through a circuit having impedance of $(5 + j\,12)\ \Omega$ connected across 230 V, 50 Hz supply.
[**Ans.** $17.69\ \angle -67.38^\circ$ A]

7. A coil of resistance 20 Ω and inductance 0.1 H is connected in series with 50 μ F capacitor across a 100 V, 50 Hz supply. Estimate (*i*) current in the circuit (*ii*) power (*iii*) power factor (*iv*) voltage across the capacitor.
[**Ans.** (*i*) 2.636 A (*ii*) 138.9 W (*iii*) 0.5271 leading (*iv*) 167.8 V]

8. A resistance of 50 Ω, inductance of 100 mH and a capacitor of 100 μF are connected in series across 200 V, 50 Hz supply. Determine the impedance, current, phase angle, pf, active and reactive powers.
[**Ans.** $50\ \angle -0.5^\circ\ \Omega$; $4 \angle 0.5^\circ$ A ; 0.5° (lead), 0.99997 (lead); 799.97 W; 7 VARs (capacitive)]

9. A coil of resistance 10 Ω and inductance 0.1 H is connected in series with a capacitor of 150 μ F across 200 V, 50 Hz, supply. Determine the impedance, circuit current, active and reactive powers.
[**Ans.** $14.28\ \angle 45.57^\circ\ \Omega$; $14\ \angle -45.57^\circ$ A; 1,960 W; 1,960 VARs (inductive)]

10. In an ac circuit the impressed voltage is $(100 + j\,50)$ V and the current flowing through the circuit is $(4 - j\,3)$ A. Determine the real and reactive power in the circuit. [**Ans.** 250 W; 500 VARs (inductive)]

❑❑❑

Chapter 8

Single Phase Parallel and Series-Parallel Circuits

8.1. PARALLEL CIRCUITS

In practice we come across the parallel circuits more often because multiple system of transmission and distribution is extensively used nowadays.

Practically all lighting and power circuits are constant voltage circuits with the loads connected in parallel. As in parallel dc circuits, the voltage is the same across each branch of a parallel ac circuit. The total line current supplied to the circuit is equal to the phasor sum of the branch currents. The total line current supplied to a parallel circuit may be determined by determining the currents in each branch and adding them, taking into account their phase relations, or by determining an equivalent impedance and dividing the applied voltage by the equivalent impedance.

Parallel ac circuits can be solved by the following methods

1. Phasor method
2. Admittance method
3. Symbolic method or *j*-method

The application of a particular method will depend upon the conditions of the problem. However, in general, that method should be used which gives quick results.

8.2. PHASOR METHOD

Consider a parallel circuit consisting of three branches of impedances Z_1 (R_1, L_1 C_1), Z_2 (R_2, L_2, C_2) and Z_3 (R_3, L_3, C_3) respectively and connected in parallel across an alternating voltage of *V* volts (rms), as shown in Fig. 8.1 (*a*).

Since all the three branches are connected in parallel, the voltage across each branch is the same and equal to supply voltage V, but currents through them will be different.

Branch I

$$\text{Impedance, } Z_1 = \sqrt{(R_1)^2 + \left(X_{C_1} - X_{L_1}\right)^2}$$

$$\text{Current, } I_1 = \frac{V}{Z_1}$$

$$\text{Phase angle, } \Phi_1 = \text{Tan}^{-1} \frac{X_{C_1} - X_{L_1}}{R_1}$$

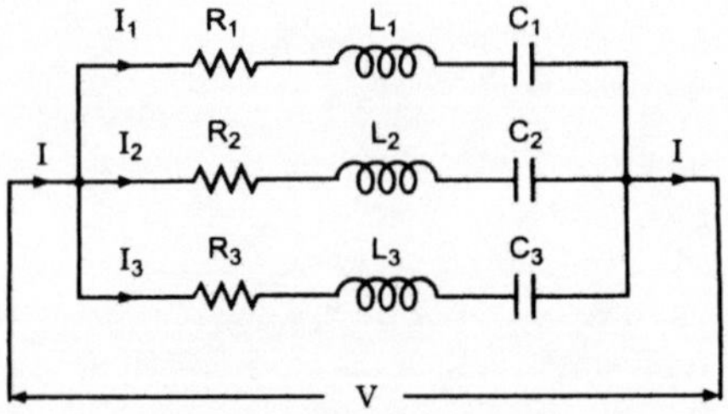

Fig. 8.1 (*a*) *Circuit Diagram*

Current I_1 will lead or lag behind the applied voltage if Φ_1 is + ve or – ve respectively.

Similarly in Branch 2

$$\text{Impedance, } Z_2 = \sqrt{(R_2)^2 + (X_{C_2} - X_{L_2})^2}$$

$$\text{Current, } I_2 = \frac{V}{Z_2}$$

$$\text{Phase angle, } \Phi_2 = \text{Tan}^{-1} \frac{X_{C_2} - X_{L_2}}{R_2}$$

and in Branch 3

$$\text{Impedance, } Z_3 = \sqrt{(R_3)^2 + (X_{C_3} - X_{L_3})^2}$$

$$\text{Current, } I_3 = \frac{V}{Z_3}$$

$$\text{Phase angle, } \Phi_3 = \text{Tan}^{-1} \frac{X_{C_3} - X_{L_3}}{R_3}$$

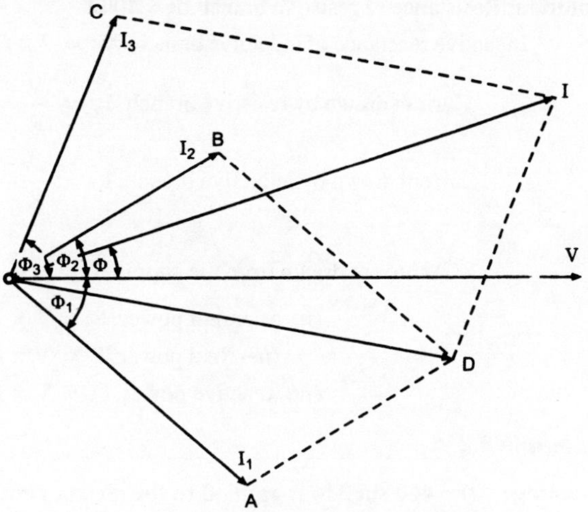

(*b*) *Phasor Diagram*

Fig. 8.1 *Single Phase Parallel Circuit*

Resultant current, I, which is phasor sum of I_1, I_2 and I_3, can be determined either by using parallelogram law of phasors, as shown in Fig. 8.1 (*b*) or by resolving branch currents I_1, I_2, and I_3 along X-axis and Y-axis and then determining the resultant of these components analytically.

Component of resultant current I along X-axis

= Sum of components of branch currents I_1, I_2 and I_3 along X-axis

or $I \cos \Phi = I_1 \cos \Phi_1 + I_2 \cos \Phi_2 + I_3 \cos \Phi_3$

Similarly component of resultant current I along Y-axis

= Sum of components of branch currents I_1, I_2 and I_3 along Y-axis

or $I \sin \Phi = I_1 \sin \Phi_1 + I_2 \sin \Phi_2 + I_3 \sin \Phi_3$

$$\text{Resultant current, } I = \sqrt{(I \cos \phi)^2 + (I \sin \phi)^2}$$

$$= \sqrt{(I_1 \cos \Phi_1 + I_2 \cos \Phi_2 + I_3 \cos \Phi_3)^2 + (I_1 \sin \Phi_1 + I_2 \sin \Phi_2 + I_3 \sin \Phi_3)^2}$$

$$\text{and Tan } \Phi = \frac{I \sin \Phi}{I \cos \Phi}$$

$$\text{or } \Phi = \text{Tan}^{-1} \frac{I_1 \sin \Phi_1 + I_2 \sin \Phi_2 + I_3 \sin \Phi_3}{I_1 \cos \Phi_1 + I_2 \cos \Phi_2 + I_3 \cos \Phi_3}$$

If Φ is + ve current I will lead the applied voltage V, if Φ is – ve current I will lag behind the applied voltage.

Example 8.1

The following circuit (Fig. 8.2) shows a parallel R-L arrangement connected across 200 volts, 50 Hz ac supply. Calculate:

(*i*) the current drawn from the supply,
(*ii*) apparent power,
(*iii*) real power, and
(*iv*) reactive power.

[*U.P. Technical Univ. Electrical Engineering Second Semester 2002-03*]

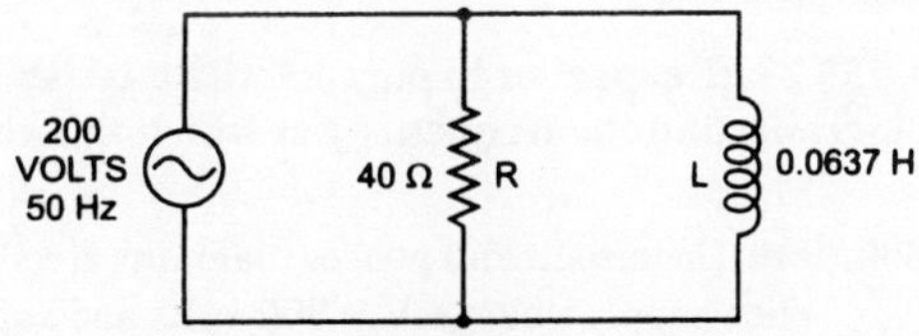

Fig. 8.2

Solution: Resistance of resistive branch, R = 40 Ω

Inductive reactance of inductive branch, $X_L = 2\pi f L = 2\pi \times 50 \times 0.0637 = 20\ \Omega$

Current drawn by resistive branch, $I_R = \frac{V}{R} = \frac{200}{40} = 5$ A

Current drawn by inductive branch, $I_L = \frac{V}{X_L} = \frac{200}{20} = 10$ A

(*i*) Current drawn from the supply, $I = \sqrt{I_R^2 + I_L^2} = \sqrt{5^2 + 10^2} = 11.18$ A **Ans.**

(*ii*) Apparent power, $S = V \times I = 200 \times 11.18 = 2{,}236$ VA or 2.236 kVA **Ans.**

(*iii*) Real power, $P = V \times I_R = 200 \times 5 = 1{,}000$ W or 1.0 kW **Ans.**

(*iv*) Reactive power, $Q = V \times I_L = 200 \times 10 = 2{,}000$ VAR or 2.0 kVAR **Ans.**

Example 8.2

A voltage $v(t) = 400 \sin 314$ is applied to the circuit shown (Fig. 8.3). Find the currents and their phase angles w.r.t. the voltages for the three branches.

$$C = 50\ \mu F\ ;\ R_1 = 100\ \Omega\ ;\ R_2 = 50\ \Omega,\ L = 0.1\ H$$

[*Pb. Technical Univ. Electrical Engineering May 2002*]

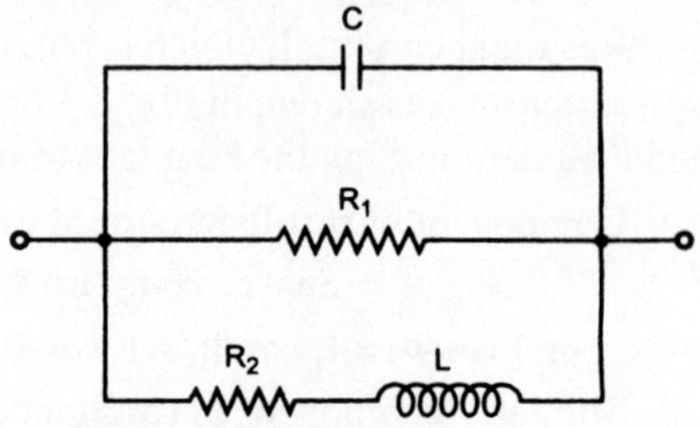

Fig. 8.3

Solution: Maximum value of applied voltage,

V_{max} = Coefficient of the sine of time angle = 400 V

RMS value of applied voltage, $V_{rms} = \frac{V_{max}}{\sqrt{2}} = \frac{400}{\sqrt{2}} = 282.84$ V

Supply frequency, $f = \frac{\text{Coefficient of time } t}{2\pi}$

$= \frac{314}{2\pi} = 50$ Hz

Current in capacitive branch, $I_1 = \frac{V}{X_C} = 2\pi f C V = 2\pi \times 50 \times 50 \times 10^{-6} \times 282.84$ A

$= 4.44$ A leading applied voltage by $\frac{\pi}{2}$ radians **Ans.**

i.e. Phase angle = 90° (leading)

Current in resistive branch, $I_2 = \frac{V}{R_1} = \frac{282.84}{100} = 2.83$ A in phase with applied voltage **Ans.**

i.e. phase angle = 0°

Current in R-L branch, $I_3 = \frac{V}{Z} = \frac{V}{\sqrt{R_2^2 + (2\pi f L)^2}} = \frac{282.84}{\sqrt{50^2 + (2\pi \times 50 \times 0.1)^2}} = \frac{282.84}{59.05} = 4.79$ A **Ans.**

Phase difference between voltage and current I_3,

$$\phi_3 = \text{Tan}^{-1} \frac{2\pi f L}{R_2} = \text{Tan}^{-1} \frac{2\pi \times 50 \times 0.1}{50} = 32.14° \text{ (lag)} \textbf{ Ans.}$$

Example 8.3

A 159.23 μF capacitor in parallel with a resistance R draws a current of 25 A from 300 V, 50 Hz mains. Using phasor diagrams find the frequency *f* at which this combination draws the same current from a 360 V mains.

[*GATE 1994*]

Solution: The circuit and phasor diagrams are shown in Figs. 8.4 (*a*), (*b*) and (*c*) respectively.

When supply voltage V = 300 volts and supply frequency *f* = 50 Hz

Capacitive reactance,

$$X_C = \frac{1}{2\pi f C} = \frac{1}{2 \times \pi \times 50 \times 159.23 \times 10^{-6}} = 20\ \Omega$$

Supply current, I = 25 A

$$\text{Capacitive current, } I_C = \frac{V}{X_C} = \frac{300}{20}$$

= 15 A, leading supply voltage by 90°

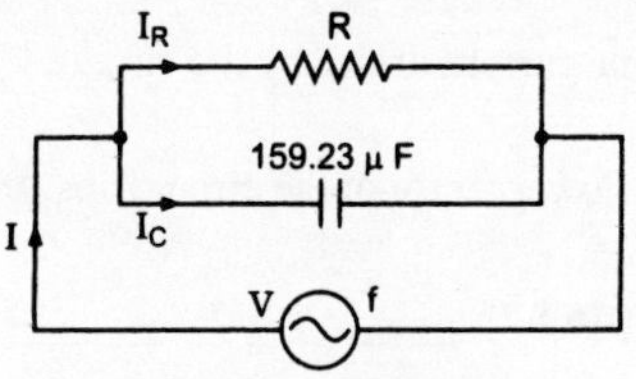

(*a*) *Circuit Diagram*

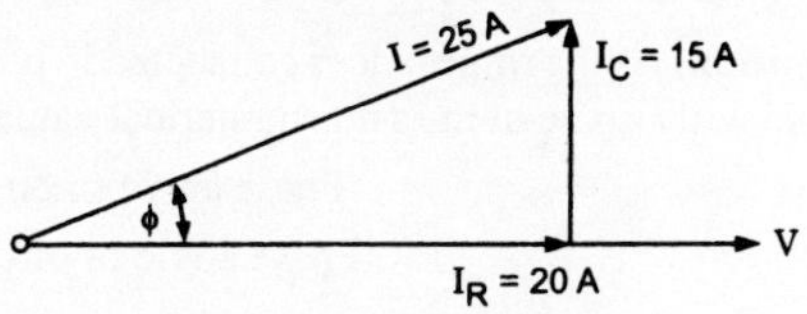

(*b*) *Phasor Diagram*

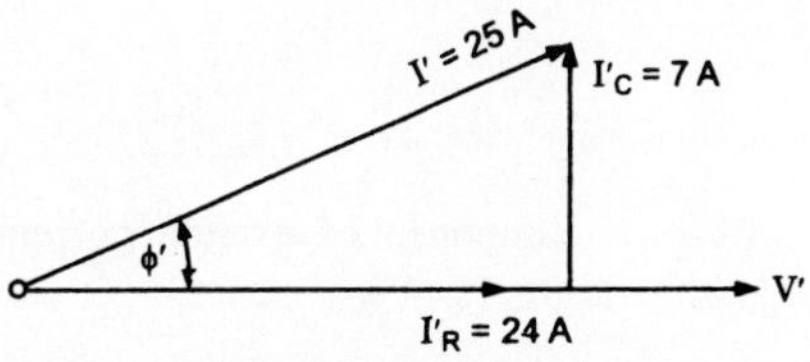

(*c*) *Phasor Diagram*

Fig. 8.4

From phasor diagram, shown in Fig. 8.4 (*b*), drawn for supply voltage of 300 V

$$\text{Resistive current, } I_R = \sqrt{I^2 - I_C^2} = \sqrt{25^2 - 15^2} = 20\ A$$

$$\text{Circuit resistance, } R = \frac{V}{I_R} = \frac{300}{20} = 15\ \Omega$$

When the supply voltage is 360 V and the supply frequency is *f'* (say)

Supply current, I' = I = 25 A, as before

$$\text{Resistive current, } I'_R = \frac{V'}{R} = \frac{360}{15} = 24\ A$$

From phasor diagram, shown in Fig. 8.4 (*c*), drawn for supply voltage of 360 V

$$I'_C = \sqrt{I'^2 - I'^2_R} = \sqrt{25^2 - 24^2} = 7\ A$$

$$\text{Capacitive reactance, } X_C = \frac{V'}{I'_C} = \frac{360}{7} = 51.43\ \Omega$$

$$\text{Supply frequency, } f' = \frac{1}{2\pi C X_C} = \frac{1}{2\pi \times 159.23 \times 10^{-6} \times 51.43} = 19.4\ \text{Hz}$$ **Ans.**

Example 8.4

An inductive coil of resistance 15 Ω and inductive reactance 42 Ω is connected in parallel with a capacitive reactance of 47.6 Ω. The combination is energized from a 200 V, 33.5 Hz ac supply. Find the total current drawn by the circuit and its power factor. Draw to the scale the phasor diagram of the circuit.

[*Bombay University, 2000*]

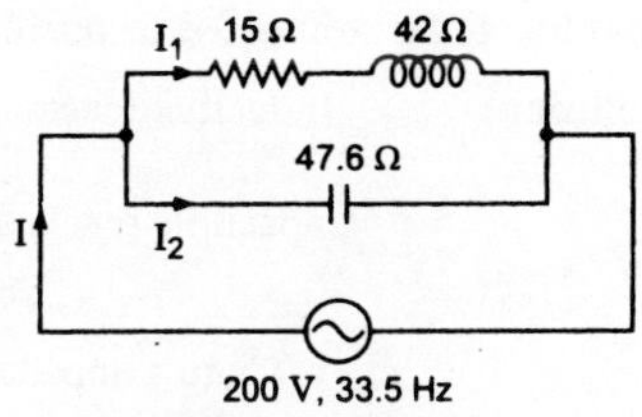

(*a*) *Circuit Diagram*

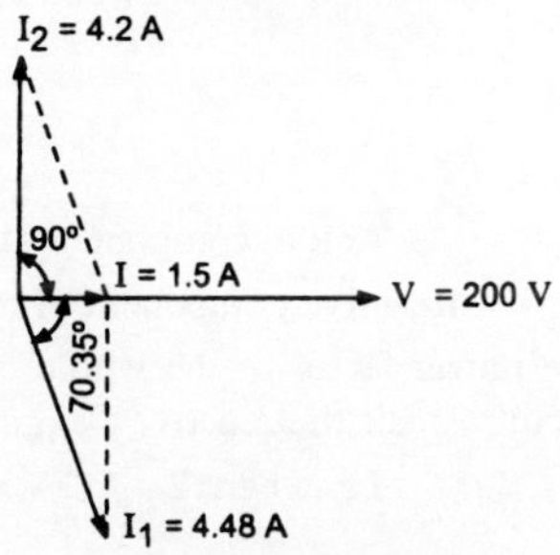

(*b*) *Phasor Diagram*

Fig. 8.5

Solution: Impedance of coil, $Z = \sqrt{R^2 + X_L^2} = \sqrt{15^2 + 42^2} = 44.6\ \Omega$

$$\text{Current in inductive coil, } I_1 = \frac{V}{Z} = \frac{200}{44.6}\ A = 4.48\ A$$

$$\text{Phase angle, } \phi_1 = \text{Tan}^{-1}\frac{X_L}{R} = \text{Tan}^{-1}\frac{42}{15} = 70.35° \text{ (lagging)}$$

$$\text{Current through capacitance, } I_2 = \frac{V}{X_C} = \frac{200}{47.6} = 4.2\ A$$

Phasor diagram is given in Fig. 8.5(*b*).

Resolving current in in-phase and quadrature components we have

$$I \cos\phi = I_1 \cos 70.35° + I_2 \cos 90°$$
$$= 4.48 \times 0.3363 + 4.2 \times 0 = 1.5\ A$$
$$I \sin\phi = I_1 \sin\phi_1 + I_2 \sin\phi_2$$
$$= 4.48 \sin(-70.35°) + 4.2 \sin 90° = -4.2 + 4.2 = 0$$

Total current drawn by the circuit, $I = \sqrt{(I\cos\phi)^2 + I(\sin\phi)^2} = \sqrt{(1.5)^2 + 0^2} = 1.5$ A **Ans.**

Power factor of the circuit, $\cos\phi = \dfrac{I\cos\phi}{I} = \dfrac{1.5}{1.5} = 1.0$ **Ans.**

Example 8.5

Two circuits having the same numerical ohmic impedance are joined in parallel. The power factor of one circuit is 0.8 (lag) and that of the other is 0.6 (lag). What is the power factor of the combination?

Solution: As the impedances connected in parallel are of the same numerical value, therefore, the currents flowing through them will also be of the same numerical value, say, I amperes. However, their phase angles will be different as below:

Phase angle of circuit I, $\Phi_1 = \text{Cos}^{-1}\,0.8 = 36.87°$

Phase angle of circuit II, $\Phi_2 = \text{Cos}^{-2}\,0.6 = 53.13°$

Resolving currents in active and reactive components we have

Active component of resultant current, $I_r\cos\Phi_r = I\cos\Phi_1 + I\cos\Phi_2$

$= I \times 0.8 + I \times 0.6$

$= 1.4\,I$ ...(*i*)

Reactive component of resultant current, $I_r\sin\Phi_r = I\sin\Phi_1 + I\sin\Phi_2$

$= I \times 0.6 + I \times 0.8$

$= 1.41$ ...(*ii*)

Dividing Eq. (*ii*) by Eq. (*i*) we have $\text{Tan}\,\Phi_r = \dfrac{1.4I}{1.4I} = 1.0$ or Φ_r

$= \text{Tan}^{-1}\,1.0 = 45°$

Power factor of the combination $= \text{Cos}\,\Phi_r = \text{Cos}\,45° = 0.707$ (lag)

Phasor diagram is shown in Fig. 8.6.

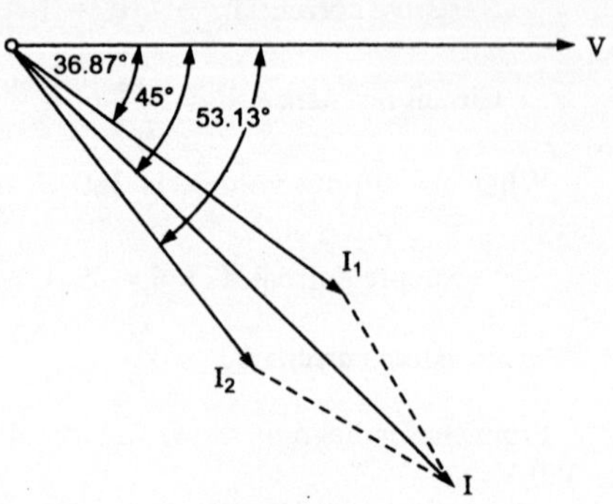

Fig. 8.6

Example 8.6

Find the active and reactive components of current taken by a series circuit consisting of a coil of inductance 0.1 H, resistance 8 Ω and a capacitor of 120 μ F connected to a 240 V, 50 Hz supply mains. Find the value of the capacitor that has to be connected in parallel with the above series circuit so that the pf of the entire circuit is unity.

Solution: Inductive reactance, $X_L = 2\pi f L = 2\pi \times 50 \times 0.1 = 31.416\ \Omega$

Capacitive reactance, $X_C = \dfrac{1}{2\pi f C} = \dfrac{1}{2\pi \times 50 \times 120 \times 10^{-6}} = 26.526\ \Omega$

Circuit impedance, $Z = \sqrt{R^2 + (X_L - X_C)^2} = \sqrt{8^2 + (31.416 - 26.526)^2} = 9.376\ \Omega$

Circuit current, $I = \dfrac{V}{Z} = \dfrac{240}{9.376} = 25.6$ A

Phase angle, $\phi = \text{Tan}^{-1}\dfrac{X_L - X_C}{R} = \text{Tan}^{-1}\dfrac{31.416 - 26.526}{8} = 31.435°$

Active component of current $= I\cos\phi = 25.6\cos 31.435° = 21.84$ A **Ans.**

Reactive component of current $= I\sin\phi = 25.6\sin 31.435° = 13.35$ A (lagging) **Ans.**

The power factor of the whole circuit will be unity if the current drawn by the capacitor, connected across the series circuit, equals the reactive (lagging) component of the series circuit.

i.e. when $2\pi f C V = I\sin\phi$

or $C = \dfrac{I\sin\phi}{2\pi f V} = \dfrac{13.35}{2\pi \times 50 \times 240} = 177\ \mu F$ **Ans.**

8.3. ADMITTANCE METHOD

In the circuit shown in Fig. 8.1 (*a*), the branch currents

$$I_1 = \frac{V}{Z_1} = \frac{V}{\sqrt{R_1^2 + (X_{L_1} - X_{C_1})^2}}$$

$$I_2 = \frac{V}{Z_2} = \frac{V}{\sqrt{R_2^2 + (X_{L_2} - X_{C_2})^2}}$$

$$I_3 = \frac{V}{Z_3} = \frac{V}{\sqrt{R_3^2 + (X_{L_3} - X_{C_3})^2}}$$

If **Z** is equivalent impedance of the circuit then total current flowing through the circuit,

$$I = \frac{V}{Z}$$

$$\text{Since } I = I_1 + I_2 + I_3 \text{ (phasor sum)}$$

$$\therefore \frac{V}{Z} = \frac{V}{Z_1} + \frac{V}{Z_2} + \frac{V}{Z_3} \text{ (phasor sum)}$$

$$\text{or } \frac{1}{Z} = \frac{1}{Z_1} + \frac{1}{Z_2} + \frac{1}{Z_3} \text{ (phasor sum)}$$

The reciprocal of impedance is known as admittance and is represented by English capital letter **Y** and expressed in siemens (S)

$$\therefore Y = \frac{1}{Z}; Y_1 = \frac{1}{Z_1}; Y_2 = \frac{1}{Z_2} \text{ and } Y_3 = \frac{1}{Z_3}$$

$$\text{and } Y = Y_1 + Y_2 + Y_3 \text{ (phasor sum)}$$

$$\text{Resultant current, } I = V \times \frac{1}{Z} = V \times Y$$

Just as impedance has two components, resistance R and reactance X similarly the admittance has two components known as *conductance*, G and *susceptance*, B respectively. The unit of conductance and susceptance is siemens (S).

From admittance and impedance triangles (Fig. 8.7)

$$\text{Conductance, } G = Y \cos \Phi = \frac{1}{Z} \cdot \frac{R}{Z} = \frac{R}{Z^2} = \frac{R}{R^2 + X^2}$$

$$\text{Susceptance, } B = Y \sin \Phi = \frac{1}{Z} \cdot \frac{X}{Z} = \frac{X}{Z^2} = \frac{X}{R^2 + X^2}$$

$$\text{Admittance, } Y = \sqrt{(Y \cos \Phi)^2 + (Y \sin \Phi)^2} = \sqrt{G^2 + B^2}$$

$$\text{Phase angle, } \Phi = \tan^{-1} \frac{B}{G}$$

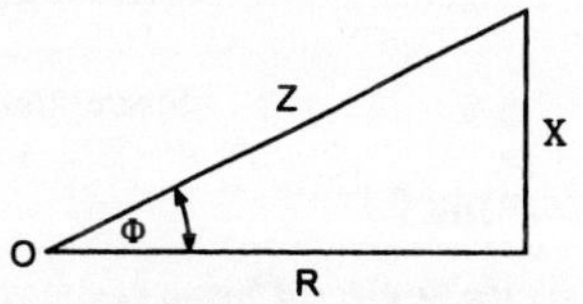

(a) Impedance Triangle

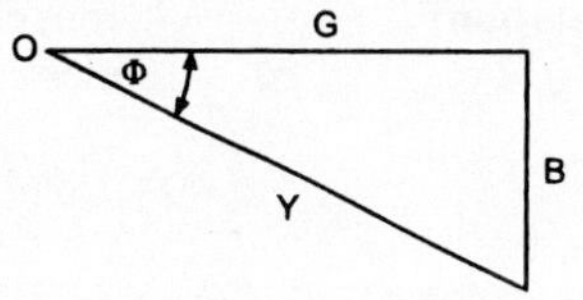

(b) Admittance Triangle

Fig. 8.7

8.3.1 Application of Admittance Method in Solution of Single Phase Parallel Circuits

Determine conductance and susceptance of individual branches from the relation

$$G = \frac{R}{Z^2} \text{ and } B = \frac{X}{Z^2}$$

taking B as + ve if X is capacitive and as –ve if X is inductive. Let the conductances of the three branches of circuit shown in Fig. 8.1 (*a*) be G_1, G_2 and G_3 respectively and susceptances be B_1, B_2 and B_3 respectively.

Find the equivalent conductance and susceptance of the circuit which will be equal to algebraic sum of conductances and susceptances of individual branches respectively.

For the circuit shown in Fig. 8.1 (*a*)

$$\text{Total conductance, } G = G_1 + G_2 + G_3$$

$$\text{and total susceptance, } B = B_1 + B_2 + B_3$$

$$\text{Total admittance of the circuit, } Y = \sqrt{G^2 + B^2}$$

The current in the circuit may be determined from the relation

$$I = V \times Y \text{ having phase angle } \Phi = \text{Tan}^{-1}\frac{B}{G}$$

Note: Capacitive susceptance will be taken as + ve and inductive susceptance as – ve though capacitive reactance is taken as – ve and inductive reactance as + ve. If Φ is + ve current will lead and if Φ is – ve current will lag behind the applied voltage.

Example 8.7

The active and lagging reactive components of current taken by an ac circuit from a 250 V supply are 50 A and 25 A respectively. Calculate the conductance, susceptance, admittance and power factor of the circuit.

Solution: Active component of current, $I\cos\phi = 50$ A

Lagging reactive component of current, $I\sin\phi = 25$ A

$$\text{Circuit current, } I = \sqrt{(I\cos\phi)^2 + (I\sin\phi)^2} = \sqrt{50^2 + 25^2} = 55.9 \text{ A}$$

$$\text{Admittance of the circuit, } Y = \frac{I}{V} = \frac{55.9}{250} = 0.224 \text{ S} \quad \textbf{Ans.}$$

$$\text{Power factor of the circuit, } \cos\phi = \frac{I\cos\phi}{I} = \frac{50}{55.9} = 0.8944 \text{ (lagging)} \quad \textbf{Ans.}$$

$$\text{Conductance of the circuit, } G = Y\cos\phi = 0.224 \times 0.8944 = 0.2 \text{ S} \quad \textbf{Ans.}$$

$$\text{Susceptance of the circuit, } B = \sqrt{Y^2 - G^2} = \sqrt{(0.224)^2 - 0.2^2} = 0.1 \text{ S (inductive)} \quad \textbf{Ans.}$$

Example 8.8

A series ac circuit has a resistance of 15 Ω and inductive reactance of 10 Ω. Calculate the value of a capacitor which is connected across this series combination so that system has unity power factor. The frequency of ac supply is 50 Hz. [*U.P. Technical Univ. Electrical Engineering Second Semester 2005-06*]

Solution: Series circuit resistance, $R = 15\ \Omega$

Series circuit reactance, $X = 10\ \Omega$

$$\text{Series circuit conductance, } G = \frac{R}{R^2 + X^2} = \frac{15}{(15)^2 + (10)^2} = \frac{15}{325} \text{ S}$$

$$\text{Series circuit susceptance, } B = \frac{X}{R^2 + X^2} = \frac{10}{15^2 + 10^2} = \frac{10}{325} \text{ S (inductive)}$$

The power factor of the whole circuit will be unity only if capacitive susceptance of the capacitor to be connected across the series combination is equal to the equivalent inductive susceptance of the series combination *i.e.* equal to $\frac{10}{325}$ S.

Reactance of the capacitance connected across the series combination,

$$X_C = \frac{1}{B_C} = \frac{1}{10/325} = 32.5\ \Omega$$

$$\text{Capacitance of the required capacitor, } C = \frac{1}{2\pi f X_C} = \frac{1}{2\pi \times 50 \times 32.5} = 98\ \mu\text{F} \quad \textbf{Ans.}$$

Example 8.9

A resistor of 12 Ω and an inductance of 0.025 H are connected in series across a 50 Hz supply. What values of resistance and inductance when connected in parallel will have the same resultant impedance and pf. Find the current in each case when the supply voltage is 230 V. *[Nagpur University, November 1996]*

Solution: Series circuit resistance, $R_S = 12\ \Omega$

Series circuit reactance, $X_S = 2\pi f L = 2\pi \times 50 \times 0.025 = 7.854\ \Omega$

$$\text{Series circuit conductance, } G_S = \frac{R_S}{R_S^2 + X_S^2} = \frac{12}{12^2 + 7.854^2} = 0.05834\ \text{S}$$

$$\text{Series circuit susceptance, } B_S = \frac{X_S}{R_S^2 + X_S^2} = \frac{7.854}{12^2 + 7.854^2} = 0.03818\ \text{S (inductive)}$$

The circuits shown in Figs. 8.8 (*a*) and (*b*) will be equivalent (the resultant impedance and pf will be same) if Conductance of parallel circuit,

G_p = Conductance of series circuit, G_S

i.e. $\frac{1}{R_p} = 0.05834$ S

or R = 17.14 Ω **Ans.**

Susceptance of parallel circuit,

B_p = Susceptance of series circuit, B_S

i.e. $\frac{1}{X_p} = 0.03818$ S or $X_p = 26.192\ \Omega$

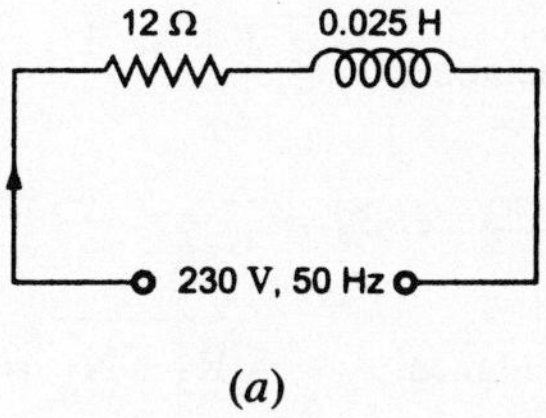

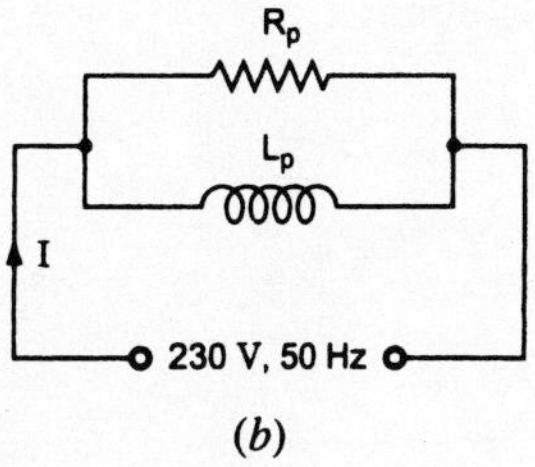

Fig. 8.8

$$\text{Inductance of equivalent parallel circuit, } L = \frac{X_p}{2\pi f} = \frac{26.192}{2\pi \times 50} = 83.4\ \text{mH} \quad \textbf{Ans.}$$

Current will be same in both cases and is given by

$$I = \frac{V}{Z} = \frac{230}{\sqrt{R_S^2 + X_S^2}} = \frac{230}{\sqrt{12^2 + 7.854^2}} = 16.04\ \text{A} \quad \textbf{Ans.}$$

$$\text{Phase angle, } \phi = \text{Tan}^{-1}\frac{X_S}{R} = \text{Tan}^{-1}\frac{7.854}{12} = 33.2^\circ \text{ (lagging)} \quad \textbf{Ans.}$$

Example 8.10

A variable capacitor is connected in series with a circuit consisting of a non-inductive resistance of 50 Ω in parallel with a coil across a 200 V, 50 Hz supply. The coil has an inductance of 0.2 H and negligible resistance.

Calculate (*a*) The capacitance of the capacitor when the pf of the whole circuit is unity. (*b*) The corresponding pd across the capacitor.

Solution:

$$\text{Conductance of parallel circuit, } G = \frac{1}{R} = \frac{1}{50} = 0.02\ \text{S}$$

$$\text{Susceptance of parallel circuit, } B = \frac{1}{X_L} = \frac{1}{2\pi f L} = \frac{1}{2\pi \times 50 \times 0.2} = 0.0159\ \text{S (inductive)}$$

$$\text{Equivalent resistance of parallel circuit, } R_{eq} = \frac{G}{G^2 + B^2} = \frac{0.02}{(0.02)^2 + (0.0159)^2} = 30.64\ \Omega$$

$$\text{Equivalent reactance of parallel circuit, } X_{eq} = \frac{B}{G^2 + B^2} = \frac{0.0159}{(0.02)^2 + (0.0159)^2} = 24.36\ \Omega \text{ (inductive)}$$

(*a*) The power factor of the whole circuit will be unity only if capacitive reactance is equal in magnitude to the equivalent inductive reactance of the parallel circuit.

i.e. $\frac{1}{\omega C} = 24.36\ \Omega$

or $C = \frac{1}{\omega \times 24.36} = \frac{1}{2\pi \times 50 \times 24.36} = 130.7\ \mu F$ **Ans.**

$I = \frac{V}{R_{eq}} = \frac{200}{30.64}$ A

PD across the capacitor

$= IX_C = \frac{200}{30.64} \times 24.36 = 159$ volts **Ans.**

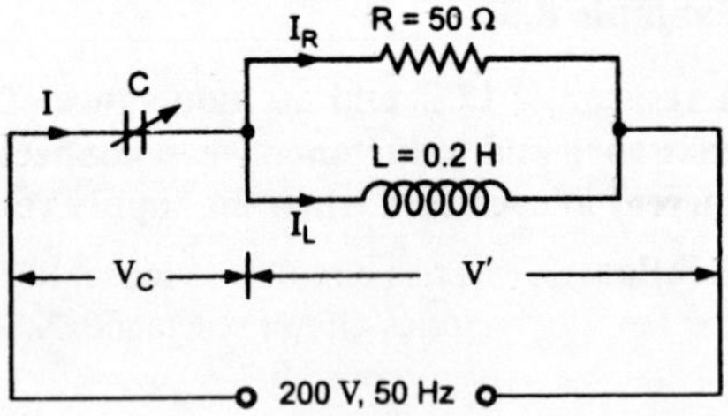

Fig. 8.9

Example 8.11

For the circuit shown in Fig. 8.10, calculate I_1, I_2 and I_3. The values marked on the inductance and capacitance give their reactances. [*Allahabad Univ. Electrical Science 1992*]

Solution:

Impedance, $Z_2 = \sqrt{R_2^2 + X_2^2} = \sqrt{4^2 + 2^2} = \sqrt{20}\ \Omega$

Conductance, $G_2 = \frac{R_2}{Z_2^2} = \frac{4}{20} = 0.2$ S

Susceptance, $B_2 = \frac{X_2}{Z_2^2} = \frac{2}{20} = 0.1$ S (inductive)

Impedance, $Z_3 = \sqrt{R_3^2 + X_3^2} = \sqrt{1^2 + 5^2} = \sqrt{26}\ \Omega$

Conductance, $G_3 = \frac{R_3}{Z_3^2} = \frac{1}{26} = 0.03846$ S

Susceptance, $B_3 = \frac{X_3}{Z_3^2} = \frac{5}{26} = 0.1923$ S (capacitive)

Total conductance of parallel combination, $G_P = G_2 + G_3 = 0.2 + 0.03846 = 0.23846$ S

Total susceptance of parallel combination, $B_P = B_2 + B_3 = (-0.1) + 0.1923 = 0.09232$ S (capacitive)

Equivalent resistance of parallel combination, $R_P = \frac{G_P}{Y_P^2} = \frac{G_P}{G_P^2 + B_P^2} = \frac{0.23846}{(0.2384)^2 + (0.0923)^2} = 3.65\ \Omega$

Equivalent reactance of parallel combination, $X_P = \frac{0.0923}{(0.2384)^2 + (0.0923)^2} = 1.41\ \Omega$ (capacitive)

Equivalent resistance of whole circuit,

$R = R_1 + R_{BC} = R_1 + R_P = 2 + 3.65 = 5.65\ \Omega$

Equivalent reactance of whole circuit,

$X = X_1 + X_P = 3 + (-1.141) = 1.59\ \Omega$

Equivalent impedance of whole circuit,

$Z = \sqrt{R^2 + X^2} = \sqrt{(5.65)^2 + (1.59)^2} = 5.87\ \Omega$

Current, $I_1 = \frac{V}{Z} = \frac{10}{5.87} = 1.7$ A **Ans.**

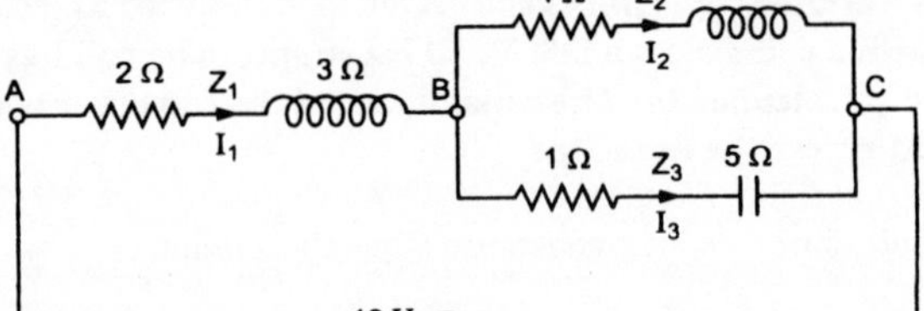

Fig. 8.10

Voltage drop across parallel circuit, $V_P = I \times Z_P = 1.7 \times \sqrt{(3.65)^2 + (1.41)^2} = 6.652$ V

Current through impedance Z_2, $I_2 = \frac{V_P}{Z_2} = \frac{6.652}{\sqrt{20}} = 1.49$ A **Ans.**

Current through impedance Z_3, $I_3 = \frac{V_P}{Z_3} = \frac{6.652}{\sqrt{26}} = 1.3$ A **Ans.**

8.4. SYMBOLIC OR j-METHOD

When two or more circuits are connected in parallel, the current in each branch may be found in complex form and then total current in the circuit may be determined by addition of complex numbers, as detailed in Art. 5.7.

In the circuit shown in Fig. 8.11.

Current in branch 1, $\mathbf{I}_1 = \dfrac{\mathbf{V}}{R_1 + jX_{L_1}} = \dfrac{\mathbf{V}}{R_1 + jX_{L_1}} \times \dfrac{R_1 - jX_{L_1}}{R_1 - jX_{L_1}} = \mathbf{V}\left(\dfrac{R_1}{R_1^2 + X_{L_1}^2} - j\dfrac{X_{L_1}}{R_1^2 + X_{L_1}^2}\right)$

Similarly current in branch 2,

$$\mathbf{I}_2 = \mathbf{V}\left(\frac{R_2}{R_2^2 + X_{L_2}^2} - j\frac{X_{L_2}}{R_2^2 + X_{L_2}^2}\right)$$

and current in branch 3, $\mathbf{I}_3 = \mathbf{V}\left(\dfrac{R_3}{R_3^2 + X_C^2} + j\dfrac{X_C}{R_3^2 + X_C^2}\right)$

Total circuit current, $\mathbf{I} = \mathbf{I}_1 + \mathbf{I}_2 + \mathbf{I}_3$

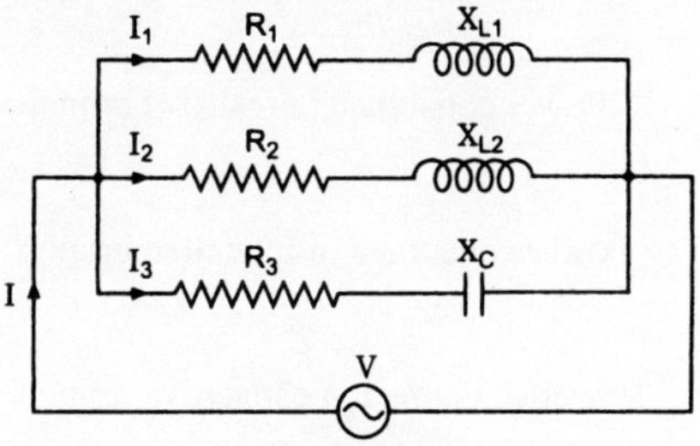

Fig. 8.11

8.4.1. Equivalent Parallel Impedance.

The ac parallel circuit, may be solved in the same way as dc parallel circuit except that complex impedances rather than simple resistances are involved. With several impedances connected in parallel the equivalent impedance is expressed as

$$\frac{1}{\mathbf{Z}} = \frac{1}{\mathbf{Z}_1} + \frac{1}{\mathbf{Z}_2} + \frac{1}{\mathbf{Z}_3} + \ldots\ldots$$

Since admittance is reciprocal of impedance, therefore,

$$\mathbf{Y} = \frac{1}{\mathbf{Z}} = \mathbf{Y}_1 + \mathbf{Y}_2 + \mathbf{Y}_3 + \ldots..$$

After finding equivalent impedance or admittance, current in the circuit may be determined from the relation

$$\mathbf{I} = \frac{\mathbf{V}}{\mathbf{Z}} = \mathbf{V} \times \mathbf{Y}$$

where $\mathbf{Y} = \dfrac{1}{\mathbf{Z}} = \dfrac{1}{(R \pm jX)} = \dfrac{R \mp jX}{(R \pm jX)(R \mp jX)} = \dfrac{R}{R^2 + X^2} \mp j\dfrac{X}{R^2 + X^2} = G \mp B$

The negative sign is used for inductive susceptance and positive sign for capacitive susceptance.

8.4.2. Symbolic Method For Solution of Single Phase Series-Parallel Circuits.

Single phase ac series-parallel circuits may be solved in the same manner as dc series-parallel circuits are solved except that complex impedances instead of simple resistances are involved. The equivalent impedance of a parallel circuit is determined first from the relation

$$\frac{1}{\mathbf{Z}_{eq}} = \frac{1}{\mathbf{Z}_1} + \frac{1}{\mathbf{Z}_2} + \frac{1}{\mathbf{Z}_3} + \ldots$$

then it is added (in complex form) to the impedance of the circuit connected in series with the combined parallel circuit and thus total impedance of the whole circuit is determined. The problems can be solved as discussed before.

Example 8.12

Consider an electric circuit shown in Fig. 8.12. Determine *(i)* the current and power consumed in each branch, *(ii)* the supply current and power factor. [*U.P. Technical Univ. Electrical Engineering, February, 2001*]

Solution: Current in 10 Ω branch, $\mathbf{I}_1 = \dfrac{\mathbf{V}}{\mathbf{R}} = \dfrac{100\angle 45^\circ}{10\angle 0^\circ} = 10\angle 45^\circ$ A **Ans.**

Current in inductive branch, $\mathbf{I}_2 = \dfrac{\mathbf{V}}{\mathbf{Z}_2} = \dfrac{100\angle 45^\circ}{5 + j5\sqrt{3}} = \dfrac{100\angle 45^\circ}{10\angle 60^\circ} = 10\angle -15^\circ$ A **Ans.**

Current in capacitive branch, $\mathbf{I}_3 = \dfrac{\mathbf{V}_3}{\mathbf{Z}_3} = \dfrac{100\angle 45^\circ}{5 - j5\sqrt{3}} = \dfrac{100\angle 45^\circ}{10\angle -60^\circ} = 10\angle 105^\circ$ A **Ans.**

Power consumed in resistive branch, $P_1 = I_1^2 R = 10^2 \times 10$

$= 1{,}000$ W **Ans.**

Power consumed in inductive branch, $P_2 = I_2^2 R_2 = 10^2 \times 5$

$= 500$ W **Ans.**

Power consumed in capacitive branch, $P_3 = I_3^2 R_3 = 10^2 \times 5$

$= 500$ W **Ans.**

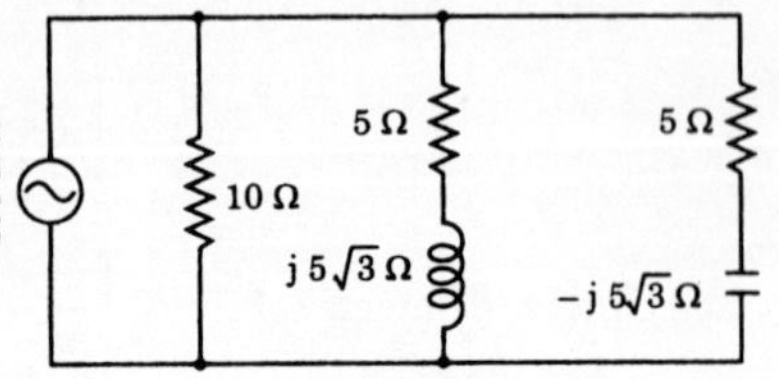

Fig. 8.12

(*ii*) Supply current, $\mathbf{I} = \mathbf{I}_1 + \mathbf{I}_2 + \mathbf{I}_3$

$= 10\angle 45^\circ + 10\angle -15^\circ + 10\angle 105^\circ$

$= (7.071 + j\,7.071) + (9.659 - j\,2.588) + (-2.588 + j\,9.659)$

$= 14.142 + j\,14.142 = 20\angle 45^\circ$ amperes **Ans.**

Power factor, $\cos\phi = \text{Cos}\,(45^\circ - 45^\circ) = 1.0$ **Ans.**

Example 8.13

Two impedances $Z_1 = (10 + j5)$ and $Z_2 = (8 + j6)$ are connected in parallel across a voltage of $V = 200 + j0$. Calculate the circuit current, power factor and reactive power.

[U.P. Technical Univ. Elec. Engineering First Semester 2004-05]

Solution : Equivalent impedance of parallel combination,

$$Z = \frac{Z_1 Z_2}{Z_1 + Z_2} = \frac{(10 + j5)(8 + j6)}{(10 + j5) + (8 + j6)} = \frac{(10 + j5)(8 + j6)}{(18 + j11)}$$

$$= \frac{11.18\angle 26.56^\circ \times 10\angle 36.87^\circ}{21.095\angle 31.43^\circ} = 5.3\angle 32^\circ\ \Omega$$

Circuit current, $\mathbf{I} = \dfrac{\mathbf{V}}{\mathbf{Z}} = \dfrac{200\angle 0^\circ}{5.3\angle 32^\circ} = 37.736\angle -32^\circ$ A

The circuit current, I = 37.736 A **Ans.**

Phase angle, $\phi = -32^\circ$

Power factor $= \text{Cos}\,\phi = \text{Cos}(-32^\circ) = 0.848$ (lagging) **Ans.**

Reactive power, $Q = VI\sin\phi = 200 \times 37.736 \times \sin(-32^\circ) = -4{,}000$ watts or 4 kW (lagging) **Ans.**

Example 8.14

Differentiate among real, reactive and apparent power.

Obtain the power factor of a two branch parallel circuit where the first branch has $Z_1 = (2 + j\,4)$ and second $Z_2 = (6 + j\,0)$. To what value must the 6 Ω resistor be changed to result in the overall power factor 0.9 lagging ?

[U.P. Technical Univ. Electrical Engineering June 2001]

Solution: Admittance of branch I, $\mathbf{Y}_1 = \dfrac{1}{\mathbf{Z}_1} = \dfrac{1}{2 + j4} = \dfrac{2 - j4}{2^2 + 4^2} = (0.1 - j\,0.2)$ S

Admittance of branch II, $\mathbf{Y}_2 = \dfrac{1}{6 + j0} = (0.167 + j\,0)$ S

Equivalent admittance of the circuit, $\mathbf{Y} = \mathbf{Y}_1 + \mathbf{Y}_2 = (0.1 - j\,0.2) + (0.167 + j\,0) = (0.267 - j\,0.2)$ S

The magnitude of equivalent admittance, $Y = \sqrt{(0.267)^2 + (0.2)^2} = 0.333$ S

Power factor of the circuit, $\text{Cos}\,\phi = \dfrac{0.267}{0.333} = 0.8$ (lagging) **Ans.**

Let the resistance of the resistor of Z_2 be changed to R ohms so as to make the power factor of whole circuit 0.9 (lagging)

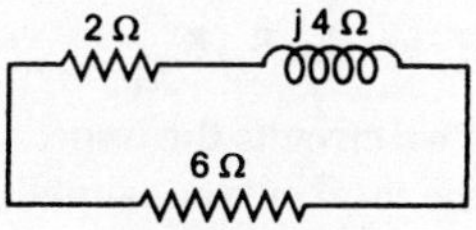

Fig. 8.13

Now total conductance of the circuit, $G = \left(\dfrac{1}{R} + 0.1\right)$ S

Total susceptance of the circuit, $B' = B = -0.2$ S

Phase angle of the circuit, $\phi' = \text{Cos}^{-1}\,0.9 = -25.84$, being lagging

Phase angle of the circuit is given as $\phi' = \text{Tan}^{-1}\dfrac{B'}{G'}$

or $\text{Tan}^{-1}\dfrac{-0.2}{\dfrac{1}{R} + 0.1} = -25.84°$ or $R = 3.2\ \Omega$ **Ans.**

Example 8.15

An inductive coil of (6 + *j* 8) Ω impedance is connected to 100 V, 50 Hz ac supply. It is desired to improve power factor of supply current to 0.8 lagging by connecting a capacitor (*i*) is in series of the coil (*ii*) in parallel of the coil. Determine the value of the capacitance in each case. *[U.P. Technical Univ. Electrical Engineering February 2002]*

Solution: Let a capacitor of capacitance C be connected in series with the coil so as to raise the pf of supply current to 0.8 lagging, as shown in Fig. 8.14.

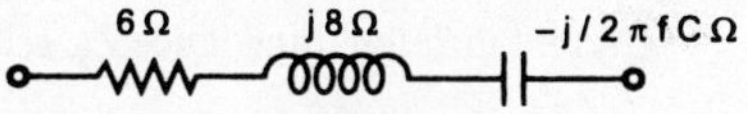

Fig. 8.14

Resistance of the circuit, $R = 6\ \Omega$

Inductive reactance of the series circuit, $X = 8 - \dfrac{1}{2\pi f C}$

Improved pf of the circuit, $\cos\phi = 0.8$ lagging

$\text{Tan}\,\phi = \text{Tan}\cos^{-1} 0.8 = 0.75$

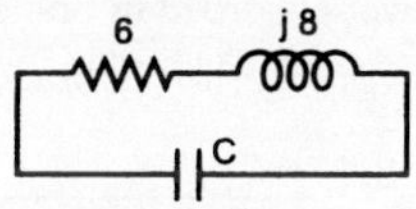

Fig. 8.15

Since $\dfrac{X}{R} = \text{Tan}\,\phi$

So $\dfrac{8 - \dfrac{1}{2\pi f C}}{6} = 0.75$

or $\dfrac{1}{2\pi f C} = 8 - 6 \times 0.75 = 3.5\ \Omega$

or $C = \dfrac{1}{2\pi f \times 3.5} = \dfrac{1}{2\pi \times 50 \times 3.5} = 909.4568\ \mu$F **Ans.**

(*ii*) Let a capacitor of capacitance C be connected across the coil so as to raise the pf of supply current to 0.8 lagging, as shown in Fig. 8.15.

Conductance of coil, $G = \dfrac{R}{R^2 + X^2} = \dfrac{6}{6^2 + 8^2} = 0.06$ S

Susceptance of coil, $B_L = \dfrac{X}{R^2 + X^2} = \dfrac{8}{6^2 + 8^2} = 0.08$ S (inductive)

Capacitive susceptance of capacitor C, $B_C = 2\pi f C$

Since $\dfrac{B}{G} = \text{Tan}\,\phi$

$$\text{or } \frac{B_L - B_C}{G} = 0.75$$

$$\text{or } B_L - B_C = 0.75 \times G = 0.75 \times 0.06 = 0.045 \text{ S}$$

$$\text{or } B_C = B_L - 0.045 = 0.08 - 0.045 = 0.035 \text{ S}$$

$$\text{or } 2\pi f C = 0.035$$

$$\text{or } C = \frac{0.035}{2\pi \times 50} = 111.4\ \mu\text{F} \textbf{ Ans.}$$

Example 8.16

Two circuits the impedances of which are given by Z_1 (10 + j 15) Ω and Z_2 = (6 – j 8) Ω are connected in parallel. If the total current supplied is 15 A. What is the power taken by each branch ?

[R.G.P.V. Bhopal Basic Electrical Engineering December 2001]

Solution: Equivalent impedance of parallel combination,

$$\mathbf{Z} = \frac{\mathbf{Z}_1 \mathbf{Z}_2}{\mathbf{Z}_1 + \mathbf{Z}_2} = \frac{(10 + j15)(6 - j8)}{(10 + j15) + (6 - j8)} = \frac{(10 + j\,15)(6 - j\,8)}{(16 + j\,7)}$$

Total current supplied, I = 15 A

Taking supply current I as reference phasor, **I** = (15 + j 0) A

$$\text{Current through impedance } Z_1, \mathbf{I}_1 = \frac{\mathbf{IZ}}{\mathbf{Z}_1} = \frac{15(10 + j15)(6 - j8)}{(16 + j7)(10 + j15)} = \frac{15(6 - j8)}{(16 + j7)} = \frac{15 \times 10 \angle -53.13^\circ}{17.464 \angle -23.63^\circ}$$

$$= 8.589 \angle -76.76^\circ = (1.967 - j\,8.361) \text{ A}$$

$$\text{Current through impedance } Z_2, \mathbf{I}_2 = \frac{\mathbf{I} \times \mathbf{Z}}{\mathbf{Z}_2} = \frac{15(10 + j15)(6 - j8)}{(16 + j7)(6 - j8)} = \frac{15(10 + j15)}{(16 + j7)} = \frac{15 \times 18.028 \angle 56.31^\circ}{17.464 \angle 23.63^\circ}$$

$$= 15.484 \angle 32.68^\circ = (13.033 + j\,8.361) \text{ A}$$

Power absorbed in branch 1, $P_1 = I_1^2 R_1 = (8.589)^2 \times 10 = 738$ W **Ans.**

Power absorbed in branch 2, $P_2 = I_2^2 R_2 = (15.484)^2 \times 6 = 1{,}438$ W **Ans.**

Example 8.17

The parallel circuit shown in Fig. 8.16 is connected across a single phase 100 V, 50 Hz ac supply. Calculate (*i*) the branch currents (*ii*) the total current (*iii*) the supply power factor (*iv*) the active and reactive power supplied by the supply.

[U.P. Technical Univ. Electrical Engineering First Semester 2006-07]

Solution: Admittance of branch 1,

$$\mathbf{Y}_1 = \frac{1}{\mathbf{Z}_1} = \frac{1}{(8 + j\,6)} = (0.08 - j\,0.06) \text{ S}$$

Admittance of branch 2,

$$\mathbf{Y}_2 = \frac{1}{\mathbf{Z}_2} = \frac{1}{(6 - j\,8)} = (0.06 + j\,0.08) \text{ S}$$

Total admittance of the circuit,

$$\mathbf{Y} = \mathbf{Y}_1 + \mathbf{Y}_2 = (0.08 - j\,0.06) + (0.06 + j\,0.08)$$
$$= (0.14 + j\,0.02) \text{ S}$$

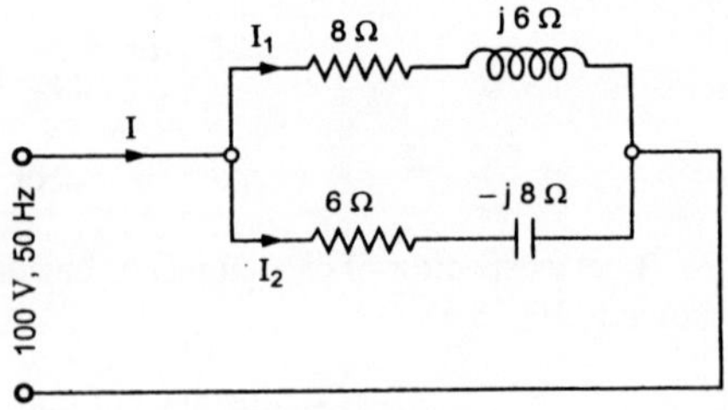

Fig. 8.16

(*i*) Branch currents, $\mathbf{I}_1 = \mathbf{V} \times \mathbf{Y}_1 = (100 + j\,0)(0.08 - j\,0.06) = (8 - j\,6)$ A or $10 \angle -36.87^\circ$ A **Ans.**

$\mathbf{I}_2 = \mathbf{V} \times \mathbf{Y}_2 = (100 + j\,0)(0.06 + j\,0.08) = (6 + j\,8)$ A or $10 \angle 53.13^\circ$ A **Ans.**

(*ii*) Total current, $\mathbf{I} = \mathbf{V} \times \mathbf{Y} = (100 + j\,0)\,(0.14 + j\,0.02) = (14 + j\,2)$ A or $14.42 \angle 8.13°$ A **Ans.**

(*iii*) Supply power factor $= \text{Cos}\,\phi = \text{Cos}\,8.13° = 0.99$ (leading) **Ans.**

(*iv*) Active power supplied, $P = V\,I\cos\phi = 100 \times 14.142 \times 0.99 = 1400$ W **Ans.**

Reactive power supplied, $Q = V\,I\sin\phi = 100 \times 14.142 \sin 8.13° = 200$ VAR (capacitive) **Ans.**

Example 8.18

A 15 mH inductor is in series with a parallel combination of an 80 Ω resistor and 20 μ F capacitor. If the angular frequency of the applied voltage is ω = 1,000 radians/second, find the admittance of the network.

[*Osmania Univ. Basic Circuit Analysis Jan/Feb. 1992*]

Solution: Inductive reactance, $X_L = \omega L = 1{,}000 \times 15 \times 10^{-3} = 15\ \Omega$

Capacitive reactance, $X_C = \dfrac{1}{\omega C} = \dfrac{1}{1{,}000 \times 20 \times 10^{-6}} = 50\ \Omega$

$R = 80\ \Omega$

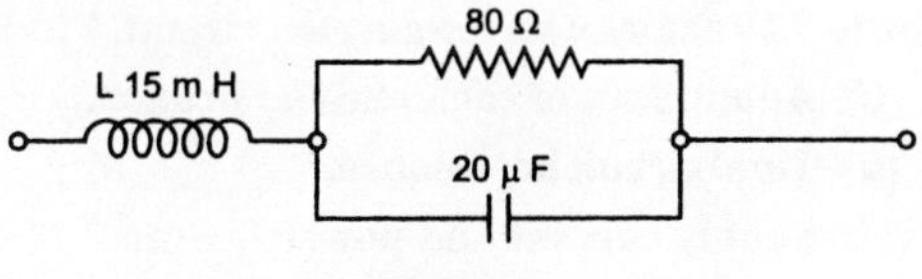

Fig. 8.17

Admittance of parallel circuit, $\mathbf{Y_P} = \dfrac{1}{80} + j\dfrac{1}{50}$

$= (0.0125 + j\,0.02)$ S

Impedance of parallel circuit, $\mathbf{Z_P} = \dfrac{1}{\mathbf{Y_P}} = \dfrac{1}{(0.0125 + j\,0.02)} = (22.48 - j\,35.96)\ \Omega$

Total impedance of the circuit, $\mathbf{Z} = (0 + j\,X_L) + \mathbf{Z_P} = (0 + j\,15) + (22.48 - j\,35.96) = (22.48 - j\,20.96)\ \Omega$

Admittance of the circuit, $\mathbf{Y} = \dfrac{1}{\mathbf{Z}} = \dfrac{1}{22.48 - j20.96} = (0.238 + j\,0.022)$ siemens **Ans.**

Example 8.19

In the circuit shown in Fig. 8.18 (*a*), the reactance of the capacitor C_1 is 4 Ω, the reactance of capacitor C_2 is 8 Ω and the reactance of L is 8 Ω. A sinusoidal voltage of 120 V is applied to the circuit. Find (*i*) current in each branch (*ii*) power loss in the circuit.

[*U.P. Technical Univ. Electrical Engineering February 2002*]

Solution: The circuit given in Fig. 8.18 (*a*) is redrawn with given values marked on it and is shown in Fig. 8.18 (*b*).

Impedance of parallel branch, $\mathbf{Z}_{bc} = \dfrac{1}{\dfrac{1}{-j8} + \dfrac{1}{(4 + j8)}} = \dfrac{-j\,8 \times (4 + j\,8)}{-j\,8 + 4 + j\,8}$

$= \dfrac{64 - j\,32}{4} = (16 - j\,8)$

$= 17.89 \angle -26.565°\ \Omega$

Total impedance of circuit, $\mathbf{Z} = \mathbf{Z}_{ab} + \mathbf{Z}_{bc} = (-j\,4) + (16 - j\,8)$

$= (16 - j\,12) = 20 \angle -36.87°\ \Omega$

Circuit current, $\mathbf{I} = \dfrac{V}{\mathbf{Z}} = \dfrac{120 \angle 0°}{20 \angle -36.87°} = 6 \angle 36.87°$ A

$= (4.8 + j\,3.6)$ A

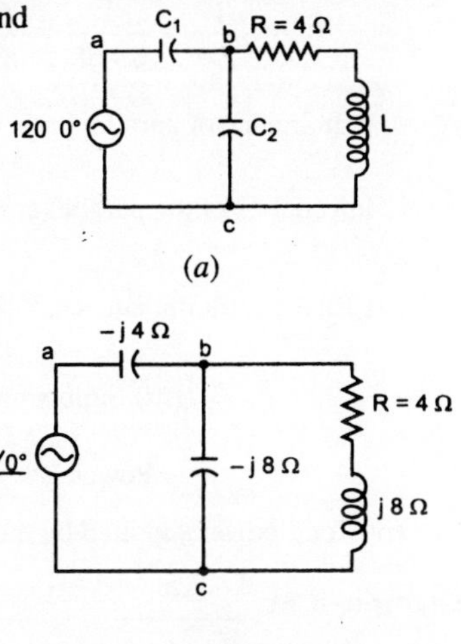

Fig. 8.18

Current in branch *ab*, $\mathbf{I}_{ab} = \mathbf{I} = (4.8 + j\,3.6)$ A **Ans.**

Voltage, $\mathbf{V}_{ab} = \mathbf{I}_{ab}\,\mathbf{Z}_{ab} = 6 \angle 36.87° \times 4 \angle -90°$ V $= 24 \angle -53.13°$ V $= (14.4 - j\,19.2)$ V **Ans.**

$$\text{Voltage, } \mathbf{V}_{bc} = \mathbf{I} \times \mathbf{Z}_{bc} = 6\angle 36.87^\circ \times 17.89\angle -26.565^\circ \text{ V}$$

$$= 107.33\angle 10.305^\circ \text{ V} = (105.6 + j\,19.2) \text{ V } \textbf{Ans.}$$

$$\text{Current in 8 } \Omega \text{ capacitor} = \frac{\mathbf{V}_{bc}}{-j8} = \frac{107.33\angle 10.305^\circ}{8\angle -90^\circ} = 13.416\angle 100.305^\circ \text{ A} = (-2.4 + j\,13.2) \text{ A } \textbf{Ans.}$$

Current in 4 Ω resistor = Current in 8 Ω inductor

$$= \frac{\mathbf{V}_{bc}}{(4 + j8)} = \frac{107.33\angle 10.305^\circ}{8.944\angle 63.435^\circ} \text{ A} = 12\angle -53.13^\circ \text{ A} = (7.2 - j\,9.6) \text{ A } \textbf{Ans.}$$

$$\text{Power loss in the circuit, P} = \text{V} \times \text{I} \times \cos\phi = 120 \times 6 \times \cos 36.87^\circ = 576 \text{ W } \textbf{Ans.}$$

Example 8.20

Figure 8.19 shows a series-parallel circuit. Find:

(*i*) Admittance of each parallel branch.

(*ii*) Total circuit impedance.

(*iii*) Supply current and power factor.

(*iv*) Total power supplied by the source.

[U.P. Technical Univ. Elec. Engineering First Semester 2005-06]

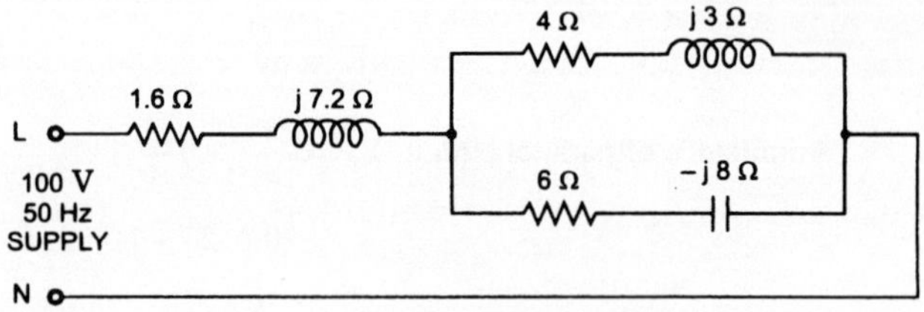

Fig. 8.19

Solution: Admittance of parallel branches,

$$\mathbf{Y}_1 = \frac{1}{\mathbf{Z}} = \frac{1}{4 + j3}$$

$$= \frac{4 - j3}{4^2 + 3^2}$$

$$= (0.16 - j\,0.12)\text{S}$$

$$= 0.2\angle -36.87^\circ \text{ S } \textbf{Ans.}$$

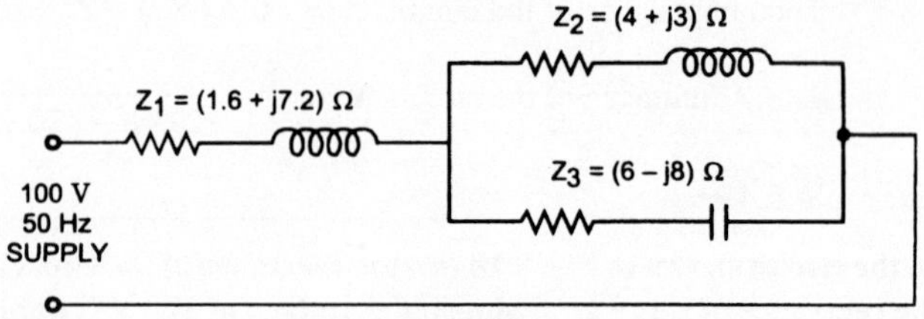

Fig. 8.20

$$\mathbf{Y}_2 = \frac{1}{\mathbf{Z}_2} = \frac{1}{6 - j8} = \frac{6 + j8}{6^2 + 8^2} = (0.06 + \text{j}0.08)\text{S or } 0.1\angle 53.13^\circ \text{ S } \textbf{Ans.}$$

$$\text{Admittance of parallel branches, } \mathbf{Y}_p = \mathbf{Y}_1 + \mathbf{Y}_2 = (0.16 - j\,0.12) + (0.06 + j\,0.08) = (0.22 - j\,0.04) \text{ S}$$

$$\text{Impedance of the parallel circuit, } \mathbf{Z}_p = \frac{1}{\mathbf{Y}_p} = \frac{1}{0.22 - j0.04} = \frac{0.22 + j0.04}{(0.22)^2 + (0.04)^2} = (4.4 + \text{j}0.8)\ \Omega$$

(*ii*) Total circuit impedance, $Z = \mathbf{Z}_s + \mathbf{Z}_p = (1.6 + j7.2) + (4.4 + j0.8) = (6 + j8)\ \Omega = 10\angle 53.13^\circ\ \Omega$ **Ans.**

(*iii*) Supply current, $I = \frac{V}{Z} = \frac{100}{10} = 10$ A **Ans.**

Power factor, $\cos\phi = \text{Cos } 53.13^\circ = 0.6$ (lagging) **Ans.**

(*iv*) Total power supplied by the source, $P = VI\cos\phi = 100 \times 10 \times 0.6 = 600$ W **Ans.**

Example 8.21

Determine the following in the circuit shown in Fig. 8.21.

(*i*) The current phasors I, $\mathbf{I}_1$ and $\mathbf{I}_2$.

(*ii*) Active power dissipated in the three resistive branches.

(*iii*) Power factor of the circuit.

[U.P. Technical Univ. Electrical Engineering Second Semester 2007-08]

Solution: Impedance of branch I of parallel circuit,

$$\mathbf{Z}_1 = R_1 + jX = (3 + j\,4)\ \Omega = 5\angle 53.13^\circ\ \Omega$$

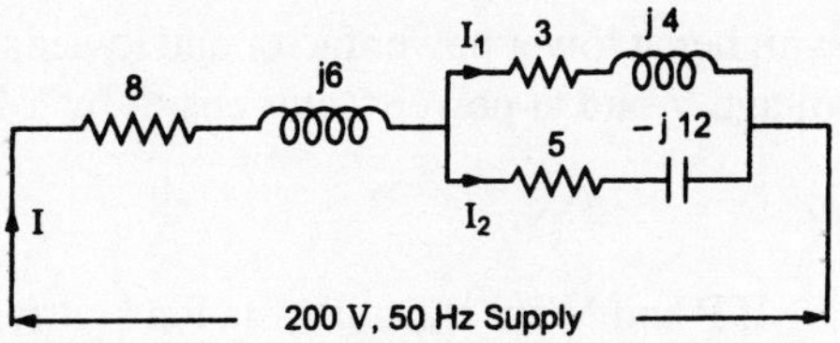

Fig. 8.21

Impedance of branch II of parallel circuit, $\mathbf{Z}_2 = (5 - j\,12)\ \Omega = 13\angle -67.38^\circ\ \Omega$

Impedance of series branch, $\mathbf{Z}_s = (8 + j\,6)\ \Omega = 10\angle 36.87^\circ\ \Omega$

Equivalent impedance of parallel circuit,

$$\mathbf{Z}_p = \frac{\mathbf{Z}_1\,\mathbf{Z}_2}{\mathbf{Z}_1 + \mathbf{Z}_2} = \frac{5\angle 53.13^\circ \times 13\angle -67.38^\circ}{(3 + j\,4) + (5 - j\,12)}$$

$$= \frac{65\angle -14.25^\circ}{(8 - j\,8)} = \frac{65\angle -14.25^\circ}{11.314\angle -45^\circ} = 5.745\angle 30.75^\circ\ \Omega$$

$$= (4.937 + j\,2.937)\ \Omega$$

Total impedance of the circuit, $\mathbf{Z} = \mathbf{Z}_s + \mathbf{Z}_p = (8 + j\,6) + (4.937 + j\,2.937)\ \Omega$

$$= (12.937 + j\,8.937)\ \Omega = 15.7237\angle 34.64^\circ\ \Omega$$

Let the supply voltage be the reference phasor i.e. $\mathbf{V} = 200\angle 0^\circ$ V

(*i*) Circuit current, $\mathbf{I} = \dfrac{\mathbf{V}}{\mathbf{Z}} = \dfrac{200\angle 0^\circ}{15.7237\angle 34.64^\circ} = 12.72\angle -34.64^\circ$ A **Ans.**

(*iii*) Power factor of the circuit, $\cos\phi = \cos(-34.64^\circ) = 0.8228$ (lagging) **Ans.**

$$\text{Current } \mathbf{I}_1 = \mathbf{I}\frac{\mathbf{Z}_2}{\mathbf{Z}_1 + \mathbf{Z}_2} = \frac{12.72\angle -34.64^\circ \times 13\angle -67.38^\circ}{(3 + j\,4) + (5 - j\,12)}$$

$$= \frac{165.36\angle -102.02^\circ}{11.314\angle -45^\circ}$$

$$= 14.616\angle -57.02^\circ \text{ A} \quad \textbf{Ans.}$$

$$\text{Current } \mathbf{I}_2 = \mathbf{I}\frac{\mathbf{Z}_1}{\mathbf{Z}_1 + \mathbf{Z}_2} = \frac{12.72\angle -34.64 \times 5\angle 53.13^\circ}{11.314\angle -45^\circ}$$

$$= 5.62\angle 63.49^\circ \text{ A} \quad \textbf{Ans.}$$

(*ii*) Active power dissipated in

Series branch $= I^2R_s = 12.72^2 \times 8 = 1{,}294.4$ W

Parallel branch 1 $= I_1^2R_1 = 14.616^2 \times 3 = 641$ W

Parallel branch 2 $= I_2^2R_2 = 5.62^2 \times 5 = 158$ W

Total power loss, $P = 1{,}294.4 + 641 + 158 = 2{,}093.4$ W

(*iii*) Power factor of the circuit, $\cos\phi = \dfrac{P}{VI} = \dfrac{2{,}093.4}{200 \times 12.72} = 0.8228$ (lagging) **Ans.**

8.5. POWER FACTOR IMPROVEMENT

The power factor for lighting and heating loads supplied from 3-phase power supply ranges from 0.95 to unity but the power factor for motor loads ranges from 0.5 to 0.9. Power factor for single phase motors may be as low as 0.4 (lagging) and for electric welding may be even 0.2 or 0.3 (lagging).

8.5.1. Disadvantages of Low Power Factor. The current for a given load supplied at a constant voltage will be higher at lower power factor and lower at higher power factor. For example if load P is supplied at terminal voltage V and at power factor cos Φ by a 1-phase system then load current is given by

$$I = \frac{P}{V \cos\Phi}$$

If P and V are constant, the load current I is inversely proportional to power factor, cos Φ *i.e.* lower the power factor, higher the current and vice-versa. The higher current due to poor power factor affects the system and results in the following disadvantages:

(*i*) Rating of generatoıs and transformers are proportional to their output current hence inversely proportional to power factor, therefore, large generators and transformers are required to deliver same load but at a low power factor.

(*ii*) The x-sectional area of the bus-bar and the contact surface of switch-gear is required to be enlarged for the same power to be delivered but at a low power factor.

(*iii*) For the same power to be transmitted but at low power factor, the transmission line or the distributor or cable will have to carry more current. Thus more conductor material is required for transmission lines, distributors and cables to deliver the same load but at a low power factor.

(*iv*) Copper losses are proportional to the square of the current hence inversely proportional to the square of the power factor *i.e.* more copper losses incur at low power factor, which results in poor efficiency.

(*v*) Low lagging power factor results in large voltage drop in generators, transformers, transmission lines and distributors which results in poor regulation. Hence extra regulating equipment is required to keep the voltage drop within permissible limits.

Thus, we see that the low power factor leads to a high capital cost for the alternators, switchgears, transformers, transmission lines, distributors and cables etc.

8.5.2. Causes of Low Power Factor.

(*i*) All ac motors (except overexcited synchronous motors and certain types of commutator motors) and transformers operate at lagging power factor. The power factor decreases with the decrease in load. For example an induction motor has a reasonable higher power factor of 0.85 at full load, 0.8 at 75% of full-load, 0.7 at half-full load, 0.5 at 25% of full-load and as low as 0.1 on no-load.

(*ii*) Arc lamps operate at low power factor (lagging) due to typical characteristic of the arc.

(*iii*) Industrial heating furnaces such as arc and induction furnaces operate on very low lagging power factor.

(*iv*) Due to increase in supply main voltage, which usually occurs during low load periods, such as lunch hours, night hours etc. the magnetizing current of inductive reactances increases and the power factor of the electrical plant as a whole comes down.

The power factor can be improved by (i) using induction motors with phase advancers (ii) connecting the static capacitors in parallel with the equipment operating at lagging power factor such as induction motors, fluorescent tubes etc or (iii) by using synchronous condensers* like shunt capacitors connected across the supply.

The capacitors draw current leading the supply voltage V by π/2 and neutralize the quadrature or wattless component of current drawn by the equipment across which they are connected.

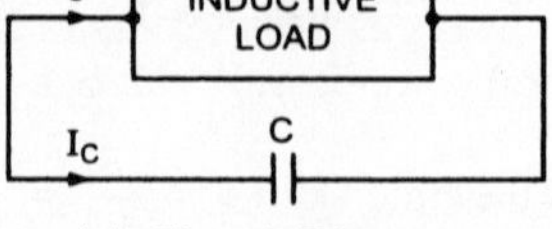

(*a*) Circuit Diagram

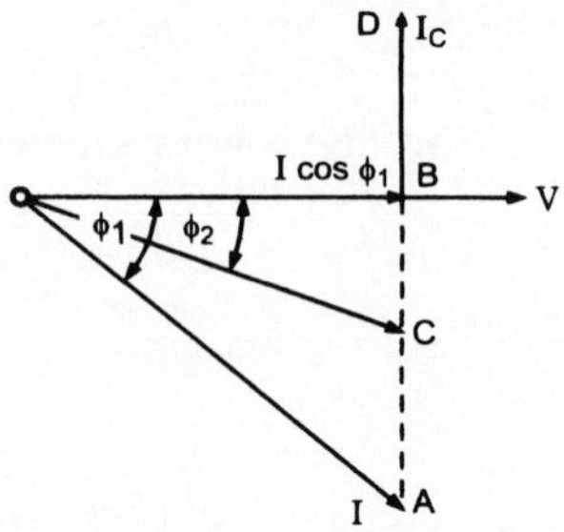

(*b*) Phasor Diagram

Fig. 8.22

* An overexcited synchronous motor operating on no load is called the *synchronous condenser* or *synchronous phase advancer* and behaves like a capacitor.

Let the current drawn by an inductive load be of I amperes lagging behind the applied voltage V by an angle ϕ_1 and pf be improved to cos ϕ_2 (lagging) by connecting a capacitor across the inductive load, as illustrated in Fig. 8.22 (*a*).

Active component of current drawn by load = I cos ϕ_1 represented by phasor OB in phasor diagram.

Reactive (lagging) component of current drawn by load = I sin ϕ_1 = I cos ϕ_1 × tan ϕ_1, represented by phasor AB in phasor diagram.

Let the current drawn by capacitor C be of I_C amperes leading the applied voltage V by $\pi/2$, as represented by phasor BD in Fig. 8.22 (*b*).

Lagging component of current drawn from supply mains, as represented by phasor BC in phasor diagram = OB tan ϕ_2 = I cos ϕ_1 tan ϕ_2.

The current drawn by capacitor C, I_C should neutralize the reactive component (lagging), represented by phasor A C in phasor diagram i.e. I_C should be equal to current represented by phasor AB less current represented by phasor BC. So

$$I_C = \text{Current represented by phasor AB} - \text{current represented by phasor BC}$$
$$= I\cos\phi_1\tan\phi_1 - I\cos\phi_1\tan\phi_2 = I\cos\phi_1(\tan\phi_1 - \tan\phi_2)$$

The value of capacitance required, $C = \dfrac{I_C}{2\pi f V}$ where V is supply voltage and *f* is the supply frequency.

8.5.4. Advantages of Power Factor Improvement. If the power factor of the supply or power station is raised to unity, the current for the same amount of power to be supplied is reduced to minimum. This results in reduction of transmission line copper losses and reduction of voltage drop in transmission line and in alternator windings as copper losses are directly proportional to the square of supply current and voltage drop is directly proportional to the supply current.

The advantages of good power factor are that (*i*) load output of a given plant is better utilized (*ii*) line losses are reduced to minimum, therefore, efficiency of the plant is improved (*iii*) voltage drop or voltage regulation is considerably reduced.

Example 8.22

A single phase load of 5 kW operates at a power factor of 0.6 lagging. It is proposed to improve the power factor to 0.95 lagging by connecting a capacitor across a load. Calculate the kVAR rating of the capacitor.

Solution: Load input, P = 5 kW

Initial power factor, cos ϕ_1 = 0.6; Tan ϕ_1 = 1.333

Power factor after improvement,

Cos ϕ_2 = 0.95;

Tan ϕ_2 = 0.328684

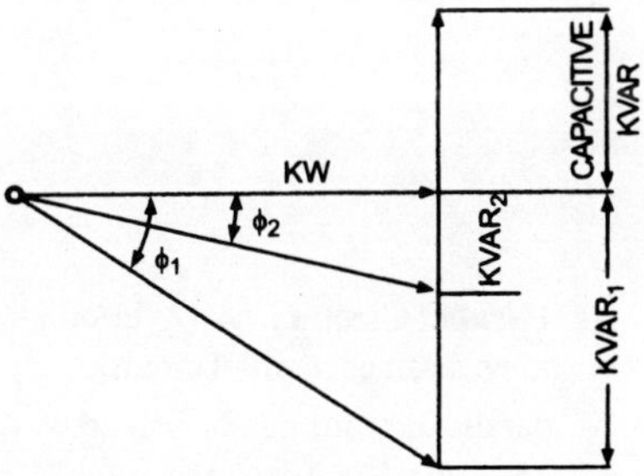

Fig. 8.23

Leading kVAR to be supplied by the capacitor to be connected across the load for power factor improvement

$$= kVAR_1 - kVAR_2 = P(\tan\phi_1 - \tan\phi_2) = 5(1.333 - 0.328684) = 5.023 \text{ kVAR } \textbf{Ans.}$$

Example 8.23

A circuit consisting of a single resistor R and an inductor L in series is driven by a 25 V_{rms} 50 Hz sinusoidal voltage source. A capacitor is to be placed in parallel with the source to improve the power factor. Given that the average power dissipated in the R is 100 W and the reactive power delivered to the L is 75 VAR, what value of C will yield a 0.9 pf lagging as seen by the source ? [*GATE 1998*]

Solution: Total power drown, P = 100 W

Reactive power drawn, Q_1 = 75 VAR

Improved pf, $\cos \phi_2 = 0.9$ (lagging)

$\phi_2 = \text{Cos}^{-1}\ 0.9$ (lagging) = 25.842° (lagging)

Reactive power drawn with improved pf,

$$Q_2 = P \tan 25.842° = 100 \times 0.4843 = 48.43 \text{ VAR}$$

Leading VAR to be supplied by the capacitor C to be connected across the load for power factor improvement

$$= VI_C = V \times 2\pi f C\, V = 2\pi f C\, V^2 = Q_1 - Q_2$$

$$\text{or} \quad C = \frac{Q_1 - Q_2}{2\pi f V^2} = \frac{75 - 48.43}{2\pi \times 50 \times 25^2} = 135.32\ \mu\text{F} \quad \textbf{Ans.}$$

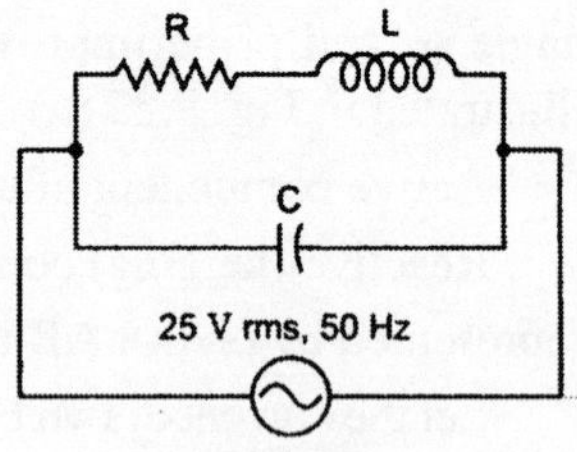

Fig. 8.24

Example 8.24

An industrial consumer is operating 3-phase, 1.0 kW induction motor at a lagging pf of 0.8 and a source voltage of 400 V_{rms}. He wishes to raise the pf to 0.95 lagging by connecting a circuit element in parallel to the load (motor). Indicate the type of circuit element (inductive or capacitive) and find its value if the operating frequency is 50 Hz. Give your comments on the new line and motor currents. *[U.P.S.C. I.E.S. Electrical Engineering-I, 2005]*

Solution: Motor input, P = 1 kW = 1,000 W

Power factor, $\cos \phi_1 = 0.8$ (lagging),

$\phi_1 = \text{Cos}^{-1}\ 0.8 = 36.87°$ (lagging)

Improved pf, $\cos \phi_2 = 0.95$ (lagging),

$\phi_2 = \text{Cos}^{-1}\ 0.95 = 18.195°$ (lagging)

Reactive power to be supplied by the pf improving device

$$= P(\tan \phi_1 - \tan \phi_2) = 1{,}000(\tan 36.87° - \tan 18.195°)$$

$$= 1{,}000(0.75 - 0.3287) = 421.3 \text{ VAR}$$

Since leading VARs are to be supplied by the power factor improving device, it must be capacitive.

If C is the capacitance of each capacitor then leading VARs supplied will be equal to $3 \times 2\pi f V_P C \times V_P = 6\pi f C V_P^2 = 6\pi f C V_L^2$ (assuming delta-connection)

$$\text{or} \quad C = \frac{421.3}{6\pi f V_L^2} = \frac{421.3}{6\pi \times 50 \times (400)^2} = 2.794\ \mu\text{F} \quad \textbf{Ans.}$$

HIGHLIGHTS

AC Parallel Circuits. An ac circuit having two or more branches, each branch having different components in series, is known as an ac parallel circuit.

AC parallel circuits can be solved by the following methods:

1. Phasor method 2. Admittance method 3. Symbolic or *j* method.

In *phasor method*, the magnitude and direction of currents flowing through different branches are determined and represented as phasors. The phasor sum of the branches currents gives the resultant current. In *admittance method*, the component values are determined as per admittance triangle which contains conductance and susceptance.

$$\text{Conductance, } G = Y \cos \phi = \frac{R}{Z^2}$$

$$\text{Susceptance, } B = Y \sin \phi = \frac{X}{Z^2}$$

$$\text{Admittance, } Y = \sqrt{G^2 + B^2}, \quad \text{where } G = G_1 + G_2 + \ldots + G_n \text{ and } B = B_1 + B_2 + \ldots + B_n$$

$$\text{Current, } I = V \times Y$$

EXERCISES

1. Define the terms: Impedance, Admittance, Power factor, Active power and Reactive power.
[Pb. Technical Univ. Elec. Engineering June 2000 ; December 2000; May 2001]
2. Write the expression of currents in different branches and the total current of a circuit consisting of an inductance, a capacitance and a non-inductive resistance connected in parallel. Draw also the phasor diagram.
3. Derive the expression of impedance, admittance, conductance and susceptance of R-L series circuit.
[R.G.P.V. Bhopal Basic Electrical Engineering Jan-Feb 2006]
4. What is admittance triangle ? Explain.
5. Define power factor and explain why it should be kept as high as possible in power supply systems ?
6. Show with the aid of phasor diagram, how can the pf of load be improved by connecting a capacitor in parallel with it.
7. What are the causes of low power factor in supply system ? Discuss its harming effects. How can the power factor be improved ? **[U.P. Technical Univ. Electrical Engineering January 2003]**

PROBLEMS

1. A circuit containing a resistance of 40 Ω in parallel with an inductive reactance of 30 Ω is connected across 240 V ac mains. Calculate (*i*) the current in each branch and the total current drawn from the mains (*ii*) the power factor of each branch and power factor of the whole circuit (*iii*) the power drawn by each branch and the total power drawn from the mains. **[Ans.** (*i*) 6A, 8A, 10A, (*ii*)1.0, zero, 0.6 (lag) (*iii*)1,440 watts, zero 1,440 watts**]**
2. An inductance of 0.6 H, a resistor of 100 Ω and a 30 μF capacitor are connected in parallel across a 230 V, 50 Hz supply. Calculate the line current, circuit phase angle, power dissipated, total circuit impedance and parameters of equivalent series circuit. **[Ans.** 2.49 A, 22.39° (lead), 529 W, 92.53 Ω, 80.51°, 35.25°**]**
3. A resistance of 100 Ω, an inductance of 0.1H and a capacitance of 150 μF are connected in parallel across 200 V, 50 Hz ac supply. Determine the branch currents, total current and the total power taken from the supply.
[Ans. 2A, 6.366 A, 9.42A, 3.655 A; 400 watts**]**
4. A series combination of R and C is in parallel with a 20 Ω resistor across 50 Hz source results in a total current of 7 A, a current through the 20 Ω resistor of 5 A and a current in the R-C branch of 3 A. Determine R and C
[Pb. Univ. Electrical Circuits & Systems December 1988]
[Ans. R = 16.667Ω; X_C = 28.8675 Ω**]**
5. An inductor of 0.5 H inductance and 90 Ω resistance is connected in parallel with a 2 μF capacitor. A voltage of 230 V at 50 Hz is maintained across the circuit. Determine the total power taken form the source **[Ans.**146 watts**]**
6. Two parallel circuits comprise respectively of (*i*) a coil of resistance 20 Ω and inductance 0.07 H and (*ii*) a capacitor or capacitance 60 μF in series with resistance of 50 Ω. Calculate the current in each branch and the current taken from the supply and power factor when connected to 220 V, 50 Hz supply. **[Ans.** 7.4 A, 3A, 7.8 A, 0.907 (lag)**]**
7. Two circuits having the same numerical ohmic impedances are joined in parallel. The power factor of one circuit is 0.8 (lag) and other is 0.6 (lead). What is the power factor of the combination ? **[Ans.** $1.4/\sqrt{2}$ lead**]**
8. An impedance $\mathbf{Z}_1 = (8 - j\,5)$ Ω is in parallel with an impedance $\mathbf{Z}_2 = (3 + j\,7)$ Ω. If 100 V are impressed on parallel combination, find the branch currents I_1, I_2 and the resultant current. Draw the corresponding phasor diagram showing each current and the voltage drop across each parameter. Calculate also the equivalent resistance, reactance and impedance of the whole circuit. **[Gwalior Univ. Electrical Technology-I, 1998]**
[Ans. $10.6\angle 32^\circ$ A; $13.16\angle 66.86^\circ$ A; $15.56\angle -24.5^\circ$ A; 5.38 Ω, 2.68 Ω**]**
9. Two circuits, the impedances of which are given by $\mathbf{Z}_1 = (15 + j\,12\ \Omega)$ and $\mathbf{Z}_2 = (8 - j\,5)$ Ω are connected in parallel. If the potential difference across one of the impedances is (250 + *j*) V, calculate (*i*) total current and branch currents (*ii*) total power and power consumed in each branch and (*iii*) overall power factor and power factor of each branch.
[Nagpur University November, 1998]
[Ans. (*i*) $33.15\angle 10.36^\circ$ A; $13\angle -38.6^\circ$ A; $26.5\angle 32^\circ$ A (*ii*) 8.153 kW; 2.535 kW, 5.618 kW, (*iii*) 0.984 lead; 0.78 lag; 0.848 lead**]**

10. Two impedances $\mathbf{Z}_1 = (6 - j\,8)$ ohms and $\mathbf{Z}_2 = (8 - j\,6)$ ohms are connected in parallel across 100 V supply. Determine (*i*) current and power factor of each branch (*ii*) overall current and power factor (*iii*) Power consumed by each branch and total power. **[Nagpur University Winter 2003]**

[**Ans.** (*i*) $10 \angle 53.13°$ A; 0.6 (leading); $10 \angle 36.87°$ A; 0.8 (leading) (*ii*) $14\sqrt{2} \angle 45°$ A, 0.707 (leading) (*iii*) 600 W, 800 W; 1,400 W]

11. Two impednaces $\mathbf{Z}_1 = (6 - j\,8)$ ohms and $\mathbf{Z}_2 = (16 + j12)$ ohms are in parallel. If the total current of the combination is $(20 + j10)$ amperes, find the complexor power taken by each impedance. Draw and explain the complete phasor diagram. **[Bombay Univ. Basic Electricity October 1971]**

[**Ans.** (2,400 + j 3,200), (1,600 – j 1,200)]

12. Two impedances $\mathbf{Z}_1 = (8 + j\,6)$ ohms and $Z_2 = (3 - j\,4)$ ohms are in parallel. If the total current of this combination is 25 A, find the power taken by each impedance.Draw and explain the complete phasor diagram. **[Bombay Univ. Basic Electricity October 1970]**

[**Ans.** 1000 watts, 1500 watts]

13. Two currents in each branch of a two branched parallel circuit is given as

$$i_a = 8.07 \sin\left(314t - \frac{\pi}{4}\right);\; i_b = 21.2 \sin\left(314t - \frac{\pi}{3}\right)$$

and supply voltage is 354 sin 314 t. Calculate (*i*) total current in the same form (*ii*) ohmic values of components in each branch. **[Nagpur University Summer 2004]**

[**Ans.** 29.07 sin (314 t – 55.88°) A (*ii*) 43.866 Ω; 43.866 Ω; 11.8 Ω; 20.45 Ω]

14. Three impedances $\mathbf{Z}_1 = (8 + j\,6)$ Ω, $\mathbf{Z}_2 = (2 - j\,1.5)$ Ω and $\mathbf{Z}_3 = 2$ Ω are connected in parallel across a 50 Hz supply. If the current through $\mathbf{Z}_1$ is $(3 + j\,4)$ A, calculate the current through the other impedances and also the power absorbed by this parallel circuit. **[V.T.U. Belgaum Karnataka Univ. Winter 2004]**

[**Ans.** (– 12 + j 16) A, (0 + j 25) A, 2.25 kW]

15. For the circuit shown in Fig. 1, find (*i*) total impedance (*ii*) total current (*iii*) total power absorbed and power factor. Draw a phasor diagram. **[Osmania Univ. Electrical Technology Jan./Feb. 1992]**

[**Ans.** (*i*) $14.12 \angle 29.2°$ Ω (*ii*) $7.08 \angle -29.2°$ (*iii*) 618.2 W (*iv*) 0.873 (lag)]

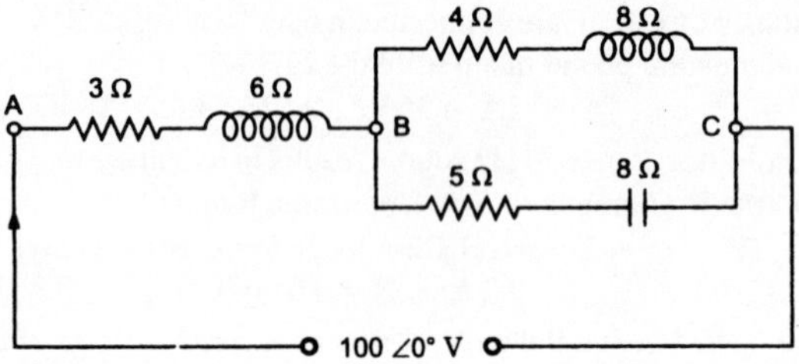

Fig. 1

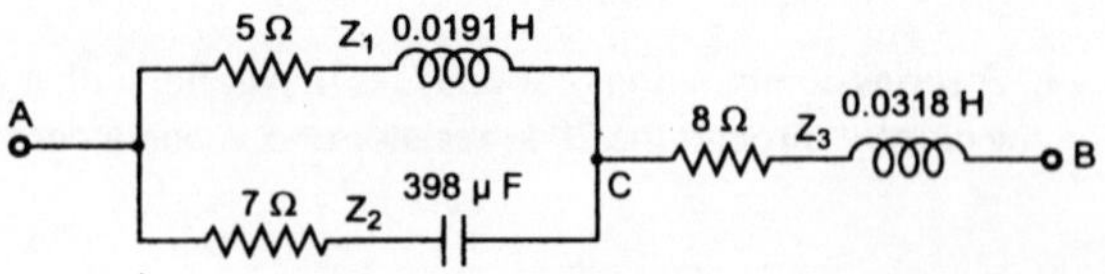

Fig. 2

16. In the circuit shown in Fig. 2, determine the voltage at 50 Hz to be applied across AB in order that a current of 10 A flows in the capacitor. **[U.P. Technical Univ. Electrical Engineering June 2001]**

[**Ans.** $288.62 \angle 22.15°$ V]

❑❑❑

Chapter 9

Resonance in R-L-C Circuits

9.1. INTRODUCTION

Resonance is identified with engineering situations which involve energy-storing elements subjected to a forcing function of varying frequency. Specifically, *resonance* is the term employed for describing the steady-state operation of a circuit or system at that frequency for which the resultant response is in time phase with the source function despite the presence of energy-storing elements. Resonance cannot take place when only one type of energy-storing element is present, *e.g.* capacitance or spring. There must exist two types of independent energy-storing elements capable of interchanging energy between them—for example, inductance and capacitance or mass and spring. Thus resonance is a phenomenon found in any system involving two independent energy-storing elements, be they electrical, mechanical, pneumatic, hydraulic or whatever.

If we have an ac circuit having a resistance R, an inductance L, and a capacitance C, connected in series (Fig. 9.1) and apply a small voltage V from a source that can keep the magnitude of V constant but can vary its frequency, we find that the magnitude of current drawn from the source of supply varies with the variation in frequency of supply source. There will be a value of frequency at which the current is maximum. *Electrical resonance* is said to exist when this condition is reached.

In this chapter we will discuss this phenomenon (more precisely called the *series* or *voltage resonance*) and also the situation when a constant voltage of variable frequency is applied to a *parallel circuit*. The simplest parallel circuit encountered in practice is a coil having resistance R and inductance L connected in parallel with a capacitor C. The resonant condition in this case is called *parallel resonance,* and sometimes *anti-resonance*. The latter name is prompted due to the fact that at resonance the input current to the parallel circuit is *minimum*.

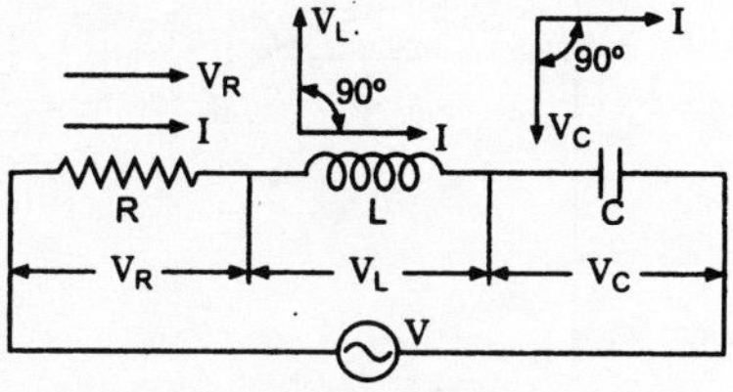

Fig. 9.1

Under the conditions of resonance, such a network becomes purely resistive in its effects, and the voltage and the current in the network are in phase. For this to occur, the inductive reactance and the capacitive reactance should be made equal.

9.2. SERIES OR VOLTAGE RESONANCE

Consider an ac circuit containing a resistance R, an inductance L and a capacitance C connected in series, as shown in Fig. 9.1.

Impedance of the circuit, $Z = \sqrt{R^2 + (X_L - X_C)^2} = \sqrt{R^2 + \left(\omega L - \dfrac{1}{\omega C}\right)^2}$

If for some frequency of applied voltage, $X_L = X_C$ in magnitude then

(*i*) Net reactance is zero *i.e.* $X = 0$

(*ii*) Impedance of the circuit, $Z = R$

(*iii*) The current flowing through the circuit is maximum and in phase with the applied voltage. The magnitude of the current will be equal to V/R

(*iv*) The voltage drop across the inductance is equal to the voltage drop across capacitance and is maximum

(*v*) The power factor is unity and

(*vi*) The power expended = VI watts.

When this condition exists, the circuit is said to be in resonance and the frequency at which it occurs is known as *resonant frequency.*

If resonant frequency is denoted by f_r, then

$$X_L = \omega L = 2\pi f_r L$$

$$\text{and } X_C = \frac{1}{2\pi f_r C}$$

$$\text{Since for resonance } X_L = X_C$$

$$\therefore\ 2\pi f_r L = \frac{1}{2\pi f_r C}$$

$$\text{or } f_r = \frac{1}{2\pi\sqrt{LC}}$$

$$\text{or } \omega_r = \sqrt{1/LC} \qquad ...(9.1)$$

From the above equation it is obvious that the value of resonance frequency depends on the parameters of the two energy-storing elements.

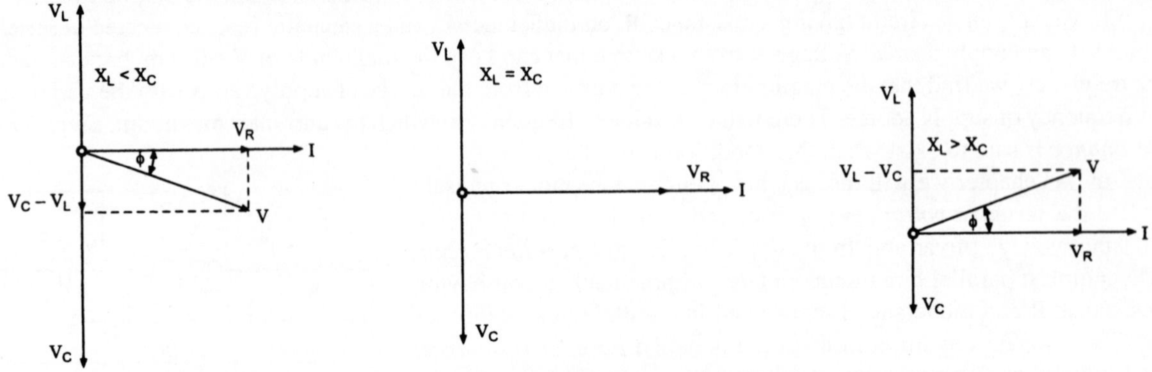

(*a*) (*b*) (*c*)

Fig. 9.2

Phasor diagrams for the series R-L-C circuit shown in Fig. 9.1 at three different frequencies *i.e.* (*a*) $f < f_r$ (*b*) $f = f_r$ and (*c*) $f > f_r$ with L and C kept constant are shown in Figs. 9.2 (*a*), 9.2 (*b*) and 9.2 (*c*) respectively.

For any frequency lower than resonant frequency f_r, inductive reactance X_L is lesser than the capacitive reactance X_C and so the circuit behaves as a capacitive circuit. Similarly for any frequency higher than resonant frequency inductive reactance is larger than capacitive reactance and so the circuit behaves as an inductive circuit. When the frequency of applied voltage is equal to the resonant frequency, the inductive reactance is equal to the capacitive reactance, the voltage drop across inductor is equal to voltage drop across capacitor in magnitude, but opposite in phase and, therefore, the circuit current I is in phase with the applied voltage *i.e.* the circuit behaves as a resistive circuit.

When the circuit is in resonance, the current is too large and will produce large voltage drop across inductance and capacitance, which will be equal in magnitude but opposite in phase and each may be several times greater than the applied voltage. If resistance R would have not been present in the circuit, such a circuit, would act like a *short-circuit* to currents of frequency to which it resonates.

Since in this resonance the voltage is maximum, it is called the *voltage resonance*. The series resonance is also called an *acceptor circuit* because such a circuit accepts currents at one particular frequency but rejects currents of other frequencies. Such circuits are used in radio-receivers.

9.2.1. Graphical Representation of Resonance in an R-L-C Series Circuit. The circuit can be made resonant in two different ways namely (*i*) by varying L and C parameters (either one or both) at a constant supply frequency or (*ii*) by varying the supply frequency *f* with parameters L and C constant. In our study of series resonance phenomenon we shall keep the applied voltage to the circuit and parameters L and C constant and allow the frequency to vary.

The circuit *resistance* R is independent of supply frequency and, therefore, remains constant. It has been represented by a straight line, parallel to the X-axis (or frequency axis) in Figure (Fig. 9.3). *Inductive reactance* X_L, being equal to ωL, increases in direct proportion to the supply frequency and is represented by a straight line passing through the origin (as X_L is regarded positive, so it lies in the first quadrant). *Capacitive reactance*, being equal to 1/ωC decreases inversely with the increase in frequency and is represented by a rectangular hyperbola lying in the fourth quadrant, below the frequency axis (capacitive reactance is considered negative).

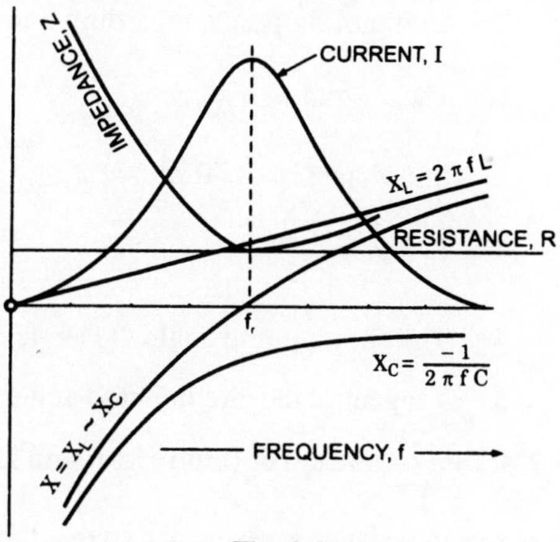

Fig. 9.3

The *net reactance* is the difference of inductive reactance X_L and capacitive reactance X_C and the curve drawn between the net reactance ($X_L \sim X_C$) and frequency will be a hyperbola (not rectangular) as shown in Fig. 9.3. The frequency at which the reactance curve crosses the frequency axis is called the *resonant frequency*, f_r (or f_o).

The impedance of the circuit, Z, being equal to $\sqrt{R^2 + (X_L \sim X_C)^2}$ is minimum at resonant frequency f_r.

At frequencies lower than resonant frequency f_r the impedance Z is large and capacitive as $X_C > X_L$ and the power factor is leading and at frequencies higher than resonant frequency f_r, the impedance Z is again large but inductive as $X_L > X_C$ and the power factor is lagging. The power factor has the maximum value of unity at resonant frequency.

9.2.2. Resonance Curve. Current varies inversely with the variation in impedance and, therefore, it is maximum at resonant frequency when impedance is minimum and decreases with the variation in frequency on both sides of the resonant frequency (as impedance Z is large), as illustrated in Fig. 9.4. The curve drawn between circuit current and the frequency of the applied voltage is called the *resonance curve* and its shape depends upon the value of circuit resistance R, as shown in the figure For smaller values of R, the resonance curve is sharply peaked, but for larger values of R, the curve is flat (Fig. 9.4).

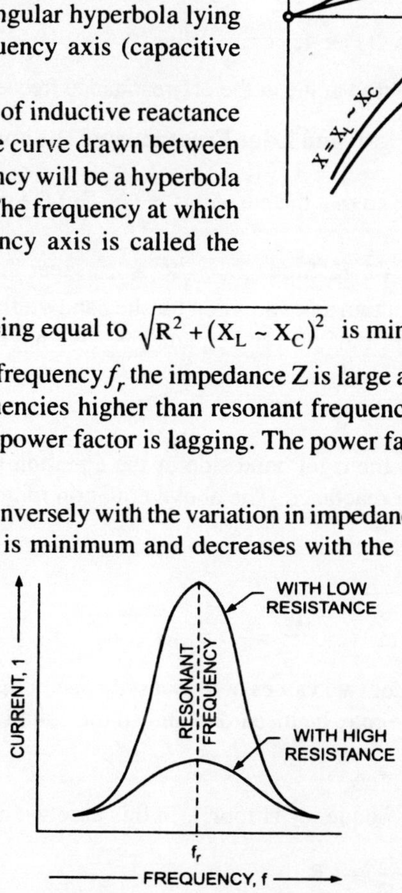

Fig. 9.4 *Resonance Curve*

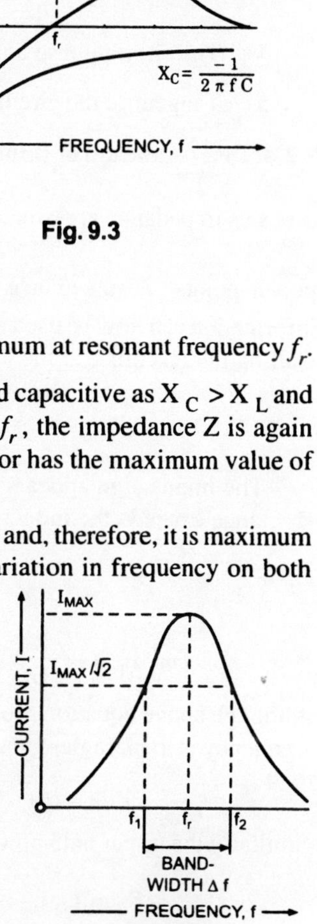

Fig. 9.5

9.2.3. Selectivity and Bandwidth. We have seen that for low resistance circuit the resonance curve is sharply peaked and such a circuit is said to be *sharply-resonant* or *highly- selective*. On the other hand, high resistance circuit have flat resonance curve and is said to have poor selectivity. Selectivity of different resonant circuits are compared in terms of their bandwidths, which are given by the bands of frequencies which lie between two points on either side of the resonant frequency where current is $1/\sqrt{2}$ times of maximum current I_{max}

$$\text{Bandwidth}, \Delta f = f_2 - f_1 \qquad ...(9.2)$$

The actual power input at frequencies f_1 and f_2

$$P = I^2 R = \left(\frac{I_{max}}{\sqrt{2}}\right)^2 R = \frac{I_{max}^2 R}{2} = \frac{1}{2} P_{max} = \frac{1}{2} \times \text{power input at resonance} \qquad ...(9.3)$$

That is why the frequencies f_1 and f_2 at the limits of the bandwidth are called the *half-power points* on the frequency scale, and the corresponding value of the bandwidth is termed as *half-power bandwidth* (B_{hp}) or – 3 dB bandwidth.

The worthnoting points regarding half-power points are

1. The current is $\frac{I_{max}}{\sqrt{2}}$;
2. Impedance is $\sqrt{2}$ R or $\sqrt{2}\, Z_{min}$;
3. $P_1 = P_2 = \frac{P_{max}}{2}$;
4. The circuit phase angle $\Phi = \pm 45°$ or $\frac{\pi}{4}$ radian;
5. Tangent of the circuit phase angle at the off-resonance frequencies f_1 and f_2, $Q = \text{Tan}\ \theta = \text{Tan}\ 45° = 1$.

9.2.4. Determination of Bandwidth and Edge Frequencies. The impedance of the tuned circuit must be $\sqrt{2}$ times its impedance at resonance so that the current is $\frac{I_{max}}{\sqrt{2}}$. But the impedance at resonance, Z = R so at half-power points the impedance is $\sqrt{2}$ R. Since $Z = \sqrt{R^2 + X^2}$, therefore, $\sqrt{2}\, R = \sqrt{R^2 + X^2}$ or X = R. This information can now be used to obtain an expression for the bandwidth in terms of the series circuit parameters. Calling the reactance at the lower half-power frequency X_1, we have

$$X_1 = \omega_1 L - \frac{1}{\omega_1 C} = -R$$

The minus sign appears on the right hand side of the equation because below resonance the capacitive reactance exceeds the inductive reactance. The above equation may also be rewritten as

$$\omega_1^2 + \frac{R}{L}\omega_1 - \frac{1}{LC} = 0$$

$$\text{or } \omega_1 = \frac{-R}{2L} \pm \sqrt{\left(\frac{R}{2L}\right)^2 + \frac{1}{LC}} = -\alpha \pm \sqrt{\alpha^2 + \omega_r^2} \qquad \text{where } \alpha = \frac{R}{2L} \because \text{From Eq. (9.1) } \omega_r = \sqrt{\frac{1}{LC}}$$

Although above equation provides two values of ω_r but only one of these is physically realizable. The negative frequency is meaningless and so may be discarded and so the equation for the lower half-power frequency is

$$\omega_1 = -\alpha + \sqrt{\alpha^2 + \omega_r^2} \qquad ...(9.4)$$

Similarly the upper half-power frequency is found. In this case we have

$$X_2 = \omega_2 L - \frac{1}{\omega_2 C} = +R$$

$$\text{or } \omega_2^2 - \frac{R}{L}\omega_2 - \frac{1}{LC} = 0$$

$$\text{or } \omega_2 = \alpha + \sqrt{\alpha^2 + \omega_r^2} \qquad \text{...(9.5)}$$

So the equation for the bandwidth becomes

$$\omega_{b\omega} = \omega_2 - \omega_1 = \alpha + \sqrt{\alpha^2 + \omega_r^2} - \left[-\alpha + \sqrt{\alpha^2 + \omega_2^2}\right] = 2\alpha = \frac{R}{L}$$

$$\text{and bandwidth, } \Delta f = f_2 - f_1 = \frac{\omega_2 - \omega_1}{2\pi} = \frac{R}{2\pi L}\text{ Hz} \qquad \text{...(9.6)}$$

$$\text{Lower half-power frequency, } f_1 = f_r - \frac{\Delta f}{2} = f_r - \frac{R}{4\pi L}\text{ Hz} \qquad \text{...(9.7)}$$

$$\text{Upper half-power frequency, } f_2 = f_r + \frac{\Delta f}{2} = f_r + \frac{R}{4\pi L}\text{ Hz} \qquad \text{...(9.8)}$$

Equation (9.6) reveals that the bandwidth of the series R-L-C circuit depends solely upon the R/L ratio. Note that it is not the individual values of R or L but rather their ratio that is important. Note too that the bandwidth is independent of the circuit capacitance C.

If Eqs. (9.4) and (9.5) are multiplied we have

$$\omega_1\,\omega_2 = \left(-\alpha + \sqrt{\alpha^2 + \omega_r^2}\right)\left(\alpha + \sqrt{\alpha^2 + \omega_r^2}\right) = \alpha^2 + \omega_r^2 - \alpha^2 = \omega_r^2$$

$$\text{or } \frac{\omega_1}{2\pi}\cdot\frac{\omega_2}{2\pi} = \frac{\omega_r^2}{(2\pi)^2}$$

$$\text{or } f_1 f_2 = f_r^2 \qquad \text{...(9.9)}$$

i.e. the resonant frequency f_r or f_o is the root of the product of the half-power frequencies of a series resonant circuit.

9.2.5. Quality Factor of a Series Resonant Circuit. The Q-factor of an R-L-C series circuit can be defined in any of the following ways.

It may be given as the voltage magnification that the circuit produces at resonance.

We have seen that at resonance, the circuit current is maximum and is equal to V/R or

$$\text{supply voltage, } V = I_{max} R$$

$$\text{and Voltage across inductance or capacitance} = I_{max} X_L = I_{max} X_C$$

$$\therefore \text{Voltage magnification} = \frac{I_{max} X_L}{I_{max} R} = \frac{X_L}{R} = \frac{\omega_r L}{R} = \frac{1}{\omega_r RC}$$

$$\text{or Q-factor at resonance, } Q_r = \frac{\omega_r L}{R} = \frac{2\pi f_r L}{R} = \frac{2\pi L}{R}\times\frac{1}{2\pi\sqrt{LC}} = \frac{1}{R}\sqrt{\frac{L}{C}} \qquad \text{...(9.10)} \quad \because f_r = \frac{1}{2\pi\sqrt{LC}}$$

$$\text{Also } Q_r = \frac{2\pi f_r L}{R} = \tan\Phi \qquad \text{...(9.11)}$$

where Φ is the power factor angle of the coil.

In case of series resonance, higher value of Q-factor means not only higher voltage magnification, but also a higher selectivity of the tuning coil and that is why it is necessary that the coil be of high inductance and low resistance. In fact, Q-factor of a series resonant circuit may be defined as the ratio of resonant frequency to bandwidth

$$\text{i.e. } Q_r = \frac{f_r}{\Delta f} = \frac{\frac{1}{2\pi\sqrt{LC}}}{R/2\pi L} = \frac{1}{R}\sqrt{\frac{L}{C}}$$

as before [see Eq. (9.10)]

Example 9.1

A coil is connected in series with a pure capacitor. The combination is fed from a 10 V of 10,000 Hz. It was observed that the maximum current of 2 A flows in the circuit when the capacitor is of the value of 1 μF. Find the parameters (R and L) of the coil *[Nagpur University April 1996]*

Solution: Maximum current in a series R-L-C circuit will flow under resonant condition and current under resonant condition is given as

$$I_{max} = \frac{V}{R}$$

So circuit resistance, $R = \frac{V}{I_{max}} = \frac{10}{2} = 5\ \Omega$ **Ans.**

Under resonant condition, also $\omega L = \frac{1}{\omega C}$

$$\text{So, } L = \frac{1}{\omega^2 C} = \frac{1}{(2\pi f)^2 C} = \frac{1}{(2\pi \times 10{,}000)^2 \times 1 \times 10^{-6}} = 0.2533 \text{ mH } \textbf{Ans.}$$

Example 9.2

A coil of resistance 40 Ω and inductance 0.75 H forms part of a series circuit for which resonant frequency is 55 Hz. If the supply is 250 V, 50 Hz, find (*i*) line current (*ii*) power factor (*iii*) power consumed and (*iv*) voltage across the coil. *[U.P. Technical Univ. Electrical Engineering June 2001]*

Solution: Resonant frequency, $f_r = 55$ Hz

At resonant frequency

Inductive reactance = Capacitive reactance

$$\text{or } 2\pi f_r L = \frac{1}{2\pi f_r C}$$

$$\text{or } C = \frac{1}{(2\pi f_r)^2 L} = \frac{1}{(2\pi \times 55)^2 \times 0.75} = 11.165\ \mu F$$

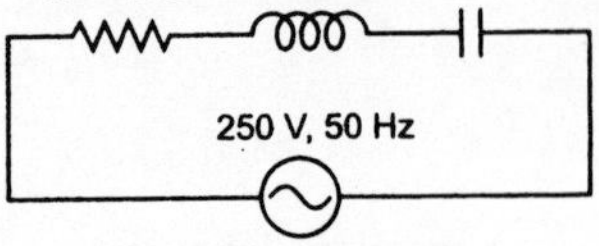

Fig. 9.6

AT 250 V, 50 Hz

Reactance of circuit, $X = \omega L - \frac{1}{\omega C}$

$$= 2\pi \times 50 \times 0.75 - \frac{1}{2\pi \times 50 \times 11.165 \times 10^{-6}} = 235.62 - 285.1 = -49.48\ \Omega$$

Impedance of the circuit, $Z = \sqrt{R^2 + X^2} = \sqrt{40^2 + 49.48^2} = 63.63\ \Omega$

(*i*) Line current, $I = \frac{V}{Z} = \frac{250}{63.63} = 3.93$ A **Ans.**

(*ii*) Power factor, $\cos\phi = \frac{R}{Z} = \frac{40}{63.63} = 0.629$ (leading) ∵ because the circuit is capacitive, X being negative

(*iii*) Power consumed, $P = I^2 R = (3.93)^2 \times 40 = 618$ W **Ans.**

(*iv*) Voltage across the coil, $V_{R-L} = I Z_{R-L} = 3.93 \sqrt{40^2 + (2\pi \times 50 \times 0.75)^2} = 939.2$ V **Ans.**

Example 9.3

A series R-L-C circuit with R = 10 Ω, $L = \frac{100}{314}$ H, $C = \frac{1}{31{,}400}$ F is excited from a 10 V, 50 Hz (ω = 314 rad/sec) source. Determine the rms values of voltages across
(*i*) Resistance (*ii*) Inductance (*iii*) Capacitance.

Give the explanation if any of the answer is more than the source voltage.

[R.G.P.V. Bhopal Basic Electrical Engineering December 2004]

Solution : Angular frequency, $\omega = 314$ radians/second

Circuit resistance, $R = 10\ \Omega$

$$\text{Inductive reactance of circuit, } X_L = \omega L = 314 \times \frac{100}{314} = 100\ \Omega$$

$$\text{Capacitive reactance of circuit, } X_C = \frac{1}{\omega C} = \frac{1}{314 \times \frac{1}{31,400}} = 100\ \Omega$$

Since $X_L = X_C$, the series R-L-C circuit is under resonant condition,

At resonance, circuit impedance, $Z = R = 10\ \Omega$

$$\text{Circuit current, } I = \frac{V}{R} = \frac{10}{10} = 1\text{ A}$$

Voltage across resistance, $V_R = I \times R = 1 \times 10 = 10$ V **Ans.**

Voltage across inductance, $V_L = I \times \omega L = 1 \times 100 = 100$ V **Ans.**

$$\text{Voltage across capacitance, } V_C = I \times \frac{1}{\omega C} = 1 \times 100 = 100\text{ V} \quad \textbf{Ans.}$$

Voltage across the inductance and capacitance is more than the source voltage because the series R-L-C circuit is under resonant condition.

Example 9.4

Voltages across resistance, inductance and capacitance connected in series are 3 V, 4 V and 5 V respectively. If supply voltage has 50 Hz frequency what is the magnitude of supply voltage. Find the resonant frequency of this series RLC circuit. *[U.P. Technical Univ. Electrical Engineering Second Semester 2004-05]*

Solution : Voltage across resistance, $V_R = 3$ V

Voltage across inductance, $V_L = 4$ V

Voltage across capacitance, $V_C = 5$ V

$$\text{Supply voltage, } V = \sqrt{V_R^2 + (V_C - V_L)^2} = \sqrt{3^2 + (5-4)^2} = \sqrt{10} = 3.162\text{ V} \quad \textbf{Ans.}$$

Voltage drop across an inductor is given as $2\pi f LI$ where f is supply frequency, L is the inductance of the inductor in Henries and I is the current flowing through it in amperes.

$$\text{So } 2\pi f LI = 4 \qquad ...(i)$$

Voltage drop across a capacitor is given as $I/2\pi fC$ where f is supply frequency, C is the capacitance of capacitor in farads and I is the current flowing through it in amperes.

$$\text{So } \frac{I}{2\pi fC} = 5 \qquad ...(ii)$$

Dividing Eq. (*i*) by Eq. (*ii*) we have

$$2\pi f LI \times \frac{2\pi fC}{I} = \frac{4}{5}$$

$$\text{or } 4\pi^2 f^2 LC = 0.8$$

$$\text{or } LC = \frac{0.8}{4\pi^2 \times 50^2} = 8.1051 \times 10^{-6} \qquad \because f = 50\text{ Hz}$$

$$\text{and resonant frequency is given as } f_r = \frac{\omega_r}{2\pi} = \frac{1}{2\pi}\sqrt{\frac{1}{LC}} = \frac{1}{2\pi}\sqrt{\frac{1}{8.1051 \times 10^{-6}}} = 55.9\text{ Hz} \quad \textbf{Ans.}$$

Example 9.5

A 10 mH coil is connected in series with a loss free capacitor to a variable frequency source of 20 V. The current in the circuit has maximum value of 0.2 A at a frequency of 100 kHz. Calculate: (*i*) the value of capacitance (*ii*) the Q-factor of the coil (*iii*) the half-power frequencies

[*U.P. Technical Univ. Electrical Engineering First Semester 2006-07*]

Solution: Supply voltage, V = 20 V

Supply frequency, f = 100 kHz = 10^5 Hz

Maximum current in series R-L-C circuit will flow under resonant condition and under resonant condition,

$$\text{Current, } I_{max} = \frac{V}{R},$$

$$\text{so circuit resistance, } R = \frac{V}{I_{max}} = \frac{20}{0.2} = 100\ \Omega$$

$$\text{Resonant frequency, } f_r = \text{Supply frequency, } f = 10^5 \text{ Hz}$$

$$\text{and } \omega L = \frac{1}{\omega C},$$

$$\text{so } C = \frac{1}{\omega^2 L} = \frac{1}{(2\pi \times 10^5)^2 \times 10 \times 10^{-3}} = 253.3 \text{ pF } \textbf{Ans.}$$

$$(ii) \text{ Q-factor of the coil} = \frac{\omega_r L}{R} = \frac{2\pi \times 10^5 \times 10 \times 10^{-3}}{100} = 62.83 \quad \textbf{Ans.}$$

$$(iii) \text{ Lower half-power frequency, } f_1 = f_r - \frac{R}{4\pi L} = 10^5 - \frac{100}{4\pi \times 0.01} \text{ Hz} = 99.204 \text{ kHz } \textbf{Ans.}$$

$$\text{Upper half-power frequency, } f_2 = f_r + \frac{R}{4\pi L} = 10^5 + \frac{100}{4\pi \times 0.01} \text{ Hz} = 100.796 \text{ kHz} \quad \textbf{Ans.}$$

Example 9.6

A series R-L-C circuit has R = 10 Ω, L = 0.1 H and C = 8 μF. Determine

(*i*) resonant frequency
(*ii*) Q-factor of the circuit at resonance
(*iii*) the half power frequencies.

[*U.P. Technical Univ. Elec. Engineering First Semester 2005-06*]

Solution:

$$(i) \text{ Resonant frequency, } f_r = \frac{1}{2\pi}\sqrt{\frac{1}{LC}} = \frac{1}{2\pi}\sqrt{\frac{1}{0.1 \times 8 \times 10^{-6}}} = 177.94 \text{ Hz } \textbf{Ans.}$$

$$(ii) \text{ Q-factor of circuit at resonance} = \frac{1}{R}\sqrt{\frac{L}{C}} = \frac{1}{10}\sqrt{\frac{0.1}{8 \times 10^{-6}}} = 11.18 \ \textbf{Ans.}$$

$$(iii) \text{ Half-power frequencies } f_1 = f_r - \frac{R}{4\pi L} = 177.94 - \frac{10}{4\pi \times 0.1} = 169.98 \text{ Hz } \textbf{Ans.}$$

$$f_2 = f_r + \frac{R}{4\pi L} = 177.94 + \frac{10}{4\pi \times 0.1} = 185.9 \text{ Hz } \textbf{Ans.}$$

Example 9.7

If the bandwidth of a resonant circuit is 10 kHz and the lower half power frequency is 120 kHz, find out the value of the upper half power frequency and the quality factor of the circuit.

[*U.P. Technical Univ. Electrical Engineering Second Semester 2003-04*]

Solution: Bandwidth of the resonant circuit, Δf = 10 kHz

Lower half power frequency, f_1 = 120 kHz

$$\text{Since lower half-power frequency, } f_1 = f_r - \frac{\Delta f}{2}$$

$$\text{Resonant frequency, } f_r = f_1 + \frac{\Delta f}{2} = 120 + \frac{10}{2} = 125 \text{ kHz}$$

$$\text{Upper half-power frequency, } f_2 = f_r + \frac{\Delta f}{2} = 125 + \frac{10}{2} = 130 \text{ kHz} \textbf{ Ans.}$$

$$\text{Quality factor of the resonant circuit, } Q_r = \frac{f_r}{\Delta f} = \frac{125}{10} = 12.5 \textbf{ Ans.}$$

Example 9.8

A 20 Ω resistor is connected in series with an inductor and a capacitor, across a variable frequency 25 V supply. When the frequency is 400 Hz, the current is at its maximum value of 0.5 A and the potential difference across the capacitor is 150 V. Calculate the resistance and inductance of the inductor.

[*U.P. Technical Univ. Electrical Engineering Junuary 2003*]

Solution: Supply voltage, V(rms) = 25 V

Frequency, f = 400 Hz

Maximum current in a series R-L-C circuit will flow under resonant condition and current under resonant condition is given as

$$I_{max} = \frac{V}{R}$$

$$\text{So circuit resistance, } R = \frac{V}{I_{max}} = \frac{25}{0.5} = 50\ \Omega \textbf{ Ans.}$$

$$\text{Voltage across the capacitor, } V_C = \frac{I_{max}}{2\pi f C}$$

$$\text{So circuit capacitance, } C = \frac{I_{max}}{2\pi f V_C} = \frac{0.5}{2\pi \times 400 \times 150} = 1.3263\ \mu F$$

$$\text{Since under resonant condition, } X_L = X_C$$

$$\text{or } 2\pi f L = \frac{1}{2\pi f C}$$

$$\therefore \text{ Inductance of inductor, } L = \frac{1}{4\pi^2 f^2 C} = \frac{1}{4\pi^2 \times (400)^2 \times 1.3263 \times 10^{-6}} = 0.119 \text{ H} \textbf{ Ans.}$$

Example 9.9

A large coil of inductance 1.405 H and resistance 40 Ω is connected in series with a capacitor of 20 μF. Calculate the frequency at which the circuit resonates. If a voltage of 100 V at the corresponding frequency is applied to the circuit, calculate the current drawn from the supply and voltages across the coil and across the capacitor.

[*Nagpur University November 1999*]

Solution:

$$\text{Resonant frequency, } f_r = \frac{1}{2\pi\sqrt{LC}} = \frac{1}{2\pi\sqrt{1.405 \times 20 \times 10^{-6}}} = 30 \text{ Hz} \textbf{ Ans.}$$

$$\text{Current drawn from supply mains, } I = \frac{V}{R} = \frac{100}{40} = 2.5 \text{ A at unity pf} \textbf{ Ans.}$$

$$\text{Voltage across the coil, } V_L = I\sqrt{R^2 + (2\pi f L)^2} = 2.5\sqrt{(40)^2 + (2\pi \times 30 \times 1.405)^2} = 670 \text{ V} \textbf{ Ans.}$$

$$\text{Voltage across the capacitor, } V_C = IX_C = \frac{I}{2\pi f C} = \frac{2.5}{2\pi \times 30 \times 20 \times 10^{-6}} = 662.6 \text{ V} \textbf{ Ans.}$$

Example 9.10

Determine the frequency at which the voltage V_{out} is zero in Fig. 9.7 shown below.

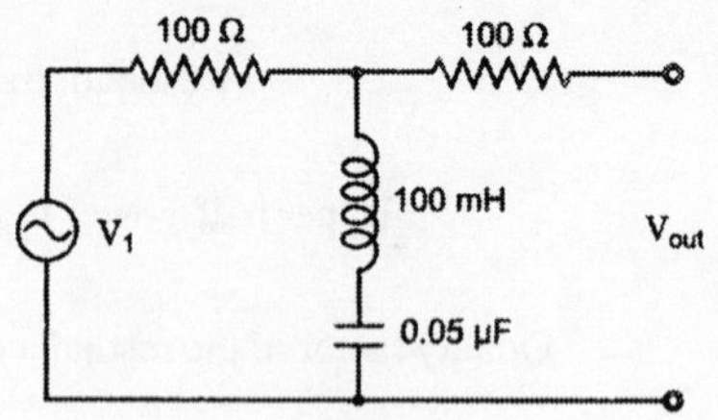

Fig. 9.7

Solution: The voltage V_{out} will be zero in the circuit when voltage drop across 100 mH inductor becomes equal to voltage drop across 0.05 μF capacitor *i.e.* when the frequency becomes equal to resonant frequency given as

$$f_r = \frac{1}{2\pi}\sqrt{\frac{1}{LC}} = \frac{1}{2\pi}\sqrt{\frac{1}{100\times10^{-3}\times0.05\times10^{-6}}}$$

$$= 2{,}250 \text{ Hz } \textbf{Ans.}$$

Example 9.11

A constant voltage at a frequency of 1 MHz is applied to an inductor in series with a variable capacitor. When the capacitor is set 500 pF, the current has its maximum value while it is reduced to one-half when the capacitance is 600 pF. Find (*i*) resistance (*ii*) the inductance, and (*iii*) the Q-factor of the inductor.

[*Bombay University 1996*]

Solution: Since circuit current has maximum value at a resonant frequency,

resonant frequency, $f_r = 1 \text{ MHz} = 1\times10^6 \text{ Hz for } C = 500\times10^{-12} \text{ F}$

$$\text{Inductance, } L = \frac{1}{(2\pi f)^2 C} = \frac{1}{(2\pi\times1\times10^6)^2\times500\times10^{-12}} = 50.66\ \mu\text{H } \textbf{Ans.}$$

Impedance of circuit with C = 500 *p*F, $Z_1 = R$ ∵ At resonance Z = R

Impedance of circuit with C = 600 pF,

$$Z_2 = \sqrt{R^2 + \left(\omega L - \frac{1}{\omega C}\right)^2} = 2R \text{ since current is halved}$$

$$\text{or } \left(\omega L - \frac{1}{\omega C}\right)^2 = 3R^2$$

$$\text{or } R = \frac{1}{\sqrt{3}}\left(\omega L - \frac{1}{\omega C}\right) = \frac{1}{\sqrt{3}}\left(2\pi\times10^6\times50.66\times10^{-6} - \frac{1}{2\pi\times10^6\times600\times10^{-12}}\right)$$

$$= 30.62\ \Omega \textbf{ Ans.}$$

$$\text{Q-factor of inductor} = \frac{\omega_r L}{R} = \frac{2\pi\times1\times10^6\times50.66\times10^{-6}}{30.62} = 10.4 \textbf{ Ans.}$$

9.3. PARALLEL OR CURRENT RESONANCE

When an inductive reactance and a capacitive reactance are connected in parallel, as shown in Fig. 9.8, conditions may reach under which *current resonance* (also known as *parallel or anti-resonance*) will take place.

Consider a practical case of a coil in parallel with a condenser, as shown in Fig. 9.8. Let the coil be of resistance R ohms and inductance L henrys and the condenser of resistance R ohms and capacitance C farads.

Such a circuit is said to be in electrical resonance when the reactive (or wattless) component of line current becomes zero. The frequency at which this happens is known as *resonant frequency.*

Circuit will be in electrical resonance if

reactive component of R-L branch current, $I_{R-L}\sin\phi_{R-L}$ = Reactive component of R-C branch current, $I_{R-C}\sin\Phi_{R-C}$

$$\text{Now since } I_{R-L} = \frac{V}{\sqrt{R^2+(\omega_r L)^2}}$$

$$\text{and } \sin\phi_{R-L} = \frac{X_L}{Z_{R-L}} = \frac{\omega_r L}{\sqrt{R^2 + (\omega_r L)^2}}$$

$$I_{R-C} = \frac{V}{Z_{R-C}} = \frac{V}{\sqrt{R_1^2 + \left(\frac{1}{\omega_r C}\right)^2}}$$

$$\text{and } \sin\phi_{R-C} = \frac{X_C}{Z_{R-C}} = \frac{1/\omega_r C}{\sqrt{R_1^2 + \left(\frac{1}{\omega_r C}\right)^2}}$$

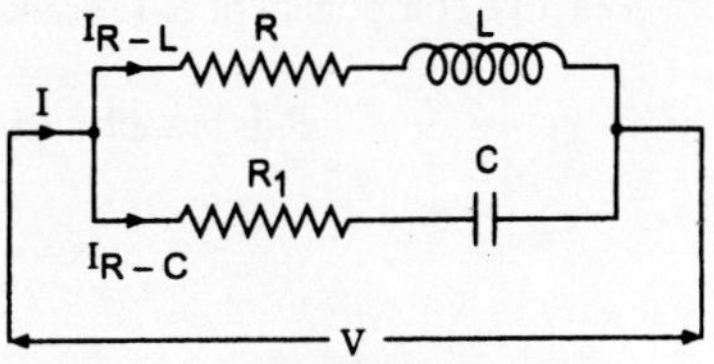

Fig. 9.8

$$\therefore \frac{V}{\sqrt{R^2 + (\omega_r L)^2}} \times \frac{\omega_r L}{\sqrt{R^2 + (\omega_r L)^2}} = \frac{V}{\sqrt{R_1^2 + \left(\frac{1}{\omega_r C}\right)^2}} \times \frac{1/\omega_r C}{\sqrt{R_1^2 + \left(\frac{1}{\omega_r C}\right)^2}}$$

$$\text{or} \quad \frac{\omega_r L}{R^2 + (\omega_r L)^2} = \frac{1/\omega_r C}{R_1^2 + (1/\omega_r C)^2}$$

$$\text{or} \quad \frac{\omega_r L}{R^2 + \omega_r^2 L^2} = \frac{\omega_r C}{\omega_r^2 R_1^2 C^2 + 1}$$

$$\text{or} \quad L\left(\omega_r^2 R_1^2 C^2 + 1\right) = C\left(R^2 + \omega_r^2 L^2\right)$$

$$\text{or} \quad \omega_r^2 L C\left(R_1^2 C - L\right) = CR^2 - L$$

$$\text{or } \omega_r = \frac{1}{\sqrt{LC}}\sqrt{\frac{CR^2 - L}{CR_1^2 - L}}$$

$$\text{or Resonant frequency}, f_r = \frac{1}{2\pi}\omega_r$$

$$= \frac{1}{2\pi\sqrt{LC}}\sqrt{\frac{CR^2 - L}{CR_1^2 - L}} \quad ...(9.12)$$

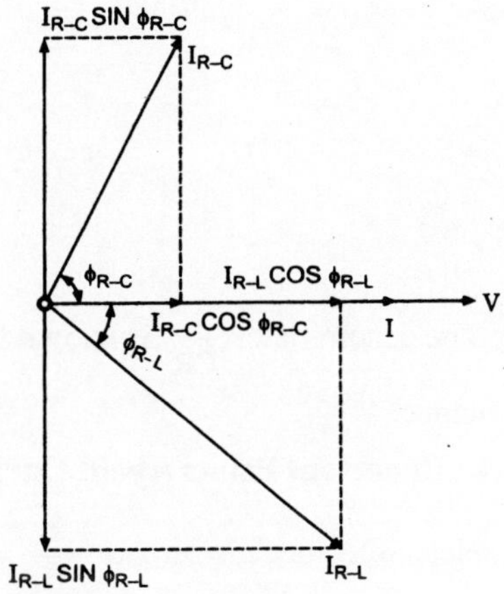

Fig. 9.9

If resistance of capacitor is negligible *i.e.* $R_1 = 0$, as is usually the case,

$$\text{Resonant frequency}, f_r = \frac{1}{2\pi\sqrt{LC}}\sqrt{1 - \frac{CR^2}{L}} = \frac{1}{2\pi}\sqrt{\frac{1}{LC} - \frac{R^2}{L^2}} \quad ...(9.13)$$

If resistance of coil R is zero

$$\text{Resonant frequency}, f_r = \frac{1}{2\pi}\sqrt{\frac{1}{LC}} \quad ...(9.14)$$

Current at Resonance.

Assuming $R_1 = 0$, as usually, at resonant frequency

$$I_{R-C} = \frac{V}{Z} = \frac{V}{\sqrt{\left(R_1^2\right)^2 + \frac{1}{(2\pi f_r C)^2}}} = 2\pi f_r C V \quad ...(9.15)$$

$$\Phi_{R-C} = \text{Cos}^{-1}\frac{R}{Z} = 90^\circ \text{ or } \frac{\pi}{2} \text{ radians}$$

Active component of R-C branch current, $I_{R-C} \cos\Phi = 0$

Reactive component of R-C branch current, $I_{R-C} \sin \phi = 2\pi f_r CV$

$$\text{R-L branch current, } I_{R\text{-}L} = \frac{V}{Z} = \frac{V}{\sqrt{R^2 + (2\pi f_r L)^2}}$$

$$\Phi_{R-L} = \text{Cos}^{-1} \frac{R}{Z}$$

$$\text{Active component of R-L branch current, } = I_{R-L} \cos \Phi_{R-L} = \frac{V}{Z} \cdot \frac{R}{Z} = \frac{VR}{Z^2}$$

$$\text{Reactive component of R-L branch current } = I_{R-L} \sin \Phi_{R-L} = \frac{V}{Z} \cdot \frac{2\pi f_r L}{Z}$$

$$\text{At resonance, } I_{R-L} \sin \Phi_{R-L} = I_{R-C} \sin \Phi_{R-C}$$

$$\text{or } \frac{V}{Z} \cdot \frac{2\pi f_r L}{Z} = 2\pi f_r CV$$

$$\text{or } \frac{L}{Z^2} = C$$

$$\text{or } Z = \sqrt{\frac{L}{C}} \quad \text{...(9.16)}$$

$$\text{Line current, } I = I_{R-L} \cos \Phi_{R-L} + I_{R-C} \cos \Phi_{R-C}$$

$$= \frac{V}{Z} \cdot \frac{R}{Z} + 0 = \frac{VR}{Z^2} = VR \times \frac{C}{L} = \frac{V}{L/CR} \quad \text{...(9.17)}$$

The denominator $\frac{L}{CR}$ is known as *effective or equivalent* or *dynamic impedance* of the parallel circuit at resonance.

9.3.1. Important Points About Current or Parallel Resonance

1. Net susceptance is zero *i.e.* $\frac{1}{X_C} = \frac{X_L}{Z^2}$

$$\text{or } \omega_r C = \frac{\omega_r L}{Z^2}$$

$$\text{or } Z = \sqrt{\frac{L}{C}}$$

2. The admittance is equal to conductance.
3. Reactive or wattless component of line current is zero, hence circuit power factor is unity.
4. Impedance is purely resistive, maximum in magnitude and is equal to L/CR.
5. Line current is minimum and is equal to $\frac{V}{L/CR}$ in magnitude and is in phase with the applied voltage.
6. Frequency is equal to $\frac{1}{2\pi}\sqrt{\frac{1}{LC} - \frac{R^2}{L^2}}$ Hz.

Note: *Parallel resonant circuit is sometimes called the rejector circuit because at resonant frequency the line current is minimum or it almost rejects it.*

Since in parallel resonant circuits circulating current between the branches is many times the line current, such type of resonance is sometimes known as current resonance.

An inductive coil of inductance L connected in parallel with a capacitance C is called the tank circuit.

9.3.2. Graphical Representation of Current or Parallel Resonance. Now we will discuss the effect of variation in frequency on the susceptances of the two parallel branches. The variations are illustrated in Fig. 9.10.

Inductive susceptance, being equal to $\frac{1}{\omega L}$ or $\frac{1}{2\pi f L}$, decreases inversely with the increase in frequency and is represented by a rectangular hyperbola lying in the fourth quadrant, below the frequency axis (inductive susceptance being considered negative). *Capacitive susceptance,* being equal to ωC or $2\pi f C$, increases in direct proportion to the supply frequency and is represented by a straight line passing through origin. As capacitive susceptance is considered positive, so it lies in first quadrant. The net susceptance B is the difference of capacitive susceptance and inductive susceptance and the curve drawn between net susceptance and frequency of the applied voltage is a hyperbola (not rectangular) as illustrated in Fig. 9.10.

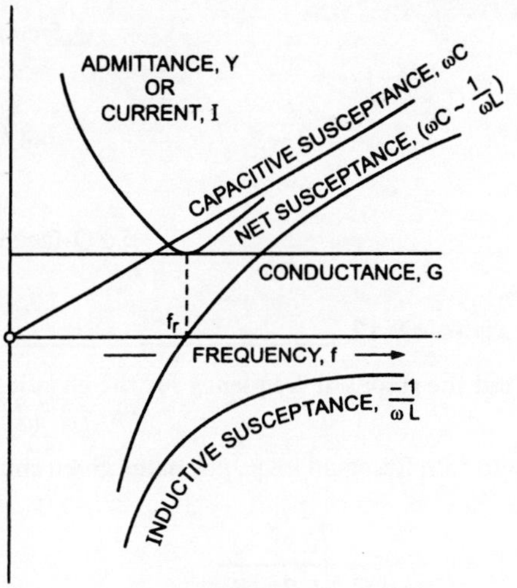

Fig. 9.10

The frequency at which the net susceptance curve crosses the frequency axis is called the *resonant frequency*. At this point impedance is maximum or admittance is minimum and is equal to G, consequently line current is minimum. Obviously at frequency lower than resonant frequency the inductive susceptance is more than capacitive susceptance, therefore, the circuit is inductive and the line current lags behind the applied voltage. But for frequencies exceeding resonant frequency capacitive susceptance predominates, therefore, the circuit is capacitive and the line current leads the applied voltage.

If the resistance is relatively low the current will fall considerably at resonant frequency and if the resistance is high, decrease in current will be less pronounced, as illustrated in Fig. 9.11.

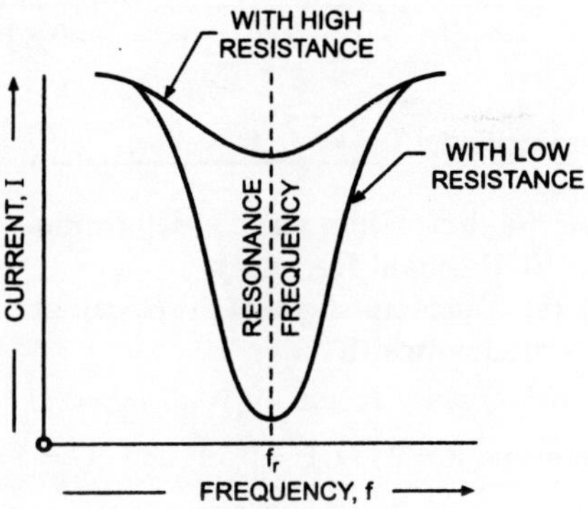

Fig. 9.11

9.3.3. Bandwidth in Case of a Parallel Resonant Circuit. The bandwidth in case of a parallel circuit is defined in the same way as in case of a series circuit. In this case also there are upper and lower half-power points where power absorbed is half of that at resonant frequency.

At bandwidth frequencies, the net susceptance B is equal to the conductance G. So at freqµency f_1 the net susceptance $B_{L_1} - B_{C_1} = G$ and at frequency f_2, $B_{C_2} - B_{L_2} = G$. Thus admittance $Y = \sqrt{G^2 + B^2} = \sqrt{2}\,G$ and phase angle $\Phi = \tan^{-1}\frac{B}{G} = \tan^{-1} 1 = 45°$ or $\frac{\pi}{4}$ radian.

9.3.4. Q-Factor or Current Magnification Factor. Q-factor of a parallel circuit is defined as the ratio of the circulating current to the line current or as the current magnification.

$$\therefore \text{ Q-factor} = \frac{\text{Circulating current between L and C}}{\text{Line current}}$$

Since from Eq. (9.15), the circulating current, $I_C = 2\pi f_r C V$

and from Eq. (9.17), the line current, $I = \frac{V}{L/CR} = VCR/L$

$$\therefore \text{ Q-factor } = \frac{I_C}{I} = \frac{2\pi f_r CV}{VCR/L} = \frac{2\pi f_r L}{R}$$

$$\text{but } f_r = \frac{1}{2\pi}\sqrt{\frac{1}{LC}} \text{ neglecting R}$$

$$\text{So Q-factor } = 2\pi \times \frac{1}{2\pi}\sqrt{\frac{1}{LC}} \times \frac{L}{R} = \frac{1}{R}\sqrt{\frac{L}{C}}$$, It is the same as for series resonance.

Example 9.12

Find the resonant frequency for the circuit shown in Fig. 9.12.

[*Pb. Technical Univ. Basic Electronics and Elec. Engineering Second Semester 2004-05*]

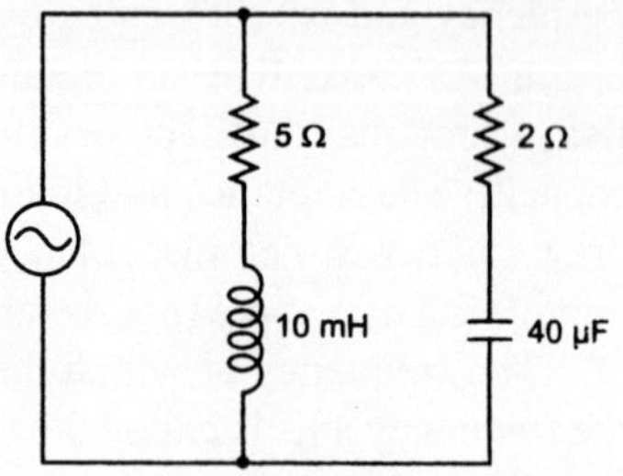

Fig. 9.12

Solution: Resonant frequency of the given circuit

$$f_r = \frac{1}{2\pi\sqrt{LC}}\sqrt{\frac{CR^2 - L}{CR_1^2 - L}}$$

$$= \frac{1}{2\pi\sqrt{10\times10^{-3}\times40\times10^{-6}}} \times \sqrt{\frac{40\times10^{-6}\times5^2 - 10\times10^{-3}}{40\times10^{-6}\times2^2 - 10\times10^{-3}}} = 263 \text{ Hz } \textbf{Ans.}$$

Example 9.13

For the circuit shown below, determine

(*i*) Resonant frequency.

(*ii*) Total impedance of the circuit at resonance.

(*iii*) Bandwidth

(*iv*) Quality factor. [*U.P. Technical Univ. Elec. Engineering First Semester 2005-06*]

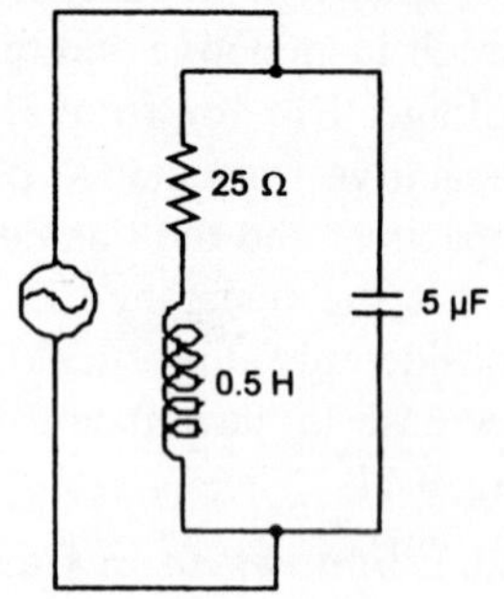

Fig. 9.13

Solution: R = 25 Ω, L = 0.5 H and C = 5 × 10^{-6} F

$$(i) \text{ Resonant frequency, } f_r = \frac{1}{2\pi}\sqrt{\frac{1}{LC} - \frac{R^2}{L^2}} \text{ Hz}$$

$$= \frac{1}{2\pi}\sqrt{\frac{1}{0.5\times5\times10^{-6}} - \frac{25^2}{(0.5)^2}} \text{ Hz} = 100.34 \text{ Hz } \textbf{Ans.}$$

$$(ii) \text{ Impedance of the circuit at resonance, } Z = \frac{L}{CR} = \frac{0.5}{5\times10^{-6}\times25} = 4{,}000\ \Omega \textbf{ Ans.}$$

$$(iii) \text{ Bandwidth } = \frac{R}{2\pi L} = \frac{25}{2\pi\times0.5} = 7.958 \text{ Hz } \textbf{Ans.}$$

$$(iv) \text{ Quality factor, } Q = \frac{1}{R}\sqrt{\frac{L}{C}} = \frac{1}{25}\sqrt{\frac{0.5}{5\times10^{-6}}} = 12.65 \textbf{ Ans.}$$

Example 9.14

In the circuit shown in Fig. 9.14, find the value of R such that the impedance of the whole circuit should be independent of the frequency of the supply. If voltage = 200 V, L = 0.16 H and C = 100 μ F, calculate the power loss in the circuit. [*U.P. Technical Univ. Electrical Engineereing February 2002*]

Solution: Impedance of inductive branch, $Z_L = (R + j\omega L)$

Impedance of capacitive branch, $Z_C = \left(R - \frac{j}{\omega C}\right)$

Impedance of the whole circuit shown in Fig. 9.14,

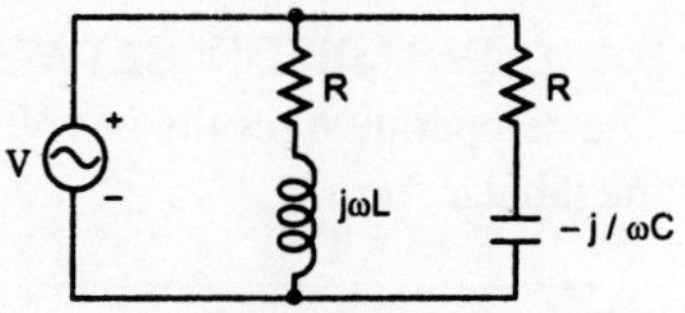

Fig. 9.14

$$\mathbf{Z} = \frac{\mathbf{Z_L\, Z_C}}{\mathbf{Z_L + Z_C}} = \frac{(R + j\omega L)\left(R - \frac{j}{\omega C}\right)}{(R + j\omega L) + \left(R - \frac{j}{\omega C}\right)}$$

$$= \frac{R^2 + \frac{L}{C} + jR\left(\omega L - \frac{1}{\omega C}\right)}{2R + j\left(\omega L - \frac{1}{\omega C}\right)}$$

We note that *j* term in the numerator of the above expression is R times the denominator so if the real term *i.e.* $\left(R^2 + \frac{L}{C}\right)$ is also R times of the real term in denominator *i.e.* R × 2 R, then the term consisting of ω will disappear. Thus the impedance of the circuit will be independent

$$\text{if } R^2 + \frac{L}{C} = R \times 2R \quad \text{or } R^2 = \frac{L}{C} \quad \text{or} \quad R = \sqrt{\frac{L}{C}} \quad \textbf{Ans.}$$

Substituting L = 0.16 H and C = 100×10^{-6} F in above expression we have

$$R = \sqrt{\frac{0.16}{100 \times 10^{-6}}} = 40\ \Omega\ \textbf{Ans.}$$

$$\text{Power loss in the circuit, } P = \frac{V^2}{R} = \frac{(200)^2}{40} = 1{,}000 \text{ W or } 1 \text{ kW} \ \textbf{Ans.}$$

Example 9.15

What are the features of resonance in parallel circuits ? Calculate the value of C which results in resonance for the circuit shown in Fig. 9.15 when frequency is 1,000 Hz and find Q-factor for each branch.

[U.P. Technical Univ. Electrical Engineering June 2001]

Solution: Admittance of inductive branch, $\mathbf{Y}_{R-L} = \frac{1}{4 + j8} = (0.05 - j\,0.1)$ Siemen

Admittance of capacitive branch, $\mathbf{Y}_{R-C} = \frac{1}{5 - jX_C} = \left(\frac{5}{25 + X_C^2} + j\frac{X_C}{25 + X_C^2}\right)$ Siemen

For resonance, *j* parts of admittances must be equal in magnitude i.e.

$$\frac{X_C}{25 + X_C^2} = 0.1$$

$$\text{or} \quad 0.1\, X_C^2 - X_C + 2.5 = 0$$

$$\text{or} \quad X_C = 5\ \Omega$$

$$\text{and } C = \frac{1}{2\pi f X_C} = \frac{1}{2\pi \times 1{,}000 \times 5} = 31.83\ \mu\text{F} \quad \textbf{Ans.}$$

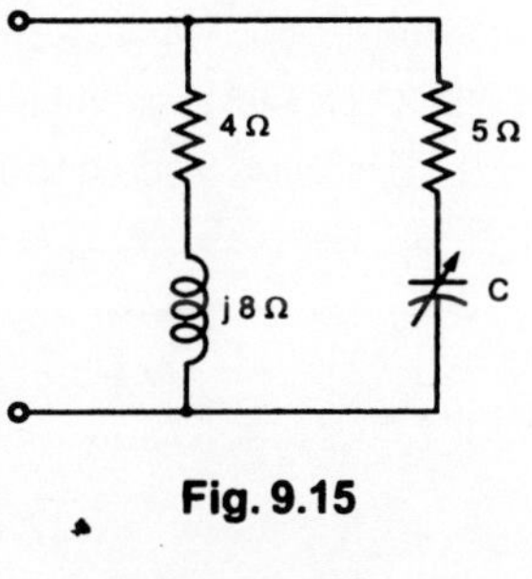

Fig. 9.15

$$\text{Q factor for R-L branch} = \frac{\omega L}{R} = \frac{8}{4} = 2 \ \textbf{Ans.}$$

$$\text{Q factor for R-C branch} = \frac{1}{\omega RC} = \frac{1}{\omega CR} = \frac{5}{5} = 1.0 \ \textbf{Ans.} \qquad \left[\because \frac{1}{\omega C} = X_C = 5\ \Omega\right]$$

9.4. COMPARISON BETWEEN VOLTAGE RESONANCE AND CURRENT RESONANCE

The comparative results of voltage resonance and current resonance studied so far are summarized below in the tabular form.

S.No.	*Particulars*	*Voltage Resonance*	*Current Resonance*
1.	Impedance under *Resonance* conditions	Z = R, the minimum	Z = L/CR, the maximum
2.	Magnitude of current under resonant condition	I = V/R, the maximum	I = V/(L/CR), the minimum
3.	Power factor at resonance	Unity	Unity
4.	Resonant frequency	$f_r = \frac{1}{2\pi\sqrt{LC}}$	$f_r = \frac{1}{2\pi}\sqrt{\frac{1}{LC} - \frac{R^2}{L^2}}$
5.	Magnification occurs of	voltage	current
6.	Magnification	ω L/R	ω L/R
7.	Bandwidth	R/2πL Hz	R/2πL Hz

HIGHLIGHTS

1. An R-L-C series circuit is said to be in electrical resonance when

$$X_L = X_C \textit{ i.e. } \text{the net reactance } X = 0$$

Under resonant conditions

(*i*) Circuit impedance, Z = R

(*ii*) The current flowing through the circuit is maximum and is in phase with the applied voltage and will be equal to V/R

(*iii*) The voltage drop across inductance is equal to the voltage drop across capacitance and is maximum.

(*iv*) The power factor is unity.

(*v*) The power expended, P = VI watts and

(*vi*) The resonant frequency, $f_r = \frac{1}{2\pi\sqrt{LC}}$ Hz

Q-factor of a series resonant circuit is given as

$$\text{Q-factor} = \frac{\omega_r L}{R} = \frac{2\pi f_r L}{R} = \frac{1}{R}\sqrt{\frac{L}{C}}$$

$$\text{Bandwidth, BW} = \frac{R}{2\pi L} = \frac{f_r}{\text{Q-factor}}$$

2. A parallel ac circuit is said to be in resonance when it draws no reactive current.

Under resonant condition of parallel ac circuit

$$\text{Impedance, Z} = \frac{L}{CR}, \text{ the maximum}$$

$$\text{Current, I} = \frac{V}{L/CR}, \text{ the minimum}$$

$$\text{Power factor} = \text{Unity}$$

$$\text{Resonant frequency, } f_r = \frac{1}{2\pi}\sqrt{\frac{1}{LC} - \frac{R^2}{L^2}} \simeq \frac{1}{2\pi\sqrt{LC}} \quad \text{if R is negligibly small}$$

$$\text{Q-factor} = \frac{1}{R}\sqrt{L/C}$$

$$\text{Bandwidth} = \frac{R}{2\pi L} \text{ Hz}$$

EXERCISES

1. Discuss the resonance in R-L-C circuit. What is condition for it ? **[M.D. Univ. Rohtak Electrical Technology May 2002]**
2. Derive the conditions of resonance in ac R-L-C series circuits.
[Pb. Technical Univ. Basic Electrical and Electronics Engineering, December 2006]
3. Give the properties of R-L-C series resonance and also draw graph showing the variation of resistance, inductive reactance and capacitive reactance with frequency. **[M.D. Univ. Rohtak Electrical Technology May 2003]**
4. Define the Q-factor for the series resonant circuit and express it in terms of the circuit parameters.
[M.D. Univ. Rohtak Electrical Technology May 2002]
5. Explain the concept of bandwidth and quality factor for a series R-L-C circuit. Derive their expressions.
[U.P. Technical Univ. Electrical Engineering Second Semester 2007-08]
6. Derive expressions for lower and upper power frequencies for an R-L-C circuit.
[G.G.S.I.P. Univ. Delhi Electrical Science May 2005]
7. For an LCR series circuit derive an expression for resonant frequency, bandwidth and quality factor.
[U.P. Technical Univ. Electrical Engineering First Semester, 2004-05]
8. Explain series resonance in R-L-C circuit. What are bandwidth and quality factor of the circuit ?
[U.P. Technical Univ. Electrical Engineering Second Semester, 2006-07]
9. Explain the selectivity of a series RLC resonance circuit. Prove that selectivity is reciprocal of quality of the circuit.
[U.P. Technical Univ. Electrical Engineering Second Semester, 2004-05]
10. Show that for an RLC series circuit the resonant frequency $\omega_0 = \sqrt{\omega_1\omega_2}$, where ω_1 and ω_2 are the lower and upper half power frequencies respectively.
[U.P. Technical Univ. Electrical Engineering Second Semester, 2004-05]
11. Explain in brief parallel resonance.
[M.D. Univ. Electrical Technology May 2003]
12. Find out the resonant frequency of the circuit shown in Fig. 1 in terms of circuit parameters. **[U.P.S.C. I.E.S. Electrical Engineering-I, 2006]**
13. Derive an expression for parallel resonance and mention its salient features.
[U.P. Technical Univ. Electrical Engineering First Semester 2006-07]
14. Explain parallel resonance. Why is parallel resonance called the current resonance? Show the graphical representation of current in parallel resonance.
[U.P. Technical Univ. Electrical Engineering Second Semester, 2005-06]
15. Define parallel resonance. Calculate at resonance the resultant current and quality factor in terms of the parameters of a circuit. **[U.P. Technical Univ. Electrical Engineering First Semester, 2007-08]**
16. Compare the series and parallel resonance of R-L-C series and R-L-C parallel circuits. **[Gujarat Univ. June/July 2003]**

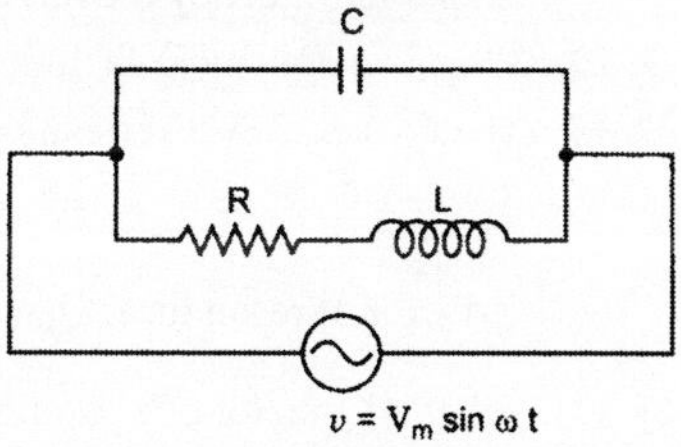

Fig. 1

SHORT ANSWER TYPE QUESTIONS WITH ANSWERS

Q. 1. What is meant by resonant frequency ?

Ans. The frequency of supply at which two reactances (inductive reactance and capacitive reactance) are equal is called resonant frequency.

Q. 2. What is series resonance ?

Ans. When in an R-L-C series circuit the inductive reactance equals capacitive reactance, the circuit is said to be in resonance, called the series resonance.

Q. 3. Why is the series resonance called the voltage resonance?

Ans. Since in series resonance the voltage across inductance and capacitance is maximum, it is called the voltage resonance.

Q. 4. Why is the series R-L-C circuit also called the acceptor circuit ?

Ans. Series R-L-C circuit accepts currents at one particular frequency but rejects currents of other frequencies and therefore, series R-L-C circuit is called the acceptor circuit.

Q. 5. In a series RLC circuit, the voltage across L and C at resonance may exceed even the supply voltage. Why ?

[U.P. Technical Univ. First Semester 2005-06]

Ans. When the series circuit is in resonance, the voltage drops across inductance and capacitance will be equal in magnitude but opposite in phase and so nullify each other and supply voltage will be equal to voltage drop across resistance. So the current will be too large and the voltage drops may exceed the supply voltage in case the inductive reactance (or capacitive reactance) exceed the resistance of the circuit.

Q. 6. What do you understand by the term bandwidth of a resonant circuit ?

Ans. The frequency band over which the magnitude of current is equal to $1/\sqrt{2}$ times the peak current is referred to as bandwidth of the circuit.

Q. 7. What is a resonance in a parallel circuit ?

Ans. A parallel electrical circuit (Fig. 1) is said to be in electrical resonance when the reactive or wattless component of line current is zero i.e. when reactive component of R-L branch current is equal to current drawn by capacitance.

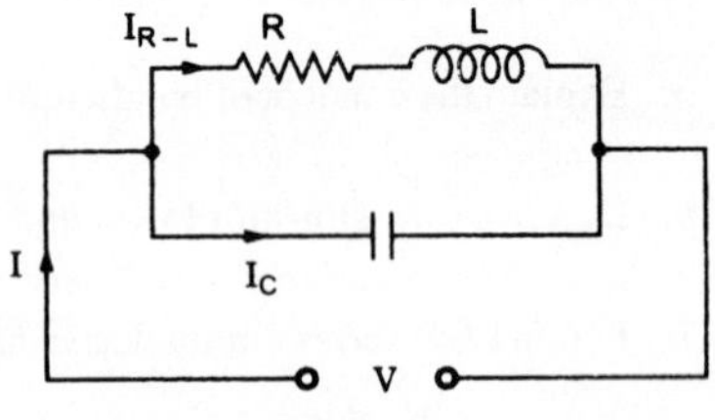

Fig. 1

Q. 8. Why is parallel resonance called the current resonance ?

Ans. Since in parallel resonant circuit, circulating current between the branches is many times the line current, such type of resonance is called the current resonance.

Q. 9. Why a series resonant circuit is called as an acceptor circuit and a parallel resonant circuit as a rejector circuit ?

[U.P. Technical Univ. First Semester 2000-01]

Ans. The series resonant circuit is called an acceptor circuit because such a circuit accepts currents at one particular frequency but rejects currents of other frequencies. Parallel resonant circuit is called the rejector circuit because at resonant frequency the line current is minimum or it almost rejects it.

Q. 10. What is meant by Q-factor of series resonant circuit ?

Ans. The Q-factor of a series resonant circuit may be given as the voltage magnification that the circuit produces at resonance.

$$\text{Q-factor at resonance, } Q_r = \frac{\omega_r L}{R} = \frac{2\pi f_r L}{R} = \frac{1}{R}\sqrt{\frac{L}{C}}$$

Q. 11. What is Q-factor of a parallel resonant circuit ?

Ans. Q-factor of a parallel resonant circuit is defined as the ratio of the circulating current to the line current at resonance or as the current magnification.

Q-factors for series resonance and parallel resonance are the same.

Q. 12. What is the power factor of an R-L-C circuit under resonant condition ?

Ans. Unity.

Q. 13. At resonance the current is maximum in a series circuit and minimum in parallel circuit. Why ?

[U.P. Technical Univ. First Semester 2005-06]

Ans. At resonance in series circuit the inductive reactance is equal to capacitive reactance in magnitude but opposite in phase and so nullify each other and thus impedance is minimum and equal to the resistance of the circuit and, therefore, current is maximum.

But at resonance in parallel circuit, net susceptance is zero, the admittance is, therefore, minimum and is equal to conductance of the circuit and the current, being the product of supply voltage and circuit admittance is minimum.

Q. 14. The shape of resonance curve depends on Q of the coil. Why ? **[U.P. Technical Univ. First Semester 2005-06]**

Ans. Bandwidth of the series RLC circuit, being equal to f_r/Q_r, depends upon the Q of the coil i.e. with high-Q coil the resonance curve will be of smaller bandwidth (i.e. sharply peaked) while with small-Q coil the resonance curve will be of larger bandwidth (flat). Thus the shape of the resonance curve depends upon the Q of the coil.

Q. 15. Draw the phasor diagram showing voltages across each element and current through each element for:

(A) series R-L-C circuit.

(B) a capacitor in parallel with R-L series combination under resonance in both cases.

[R.G.P.V. December 2002]

Ans. Phasor diagram showing voltages across each element and current flowing through each element for (A) series R-L-C circuit and (B) a capacitor in parallel with R-L series combination under resonant conditions are shown in Figs. 2 and 3 respectively.

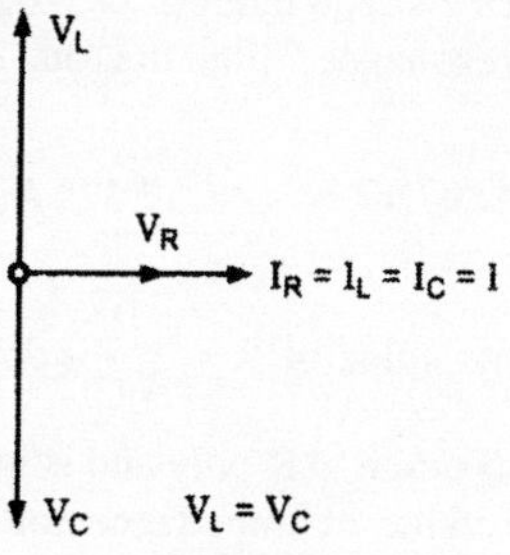

Fig. 2

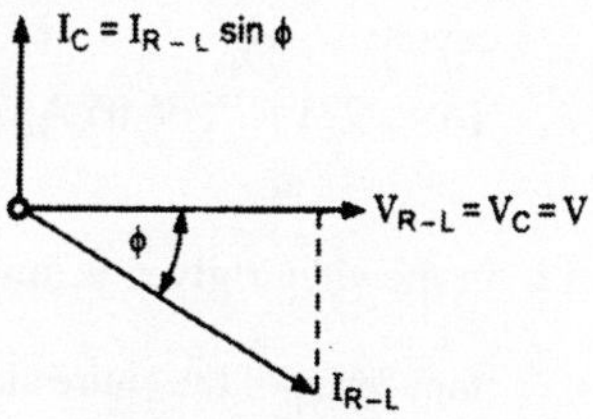

Fig. 3

PROBLEMS

1. A resistance of 15 Ω, choke coil of 0.05 H and the capacitor of 300 μF are connected in series. If the circuit is connected to 250 V, 50 Hz, power supply, find (*i*) circuit impedance (*ii*) power factor (*iii*) absorbed power (*iv*) resonant frequency and (*v*) Q-factor of the coil.

   [**Ans.** (*i*) 15.84 Ω (*ii*) 0.947 (lag) (*iii*) 3,735 W (*iv*) 41 Hz (*v*) 1.056]

2. A coil of 1.5 H inductance and 5 Ω resistance and a capacitor of 5 μ F are connected in series. Calculate the resonant frequency and the power consumed if it is connected to a 200 V ac mains. [**Ans.** 58 Hz; 8,000 W]

3. A series R-L-C circuit has R = 10 Ω, L = 0.1 H and C = 8 μ F. Determine (*i*) resonant frequency (*ii*) Q-factor of the circuit at resonance and (*iii*) the half-power frequencies. **[U.P. Technical Univ. Electrical Engineering September 2001]**

   [**Ans.** (*i*) 177.94 Hz (*ii*) 11.18 (*iii*) 170 Hz, 185.8 Hz]

4. A 20 Ω resistor is connected in series with an inductor and a capacitor, across a variable frequency 25 V supply. When the frequency is 400 Hz, the current is at its maximum value of 0.5 A and the potential difference across the capacitor is 150 V. Calculate the resistance and inductance of the inductor.

   [U.P. Technical Univ. Electrical Engineering Junuary 2003]

   [**Ans.** 50 Ω; 0.119 Ω]

5. An inductive coil of resistance 10 Ω and inductance 0.1 H is connected in parallel with a 150 μF capacitor to a variable frequency, 200 V supply. Find the resonant frequency at which the total current taken from the supply is in phase with the supply voltage. Also find the value of this current. **[G.G.S.I.P. Univ. Delhi Electrical Science June 2006]**

   [**Ans.** 37.886 Hz; 3 A]

6. A series RLC circuit with R = 250 Ω, and L = 0.6 H results in a leading phase angle of 60° at a frequency of 40 Hz. At what frequency will the circuit resonate ? **[Pb. Technical Univ. Basic Electrical Engineering December 2006]**

   [**Ans.** 81.2 Hz]

7. An R-L-C series circuit consists of a resistance of 1,000 Ω, an inductance of 100 mH and a capacitance of 10 μμF. If the voltage of 100 V is applied across the combination, find (*i*) the resonance frequency (*ii*) Q-factor of the circuit and (*iii*) the half-power points. **[Bombay Univ. Nov. 1971]**

   [**Ans.** (*i*) $10^6/2\pi$ Hz; (*ii*) 100; (*iii*) 158.4 kHz and 160 kHz]

8. A circuit comprises an inductance of 26 μ H associated with resistance of 0.1 Ω in parallel with a capacitance of 0.16 μ F. Find the ratio of the circulating current to the supply current at resonant frequency. [**Ans.** 127.5]

9. An inductive coil when connected across 500 V, 50 Hz supply draws 1.0 A at 0.8 p f. What capacitance is connected in parallel with the coil to make the power factor of the combination unity ? What will be the current (*i*) drawn from the supply; and (*ii*) in each branch with this new capacitor. [**Ans.** 3.82 μ F; (*i*) 0.8 A (*ii*) 1.0 A, 0.6 A]

10. An a c voltage of 200 V and variable frequency is applied to a coil of 20 Ω resistance and 0.3 H inductance connected in parallel with a capacitance of 200 μ F. Determine the parallel resonant frequency and the input current at this frequency. [**Ans.** 20.5 Hz; 2.667 A]

11. An inductive circuit of resistance 2 Ω and inductance 0.01 H is connected to 250-V, 50 Hz supply. What capacitance placed in parallel will produce resonance ? Find the total current taken from the supply and the current in the branch circuits. **[Kerala Univ. Electrical Engineering 1987]**

[**Ans.** 721 μ F; 36.05 A; 125 A; 67.13 A $\angle -57.52°$ A; 56.63 $\angle 90°$ A

12. In the circuit given below, if the value of $R = \sqrt{\frac{L}{C}}$, then prove that the impedance of the entire circuit is equal to R only and is independent of the frequency of supply. Find the value of impedance for L = 0.02 H and C = 100 μ F. **[Hyderabad Univ. Communication System 1991]**

[Ans. 14.142 Ω]

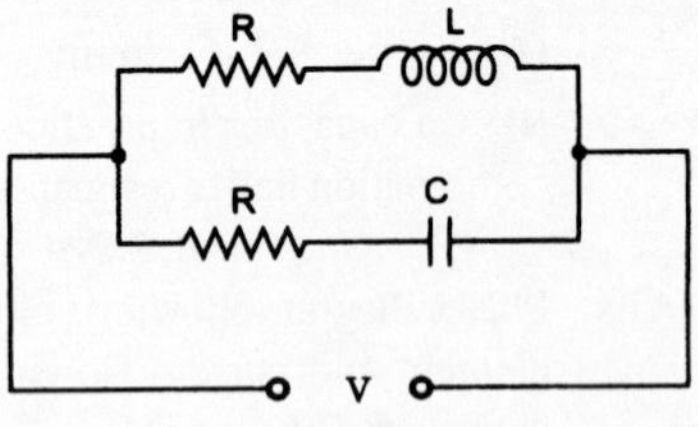

Fig. 1

❑❑❑

UNIT-V
Polyphase Circuits

Contents in This Unit

Polyphase Circuits: Generation of polyphase voltages; 3-Phase system; Phase sequence; Interconnection of 3-phases; Voltage, current and power relationships in balanced three phase circuits; Power measurement in Single phase and 3-phase circuits.

Chapter 10: Polyphase Circuits

Chapter

10 Polyphase Circuits

10.1. INTRODUCTION

The type of alternating currents and voltages discussed so far in the book are termed as *single phase currents and voltages* as they consist of a single alternating current and voltage waves. Single phase systems involving single phase currents and voltages are quite satisfactory for domestic applications. Even the motors employed in domestic applications are mostly single phase, for example, motors for mixers, coolers, fans, air-conditioners, refrigerators. However, the single phase system has its own limitations and, therefore, has been replaced by polyphase system. For supplying power to electric furnaces 2-phase system is generally employed. Six phase system is usually employed in connection with converting machinery and apparatus. For general supply three-phase system is universally used. For generation, transmission and distribution of electric power 3-phase system has been universally adopted. The two-phase supply and six-phase supply are obtained from 3-phase supply.

Polyphase system means the system which consists of numerous (poly means numerous or multiple) windings or circuits (phase* means winding or a circuit).

A polyphase system is essentially a combination of several single phase voltages having same magnitude and frequency but displaced from one another by equal angle (electrical), which depends upon the number of phases and can be determined from the following relation

$$\text{Electrical displacement} = \frac{\text{360 electrical degrees}}{\text{Number of phases}} \quad ...(10.1)$$

The above relation does not hold good for two phase windings, which are displaced by 90 electrical degrees apart.

A supply system is said to be symmetrical when several voltages of the same frequency have equal magnitude and are displaced from one another by equal time angle. 3-phase, 3-wire or 4-wire supply system will be symmetrical when the line to line voltages are equal in magnitude and displaced in phase by 120 electrical degrees with respect to each other. Further in a four-wire system the voltage with respect to neutral of the three phase wires should be equal to one another in magnitude and displaced in phase by 120° with respect to each other.

A 3-phase supply will be unbalanced when either of the three-phase voltages are unequal in magnitude or the phase angle between these phases is not equal to 120°.

A load circuit is said to be *balanced* when the loads (impedances) connected in various phases are same in magnitude as well as in phase. Any three phase load in which the impedances in one or more phases differ from the impedances of other phases is called an *unbalanced three-phase load.*

---

* Each emf of a polyphase system is referred to as a *phase*.

In case one phase out of three phases of a 3-phase supply connected to the 3-phase load is not available, then such a condition is called the *single phasing*.

10.1.1. Merits and Demerits of Polyphase System Over Single Phase System. The advantages of a polyphase system over single phase system are enumerated below:

(*i*) In a single phase circuit the power delivered is pulsating. Even when current and voltage are in phase, the power is zero twice in each cycle, and when the current leads or lags behind the voltage, the power is negative twice and zero four times during each cycle. This is not objectionable for lighting and small motors but with large motors it causes excessive vibrations. In polyphase system the total power delivered is constant if loads are balanced though the power of any one phase or circuit may be negative. So polyphase system is highly desirable particularly for power loads.

(*ii*) The rating of a given machine increases with the increase in number of phases. For example output of a 3-phase motor is 1.5 times the output of a single phase motor of same size.

(*iii*) Single phase induction motors have no starting torque and so it is necessary to provide these motors with an auxiliary means of starting , but in case of three-phase motors except synchronous motors, there is no need of providing an auxiliary means for starting.

(*iv*) Power factor of a single-phase motor is lower than that of a polyphase motor of the same rating (output and speed). The efficiency of a polyphase motor is also higher than that of a single-phase motor.

(*v*) Three phase-system requires 3/4th weight of copper of that required by single-phase system to transmit the same amount of power at a given voltage and over a given distance.

(*vi*) Rotating magnetic field can be set up by passing polyphase currents through stationary coils.

(*vii*) Polyphase system is more capable and reliable than single-phase system .

(*viii*) Parallel operation of polyphase alternators is simple as compared to that of single-phase alternators because of pulsating reaction in single phase alternators.

However, three-phase operation is not as practical for domestic applications where motors are usually smaller than 1 kW and where lighting circuits supply most of the load.

10.1.2. Commonly Used Polyphase System. Although there are several polyphase systems such as two-phase, three-phase but 3-phase system is invariably adopted because of its inherent advantages over all other polyphase systems .

The demand for two-phase system has almost disappeared because that does not have any advantage that is not equalled or surpassed by three-phase system in generation, transmission or utilization. Three-phase system is universally employed for generation, transmission and distribution of electric power. Two-phase supply and six-phase supply, when required, is obtained from 3-phase supply.

Systems with number of phases, more than three, increase complexity and cost of transmission and utilisation hardware and becomes uneconomical.

A knowledge of 3-phase power systems is, therefore, essential for understanding of power technology. Fortunately, the basic circuit technique used in solving single phase circuits are directly applicable to 3-phase circuits because three phases are identical and one phase represents the behaviour of all. In this chapter we will discuss only 3-phase systems.

10.2. GENERATION OF EMFs

10.2.1. Generation of Single Phase EMF. As already discussed when a coil rotates in a uniform magnetic field, an alternating emf is induced in the coil. At the instant the coil is at right angles to the direction of field, as in position shown in Fig. 10.1 (*a*) the linking flux in the coil is maximum, but the rate of change of flux is zero, therefore induced emf is zero. But when the plane of the coil is in the plane of the field as in position CD shown in Fig. 10.1 (*a*) the linking flux is zero but the rate of change of flux is maximum so emf induced in the coil at this instant is maximum. The emf induced in the coil in various *positions of the coil with respect to the field is shown in Fig. 10.1 (b).*

10.2.2. Generation of Two Phase EMF. When two coils fastened rigidly together 90 degrees (electrical) apart from each other rotate in a uniform magnetic field, the emf induced in each coil will have phase difference of 90º or $\pi/2$ radians.

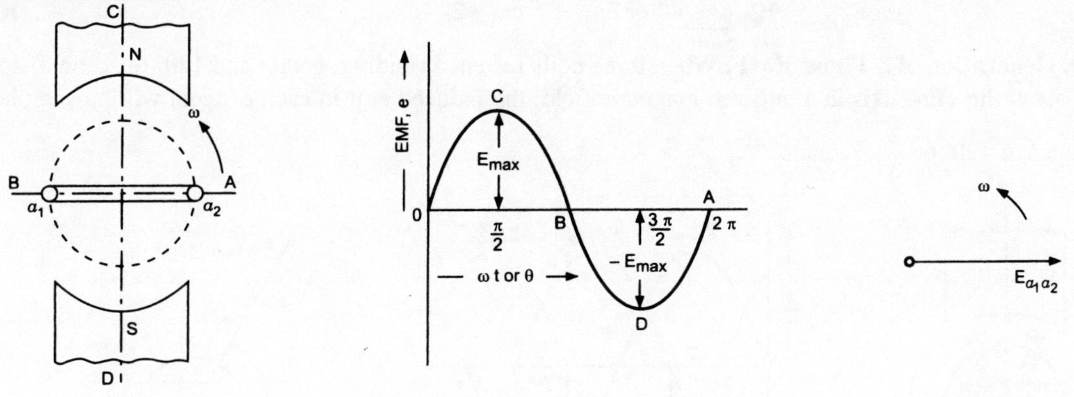

(a) (b) (c)

Fig. 10.1 *Generation of Single Phase EMF*

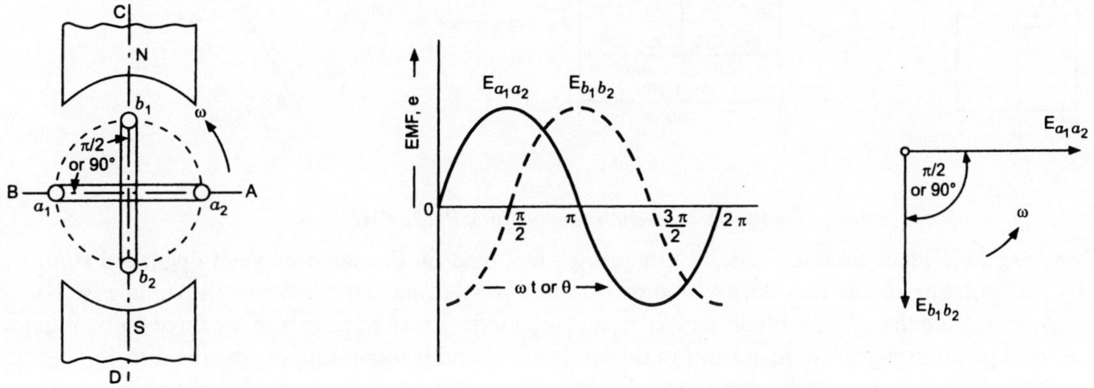

(a) (b) (c)

Fig. 10.2 *Generation of Two Phase EMF*

Consider two identical coils a_1a_2 and b_1b_2 fastened together at right angles to each other, rotating in counter-clockwise direction in a bipolar field, as shown in Fig. 10.2 (*a*). The terminal a_1 of coil a_1a_2 leads the terminal b_1 of coil b_1b_2 by 90º or $\pi/2$ radians. At the instant, the coil a_1a_2 is in position AB, the coil b_1b_2 is in position CD and, therefore, at this instant the induced emfs in coil a_1a_2 and b_1b_2 are zero and maximum (in negative direction)respectively. When the coils rotate through $\pi/2$ radians , the coil a_1a_2 and b_1b_2 come in position CD and AB respectively, therefore, emfs induced in them will be maximum (in positive direction) and zero respectively.

Since each coil has an equal number of turns and is wound with the wire of the same type and same x-section, the induced emfs in both of the coils are, therefore, of equal magnitude. The induced emf in each coil is also of same frequency and same waveform (sinusoidal in this case) but is displaced from each other by $\pi/2$ radians or 90º, as illustrated in Fig. 10.2(*b*) by waveforms. The emf induced in coil a_1a_2 leads the emf induced in coil b_1b_2 by $\pi/2$ radians or 90º. If instantaneous induced emf in coil a_1a_2 is given as

$e_{a_1a_2} = E_{max} \sin \omega t$ then instantaneous induced emf in coil b_1b_2 will be given as

$$e_{b_1b_2} = E_{max} \sin\left(\omega t - \frac{\pi}{2}\right) \text{ or } E_{max} \sin(\omega t - 90°) \qquad ...(10.2)$$

In symbolic notation, the real axis is usually made to coincide with the voltage or emf of coil a_1a_2 (say phase A). Then the effective values of the system are represented by the complex numbers as

$$\mathbf{E}_{a_1a_2} = \mathrm{E_P}$$

$$\mathbf{E}_{b_1b_2} = \mathrm{E_P} \angle -90^\circ \text{ or } \mathrm{E_P}\, e^{-j\,90^\circ} \text{ or } -j\,\mathrm{E_P} \qquad ...(10.3)$$

10.2.3. Generation of 3-Phase EMF. When three coils fastened rigidly together and 120° (electrical) apart rotate about the same axis in a uniform magnetic field, the induced emf in each of them will have a phase difference of 120° or $\frac{2}{3}\pi$ radians.

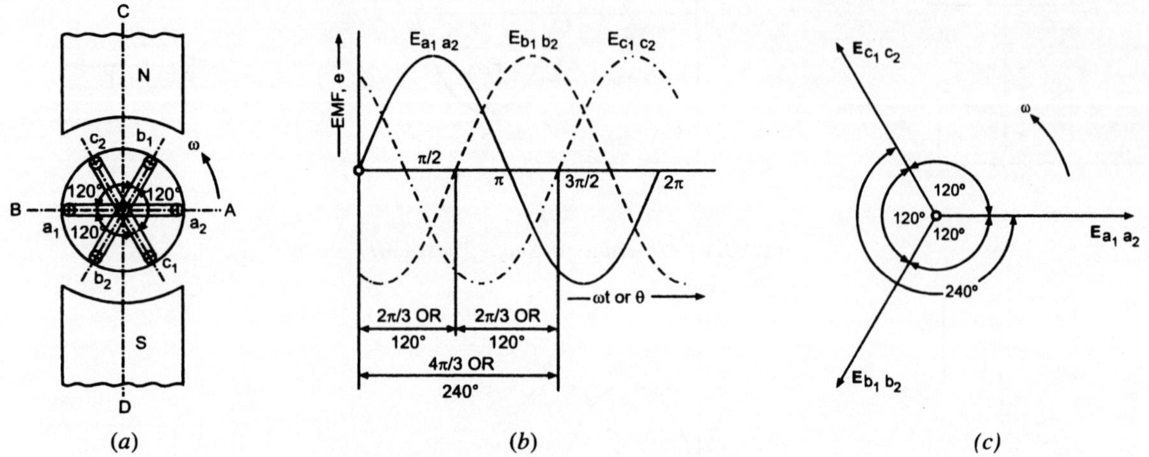

Fig. 10.3 *Generation of Three Phase EMF*

Consider three identical coils a_1a_2, b_1b_2 and c_1c_2 mounted on the same axis but displaced from each other by 120° rotating in counter-clockwise direction in a bipolar magnetic field, as shown in Fig. 10.3(*a*). Here a_1, b_1 and c_1 are the start terminals and a_2, b_2 and c_2 are the finish terminals of the three coils. When the coil a_1a_2 is in position AB the induced emf in this coil is zero and is increasing in positive direction, the coil b_1b_2 is 120° behind coil a_1a_2 so the emf induced in this coil is approaching its maximum negative value and the coil c_1c_2 is 240° behind coil a_1a_2 so as the emf induced in the coil has passed its positive maximum value and is decreasing.

Since each coil being indentical has an equal number of turns and is wound with the wire of the same type and same x-section, the induced emfs in each of the coils are of same magnitude. The induced emf in each coil is also of same frequency and same waveform (sinusoidal in this case)but displaced from each other by $\frac{2\pi}{3}$ radians or 120°, as illustrated in Fig. 10.3 (*b*) by waveforms .

Accordingly, the instantaneous values of the emfs induced in coils a_1a_2, b_1b_2 and c_1c_2 may be given as

$$\left.\begin{aligned} e_{a_1a_2} &= \mathrm{E_{max}} \sin \omega t \\ e_{b_1b_2} &= \mathrm{E_{max}} \sin\left(\omega t - \frac{2\pi}{3}\right) \text{ or } \mathrm{E_{max}} \sin(\omega t - 120^\circ) \\ e_{c_1c_2} &= \mathrm{E_{max}} \sin\left(\omega t - \frac{4\pi}{3}\right) \text{ or } \mathrm{E_{max}} \sin(\omega t - 240^\circ) \\ &\quad \text{or } \mathrm{E_{max}} \sin(\omega t + 120^\circ) \end{aligned}\right\} \qquad ...(10.4)$$

if $t = 0$ corresponds to the instant when the voltage or emf of coil a_1a_2 passes through zero and increases in positive direction.

In symbolic notation, the real axis is usually made to coincide with the emf (or voltage) of coil a_1a_2 (say phase A). Then the effective values of the symmetrical system are given in complex numbers as

$$\left.\begin{aligned} E_{a_1a_2} &= E_P \\ E_{b_1b_2} &= E_P \angle -120^\circ \text{ or } E_P\, e^{-j\,120^\circ} \text{ or } E_P\,(-0.5 - j\,0.866) \\ E_{c_1c_2} &= E_P \angle +120^\circ \text{ or } E_P\, e^{+j\,120^\circ} \text{ or } E_P\,(-0.5 + j\,0.866) \end{aligned}\right\} \quad \text{...(10.5)}$$

It is easy to prove that, at any instant, the algebraic sum of the three emfs is zero.

This result can be established analytically with the aid of the relations in Eq. (10.4)

$$\begin{aligned} \text{Resultant} &= e_{a_1a_2} + e_{b_1b_2} + e_{c_1c_2} \\ &= E_m\ [\sin \omega t + \sin (\omega t - 120^\circ) + \sin (\omega t - 240^\circ)] \\ &= E_m\ [\sin \omega t - 2 \sin (\omega t - 180^\circ) \cos 60^\circ] \\ &= E_m\ [\sin \omega t - 2 \sin \omega t \cos 60^\circ] = 0 \end{aligned}$$

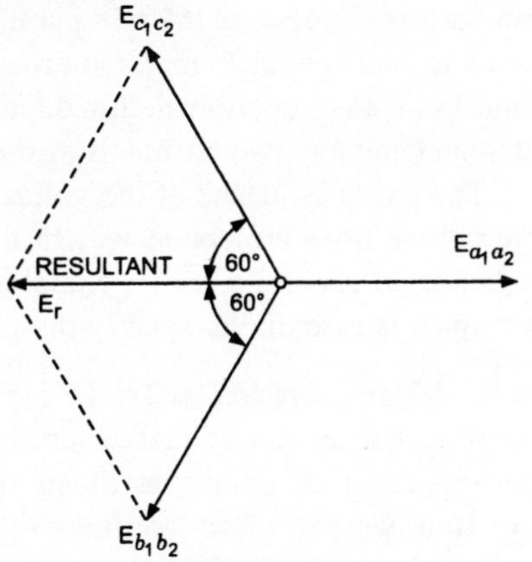

Fig. 10.4

It is easier to establish the result by phasor diagram shown in Fig. 10.4.

Indeed, combining the phasors $E_{b_1b_2}$ and $E_{c_1c_2}$ we find that, due to symmetry, the phasor $\mathbf{E}_{b_1b_2} + \mathbf{E}_{c_1c_2}$ is directed horizontally to the left, *i.e.* it is opposite to phasor $E_{a_1a_2}$.

$$i.e.\ \mathbf{E}_{a_1a_2} + \mathbf{E}_{b_1b_2} + \mathbf{E}_{c_1c_2} = \mathbf{E}_{a_1a_2} - (\mathbf{E}_{a_1a_2}) = 0$$

The result can be established by using complex notation– relations (10.5)

$$\mathbf{E}_{a_1a_2} + \mathbf{E}_{b_1b_2} + \mathbf{E}_{c_1c_2} = E_P + E_P\,(-0.5 - j\,0.866) + E_P\,(-0.5 + j\,0.866) = 0$$

The result is also obvious from wave diagram shown in Fig. 10.3 (*b*).

10.3. DOUBLE SUBSCRIPT NOTATION

The solution of problems involving circuits and systems containing a number of voltages and currents is simplified and less susceptible to error if the voltage and current phasors are designated by some systematic notation. The double subscript notation is a very useful concept from this point of view. In this notation two letters are placed at the foot of the symbol for voltage or current, the order in which the subscripts are written, indicates the direction in which the voltage acts or current flows. For example, if a voltage in a circuit acts in such a direction as to cause a current to flow from A to B, the positive direction of voltage is from A to B, and the voltage may be represented by V_{AB} or E_{AB}, the order of the subscripts denoting that the voltage or emf is acting from A to B. If the voltage is indicated by V_{BA} or E_{BA} it means that point B is positive w.r.t. point A (during its positive half-cycle) and the voltage causes the current to flow from B to A *i.e.* V_{BA} or E_{BA} indicates that voltage or emf acts in a direction opposite to that in which V_{AB} or E_{AB} acts.

$$\text{So } \mathbf{V}_{BA} = -\mathbf{V}_{AB} \quad \text{...(10.6)}$$

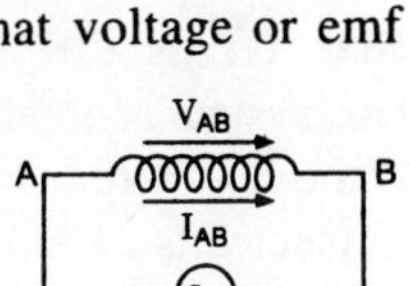

Fig. 10.5

Similarly I_{AB} indicates that current is flowing from A to B but I_{BA} indicates that current is flowing in direction from B to A

$$i.e.\ \mathbf{I}_{BA} = -\mathbf{I}_{AB}$$

10.4. PHASE SEQUENCE

Phase sequence is the order or sequence in which the currents or voltages in different phases attain their maximum values one after the other.

In Fig. 10.3 (*a*) three coils a_1a_2, b_1b_2 and c_1c_2 rotating in anti-clockwise direction are shown. Since the coil b_1b_2 is 120° behind coil a_1a_2 and coil c_1c_2 is 240° behind coil a_1a_2, therefore, first coil a_1a_2 attains maximum or peak value of induced emf, the coil b_1b_1 attains maximum or peak value of induced emf, when the coils rotate further 120° (electrical) and coil c_1c_2 attains peak value of induced emf when the coils rotate through 240° (electrical). Since the induced emfs in three coils a_1a_2, b_1b_2 and c_1c_2 attain maximum values in order *a, b, c*, phase sequence is *a b* c.

If the direction of rotation of the coils is reversed *i.e.* clockwise, the three-phase emfs would attain their maximum values in order *a, c, b* and, therefore, phase sequence would be *acb*. Since the coils can be rotated in clockwise or in counter-clockwise direction, there can be only two possible phase sequences.

The knowledge of phase sequence of a 3-Φ supply is essential for making connections for alternators and transformers to operate them in parallel. Reversal of phase sequence of a 3-phase alternator which is to be paralleled with another similar alternator may cause extensive damage to both the machines. The direction of rotation of an induction motor depends on the phase sequence. If the phase sequence is reversed by interchanging any two terminals of the 3-Φ supply, the motor would rotate in the opposite direction.

The phase sequence of the voltage applied to the load, in general, is determined by the order in which three phase lines are connected. In the case of 3-phase unbalanced loads, the effect of reversal of phase sequence, in general causes a completely different set of values of the currents. So while working on such systems it is essential to specify the phase sequence clearly to avoid unnecessary confusion.

10.5. INTERCONNECTION OF THREE PHASES

Each coil of three phases has two terminals (one 'start' and another 'finish') and if individual phase is connected to a separate load circuit, as illustrated in Fig. 10.6, we get a non-interlinked three-phase, six-wire system. Then, by Ohm's law, the current in individual phase will be

$$I_P = E_P/Z_P \qquad ...(10.7)$$

where I_P is the phase current, E_P is the phase emf or voltage.

In such a system, each circuit will require two conductors, therefore, 6 conductors in all.

Hence such a system will be very complicated and expensive.

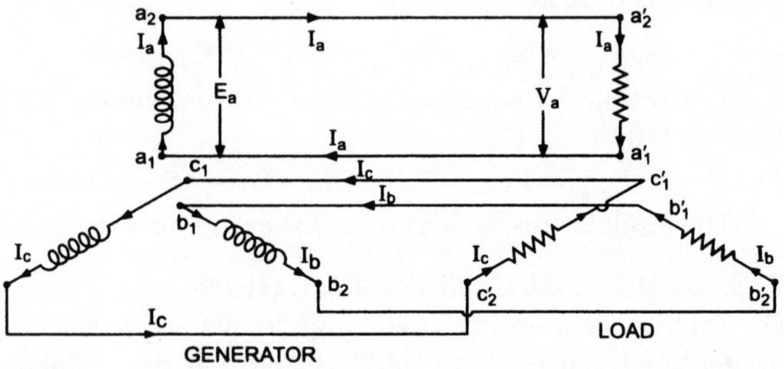

Fig. 10.6 *Non-interlinked Three-Phase System*

In practice non-interlinked 3-phase, 6-wire system is not employed but the phases are interconnected to give 3 or 4-wire, 3-phase system. The two principal methods of connecting the separate phases of three-phase system are the star (or wye) connections and the delta (or mesh) connections.

The interconnections must be carried out in such a way that if closed circuits are formed the sum of instantaneous emfs in them must be zero in order that there shall be no circulating current in these circuits.

10.6. STAR OR WYE (Y) CONNECTED SYSTEM

This system is obtained by joining together similar ends, either the start or the finish, the other ends are joined to the line wire, as shown in Fig. 10.7 (*a*). The common point N at which similar (start or finish) ends are connected is called the *neutral or star point*. Ordinarily only three wires are carried to the external circuit giving 3-phase 3-wire star-connected system but sometimes a fourth wire, known as *neutral wire* is carried to the neutral point of the external load circuit giving 3-phase, 4-wire star-connected system. The voltage between any line and the neutral point *i.e.* voltage across the phase winding is called the *phase voltage*; while the voltage between any two outers is called the *line voltage*. The neutral point is usually earth-connected.

In Fig. 10.7 (*a*) positive directions of emfs are taken from star point outwards. The arrowheads on emfs and currents indicate the positive direction and not their directions at any instant. The three phases be either numbered as 1, 2, 3, or a, b, c or R, Y, B. In Fig. 10.7 (*a*) the three-phases are numbered as usually done; R,Y and B indicate the three natural colours red, yellow and blue respectively.

It is worthwhile to state here that there are only two possible phase sequences, namely RYB and RBY. By convention sequence RYB is taken as positive and RBY as negative.

In Fig. 10.7 (*b*) the emfs induced in 3-phases are shown by phasors. In star-connection, there are two windings between each pair of outers and due to joining of similar ends together the emf induced in them are in opposition.

Hence the potential difference between the two outers, known as *line voltage,* is the phasor difference of phase emfs of the two phases concerned.

For example, the potential difference between outers R and Y or line voltage E_{RY} is the phasor difference of phase emfs E_R and E_Y or phasor sum of phase emfs E_R and $(-E_Y)$.

i.e. $\mathbf{E}_{RY} = \mathbf{E}_R - \mathbf{E}_Y$ (phasor difference)

or $\mathbf{E}_{RY} = \mathbf{E}_R + (-\mathbf{E}_Y)$ (phasor sum)

Since phase angle between phasors E_R and $(-E_Y)$ is 60°,

From phasor diagram shown in Fig. 10.7 (*b*)

$$E_{RY} = \sqrt{E_R^2 + E_Y^2 + 2\,E_R\,E_Y \cos 60°}$$

Let $E_R = E_Y = E_B = E_P$ (phase voltage) (Assuming balanced system) then line voltage

$$E_{RY} = \sqrt{E_P^2 + E_P^2 + 2E_P\,E_P \times (1/2)} = \sqrt{3}\,E_P$$

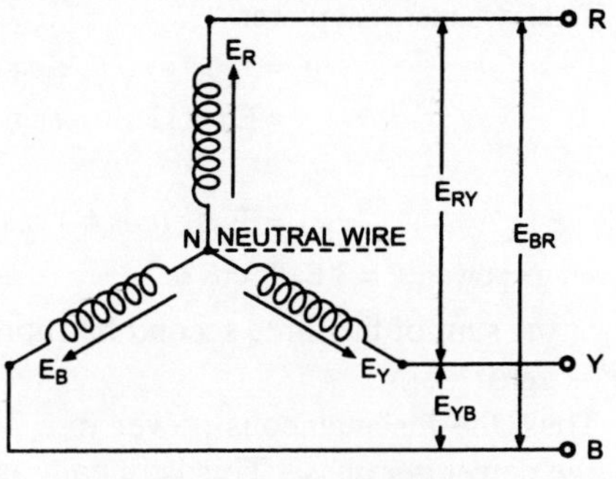

(*a*) *Connection Diagram*

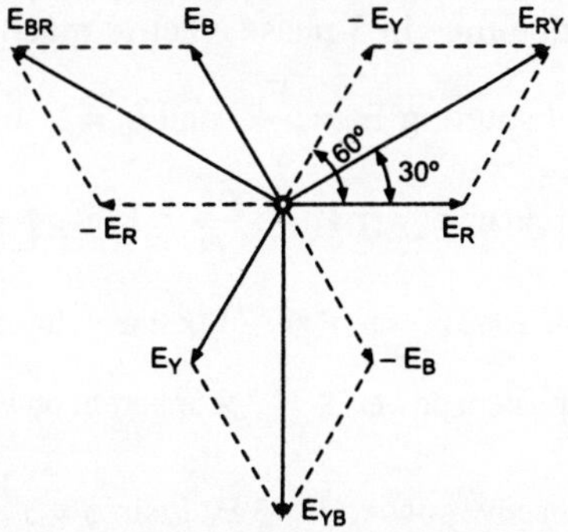

(*b*) *Phasor Diagram of Line and Phase Voltages*

Fig. 10.7 *Three-Phase Star-Connected System*

Similarly potential difference between outers Y and B

or line voltage $\mathbf{E}_{YB} = \mathbf{E}_Y - \mathbf{E}_B = \sqrt{3}\,E_P$ in magnitude

and potential difference between outers B and R

or line voltage, $\mathbf{E}_{BR} = \mathbf{E}_B - \mathbf{E}_R = \sqrt{3}\,E_P$ in magnitude

In balanced star (Y) system E_{RY}, E_{YB} and E_{BR} are equal in magnitude and are known as *line voltages*. Let each of them be equal to E_L

$$\therefore\ E_L = \sqrt{3}\,E_P \text{ (in balanced star-connected system)} \qquad ...(10.8)$$

Since in star-connected system each line conductor is connected to separate phase, current flowing through the lines and phases are same

i.e. Line current, I_L = Phase current, I_P ...(10.9)

10.6.1. 3-Phasor Power. Phase voltages and currents in a balanced three-phase circuit (star or delta) may be written in the instantaneous form as below :

$$e_R = \sqrt{2}\,E_P \sin \omega t \qquad ...(10.10\ a)$$

$$e_Y = \sqrt{2}\,E_P \sin (\omega t - 120°) \qquad ...(10.10\ b)$$

$$e_B = \sqrt{2}\,E_P \sin (\omega t + 120°) \qquad ...(10.10\ c)$$

$$\text{and } i_R = \sqrt{2}\,I_P \sin (\omega t - \phi) \qquad ...(10.11\ a)$$

$$i_Y = \sqrt{2}\,I_P \sin (\omega t - \phi - 120°) \qquad ...(10.11\ b)$$

$$i_B = \sqrt{2}\,I_P \sin (\omega t - \phi + 120°) \qquad ...(10.11\ c)$$

where ϕ is the phase angle between phase voltage and phase current.

Total instantaneous power,

$$p = e_R i_R + e_Y i_Y + e_B i_B$$
$$= E_P I_P [2 \sin \omega t \sin (\omega t - \phi)] + E_P I_P [2 \sin (\omega t - 120°) \sin (\omega t - \phi - 120°)$$
$$+ E_P I_P [2 \sin (\omega t + 120°) \sin (\omega t - \phi + 120°]$$
$$= E_P I_P [\cos \phi - \cos (2 \omega t - \phi) + \cos \phi - \cos (2 \omega t - \phi - 240°) + \cos \phi - \cos (2 \omega t - \phi + 240°)]$$

or Average power, $P = 3 E_P I_P \cos \phi$ ...(10.12)

∵ the sum of the three second harmonic oscillating terms which have a progressive phase difference of 120° is zero.

Thus, the instantaneous power in a 3-phase balanced system is constant and equal to three times the average power per phase. This is in contrast to power in a single phase system which has a second harmonic oscillating component. Constancy of power in a balanced 3-phase system affords the unique advantage of uniform torque in 3-phase electric machines.

Substituting $E_P = \frac{E_L}{\sqrt{3}}$ and $I_P = I_L$ for a 3-phase star-connected balanced system in Eq. (10.12) we have

Total circuit power, $P = 3 \frac{E_L}{\sqrt{3}} \times I_L \cos \phi = \sqrt{3} E_L I_L \cos \phi$ ...(10.13)

i.e. Total power, $P = \sqrt{3}$ × line voltage × line current × power factor.

Total apparent power, $S = 3 \times$ apparent power per phase $= 3 E_P I_P = 3 \frac{E_L}{\sqrt{3}} I_L = \sqrt{3} E_L I_L$ ...(10.14)

Total reactive power, $Q = 3 E_P I_P \sin \phi = 3 \frac{E_L}{\sqrt{3}} I_L \sin \phi = \sqrt{3} E_L I_L \sin \phi = \frac{P}{\cos \phi} \sin \phi = P \tan \phi$ ...(10.15)

It should be noted that in a balanced star- connected system

(*i*) Line voltages are 120° apart.

(*ii*) Line voltages are 30° ahead of the respective phase voltages.

(*iii*) Line voltages are $\sqrt{3}$ times of phase voltages.

(*iv*) Line currents are equal to phase currents.

(*v*) The angle between line currents and the corresponding line voltages is (30° ± Φ); + ve for lagging currents and –ve for leading currents.

(*vi*) True power output = $\sqrt{3} E_L I_L \cos \Phi$, where Φ is the angle between the respective phase current and phase voltage (not between line current and line voltage).

(*vii*) Apparent power = $\sqrt{3} E_L I_L$

(*viii*) In balanced system, the potential of neutral or star point is zero because potential at neutral or star point,

$$E_{NR} + E_{NY} + E_{NB} = 0$$

Example 10.1

Phase voltages of a star-connected alternator are $E_R = 231 \angle 0°$ V; $E_Y = 231 \angle -120°$ V; and $E_B = 231 \angle +120°$ V. What is the phase sequence of the system ? Compute the line voltages E_{RY} and E_{YB}.

Solution: The phase voltages of the star- connected alternator are given as $E_R = 231 \angle 0°$; $E_Y = 231 \angle -120°$ V and

$$E_B = 231 \angle +120° \text{ V or } 231 \angle -240° \text{ V}$$

From the above expressions of phase voltages in polar form, it is obvious that E_R is the reference voltage, E_Y lags behind it by 120° and E_B lags behind E_R by 240° and E_Y by 120°. So phase sequence is RYB.

Since phase voltages are of equal magnitude and are displaced from one another by equal time angle, it is a symmetrical 3-phase voltage system and so the line voltage

$$E_{RY} = E_{YB} = \sqrt{3} \times 231 = 400 \text{ V } \textbf{Ans.}$$

Example 10.2

If the phase voltage of a 3-phase star-connected system is 200 V, what will be the line voltages (a) when the phases are correctly connected (b) when connections to one of phases are reversed ?

[U.P. Technical Univ. Electrical Engineering First Semester 2000-01; First Semester 2003-04]

Solution: (*i*) When the phases are correctly connected *i.e.* similar ends are connected together, as shown in Fig. 10.7 (*a*) all line voltages will be equal in magnitude and will be given as

Each line voltage, $E_L = \sqrt{3}\ E_P = \sqrt{3} \times 200 = 346.41$ V **Ans.**

(*ii*) When the connections to one of the phases are reversed, as shown in Fig. 10.8 (*a*), voltage in phase R is in reverse direction. The voltage phasor diagram for such a case is shown in Fig. 10.8 (*b*).

It should be noted that E_R has been drawn in reverse direction, so the angles between the three phase voltages are of 60°, as shown in Fig. 10.8(*b*)

So $\mathbf{E_{RY} = E_R - E_Y}$
$= 2\ E_P \cos 60° = E_P = 200$ V **Ans.**

$\mathbf{E_{YB} = E_Y - E_B}$
$= 2E_P \times \cos 30°$

$= 2 \times 200 \times \frac{\sqrt{3}}{2} = 346.41$ V **Ans.**

and $\mathbf{E_{BR} = E_B - E_R} = 2\ E_P \cos 60° = 200$ V **Ans.**

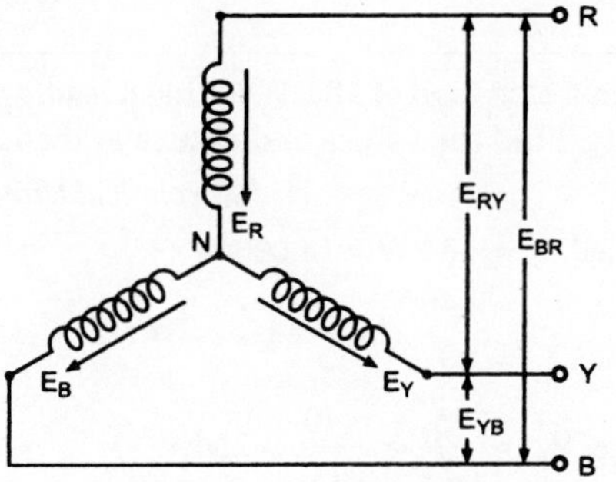

(*a*) Connection Diagram

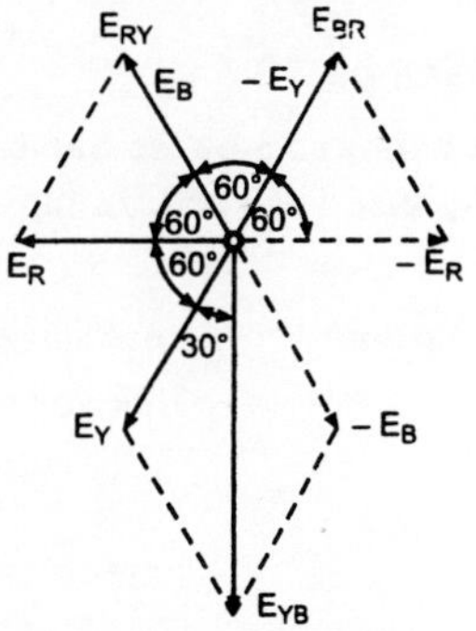

(*b*) Phasor Diagram of Line and Phase Voltages

Fig. 10.8

Example 10.3

A balanced star-connected load of (8 + j6) Ω per phase is connected to a balanced 3-phase, 400 V supply. Find the line current, power factor, power and total volt-amperes.

[U.P. Technical Univ. Electrical Engineering Second Semester 2005-06]

Solution:

$$\text{Phase voltage, } V_P = \frac{\text{Line voltage, } V_L}{\sqrt{3}} = \frac{400}{\sqrt{3}} = 231 \text{ V}$$

$$\text{Impedance per phase, } Z_P = \sqrt{R_P^2 + X_P^2} = \sqrt{8^2 + 6^2} = 10\ \Omega$$

$$\text{Phase current, } I_P = \frac{V_P}{Z_P} = \frac{231}{10} = 23.1 \text{ A}$$

$$\text{Line current, } I_L = I_P = 23.1 \text{ A } \textbf{Ans.}$$

$$\text{Power factor, } \cos\phi = \frac{R_P}{Z_P} = \frac{8}{10} = 0.8 \text{ (lagging) } \textbf{Ans.}$$

$$\text{Total power, } P = \sqrt{3}\ V_L I_L \cos\phi = \sqrt{3} \times 400 \times 23.1 \times 0.8 = 12{,}800 \text{ W } \textbf{Ans.}$$

$$\text{Total volt-amperes, } S = \sqrt{3}\ V_L I_L = \sqrt{3} \times 400 \times 23.1 = 16{,}000 \text{ VA } \textbf{Ans.}$$

$$\text{Also total power, } P = 3\ I_P^2 R_P = 3 \times (23.1)^2 \times 8 = 12{,}800 \text{ W } \textbf{Ans.}$$

Example 10.4

For star-connected system in a 3-phase circuit prove that $V_L = \sqrt{3}\ V_{ph}$ and $I_L = I_{ph}$.

A 3-phase, 400 V supply is connected to a 3-phase star connected balanced load. The line current is 20 A and the power consumed by the load is 12 kW. Calculate the impedance of the load, phase current and power factor.

[U.P. Technical Univ. Electrical Engineering Second Semester 2006-07]

Solution:

Load supplied, $P = 12$ kW $= 12{,}000$ W

Line voltage, $V_L = 400$ V

Line current, $I_L = 20$ A

$$\text{Phase voltage, } V_P = \frac{V_L}{\sqrt{3}} = \frac{400}{\sqrt{3}} = 230.94 \text{ V}$$

$$\text{Phase current, } I_P = I_L = 20 \text{ A } \textbf{Ans.}$$

$$\text{Load impedance, } Z_L = \frac{V_P}{I_P} = \frac{230.94}{20} = 11.547\ \Omega \textbf{ Ans.}$$

$$\text{Load power factor, } \cos\phi = \frac{P}{\sqrt{3}\, V_L I_L} = \frac{12{,}000}{\sqrt{3} \times 400 \times 20} = 0.866 \textbf{ Ans.}$$

Example 10.5

A balanced 3-phase star-connected load of 18 kW taking a leading current of 60 amperes when connected across a 3-phase 440 V, 50 Hz supply. Find the values and nature of the load.

[U.P. Technical Univ. Electrical Engineering Second Semester 2007-08]

Solution:

$$\text{Load supplied, } P = 18 \text{ kW} = 18{,}000 \text{ W}$$

$$\text{Line voltage, } V_L = 440 \text{ V}$$

$$\text{Line current, } I_L = 60 \text{ A}$$

$$\text{Phase voltage, } V_P = \frac{V_L}{\sqrt{3}} = \frac{440}{\sqrt{3}} = 254 \text{ V}$$

$$\text{Phase current, } I_P = I_L = 60 \text{ A}$$

$$\text{Load impedance, } Z = \frac{V_P}{I_P} = \frac{254}{60} = 4.234\ \Omega$$

$$\text{Load power factor, } \cos\phi = \frac{P}{\sqrt{3}\, V_L I_L} = \frac{18{,}000}{\sqrt{3} \times 440 \times 60} = 0.39365 \text{ (leading) } \textbf{Ans.} \quad \because \text{Current is leading}$$

$$\text{Resistance of load, } R = Z\cos\phi = 4.234 \times 0.39365 = 1.6667\ \Omega \textbf{ Ans.}$$

$$\text{Reactance of load, } X = Z\sin\phi = 4.234\sqrt{1 - 0.39365^2} = 4.234 \times 0.9193 = 3.89242\ \Omega \text{ (capacitive) } \textbf{Ans.}$$

$\because$ Current is leading

Example 10.6

A 3-ϕ, 3-wire Y-connected system has 150 V between two phases. Each phase has Z = 5 $\angle -30°$ Ω. Find: (i) Current in each phase (ii) total power (iii) Draw phasor diagram.

[U.P. Technical Univ. Electrical Engineering First Semester 2007-08]

Solution:

$$\text{Line voltage, } V_L = \text{Phase-to-phase voltage} = 150 \text{ V}$$

$$\text{Phase voltage, } V_P = \frac{V_L}{\sqrt{3}} = \frac{150}{\sqrt{3}} = 86.6 \text{ V}$$

$$(i) \text{ Current in each phase, } I_P = \frac{V_P}{Z_P} = \frac{86.6}{5} = 17.32 \text{ A } \textbf{Ans.}$$

$$\text{Power factor of the circuit, } \cos\phi = \text{Cosine of phase angle} = \cos 30° = 0.866 \text{ (leading)}$$

$$(ii) \text{ Total power, } P = 3 V_P I_P \cos\phi = 3 \times 86.6 \times 17.32 \times 0.866 = 3{,}897 \text{ W or } 3.897 \text{ kW } \textbf{Ans.}$$

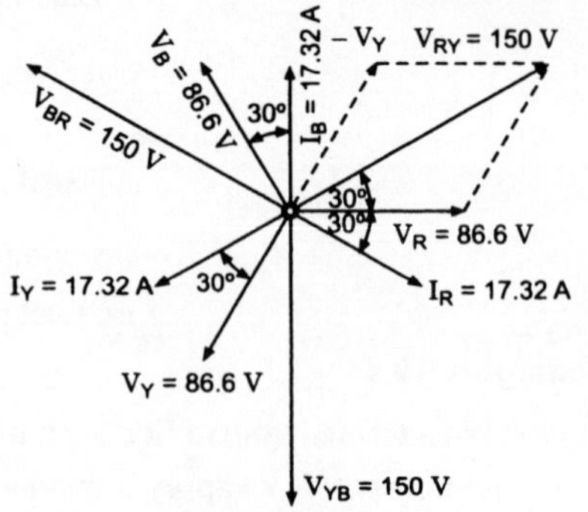

Fig. 10.9

Phasor diagram is shown in Fig. 10.9.

Example 10.7

A balanced star-connected load of (8 + *j*6) Ω per phase is connected to a 3-phase, 230 V, 50 Hz supply. Find the line current, pf, power, volt-amperes and reactive power. Draw the phasor diagram for the above circuit.

[Pb. Technical Univ. Electrical Engineering, June 1999]

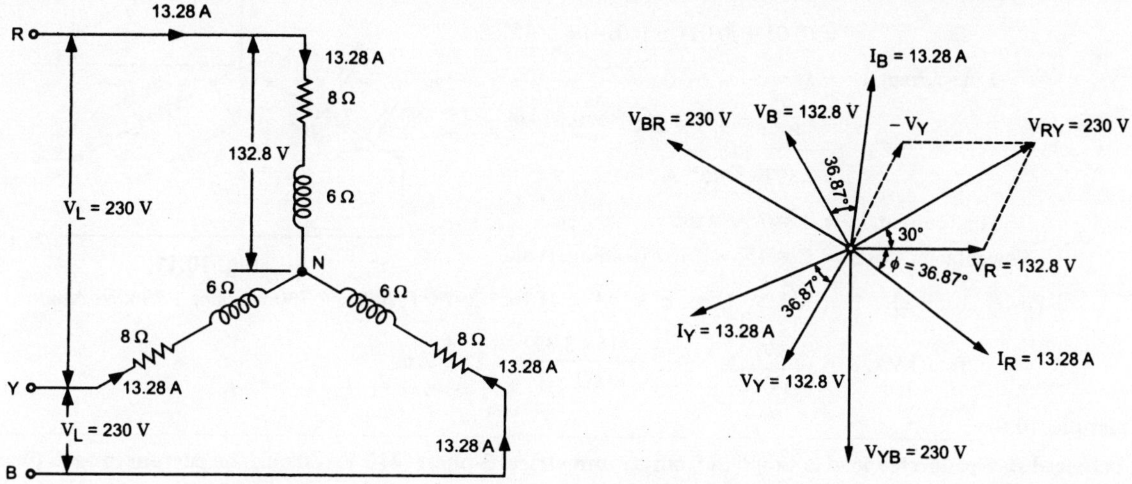

(a) Circuit Diagram *(b) Phasor Diagram For Currents and Voltages*

Fig. 10.10

Solution:

$$\text{Phase voltage, } V_P = \frac{\text{Line voltage, } V_L}{\sqrt{3}} = \frac{230}{\sqrt{3}} = 132.8 \text{ V}$$

$$\text{Impedance per phase, } Z_P = \sqrt{R_P^2 + X_P^2} = \sqrt{8^2 + 6^2} = 10\ \Omega$$

$$\text{Phase current, } I_P = \frac{V_P}{Z_P} = \frac{132.8}{10} = 13.28 \text{ A}$$

$$\text{Line current, } I_L = I_P = 13.28 \text{ A } \textbf{Ans.}$$

$$\text{Power factor, } \cos\phi = \frac{R_P}{Z_P} = \frac{8}{10} = 0.8 \text{ (lagging) } \textbf{Ans.}$$

$$\text{Power, } P = \sqrt{3}\, V_L I_L \cos\phi = \sqrt{3} \times 230 \times 13.28 \times 0.8 = 4{,}232 \text{ watts or } 4.232 \text{ kW } \textbf{Ans.}$$

$$\text{Total volt-amperes, } S = \sqrt{3}\, V_L I_L = \sqrt{3} \times 230 \times 13.28 = 5{,}290 \text{ VA } \textbf{Ans.}$$

$$\text{Reactive power, } Q = \sqrt{3}\, V_L I_L \sin\phi = \sqrt{3} \times 230 \times 13.28 \sin\cos^{-1}(0.8)$$

$$= \sqrt{3} \times 230 \times 13.28 \times 0.6 = 3{,}174 \text{ VAR or } 3.174 \text{ kVAR } \textbf{Ans.}$$

Phasor diagram for currents and voltages is shown in Fig. 10.10 (*b*).

Example 10.8

Each phase of a star-connected load consists of a resistance of 100 Ω in parallel with a capacitance 31.8 μF. Calculate the line current, power absorbed, total kVA and power factor of the load when connected to a 416 V, 3ϕ, 4 wire 50 Hz supply.

[Pb. Technical Univ. Basic Electrical Engineering December 2003]

Solution:

$$\text{Phase voltage, } V_P = \frac{V_L}{\sqrt{3}} = \frac{416}{\sqrt{3}} = 240.2 \text{ V}$$

$$\text{Taking phase voltage as reference phasor, } \mathbf{V_P} = 240.2 \angle 0^\circ \text{ V}$$

Admittance per phase, $\mathbf{Y_P} = \frac{1}{R} + j\,\omega\,C$

$= \frac{1}{100} + j\,2\,\pi \times 50 \times 31.8 \times 10^{-6}$

$= (0.01 + j0.01)$ or $0.01414 \angle 45°$ S

Line current, $\mathbf{I_L} = \mathbf{I}$

$= \mathbf{V_P}\,\mathbf{Y_P} = 240.2 \angle 0° \times 0.01414 \angle 45°$

$= 3.397 \angle 45°$ A

So line current, $I_L = 3.397$ A **Ans.**

Power factor, $\cos\phi = \text{Cos } 45° = 0.707$ (leading) **Ans.**

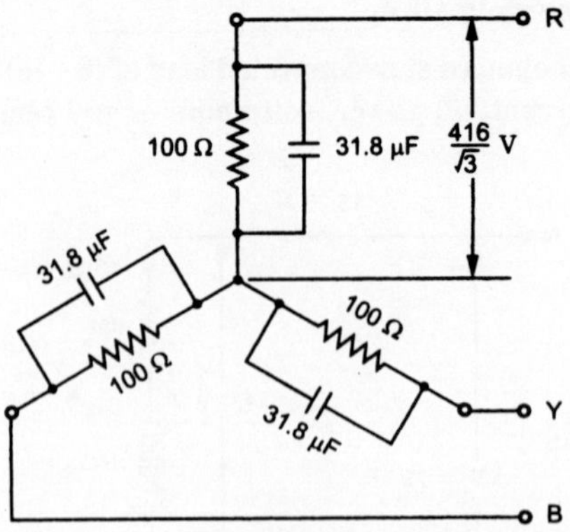

Fig. 10.11

Power absorbed, $P = \sqrt{3}\,V_L\,I_L\cos\phi = \sqrt{3} \times 416 \times 3.397 \times 0.707 = 1{,}730$ watts or 1.73 kW **Ans.**

Total kVA, $S = \frac{\sqrt{3}V_L\,I_L}{1{,}000} = \frac{\sqrt{3}\times 416 \times 3.397}{1{,}000} = 2.45$ **Ans.**

Example 10.9

A balanced star-connected load is supplied from a symmetrical 3-phase, 410 V system. The current in each phase is 30 A and lags 30° behind the phase voltage. Find (*i*) phase voltage (*ii*) total power (*iii*) reactive power drawn by load. [*R.G.P.V. Bhopal Basic Electrical Enginnering 2001*]

Solution: (*i*) Phase voltage, $V_P = \frac{V_L}{\sqrt{3}} = \frac{410}{\sqrt{3}} = 236.7$ V **Ans.**

Phase current, $I_P = 30$ A

Power factor, $\cos\Phi = \text{Cos } 30° = 0.866$ (lag)

(*ii*) Power drawn by load, $P = 3\,V_P\,I_P\cos\Phi = 3 \times 236.7 \times 30 \times 0.866 = 18{,}450$ W or 18.45 kW **Ans.**

(*iii*) Reactive power drawn by load, $Q = 3\,V_P\,I_P\sin\Phi = 3 \times 23.67 \times 30 \times \frac{1}{2} = 10{,}652$ VAR or 10.652 kVAR **Ans.**

Example 10.10

Three-phase star-connected load when supplied from a 400 V, 50 Hz source takes a line current of 10 A at 66.86° w.r.t. to its line voltage. Calculate : (*i*) impedance parameters (*ii*) pf and active power consumed. Draw the phasor diagram. [*Nagpur University, April 1998*]

Solution: As mentioned in Art 10.6, line voltages are 30° ahead of respective phase voltage *i.e.* phase voltages lag behind respective line voltages by 30°. It is also illustrated in phasor diagram shown in Fig. 10.12. Thus phase angle between phase current and its phase voltage is (66.86° – 30°) = 36.86° (lagging). This is because a current differing in phase with respect to its line voltage by 66.86° can only be lagging.

(*i*) Impedance $Z = \frac{V_P}{I_P} = \frac{V_L/\sqrt{3}}{I_L} = \frac{400}{\sqrt{3}\times 10} = 23.1\ \Omega$ **Ans.**

Resistance $R = Z\cos\phi = 23.1 \cos 36.86° = 18.48\ \Omega$ **Ans.**

Reactance $X_L = Z\sin\phi = 23.1 \sin 36.86° = 13.86\ \Omega$ **Ans.**

(*ii*) Power factor $= \text{Cos }\phi = \text{Cos } 36.86° = 0.8$ (lagging) **Ans.**

Total active power consumed

$= \sqrt{3}\,V_L\,I_L\cos\phi = \sqrt{3} \times 400 \times 10 \times 0.8$

$= 5{,}544$ W or 5.544 kW **Ans.**

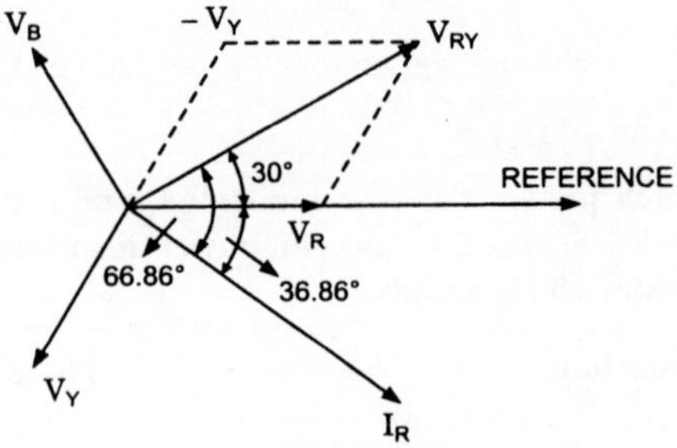

Fig. 10.12

Example 10.11

A balanced 3-phase star-connected load of 120 kW takes a leading current of 85 amp., when connected across a 3-phase 1,100 volts, 50 Hz supply. Obtain the values of the resistance, impedance and capacitance of the load per phase and also calculate the power factor of the load. [*U.P. Technical Univ. Electrical Engineering Second Semester 2003-04*]

Solution:

Supply voltage, $V_L = 1{,}100$ V

Supply current, $I_L = 85$ A

Power supplied, $P = 120$ kW or 1,20,000 W

$$\text{Power factor of the load, } \cos\phi = \frac{P}{\sqrt{3}\, V_L\, I_L} = \frac{1{,}20{,}000}{\sqrt{3}\times 1{,}100\times 85} = 0.741 \text{ (leading) } \textbf{Ans.}$$

$$\text{Impedance of the load per phase, } Z = \frac{V_P}{I_P} = \frac{V_L/\sqrt{3}}{I_L} = \frac{1{,}100/\sqrt{3}}{85} = 7.472\ \Omega\ \textbf{Ans.}$$

$$\text{Resistance of the load per phase, } R = Z\cos\phi = 7.472\times 0.741 = 5.536\ \Omega\ \textbf{Ans.}$$

$$\text{Reactance of the load per phase, } X_C = \sqrt{Z^2 - R^2} = \sqrt{(7.472)^2 - (5.536)^2} = 5.02\ \Omega$$

$$\text{Capacitance of the load per phase, } C = \frac{1}{2\pi f X_C} = \frac{1}{2\pi\times 50\times 5.02} = 634.3\ \mu\text{F}\ \textbf{Ans.}$$

Example 10.12

A 3-phase balanced load draws 10 kW power from a 400 V, 3-phase, 50 Hz, 4-wire supply at 0.8 lagging power factor.

(*i*) Determine the line current.

(*ii*) What would be the line current if power factor were raised to unity, power remaining the same?

(*iii*) How can the power factor be improved to unity? [*R.G.P.V. Bhopal Basic Electrical Engineering December 2004*]

Solution:

Power, $P = 10$ kW

Line voltage, $V_L = 400$ V

Power factor, $\cos\Phi = 0.8$ (lagging)

$$(i)\ \text{Line current, } I_L = \frac{P}{\sqrt{3}\, V_L \cos\phi} = \frac{10\times 1{,}000}{\sqrt{3}\times 400\times 0.8} = 18.04 \text{ A } \textbf{Ans.}$$

Line current when power factor is raised to unity, power remaining the same

$$I_L' = \frac{10\times 1{,}000}{\sqrt{3}\times 400\times 1} = 14.43 \text{ A } \textbf{Ans.}$$

Power factor can be improved to unity by connecting capacitors of suitable capacity across the load.

Example 10.13

A three-phase, three-wire RYB system, with effective line voltage of 150 V has a balanced Y-connected load of $Z = 5\ \angle -30°\ \Omega$ in each phase. Obtain the currents, power supplied to load and draw voltage-current phasor diagram. [*U.P. Technical Univ. Elec. Engineering June 2001*]

Solution: Phase voltage, $V_P = \dfrac{\text{Line voltage, } V_L}{\sqrt{3}}$

$$= \frac{150}{\sqrt{3}} = 86.6 \text{ V}$$

Taking phase voltage, V_P as reference phasor

$$\mathbf{V_P} = 86.6\ \angle 0°\ \text{V}$$

$$\text{Impedance per phase, } \mathbf{Z_P} = 5\ \angle -30°\ \Omega$$

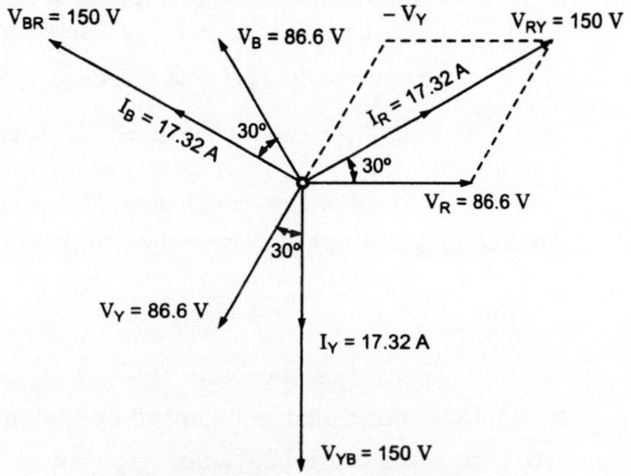

Fig. 10.13 *Phasor Diagram*

$$\text{Line current, } \mathbf{I_L} = \text{Phase current, } \mathbf{I_P} = \frac{\mathbf{V_P}}{\mathbf{Z_P}} = \frac{86.6 \angle 0^\circ}{5 \angle -30^\circ} \text{ A} = 17.32 \angle 30^\circ \text{ A } \textbf{Ans.}$$

$$\text{Power supplied to load, } P = 3V_P\, I_P \cos\phi = 3 \times 86.6 \times 17.32 \cos 30^\circ = 3{,}897 \text{ W } \textbf{Ans.}$$

Phasor diagram is shown in Fig. 10.13.

10.7. MESH OR DELTA (Δ) CONNECTED SYSTEM

When the starting end of one coil is connected to the finishing end of another coil, as shown in Fig. 10.14 (*a*), the *delta or mesh connection is obtained.* The direction of emfs in the coils have been taken as positive from start end to finish end in Fig. 10.14 (*a*).

From Fig. 10.14 (*b*) it is obvious that line current is phasor difference of phase currents of two phases concerned. For example the line current in red outer I_R will be equal to the phasor difference of phase currents I_{YR} and I_{RB}. The current phasors are shown in Fig. 10.14 (*b*).

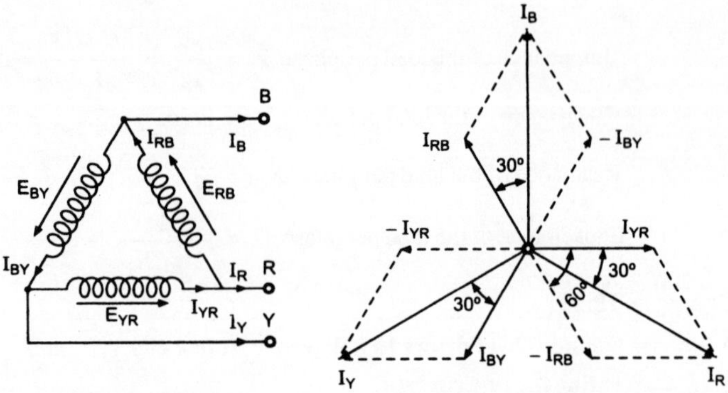

(*a*) *Connection Diagram* (*b*) *Phasor Diagram*

Fig. 10.14 *Three Phase Delta-Connected System*

From Figs. 10.14 (*a*) and 10.14 (*b*) we have

$$\text{Line current, } \mathbf{I_R} = \mathbf{I_{YR}} - \mathbf{I_{RB}}$$
$$= \mathbf{I_{YR}} + (-\mathbf{I_{RB}})$$

Since phase angle between phase current phasors $\mathbf{I_{YR}}$ and $-\mathbf{I_{RB}}$ is 60°,

$$I_R = \sqrt{I^2_{YR} + I^2_{RB} + 2\, I_{YR}\, I_{RB} \cos 60^\circ}$$

Assuming balanced load

The phase current in each winding is equal and let each be equal to I_P

$$\therefore \text{ Line current } I_R = \sqrt{I^2_P + I^2_P + 2\, I_P\, I_P \times \frac{1}{2}} = \sqrt{3}\, I_P \qquad \text{...(10.16)}$$

$$\text{Similarly line current } \mathbf{I_Y} = \mathbf{I_{BY}} - \mathbf{I_{YR}} = \sqrt{3}\, I_P \text{ in magnitude}$$

$$\text{and line current } \mathbf{I_B} = \mathbf{I_{RB}} - \mathbf{I_{BY}} = \sqrt{3}\, I_P \text{ in magnitude}$$

Since in delta-connected system, only one phase is included between any pair of line outers, potential difference between the line outers, called the *line voltage*, is equal to phase voltage.

$$\textit{i.e.} \text{ Line voltage, } E_L = \text{Phase voltage, } E_P \qquad \text{...(10.17)}$$

$$\text{Power output per phase} = E_P\, I_P \cos\Phi$$ where $\cos\Phi$ is the power factor of the load

$$\text{Total power output} = 3\, E_P\, I_P \cos\Phi = 3\, E_L \frac{I_L}{\sqrt{3}} \cos\Phi = \sqrt{3}\, E_L\, I_L \cos\Phi \qquad \text{...(10.18)}$$

i.e. Total power output $= \sqrt{3} \times$ line voltage $\times$ line current $\times$ power factor

Apparent power of 3-phase delta-connected system = 3 × apparent power per phase

$$= 3\, E_P\, I_P = 3\, E_L \frac{I_L}{\sqrt{3}} = \sqrt{3}\, E_L\, I_L \qquad \text{...(10.19)}$$

$$\text{Total reactive power, } Q = 3\, E_P\, I_P \sin\phi = \sqrt{3}\, E_L\, I_L \sin\phi = P \tan\phi \qquad \text{...(10.20)}$$

It should be noted that in balanced delta-connected system

(*i*) Line currents are 120° apart.

(*ii*) Line currents are 30° behind the respective phase currents.

(*iii*) Line currents are $\sqrt{3}$ times phase currents.

(*iv*) The angle between line currents and corresponding line voltage is $(30° \pm \Phi)$ as in star system.

(*v*) Line voltages are equal to phase voltages.

(*vi*) Power = $\sqrt{3}\ E_L I_L \cos\Phi$, where Φ is the phase angle between respective phase current and phase voltage.

(*vii*) Apparent power = $\sqrt{3}\ E_L I_L$

(*viii*) In balanced system the resultant emf in the closed circuit will be zero.

$$i.e.\ \mathbf{E}_{RY} + \mathbf{E}_{YB} + \mathbf{E}_{BR} = 0$$

Hence there will be no circulating current in the mesh if no load is connected to the lines.

Example 10.14

Three identical coils connected in delta across 400 V, 50-Hz, 3-phase ac supply, take a line current of 17.32 A at power factor of 0.8 lagging. Calculate (*i*) the phase current (*ii*) the resistance and inductance of each coil (*iii*) the power drawn by each coil. [*U.P. Technical Univ. Electrical Engineering February 2002*]

Solution:

Phase voltage, $V_P = V_L = 400\ \text{V}$

(*i*) Phase current, $I_P = \dfrac{I_L}{\sqrt{3}} = \dfrac{17.32}{\sqrt{3}} = 10\ \text{A}$ **Ans.**

Impedance of each coil, $Z_P = \dfrac{V_P}{I_P} = \dfrac{400}{10} = 40\ \Omega$

(*ii*) Resistance of each coil, $R_P = Z_P \cos\phi = 40 \times 0.8 = 32\ \Omega$ **Ans.**

Inductive reactance of each coil, $X_P = \sqrt{Z_P^2 - R_P^2} = \sqrt{40^2 - 32^2} = 24\ \Omega$

Inductance of each coil, $L = \dfrac{X_P}{2\pi f} = \dfrac{24}{2\pi \times 50} = 0.0764\ \text{H or } 76.4\ \text{m H}$ **Ans.**

(*iii*) Power drawn by each coil $= I_P^2 R_P = 10^2 \times 32 = 3{,}200\ \text{W}$ **Ans.**

Example 10.15

A balanced Δ–connected load of (8 + *j*6) ohm per phase is connected to 3-phase 230 V supply. Find the line current. [*U.P. Technical Univ. Electrical Engineering First Semester 2004-05*]

Solution :

Impedance in each phase, $Z_P = \sqrt{8^2 + 6^2} = 10\ \Omega$

Phase current, $I_P = \dfrac{V_P}{Z_P} = \dfrac{V_L}{Z_P} = \dfrac{230}{10} = 23\ \text{A}$

Line current, $I_L = \sqrt{3}\ I_P = \sqrt{3} \times 23 = 39.84\ \text{A}$ **Ans.**

Example 10.16

Derive the relationship between the line and phase voltages of an alternator.

Three similar coils each having a resistance of 8 ohm and an inductance of 0.0191 H in series in each phase is connected across a 400 V, three phase, 50 Hz supply. Calculate the line current, power input, kVA and kVAR taken by the load. [*U.P. Technical Univ. Electrical Engineering First Semester 2006-07*]

Solution:

Resistance of coil, $R = 8\ \Omega$

Inductive reactance of coil, $X_L = 2\pi f L = 2\pi \times 50 \times 0.0191 = 6\ \Omega$

Impedance of each coil, $Z = \sqrt{R^2 + X_L^2} = \sqrt{8^2 + 6^2} = 10\ \Omega$

Power factor of coil, $\cos\phi = \dfrac{R}{Z} = \dfrac{8}{10} = 0.8$ (lagging) $\quad\because$ Circuit is inductive

When coils are connected in star

$$\text{Line current, } I_L = \text{Phase current, } I_P = \frac{V_L/\sqrt{3}}{Z} = \frac{400}{\sqrt{3}\times 10} = 23.094 \text{ A}$$

$$\text{Power input, } P = \frac{\sqrt{3}\, V_L I_L \cos\phi}{1{,}000} = \frac{\sqrt{3}\times 400\times 23.094\times 0.8}{1{,}000} = 12.8 \text{ kW } \textbf{Ans.}$$

$$\text{KVA taken by load} = \frac{\sqrt{3}\, V_L I_L}{1{,}000} = \frac{\sqrt{3}\times 400\times 23.094}{1{,}000} = 16 \textbf{ Ans.}$$

$$\text{KVAR taken by load} = \frac{\sqrt{3}\, V_L I_L \sin\phi}{1{,}000} = \frac{\sqrt{3}\times 400\times 23.094\times 0.6}{1{,}000} = 9.6 \text{ (lagging) } \textbf{Ans.}$$

When coils are connected in delta

$$\text{Line current, } I_L = \sqrt{3} \times \text{phase current} = \frac{\sqrt{3}\, V_L}{Z} = \frac{\sqrt{3}\times 400}{10} = 69.282 \text{ A } \textbf{Ans.}$$

$$\text{Power input, } P = \frac{\sqrt{3}\, V_L I_L \cos\phi}{1{,}000} = \frac{\sqrt{3}\times 400\times 69.282\times 0.8}{1{,}000} = 38.4 \text{ kW } \textbf{Ans.}$$

$$\text{KVA taken by load} = \frac{\sqrt{3}\, V_L I_L}{1{,}000} = \frac{\sqrt{3}\times 400\times 69.282}{1{,}000} = 48 \textbf{ Ans.}$$

$$\text{KVAR taken by load} = \frac{\sqrt{3}\, V_L I_L \sin\phi}{1{,}000} = \frac{\sqrt{3}\times 400\times 69.282\times 0.6}{1{,}000} = 28.8 \text{ (lagging) } \textbf{Ans.}$$

Example 10.17

Each phase of a delta-connected load has a resistance of 25 Ω, an inductance of 0.15 H and a capacitance of 120 μF in series. The load is connected across a 400 V, 50 Hz, 3-phase supply. Determine the line current, active power and reactive volt-amperes. *[U.P. Technical Univ. Electrical Engineering July 2002]*

Solution: Resistance per phase, $R_P = 25\ \Omega$

$$\text{Inductive reactance per phase, } X_{LP} = 2\pi f L = 2\,\pi \times 50 \times 0.15 = 47.124\ \Omega$$

$$\text{Capacitive reactance per phase, } X_{CP} = \frac{1}{2\,\pi f C} = \frac{1}{2\,\pi\times 50\times 120\times 10^{-6}} = 26.526\ \Omega$$

$$\text{Net reactance per phase, } X_p = X_{LP} - X_{CP} = 47.124 - 26.526 = 20.6\ \Omega$$

$$\text{Impedance per phase} = Z_P = \sqrt{R_P^2 + X_P^2} = \sqrt{25^2 + 20.6^2} = 32.4\ \Omega$$

$$\text{Phase angle, } \phi = \text{Tan}^{-1}\frac{X_p}{R_p} = \text{Tan}^{-1}\frac{20.6}{25} = 39.5^\circ \text{ (lag)}$$

$$\text{Phase current, } I_P = \frac{V_P}{Z_P} = \frac{V_L}{Z_P} = \frac{400}{32.4} = 12.35 \text{ A}$$

$$\text{Line current, } I_L = \sqrt{3}\, I_p = \sqrt{3} \times 12.35 = 21.38 \text{ A } \textbf{Ans.}$$

$$\text{Active power, } P = \sqrt{3}\ V_L I_L \cos\phi = \sqrt{3} \times 400 \times 21.38 \times \cos 39.5^\circ = 11{,}430 \text{ W or } 11.43 \text{ kW } \textbf{Ans.}$$

$$\text{Reactive volt-amperes} = \sqrt{3}\ V_L I_L \sin\phi = \sqrt{3} \times 400 \times 21.38 \times \sin 39.5^\circ = 9{,}420 \text{ VAR or } 9.42 \text{ kVAR } \textbf{Ans.}$$

Example 10.18

A 400 V, 3-ϕ voltage is applied to a balanced 3-ϕ Δ-connected load of phase impedance (15 + j 20) Ω. Find (*i*) the phasor current in each line. (*ii*) What is the power consumed per phase ? (*iii*) What is the phasor sum of three line currents ? Why does it have this value ? Explain. Also solve this by symbolic notation.

Solution: The circuit is given in Fig. 10.15.

Phase voltage, $V_P = V_L = 400$ V

Phase impedance, $Z_P = \sqrt{R_P^2 + X_P^2} = \sqrt{15^2 + 20^2} = 25\ \Omega$

Phase current, $I_P = \dfrac{V_P}{Z_P} = \dfrac{400}{25} = 16$ A

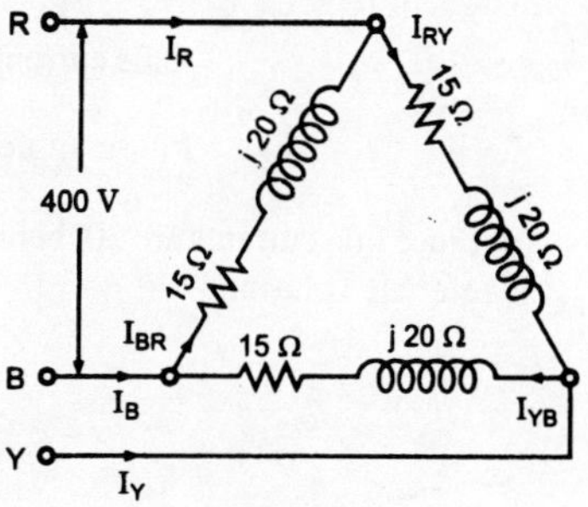

Fig. 10.15

(*i*) Line current, $I_L = \sqrt{3}\ I_P = \sqrt{3} \times 16 = 27.7$ A **Ans.**

(*ii*) Power consumed per phase

$= I_P^2\,R_P = 16^2 \times 15 = 3{,}840$ W **Ans.**

(*iii*) Phasor sum of three line currents would be zero because the three currents are equal in magnitude and have a mutual phase difference of 120°.

Solution By Symbolic Notation: Taking pd across phase RY as reference phasor and phase sequence RYB we have

$$\mathbf{V}_{RY} = 400\angle 0^\circ \text{ V};\ \mathbf{V}_{YB} = 400\angle -120^\circ \text{ V and } \mathbf{V}_{BR} = 400\angle 120^\circ \text{ V}$$

Impedance per phase, $\mathbf{Z}_P = (15 + j\,20) = 25\angle 53.13^\circ\ \Omega$

Phase current, $\mathbf{I}_{RY} = \dfrac{\mathbf{V}_{RY}}{\mathbf{Z}_P} = \dfrac{400\angle 0^\circ}{25\angle 53.13^\circ} = 16\angle -53.13^\circ$ or $(9.6 - j\,12.8)$ A

Phase current, $\mathbf{I}_{YB} = \dfrac{\mathbf{V}_{YB}}{\mathbf{Z}_P} = \dfrac{400\angle -120^\circ}{25\angle 53.13^\circ} = 16\angle -173.13^\circ$ or $(-15.88 - j\,1.91)$ A

Phase current, $\mathbf{I}_{BR} = \dfrac{\mathbf{V}_{BR}}{\mathbf{Z}_P} = \dfrac{400\angle 120^\circ}{25\angle 53.13^\circ} = 16\angle 66.87^\circ$ or $(6.28 + j\,14.71)$ A

(*i*) Line current in phase R, $\mathbf{I}_R = \mathbf{I}_{RY} - \mathbf{I}_{BR} = (9.6 - j\,12.8) - (6.28 + j\,14.71) = (3.32 - j\,27.51)$ or $27.7\angle -83.14^\circ$ A **Ans.**

Line current in phase Y, $\mathbf{I}_Y = \mathbf{I}_{YB} - \mathbf{I}_{RY} = (-15.88 - j\,1.91) - (9.6 - j\,12.8) = (-25.48 + j\,10.89)$ or $27.7\angle 156.86^\circ$ A

Line current in phase B, $\mathbf{I}_B = \mathbf{I}_{BR} - \mathbf{I}_{YB} = (6.28 + j\,14.71) - (-15.88 - j\,1.91) = (22.16 + j16.62)$ or $27.7\angle 36.86^\circ$ A **Ans.**

(*ii*) For R-Y phase, $S = \mathbf{V}_{CRY} \times \mathbf{I}_{RY} = (400 - j\,0)\ (9.6 - j\,12.8) = (3{,}840 - j\,5{,}120)$ VA

So true power per phase $= 3{,}840$ W **Ans.**

(*iii*) Phasor sum of three line currents $= \mathbf{I}_R + \mathbf{I}_Y + \mathbf{I}_B = (3.32 - j\,27.51) + (-25.48 + j\,10.89) + (22.16 + j\,16.62) = 0$

As expected, phasor sum of three line currents drawn by a balanced load is zero.

Example 10.19

A 220-V 3-Φ voltage is applied to a balanced delta-connected 3-Φ load of phase impedance (15 + *j* 20) Ω.

(*a*) Find the phasor current in each line.

(*b*) What is the power consumed per phase ?

(*c*) What is the phasor sum of the three line currents ? Why does it have this value ?

[*Banaras Hindu Univ. Elec. Circuits & Instruments 1985*]

Solution: Phase voltage, $V_P = V_L = 220$ V

Phase impedance, $Z_P = \sqrt{R_P^2 + X_P^2} = \sqrt{15^2 + 20^2} = 25\ \Omega$

Phasor current, $I_P = \dfrac{V_P}{Z_P} = \dfrac{220}{25} = 8.8$ A

$$\text{Line current, } I_L = \sqrt{3}\, I_P = \sqrt{3} \times 8.8 = 15.24 \text{ A}$$

$$\text{Phase angle, } \Phi = \cos^{-1}\frac{R_P}{Z_P} = \cos^{-1}\frac{15}{25} = \cos^{-1} 0.6 = 53.13° \text{ (lag)}$$

(*a*) Since line currents are 30° behind their respective phase currents, line currents are expressed as, taking phase voltage V_{RY} as reference phasor,

$$\mathbf{I_R} = 15.24 \angle -83.13° \text{ A}, \ \mathbf{I_Y} = 15.24 \angle -203.13° \text{ A}$$

$$\text{or } \mathbf{I_Y} = 15.24 \angle 156.87° \text{ and } \mathbf{I_B} = 15.24 \angle 36.87° \text{ A } \textbf{Ans.}$$

(*b*) Power consumed per phase $= V_P I_P \cos\Phi = 220 \times 8.8 \times 0.6 = 1{,}161.6$ watts **Ans.**

(*c*) Phasor sum of the three line currents $= \mathbf{I_R} + \mathbf{I_Y} + \mathbf{I_B} = 15.24 \angle -83.13° + 15.24 \angle 156.87° + 15.24 \angle 36.87°$

$$= (1.823 - j\,15.131) + (-14.015 + j\,5.987) + (12.192 + j\,9.144)$$

$$= 0 \ \textbf{Ans.}$$

As expected, phasor sum of three line currents drawn by a balanced load is zero because they are equal in magnitude and have a phase difference of 120° amongst themselves.

10.8. COMPARISON BETWEEN STAR AND DELTA SYSTEMS

1. In star-connected system similar ends are joined together whereas in delta-connected system dissimilar ends are joined together.
2. In a star-connected system phase voltage is equal to $\frac{1}{\sqrt{3}}$ times the line voltage, but in delta-connected system phase voltage is equal to line voltage, therefore, star-connected alternator needs less number of turns per phase and less insulation than a delta-connected alternator for the same line voltage. These are the reasons that 3-phase alternators are usually star-connected.
3. In star-connected system, phase current is equal to line current *i.e.* $I_P = I_L$ but in delta-connected system, the phase current is $\frac{1}{\sqrt{3}}$ times of line current *i.e.* $I_P = \frac{I_L}{\sqrt{3}}$
4. It is possible to carry neutral wire to the load in star-connected system only *i.e.* only star-connected system can give 3-phase 4-wire arrangement and hence only star-connected system can be used for lighting as well as power loads *i.e.* for an unbalanced load. Lighting loads are connected between neutral and on any one of the outers and power circuits can be connected to the lines.
5. Neutral point of a star-connected system can be connected to earth, so relays and protective devices can be provided in star-connected system for protection of the system against ground faults and it is usually done in practice.
6. Delta-connections are mostly used in transformers for operating small low-voltage 3-phase motors and are best suited for rotary convertors. Most of the three-phase induction motors are delta-connected, 3-phase loads are also usually delta-connected as it provides flexibility for addition or removal of a load on a single phase in case of an unbalanced load.

10.9. CONVERSION OF BALANCED LOAD SYSTEM FROM STAR TO DELTA AND VICE-VERSA

Any balanced star-connected system can be completely replaced by an equivalent delta-connected system or vice-versa because of their relationships between phase and line voltages and currents. For example, a balanced star-connected load having an impedance of magnitude Z with a power factor cos Φ (or $Z \angle \Phi$) in each phase can be replaced by an equivalent delta-connected load having an impedance of magnitude 3Z and power factor cos Φ (*i.e.* $3Z \angle \Phi$) in each phase.

This may be established as follows:

For a balanced star-connected load, let the line voltage be V_L, line current be I_L and impedance per phase be Z

$$\text{Phase voltage, } V_P = \frac{V_L}{\sqrt{3}}$$

$$\text{Phase current, } I_P = I_L$$

$$\text{So impedance per phase, } Z_S = \frac{V_P}{I_P} = \frac{V_L/\sqrt{3}}{I_L} = \frac{1}{\sqrt{3}}\frac{V_L}{I_L} \quad ...(10.21)$$

Now in equivalent delta-connected system for the same line values of voltage and current as in case of star-connected system

$$\text{Phase voltage, } V_P = V_L$$

$$\text{Phase current, } I_P = \frac{I_L}{\sqrt{3}}$$

$$\text{So impedance per phase, } Z_D = \frac{V_P}{I_P} = \frac{V_L}{I_L/\sqrt{3}} = \sqrt{3}\frac{V_L}{I_L} \quad ...(10.22)$$

Comparing Eqs. (10.21) and (10.22) we have

$$Z_S = \frac{1}{3}Z_D \text{ or } Z_D = 3\,Z_S \quad ...(10.23)$$

10.10. BALANCED PARALLEL LOADS

A combination of balanced 3-Φ loads connected in parallel may be solved by any one of the following methods.

1. All the given loads may be converted into either equivalent Y or Δ-connected loads and then combined together according to the law governing parallel circuits.
2. Alternative method is of working out volt-amperes. The real power and reactive power of various loads may be added arithmetically and algebraically respectively in order to give total volt-amperes according to expression

$$VA = \sqrt{W^2 + VAR^2}$$

$$\text{or } S = \sqrt{P^2 + Q^2}$$

where P is the power is watts (or kW), Q is the reactive power in reactive volt-amperes (or kVAR) and S is the volt-amperes (or kVA)

Example 10.20

A 3-phase star-connected 1,000 V alternator supplies power to 500 kW delta-connected induction motor. If the motor power factor is 0.8 lagging and its efficiency 0.9, find the current in each alternator and motor phase.

Solution:

$$\text{Motor input, } P = \frac{\text{Motor output}}{\text{Motor efficiency}} = \frac{500}{0.9} = 555.55 \text{ kW}$$

$$\text{Motor power factor, } \cos\Phi = 0.8 \text{ (lagging)}$$

$$\text{Line current, } I_L = \frac{\text{P in watts}}{\sqrt{3}\,V_L\cos\Phi} = \frac{555.55\times 1{,}000}{\sqrt{3}\times 1{,}000\times 0.8} = 401 \text{ A}$$

Current in star-connected alternator phase, $I_P = I_L = 401$ A **Ans.**

Current in delta-connected motor phase, $I_P = \frac{I_L}{\sqrt{3}} = \frac{401}{\sqrt{3}} = 231.5$ A **Ans.**

Example 10.21

Three similar coils each having a resistance of 20 Ω and an inductance of 0.05 H are connected in star to a 3-phase, 50 Hz supply with 400 V between lines. Calculate power factor, total power absorbed and line current. If the same coils are reconnected in delta across the same supply what will be the power factor, total power absorbed and line current ?

[Pune University 2003; Nagpur University Winter 2003]

Solution:

$$\text{Load resistance, } R = 20\ \Omega$$

$$\text{Load reactance, } X_L = 2\pi f L = 2\pi \times 50 \times 0.05 = 15.7\ \Omega$$

$$\text{Load impedance, } Z = \sqrt{R^2 + X_L^2} = \sqrt{20^2 + 15.7)^2} = 25.43\ \Omega$$

$$\text{Power factor, } pf = \frac{R}{Z} = \frac{20}{25.43} = 0.78644 \text{ (lagging) } \textbf{Ans.}$$

In Star-connection

$$\text{Line current, } I_L = \frac{V_L/\sqrt{3}}{Z} = \frac{400}{\sqrt{3} \times 25.43} = 9.08 \text{ A } \textbf{Ans.}$$

$$\text{Total power absorbed, } P = 3\, I_P^2\, R = 3 \times (9.08)^2 \times 20 = 4{,}948 \text{ W } \textbf{Ans.}$$

In Delta Connection:

$$\text{Phase current, } I_P = \frac{V_P}{Z} = \frac{400}{25.43} = 15.73 \text{ A}$$

$$\text{Line current, } I_L = \sqrt{3}\, I_P = \sqrt{3} \times 15.73 = 27.24 \text{ A } \textbf{Ans.}$$

i.e. three times the line current in star-connection

$$\text{Total power absorbed, } P = 3\, I_P^2\, R = 3 \times (15.73)^2 \times 20 = 14{,}844 \text{ W } \textbf{Ans.}$$

i.e. three times the power absorbed in star-connection.

Example 10.22

Three similar resistors connected in star draw a line current of 5 A from a 400 V, 3-phase mains. To what value should the line voltage be changed to obtain the same line current with the resistors connected in delta ?

Solution: When star-connected

$$\text{Phase voltage, } V_P = \frac{V_L}{\sqrt{3}} = \frac{400}{\sqrt{3}} = 231 \text{ V}$$

$$\text{Phase current, } I_P = I_L = 5 \text{ A}$$

$$\text{Phase resistors, } R_P = \frac{V_P}{I_P} = \frac{231}{5} = 46.2\ \Omega$$

When delta-connected

$$\text{Line current, } I_L = 5 \text{ A, the same as in star-connection}$$

$$\text{Phase current, } I_P = \frac{I_L}{\sqrt{3}} = \frac{5}{\sqrt{3}} = 2.887 \text{ A}$$

$$\text{Phase voltage, } V_P = I_P R_P = 2.887 \times 46.2 = 133.33 \text{ V}$$

$$\text{Line voltage, } V_L = V_P = 133.33 \text{ V,}$$

one-third of line voltage for star-connection **Ans.**

Example 10.23

A 3-phase delta-connected load, each phase of which has an inductive reactance of 40 Ω and a resistance of 25 Ω, is fed from the secondary of a 3-phase star-connected transformer, which has phase voltage of 240 V. Draw the circuit diagram of the system and calculate:

(*a*) The current in each phase of the load;
(*b*) The voltage across each phase of the load;
(*c*) The current in the transformer secondary windings;
(*d*) The total power taken from the supply and its power factor.

Solution:

$$\text{Resistance, } R_P = 25\ \Omega$$

$$\text{Reactance, } X_P = 40\ \Omega$$

$$\text{Impedance, } Z_P = \sqrt{R_P^2 + X_P^2} = \sqrt{(25)^2 + (40)^2} = 47.2\ \Omega$$

Phase voltage across transformer secondary = 240 V

Line voltage across delta-connected load = Line voltage across transformer secondary = $\sqrt{3} \times 240 = 415.7$ V

(*b*) Voltage across each phase of the load,

$$V_P = V_L = 415.7 \text{ V } \textbf{Ans.}$$

(*a*) Current in each phase of the load,

$$I_P = \frac{V_P}{Z_P} = \frac{415.7}{47.2} = 8.8 \text{ A } \textbf{Ans.}$$

(*c*) Current in the transformer secondary

= Line current of the load

$= \sqrt{3} \times 8.8 = 15.3$ A **Ans.**

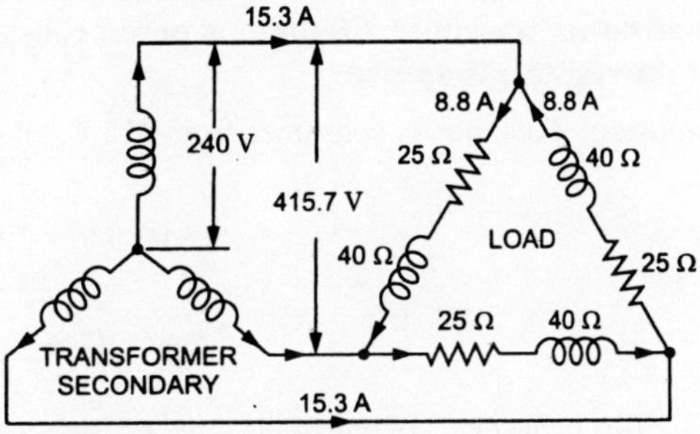

Fig. 10.16

(*d*) Power factor, $\cos \Phi = \frac{R_p}{Z_p} = \frac{25}{47.2} = 0.53$ (lag) **Ans.**

Total power drawn, $P = \sqrt{3}\, V_L I_L \cos \Phi = \sqrt{3} \times 415.7 \times 15.3 \times 0.53 = 5{,}825$ watts **Ans.**

Example 10.24

A 3- wire, 3-phase supply feeds a load consisting of three equal resistors. By how much is the load reduced if one of the resistor be removed ? (*i*) When the load is star-connected (*ii*) When the load is delta-connected.

[R.G.P.V. Bhopal Basic Electrical Engineering June 2002]

Solution: Let the line voltage of 3-phase supply be V_L volts and resistance of each phase be of R ohms.

(*i*) **In Star**

$$\text{Power consumed, } P = 3 \times \frac{V_P^2}{R} = 3 \times \left(\frac{V_L}{\sqrt{3}}\right)^2 \times \frac{1}{R} = \frac{V_L^2}{R}$$

After disconnection or removal of one of the resistances, the circuit is no longer 3-phase but converted into single phase circuit, having two resistors, each of R ohms connected in series across supply of V_L volts, as shown in Fig. 10.17(*b*).

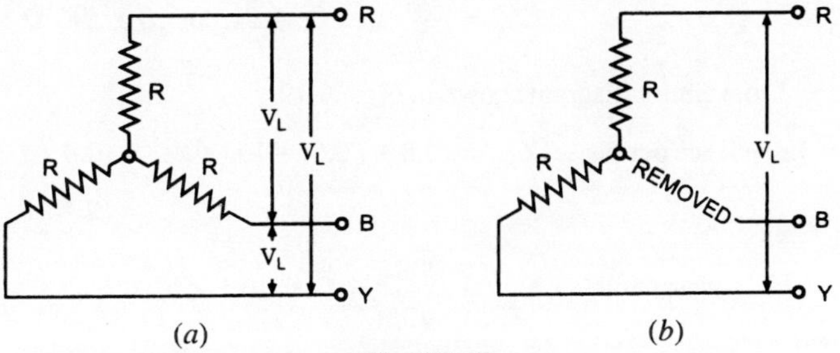

Fig. 10.17

$$\text{Hence total power consumed} = 2\,\frac{\left(\frac{V_L}{2}\right)^2}{R} = \frac{V_L^2}{2R}$$

$$\therefore \text{ Reduction in load} = \frac{\frac{V_L^2}{R} - \frac{V_L^2}{2R}}{\frac{V_L^2}{R}} \times 100 = 50\% \textbf{ Ans.}$$

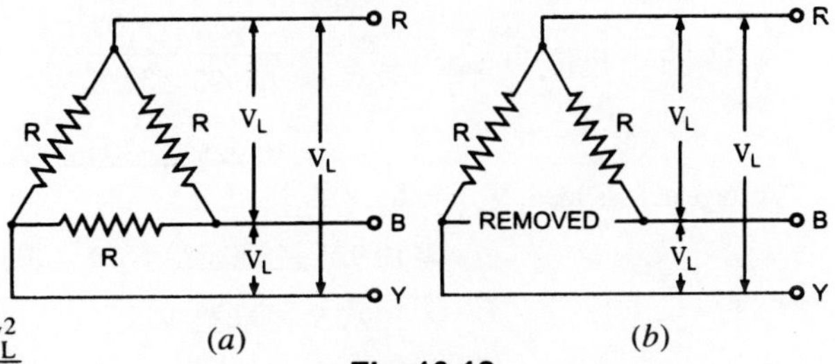

Fig. 10.18

(*ii*) **In Delta**

$$\text{Power consumed} = 3 \times \frac{V_P^2}{R} = 3 \times \frac{V_L^2}{R} = 3\,\frac{V_L^2}{R}$$

After removal of one of the resistance, each of the remaining two resistances is connected across supply of V_L volts.

$$\therefore \text{ Power consumed} = 2 \times \frac{V_L^2}{R} = \frac{2V_L^2}{R}$$

$$\text{So reduction in load} = \frac{\frac{3V_L^2}{R} - \frac{2V_L^2}{R}}{\frac{3V_L^2}{R}} \times 100 = 33.33\% \textbf{ Ans.}$$

Example 10.25

Three identical resistors of 20 Ω each are connected in star to a 415 V, 50 Hz, three-phase supply. Calculate (*i*) the total power consumed, (*ii*) the total power consumed, if they are connected in delta (*iii*) the power consumed, if one of the resistors is opened.
[U.P. Technical Univ. Electrical Engineering January 2003]

Solution: Total power consumed from 415 V, 50 Hz 3-phase supply, when three 20 Ω resistors are connected in

$$(i)\ \text{star},\ P_S = \frac{3\times\left(V_L/\sqrt{3}\right)^2}{R} = \frac{V_L^2}{R} = \frac{415^2}{20} = 8{,}611.25\ \text{W or } 8.61125\ \text{kW}\ \textbf{Ans.}$$

$$(ii)\ \text{delta},\ P_D = 3\times\frac{V_L^2}{R} = 3\times\frac{415^2}{20} = 25{,}833.75\ \text{W or } 25.83375\ \text{kW}\ \textbf{Ans.}$$

(*iii*) In the event of one of the three resistors getting open,

$$(a)\ \text{Power drawn in star-connections} = \frac{2\times(V_L/2)^2}{R} = \frac{V_L^2}{2R} = \frac{415^2}{2\times 20} = 4{,}305.625\ \text{W or } 4.306\ \text{kW}\ \textbf{Ans.}$$

$$(b)\ \text{Power drawn in delta-connections} = \frac{2\times V_L^2}{R} = \frac{2\times 415^2}{20} = 17{,}222.5\ \text{W or } 17.2225\ \text{kW}\ \textbf{Ans.}$$

Example 10.26

Three identical impedances of 30 $\angle 30°$ ohms are connected in delta to a 3-phase, 3-wire, 208 V abc system by conductors which have impedances of (0.8 + *j* 0.6) ohm. Find the magnitude of the voltage at the load end.
[Pb Univ. Electrical Engineering May 1990]

Solution: The equivalent star load of the delta-connected load is given as

$$\mathbf{Z}_S = \frac{\mathbf{Z}_D}{3}\ \textit{i.e.}\ \frac{30\angle 30°}{3}\ \text{or}\ 10\angle 30°\ \Omega \qquad \text{(Art. to 10.9)}$$

From circuit diagram shown in Fig. 10.19.

$$\text{Impedance per phase, } \mathbf{Z}_{an} = (0.8 + j\,0.6) + 10\angle 30° = (0.8 + j\,0.6) + (8.66 + j\,5)$$
$$= (9.46 + j\,5.6)\ \text{or}\ 10.993\angle 30.62°\ \Omega$$

$$\text{Phase voltage, } \mathbf{V}_{an} = \frac{208}{\sqrt{3}}\angle 0° = 120.1\angle 0°\ \text{V}$$

taking $\mathbf{V}_{an}$ as reference phasor

$$\text{So phase current, } \mathbf{I}_{an} = \frac{\mathbf{V}_{an}}{\mathbf{Z}_{an}} = \frac{120.1\angle 0°}{10.993\angle 30.62°}$$
$$= 10.925\angle -30.62°\ \text{A}$$

$$\text{Voltage across load, } V_{a'n} = \mathbf{I}_{an}\times\mathbf{Z}_L$$
$$= 10.925\angle -30.62°\times 10\angle 30°$$
$$= 109.25\angle 0.62°\ \text{V}$$

$$\text{Line voltage at the load} = \sqrt{3}\,V_{a'n} = \sqrt{3}\times 109.25 = 189.2\ \text{V}\ \textbf{Ans.}$$

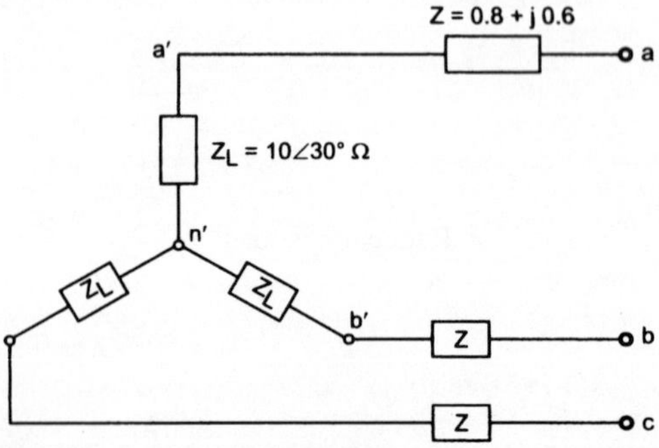

Fig. 10.19

10.11. POWER MEASUREMENT IN SINGLE PHASE CIRCUITS

Power may be defined as the rate at which energy is transformed or made available. The power in a circuit at any instant is equal to the product of the current in the circuit and the voltage across its terminals at that instant. In a dc circuit, if the current and voltage are constant, $P = VI$ so that it is necessary only to determine the current and voltage and to take their product in order to obtain the value of the power in the circuit. Alternatively, if the circuit resistance is known, power may be computed from one of the equivalent formulae, $P = V^2/R = I^2R$.

In almost all cases the power in a dc circuit is best measured by separately measuring quantities, V and I and by computing power by the formula P = VI rather than by measuring the power directly with a wattmeter. If voltage and current are measured simultaneously, allowance for power required for operation of voltmeter and ammeter must be made. Of course the power loss in the ammeter or in the voltmeter is often very small compared to the load power and may be safely neglected.

In the case of ac circuits the instantaneous power varies continuously as the current and voltage go through a cycle of values. The cyclic variation of power has a period so short that it can be followed only by special instruments such as oscillographs. However, we are not usually interested in the instantaneous power (except where transient phenomena are being studied) but in its time average. If the voltage and current are both sinusoidal the average power over a cycle is given by the expression P = VI cos Φ watts, where V and I are the rms values of voltage and current and Φ is the phase angle by which current lags behind or leads the voltage.

The fact that the power factor (cos Φ) is involved in the expression for the power means that a wattmeter must be used instead of merely an ammeter and a voltmeter, since the latter method takes no account of power factor.

A wattmeter is essentially an inherent combination of an ammeter and a voltmeter and, therefore, consists of two coils known as *current coil* and *pressure coil*. The operating torque is produced due to interaction of fluxes on account of currents in current and pressure coils. The current coil is inserted in series with the line carrying current to be measured and the pressure coil in series with a high non-inductive resistance R is connected across the load or supply terminals, as shown in Fig. 10.20.

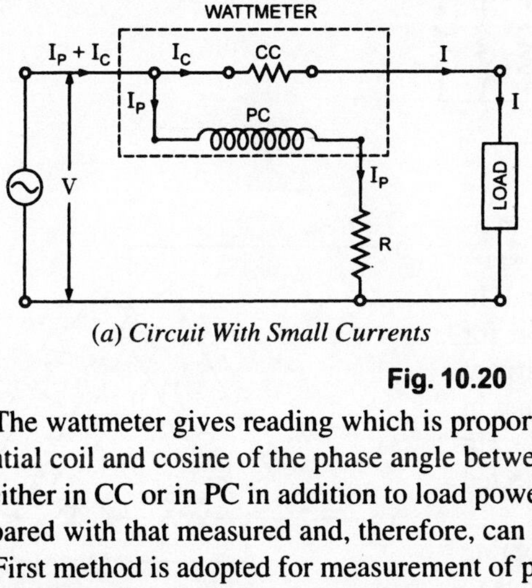

(*a*) *Circuit With Small Currents* (*b*) *Circuit With Large Currents*

Fig. 10.20 *Wattmeter Connections*

The wattmeter gives reading which is proportional to current flowing through its current coil, pd across potential coil and cosine of the phase angle between voltage and current. The wattmeter indicates the power lost either in CC or in PC in addition to load power. Normally the power lost in CC or in PC is very small as compared with that measured and, therefore, can be neglected.

First method is adopted for measurement of power in the circuits carrying small currents while the latter one is adopted for measurement of power in the circuits carrying large currents.

Example 10.27

A dynamometer type voltmeter with its voltage coil connected across the load side of the instrument reads 250 W. If the load voltage be 200 V, what power is being taken by the load? The voltage coil branch has a resistance of 2,000 Ω. [*Madras Univ. Elec. Engineering, 1985*]

Solution:

Wattmeter reading = 250 watts

Pressure coil circuit resistance, R_p = 2,000 Ω

Load voltage, V_L = 200 V

$$\text{Power lost in pressure coil} = \frac{V_L^2}{R_p} = \frac{(200)^2}{2{,}000} = 20 \text{ watts}$$

Since voltage coil is connected across the load side of the wattmeter as shown in Fig. 10.20 (*b*), the power consumed by it is also included in the reading of the wattmeter and so power taken by the load is equal to wattmeter reading less power lost in the pressure coil circuit *i.e.*

250 – 20 = 230 watts **Ans.**

10.12. MEASUREMENT OF POWER IN THREE PHASE CIRCUITS

10.12.1. Three Wattmeter Method. This method is employed for measurement of power in 3-phase, 4-wire circuits, the connections being as shown in Fig. 10.21 (*a*). As the neutral wire is common to the three phases, each wattmeter reads power in its own phase, and the total power of the load circuit is given by the sum of the readings of the three watt-meters

$$i.e. \text{ Total power of load circuit, } P = W_1 + W_2 + W_3$$

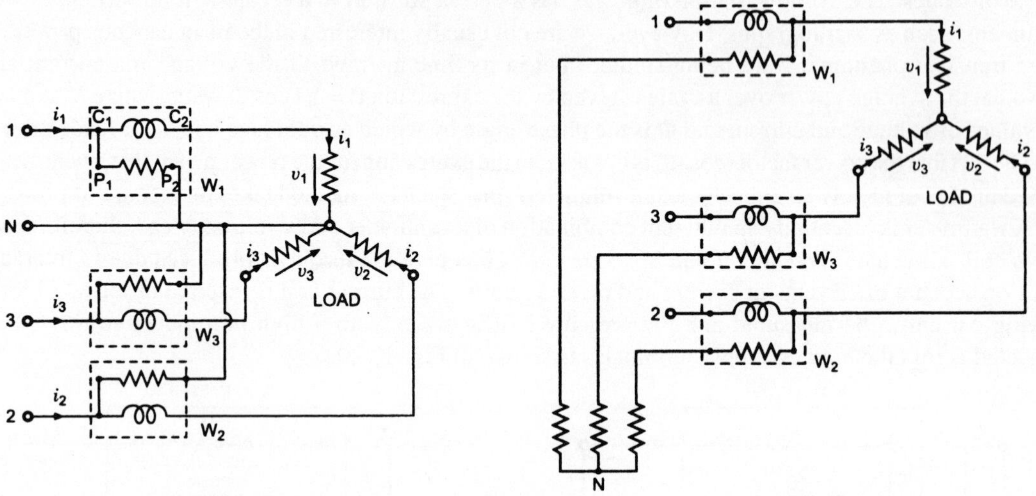

(*a*) (*b*)

Fig. 10.21

The above method is only useful for measuring power in 3-phase, 4-wire load circuits. In case of 3-phase, 3-wire star-connected circuits difficulty is experienced in getting neutral. In special cases when it is necessary to employ this method for measurement of power in 3-phase, 3-wire circuits an "artificial star" can be formed by connecting three equal high resistances in star to the three line conductors, as shown in Fig. 10.21 (*b*). (In case of low-voltage circuits three potential coils may be connected to form the common star).

In case of delta-connected circuits the difficulty in adopting the above mentioned method for measurement of power is due to the fact that the phase coils are required to be broken for inserting current coils of wattmeters, as shown in Fig. 10.21 (*c*).

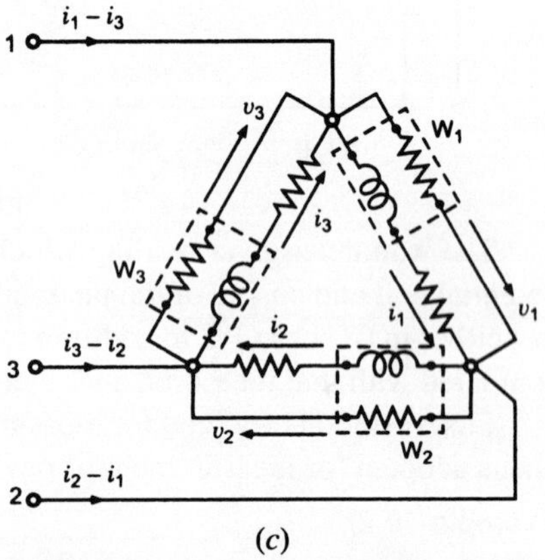

(*c*)

Fig. 10.21 *Three Wattmeter Method of Measuring 3-Phase Power*

10.12.2. One Wattmeter Method. In a balanced 3-wire, 3-phase load circuit the power in each phase is equal and, therefore, total power of the circuit can be determined by multiplying the power measured in any one phase by three.

Figure 10.22 (*a*) shows connections for measurement of power in one phase of the 3-phase, 3-wire star-connected balanced load circuit. Total power of the load circuit will be 3 times the reading of wattmeter.

Figure 10.22 (*b*) shows connections for measurement of power in 3-phase, 3-wire delta-connected balanced load circuits by one wattmeter method. In this method of measurement of power, resistance coils of resistance equal to that of pressure coil is connected in each of remaining two phases, as shown in Fig. 10.22 (*b*). The pressure coil and resistance coils form a balanced star-connection.

It should be noted that the one wattmeter method has the disadvantage that even a slight degree of unbalance in the loading produce a large error in the measurement.

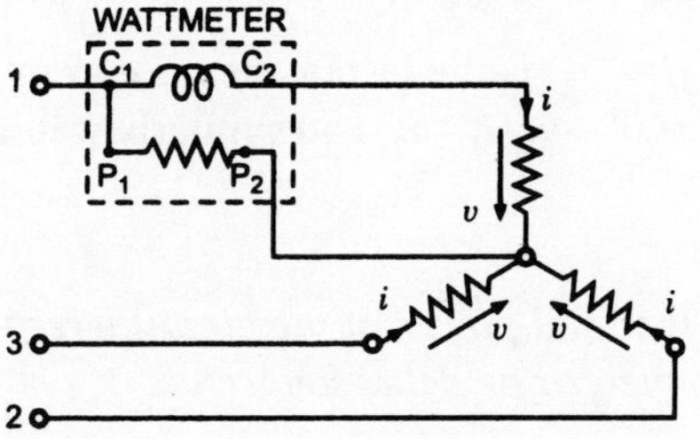

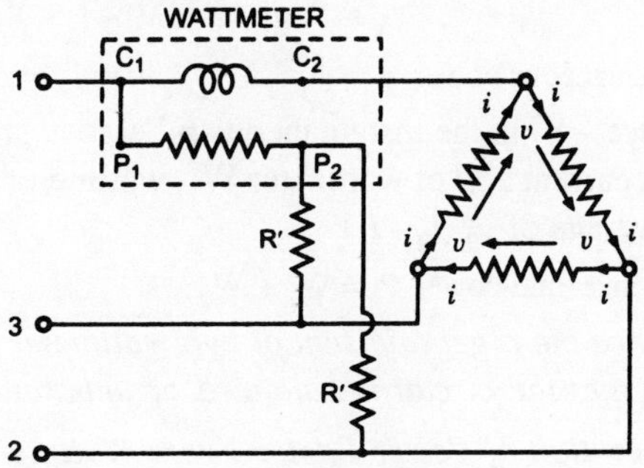

Fig. 10.22 *One Wattmeter Method of Measuring Power in 3-Phase 3-Wire Balanced Load Circuits*

10.12.3. Two Wattmeter Method. This is the generally used method for measurement of power in 3-phase, 3-wire load circuits. The current coils of two wattmeters are inserted in any two lines and pressure coil is connected from its own current coil to the line without a current coil. The connections are shown in Fig. 10.23.

Let v_1 , v_2 , v_3 and i_1 , i_2 , i_3 be the voltages and currents of the three loads connected across three different phases at any particular instant. These being instantaneous values, the power at the instant under consideration is equal to the sum of their products regardless of power factor

i.e. instantaneous power,

$$p = v_1 i_1 + v_2 i_2 + v_3 i_3 \text{ watts} \quad ...(10.24)$$

(*a*) **Star-Connected System.** Since all the three phases meet at a star point so according to Kirchhoff's first law, the algebraic sum of three instantaneous currents is zero.

$$i.e.\ i_1 + i_2 + i_3 = 0 \text{ or } i_3 = -(i_1 + i_2)$$

Substituting $i_3 = -(i_1 + i_2)$ in Eq. (10.24) we get

$$\text{Instantaneous power, } p = v_1 i_1 + v_2 i_2 - v_3 (i_1 + i_2)$$
$$= i_1 (v_1 - v_3) + i_2 (v_2 - v_3)$$

Since i_1 is the instantaneous current flowing through the current coil and $(v_1 - v_3)$ is the instantaneous potential difference across pressure coil of wattmeter W_1, $(v_1 - v_3)\, i_1 = w_1$, the instantaneous power measured by wattmeter W_1.

Similarly i_2 is the instantaneous current flowing through the current coil and $(v_2 - v_3)$ is the instantaneous potential difference across pressure coil of wattmeter W_2, therefore, $(v_2 - v_3)\, i_2 = w_2$ the instantaneous power measured by wattmeter W_2.

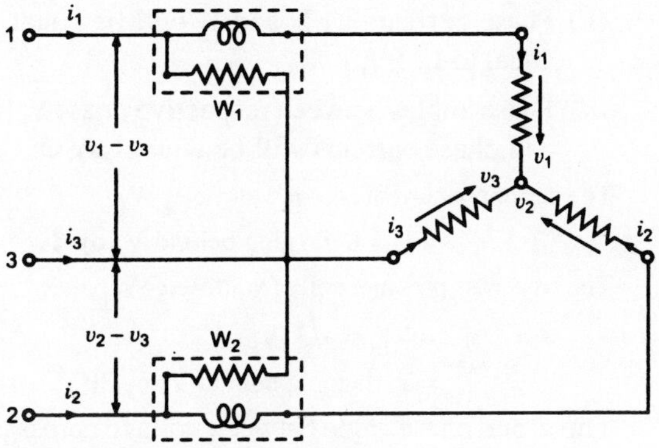

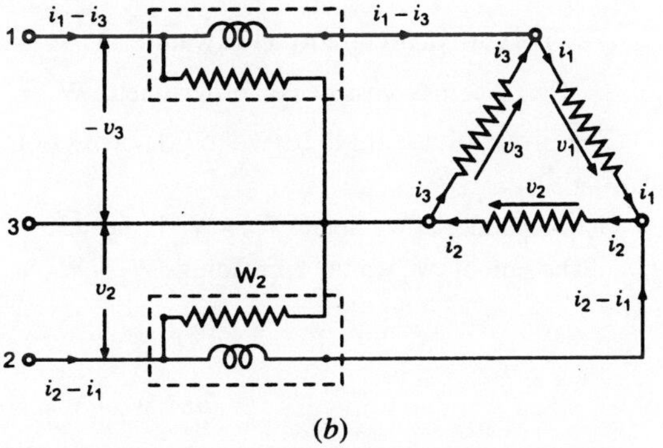

Fig. 10.23 *Two-Wattmeter Method of Measuring 3-Phase 3-Wire Power*

$$\text{Hence total instantaneous power, } p = w_1 + w_2$$

$$\text{or Total average power, } P = W_1 + W_2 \quad ...(10.25)$$

Hence the algebraic sum of two wattmeter readings gives the total power in the 3-phase, 3-wire star-connected load circuits whether the load is balanced or unbalanced.

(*b*) **Delta-Connected System.** In delta-connected system the three phases form a closed loop and according to Kirchhoff's second law.

$$v_1 + v_2 + v_3 = 0 \text{ or } v_1 = -(v_2 + v_3)$$

Instantaneous power, $p = v_1 i_1 + v_2 i_2 + v_3 i_3 = -(v_2 + v_3) i_1 + v_2 i_2 + v_3 i_3 = -v_3 (i_1 - i_3) + v_2 (i_2 - i_1)$ ...(10.26)

Since $-v_3$ is the instantaneous pd across pressure coil and $(i_1 - i_3)$ is the instantaneous current flowing through current coil of wattmeter W_1, wattmeter W_1 reads average of $-v_3 (i_1 - i_3)$ and similarly wattmeter W_2 reads average of $v_2 (i_2 - i_1)$

Hence total power, $P = W_1 + W_2$ ...(10.27)

Hence the algebraic sum of two wattmeter readings gives the total power of the circuit irrespective of the fact that the circuit is balanced or unbalanced and star-connected or delta-connected.

Determination of Power Factor From Wattmeter Readings. If load is balanced then pf of the load can also be determined from the wattmeter readings.

The phasor diagram for a balanced star-connected inductive load is shown in Fig. 10.24. Let V_1, V_2, and V_3 be the rms values of phase voltages and I_1, I_2 and I_3 be the rms values of phase currents.

Since load is balanced,

(*i*) Phase voltages V_1, V_2 and V_3 will be equal (say, equal to V_P)

(*ii*) Phase currents I_1, I_2 and I_3 will be equal (say, equal to I_P or I_L)

(*iii*) Phase angles between respective phase voltages and phase currents will be equal, say Φ

The current in current coil of wattmeter W_1

$= I_1 = I_L$ lagging behind V_1 by Φ.

The pd across pressure coil of wattmeter W_1

$= V_{13} = \sqrt{3}\ V_P$

$= V_L$ lagging behind V_1 by 30°.

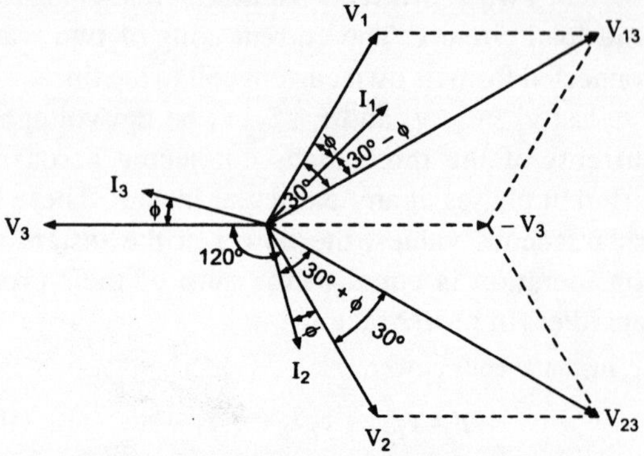

Fig. 10.24 *Phasor Diagram For Balanced Star-connected Inductive Load*

Therefore, phase angle between voltage across potential coil and current through current coil of wattmeter W_1 is $(30° - \Phi)$ and reading of wattmeter

$$W_1 = V_L I_L \cos(30° - \Phi) \quad ...(10.28)$$

The current in current coil of wattmeter $W_2 = I_2 = I_L$ lagging behind V_2 by Φ.

The pd across pressure coil of wattmeter $W_2 = V_{23} = \sqrt{3}\ V_P = V_L$ leading V_2 by 30°.

Therefore, phase angle between pd across potential coil and current through current coil of wattmeter W_2 is $(30° + \Phi)$.

Hence reading of wattmeter $W_2 = V_L I_L \cos(30° + \Phi)$ ...(10.29)

The sum of two wattmeter readings $W_1 + W_2 = V_L I_L \cos(30° - \Phi) + V_L I_L \cos(30° + \Phi)$

$= V_L I_L \times 2 \cos 30° \cos \Phi = \sqrt{3}\ V_L I_L \cos \Phi$

$=$ True power of load ...(10.30)

and $W_1 - W_2 = V_L I_L \cos(30° - \Phi) - V_L I_L \cos(30° + \Phi)$

$= V_L I_L \times 2 \sin 30° \sin \Phi = V_L I_L \sin \Phi$ ...(10.31)

Dividing Eq. (10.31) by Eq. (10.30) we get

$$\frac{W_1 - W_2}{W_1 + W_2} = \frac{\text{Tan}\,\Phi}{\sqrt{3}}$$

$$\text{or } \Phi = \text{Tan}^{-1}\left(\frac{W_1 - W_2}{W_1 + W_2}\right) \times \sqrt{3}$$

and pf of load circuit, $\cos\Phi = \text{Cos}\tan^{-1}\dfrac{\sqrt{3}(W_1 - W_2)}{W_1 + W_2}$ ...(10.32)

Hence phase angle Φ and pf cos Φ can be determined from readings of two wattmeters.

Alternative Method of Determination of Power Factor From Wattmeter Readings. This is convenient method for determining the power factor from wattmeter readings for a balanced load. In this method of determination of power factor from the wattmeter readings, for various values of phase angles the ratio of two wattmeter readings, $\dfrac{\text{Cos } 30° + \Phi}{\text{Cos } 30° - \Phi}$ $\left(\textit{i.e.} \dfrac{\text{smaller reading}}{\text{larger reading}}\right)$ and the corresponding power factor values are determined and plotted as shown in Fig. 10.25. Now by using the watt-ratio curve (Fig. 10.25) the power factor is read directly, the ratio, $\dfrac{W_2}{W_1}$ being known.

It is seen that, when $\dfrac{W_2}{W_1} = 1.0$, the power factor is unity; when $\dfrac{W_2}{W_1} = 0$, the power factor is 0.5; when $\dfrac{W_2}{W_1}$ is negative i.e. when it becomes necessary to reverse W_2, the power factor is less than 0.5.

Variation in Wattmeter Readings. Wattmeter readings for inductive loads

$$W_1 = V_L I_L \cos(30° - \Phi)$$
$$W_2 = V_L I_L \cos(30° + \Phi)$$

(*i*) When pf of load is unity *i.e.* Φ = 0 then

$$W_1 = V_L I_L \cos 30° = \frac{\sqrt{3}}{2} V_L I_L = \text{Half of total power}$$

$$W_2 = V_L I_L \cos 30° = \frac{\sqrt{3}}{2} V_L I_L = \text{Half of total power}$$

Therefore, readings of both wattmeters are same, positive and equal to half the total power of the circuit.

(*ii*) When pf of load is 0.5 *i.e.* Φ = 60° then

$$W_1 = V_L I_L \cos(30° - 60°) = \frac{\sqrt{3}}{2} V_L I_L = \text{Total power}$$

$$W_2 = V_L I_L \cos(30° + 60°) = 0$$

Hence reading of wattmeter W_1 will give total power of the load.

When the pf of load is zero *i.e.* Φ = 90° then

$$W_1 = V_L I_L \cos(30° - 90°) = \frac{1}{2} V_L I_L$$

$$W_2 = V_L I_L \cos(30° + 90°) = -\frac{1}{2} V_L I_L$$

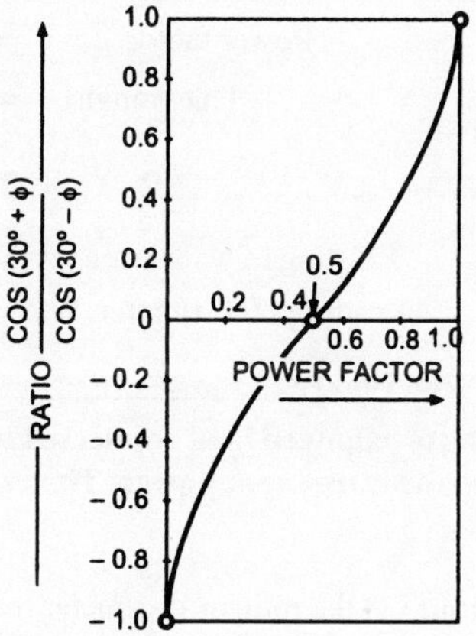

Fig. 10.25 *The Watt-ratio Curve*

Hence the readings of two wattmeters are equal but of opposite sign.

So wattmeter W_2 gives –ve reading when phase angle Φ varies from 60° to 90°. For obtaining the reading of wattmeter W_2 either the connection of current coil or pressure coil should be changed and readings obtained after the reversal of connections should be subtracted from the other wattmeter reading in order to get the total power.

Wattmeter Readings For Capacitive Loads. In the above discussion inductive load has been considered. If the circuit is capacitive the expression for readings of two meters are obtained by substituting in Φ by (– Φ) in the above expressions for wattmeter readings.

$$W_1 = V_L I_L \cos[30° - (-\Phi)] = V_L I_L \cos(30° + \Phi)$$
$$W_2 = V_L I_L \cos[30° + (-\Phi)] = V_L I_L \cos(30° - \Phi)$$

i.e. readings of wattmeters are interchanged.

Example 10.28

In the two wattmeter method of power measurement in a 3 phase circuit, the readings of the wattmeters are 1,200 W and 300 W. What is the pf of the load : Prove the formula used. [*M.D. Univ. Rohtak Electrical Technology May 2003*]

Solution: $W_1 = 1{,}200$ watts and $W_2 = 300$ watts

$$\text{Power factor of load, } \cos\phi = \text{Cos}\tan^{-1}\frac{W_1 - W_2}{W_1 + W_2}\sqrt{3} = \text{Cos}\tan^{-1}\frac{1{,}200 - 300}{1{,}200 + 300}\sqrt{3} = 0.6934 \text{ (lag) } \textbf{Ans.}$$

Example 10.29

Two wattmeters connected to measure the total power in a 3-phase balanced circuit. One measures 4,800 W, while the other reads backwards. On reversing the latter it is found to read 400 W. What is the total power and power factor ? Draw the connection diagram and phasor diagram of the circuit.

[*R.G.P.V. Bohpal Basic Electrical Engineering 2001*]

Solution: Wattmeter reading, $W_1 = 4{,}800$ W

Wattmeter reading $W_2 = -400$ W

Total power of the circuit, $P = W_1 + W_2 = 4{,}800 + (-400) = 4{,}400$ W **Ans.**

$$\text{Power factor of the circuit} = \text{Cos}\tan^{-1}\frac{W_1 - W_2}{W_1 + W_2}\sqrt{3} = \text{Cos}\tan^{-1}\frac{4{,}800 - (-400)}{4{,}800 + (-400)}\sqrt{3} = \text{Cos}\tan^{-1} 2.047$$

$$= 0.438 \text{ (lagging) } \textbf{Ans.}$$

Connection diagram is shown in Fig. 10.23 and phasor diagram is shown in Fig. 10.24.

Example 10.30

In a 2 wattmeter method power measured was 30 kW at 0.7 pf lagging. Find the reading of each wattmeter.

[*U.P. Technical Univ. Electrical Engnieering Second Semester 2006-07*]

Solution: Total load supplied, P = 30 k W

Power factor, $\cos\phi = 0.7$ (lagging)

Phase angle, $\phi = \text{Cos}^{-1}\,0.7 = 45.57°$ (lagging)

$$\text{So } V_L I_L = \frac{\text{P in kW} \times 1{,}000}{\sqrt{3}\cos\phi} = \frac{30 \times 1{,}000}{\sqrt{3} \times 0.7} = 24{,}743.6 \text{ V A}$$

Reading of wattmeter $W_1 = V_L I_L \cos(30° - \phi) = 24{,}743.6\cos(30° - 45.57°) = 23{,}835$ W or 23.835 k W **Ans.**

Reading of wattmeter, $W_2 = V_L I_L \cos(30° + \phi) = 24{,}743.6\cos(30° + 45.57°) = 6{,}165$ W or 6.165 kW **Ans.**

Example 10.31

A 3-phase balanced load connected across a 3-ϕ, 400 V ac supply draws a line current of 10 A. Two wattmeters are used to measure input power. The ratio of two wattmeter readings is 2 : 1. Find the readings of the two wattmeters.

[*U.P. Technical Univ. Electrical Engineering February 2002*]

Solution: Let the ratio of wattmeter readings be *r i.e.* $\frac{W_2}{W_1} = r$

$$\text{and } \tan\Phi = \sqrt{3}\left(\frac{W_1 - W_2}{W_1 + W_2}\right) = \sqrt{3}\left(\frac{1 - \frac{W_2}{W_1}}{1 + \frac{W_2}{W_1}}\right) = \sqrt{3}\left(\frac{1-r}{1+r}\right)$$

$$\text{and power factor, } \cos\Phi = \frac{1}{\sec\phi} = \frac{1}{\sqrt{1 + \tan^2\phi}} = \frac{1}{\sqrt{1 + 3\left(\frac{1-r}{1+r}\right)^2}}$$

$$\text{Substituting } r = \frac{W_2}{W_1} = \frac{1}{2} = 0.5 \text{ in above expression we get}$$

$$\text{Power factor of the load, } \cos\phi = \frac{1}{\sqrt{1+3\left(\frac{1-0.5}{1+0.5}\right)^2}} = 0.866$$

$$\text{or phase angle, } \phi = \text{Cos}^{-1}\ 0.866 = 30^\circ$$

$$\text{Wattmeter reading, } W_1 = V_L I_L \cos(30^\circ - 30^\circ) = 400 \times 10 \cos 0^\circ = 4{,}000 \text{ W} \textbf{ Ans.}$$

$$\text{Wattmeter reading, } W_2 = V_L\, I_L \cos(30^\circ + 30^\circ) = 400 \times 10 \times \cos 60^\circ = 2{,}000 \text{ W} \textbf{ Ans.}$$

Example 10.32

Three equal impedances, each consisting of R and L in series are connected in star and are supplied from a 400 volts, 50 Hz, 3-phase, 3-wire balanced supply system. The power input to the load is measured by 2-wattmeter method and the two wattmeters read 3 kW and 1 kW. Determine the values of R and L connected in each phase.

[U.P. Technical Univ. Electrical Engineering Second Semester 2002-03]

Solution:

$$\text{Reading of wattmeter } W_1 = 3 \text{ kW}$$

$$\text{Reading of wattmeter } W_2 = 1 \text{ kW}$$

$$\text{Total power, } P = W_1 + W_2 = 3 + 1 = 4 \text{ kW}$$

$$\text{Power factor of the circuit, } \cos\phi = \text{Cos}\tan^{-1}\frac{W_1 - W_2}{W_1 - W_2}\sqrt{3} = \text{Cos}\tan^{-1}\frac{3-1}{3+1}\sqrt{3}$$

$$= \text{Cos}\tan^{-1}\frac{\sqrt{3}}{2} = \text{Cos } 40.89^\circ = 0.7559 \text{ (lagging)}$$

$$\text{Line current, } I_L = \frac{P}{\sqrt{3}\, V \cos\phi} = \frac{4 \times 1{,}000}{\sqrt{3} \times 400 \times 0.7559} = 7.6376 \text{ A}$$

$$\text{Impedance of the circuit per phase, } Z_P = \frac{V_P}{I_P} = \frac{400/\sqrt{3}}{7.6376} = 30.237\ \Omega$$

$$\text{Resistance per phase, } R = Z_P \cos\phi = 30.237 \times 0.7559 = 22.856\ \Omega \textbf{ Ans.}$$

$$\text{Reactance per phase, } X_L = \sqrt{Z_P^2 - R^2} = \sqrt{(30.237)^2 - (22.856)^2} = 19.796\ \Omega$$

$$\text{Inductance per phase, } L = \frac{X_L}{2\pi f} = \frac{19.796}{2\pi \times 50} = 0.063 \text{ H or } 63 \text{ mH} \textbf{ Ans.}$$

Example 10.33

A star-connected three-phase load has a resistance of 8 Ω and an inductive reactance of 6 Ω in each phase. It is fed from a 400 V, 3-phase balanced supply. Determine the line current, power factor, active and reactive powers. Draw phasor diagram showing phase and line voltages and currents. If power measurement is made using two-wattmeter method, what will be the readings of both wattmeters ? *[U.P. Technical Univ. Electrical Engineering First Semester 2005-06]*

Solution:

$$\text{Phase voltage, } V_P = \frac{\text{Line voltage, } V_L}{\sqrt{3}}$$

$$= \frac{400}{\sqrt{3}} = 231 \text{ V}$$

$$\text{Impedance per phase, } Z_P = \sqrt{R_P^2 + X_P^2} = \sqrt{8^2 + 6^2} = 10\ \Omega$$

$$\text{Line current, } I_L = \text{Phase current, } I_P$$

$$= \frac{V_P}{Z_P} = \frac{231}{10} = 23.1 \text{ A} \textbf{ Ans.}$$

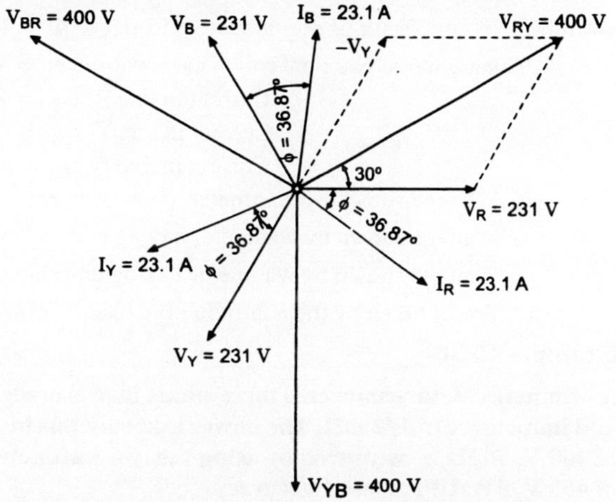

Fig. 10.26 *Phasor Diagram For Current and Voltages*

$$\text{Power factor, } \cos\phi = \frac{V_P}{Z_P} = \frac{8}{10} = 0.8 \text{ (lagging) } \textbf{Ans.}$$

$$\text{Phase angle, } \phi = \text{Cos}^{-1}\, 0.8 = 36.87° \text{ (lagging)}$$

$$\text{Active power, P} = \sqrt{3}\, V_L I_L \cos\phi = \sqrt{3} \times 400 \times 23.1 \times 0.8 = 12{,}800 \text{ W } \textbf{Ans.}$$

$$\text{Reactive power, Q} = \sqrt{3}\, V_L I_L \sin\phi = \sqrt{3} \times 400 \times 23.1 \times 0.6 = 9{,}600 \text{ VA (reactive) } \textbf{Ans.}$$

$$\text{Wattmeter readings, } W_1 = V_L I_L \cos(30° - \phi) = 400 \times 23.1 \cos(30° - 36.87°) = 9{,}171.3 \text{ W } \textbf{Ans.}$$

$$W_2 = V_L I_L \cos(30° + \phi) = 400 \times 23.1 \cos(30° + 36.87°) = 3{,}628.7 \text{ W } \textbf{Ans.}$$

Phasor diagram for currents and voltages is shown in Fig. 10.26.

Example 10.34

The power input to a synchronous motor is measured by two wattmeters both of which indicate 50 kW. If the pf of the motor be changed to 0.866 leading determine the readings of the wattmeters, the total input power remaining the same. Draw the phasor diagram for the second condition of the load. [*Nagpur Univ. Elec. Technology 1992*]

Solution: Let the readings of wattmeters in second conditions be W_1 and W_2

Now $W_1 + W_2$ = Total power of the load = 100 kW ...(*i*)

as before change of pf of operation

$$\text{Phase angle, } \Phi = \text{Cos}^{-1}\text{ pf} = \text{Cos}^{-1}\, 0.866 = 30° \text{ (lead)}$$

$$\text{Since } \Phi = \text{Tan}^{-1}\left(-\sqrt{3}\,\frac{W_1 - W_2}{W_1 + W_2}\right)$$

$$\text{So } \tan\Phi = -\sqrt{3}\,\frac{W_1 - W_2}{W_1 + W_2}$$

$$\text{or } W_1 - W_2 = (W_1 + W_2) \times \frac{\tan 30°}{-\sqrt{3}}$$

$$= 100 \times \frac{1/\sqrt{3}}{-\sqrt{3}} = -33.34 \text{ kW} \quad ...(ii)$$

From Eqs. (*i*) and (*ii*) we get

W_1 = 33.33 kW **Ans.** and W_2 = 66.67 kW **Ans.**

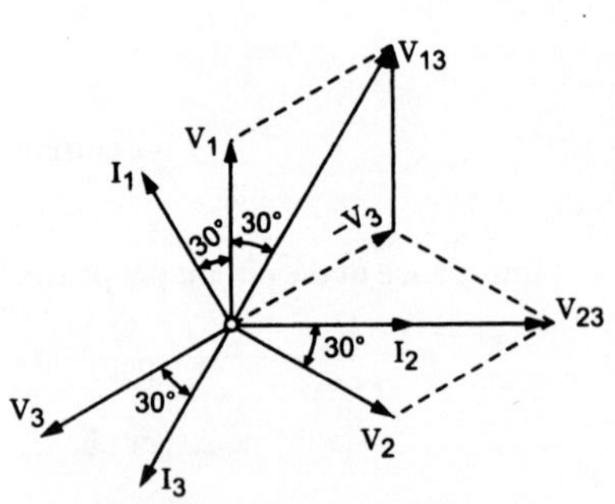

Fig. 10.27 *Phasor Diagram*

Example 10.35

A 3-phase, 4-wire star-connected system supplies only non-inductive load. The current in line R is 8 A, the current in line Y is 10 A and the current in line B is 6 A. The voltage from line to neutral is 120 V (*i*) Find wattage shown by each of the three wattmeters (*ii*) power taken by the load.

Solution: Power factor of non-inductive load, $\cos\Phi = 1.0$

Voltage across potential coil of each wattmeter = Voltage from each line to neutral *i.e.* $V_P = 120$ V

Current in line R, I_{PR} = 8 A

Current in line Y, I_{PY} = 10 A

Current in line B, I_{PB} = 6 A

(*i*) Wattage shown by wattmeter having current coil in line R = $V_P \times I_{PR} \cos\Phi = 120 \times 8 \times 1 = 960$ watts **Ans.**

Wattage shown by wattmeter having current coil in line Y = $V_P \times I_{PY} \cos\Phi = 120 \times 10 \times 1 = 1{,}200$ watts **Ans.**

Wattage shown by wattmeter having current coil in line B = $V_P \times I_{PB} \cos\Phi = 120 \times 6 \times 1 = 720$ watts **Ans.**

(*ii*) Power taken by the non-inductive load = 960 + 1, 200 + 720 = 2,880 watts **Ans.**

Example 10.36

A symmetrical star-connected three-phase load is made of three identical coils having an internal resistance of 12 Ω and inductance of 142 mH. The power taken by this load when connected to a balanced 3-phase, three wire supply of 400 V, 50 Hz in measured by using the two wattmeter method. The wattmeters used are similar and are of type of 400 V, 5 A, UPF, class 1 accuracy.

(*i*) Determine expected readings of W_1 and W_2 (*ii*) Compute the total power. [*GATE 1995*]

Solution:

Line voltage, $V_L = 400$ V

Phase voltage, $V_P = \frac{V_L}{\sqrt{3}} = \frac{400}{\sqrt{3}} = 231$ V

Coil resistance, $R_P = 12\ \Omega$

Coil Inductive reactance, $X_P = 2\pi f L_P = 2\pi \times 50 \times 142 \times 10^{-3} = 44.61\ \Omega$

Coil impedance, $Z_P = \sqrt{R_P^2 + X_P^2} = \sqrt{12^2 + 44.61^2} = 46.2\ \Omega$

Coil phase angle, $\Phi = \text{Cos}^{-1} \frac{R_P}{Z_P} = \text{Cos}^{-1} \frac{12}{46.2} = 74.94°$ (lagging)

Phase current, $I_P = \frac{V_P}{Z_P} = \frac{231}{46.2} = 5$ A

Line current, $I_L = I_P = 5$ A

Reading of wattmeter $W_1 = V_L I_L \cos(30° - \Phi) = 400 \times 5.0 \cos(30° - 74.94°) = 1{,}415.7$ W

Reading of wattmeter $W_2 = V_L I_L \cos(30° + \Phi) = 400 \times 5.0 \cos(30° + 74.94°) = -515.7$ W **Ans.**

Total power, $P = W_1 + W_2 = 1{,}415.7 + (-515.7) = 900$ W **Ans.**

HIGHLIGHTS

1. *Polyphase system* means the system which consists of numerous windings or circuits.
2. A supply system is said to be *symmetrical* when different voltages of the same frequency have equal magnitude and are displaced from one another by equal time angle. 3-phase, 3-wire supply system will be symmetrical when line-to line voltages are equal in magnitude and displaced by 120 electrical degrees w.r.t. each other.
3. A 3-phase supply will be *unbalanced* when either of three phase voltages are unequal in magnitude or the phase angles between these phases is not equal to 120°.
4. A load circuit is said to be balanced when the loads (impedances) connected in different phases are same in magnitude as well as in phase.
5. For a balanced system, the phasor sum of the three phase voltages is always zero.
6. In case one phase out of three phases of a 3-phase supply connected to the three phase load is not available, then such a condition is called the single phasing.
7. Commonly used polyphase system is 3-phase system.
8. *Phase sequence* is the order in which the currents or voltages in different phases attain their maximum values.
9. The three phase system is connected in two ways namely star connection and delta connection.
10. Star-connected system is obtained by joining together similar ends, either the start or the finish, the other ends are joined to the line wires. The common point at which similar ends are connected is called the neutral or star point.
11. In a balanced star-connected system
    (*i*) Line voltages are 120° apart.
    (*ii*) Line voltages are 30° ahead of the respective phase voltages.
    (*iii*) Line voltages are $\sqrt{3}$ times of phase voltages.
    (*iv*) Line currents are equal to phase currents.
    (*v*) The angle between line currents and the corresponding line voltages is $(30° \pm \Phi)$; + ve for lagging currents and –ve for leading currents.
    (*vi*) True power output = $\sqrt{3}\ E_L I_L \cos \Phi$, where Φ is the angle between the respective phase current and phase voltage (not between line current and line voltage).
    (*vii*) Apparent power = $\sqrt{3}\ E_L I_L$
    (*viii*) In balanced system, the potential of neutral or star point is zero because potential at neutral or star point,

$$E_{NR} + E_{NY} + E_{NB} = 0$$

12. When the starting end of one coil is joined to the finishing end of another coil, the delta or mesh connection is obtained. The direction of emf in the coils are taken as positive from start to finished ends.

13. In balanced delta connected system

(*i*) Line currents are 120° apart.

(*ii*) Line currents are 30° behind the respective phase currents.

(*iii*) Line currents are $\sqrt{3}$ times phase currents.

(*iv*) The angle between line currents and corresponding line voltage is (30° ± Φ) as in star system.

(*v*) Line voltages are equal to phase voltages.

(*vi*) Power = $\sqrt{3}\ E_L I_L \cos\Phi$, where Φ is the phase angle between respective phase current and phase voltage.

(*vii*) Apparent power = $\sqrt{3}\ E_L I_L$

(*viii*) In balanced system the resultant emf in the closed circuit will be zero.

$$i.e.\ \mathbf{E}_{RY} + \mathbf{E}_{YB} + \mathbf{E}_{BR} = 0$$

Hence there will be no circulating current in the mesh if no load is connected to the lines.

14. Two wattmeter method is generally used to measure 3-phase power. In this method, the current coils of the two wattmeters are connected in any two lines and potential coil is connected from its own current coil to the line without a current coil. The sum of the two wattmeter readings gives the total power of the circuit.

This method of power measurement is applicable to the 3-phase, 3-wire circuit irrespective of the fact that the circuit is balanced or unbalanced, and star-connected or delta connected.

If load is balanced, then its power factor can also be calculated from the readings of the two wattmeters, which are given as

$$W_1 = V_L I_L \cos(30° - \Phi)$$

$$\text{and } W_2 = V_L I_L \cos(30° + \Phi) \text{ in case of inductive loads}$$

$$\text{and power factor } \cos\Phi = \text{Cos}\left[\tan^{-1}\frac{\sqrt{3}(W_1 - W_2)}{W_1 + W_2}\right]$$

$$W_1 = V_L I_L \cos(30° + \Phi)$$

$$\text{and } W_2 = V_L I_L \cos(30° - \Phi) \text{ in case of capacitive circuits.}$$

EXERCISES

1. What do you understand by the following terms.

(*i*) Three phase balanced supply.

(*ii*) Three phase balanced load.

(*iii*) Three phase unbalanced load.

(*iv*) Advantages of 3-phase over single phase. **[M.D. Univ. Rohtak Electrical Technology May 2002]**

2. Compare single phase and 3-phase systems. **[M.D. Univ. Rohtak Electrical Technology May 2002]**

3. State the advantages of three phase system over single phase system.
[Pb. Technical Univ. Basic Electrical Engineering June 2003]

4. Deduce the relationship between the phase and line voltages and currents in a 3-phase star-connected circuit. Draw phasor diagram to establish it. **[M.D. Univ. Rohtak Electrical Technology May 2002]**

5. Derive the relationship between the line and phase voltages of an alternator.
[U.P. Technical Univ. Electrical Engineering First Semester 2006-07]

6. For star-connected system in a 3-phase circuit prove that $V_L = \sqrt{3}\ V_{ph}$ and $I_L = I_{ph}$.
[U.P. Technical Univ. Electrical Engineering Second Semester 2006-07]

7. Prove that the average power in an ac circuit is given by W = VI cos ϕ where symbols have their usual meaning.
[U.P.S.C. I.E.S. Electrical Engineering-I, 2002]

8. Derive the relationship between line current and phase current for delta connected 3-phase load when supplied from 3-phase balanced supply. **[Nagpur University Summer 2004]**

9. Derive the relationship between line and phase voltages and currents for a delta-connected 3-phase balanced system.
[U.P. Technical Univ. Electrical Engineering First Semester 2004-05]

10. Derive expressions for power in 3-phase star and delta connections.
[U.P. Technical Univ. Electrical Engineering Second Semester 2005-06]

11. Establish relationship between line and phase voltages and currents in a balanced delta-connection. Draw complete phasor diagram of voltages and currents. **[R.G.P.V. Bhopal Basic Electrical Engineering December 2003]**

12. Derive the expression for the active power and reactive power consumed by a balanced three-phase-connected load in terms of line voltage, line current and power factor. **[R.G.P.V. Bhopal Basic Electrical Engineering February 2005]**

13. Explain star and delta connections in 3-phase circuits. State the advantages of each type.
[R.G.P.V. Bhopal Basic Electrical Engineering June 2007]

14. Derive the relationship between line voltage, phase voltage, line current and phase current in a 3-phase star-connected and delta-connected circuits. **[Gujarat University June/July 2003]**

15. Show that the power consumed by three identical phase loads connected in delta is equal to three times the power consumed when the phase loads are connected in star. **[Nagpur University Summer 2002]**

16. Discuss the concept of balanced 3-phase circuits. What is the difference between three-phase star and three-phase delta connections. **[M.D. Univ. Electrical Technology Ist Semester E. Scheme New Examination, 2006-07]**

17. Explain two wattmeter method to measure 3-ϕ power in Y-connected load. Derive the phase angle in terms of wattmeter readings. Draw phasor diagram also. **[U.P. Technical Univ. Electrical Engineering First Semester 2007-08]**

18. Derive and explain the two wattmeter method of measurement of three phase power for a balanced star connected load. How is the three phase power determined ? Discuss the variations in readings for different power factors of loads from unity to zero. **[U.P. Technical Univ. Electrical Engineering Second Semester 2007-08]**

19. Explain the methods to measure power in 3-phase circuits.
[U.P. Technical Univ. Electrical Engineering Second Semerter 2006-07]

20. With the help of connection and phasor diagrams, show that the power factor of a balanced three-phase load can be determined, using two wattmeters. **[M.D. Univ. Rohtak Electrical Technology May 2002]**

21. Show how the active power in 3-phase ac circuits can be measured with the help of two wattmeters. Deduce the expression for power factor in terms of two wattmeter readings.
[R.G.P.V. Bhopal Basic Electrical Engineering June 2003]

22. Prove that the power in a three-phase balanced circuit can be deduced from the readings of two wattmeters with the help of suitable vector diagrams. Discuss the nature of power factor when, (*i*) the two wattmeters read equal and positive values, (*ii*) the two wattmeters read equal and opposite values, (*iii*) the one wattmeter reads zero value.
[U.P. Technical Univ. Elec. Engineering February 2001]

SHORT ANSWER TYPE QUESTIONS WITH ANSWERS

Q. 1. What do you mean by three-phase balanced load ?

Ans. Any three-phase load will be balanced when the loads (impedances) connected in three phases are same in magnitude as well as in phase.

Q. 2. What is three-phase unbalanced load ?

Ans. Any three-phase load will be unbalanced if impedances in one or more phases differ from the impedance (s) of the remaining phase (s).

Q. 3. Why is an unbalanced load not normally used on a 3-phase 3-wire system ? Any line/phase voltages equal in such situations ? **[P.T.U. 1999]**

Ans. Normally, unbalanced load is not employed on 3-phase, 3-wire ungrounded star-connected system because unbalanced loading may cause different voltage drops in lines. Consequently the three phase voltages are different (or unbalanced) in magnitude as well as in phase and it is quite possible that one phase voltage may exceed the line voltage. Such a condition is undesirable, since some loads may operate inefficiently due to lowering of the voltage and the other equipment may get damaged due to overvoltage.

Q. 4. What is phase sequence in a 3-phase system ? How is given phase sequence reversed ? **[P.T.U. 1999]**

Ans. Phase sequence is the order or sequence in which the currents or voltages in different phases attain their maximum values one after the other.

The given phase sequence can be reversed by interchanging any two terminals of the supply.

Q. 5. Why interconnection of 3-phase system is necessary ?

Ans. Non-interlinked three-phase system will require 6 wires while interconnected 3-phase system will require 3 or 4 wires. So non-interlinked system will be very complicated and expensive.

Q. 6. Differentiate between star and delta connections. **[P.T.U. June 2003]**

Ans. In star-connected system similar (either start or finish) terminals of the three phases are connected together to provide star or neutral point [refer to Fig. 10.7 (*a*)] while in delta-connected system start terminal of one phase is connected to the finish terminal of the second phase and start terminal of second phase is connected to finish terminal of the third phase and so on to provide a closed circuit in 3-phase system [refer Fig. 10.14 (*a*)].

Q. 7. Discuss, why the stator windings of alternator are generally star-connected ? **[P.T.U. December 2000]**

Ans. In a star-connected system phase voltage is equal to $1/\sqrt{3}$ times the line voltage, but in a delta-connected system phase voltage is equal to line voltage, therefore, star-connected alternator needs less number of turns per phase and less insulation than a delta-connected alternator for the same line voltage.

Neutral point of a star-connected system can be connected to earth, so relays and protective devices can be provided in star-connected system for protection of the system against ground faults and it is usually done in practice. These are the reasons that 3-phase alternators are usually star-connected.

Q. 8. Why does star-connected alternator need less number of turns per phase as compared to that required by delta-connected one to provide the same line voltage ?

Ans. In a star-connected system phase voltage is equal to $1/\sqrt{3}$ times the line voltage, but in a delta-connected system phase voltage is equal to line voltage, therefore, star-connected alternator needs less number of turns per phase than a delta-connected alternator for the same line voltage.

Q. 9. Why delta-connected system cannot be employed for supplying lighting as well as power loads ?

Ans. Delta-connected system cannot provide 3-phase 4-wire system (i.e. neutral wire cannot be carried) which is essential for supplying lighting as well as power loads.

Q. 10. Name two electrical equipments in each case having unity power factor and lagging power factor. **[P.T.U. June 2003]**

Ans. Electrical equipments operating

at unity power factor: Incandescent lamps, electric heater.

at lagging power factor: Arc lamps, induction furnaces.

Q. 11. Draw the circuit diagram for 3-phase power measurement by two wattmeters. Under what operating conditions, one of the wattmeters will show a negative reading ? **[P.T.U. December 2001]**

Ans. The circuit diagrams for 3-phase power measurement by two wattmeters are shown in Figs 10.23 (*a*) and 10.23 (*b*). One of the wattmeters will show a negative reading when the power factor of the load circuit is below 0.5.

Q. 12. While measuring power in a balanced three-phase circuit, one of the wattmeters keeps showing zero as the load is varied. How do you account for it ? **[P.T.U. May 2002]**

Ans. One of the two wattmeters, in a two-wattmeter method of measuring power in a balanced 3-phase circuit, will keep showing zero as the load is varied if and only if the power factor of the load remains constant at 0.5. The reading of other wattmeter will vary with the variation in load.

PROBLEMS

1. Three similar coils each of resistance 20 Ω and an inductance of 0.5 H are connected in star across a three-phase supply source of 400 V, 50 Hz. Calculate the line current and power absorbed by the circuit.
[Pb. Technical Univ. Basic Electrical Engineering, December 2002]
[**Ans.** 1.458 A; 127.6 W]

2. A star-connected load has impedance of $(6 + j\,8)$ Ω in each phase and is connected across a balanced 400 V, 3-phase supply. Obtain the line currents, total real power and reactive power consumed by the load.
[R.G.P.V. Bhopal Basic Electrical Engineering December 2003]
[**Ans.** 23.1 A; 9.6 kW, 12.8 kVAR]

3. Given a balanced 3-Φ, 3-wire system with Y-connected load for which line voltage is 230 V and impedance of each phase is $(6 + j\,8)$ Ω. Find the line current and power absorbed. **[Pune Univ. Elec. Engineering II, 1991]**
[**Ans.** 13.28 A; 1,058 W]

4. Three similar coils each of impedance $\mathbf{Z} = (8 + j\,10)\ \Omega$ are connected in star and supplied from 3-phase, 400 V, 50 Hz supply. Find the line current, power factor, power and total volt-amperes. **[Nagpur University Summer 2003]**
[**Ans.** 18.03 A; 0.6247 (lagging); 7.8 kW; 12.5 kVA]

5. A balanced star-connected load of $(8 + j\,6)\ \Omega$ is connected to a 3-phase 230 V supply. Find the line current, power factor, power, reactive volt-amperes and total volt-amperes. **[V.T.U. Belgaum Karnataka Univ., July/August 2003]**
[**Ans.** 13.28 A, 0.8 (lagging), 4.232 kW, 3.174 kVAR, 5.29 kVA]

6. A balanced three-phase, star-connected load of 100 kW takes a leading current of 80 A, when connected across a 3-phase 1100 V, 50 Hz supply. Find the circuit constants of the load per phase. **[Nagpur University April 1996]**
[**Ans.** 7.94 Ω; 5.21 Ω; 530.8 μF]

7. A 3-phase 230 V system supplies a total load of 2,000 W at a line current of 6 A when three identical impedances are in star-connection across the line terminals of the system : Determine the resistive and reactive components of each impedance. **[Nagpur University Winter 2002]**
[**Ans.** 18.52 Ω; 12.12 Ω]

8. A balanced 3-phase star-connected load of 100 kW takes a leading current of 100 A when connected across a 3-phase 1,100 V, 50 Hz supply. Calculate the circuit constants of the load per phase.
[Mumbai University 2003; Gujarat University June/July 2003]
[**Ans.** 3.334 Ω; 589.5 μF]

9. A 3-phase star-connected load when supplied from 440 V, 50 Hz source takes a line current of 12 A lagging w.r.t. line voltage by 70°. Calculate (*i*) impedance parameters (*ii*) power factor and its nature (*iii*) Draw phasor diagram indicating all voltages and currents. **[Nagpur University Summer 2004]**
[**Ans.** (*i*) 16.217 Ω; 13.6 Ω; (*ii*) 0.766 (lagging)]

10. A delta-connected 3-phase load has a resistance of 6 Ω and an inductive reactance of 8 Ω in each branch line. Line voltage is 230 V, 50 Hz. What are the rms values of current and voltage in each branch ? Calculate the total power consumed by the circuit and the power factor. **[Pb. Technical Univ. Electrical Engineering December 2001]**
[**Ans.** 230 V, 23 A, 0.6 (lag), 9.522 kW]

11. Define phase sequence in a 3-phase system.
Three 100 Ω resistors are connected in (*a*) star and (*b*) delta across 440 V, 50 Hz, 3-phase lines : Calculate the line currents and phase currents and power taken from supply man's in each case.
[Pb. Technical Univ. Basic Electrical and Electronics Engineering Second Semester 2004-05]
[**Ans.** (*a*) 2.54 A, 2.54 A, 1,936 W (*b*) 4.4 A, 7.62 A, 5808 W]

12. Three-phase 400 V, 50 Hz ac supply is feeding a three-phase delta-connected load with each phase having a resistance of 25 Ω, an inductance of 0.15 H and a capacitor of 120 μF in series. Determine the line current, volt-amperes, active power and reactive volt-amperes. **[Nagpur University November 1999]**
[**Ans.** 21.4 A, 14.82 kVA, 11.43 kW, 9.42 kVAR (lagging)]

13. Three coils each of impedance $20\angle 60°\ \Omega$ are connected in delta across a 400 V, 3-phase, 50 Hz ac supply. Calculate the line current and total power. **[V.T.U. Belgaum Karnataka Univ. January/February 2003]**
[**Ans.** 34.64 A; 12 kW]

14. A star-connected load having R = 42.6 Ω/phase and X_L = 32 Ω/phase is connected across 400 V, 3-phase supply. Calculate line current, reactive power and power loss (*ii*) line current when one of the load becomes open-circuited.
[Nagpur University, Summer 2001]
[**Ans.** (*i*) 4.334 A, 1.802 kVAR; 2.402 kW, (*iii*) 3.754 A]

15. Three non-inductive resistances each of 100 Ω are connected in star to a 3-phase, 440 V supply. Three inductive coils, each of reactance 100 Ω connected in delta are also connected to the supply. Calculate (*i*) line currents and (*ii*) pf of the system. **[Nagpur University November 1998]**
[**Ans.** (*i*) $8.032\angle -71.57°$ A; 0.3162 (lagging)]

16. In the two-wattmeter method of power measurement in a three-phase, the readings of the wattmeter are 1,000 W and 550 W. What is the power factor of the load ? **[U.P. Technical Univ. Electrical Engineering September 2001]**
[**Ans.** 0.893 (lagging)]

17. For a certain load, one of the wattmeter reads 20 kW and the other 5 kW after the voltage of this wattmeter has been reversed. Calculate the power and power factor of the load. **[U.P. Technical Univ. Electrical Engineering July 2002]**
[**Ans.** 15 kW, 0.3273 (lagging)]

18. The power readings of two wattmeters are + 15 kW and – 4 kW for a 3-phase balanced load. If the supply voltage is balanced 440 V, find the true power drawn by the load, the power factor and line current.
[R.G.P.V. Bhopal Basic Electrical Engineering June 2005]
[**Ans.** 11 kW, 0.317 (lagging), 45.53 A]

19. The power input to a 3-phase circuit was measured by two wattmeter method and the readings were 3400 and – 1200 W respectively. Calculate the total power and power factor. **[V.T.U. Belgaum Karnataka Univ.; July/August 2002]**
[**Ans.** 2200 W; 0.2662 (lagging)]

20. Two wattmeters connected to measure power in a 3-phase circuit measure 5 kW and 1 kW, the latter reading being obtained after reversing current coil connections. Calculate pf of the load and the total power consumed.
[V.T.U. Belgaum Karnataka Univ. January/February 2003]
[**Ans.** 0.3592 (lagging); 4 kW]

21. A 3-phase balanced load power was measured by two wattmeter method. If the readings of the two wattmeters so connected are 5.0 kW and 0.5 kW, the latter reading being obtained after reversal of current coil connections, calculate the power factor of the load. **[Pb. Univ. Elec. Measurement-I Dec; 1992]**
[**Ans.** 0.427 (lagging)]

22. Phase voltage and current of a star-connected inductive load is 150 V and 25 A. Power factor of the load is 0.707 (lag). Assuming that the system is 3-wire and power is measured using two wattmeters , find the readings of wattmeters.
[Nagpur Univ. Elec. Instruments & Measurements 1993]
[**Ans.** 6,247 W, 1,681 W]

23. Each of the two wattmeters connected to measure the input to a three-phase circuit reads 20 kW. What does each instrument reads, when the load pf is 0.866 lagging with the total three phase power remaining unaltered in the altered condition ? **[V.T.U. Belgaum Karnataka Univ. January/February 2003]**
[**Ans.** 26.667 kW; 13.333 kW]

24. Three identical coils, each having a reactance of 20 Ω and resistance of 20 Ω are connected in (*a*) star (*b*) delta across a 440 V, 3-phase line. Calculate for each method of connection the line current and readings on each of two wattmeters connected to measure the power. **[Allahabad University Electro-mechanics 1992]**
[**Ans.** (*a*) 8.98 A; 3,817 W; 1,023 W (*b*) 26.94 A, 11,451; 3,069 W]

25. Three inductive coils each having a resistance of 20 Ω and reactance of 15 Ω are connected in star across a 3-phase, 400 V, 50 Hz supply. Calculate the readings of two wattmeters connected in the circuit to measure the power input.
[R.G.P.V. Bhopal Basic Electrical Engineering February 2005]
[**Ans.** 3.6685 kW, 1.4515 kW]

26. A star-connected balanced 3-balanced 3-phase circuit containing $Z = 20 \angle 30°$ Ω in each branch is connected across 400 V, 3-phase, 50 Hz, 4-wire supply. Calculate
(*i*) current flowing through each line, (*ii*) total power drawn,
(*iii*) readings of two wattmeters connected for the measurement of total power.
[R.G.P.V. Bhopal Basic Electrical Engineering December 2002]
[**Ans.** 11.55 A, 6.93 kW, 4.62 kW, 2.31 kW]

27. A 3-phase, 3-wire, 415 V system supplies a current of 20 A at a power factor of 0.8 to a star-connected balanced load. Two wattmeters are used to measure total power. The current coil of wattmeter 1 is in the phase R and its coil is across R and Y. The current coil of wattmeter 2 is in the phase B and its voltage coil is across B and Y. Draw the relevant connection and vector diagrams, and (*b*) Calculate the reading on wattmeter W_1 and the reading on wattmeter W_2.
[**Ans.** 8,240 W, 3,260 W]

28. Power in a balanced 3-phase system is measured by the two wattmeter method and it is found that the ratio of the two readings is 2 to 1. What is the power factor of the system ? **[Allahabad Univ. Electrical Science-I 1991]**
[**Ans.** 0.866 (lag)]

29. The power input to a 2,000 V, 50 Hz, 3-phase motor running on full load at an efficiency of 90% is measured by two wattmeters which indicate 300 kW and 100 kW respectively. Calculate (*i*) input (*ii*) power factor (*iii*) line current and (*iv*) hp output. [**Ans.** (*i*) 400 kW (*ii*) 0.756 (lag) (*iii*) 152 A (*iv*) 490 (metric)]

30. A 3-phase motor draws a line current of 50 A from 220 V source while starting. The pf is 0.4. Find the readings of the two wattmeters connected to measure power. **[Pb. Univ. Elec. Measurements-I, June 1993]**
[**Ans.** 8.85 kW ; (–) 1.23 kW]

❑❑❑

UNIT-VI
Batteries and Cables

Contents in This Unit

Batteries: Battery; Life of batteries; Charging and discharging of battery.

Cables: 2, 2½, 3 and 4 core armored and unarmored cables.

Chapter 11

Batteries

11.1. INTRODUCTION

As the world increasingly comes to rely on electrical and electronic systems for its daily functions, both vital and mundane, we are ever more dependent on batteries. As electrical and electronic systems become smaller and more efficient, batteries provide the key to portability. And, even with stationary systems, batteries are becoming increasingly important to provide memory backup or keep essential functions operational when the main power is not available.

Not only has the world become more dependent on batteries in general, it has also come to appreciate the economy and reliability of rechargeable batteries. It is only due to rechargeable batteries that travellers can now work at their laptop computer throughout a transcontinental airline flight and then recharge their batteries overnight using only a few rupees worth of electricity. At home, rechargeable appliances kept on their chargers are always fully charged and ready for use. Rechargeable flashlights provide the security of knowing they will respond when needed–eliminating the mystery of when the last time the batteries were charged and guessing at how much they have been used since then.

Now, improvements in battery technology are opening many new applications for batteries. The rapid development of markets such as portable electronics and cordless appliances and power tools have relied heavily on today's improved batteries. In a diversity of applications, creative use of batteries is translating directly into improved sales and increased profits. But, creative use of batteries is aided greatly by knowing how batteries operate.

Interest in new and improved batteries remains strong today. The demand for versatile, clean, high-power energy sources grows as electronics becomes an increasingly essential part of both consumer and industrial products. To date, the results of the battery industry's development efforts have been most evident in the dramatic improvements in existing battery types. However, the newly developed batteries, such as nickel-hydrogen cells, silver zinc batteries, nickel-metal hydride batteries, using different materials and technology are, exhibiting the greatest application in commercial products.

11.1.1. Primary Vs Secondary Batteries. Batteries are either primary or secondary. Primary batteries can be used only once because the chemical reactions that supply the current are irreversible. Secondary batteries, sometimes called *storage batteries* or *accumulators*, can be used, recharged and reused. In these batteries, the chemical reactions that provide current from the battery are readily reversed when current is supplied to the battery. The process of inducing or storing energy in an accumulator is called the *charging*, and the process of giving out energy in the form of an electric current, the *discharging*. Accumulators or storage batteries owe their name "secondary" due to the fact that they can supply electrical energy only after they have been charged.

Primary batteries are the most common batteries available today because of their low cost and simplicity in use. Carbon-zinc dry cells and alkaline cells dominate portable consumer battery applications where currents are low and usage is sporadic. Other primary batteries, such as those using mercury or lithium-based chemistries, may be used in applications when high energy densities, small sizes, or long shelf life are especially important. In general, primary batteries have dominated two areas: consumer products where the initial cost of battery is very important and electronic products (such as watches, hearing aids and pacemakers) where drains are low or recharging is not feasible.

Secondary batteries, which are rechargeable, have traditionally been most widely used in industrial and automotive applications. Here users are willing to trade higher initial cost and additional handling and care requirements for high current delivery and the economies of a rechargeable product. Only two rechargeable battery chemistries, lead acid and nickel-cadmium, have, to-date, achieved significant commercial success. The recently introduced nickel-metal hydride couple currently shows promise of supplementing nickel-cadmium cells in many commercial applications. Other couples, notably nickel-hydrogen and silver-zinc, are in use in special applications where performance requirements are more important than cost.

Recently, the battery market has changed significantly with consumers indicating that, for many products, they are willing to pay the initial premium necessary to have the long-term benefits of rechargeable batteries. Wet (flooded) versions of rechargeable batteries had achieved only limited penetration of the consumer and portable markets. Development of sealed high-performance forms of both lead-acid and nickel-cadmium batteries has allowed secondary batteries to make substantial inroads into traditional primary battery markets such as consumer products.

Recent improvements in secondary battery technology have improved performance and reduced costs. These improvements have spurred sustained growth in applications, both consumer and industrial, relying on rechargeable battery power.

11.1.2. Batteries Vs Cells. An electric cell is a device in which difference of electric potential is established between the two electrodes as a result of chemical reactions between the electrodes and the electrolyte. Single unit is known as a *cell* whereas a combination of several such units, connected electrically (usually in series) is known as a *battery*. In many batteries, such as those used in automobiles, all intercell connections are made internally. The number of cells within the battery may not even be apparent from the finished package.

Cell voltage is determined by the electrochemistry involved. Nickel-cadmium cells nominally develop about 1.2 V per cell while lead-acid batteries develop about 2 V per cell. Battery voltages then must be multiples of basic unit. For example, nickel-cadmium battery voltages may be 1.2, 2.4, or 3.6 volts, but not 2.8 or 4.2 volts. The voltage obtained on discharge or the voltage required on charge is usually just the appropriate single cell voltage multiplied by the number of cells in the battery. For example a 6-cell lead-acid battery will develop about 12 V.

Often in discussions, the term battery and cell are used interchangeably. The meaning is normally apparent from the context.

11.2. CLASSIFICATION OF SECONDARY BATTERIES

The secondary batteries, on the basis of their uses, may be classified as (*i*) automotive (or SLI/portable) batteries (*ii*) motive-power (or vehicle traction/industrial) batteries and (*iii*) stationary batteries.

Automotive batteries, such as lead-acid batteries, nickel cadmium batteries, are employed for starting, lighting and ignition (SLI) in internal combustion (IC)-engined vehicles.

Motive-power batteries, such as lead-acid batteries, nickel-iron batteries, silver-zinc batteries, are employed as a motive power source for various types of vehicles.

Stationary batteries, such as lead-acid accumulators, are employed in power stations as (*i*) standby power system and (*ii*) load-levelling system. In the former case the batteries are employed intermittently whereas in the latter case the batteries store energy during light load hours and supply energy during peak load hours to meet the peak demand.

11.3. LEAD-ACID BATTERIES

11.3.1. Classification. Lead-acid batteries, according to service rendered by them, are classified into automotive, motive-power and stationary batteries.

Lead-acid automotive batteries are used to supply power for starting, lighting and ignition of IC engines employed to propel auto-vehicles (motor cycles, motor cars, buses, trucks etc.). Normally these batteries consist of 6 cells, connected in series, with capacity of the order of 100 ampere-hours and provide power at 12 V.

Nowadays maintenance free (M F) batteries are available for this purpose. Such batteries are constructed of such material that no gassing occurs during charging, the electrolyte is either absorbed within the microporous separators and the plates or immobilized with suitable gelling agents and addition of water is not required throughout their normal service life of 2–5 years.

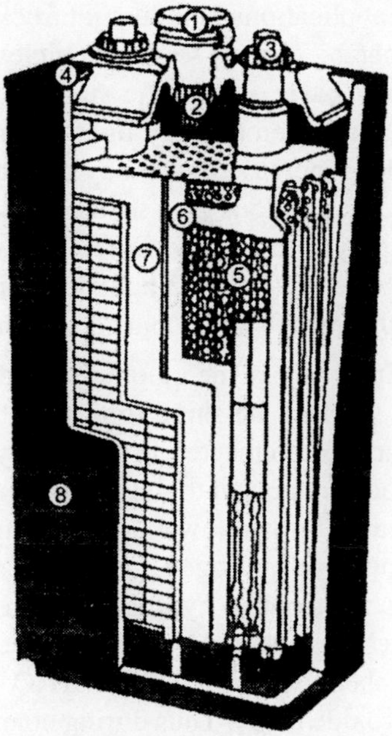

(a) Constructional Features of Lead-Acid SLI Battery

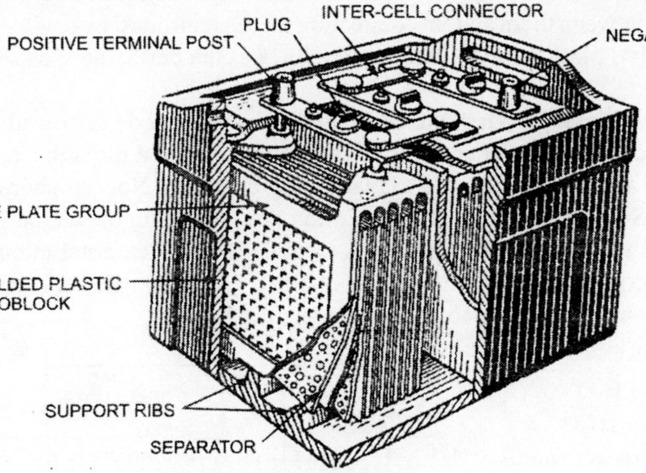

(b) Exploded View of Exide HSP Classic Motive Power Battery

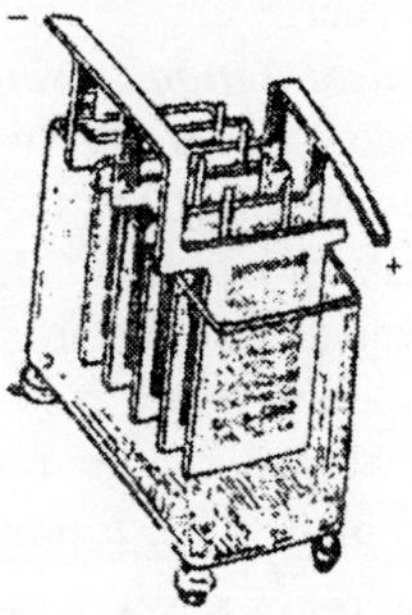

(c) Lead-acid Stationary Accumulator

Fig. 11.1

Automotive batteries are charged from ac supply (or alternator) through a full-wave bridge rectifier employing semiconductor diodes. The alternator output voltage is controlled by the transistorised voltage controller (not by cut-out) to meet the requirements of batteries under charge. The batteries are charged at constant voltage.

Lead-acid automotive (or SLI) battery is shown in Fig. 11.1 (*a*).

Lead-acid motive-power batteries are of better quality in comparison to automotive batteries and provide constant output voltage, high volumetric capacity, good resistance to vibration and a long service life (1,000–1,500 cycles). Such batteries are capable to withstand prolonged and deep discharges and deep recharges on a daily basis. These batteries supply power at voltage ranging from 12 V to 240 V.

These batteries find applications in industrial trucks and commercial vehicles of various types plying on roads, mining, airport tractors, aircraft service vehicles, electric cars and in robotics and guide vehicles.

Lead-acid stationary batteries, in the standby applications, are used to supply power to essential services or to provide alarms or emergency lighting, in case of breakdown of main power supply. In recent years, there is a tremendous increase in standby applications due to increase in demand for uninterruptable power systems (UPS) and tremendous growth in new telecommunication networks. Banks of sealed lead-acid standby batteries are employed these days in telecommunication systems and for UPS applications.

Advanced lead-acid batteries (100 MWh capacity) have been developed for use in electric power generating stations to store energy during off-hours and supply the stored energy during peak-load hours.

11.3.2. Working Principle. When the sulphuric acid is dissolved, its molecules break up into hydrogen positive ions ($2H^+$) and sulphate negative ions (SO_4^{--}) and move freely. Now if two lead electrodes are immersed in this solution and connected to dc supply mains, the hydrogen ions being positively charged move towards the electrode connected to the negative terminal of the supply mains (*i.e.* cathode) and SO_4^{--} ions being negatively charged move towards the electrode connected to the positive terminal of the supply main (*i.e.* anode). Each positive hydrogen ion (H^+) reaching the cathode gets one electron from it, becomes atom of hydrogen and escapes out as hydrogen gas in the form of bubbles. Each of the negatively charged sulphate ion (SO_4^{--}) reaching the anode gives two electrons to it, reacts with water and forms sulphuric acid and oxygen according to chemical reaction $SO_4 + H_2O \longrightarrow H_2SO_4 + O$. The oxygen produced attacks the lead anode and forms lead peroxide PbO_2. Thus during charging lead cathode remains as lead but lead anode gets converted into lead peroxide (PbO_2), chocolate in colour.

Now if the cell is disconnected from dc source of supply and voltmeter is connected between the electrodes, it will show potential difference between them and on connecting these electrodes, by a wire, current will flow from positive plate to the negative plate through external circuit, *i.e.* the cell is now capable of supplying electrical energy.

Chemical Action During Discharging. When the cell is fully charged anode is of lead peroxide (PbO_2) and cathode is of sponge metallic lead (Pb). When the electrodes are connected through a resistance, the cell discharges and electrons flow in a direction opposite to that during charging. Now sulphuric acid molecules again break up into ($2H^+$) and (SO_4^{--}) ions. Each hydrogen ion now moves to the anode and reaching the anode receives one electron from the anode (coming from cathode through external circuit) and becomes hydrogen atom. Since it is directly in contact with anode of PbO_2, so it attacks and forms lead sulphate ($PbSO_4$), whitish in colour and water according to chemical equation

$$PbO_2 + 2H = PbO + H_2O$$
$$PbO + H_2SO_4 = PbSO_4 + H_2O$$
$$\overline{PbO_2 + H_2SO_4 + 2H = PbSO_4 + 2H_2O}$$

Each sulphate ion (SO_4^{--}) moves towards the cathode and reaching there gives up two electrons (moving to the anode through external circuit) becomes radical SO_4, attacks the metallic lead cathode and forms lead sulphate, whitish in colour according to chemical equation $Pb + SO_4 = PbSO_4$.

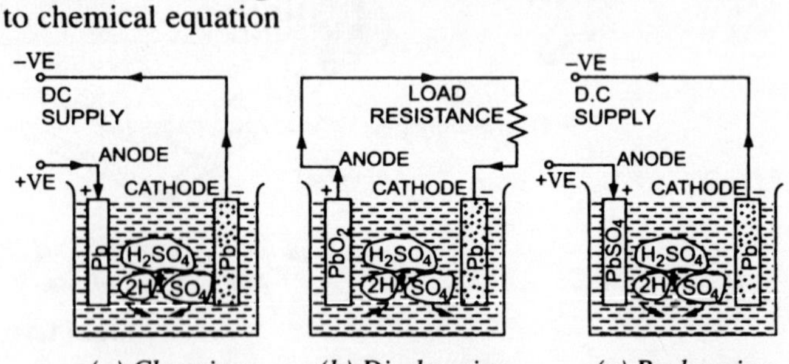

(a) Charging *(b) Discharging* *(c) Recharging*

Fig. 11.2 *Charging and Discharging of Lead Acid Cells*

Thus during discharging (*i*) both of the electrodes are converted into lead sulphate, whitish in colour (*ii*) terminal potential of the cell falls (*iii*) density of the electrolyte decreases due to formation of water during chemical action and (*iv*) chemical energy stored is converted into electrical energy and is supplied to an external circuit.

Chemical Action During Recharging. When the cell gets discharged as explained above both of the electrodes are converted into lead sulphate ($PbSO_4$), whitish in colour. On recharging of the cell, the anode and cathode are connected to the positive and negative terminals, of the dc supply mains respectively. The molecules of sulphuric acid (H_2SO_4) in solution again break up into ($2H^+$) and (SO_4^{--}) ions. Hydrogen ions ($2H^+$) being positively charged move to the cathode, receives two electrons from there (coming from anode through external circuit), form hydrogen atom and react with lead sulphate cathode forming lead and sulphuric acid according to chemical equation,

$$PbSO_4 + 2H = H_2SO_4 + Pb$$

SO_4^{--} ion moves to the anode, gives up its two additional electrons, (moving to the cathode through external circuit) becomes radical SO_4, reacts with lead sulphate anode and forms lead peroxide (PbO_2) and sulphuric acid according to chemical equation

$$PbSO_4 + 2H_2O + SO_4 = PbO_2 + 2H_2SO_4$$

Hence during recharging

(*i*) Lead sulphate anode gets converted into lead per-oxide (PbO_2) dark chocolate brown in colour.

(*ii*) Lead sulphate cathode gets converted into lead, grey in colour.

(*iii*) Terminal potential of the cell increases.

(*iv*) Density (concentration) of the electrolyte (H_2SO_4) increases (due to formation of H_2SO_4 during chemical reaction).

(*v*) Electrical energy received from the external source is stored in the form of chemical energy.

The charging and discharging of cell can be represented by a single reversible equation given below :

$$\underset{\text{Positive plate}}{PbO_2} + 2H_2SO_4 + \underset{\text{Negative plate}}{Pb} \underset{\text{Discharge}}{\overset{\text{Charge}}{\rightleftharpoons}} \underset{\text{Positive plate}}{PbSO_4} + 2H_2O + \underset{\text{Negative plate}}{PbSO_4} + \text{Electrical energy}$$

The equation should be read from left to right for discharge and from right to left for recharge.

11.3.3. Active Materials. The materials, in a cell (or battery), taking active participation in chemical reaction (absorption or evolution of electrical energy) during charging or discharging are called the *active materials* of the cell.

The active materials of a lead-acid battery are

1. **Lead peroxide** (PbO_2) dark chocolate brown in colour. It forms the positive active material.
2. **Sponge lead** (Pb) grey in colour. It forms the negative active material.
3. **Dilute sulphuric acid** (H_2SO_4) is used as electrolyte. The electrolyte of a fully charged battery contains about 31% sulphuric acid by weight and 21% by volume for specific gravity of 1.23 at 70 °C. This is the recommended value for the tropical climate. Slightly higher values can be used in winter, if desired. The acid should be made from sulphur and not from pyrites, as the latter is liable to contain injurious substances.

The mixture should be made by pouring the acid slowly into the water, *never the reverse*.

The lead peroxide and sponge lead, which form the positive and negative active materials respectively of a fully charged cell, as mentioned above, have little mechanical strength and, therefore, cannot be used alone. Supports of grids required for this purpose, in modern form, are made of either pure lead or an alloy of lead and antimony.

11.3.4. Construction. The various parts of a lead-acid battery are shown in Fig. 11.3.

1. Containers. The materials of which the battery containers are made should be resistant to sulphuric acid, should not deform or become porous, or contain impurities deterious to the electrolyte; of these iron and manganese are especially intolerable.

The containers of lead-acid cells are made of glass, lead lined wood, ebonite, hard rubber of bituminous compound, ceramic materials or moulded plastics and are sealed at the top (with three holes, one at each end for the posts and one in the middle for the vent plug and through which the electrolyte is poured and the gases escape out) to avoid spilling of electrolyte. At the bottom of the container there are four ribs (prisms), on two of them rest the positive plates, and on the other two the negative plates. These prisms serve as supports for the plates and at the same time protect them from short-circuits that would otherwise occur as a result of fall of the active material from the plates on to the bottom of the container.

2. Plates. The lead per oxide and spongy lead, which form the positive and negative active materials respectively of a fully charged cell have little mechanical strength and, therefore, cannot be used alone. Supports of grids required for this purpose, in modern form, are made of either pure lead or an alloy of lead and antimony.

There are two ways of forming the plates.

(*a*) *Formed Plates or Plante Process.* In this method the plates are formed electro-chemically (by repeatedly charging and discharging) out of the lead plates itself. The plates are grooved or made in the form of grill so as to have a large exposed surface. Plante plates are used largely for stationary batteries as these are heavier in weight and more costly than pasted plates but are more durable and less liable to lose active material by rapid charging and discharging.

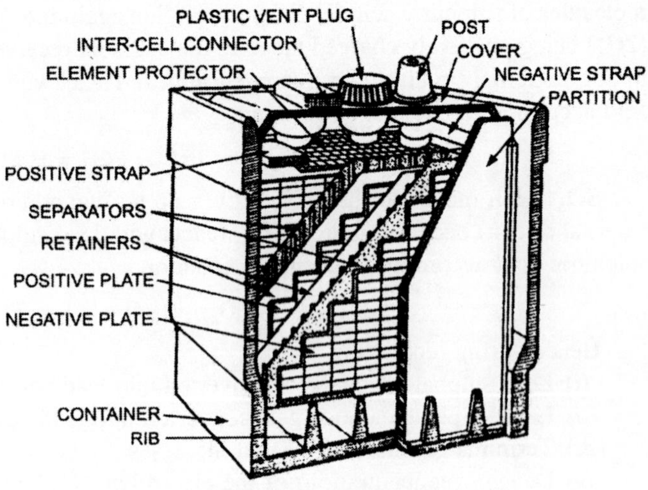

Fig. 11.3 *Lead Acid Battery*

(*b*) *Pasted or Faure Plates.* These plates are formed by spreading a paste of the active material on to a sporting grid of lead hardened with a small quantity of antimony. These plates being lighter in weight and cheaper in cost are usually employed for automobile and motor truck service.

It is customary to make positive plates of the plante type and negative plates of the Faure type.

In order to obtain large capacity from a battery, a large surface must be exposed to the electrolyte, and since the size of a single plate is limited so increased capacity is obtained by connecting a number of plates in parallel to form a group. Two sets of plates are then sandwiched together, adjoining plates being separated from one another by perforated plastic, fiberglass, hard rubber or wooden separators. The –ve group of the plates contains one more plate than that of the +ve group so that when interleaved both the end plates become the –ve. This is done so that all the positive plates can work equally on both sides. During charging lead sulphate ($PbSO_4$) on the +ve plate is changed to PbO_2 and the active material of the plate expands. If a positive plate were used at one end, one of its sides would remain inactive, when the other side expands and thus the plate is forced to buckle. The arrangement of plates in a cell is illustrated in Fig. 11.4.

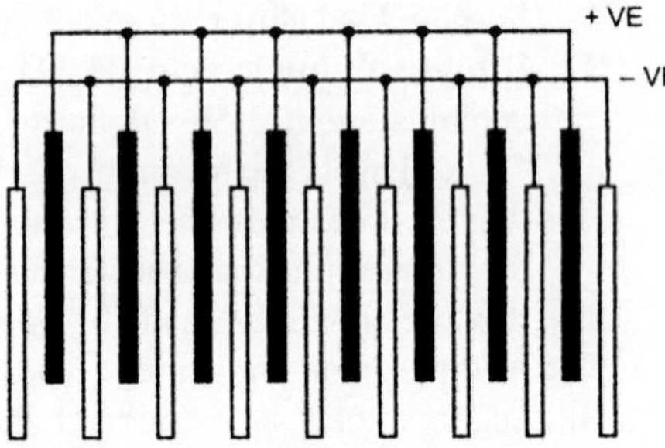

Fig. 11.4

3. Separators. The separators, which are thin sheets of non-conducting porous material made of chemically treated wood, porous rubber or mats of glass fibres, are placed between the positive and negative plates to insulate them from each other. Separators are grooved vertically on one side and are smooth on the other. The grooved side is placed facing the positive plates to provide greater acid volume next to the surface for improved efficiency, to facilitate acid circulation within the cell, and to minimise the area of contact with the positive plate which has a highly oxidising effect on wood separators.

4. Battery Terminals. A battery has two terminals — the positive and the negative. The positive terminal with a diameter of 17.5 mm at the top is slightly larger than the negative terminal which is 16 mm in diameter.

11.3.5. Battery Charging. The batteries are prepared for charging as follows :

The batteries are wiped dry, the terminal posts and cable connectors are cleaned of oxides, and the plugs in the cell covers are removed. Any protective hood provided for the terminals, and the cover (if any) are removed. The level of the electrolyte in the cells is checked and, if necessary, the cell is topped up with distilled water. The sealing washers are removed from under the plugs, and the vent tubes are removed from the vent holes. The vent holes are cleaned. The batteries are placed on the charging rack and connected in series or in parallel or in series-parallel as per needs.

For initial charging the new battery is filled with distilled water. It is kept 2 to 3 hours so that plates are soaked in water. Then water is removed and the battery is filled with the electolyte made up according to the given climatic conditions. It is kept for 24 hours. Now the battery is charged at a low rate *i.e.* rd of the rated charging rate, the battery is charged 20% more than the rated ampere-hour capacity of the battery. Then the battery is discharged at a higher rate *i.e.* 2 to 3 times the rated charging rate. The battery is then kept idle for 2 to 3 hours. The battery is then charged at normal rate *i.e.* from 3 to 10 A depending upon the capacity of the battery and also the time in which the battery is to be charged.

Charging Methods. Direct current is essential, and this may be obtained in some cases direct from the supply mains. In case the available source of supply is ac then it is converted into dc by some means such as motor-generator set, rotary convertor set or rectifier.

1. Normal Voltage DC Supply Method. An arrangement for charging the secondary batteries from d c supply mains at normal voltage is shown in Fig. 11.5. The cells to be charged are connected in series or in parallel or in series-parallel according to the requirements, but in all cases it is best to use the simplest arrangement practicable. In Fig. 11.5 all the cells connected in series are shown.

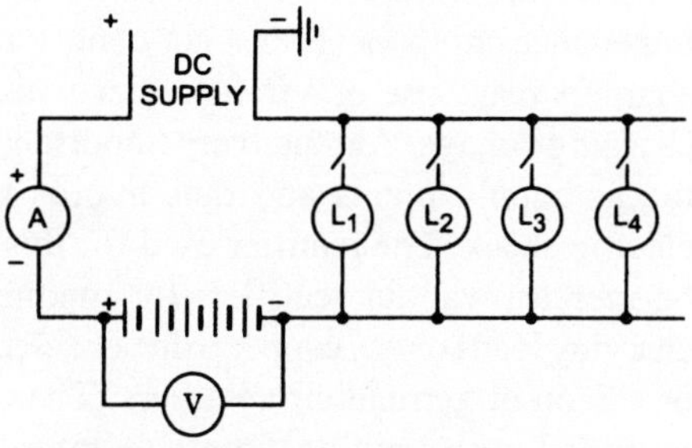

Fig. 11.5 *Charging of Batteries*

The advantage of this method is that it requires no apparatus except the lamps of variable resistance, but it has a great disadvantage of low efficiency because of the wastage of considerable amount of energy in the lamps or in controlling resistance.

The positive terminal of the battery is connected to the positive terminal of dc supply mains and negative one with negative terminal of dc supply mains. Voltmeter is connected to measure potential difference across the batteries connected in series for charging, L_1, L_2, L_3 and L_4 are the high wattage incandescent lamps connected in parallel and are used as resistance to control the current drawn from the supply mains, which is known as *charging current*. Higher the number of lamps in the circuit, higher will be the charging current. The potential across the batteries shown by voltmeter connected for this purpose must be about 5% higher than the sum of normal terminal voltages of the batteries connected in series for charging and charging current should be such that temperature of the electrolyte may not increase beyond 40 °C to 45 °C. Usually the charging current* is marked by the manufacturer so its value should be adjusted accordingly. The ammeter is connected in series with the batteries to indicate the charging current.

The charging is continued until gases are freely evolved at the electrodes indicating that batteries have been fully charged. The cell should not be continued charging for longer period after the gases start coming out, because in doing so active material is losened from the plates and gets deposited at the bottom. Hence the cells should not be overcharged. But if the cells remain undercharged, sulphations formed on the plates are never completely removed which reduces the capacity of the batteries, so the batteries should never be left undercharged.

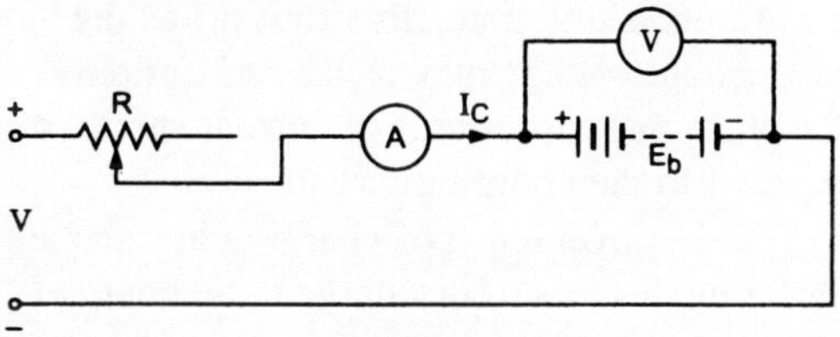

Fig. 11.6

Note: If the charging is done too rapidly, by using a current higher than that marked by the manufacturer, there is a risk of making the positive plates crumble and buckle up. It is safe, however, to charge at a smaller current than rated one.

Alternatively the circuit shown in Fig. 11.6 may be used for charging the secondary batteries from dc supply mains.

The charging current is given by the expression

$$I_c = \frac{V - E_b}{R + r}$$

where V is supply voltage, E_b is battery voltage, R is current limiting resistance and r is the internal resistance of the battery.

If the source voltage is only slightly greater than the battery voltage, small source voltage or battery voltage variations will cause large variations in charging current. It is, therefore, better, whenever possible, to select a higher source voltage and a large resistance. This provides a more uniform charging current flow despite major changes in the battery voltage.

2. Charging of Batteries From AC Power Source. The basic requirements of common ac source chargers are like those of the dc power source, namely, the source voltage must be significantly greater than the battery voltage and the impedance must limit the current. The impedance of an ac power source can be reactive, resistive or combination of these. Limiting the current by reactance has the advantage that the reactive impedance component does not contribute to rectification. Rectification is used to convert ac into dc. Most chargers make use of a transformer which permits optimized matching of the source voltage to the battery charging voltage. Another very important feature of the transformer is that it isolates the charging output from the ac supply mains and thus avoids the hazard of electric shock. The rectifier used for this purpose may be mercury vapour rectifier (for medium and heavy charging loads only), copper oxide or selenium rectifiers or silicon or germanium rectifiers. The unit is usually provided with tapping points to give different low voltages *e.g.* 6, 12, 18, 24 V. This is most economical and simple method of charging batteries. The circuit diagram is shown in Fig. 11.7.

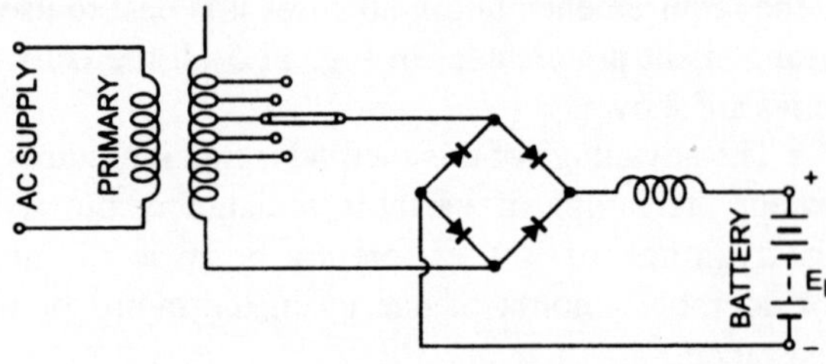

Fig. 11.7 *Full-wave Bridge Charging Circuit*

DC supply for battery charging from ac supply source can also be made available by employing a dc shunt wound generator coupled to an ac motor. In this method of charging batteries, the charging current can be controlled simply by regulating the generator excitation. In the event of failure of supply to the motor during charging, say because of blown fuses, the generator would continue to run as a shunt motor in the same direction of rotation, with no damage to battery.

Types of Charging. There are mainly two types of charging namely *constant voltage charging* and *constant current charging.*

(*a*) *Constant Voltage Charging.* In this method, the charging voltage is kept constant throughout the charging process. In this method the charging current is high in the beginning when a battery is in discharged condition, and it gradually drops off as the battery picks up charge resulting in increased back emf. Charging at constant voltage may be carried out only when the batteries have the same voltage, for example, 6, 12 or 24 V. In this case source of current should have a voltage of 7.5, 15 or 30 V; these batteries are connected in parallel to the charging circuit.

The convenience of charging at constant voltage is that it allows cells with different capacities and at different degrees of discharge to be charged. The large charging current at the beginning of the charge is of relatively short duration and will not harm the cells. At the end of the charge the charging current drops to almost zero because the voltage of the battery becomes nearly equal to the voltage of the supply circuit. This method, is however, not very suitable for old, badly sulphated batteries which need prolonged charging at a slow rate.

This method is the most common method of charging lead-acid batteries and has been used successfully for over 50 years for different types of lead-acid batteries. With this method of charging, the charging time is almost reduced to half, capacity is increased by approximately 20% but efficiency is reduced by approximately 10%.

(*b*) *Constant Current Charging.* In this method of charging of batteries, the batteries are connected in series so as to form groups and each group is charged from the dc supply mains through loading rheostats. The number of batteries in each group depends on the charging circuit voltage which should not be less than 2.7 V per cell. The charging current is kept constant throughout the charging period by reducing the resistance in the circuit as the battery voltage goes up. This method is usually employed for initial charging of lead-acid batteries and for charging portable batteries in general.

In order to avoid excessive gassing or overheating, the charging may also be carried out in two steps, an initial charging of comparatively higher current and a finishing rate of low current. In this method the charge current is kept one-eighth of its ampere-hour rating.

The excess voltage of the supply circuit is absorbed in the series resistance. However, the groups of batteries to be charged should be so connected that the series resistance absorbs as little energy as possible. The series resistance used should be of current carrying capacity equal to or greater than the charging current required, otherwise the series resistance will overheat and burn out. The groups of batteries should be selected so that they all have the same capacity. If the batteries have the different capacities the charging current will have to be set according to the battery having the least capacity. This would both retard and complicate the charging.

Precautions During Charging. (*i*) During the charging period the temperature of the electrolyte should not exceed beyond 40 to 45 °C because of the danger of plate buckling. Therefore, the electrolyte temperature is carefully watched and if the temperature approaches the danger limit before the end of the charge, the charging current should be reduced or the charging discontinued until the batteries cool down.

(*ii*) While the charging is in process the fans in the battery room should be kept operating continuously. This is specially important during the last period of the charging in order to exhaust from the premises the dangerous oxygen and hydrogen gas evolved by the batteries during charging.

(*iii*) The gases, hydrogen and oxygen evolving from the batteries under charge can explode if spark or flame is brought too near. No metal tool should be used on batteries without switching off the battery charger.

(*iv*) Before the end of the charging it is necessary to examine all the batteries and make sure that gassing begins simultaneously in all the cells and with the same intensity. It is also necessary to measure the density of the electrolyte and the voltage of the battery cells.

Example 11.1

A storage battery of 15 cells is to be charged from a 40 volt supply at constant rate of 5 A. The emf of each cell is 1.2 V at the beginning of charge and the internal resistance of each cell is 0.03 Ω. Calculate the series resistance required in the circuit at the beginning of the charging.

Solution:

Supply voltage, V = 40 volts

Battery emf, $E_b = 15 \times 1.2 = 18$ V

Internal resistance of the battery $= n \times r = 15 \times 0.03 = 0.45\ \Omega$

Charging current at start of charging, $I_c = 5$ A

Total resistance in the circuit $= \dfrac{V - E_b}{I_c}$

$$\text{or } R + nr = \frac{40 - 18}{5} = 4.4\ \Omega$$

or $R = 4.4 - 0.45 = 3.95\ \Omega$ **Ans.**

Example 11.2

A battery of 60 cells is charged from a supply of 250 V. Each cell has an emf of 2 V at the start of charge and 2.5 V at the end. If the internal resistance of each cell is 0.1 Ω and if there is an external resistance of 19 Ω in the circuit, calculate (*a*) the initial charging current; (*b*) the final charging current; and (*c*) the additional resistance which must be added to give a finishing charge of 2 A rate.

Solution:

Supply voltage, $V = 250$ V

EMF of battery at the start of charging, $E_{b1} = 60 \times 2 = 120$ V

EMF of battery at the end of the charging, $E_{b2} = 60 \times 2.5 = 150$ V

Internal resistance of battery, $r = 60 \times 0.1 = 6\ \Omega$

External resistance, $R = 19\ \Omega$

(*a*) Initial charging current, $I_{C1} = \frac{V - E_{b1}}{R + r} = \frac{250 - 120}{19 + 6} = 5.2$ A **Ans.**

(*b*) Final charging current, $I_{C2} = \frac{V - E_{b2}}{R + r} = \frac{250 - 150}{19 + 6} = 4$ A **Ans.**

(*c*) Total resistance in the circuit to give charging current of 2 A at the end of charging,

$$R' + R + r = \frac{V - E_{b2}}{I'_{C2}} = \frac{250 - 150}{2} = 50\ \Omega$$

or additional resistance required, $R' = 50 - (R + r) = 50 - (19 + 6) = 25\ \Omega$ **Ans.**

Example 11.3

A storage battery consists of 55 cells in series. Each has an emf of 2 V and an internal resistance of 0.001 Ω. The full load current per cell is 0.04 A/cm² of positive plate surface and each cell has 10 positive plates each 12 cm square. Find the full load terminal voltage and the power wasted in the battery if the connectors have a total resistance of 0.02 Ω. [*Pb. Univ. Applied Electricity 1973 Supp*]

Solution: Since in batteries negative group of plates contains one more plate than that of positive group so that when interleaved both the end plates become the negative and all the positive plates work equally on both sides.

∴ Total area (both sides) of ten positive plates $= 2 \times 10 \times 12 \times 12 = 2{,}880$ cm²

Full load current, $I = 0.04 \times 2{,}880 = 115.2$ A

Battery emf, $E_b = 55 \times 2 = 110$ V

Voltage drop in battery and across connections $= I\,[r + r'] = 115.2\,[55 \times 0.001 + 0.02] = 8.64$ V

Battery terminal voltage on full load, $V = E - I\,(r + r') = 110 - 8.64 = 101.36$ V **Ans.**

Power wasted in the battery $= I^2\,(r + r') = (115.2)^2 \times 0.075 = 995$ watts **Ans.**

11.3.6. Charging and Discharging Curves. Typical charge and discharge curves (variations in terminal voltage) of a lead-acid accumulator are shown in Fig. 11.8. When the cell is charged, the voltage of the cell increases from 1.8 V to 2.2 V during first two hours, then increases very slowly, rather remains almost constant for sufficient time and finally rises to 2.5 to 2.7 V.

When a charged storage cell has just been disconnected from a charging source , its terminal voltage falls rapidly to 2.2 V. On discharge, the voltage of the cell drops to 2.0 V in the beginning; remains constant for sufficient time and falls to 1.8 V finally, as illustrated in Fig. 11.8.

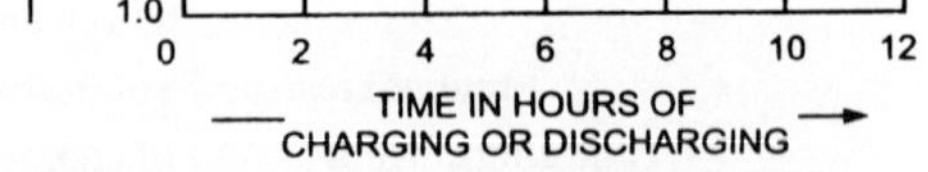

Fig. 11.8 *Typical Charge and Discharge Curves For Lead-Acid Accumulator*

Caution. The cell should never be allowed to discharge beyond 1.75 V otherwise lead sulphate will be formed on the electrodes which is hard, insoluble and increases the internal resistance of the cell. The conversion of active material into lead sulphate is termed *sulphatization.*

11.3.7. Charging Indications. Full charging of lead-acid accumulator (or cells) can be judged from the following indications :

1. *Gassing*. When the cell is fully charged, the hydrogen and oxygen gases are liberated at the cathode and anode respectively, so liberation of gases (hydrogen and oxygen), known as *gassing*, on the electrodes indicates that the cells are fully charged.

2. *Colour*. When the cell is fully charged, the lead sulphate anode gets converted into lead per oxide (PbO_2) dark chocolate brown in colour and lead sulphate cathode gets converted into lead (Pb), grey in colour. It is considered one of the best tests for ascertaining the condition of a battery.

3. *Voltage*. When the cell is fully charged its terminal potential will be approximately 2.6 volts.

4. *Specific Gravity of The Electrolyte*. When the cell is fully charged, the specific gravity of the electrolyte will be approximately 1.21. When the cell is fully discharged its value falls to 1.17.

Cells are considered to be fully charged once three successive hourly readings of cell voltage and electrolyte gravity are found to be constant. However, the minimum total ampere-hours input, as mentioned by the manufacturers, must be provided to the cells even if the voltages and specific gravities are observed to be constant before that.

11.3.8. Electrical Characteristics. There are three important characteristics of an accumulator (or storage battery) namely (*i*) voltage (*ii*) capacity and (*iii*) efficiency.

(*i*) *Voltage.* Average emf of cell is approximately 2.0 volts. The value of emf of a cell does not remain constant but varies with the change in specific gravity of electrolyte, temperature and the length of time since it was last charged.

The emf of the cell increases with the increase in specific gravity of the electrolyte and vice-versa but increase in specific gravity of the electrolyte also causes increase in internal resistance of the cell, therefore, its value should not go beyond 1.22. Best results are obtained with the electrolyte of specific gravity 1.21.

The emf of the cell, though not much, but slightly increases with the increase in temperature.

The terminal voltage of a battery is higher during charge than that during discharge due to the following reasons.

The internal voltage developed by chemical action, depends on the strength of electrolyte and increases slightly as the acid becomes stronger and the concentration of electrolyte increases due to formation of H_2SO_4 during charging and decreases due to formation of water during discharging. Moreover, since acid is formed in the pores of active material during charging and water is formed during discharging, and since it takes time for the acid or water to diffuse out, it follows that the strength of the electrolyte that is in actual contact with the active material is considerably greater during charging than the average strength of the electrolyte (acid) while during discharging it is considerably less than the average. Therefore, emf of the cell is greater during charging than that during discharging.

The terminal voltage of a battery is equal to $E + Ir$ while charging and to $E - Ir$ while discharging where I is the charging or discharging current and r is the internal resistance of the cell.

(*ii*) *Capacity.* The ability of an accumulator to last and give current is called *rated output* or the *capacity*. While the voltage of a cell is determined by its chemistry, the capacity of a cell is infinitely variable.

The capacity of a cell is essentially the number of electrons that can be obtained from it. Since the current is the number of electrons per unit time, cell capacity is the integration of current supplied by the cell over time. The capacity of the cell is, therefore, expressed in *ampere-hours* (A -h) and is equal to the product of the specified discharge current in amperes multiplied by the number of hours before the cell discharges to the specified extent. Thus, a rated output (or capacity) of 10 ampere-hours means that one ampere current can be drawn for 10 hours, or half an ampere current for 20 hours.

Thus the capacity of a battery may be defined as the useful quantity of electricity that can be drawn from a battery at the specified discharge rate before it falls to the specified value of voltage, which is equal to 1.75 V multiplied by the number of cells. The capacity of battery depends upon several factors, principal among which are area of plate surface; quantity, arrangement and porosity of the active material used in the manufacture of the plates; quantity and specific gravity of the electrolyte used; and the porosity of the separators. Rate of discharge and temperature also play important role.

The capacity of the cell increases with the increase in *plate surface area.* A rough rule for estimating the capacity of a battery is the surface area of positive plates in mm^2 multiplied by the number of such plates and divided by 1,000. For example the capacity of a battery having 5 positive plates each of dimensions of 100 mm and 50 mm, will be $\frac{100 \times 50 \times 5}{1{,}000}$ *i.e.* 25 Ah.

Since electricity is produced by chemical action taking place within the cells, the capacity of the battery depends directly upon the *kind and amount of active material used.* Theoretically, roughly 4 grams of metallic lead on either element is required to be reduced to sponge lead or to lead peroxide to produce one ampere-hour of electricity. In practice, from four to six times this amount is required. The reason for this is that it is impossible to reduce all the active material, to bring every particle in contact with the electrolyte, or to cause every part to be penetrated by the current. Experiment shows that from 15 to 22 grams of sponge lead, and from 16 to 24 grams of metallic lead converted into PbO_2, are required on their respective elements to produce a discharge of one ampere-hour at ordinary commercial rates.

The capacity of the cell depends on the *concentration* or the *specific gravity* of the electrolyte as it affects the internal resistance and vigourisity of chemical reaction taking place in a cell. It increases with the increase in specific gravity of the electrolyte.

At a particular temperature, the capacity of the cell depends on its rate of discharging. For instance, a 100 ampere-hour battery capable of giving a continuous discharge of 10 A for 10 hours should theoretically give a discharge of 20 A continuously for 5 hours or 50 A for 2 hours or 100 A for one hour, but in reality, the *ampere-hour capacity decreases with the increase in discharging rate.* With the increase in the rate of discharge, voltage of the cell falls more rapidly due to internal resistance of the cell, the chemical reactions become brisk and so weaken the plates and reduce the capacity of the cell. If the cell is discharged too rapidly it may break the plates, and in case of pasted plates, a very sudden discharge will dislodge the paste. The capacities of chloride tubular stationary lead- acid cells at various rates of discharge, expressed as a percentage of the ampere-hours available at the 10 hour rate, are given in Fig. 11.9.

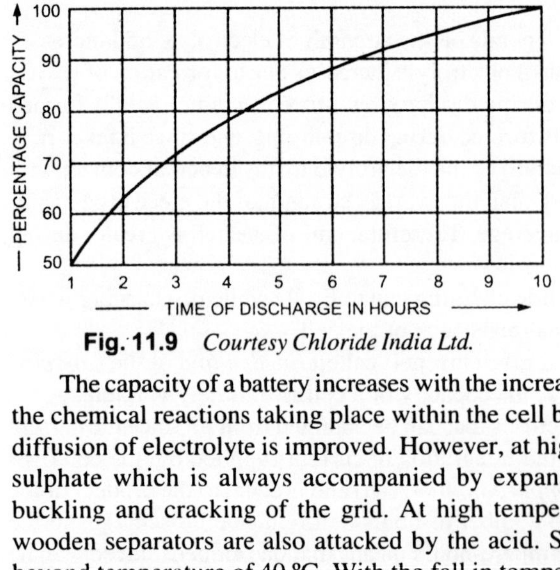

Fig. 11.9 *Courtesy Chloride India Ltd.*

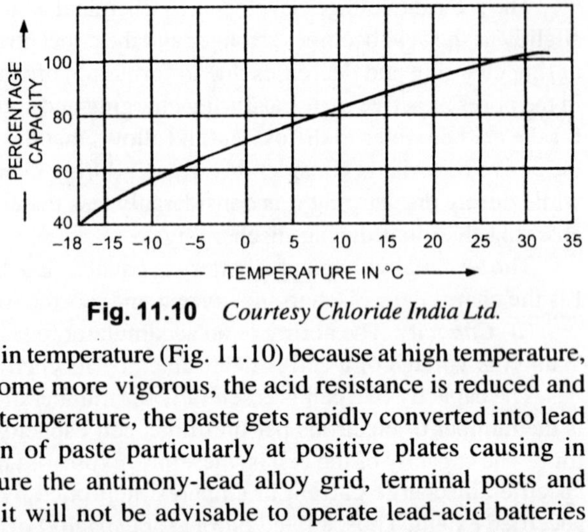

Fig. 11.10 *Courtesy Chloride India Ltd.*

The capacity of a battery increases with the increase in temperature (Fig. 11.10) because at high temperature, the chemical reactions taking place within the cell become more vigorous, the acid resistance is reduced and diffusion of electrolyte is improved. However, at high temperature, the paste gets rapidly converted into lead sulphate which is always accompanied by expansion of paste particularly at positive plates causing in buckling and cracking of the grid. At high temperature the antimony-lead alloy grid, terminal posts and wooden separators are also attacked by the acid. So, it will not be advisable to operate lead-acid batteries beyond temperature of 40 °C. With the fall in temperature, the chemical reactions become slow, cell internal resistance increases and diffusion of electrolyte becomes poor. Consequently, the capacity of the cell decreases with the fall in temperature till at freezing point (–35°C at specific gravity of 1.22 of electrolyte) the capacity is reduced to zero even though the battery otherwise be fully charged.

(*iii*) *Efficiency.* The efficiency of the cell can be given in two ways, enumerated and explained below.

1. *The Quantity or Ampere-Hour (A-H) Efficiency.* Since variations in terminal potential of the cell during charge and discharge are not taken into account while working out this efficiency and terminal potential of the cell during charge is higher than that during discharge, therefore, quantity efficiency is always higher than energy efficiency, in which variations of terminal potential of the cell are taken into account.

As generally efficiency is defined as the ratio of output to the input, similarly quantity efficiency or ampere-hour efficiency is defined as the ratio of ampere-hours of discharge and ampere-hours of charge.

$$i.e.\ \text{Ampere-hour efficiency, } \eta_{AH} = \frac{\text{Ampere-hours of discharge}}{\text{Ampere-hours of charge}} \times 100 = \frac{I_d \times T_d}{I_c \times T_c} \times 100$$

The quantity efficiency of the lead acid cell varies from 90 to 95%. It would be 100 percent if it were not for the gassing on charge, which represents a non-reversible chemical reaction. If the charging is discontinued each time as soon as the gassing becomes appreciable, the ampere-hour efficiency will be nearly 100 percent but ampere-hour capacity will be reduced and it is advisable to give the battery a full charge from time to time in order to avoid deterioration of the otherwise unused lead sulphate. The quantity efficiency also decreases due to self discharge of the plates caused by local reactions and because of current leakage caused by faulty insulation between the cells and the battery.

2. *Energy or Watt-Hour Efficiency.* Energy efficiency is defined as the ratio of energy delivered in watt-hours by the cell during discharge and the energy drawn in watt-hours during charge.

$$i.e.\ \text{Energy or watt-hour efficiency, } \eta_{W-H} = \frac{\text{Output in watt-hours}}{\text{Input in watt-hours}} \times 100$$

$$= \frac{\text{Current delivered} \times \text{time of discharge} \times \text{average pd during discharge}}{\text{Current drawn} \times \text{time of charge} \times \text{average pd during charge}} \times 100$$

$$= \frac{I_d \times T_d \times V_d}{I_c \times T_c \times V_c} \times 100 = \frac{I_d\ T_d}{I_c\ T_c} \times 100 \times \frac{V_d}{V_c} = \eta_{A-H} \times \frac{V_d}{V_c}$$

Operation at low rate of charge and discharge and at reduced ampere-hour capacity both tend to raise the watt-hour efficiency. Actual watt-hour or energy efficiency obtained in practice ranges from about 75 to 85 percent.

11.3.9. Battery Ratings. The standards adopted, both by industry and government organisations, are given below :

1. *Ampere-hour Capacity Rating.* The ampere-hour rating of a battery is usually determined from its ability to deliver current continuously for 20 hours at 27 °C. A battery that can deliver 5 amperes steadily for 20 hours then its rating will be 100 Ah.

2. *Reserve Capacity.* The reserve capacity of a battery is indicated in terms of minutes, a battery is capable of tolerating a drain of 25 A without dropping below 10.5 V (1.75 volts per cell). Higher this rating, better the battery is.

3. *Cold Rating.* The cold rating of a battery indicates the number of minutes a battery can deliver 300 A at – 18 °C (0 °F). This is applicable in relation to craft which ply in freezing weather.

4. *Cold Cranking Power Rating.* This rating is applicable to all 12 V batteries irrespective of their size. The battery is discharged at – 18 °C till its terminal voltage falls to 7.2 V. The output current is measured for 30 s. Higher the output, better the battery is.

Example 11.4

Determine the weight of lead and lead peroxide required to produce 1 ampere-hour in a lead acid cell. The gram equivalent of lead is 103.6 g and Faraday's constant is 96,500 C.

Solution: Gram equivalent of lead = 103.6

$$\text{Faraday's constant} = 96{,}500\ \text{C} = \frac{96{,}500}{3{,}600}\ \ i.e.\ 26.8\ \text{Ah} \qquad \text{since } 1\ \text{Ah} = 3{,}600\ \text{As or C}$$

$$\text{Quantity of lead required to deliver 1 A-h} = \frac{\text{Gram-equivalent of lead}}{\text{Faraday's constant}} = \frac{103.6}{26.8} = 3.865\ \text{g}\ \textbf{Ans.}$$

$$\text{Atomic weight of lead peroxide} = \text{Atomic weight of lead} + 2 \times \text{atomic weight of oxygen}$$

$$= 207.2 + 2 \times 16 = 239.2$$

$$\text{Gram-equivalent of lead peroxide} = \frac{\text{Atomic weight}}{\text{Valency}} = \frac{239.2}{2} = 119.6$$

$$\text{Quantity of lead peroxide required to deliver 1 Ah} = \frac{119.6}{26.8} = 4.46 \text{ g } \textbf{Ans.}$$

Example 11.5

An alkaline cell is discharged at a steady current of 4 A for 12 hours, the average terminal voltage being 1.2 V. To restore it to original state of voltage, a steady current of 3 A for 20 hours is required, the average terminal voltage being 1.44 V. Calculate the ampere-hour and watt-hour efficiencies in this particular case.

Solution: Charging current, I_c = 3 A

Charging average voltage, V_c = 1.44 V

Charging period, T_c = 20 hours

Discharging current, I_d = 4 A

Discharging average voltage, V_d = 1.2 V

Discharging period, T_d = 12 hours

$$\text{Ampere-hour efficiency, } \eta_{AH} = \frac{I_d\, T_d}{I_c\, T_c} \times 100 = \frac{4 \times 12}{3 \times 20} \times 100 = 80\% \textbf{ Ans.}$$

$$\text{Watt-hour efficiency, } \eta_{WH} = \eta_{AH} \times \frac{V_d}{V_c} = 80 \times \frac{1.2}{1.44} = 66.7\% \textbf{ Ans.}$$

Example 11.6

Thirty five lead acid secondary cells each of discharge capacity 120 Ah at 12 hours rate are to be fully charged at constant current for 10 hours. The dc supply is 125 volts, the ampere-hour efficiency is 80% and the emf of each cell at start and at the end of the charge is 1.9 volts and 2.6 volts respectively. Calculate the maximum and minimum values of external resistance necessary. Neglect the internal resistance of the cells.

Solution:

$$\text{Ampere-hours input per cell} = \frac{\text{Discharge capacity per cell}}{\text{Quantity efficiency, } \eta_{AH}} = \frac{120}{0.8} = 150 \text{ Ah}$$

$$\text{Charging current, } I_c = \frac{\text{Input in A}h}{Tc} = \frac{150}{10} = 15 \text{ A}$$

Combined back emf of 35 cells connected in series at the start of the charge,

$$E_{b1} = 35 \times 1.9 = 66.5 \text{ V}$$

Combined back emf of 35 cells connected in series at the end of the charge,

$$E_{b2} = 35 \times 2.6 = 91 \text{ V}$$

The resistance necessary to be added in the circuit to maintain charging current constant at 15 A will be maximum at the start of charge when back emf will be minimum and will be given by the expression

$$R_{max} = \frac{V_1 - E_{b_1}}{I_c} = \frac{125 - 66.5}{15} = 3.9\ \Omega \textbf{ Ans.}$$

The resistance necessary to be added in the circuit to maintain charging current constant at 15 A will be minimum at the end of the charge when back emf will be maximum and will be given by the expression

$$R_{min} = \frac{V - E_{b_2}}{I_c} = \frac{125 - 91}{15} = 2.267\ \Omega \textbf{ Ans.}$$

Example 11.7

A discharged battery is put on charge at 5 A for hours at a mean charging voltage of 13.5 V. It is then discharged in 6 hours at a constant terminal voltage of 12 V through a resistance of 'R' ohms. Determine (*i*) value of 'R' for an ampere-hour efficiency of 85% (*ii*) watt-hour efficiency of the battery.

Solution: Ampere-hour efficiency, η_{AH} = 85% = 0.85

$$\therefore \text{ Discharging current, } I_d = \eta_{AH} \times \frac{I_c T_c}{T_d} = 0.85 \times \frac{5 \times 3\frac{1}{2}}{6} = 2.48 \text{ A} \qquad \because \eta_{AH} = \frac{I_d T_d}{I_c T_c}$$

$$\text{Discharging resistance, } R = \frac{V_d}{I_d} = \frac{12}{2.48} = 4.84\ \Omega \textbf{ Ans.}$$

$$\text{Watt-hour efficiency, } \eta_{WH} = \eta_{AH} \times \frac{V_d}{V_c} = 85 \times \frac{12}{13.5} = 75.55\% \textbf{ Ans.}$$

11.3.10. Uses of Lead-Acid Batteries. The storage batteries are employed for a great variety and range of purposes such as to supply current for electric vehicles and gas engine ignition, and in telephone exchanges, railway trains, mines, laboratories, hospitals, broadcasting stations, telecommunication systems, emergency lightings, UPS systems, solar photovoltaic systems, power generating stations and distribution work as standby unit.

The storage batteries are employed in central power stations for (*i*) supplying the whole load during light load periods (*ii*) supplying peak load during the peak load hours (*iii*) local lighting for odd time of breakdown (*iv*) regulation of load and voltage (*v*) compensating feeder drop and as a preventive against shut down.

A very important use of storage batteries is in the providing of standby power for various electrical systems. In some electrical systems, storage batteries are connected in parallel with the generator and load. When the generator is in operation the batteries draw enough current to keep them fully charged. In case of shut down/breakdown of supply from the generator, the batteries supply the load.

Railway-car lighting systems are supplied from axle-driven generators when the train is running, with batteries supplying the system when the speed of the train is low or the train is stopped.

Automobile electrical systems are similar to the above mentioned system in that the generator, and the load are connected in parallel, the battery supplying power for starting and lighting when the generator is not in operation.

Hospitals and other places, where a continuous source of power supply is absolutely essential, often use storage batteries as an emergency (or standby) supply.

Applications in which storage batteries supply the normal current are in industrial truck or mine-locomotive propulsion portable lighting equipment, portable radios and other applications in which continuous supply from a generator is impracticable.

11.3.11. Maintenance and Care of Lead-Acid Accumulators. The following general instructions should be followed in the care and maintenance of lead-acid accumulators.

1. The accumulators must always be charged with direct current and in right direction. Connect battery positive terminal to positive charging lead and negative terminal to negative charging lead with suitable control resistance included in the circuit. The cells will get damaged if wrongly connected.
2. The level of electrolyte must be maintained at least 10 or 15 mm above the plates and the plates should not be exposed to air. In the normal operation of lead-acid accumulator, a certain amount of water is lost from the electrolyte by evaporation and because of gassing. Gassing takes place during charging when the water in the electrolyte is decomposed into hydrogen and oxygen, each of which are given in the form of gas. Evaporation of electrolyte should be made up by adding distilled water only at regular intervals. Acid is not added to a battery unless the electrolyte has been spilled. Only pure and colourless sulphuric acid should be added whenever necessary.
3. Batteries may be charged at any rate provided the electrolyte temperature does not exceed 45 °C and that excessive gassing does not take place. Although violent gassing or gassing for long periods should be avoided, a small amount of gassing for a short period at the end of a charge is desirable to ensure that no lead sulphate is formed on the plates.
4. Lead-acid batteries should not be discharged beyond a terminal voltage of about 1.75 V per cell and the specific gravity has dropped its normal value.

5. A battery should not be left in discharged condition for any length of time (more than a day) since the lead sulphate formed on the plates crystallizes so the proper reforming of the plates becomes difficult if not impossible. If a battery is to be stored for any length of time, distilled water should be added to the proper level, after which the battery should be fully charged. At intervals of 4 to 6 weeks the battery should be given a freshening charge to keep it in the fully charged state.
6. Keep naked flames away from the battery at all times. This is to avoid the risk of explosion of the gaseous mixture in the upper parts of the cells, and in their immediate vicinity.
7. The batteries should be overcharged occasionally in order to remove the whole of the sulphate formed on the plates.
8. In preparing the electrolyte never pour water into the acid.
9. The specific gravity of the electrolyte should be checked after and before charging of the battery. The specific gravity of the electrolyte when the cells are fully charged should be between 1.200 to 1.210 (1.200 ± 0.005 at 27 °C), and when fully discharged, at the 10 hour rate, about 1.170. If at any time the specific gravity rises above 1.210, it must be reduced to normal value by adding distilled water.
10. During a charge, whenever possible, open the battery compartment to provide additional ventilation. Leave the vent plugs firmly in position as they incorporate special baffle devices which return the electrolyte spray to the cells whilst allowing gases to escape.
11. Keep vent plugs and connections tight. Remove filling plugs only when topping up and taking specific gravity readings.
12. Check and clean of the vent holes in the plugs.
13. Keep the cells, trays and containers dry and clean. Should there be any corrosion of the metal work, clean the affected parts and neutralize any remaining acid with a solution of water and sodium bicarbonate or sodium carbonate or with dilute ammonia. Protect from further corrosion by covering with acid-proof paint or petroleum jelly. Wash down the trays occasionally with one of the above solutions and, when dry, give them a coat of acid proof paint.
14. Wipe off the dirt and dust which could contaminate the electrolyte and reduce the service life of the battery. Remove any electrolyte spilt on the surface of the battery with a dry cloth; dirty surfaces lead to self discharge of the battery.
15. Check externally for fit condition of the container and sealing compound cracks in the compound and seepage of electrolyte from the container should be timely detected.
16. Ensure that leads wires are reliably fixed to the terminal posts, clean of oxidized terminals and lead wire clamps. The contact surfaces should fit tightly together to avoid any increase in contact resistance which would hinder the flow of current. While cleaning or filing the terminals and clamps a minimum layer of metal should be removed to avoid weakening of the parts.
17. While cells are not in use or kept as standby, keep them on trickle charge, which will supply energy just sufficient to balance the open-circuit losses.

It is never permissible to use commercial sulphuric acid or contaminated water for making up the electrolyte because this will result in damage of plates. When the electrolyte becomes contaminated, foreign particles may get between the plates, this also occurring when particles of the active material shed off the plates. At the bottom of the cells a large quantity of various sediments may accumulate. As a result, short-circuit may occur between the plates and cause a reduction in density of the electrolyte and a drop in a battery voltage. When found to be contaminated, the electrolyte is removed from each cell and for 2 to 3 hours the cells are left filled with distilled water. Then the cells are washed out with distilled water until the water becomes clean. After this the cells are refilled with electrolyte and charged.

11.3.12. Sulphations of Plates. The formation of lead sulphate on the electrodes, due to chemical combination of active material with electrolyte, during prolonged discharge is called the *sulphatization*. The lead sulphate is an insulator, insoluble, hard, white in colour and of greater volume in proportion as compared to the active material. Sulphation is evidence by electrodes becoming lighter in colour as the sulphate lessens the active surface.

It is sometimes caused by a too weak or too strong electrolyte (acid solution), but more generally by continued over discharging, or too rapid discharging of batteries, or by allowing them to remain discharged for long period of time.

Sulphation tends to cause shedding of the active material, buckling of plates, loss of capacity, increase of internal resistance and consequent reduction of efficiency and increase of temperature with flow of current. A sufficient amount of lead peroxide and sponge lead must be retained on the plates to reduce this resistance, otherwise the charging current cannot flow through the active material and regenerate the battery.

In case of sulphation, the battery should be charged below the maximum rate, necessarily prolonging the charge, until the plates assume the proper colour. This is a tedious task and it must not be hastened, as rapid charging will cause serious buckling of plates. The charging should be done at low rate. Discharge should not be allowed below 1.8 V per cell, and charging current should be stopped as shown as soon as cell shows 2.4 V.

If the plates are in very bad condition, a little of white sulphate deposit on each of the positive plates may be removed with a stick, thus exposing a part of the good surface to the action of electrolyte.

If the positive plates cannot be restored to their proper colour as discussed, it will be cheaper to replace them by a new set, rather than to attempt their recovery by means of reversal.

11.3.13. Testing. The condition of each cell should be tested at least once a week by a cell testing voltmeter. This test should be made at the end of discharge, when the voltage per cell should not fall below 1.85 V, the current being at the 10 hour discharge rate. The voltage of each cell should also be tested at the end of charge and the readings taken about 10 minutes before charging current is stopped.

The specific gravity of electrolyte in each cell should be tested once a week at the end of charge and, at the same time, the plates in every cell should be carefully examined. If any internal contacts are found between adjacent plates, they must be removed immediately.

11.3.14. Special Charges. The following types of special charges are usually required for proper maintenance of secondary batteries.

1. *Boosting Charge.* Though batteries are usually of sufficient capacity to work for the whole day without requiring any intermediate charge during the day. But in some cases such as in battery operated vehicles, however, the batteries may require an extra charge known as "*boosting charge*" in the middle of the day.

The current for this boosting charge is usually determined by an emperical rule according to which the *boosting charge current in amperes is numerically equal to battery discharge in ampere-hours up to that time when boosting charge is to be given.* For example for an 80 ampere-hour battery when discharged to the extent of 25 ampere-hours, the boosting charge current will be 25 amperes. Due to this charging current there is no gassing.

2. *Equalising Charge.* The equalising charge is a low rate charging and is given to the cells to ensure that every cell is brought into a fully charged condition at regular intervals. This charge is given until the voltage and specific gravity of the electrolyte remains constant for three successive hourly readings. This keeps the cell healthy for a longer period of time. The usual practice of giving equalising charge is as follows :

(*i*) For batteries subjected to "trickle charge"—not required.
(*ii*) For "floating" or idle batteries—once a month.
(*iii*) For batteries charged less frequently—once a fortnight.
(*iv*) For batteries charged every day—once a week.

3. *Trickle Charge.* The trickle charge is extremely a low rate charge and is applied to standby batteries for compensating the open-circuit losses. The charge current should be of such a magnitude that there is no gassing and when it is allowed to flow continuously through the fully charged batteries, specific gravity remains constant at the specific value.

The batteries subjected to 'trickle charge' can be maintained always in a fully charged condition for several years. The service life of the battery also increases.

11.3.15. Voltage Control Methods. As explained in Art 11.3.6, the output voltage of a lead-acid accumulator varies over a considerable range—the voltage of lead-acid cell reaches about 2.6 V at the finish of charge, begins to discharge at about 2.1 volts and drops to 1.8 V at the end of discharge. It is, therefore, necessary to employ some means to control the battery output voltage so as to confine voltage variations within reasonable limits. The various methods employed for this purpose are explained below:

1. *Rheostatic Control.* This method of voltage control may be employed advantageously only with batteries of small capacity and in small lighting plants, where the space available for battery auxiliaries is limited, and where the cost of energy is so low that the energy loss in rheostat is not objectionable. Rheostatic control consists of a rheostat having a sufficient number of steps so as to insert suitable resistance in the circuit when the battery is fully charged and gradually cut out as the discharge continues.

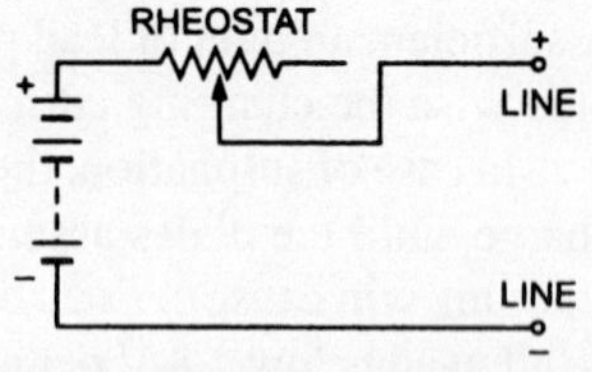

Fig. 11.11 *Rheostatic Control*

2. *End Cell Control.* Though the rheostatic control is very simple but wasteful and is not suitable in large capacity installations. Hence other economical systems have been devised and *end cell control* is one of these. In this system, suitable regulator switches are employed to take one or more of a number of cells out of the circuit when the battery is fully charged and into the circuit again as the battery discharges. To make the process of cutting cells in and out of the battery circuit simple, a group of cells selected for this purpose is kept at the end of the battery. Thus it derives the name *end cell.*

For example a battery intended for use on 110 V lighting circuit will require $\frac{110}{1.8}$ *i.e.* 62 cells since the voltage per cell at the end of discharge should not fall below 1.8 V. The voltage per cell at the end of charge is about 2.6 V, so the number of cells required to give 110 V at the end of charge will be $\frac{110}{2.6}$ *i.e.* 44. Switching arrangements are, therefore, provided to tap in one of the two lead connections at any junction of the last 18 cells at one end of the battery. A similar tapping arrangement is necessary for one of the charging connections, as otherwise the cells that are used least in the discharge will be over charged during the charging .

The end cell switch is shown in Fig. 11.12. The switch contact arm is made in two parts, A and B, which are insulated from each other through the protective resistance R. The end cell contacts are so spaced that when the main current carrying part A, of the switch arm is squarely on one end cell contact such as X, the part B, does not touch any other contact such as Y, but when the switch arm is advanced for cutting into circuit another end cell, the part B reaches the contact Y, before the part A, leaves the contact X, thus keeping the battery circuit closed, while the resistance R, limits the current in the short- circuited cell at the instant the switch arm passes from one end cell contact to the next.

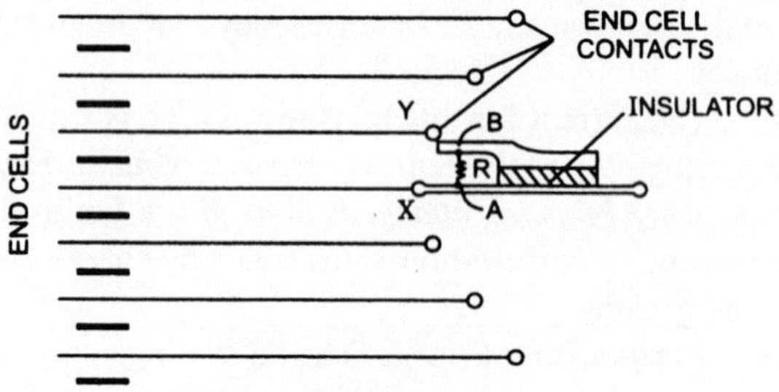

Fig. 11.12

3. *Counter* EMF *Cell Control.* The cells used for voltage control consist of nickel plates immersed in an alkaline solution. When current is driven through them, water is decomposed and a counter-emf of 2 volts per cell is setup, independent of the magnitude of discharging current flowing through them, thereby reducing the voltage of the battery, so that the line voltage depends on the number of reverse current cells in series in the battery circuit. With the drop in the voltage of battery during discharging the reverse pressure cells are cut out, successively, thus keeping the external or line voltage constant.

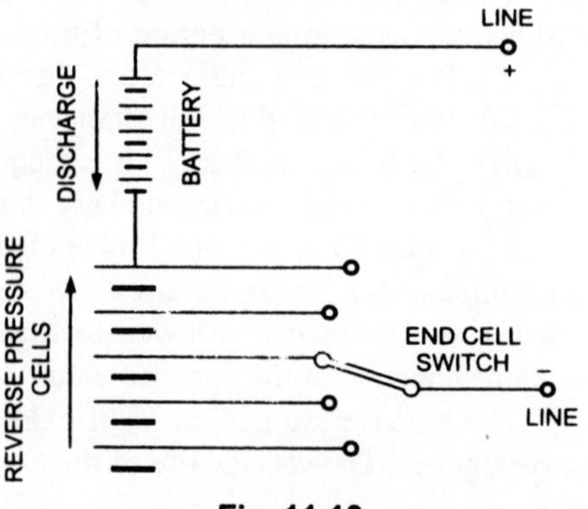

Fig. 11.13

4. *Booster Control.* In general, a booster may be defined as a low voltage *dynamo inserted* in series in a circuit, to change its voltage. The function of a booster is to add or to subtract an electric pressure derived from another source.

The average or mean voltage of a lead-acid cell is $\frac{2.6 + 1.8}{2}$ *i.e.* 2.2 V. The number of cells required in a cell to provide 110 V is $\frac{110}{2.2}$ *i.e.* 50. The undesired voltage fluctuations of the battery are at all times neutralized by the emf of the booster permanently connected in series with the battery. At the end of charge the booster emf must be $(50 \times 2.6 - 110)$ *i.e.* 20 volts opposing the emf of the battery while at the end of discharge the booster emf must again be 20 V, but assisting the battery emf. The voltage rating of the booster will thus be 20 V.

11.4. SEALED LEAD CELLS AND BATTERIES

Starved-electrolyte sealed lead batteries are the most advanced form of lead acid batteries in use today. Because of their demonstrated superior performance and reliability these batteries are essential components of a variety of products ranging from inexpensive toys and consumer products to telecommunication systems and aircraft.

11.4.1. Features and Benefits.

1. *Rugged Construction.* Starved-electrolyte sealed lead batteries are tough and of rugged construction. It is a great testimonial to ruggedness of such batteries that they are the battery of choice in a wide array of equipment.

2. *Long Life.* These batteries have demonstrated an enviable longevity whether they are employed in float or cyclic duty or just kept in storage. These batteries, when properly charged, offer up to 10 years life at room temperature before their capacity drops to 80 percent of its rated value. These batteries outperform other lead batteries when employed in cyclic service. Their self discharge rate is very low, this not only makes their storage simple, but products are much more likely to come from storage with some residual capability available. This is especially important in some consumer applications where the purchaser expects to see the product operate straight out of box, even before he has charged it.

3. *Compact in Size.* It is due to thin plates and minimum amounts of electrolyte used and results in saving in weight and volume throughout the system.

4. *Elimination of Maintenance.* Inexpensive batteries that require routine maintenance often turn out to be no bargain at all.

5. *Design Flexibility.* Conventional lead acid batteries are tolerated because of their usefulness, but disliked due to their nasty habits. Many applications require that conventional batteries be isolated in separate compartments made from acid resistant materials with independent ventilation and drainage systems. Starved-electrolyte cells and batteries do not require this special treatment, as spillage and corrosions are not the problems with these batteries. In fact even sensitive applications, such as computers and aircraft, now locate these batteries among other electronic equipment. These batteries no longer have to be mounted vertically; allowing equipment designers enhanced flexibility.

6. *Simplified Charging.* The gas recombination within the battery improves its ability to accommodate overcharge, giving designers a new freedom in tailoring the charger to the application. These batteries may be charged either by constant voltage or two step constant current methods to give maximum life while giving adequate discharge performance.

7. *Excellent Performance.* The thin-plate, low-impedance design of many sealed-lead cells and batteries gives much higher discharge currents than traditional designs of the same rating.

Power density is higher due to thin plates and minimum amounts of active material used in these batteries.

With these batteries, the output voltage characteristics are very good, especially at higher discharge rates and at temperature extremes. Many of the design compromises necessary due to voltage variation may be eliminated with these batteries.

These batteries reduce cold weather problems.

11.4.2. Applications. Starved-electrolyte batteries have a wide range of applications. These batteries are employed as standby units in computers, emergency lighting, medical equipment, security alarm systems, telecommunication equipment and uninterruptible power supplies; as portable units in cellular phones, toys, cordless appliances and portable lighting, and as alternate units in computers, consumer electronics and instrumentation.

11.5. ALKALINE BATTERIES

The alkaline batteries are principally the outcome of attempts to overcome the weight and mechanical weakness of the lead plates. The alkaline cells may be of diverse designs but, like lead-acid cells, consist of essentially positive and negative plates immersed in an electrolyte. The plates and electrolytes are placed in a suitable container.

The alkaline batteries according to the composition of the active material of the plates may be classified as nickel iron (or Edison), nickel-cadmium (or Nife), silver zinc and Alkum batteries.

By method of assembly these batteries may be classified as sealed and non-sealed cells or batteries.

By design of plates these batteries may be classified as enclosed pocket and open pocket type.

These batteries make use of ebonite sticks to prevent plates of opposite polarity from coming into direct contact; in sealed or open-pocket alkaline cells plastic (capron) separators are employed for this purpose.

The active material of enclosed-pocket type plates is enclosed in nickel-plated perforated steel pockets or packs which are pressed into the steel frames of the plates. Graphite or nickel is added to the active material so as to improve the contact and conductivity between the active material and the nickel plated pocket casing. Open-pocket type plates may be of a metal ceramic material, foil, or pressed metal plates and the active material is outside the plates and is directly in contact with the electrolyte.

The frame of *metal-ceramic plates* is nickel plated steel grid with the active material pressed in under a pressure of 8 to 9 kg/mm^2. The base of *foil plates* is nickel foil of 0.05 mm thickness and coated with a layer of nickel suspension deposited by a spray technique. The base member of *pressed plates* is a nickel plated pressed steel grid. The active material is pressed into pressed plates at a pressure of about 4 kg/mm^2.

11.6. NICKEL IRON (OR EDISON) BATTERIES

These batteries are going to become more and more popular as there is a possibility of their development into high energy density batteries for electric vehicles.

11.6.1. Active Materials. The *positive plate* consists of nickel hydroxide [$Ni(OH)_3$] or nickel peroxide (NiO_2). About 17 percent of graphite is added to the above so as to increase its conductivity and about 2 percent of activating additive, barium hydroxide, is also introduced into the active material. The additive causes increase in service life of plates and coefficient of utilization of the positive plate material.

The *negative plate* consists of powdered iron and its oxides [FeO and Fe $(OH)_2$]. Small quantities of nickel sulphate and ferrous sulphide are added to the active material so as to enhance the conditions for forming of the cell considerably and improve the performance.

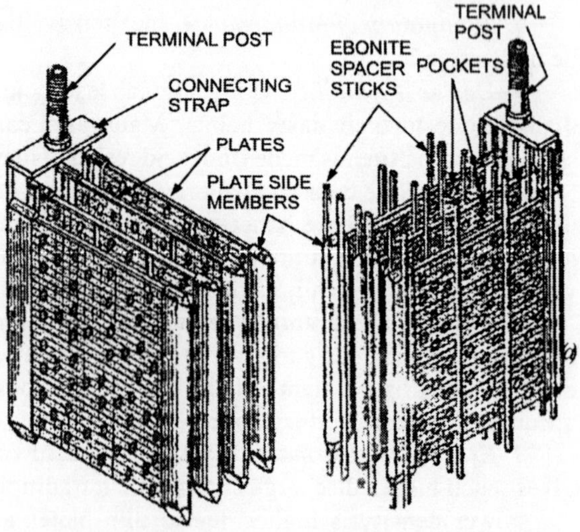

Positive Plate Group *(b) Negative Plate Group*

Fig. 11.14 *Plate Groups of An Altaline Cell*

The *electrolyte* is a 20 percent solution of potassium hydroxide (KOH) in distilled water with a small amount of lithium hydroxide added. Addition of lithium hydroxide to the electrolyte results in a more complete utilization of the nickel peroxide (NiO_2) and thus increases both the capacity and the life of the battery.

11.6.2. Construction. The vessel containing the electrolyte and electrodes is also made of nickel plated iron with a welded lid provided with holes for the terminals and for pouring in electrolyte. Ebonite sticks are placed between the plates of opposite polarities (*i.e.* positive and negative plates) to prevent them from coming into direct contact and cause short-circuiting. The number of negative plates is one more than the number of positive plates. The extreme negative plates are electrically connected to the container.

Plates of same polarity with their pockets filled, are assembled into cell groups. For this purpose, they are welded to a common strap which has a threaded post.

11.6.3. Operation. The chemistry of nickel iron cell is very complicated and cannot be entirely understood because the exact formula for positive active material is not yet well established, but the action of the cell may be explained by assuming it $Ni(OH)_3$.

In a fully charged condition, the active material in the tubes of the positive plates is nickel hydroxide $Ni(OH)_3$ while that in the pockets of the negative plate is iron Fe. As the cell delivers current to the load, the active material of the positive plate changes from $Ni(OH)_3$ to $Ni(OH)_2$ and that of negative plate changes from iron to ferrous hydroxide $Fe(OH)_2$. Charging reverses this process, namely the active material in the tubes of positive plates becomes $Ni(OH)_3$ and that in the tubes of negative plates Fe. The electrochemical process in the nickel-iron accumulators can be expressed by the equation.

$$\underset{\text{Positive plate}}{2Ni(OH)_3} + 2KOH + \underset{\text{Negative plate}}{Fe} \underset{\text{Discharge}}{\overset{\text{Charge}}{\rightleftarrows}} \underset{\text{Positive plate}}{2Ni(OH_2)} + 2KOH + \underset{\text{Negative plate}}{Fe(OH)_2} + \text{Electrical energy}$$

The equation should be read left to right for discharge and from right to left for charge. The reactions are accompanied by transfer of electrons through the external circuit to the positive plate during discharge. In the nickel-iron accumulators the electrolyte, potassium hydroxide in water, does not take part in chemical reactions, but does in fact, maintain its original concentration both on discharge and charge. This allows a comparatively small quantity of electrolyte to be used and the accumulators to be made more compact.

The state of charge of the nickel iron cell cannot be determined by the specific gravity of electrolyte as it does not change.

No corrosive fumes are given off from the electrolyte, so no special care is needed to be taken in mounting the cell.

11.6.4. Electrical Characteristics. Fig. 11.15 illustrates how the voltage of a nickel-iron cell varies during charging and discharging. The emf, when fully charged, is about 1.4 V, decreases to 1.3 V rapidly and then very slowly to 1.1 or 1.0 V of discharge. An average discharge voltage is about 1.2 V. The average charging voltage is about 1.7 V per cell. The voltage characteristics are similar to those of a lead-acid cell. There is no lower limit to the voltage of the Edison cell because in it there is nothing like sulphation, but discharge is not continued below a useful lower limit. The emf of the cell or battery increases slightly with the temperature.

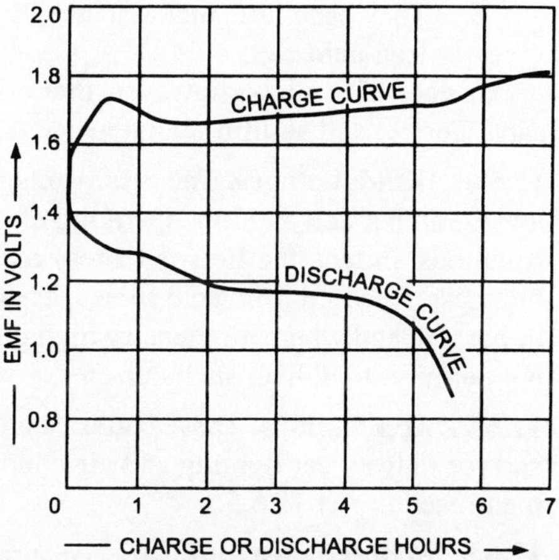

Fig. 11.15 *Typical Charge and Discharge Curves For Edison Cell*

Edison batteries are rated at a 7-hour charge rate and a 5 hour discharge rate with the same current in each case, the ampere-hour efficiency being about 80 percent at this rate and the internal heating not more than permissible. A higher rate of discharge may be employed as long as the internal temperature does not exceed about 45 °C; continual operation at higher temperatures shortens the battery service life. A longer charge rate than 7 hour should not be used because, with low currents, the iron element is not completely reduced; this, however, does not permanently injure the battery but makes it necessary to overcharge the battery at normal rate and then discharge it completely to restore it to normal condition.

The internal resistance of the nickel-iron cell is nearly 5 times that of lead acid cell and for this reason it is able to deliver less current at short-circuit or to low resistance load. But because of high internal resistance of the nickel iron cell, the difference between the charge and discharge voltages is greater than that of a lead-acid cell, as illustrated in Fig. 11.15.

Due to comparatively high internal resistance, the efficiencies of the Edison batteries are lower than those of the lead-acid batteries. The ampere-hour and watt-hour efficiencies of the Edison batteries are about 80 percent and 60 percent respectively. Average energy density is about 50 Wh per kg of cell.

When assembling batteries for the same voltage, the number of nickel-iron cells required are more than that of lead-acid cells. For example, a 12-V lead-acid battery will need 12/2 *i.e.* 6 series-connected cells, whereas the nickel-iron battery will require $\frac{12}{1.2}$ *i.e.* 10 series-connected cells to give the same voltage *i.e.* 12 V.

The capacity of an Edison cell is affected appreciably by the variations in temperature. Decrease in temperature causes the capacity to fall and becoming practically zero at 4 °C, even though the cell is fully charged. This is a serious drawback in case of electrically driven vehicles in cold weather. Care is taken to heat up the battery before starting. However, the internal losses (I^2R losses) of the battery are sufficient to keep the battery warm when running.

11.6.5. Advantages. Nickel-iron cells (or batteries) offer a number of advantages over lead-acid accumulators.

1. They are comparatively lighter in weight as they need less quantity of electrolyte and their plates are lighter. They weight only about half as much as lead-acid cells of equivalent energy storage.
2. They have rugged construction, more durable and mechanically strong, because of all steel construction. They are not affected by vibrations, jolts or shocks, heavy currents, short discharges, and do not suffer damage during short-circuits. They do not deteriorate even left discharged for long periods of time.
3. They have long service life.
4. Lead, which is expensive material, is not used in these cells.
5. They do not evolve obnoxious fumes, the cells may be placed in the same room as other machinery/ equipment without any risk of corrosion of that machinery/equipment.
6. They need little maintenance. Even when their maintenance is inadequate they fail less quickly than lead-acid cells.

It does not follow, however, that alkaline cells (or batteries) can operate satisfactorily without proper care. Correct and skillful maintenance is obligatory for both types

11.6.6. Disadvantages. The disadvantages of nickel-iron cells are high initial cost, high internal resistance, lower emf per cell and low operating efficiency. These drawbacks are sufficient to prevent their use in most situations. In fact, the Edison battery may be considered as a special-service battery to be employed where lightness, freedom from acid fumes, or ability to withstand severity of vibrations and shocks are particularly important, and where momentary high discharge currents are not required, as its high internal resistance will not permit it to deliver such currents.

11.6.7. Applications. These batteries are used for the propulsion of industrial trucks, and mine locomotives and for railway car lighting and air-conditioning because of their rugged construction and other advantages mentioned in Art 11.6.5.

11.6.8. Maintenance and Care. As already mentioned, the maintenance of Edison batteries is much more simpler than that of lead-acid accumulators. The care of alkaline batteries mainly amounts to keep them clean and add water when necessary to maintain a normal electrolyte level.

Any nickel-iron battery, when operated with the electrolyte level below the top edge of the plates, will suffer irrevocable loss in capacity and have its plates quickly damaged by washout of the active material from the bared pockets. At least once in a week the cells should be topped up with distilled water to maintain the level of electrolyte.

The electrolyte is required to be changed from time to time, whenever its specific gravity falls below 1.16. For changing the electrolyte, the battery should first be completely discharged down to zero volt and left short-circuited for 2 hours or more. It should be refilled with the new electrolyte immediately after the old is dumped out.

When it is desired to take the battery out of service for a long period, it is simply discharged down to zero volt and stored with the terminals short-circuited.

The Edison cell is not damaged by freezing.

11.7. NICKEL-CADMIUM ACCUMULATORS

The general construction of a nickel-cadmium (or Nife) cell is shown in Fig. 11.16. The nickel-cadimum alkaline accumulator is similar in construction to that of nickel-iron (or Edison) accumulator. The positive plates of both types of accumulators are similar in every details; the negative plates differ in the active material they employ—the negative plates of nickel cadmium accumulators consist of spongy cadmium, Cd whereas the negative plates of Edison accumulators consist of spongy iron, Fe. The electrolyte is the same as in Edison accumulators *i.e.* a solution of potassium hydroxide (KOH) in distilled water with a specific gravity of 1.2 approximately. The use of cadmium in the accumulators reduces the internal resistance.

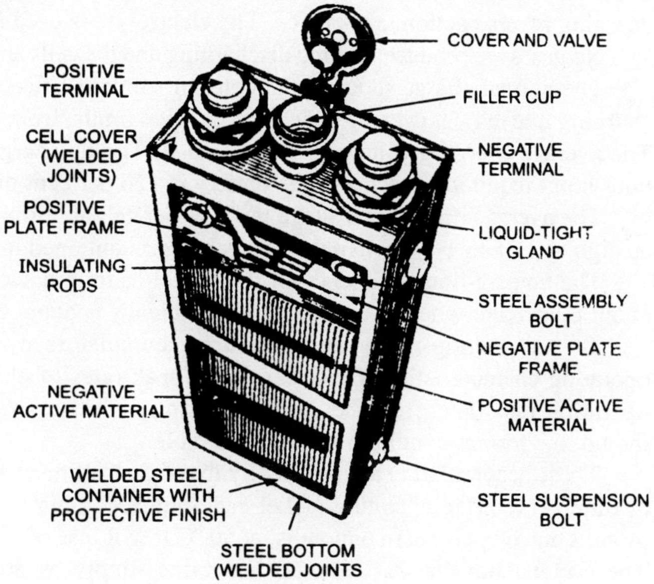

Fig. 11.16 *Nickel-Cadmium Alkaline Accumulator*

The most important difference between the nickel-cadmium accumulators and nickel-iron accumulators is the arrangement of the plates. Whereas in nickel-iron accumulators the extreme plates are negative, but in the nickel-cadmium accumulators the extreme plates are positive. The reason for this is that, for proper operation, the positive active material of the nickel-cadmium accumulators must occupy more space that the negative active material. Seen from outside, the positive plates of nickel-cadmium accumulators are little thicker than the negative plates. The extreme negative plates of nickel-iron accumulator are electrically connected to the container, the nickel-cadmium accumulator on the contrary has its container electrically connected to the positive plates. Arrangement of positive and negative plates in nickel-cadmium and nickel-iron accumulators are shown in Figs. 11.17 (*a*) and 11.17 (*b*) respectively.

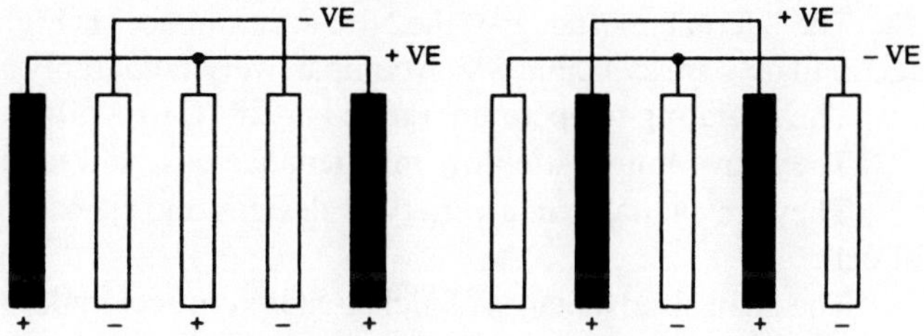

(a) Nickel-Cadmium Cell *(b) Nickel-Iron Cell*

Fig. 11.17

The chemical action is perfectly reversible. On discharge the nickel hydroxide [$Ni(OH)_3$] in the positive plates loses oxygen and is reduced to a lower form [$Ni(OH)_2$], while the cadmium in the negative plate is converted into cadmium hydroxide [$Cd(OH)_2$]. On charge the reverse action takes place, the cadmium hydroxide is converted back to plain cadmium metal atoms and lower nickel hydroxide is restored to the original condition. The electrolyte does not enter into chemical combination with the plates, the specific gravity remaining unchanged during charging and discharging of accumulators.

The electrochemical process in the nickel-cadmium accumulators for charge and discharge may be given as

$$\underset{\text{Positive plate}}{2Ni(OH)_3} + 2KOH + \underset{\text{Negative plate}}{Cd} \underset{\text{Discharge}}{\overset{\text{Charge}}{\rightleftarrows}} \underset{\text{Positive plate}}{2Ni(OH)_2} + 2KOH + \underset{\text{Negative plate}}{2Cd(OH)_2} + \text{Electrical energy}$$

The equation should be read left to right for discharge and from right to left for charge.

During discharge, two negative hydroxyl ions $(2OH)^-$ combine with cadmium (Cd), form cadmium hydroxide and release two electrons. One of the electrons is used to form hydroxyl ion $(OH)^-$ and the change a molecule of $Ni(OH)_3$ and the other electron goes to positive plate through the external circuit. During charging reverse action takes place. The electrolyte is used for conducting purpose only.

No gases are produced during discharging and the cells are sealed with vent caps to make them unspillable.

The normal charge should be at constant curent sufficient to give full charge to the cell in 6 hours. The charging rate may, however, vary between wide limits, from a heavy boosting charge down to trickle charge. The average charging voltage is about 1.45 V. Trickle charge voltage is about 1.35 V per cell. The ampere-hour input required to recharge the battery is 120 per cent of its rated capacity.

The average discharge voltage is 1.2 V per cell at normal current. However, the voltage per cell may be as high as 1.35 to 1.4 V when the cell is being maintained in the fully charged condition by a trickle charge.

The ampere-hour efficiency of the nickel-cadmium accumulator under normal operating conditions is about 80 percent whereas the watt-hour efficiency is about 65 percent.

To last so long the nickel-cadmium accumulators would be expected to have unusually favourable operating characteristics. These can stand peak rates of charge and discharge—upto 20 times the normal operating rate. These can be stored indefinitely either in discharged condition or in charged condition, though the former condition is more desirable.

These accumulators are practically free from self discharge and may be left unattended for long periods of time. At room temperature nickel-cadmium accumulator will lose 40 percent of full charge capacity in 3 months and 60 percent in 6 months; at 50 °C it will lose 60 percent in one month and 90 percent in 3 months. The accumulator can be restored to service simply by applying normal voltage. There is no permanent damage, such as sulphation, which occurs in lead-acid accumulator that stands for months in a discharge condition.

The internal resistance of the Nife accumulators is higher than that of lead-acid accumulators but special accumulators are available which compare very favourably and maintain their terminal voltage nearly constant.

The operating temperature range is – 28 °C to + 60 °C.

They have longer life, low maintenance cost and reliability under severe operating conditions.

They are available in a variety of designs and sizes. They have energy density as much as 55 Wh per kg of cell.

The main disadvantages of nickel-cadmium accumulators are that they are heavier in weight and expensive in cost in comparison to nickel-iron accumulators.

The nickel-cadmium accumulators are superior to lead-acid accumulators as well as nickel-iron accumulators for all applications except those in which weight is important criterion of selection of battery. Their use is also restricted due to their higher cost.

Like the nickel-iron batteries, the nickel-cadmium batteries are used for industrial services, where rugged construction, long life and low maintenance cost outweigh the fact that their original cost is as much as four times of that lead-acid batteries for the same ampere-hour output.

Such batteries are more suitable than Edison batteries for floating duties in conjunction with charging dynamo. This is due to the fact that there is not much difference between their charging and discharging emfs.

Small rechargeable thin sintered plate nickel-cadmium cells/batteries are used in cordless electric appliances such as shavers, photography equipment (movie cameras, photoflash and photoflood lights), and in aerospace applications.

They are also employed in commercial airlines, military aeroplanes and helicopters for starting main engines or auxiliary turbines and for emergency power supply.

11.8. FEATURES AND BENEFITS OF SEALED NICKEL-CADMIUM ACCUMULATORS

Among the significant features that lead to their position as the preferred power source are (*i*) rugged construction (*ii*) long operating life (*iii*) long storage life (*iv*) high energy (or watt-hour) density (*v*) high rate discharge capability (*vi*) consistent discharge voltage (*vii*) fast charge rate (*viii*) continuous overcharge capability (*ix*) maintenance free service (*x*) operation over a wide range of temperatures (*xi*) operation in a wide range of environments and (*xii*) operation in any orientation.

Sealed nickel-cadmium accumulators are employed in diverse applications due to their unique characteristics and capabilities. These batteries are well suited to applications where the convenience of a self contained, always readily available power source increases the value of the end product. In rechargeable battery applications requiring high power delivery or high power density, a nickel-cadmium battery is usually the only logical choice.

11.9. COMPARISON BETWEEN LEAD-ACID, NICKEL–IRON AND NICKEL CADIUM ACCUMULATORS

Particulars	*Lead-Acid*	*Nickel-Iron*	*Nickel-Cadmium*
1. Active materials			
(*a*) Positive plate	PbO_2 (lead peroxide) dark chocolate brown in colour	Nickel-hydroxide, $Ni(OH)_3$	Nickel-hydroxide, $Ni(OH)_3$
(*b*) Negative plate	Pb (lead) grey in colour	Spongy iron	Spongy cadmium
(*c*) Electrolyte	Dilute sulphuric acid (H_2SO_4) solution	20% of water solution of potassium hydroxide (KOH) with a small addition of lithium monohydrate containing not less than 50% lithium hydroxide (LiOH)	20% of water solution of patassium hydroxide (KOH) with a small addition of lithium monohydrate containing not less than 50% lithium hydroxide, (LiOH)
2. Mechanical Properties (For same A-H rating)			
(*a*) Weight	Moderate	Light	Heavy
(*b*) Volume	Varies	Small	Large
(*c*) Number of plates per cell	Few	Few	Many
(*d*) Mechanical strength	Poor	Good	Good
(*e*) High temperature damage	Yes	No	No
(*f*) Life	1,250 charges and discharges.	About 5 years	Long life
3. Maintenance Factors			
(*a*) Damage likely due to freezing	Yes	No	No
(*b*) Rate of self discharge	High	High	Low
(*c*) Trickle Charge Use	Yes	No	Yes
4. Electrical Properties			
(*a*) Average emf	2.0 V per cell	1.2 V per cell	1.2 V per cell
(*b*) Internal resistance	Comparatively low (about 0.001 to 0.002 Ω)	High (about 0.005 to 0.01 Ω)	Low

Contd...

(c) Efficiency			
(i) Quantity	90 to 95%	About 80%	80%
(ii) Energy	75 to 85%	About 60%	65%
(d) Voltage regulator control	Easy	Difficult	Easy
(e) Gas production during charge	Yes	—	—
(f) Ampere-hour capacity			
(i) Reduced by over-charge	Yes	No	No
(ii) Reduced by occasional over-discharge	No	No	No
(iii) Reduced by frequent over-discharge	Yes	No	Yes

11.10. SILVER ZINC BATTERIES

The active material for the positive plate is made of reduced silver oxide which is pressed into the plate and then heat treated. The active material of the negative plate consists of a mixture of zinc oxide and zinc powder. The negative plates are prepared as the positive plates and then also subjected to heat treatment.

The charging and discharging of silver zinc cell may be represented by a single reversible equation given below :

$$\underset{\text{Positive plate}}{Ag_2O} + \underset{\text{Negative plate}}{Zn} \underset{\text{Discharge}}{\overset{\text{Charge}}{\rightleftharpoons}} \underset{\text{Positive plate}}{2Ag} + \underset{\text{Negative plate}}{Zno} + \text{electrical energy}$$

The monovalent silver oxide (Ag_2O) is reduced to silver (Ag) at positive plate and zinc is oxidized to zinc oxide (ZnO) at negative plate during discharge. During charge reverse reactions take place.

From the point of view of expenditure of material the most economical type of cell is the silver-zinc cell, but it is also the most expensive of the cells.

Silver-zinc cells have the highest ampere-hour and watt-hour efficiencies. The specific capacity (capacity per unit weight) of a silver-zinc battery is 4 to 5 times greater than that of other types of batteries. Their self discharge is low. They have energy density as high as 150 Wh/kg weight of cell.

These batteries are designed for both short-time and continuous duty. These batteries can withstand much heavier discharge currents than the permissible for other types of batteries. They can operate over a wide range of temperatures (– 20 °C to + 60 °C).

They are mainly employed in military services in communication equipment, portable radar sets and night vision equipment. They are also employed in heavy weight torpedoes and for submarine propulsion.

11.11. FUEL CELLS

Fuel cells differ from storage batteries in that their electrode materials are not changed in chemical composition and, therefore, they do not need recharging. They possess advantages over other types of energy convertors (engines, turbines, generators)—they have no moving part except certain control components, they operate at room temperature, produce no obnoxious fumes, use fuels that are not difficult to procure, need little maintenance, are silent and more efficient in operation than any other type of energy convertor. Another reason for interest in the fuel cells is that their efficiency and cost per kW of power are independent of size (or rating) of the fuel cell. This advantage makes the prospects of fuel cells very attractive as portable power plants for space crafts, locomotives etc.

Theoretically a fuel cell should be capable of generating electrical energy very efficiently, however, the development costs are very high. It is necessary to work at high temperatures or high pressures or use only costly catalysts for the reactions to take place at a speed in order to give high current densities required for an economic plant. Other drawbacks of the fuel cells are low voltage and low service life.

The fuel cell has considerable potentialities, especially for such applications as submarines, vehicles and locomotives and in remote and rural areas.

A fuel cell normally consists of two electrodes separated by an electrolytic solution. A fuel reactant, usually hydrogen or carbon monoxide is fed into one porous electrode and oxygen or air is fed into the other porous electrode. The electrodes should be capable of passing through both fuel and electrolyte and also conduct electrons to the terminals. The electrodes must contain a chemical catalyst that breaks the fuel compound into atoms so that they are more reactive. The most commonly used catalysts are platinum and sintered nickel. The electrodes should neither have pores of too large size to cause bubbles of fuel gas nor of too small size to cause insufficient contact between the reactant and the electrolyte. The electrolyte solution must be high permeable to either H^+ or OH^- ion which is produced as an intermediate product at one of the electrodes. The same ion is transferred through the electrolyte to the other electrode where it combines with the other reactant.

The working of a fuel cell using hydrogen and oxygen is explained below :

When two permeable nickel electrodes are immersed in a well conducting electrolyte (say a solution of H_2SO_4 or KOH), negative electrode is fed with hydrogen, bubbled around it through the solution and positive electrode is fed with oxygen and the electrodes are connected together through an external circuit, then for every molecule of hydrogen consumed, two electrons pass from negative electrode to positive electrode where they react with absorbed oxygen. The operation of a fuel cell can be summarized as follows:

Reaction at negative electrode

$$2H_2 \longrightarrow 4H^+ + 4\,e^-$$

Every hydrogen molecule brought to the electrode surface is dissociated into two atoms by virtue of catalystic properties of the surface. These enter the solution as hydrogen ions leaving behind two electrons which pass through the external circuit to the positive electrode.

Reaction at positive electrode

$$4H^+ + O_2 + 4\,e^- \longrightarrow 2H_2O$$

The oxygen supplied to the positive electrode reacts with hydrogen ions from the electrolyte and the electrons to form water. Thus water is the waste product of the cell. For the above case the electrolyte is acidic and the intermediate ion is H^+.

In case of alkaline electrolyte (a typical 40% solution of KOH), the intermediate ion will be OH^- and the chemical reactions in the cell will be as follows:

At cathode $2H_2 + 4\,OH^- \longrightarrow 4\,H_2O + 4\,e^-$

At anode $2H_2O + O_2 + 4\,e^- \longrightarrow 4\,OH^-$

So, when the electrolyte is acidic, water is formed at the anode and when the electrolyte is alkaline, water is formed at the cathode.

If the electrodes are open-circuit, negative charges accumulate at hydrogen electrode. These negative charges attract potassium ion (K^+) of the electrolyte producing a double layer. Similarly the loss of electrons from the oxygen electrode results in a layer of positive charges which in turn attracts hydroxyl ions (OH^-) from the electrolyte and form a double layer. These electrical double layers build up at the electrodes until the potentials are such that they inhibit any further reaction between the electrolyte and fuel gases. An open-circuit voltage of 1.23 V at one atmospheric pressure at 25 °C is developed. If the circuit is closed through an external load resistance, the electrons flow from the hydrogen electrode to the oxygen electrode through the external circuit and take part in the reactions as mentioned above. The movement of electrons is from the hydrogen electrode to oxygen electrode. Thus hydrogen electrode serves as the *cathode* and the oxygen electrode as *anode*.

Fuel cell can be adopted to a variety of fuels by changing the catalyst, but 'hydrox' fuel cells using hydrogen and oxygen as fuels are the most efficient and most highly developed cells. Hydrogen can be obtained from natural gas by catalytic cracking of ammonia or hydro-carbons or as a by-product from some processes. Oxygen may be obtained from the air or from decomposition of oxides. A fuel cell power plant generally also contains a reformer and an invertor. The reformer uses chemical processes to convert the fuel to form that can be utilized by the cell while an invertor is used for converting output dc into ac.

A single hydrox fuel cell gives an emf of 1.23 V at 1 atmospheric pressure at 25 °C, as already mentioned. However, it is possible to develop useful potentials of 100 to 1,000 V and power level of 1 kW to 100 MW by connecting a number of cells in series, parallel combination. The current depends upon the physical size of the cell. The output of the fuel cell varies directly with pressure, so to increase the cell output, the gas pressure is raised. The optimum size of the cell is about 0.027 cubic meter per kW.

Fuel cells are particularly suited for low voltage and high current applications. Apollo Astronauts going to the Moon used fuel cells to convert hydrogen and oxygen to electricity. The power cells were located in the Apollo service module and provided the primary power source to operate life, support communication, guidance and other electrical systems.

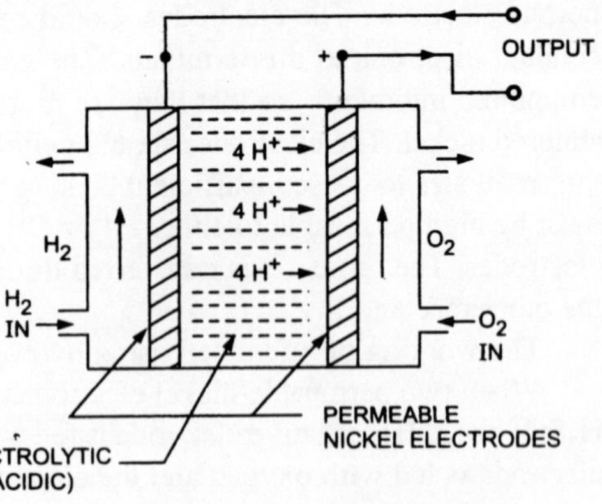

Fig. 11.18 *Fuel Cell*

11.12. NICKEL-HYDROGEN CELLS

Light-weight high-performance nickel-hydrogen cells are becoming more and more popular as the backup power source for space vehicles. The nickel-hydrogen cell is something of a hybrid between a battery and fuel cell combining many of the features of a nickel-cadmium aerospace cell with aspects of a hydrogen-oxygen fuel cell.

Nickel-hydrogen system have the advantages of (*i*) high energy density, (*ii*) a long cycle life, even in deep cycling applications, (*iii*) high tolerance for electrical abuse (over-charging and reversing) and (*iv*) state of charge indication through hydrogen pressure.

The drawbacks are high cost and complexity.

These cells employ a nickle positive electrode that is similar to those used on aerospace nickel-cadmium cells. The electrolyte is potassium hydroxide (KOH) just as with the nickel-cadmium cells. The negative electrode consists of gaseous hydrogen contained at high pressure.

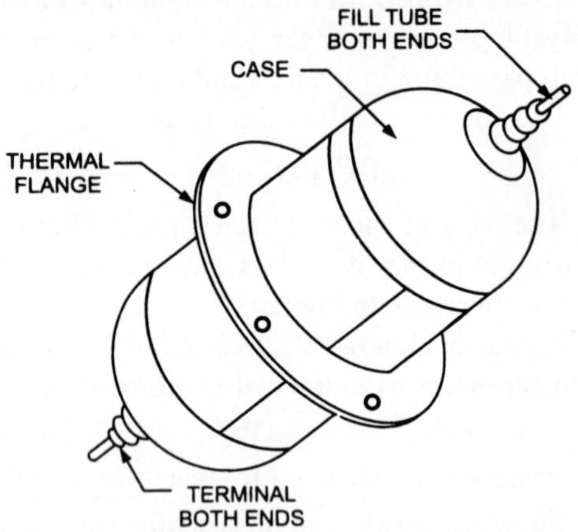

Fig. 11.19 *Typical Nickel-Hydrogen Aerospace Cell*

The pertinent reactions are

At positive electrode

$$NiOOH + H_2O + e^- \underset{\text{Charge}}{\overset{\text{Discharge}}{\rightleftharpoons}} Ni(OH)_2 + OH^- \qquad ...(0.490\ V)$$

At negative electrode

$$\frac{1}{2}H_2 + OH^- \underset{\text{Charge}}{\overset{\text{Discharge}}{\rightleftharpoons}} H_2O + e^- \qquad ...(0.828\ V)$$

Overall cell reaction

$$\frac{1}{2}H_2 + NiOOH \underset{\text{Charge}}{\overset{\text{Discharge}}{\rightleftharpoons}} Ni(OH)_2 \qquad ...(1.318\ V)$$

During overcharge oxygen generated at the nickel electrode and excess hydrogen recombine to form water using a platinum electrode to catalyze the recombination. The efficiency of this recombination process is one of the reasons that the nickel hydrogen cell is so tolerant of overcharge.

To be effective, these cells must maintain a hydrogen pressure of about 0.7 kg/mm^2. Developing light weight pressure vessels capable of containing hydrogen at such a high pressure needs innovative design and careful attention to details of material selection and manufacturing techniques.

The first aerospace nickel-hydrogen battery was launched in 1977. Since then, nickel-hydrogen battery performance has been highly satisfactory. Nickel-hydrogen batteries have been powering the INTELSAT–V family of communication satellites since 1983.

HIGHLIGHTS

1. Batteries are either primary or secondary. Primary batteries can be used only once because the chemical reactions that supply the current are irreversible. Secondary batteries, sometimes called *storage batteries* or *accumulators*, can be used, recharged and reused. In these batteries, the chemical reactions that provide current from the battery are readily reversed when current is supplied to the battery. The process of inducing or storing energy in an accumulator is called the *charging*, and the process of giving out energy in the form of an electric current, the *discharging*. Accumulators or storage batteries owe their name "secondary" due to the fact that they can supply electrical energy only after they have been charged.
2. The charging and discharging of a lead acid cell may be represented by a single reversible equation given below:

$$\underset{\text{Positive plate}}{PbO_2} + 2H_2SO_4 + \underset{\text{Negative plate}}{Pb} \underset{\text{Discharge}}{\overset{\text{Charge}}{\rightleftharpoons}} \underset{\text{Positive plate}}{PbSO_4} + 2H_2O + \underset{\text{Negative plate}}{PbSO_4} + \text{Electrical energy}$$

3. The efficiency of a cell can be given in two ways as given below:

$$\text{Quantity or ampere-hour efficiency, } \eta_{AH} = \frac{\text{Ampere-hours of discharge}}{\text{Ampere-hours of charge}} \times 100 = \frac{I_d T_d}{I_c T_c} \times 100$$

$$\text{Energy or watt-hour efficiency, } \eta_{WH} = \frac{\text{Output in watt-hours}}{\text{Input in watt-hours}} \times 100$$

$$= \frac{I_d \times T_d \times V_d}{I_c \times T_c \times V_c} \times 100 = \frac{I_d \times T_d}{I_c \times T_c} \times 100 \times \frac{V_d}{V_c} = \eta_{AH} \frac{V_d}{V_c}$$

4. Sealed lead cells and batteries are rugged in construction, compact in size, have long life, excellent performance, simplified charging, design flexibility and need little maintenance.
   Starved-electrolyte batteries have a wide range of applications. These batteries are employed as standby units in computers, emergency lights, medical equipment, security alarm systems, telecommunication equipment and uninterruptible power supplies; as portable units in cellular phones, toys, cordless appliances and portable lighting, and as alternate units in computers, consumer electronics and instrumentation.
5. The electrochemical process in the nickel-iron accumulators can be expressed by the equation

$$\underset{\text{Positive plate}}{2Ni(OH)_3} + 2KOH + \underset{\text{Negative plate}}{Fe} \underset{\text{Discharge}}{\overset{\text{Charge}}{\rightleftharpoons}} \underset{\text{Positive plate}}{2Ni(OH_2)} + 2KOH + \underset{\text{Negative plate}}{Fe(OH)_2} + \text{Electrical energy}$$

6. Nickel-iron batteries are comparatively lighter in weight, rugged in construction, more durable and mechanically stronger, they need little maintenance, they have long service life and they do not evolve obnoxious fumes. But they have the disadvantages of high initial cost, high internal resistance, lower emf per cell and lower operating efficiency. Edison battery may be considered as a special-service battery to be employed where lightness, freedom from acid fumes or ability to stand severity of vibrations and shocks are particularly important, and where momentary high discharge currents are not required.
7. The electrochemical process in the nickel-cadmium cells for charge and discharge may be given as

$$\underset{\text{Positive plate}}{2Ni(OH)_3} + 2KOH + \underset{\text{Negative plate}}{Cd} \underset{\text{Discharge}}{\overset{\text{Charge}}{\rightleftharpoons}} \underset{\text{Positive plate}}{2Ni(OH)_2} + 2KOH + \underset{\text{Negative plate}}{2Cd(OH)_2} + \text{Electrical energy}$$

8. The nickel-cadmium accumulators are superior to lead acid accumulators as well as nickel-iron accumulators for all applications except those in which weight is important criterion of selection of accumulator.
9. The charging and discharging of silver zinc cell may be represented by a single reversible equation given below:

$$\underset{\text{Positive plate}}{Ag_2O} + \underset{\text{Negative plate}}{Zn} \underset{\text{Discharge}}{\overset{\text{Charge}}{\rightleftharpoons}} \underset{\text{Positive plate}}{2Ag} + \underset{\text{Negative plate}}{Zno} + \text{Electrical energy}$$

10. Fuel cells differ from storage batteries in that their electrode materials are not changed in chemical composition and, therefore, they do not need charging.

EXERCISES

1. Distinguish between a primary and secondary cell. In what form is the energy stored in the cell ?
2. Give the classification of secondary accumulators. What for stationary batteries are used?
   Describe the construction of lead-acid battery. Write down the chemical reactions during charging and discharging process of lead-acid cells.
3. Explain with connection diagrams the two charging methods of a lead-acid cell by constant current. State the indications of a fully charged lead-acid cells. State the precautions to be taken during charging.
4. (*a*) List out the different parts of lead acid battery and mention the materials used for each part.
   (*b*) What are the indications of a fully charged cell ?
   (*c*) List factors affecting the capacity of a cell.
5. State the factors which affect the capacity of battery mentioning the nature of effects.
   Write the care and maintenance of lead acid batteries.
   What do you understand by the term trickle charging ?
   Give the features and benefits of sealed lead-acid batteries.
6. Explain the construction of nickel-iron cell. Write the chemical reactions during charging and discharging of nickel-iron cells. Give its electrical characteristics, advantages, disadvantages and applications.
7. Give the construction, characteristics, merits, demerits and applications of nickel-cadmium accumulators.
   Mention the application of (*i*) lead acid (*ii*) nickel-iron and (*iii*) nickel-cadmium cells.
8. Compare lead acid cell with nickel-iron cell in respect of (*a*) internal resistance (*b*) life (*c*) cost and (*d*) ampere-hour efficiency.
9. Write down the active materials of silver-zinc batteries. Give the merits, demerits and applications of silver-zinc batteries.
10. Give the construction and working of a fuel cell in brief. Give its merits, limitations and applications.
11. Write short notes on the following:
    (*a*) Care and maintenance of lead-acid batteries.
    (*b*) Features and benefits of sealed nickel-cadmium accumulators.
    (*c*) Silver zinc batteries.
    (*d*) Nickel hydrogen cell.

SHORT ANSWER TYPE QUESTIONS WITH ANSWERS

Q. 1. What is difference between primary and secondary batteries ?

Ans. Batteries are either primary or secondary. Primary batteries can be used only once because the chemical reactions that supply the current are irreversible. Secondary batteries, sometimes called *storage batteries* or *accumulators*, can be used, recharged and reused. In these batteries, the chemical reactions that provide current from the battery are readily reversed when current is supplied to the battery.

Q. 2. Lead acid battery can be charged where as primary cell cannot be charged. Why ?

Ans. Primary cells can be used only once because the chemical reactions that supply the electric current are irreversible but in lead acid battery, the chemical reactions that provide the current are readily reversed when current is supplied to the battery.

Q. 3. In what form is energy stored in a secondary battery ?

Ans. Energy is stored in the form of electrical energy in a secondary battery.

Q. 4. Why Faure process is much suitable for manufacturing of negative plates rather than positive plates?

Ans. Faure process is much suitable for manufacturing of negative plates because negative active material (sponge lead) is quite tough (not hard and brittle as lead peroxide is) and moreover, it undergoes a comparatively negligible change in volume during charging and discharging processes.

Q. 5. Why in grouping of lead acid cells the negative group is made to have one plate more than positive group ?

Ans. The negative group of the plates contains one more plate than that of the positive group so that when interleaved both the end plates become the negative. This is done so that all the positive plates can work equally well on both sides.

If a positive plate were used at one end, one of its sides would remain inactive, when the other side expands, and thus the plate is likely to buckle.

Q. 6. Do we use ac or dc to charge a battery ?

Ans. For charging of a battery dc is essential.

Q. 7. Why should the lead acid cells not be allowed to stand completely discharged ?

Ans. Lead acid cells shall not be allowed to stand completely discharged because the lead sulphate formed on the plates crystallizes and the proper reforming of the plates becomes difficult.

Q. 8. What are the factors that affect the battery capacity.

Ans. The capacity of battery depends upon several factors, principal among which are area of plate surface, quantity, arrangement and porosity of the active material used in the manufacture of plates, quantity and specific gravity of the electrolyte used and the porosity of the separators, rate of charge and temperature.

Q. 9. Why the lead acid battery should not be allowed to discharge beyond 1.75 V ?

Ans. The lead acid battery should never be allowed to discharge beyond 1.75 V otherwise lead sulphate will be formed on the electrodes which is hard, insoluble and increases the internal resistance of the battery.

Q. 10. Why is the watt-hour efficiency of a cell lower than its ampere-hour efficiency ?

Ans. Watt-hour efficiency is given as

$$\eta_{AH} \times \frac{\text{average pd during discharge, } V_d}{\text{average pd during charge, } V_c}$$

and V_d is always lower than V_c, therefore η_{WH} is always lower than its η_{A-H}.

Q. 11. Why the state of charge of a nickel-iron cell cannot be determined by measuring the specific gravity of the electrolyte ?

Ans. The state of charge of a nickel-iron cell cannot be determined by the specific gravity of electrolyte as it does not change during charging and discharging.

Q. 12. Why the ampere-hour and watt-hour efficiencies of an Edison cell are lower than those for lead-acid cell ?

Ans. Due to comparatively high internal resistance, the efficiencies of the Edison cell are lower than those of the lead acid cell.

Q. 13. Why and where will you recommend the use of nickel-iron batteries ?

Ans. Nickel-iron batteries are used for the propulsion of industrial trucks, and mine locomotives and for railway car lighting and air-conditioning because of their rugged construction, compact size, lighter in weight, more durability, long service life and as they need little maintenance and there is no evolution of obnoxious fumes.

Q. 14. What is the basic difference between nickel-iron and nickel- cadmium cells ?

Ans. The most important difference between the nickel-cadmium and nickel-iron cells is the arrangement of plates. Where as in nickel-iron cells the extreme plates are negative, but in the nickel-cadmium cells the extreme plates are positive.

Q. 15. Why use of nickel-cadmium accumulators is limited in spite of its numerous advantages over nickel-iron accumulators and lead acid accumulators ?

Ans. The use of nickel-cadmium accumulators is limited as they are heavier in weight and expensive in cost.

PROBLEMS

1. A battery of accumulators of emf 50 V and internal resistance 2 Ω is charged on 100 V. What series resistance will be required to give a charging current of 2 A ? **[Ans.** 23 Ω**]**
2. A battery of emf 80 V and internal resistance 2 Ω is to be charged from 200 V mains. What value of series resistance will be required to give a charging current of 5 A. If the energy costs 25 paise per kWh; what will it cost to charge the battery for 10 hours ? Also calculate the amount of energy wasted as heat in the external resistance.
[Pb. Univ. Applied Electricity Nov; 1974]
[Ans. 22 Ω; Rs 2.50; 5.5 kWh**]**
3. Determine the ampere-hour and watt-hour efficiencies of an accumulator which is charged for 12 hours at 25 A at an average voltage of 2.5 V and is discharged in 10 hours at a load of 20 A at an average voltage of 2.25 V.
[Pb. Univ. Applied Electricity 1972]
[Ans. 66.67%; 60%**]**

4. A discharged battery is put on charge at 5 A for 3.5 hours at a mean charging voltage of 13.5 V. It is then discharged in 6 hours at constant terminal voltage of 12 V through a resistance of R ohms. Calculate the value of R for an A-H efficiency of 85%. Also calculate the watt-hour efficiency. **[Ans.** 4.84 Ω ; 75.56%]
5. A battery is charged keeping the voltage constant. The current initially was 10.5 A and finally 2.5 A in a period of 8 hours. The battery was discharged at a constant current of 6 A for 7 hours. Calculate ampere-hour efficiency of this battery. **[Ans.** 80.77%]
6. A battery is charged for 24 hours at a constant current of 5 A with initial voltage of 1.8 V and final voltage of 2.0 V. Calculate the watt-hour efficiency of the battery if it is discharged at a constant current of 6 A at 1.9 V for 18 hours. **[Ans.** 81%]
7. An alkaline cell is discharged at a steady current of 4 A for 12 hours, the average terminal voltage being 1.2 V. To restore to its original state of charge, a steady current of 3 A for 20 hours is required, the average terminal voltage being 1.44 V. Calculate watt-hour efficiency. [Ans. 66.67%]

❑❑❑

Chapter 12

Cables

12.1. WIRE AND CABLE

The terms wire and cable are used more or less synonymously in house wiring. Strictly speaking, single wire, may be bare or covered with insulation, is known as a wire and several wires stranded together is known as a *cable.* But in practice bare conductors, whether single or stranded together are termed as *wires* and conductors covered with insulation are termed as cables.

The necessary requirements of a cable are that it should conduct electricity efficiently, cheaply and safely. This should neither be so small so as to have a large internal voltage drop nor be too large so as to cost too much. Its insulation should be such as to prevent leakage of current in unwanted direction and thus to minimize risk of fire and shock.

The cable consists of three parts :

(*a*) the *conductor or core;* the metal wire or strand of wires carrying the current :

(*b*) the *insulation or dielectric;* a covering of insulating material to avoid leakage of current from the conductors and

(*c*) the *protective covering* for protection of insulation from mechanical damage.

12.2.1. Conductor Materials Used in Cables. The function of conductor, usually known as a core, in a cable is to carry electrical current. Copper and aluminium are the materials used as conductors in power and lighting cables.

1. **Copper.** Though silver is the best conductor but due to its higher cost it is rarely used. The next best conductor is copper. It is cheaper comparatively.

The electrical conductivity of copper is comparatively high. The resistivity of pure copper at 20ºC is 1.786×10^{-8} Ω-m. It is mechanically strong, hard, extremely tough, durable and ductile. It is highly resistive to corrosion, oxidation and pitting. It can withstand dampness and high temperature. It can be easily soldered and welded. The specific weight of copper is 8,900 kg/m^3 at 20ºC. Its melting point is 1,083 ºC.

The copper conductors may be annealed or hard drawn. Annealed copper conductors are comparatively soft and pliable and are most suitable for indoor and outdoor wires and cables laid or fixed in position. Hard drawn copper conductors, which have a very high tensile strength (40 kg/mm^2), are used as overhead wires mainly in the bare form.

2. **Aluminium.** Aluminium is frequently used in place of copper for bare electric cables used for long distance power distribution. The electrical conductivity of aluminium is about 60% of that of copper ((resistivity being 2.87×10^{-8} Ω-m at 20ºC) so for same resistance for a given length, the aluminium required will be 1.61 times of that of copper in volume and 1.26 times of that of copper in diameter. When the cost of

insulating the large conductor is reckoned in, aluminium cables only come out cheaper in the larger sizes, and even then the film of oxide, which is always present on the surface of the metal, necessitates special precautions when connecting up. The only application of aluminium cables for wiring in buildings is for the 'continuous bus-bar" system of distribution, used some times in blocks of flats or office buildings for rising mains and submains of larger sectional area, for which this metal is some times selected instead of copper.

12.2.2. Insulating Materials. Insulating materials used in cable will be discussed in detail later on (Refer to Art 12.8).

12.1.3. Mechanical Protection. As all the insulating materials used in the manufacture of cables are mechanically weak, they require some form of protection against mechanical injury. Mechanical protection is usually provided to power cables laid direct in the ground by providing two layers of steel tape in such a way that upper layer covers the joint in the lower layer. With regard to ordinary main cables, an aluminium sheathing has been recently introduced. The cables provided with aluminium sheathing do not require armouring and are usually finished with water tight protection such as rubber sandwhich or a covering of PVC, the latter being preferred.

All cables, whether armoured or not, are provided with an external covering or serving of impregnated *braiding* or tape as an additional protection against damage and moisture.

12.2. TYPES OF CABLES USED IN INTERNAL WIRING

The wires employed for internal wiring of buildings may be divided into different groups according to (*i*) conductor used (*ii*) number of cores used (*iii*) voltage grading and (*iv*) type of insulation used.

According to the conductor material used in cables, these may be divided into two classes known as *copper conductor cables* and *aluminium conductor cables.*

According to the number of cores, the cable consists of, the cables may be divided into classes known as *single core* cables; *twin core* cables; *three core cables;* two core with ECC (earth continuity conductor) cables etc.

According to voltage grading the cables may be divided into two classes : (*i*) 250/440 volt cables and (*ii*) 650/1,100 volts cables.

According to type of insulation the cables are of the following types :

1. Vulcanized Indian-Rubber (VIR) insulated cables.
2. Tough rubber sheathed (TRS) or cab tyre sheathed (CTS) cables.
3. Lead sheathed cables.
4. Polyvinyl chloride (PVC) cables.
5. Weather proof cables.
6. Flexible cords and cables.
7. XLPE cables.

1. **Vulcanized Indian Rubber (VIR) Cables.** VIR, cables are available in 240/415 volt as well as in 650/1,100 volt grades.

VIR cable consists of either tinned copper conductor covered with a layer of vulcanized Indian rubber insulation. Over the rubber insulation cotton tape sheathed covering is provided with moisture resistant compound bitumen wax or some other insulating material for making the cables moisture proof. The thickness of rubber insulation depends upon the voltage grade for which the cable is required.

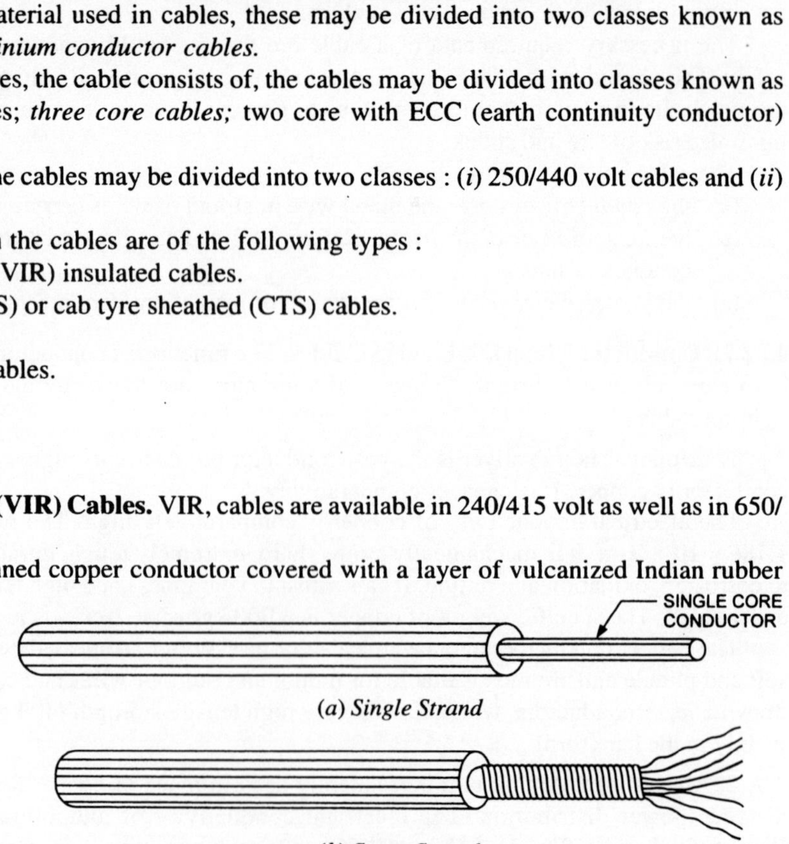

Fig. 12.1 *Single Core VIR Cables*

The copper conductor is tinned to provide protection against corrosion due to presence of traces of sulphur, zinc oxide and other mineral ingredients in the VIR.

A single core single strand VIR wire may be employed but larger cables have to be stranded.

Single core single strand VIR wire and single core seven strand VIR cables are shown in Figs 12.1(*a*) and 12.1(*b*) respectively.

2. **Tough Rubber Sheathed (TRS) or Cab Tyre Sheathed (CTS) Cables.** These cables are available in 250/440 volt and 650/1,100 volt and used in CTS (or TRS) wiring. TRS cable is nothing but a vulcanized rubber insulated conductor with an outer protective covering of tough rubber, which provides additional insulation and protection against wear and tear. These cables are water proof, hence can be used in wet conditions. These cables are available as single core, circular twin core, circular three core, flat three core, twin or three core with an earth continuity conductor (ECC). The cores are insulated from each other and covered with a common sheathing. Different types of TRS cables are shown in Fig. 12.2.

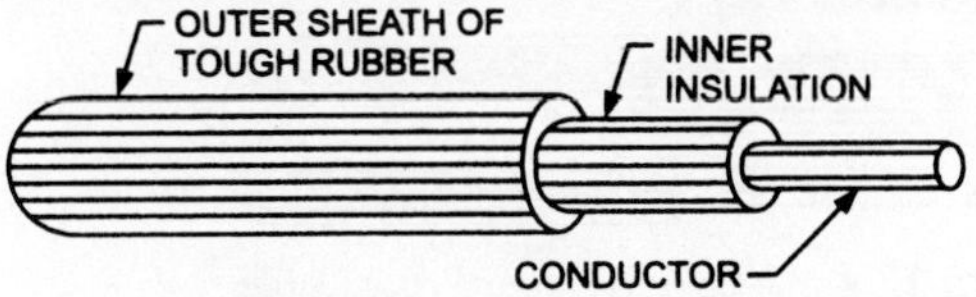

(*a*) *Single Core Single Strand TRS Wire*

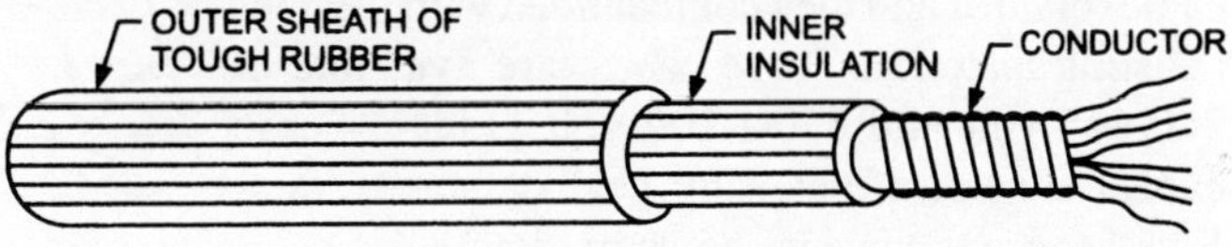

(*b*) *Single Core Seven Strand TRS Cable*

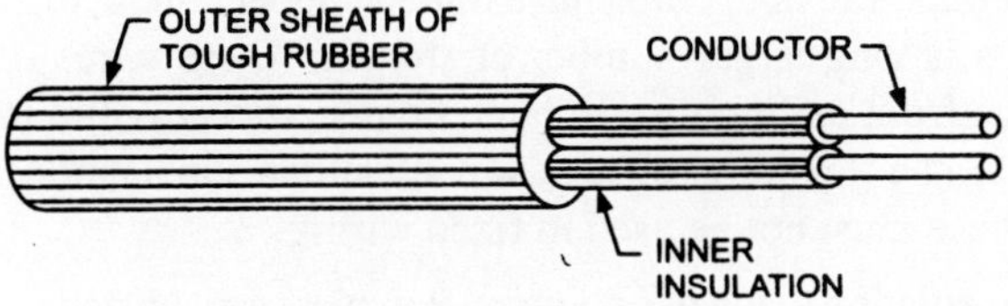

(*c*) *Twin Core Single Strand TRS Cable For Indoor Service*

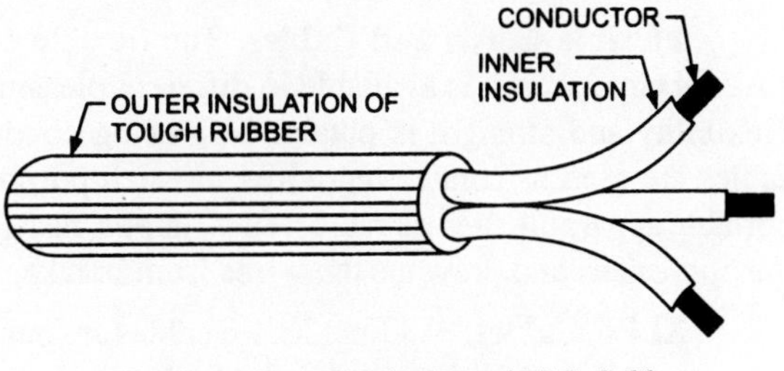

(*d*) *Three Core Single Strand TRS Cable For Indoor Service*

Fig. 12.2

In wiring of 3 pin plugs separate earth wire may be used as it is cheaper in cost and easier in installation.

These cables are cheaper in cost and lighter in weight than lead alloy sheathed cables and have the properties similar to those of lead sheathed cables and thus provide cheaper substitute to lead sheathed cables.

3. **Lead Sheathed Cables.** These cables are available in 240/415 volt grade. The lead sheathed cable is a vulcanized rubber insulated conductor covered with a continuous sheath of lead. The lead sheath provides very good protection against the absorption of moisture and sufficient protection against mechanical injury and so can be used without casing or conduit system. It is available as a single core, flat twin core, flat three core and flat twin or three core with an earth continuity conductor. Two-core lead sheathed cable is shown in Fig. 12.3.

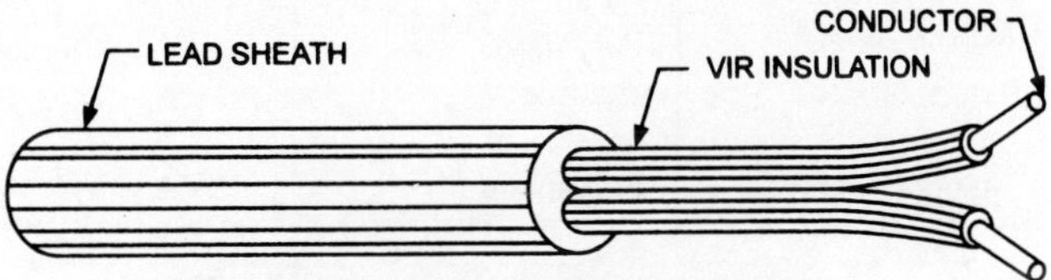

Fig. 12.3 *2-Core Lead Sheathed Cable*

4. **Polyvinyl Chloride (PVC) Insulated Cables.** These cables are available in 250/440 volt and 650/1,100 volt grades and are used in casing-capping, batten and conduit wiring system. In this type of cable conductor is insulated with PVC insulation. Since PVC is harder than rubber, PVC cable does not require cotton taping and braiding over it for mechanical and moisture protection.

PVC insulation is preferred over VIR insulation because of the following reasons:

(*i*) PVC insulation has better insulating qualities.

(*ii*) PVC insulation provides better flexibility.

(*iii*) PVC insulation has no chemical effect on metal of the wire.

(*iv*) Thin layer of PVC insulation will provide the desired insulation level.

(*v*) PVC coated wire gives smaller diameter of cable and, therefore, more number of wires can be accommodated in the conduit of a given size in comparison to VIR or CTS wires.

PVC cables are most widely used for internal wiring these days. Though the insulation resistance of PVC is lower than that of VIR but its effect is negligible for low and medium voltages, below 600 V.

5. **Weather Proof Cables.** These cables are used for outdoor wiring and for power supply or industrial supply. These cables are either PVC insulated or vulcanized rubber insulated conductors being suitably taped (only in case of vulcanized rubber insulated cable) braided and then compounded with weather resisting material. These cables are available in 240/415 volt and 650/1,100 volt grades. These cables are not affected by heat or sun or rain. Weather proof cable is shown in Fig. 12.4.

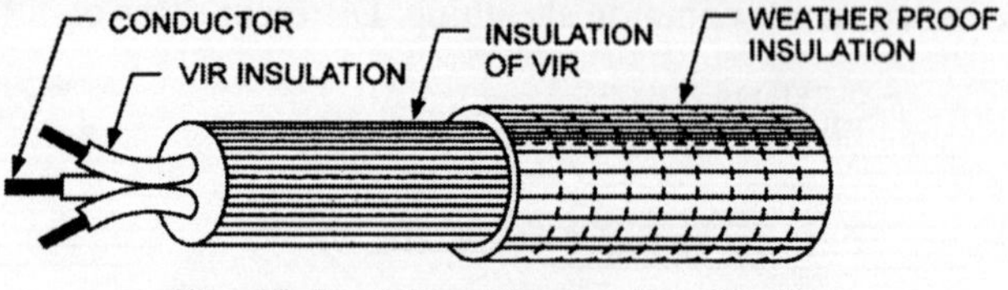

Fig. 12.4 *3-Core Weather Proof Cable*

Although TRS cables can be used for outdoor purposes but due to their higher cost, weather proof cables are generally used for outdoor services.

6. **Flexible Cords and Cables.** The flexible cords consist of wires silk/cotton/plastic covered. Plastic cover is popular as it is available in different pleasing colours. Flexible cords have tinned copper conductors. Flexibility and strength is obtained by using conductors having larger number of strands. These wires or cables are used as connecting wires for such purposes as from ceiling rose to lamp holder, socket outlet to portable apparatus such as radios, fans, lamps, heaters etc. The flexibility of such wires facilitates in handling the appliances and prevents the wires from breakage. These must not be used in fixed wiring.

7. **XLPE Cables.** PVC and XLPE cables are built of insulation made of polymers. Polymers are substances consisting of long macro molecules built up of small molecules or groups of molecules as repeated units. These are divided into homopolymers and copolymers. Homopolymers are built by reactions of identical monomers. Copolymers are built up of at least two different kinds of monomers.

TABLE 12.1
Polymers Used With Cables

	Thermo-Plastics	*Intermediate Substances*	*Elastomers*	
	PE		Natural Rubber	
Homopolymers	Polyethylene	–	XLPE	Cross-Linked Polyethylene
	PVC			
	Polyvinyl chloride	–	PCP	Poly-Chloro-Prene
			CSPE	Chloro-Sulphonated-Polyethylene
Copolymers	Polyamide	EVA Ethylene Vinyl Acetate Copolymer	SiR	Silicon Rubber
			EPM	Ethylene Propylene Rubber
			EPDM	Ethylene Propylene Terpolymer Rubber
		Polyvrethane	Butyl Rubber	
			SBR	Styrene Butadien Rubber
			NBR	Nitride Butadien Rubber

The mechanical properties of the polymers *e.g.* tensile strength, elongation elasticity and resistance against cold depend upon chemical structure. Their resistance against external chemical influences, acids, bases or oils together with their electrical and thermal characteristics are the decisive factors for the usefulness of cables insulated and sheathed with these materials.

Advantages of PVC Cables Over Other Types of Cables.

1. Non-hygroscopic insulation almost unaffected by moisture.
2. Non-migration of compound allowing vertical installation.
3. Complete protection against most forms of electrolytic/chemical corrosion.
4. Tough/Resilient sheath with excellent fire resisting qualities.
5. Good ageing characteristics.
6. Not affected by vibrations.

Advantages of XLPE Cables Over Both PVC and All Other Types of Cables.

1. Higher current rating.
2. Higher short-circuit current rating.
3. Longer service life.
4. Can withstand 130 °C (maximum) for short time and is favourable to endure short-circuit stresses.
5. It is less sensitive to the setting of network protection.
6. Because of thermosetting process taking place through cross-linking, crack resistance is increased.
7. Due to chemical cross-linking internal stresses are reduced. Consequently material is less sensitive, during manufacture, to the setting of the cooling gradient.
8. The thermal resistivity of cross-linked material is favourably low, compared to thermoplastic material.
9. Low dielectric loss (a significant advantage).
10. Excellent mechanical features of the insulation improves the protection against external effects.
11. The resistance of XLPE to acids, alkalies is outstanding and often compensates adverse environmental influences.

Presently XLPE cables are used extensively for high tension and low tension work. Even for distribution work at 3-phase 415 V industry is using XLPE cables.

Cable Specification Code :

Constituent	*Code*
Aluminium Conductor	A
XLPE Insulation	2X
Round Steel Wire	W
Flat Steel Strip Armour	F
Double Round Steel Wire Armour	WW
Double Flat Steel Strip Armour	FF
Non-magnetic (Al) Round Wire Armour	Wa
Non-magnetic (Al) Strip Armour	Fa
PVC Outer Sheath	Y

Operating Characteristics of XLPE Cables

Maximum conductor temperature for continuous operation :	90 °C
Ambient air temperature :	40 °C
Standard ground temperature :	30 °C
Thermal resistivity of soil :	150 °C cm/watt
Thermal resistivity :	350 °C cm/watt
Depth of laying for cables laid direct in ground :	75 cm
Minimum bending radius for multi-core cables : (D is diameter of cable)	12 D
Maximum conductor temperature during short-circuit :	250°C
Maximum ambient temperature :	85°C

12.3. MULTI-STRAND CABLES

Multi-strand cables have got the following advantages with respect to the single solid conductor and hence preferred.

(*i*) The multi-strand cables are more flexible and durable and, therefore, can be handled conveniently.

(*ii*) The surface area of multi-strand cable is more as compared to the surface area of equivalent single solid conductor, so heat radiating capacity, being proportional to the surface area, is more.

(*iii*) Skin effect is better as the conductors are tubular, specially in case of high frequency.

The number of strands in stranded cable must be 3, 7, 19, 37, 61, 91 and so on in order to obtain a circular contour. The section of a 3-strand cable at right angle to its length is three circles touching one another, the centres of which are the corners of an equilateral triangle. All other have a centrally disposed conductor with all the other around it. Thus a 7-strand cable has one central wire with 6 wires surrounding it; the 19-strand cable has another 12 wires surrounding the 7-strands; the 37-strand cable has another 18 wires surrounding the 19-strands and so on. It is seen that each layer of wires has always 6 more wires in it than the layer beneath it.

The various conductors are spiralled round the central conductor and when there is more than one layer, alternate layers are spiralled in opposite directions. This to is prevent "bird-aging" when the conductor is bent.

Since the length of each spiralled conductor is greater than the central strand and the current flows along the various conductors, resistance is increased and so we should be particular in the designation of the x-section.

The nominal cross-section is the area of the x-section of one conductor in a plane perpendicular to its length, multiplied by the number of conductors. The actual cross-section is the area of the oblique cross-section due to the cutting of the stranded cable by a plane perpendicular to the core of the cable, multiplied by the number of conductors. The equivalent x-section, *defined as the cross-section of a solid conductor of the same length as the cable and having the same resistance at the same temperature,* is taken less than nominal cross-section because of increase in resistance due to spiralling. Loss in lay *i.e.* the additional length due to spiralling is taken approximately 2% by the cable manufacturers and this percentage has been added in giving the equivalent x-sectional area. Actual cross-section is 2 percent more instead of 2 percent less. In calculating weight of the conductor the 2 percent loss in the lay has to be added.

The size of a cable is given in various manners, explained below.

The size of the cable can be given by a designation giving number of strands and gauge number of each strand. For example a cable having 3 strands each of gauge 20 SWG may be referred as 3/20. In the system of specifying the size of cables, numerator indicates the number of strands employed and denominator indicates the gauge number of each strand.

The size of cables, may also be given in terms of number of strands and diameter of each strand in mm. For example a cable having 19 strands, each strand of diameter 1.12 mm may be referred as 19/1.12 mm.

The cable size is often denoted in terms of total cross-sectional area of the core instead of number and diameter of strands. As 19/1.12 mm cable has a cross-section of 19.35 mm^2 so this cable is often referred to as a 19.35 mm^2 cable.

12.4. GENERAL SPECIFICATIONS OF CABLES

Cables are specified providing

(*i*) Size of the cable in metric system (*e.g.* 19/2.24, 7/1.70, 7/2.24, 7/2.50 etc.) giving the number of strands used and diameter of each strand, or giving the area of cross-section of conductor used as explained in Art 12.3.

(*ii*) Type of conductor used in cables (copper or aluminium).

(*iii*) Number of cores that cable consists of *e.g.* single core, twin core, three core, twin core with ECC etc.

(*iv*) Voltage grade (240/415 V or 650/1,100 V grade).

(*v*) Type of cable with clear description regarding insulation, shielding, armouring, braiding etc.

A few specifications of cables are given below :

(*i*) 7/1.12, aluminium conductor, flat twin core, 650/1100 V grade PVC cable conforming to the IS : 1885 (Part 32)-1972.

(*ii*) 50 mm^2 aluminium conductor 3½ core 1100 V PVC cable, PVC sheathed, armoured cable conforming to the IS : (Part 32)-1972.

12.5. GENERAL CONSTRUCTION OF UNDERGROUND CABLE

The general construction of an underground cable is given below.

The underground cable employed for transmission of power at high voltage consists of one central core or a number of cores (two, three or four) of tinned stranded copper or aluminium conductors insulated from each other by paper or varnished cambric or vulcanized bitumen or impregnated paper. A metallic sheath of lead, alloy or of aluminium is provided around the insulation to protect it against ingress of moisture, gases or other damaging liquids (acid or alkalies) in the soil and atmosphere. A metal is essential for sheath because no organic material is sufficiently impervious to moisture. The main advantages of lead sheaths are the comparative ease with which they are made in lead presses, their flexibility and high corrosion resistance while they have got drawbacks of large specific gravity, low mechanical strength, fluidity and small resistance to vibrations. The hardness, mechanical strength and resistance to vibrations of lead sheaths can be considerably improved by adding alloying mixtures (about 1 per cent of copper, tin, bismuth, etc). Aluminium sheaths being cheaper (costing 3.5 times less than the lead sheaths) and having much greater mechanical strength than lead sheaths, low weight and low fluidity are now being increasingly employed.

For the protection of metallic sheath against corrosion and from mechanical injury a layer of *bedding* consisting of paper tape compounded with a fibrous material (such as jute, cotton or hessian tape) is provided over the metallic sheath. The compound used should be of such a nature that it does not react with the material of the armouring and the lead sheath but simultaneously it is adhesive enough so as to stick both to the lead sheath and the armouring. Bedding is employed in paper insulated lead covered cables. It is not required in PVC cables.

Over the layer of bedding *armouring* consisting of one or two layers of galvanized steel wire or steel tape is provided to save the cable from mechanical injury and over the armouring a layer of fibrous material similar to that of bedding, known as *serving,* is provided in order to protect the armouring. Steel tape armour consists of two steel tapes coated with preservative compound, applied hellically in the same direction over a fibrous bedding, the outer layer covering the spaces between the turns of the inner layer. This provides sufficient protection against mechanical damage but steel wire armour is recommended where additional longitudinal stresses may occur during installation or in service. Double wire armours are employed for high degree of mechanical protection.

Single core cables for ac systems are not provided with armouring because eddy currents induced in the steel armour causes additional power losses. In such cables mechanical protection is provided by plastic wrap. Aluminium sheathed cables, owing to comparative hardness of sheath, need not to be armoured in the same way as the lead sheathed cables but for special conditions armouring is provided for these cables too.

Fig. 12.5 illustrates the various parts of an underground cable.

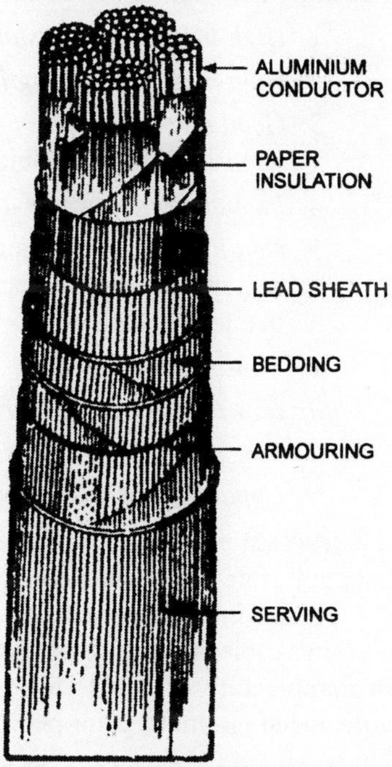

Fig. 12.5

12.6. REQUIREMENTS OF CABLE

(*i*) The conductor (usually annealed copper with about 99% purity or aluminium) used in cables should be stranded one in order to provide flexibility to the cables and should be of such x-sectional area that it may carry the desired load current without overheating and causing excessive voltage drop.

(*ii*) The insulation provided should be of such thickness that it may give high degree of safety and reliability at the working voltage for which it is designed.

(*iii*) The cables should be provided with a mechanical protection so that it may withstand the rough usage in laying it.

(*iv*) Materials used in manufacture of cables should be such as to give complete chemical and physical stability throughout.

12.7. CABLE CONDUCTORS

For all cables the conductor was, at one time, almost universally stranded copper but because of scarcity of copper, only aluminium is allowed to be used in power cables in India. Copper conductors are only used for cables used in control circuits.

The stranding is necessary to provide flexibility to cables and the conductor numbers are 3, 7, 19, 37, and so on, all numbers except 3 having a centrally disposed conductor with all the others around it. Thus a 7-strand has a central conductor with 6 around it. The various conductors are spiralled round the central conductor, and when there is more than one layer alternate layers are spiralled in opposite directions. This is to prevent 'bird-caging' when the conductor is bent.

The spiralling of the conductors obviously causes the increase in resistance, since the length of each spiralled conductor is greater than the central strand, and the flow of current is along the various conductors. It is therefore necessary to be particular in the designation of the cross-section.

12.8. INSULATING MATERIALS FOR CABLES

The satisfactory operation of a cable depends to a great extent upon the characteristics of insulation employed. So the proper choice of insulating material for cables is considerably important. The main requirements of the insulating materials used for cables are :

1. *High insulation resistance* to avoid leakage current.
2. *High dielectric strength* to avoid electrical breakdown of the cable.
3. *Good mechanical properties i.e.* tenacity and elasticity. Good tenacity is required in the material to withstand the mechanical handling of cables.
4. *Immune to attacks by acids and alkalies,* over a range of temperature of about – 18°C to 94°C
5. *Non-hygroscopic* because the dielectric strength of any material goes very much down with moisture content. In case the insulating material is hygroscopic, it must be enclosed in a water tight covering like lead sheath etc.
6. *Non-inflammable.*
7. *Low coefficient of thermal expansion.*
8. *Low permittivity.*
9. *Capability of withstanding high rupturing voltages.*

The last point but not the least is that a dielectric should not be *expensive* and it should be possible to handle it from the point of view of the manufacturer.

No one insulating material possesses all the above mentioned qualities, so the type of insulating material used in a cable depends upon the service for which the cable is required. The various insulating materials used in manufacture of cables are rubber, VIR, paper, polyvinyl chloride, varnished cambric, polyethylene, vulcanized bitumen, gutta-percha, silk, cotton, enamel etc.

1. *Rubber.* Rubber may be natural and synthetic. Natural rubber is obtained from the milky sap of tropical trees. Synthetic rubber is produced from alcohol or oil products. Its relative permittivity is between 2 and 3 and dielectric strength is 30 kV/mm. Though it possesses high insulating qualities but it absorbs moisture readily, softens when heated to a temperature of 60 or 70 °C, swells under the action of mineral oils and ages when exposed to light. Hence pure rubber cannot be used as insulating material.

2. ***Vulcanized India Rubber (VIR).*** It is prepared by mixing India rubber with mineral matter such as sulphur, zinc oxide, red lead etc. Though vulcanization process improves the non-hygroscopic and other properties like strength, durability, resilience, elasticity etc; but it slightly reduces its specific insulation resistance. It has dielectric strength of about 10-20 kV/mm and dielectric constant of about 2.5. The main drawback of VIR is that owing to sulphur content it attacks copper and therefore in cables using VIR insulation the copper conductor is tinned before providing this insulation. Sometimes a layer of pure rubber is also given on the conductor to protect it from sulphur.

As compared to impregnated paper, the performance of VIR is poor because VIR cables are expensive, less durable and have poor mechanical properties such as electrostatic capacity, elasticity and tenacity. VIR is used as insulating covering for low and medium voltage cables. Such type of cables are normally covered with cotton tape and jute braiding soaked in the compound in order to protect the cables against ingress of moisture.

3. ***Impregnated Paper.*** It is prepared from wood pulp rags or plant fibres by a suitable chemical process. It is quite cheap, has low capacitance, high dielectric strength (30 kV/mm), high insulation resistivity (of the order 10^7 Ω–m), high thermal conductivity, better thermal withstand capability and high durability. The only disadvantage is that it is hygroscopic, and even if it is impregnated with compound, it absorbs moisture, which lowers its insulation resistance. That is why paper insulated cables are always provided with some sort of protective covering and are never left unsealed. Its ends should be temporarily covered with wax or tar, even if it is required to be left unused on site while being laid. The maximum safe temperature of paper insulated cable is about 95 °C. The compound usually employed for impregnation purposes is a heavy-grade mineral oil mixed with resin, the actual proportions of oil and resin depending on the particular manufacturer.

For providing insulation over the conductor, the paper tape is wrapped on to the conductor until the required thickness* is obtained, it is then dried and impregnated with insulating compound. The insulating compounds are required to have, besides high electrical insulation properties, a high viscosity at operating temperatures (50–80 °C), a low viscosity at impregnation temperatures of 50–120 °C, a smooth change in viscosity with temperature and very low coefficient of thermal expansion. The impregnation of paper insulation increases its dielectric strength from about 3–5 to 10–80 kV/mm. The dielectric constant of impregnated paper insulation is about 3.5.

Paper insulated cables are used for conveying large blocks of power, in transmission and distribution, and particularly for distribution at low voltage in congested areas where the joints are to be provided only at the terminal apparatus or where the joints are rare, owing to cheapness and durability over VIR cables.

4. ***Polyvinyl Chloride.*** This is a synthetic compound. For obtaining this material as a cable dielectric or sheathing it is processed with certain materials known as *plasticizer* and its type will depend upon the use of the finished product *i.e.* PVC. It is inert to oxygen and almost inert to oils and to many alkalies and acids and therefore its use is preferred over VIR in extreme environments such as in cement factory or chemical factory. PVC has dielectric strength of 17 kV/mm and dielectric constant of 5. Maximum continuous temperature rating is 75 °C. The mechanical properties (*i.e.* elasticity and recovery from stretching) of PVC are not so good as those of rubber. PVC insulated cables are usually employed for low and medium voltage domestic and industrial light and power installations.

5. ***Polythene.*** Polythene is a straight chain polymer derived from ethylene. Its electrical properties are very good—low permittivity (dielectric constant), high insulation resistivity and very low power factor at all frequencies. The thermal dissipation properties are better than those of impregnated paper and the impulse strength compares favourably with an impregnated paper-insulated cable. The maximum operating temperature under short-circuits is 100 °C. It has limited use in power cables. However, a 3-core 11 kV self supporting aerial cable, which can be employed in rural distribution, has been developed.

---

* The thickness of insulation depends on the mechanical and electrical requirements which it has to meet in the service.

6. ***Cross-Linked Polythene.*** Low density polythene, when vulcanized under controlled conditions, results in cross-linking of carbon atoms and the compound produced is a new material having extremely high melting point with light weight, small dimensions, low dielectric constant and good mechanical strength. By using chemical process it has been made technically possible to cross-link polythene in conventional equipment for the manufacture of rubber. This is why the product is said to be ''vulcanized'' 'or cross-linked' polythene.

Cross-linked polythene (XLPE) cables have high maximum continuous temperature rating of 90 ºC with dielectric strength of about 20 kV/mm. Because of high temperature withstand capability and very low water absorption, such cables can be buried directly in soil. These cables have been found very suitable for all voltages up to 33 kV.

7. ***Varnished Cambric (or Empire Tape).*** This is a cotton cloth impregnated and coated with varnish. It has very smooth surface. The cambric is wrapped in the form of a tap on to the conductor and its surfaces are coated with petroleum jelly compound to give easy sliding of the surfaces. Such cables require protective covering like lead sheath because this insulating material is hygroscopic. Its dielectric strength is 4 kV/mm and its dielectric constant is 2.5 to 3.8. Such cables do not require sealing.

8. ***Gutta-percha.*** It is similar to rubber but it becomes soft at about 65ºC. It is non-hygroscopic but cannot withstand even medium voltages. It is mostly used for submarine cables for telephone and telegraph purposes.

9. ***Silk and Cotton.*** This is used in low voltage cables. The conductor may have a single layer or double layer covering depending upon the requirements of service. Silk or cotton covered wires are usually used for instrument and motor winding.

10. ***Enamel Insulation.*** *Enamelled* wires are also used for instruments and motor winding. The wires are cheaper than silk and cotton covered wires and therefore for low voltage machines and instruments enamelled wires are used. Enamel insulation is liable to crack.

12.9. CLASSIFICATION OF CABLES

The type of cable to be used at a particular location is determined by the mechanical considerations and the voltage at which it is required to operate. Usually the operating voltage determines the type of insulation and the cables are placed in various categories depending upon the voltage for which they are designed. According to voltage, the cables are classified as :

(*i*) Low voltage (LT) cables for operating voltage up to 1,000 V.

(*ii*) High voltage (HT) cables for operating voltage up to 11,000 V.

(*iii*) Super-tension (ST) cables for operating voltage up to 33,000 V.

(*iv*) Extra high tension (EHT) cables for operating voltage up to 66,000 V.

(*v*) Extra super-voltage cables for operating voltage beyond 1,32,000 V.

A cable may be a single core or multi-core depending upon the type of service for which it is intended to be used. A multi-core cable may be two-core, three-core or four core. For 3-phase, 3-wire systems either three single-core cables or three-core cables can be used depending upon the operating voltage and load demand. For 3-phase, 4-wire systems 4-core cables may be used.

1. Low Tension (or LT) Cables. These cables are meant for use up to 1,000 V. For voltages up to 6,600 V, the electrostatic stresses developed in cables are very small and thermal conductivity is also of not much importance so no special construction is required. The insulating materials used may be impregnated paper, varnished cambric, vulcanized rubber, or vulcanized bitumen.

The *lt* cables are of two types *viz* single core and multi-core cables. The former type has the advantages of simplicity of construction and availability of larger conductor section. The various sizes of *lt* cables are given below:

Single core cable with aluminium conductors	:	1.5 – 625 mm^2
Two-, three-, three-and-a-half and four core cables with aluminium conductors	:	1.5 – 625 mm^2
Control cables up to 61 cores with copper conductors	:	1.5 and 2.5 mm^2.

Single core *lt* cable consists of one circular core of tinned stranded copper (or aluminium) insulated by layers of impregnated paper or varnished cambric over it and a lead sheath over the insulation, as shown in Fig. 12.6. The lead sheath protects the cable against ingress of moisture and mechanical handling. If the cables are to be buried directly then an overall serving of compounded fibrous material or hessian tape is provided over the sheath in order to protect the metallic sheath against corrosion. Single core cables are usually not provided with armouring in order to avoid excessive loss in the armour.

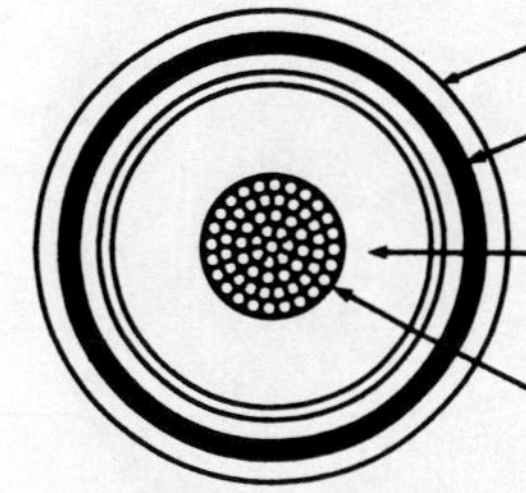

Fig. 12.6 *Single Core LT Cable*

The multicore cables for use up to 11,000 volts are of belted type. The belted type cable consists of either circular shaped or oval or sector shaped (oval or sector shaped core improve the copper space factor) cores of stranded copper (or aluminium) conductors wrapped around by impregnated paper and then wormed together. The interstices are filled with packing to make the cable of circular cross-section. Insultating belt of impregnated paper is provided surrounding the three cores. A lead sheath enclosing the whole is provided for its protection against the entry of moisture. The cable is provided with two layers of steel tape pre-dipped in compound. The serving over the armouring consists of a compound and then one layer of impregnated hessian tape. In order to prevent adhesion, a coating of lime wash is applied to the outside of the cable.

The belted type construction is not suitable for the cables used for voltages above 22 kV because of development of both the radial and tangential stresses. The tangential stresses act along the layers of insulation. The electric resistance, therefore, dielectric strength of the impregnated paper is much higher across the layers than along the layers. The leakage current on account of tangential stresses along the impregnated paper insulation causes power loss at the centre filling and local heating resulting in breakdown at any moment. Further owing to non-homogenity of dielectric in belted construction, when cables are loaded and unloaded some portions of the dielectric are less stressed whereas some portions are overstressed resulting in formation of vacuous spaces and voids. These vacuous spaces are ionised when voltage is applied and ultimately deteriorate the cable insulation.

The above difficulties have been overcome in the screened cables where leakage currents are conducted to earth through metallic sheaths.

2. Super-Tension Cables. For cables above 11,000 V a special construction is adopted. For use up to 33 kV, the cables used are screened cables where leakage currents are conducted to earth through metallic sheaths. Screened cables are of two types *viz* H-type and S L type.

(a) H-Type Cables. In the "H" type cable, invented by Hochstadter, no belt insulation is used, but each of the core is insulated with paper to the desired thickness and over this is provided a layer of metallized paper

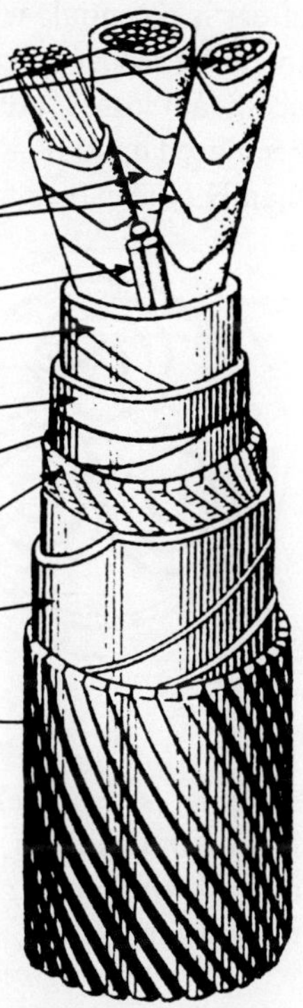

Fig. 12.7 *(a) View of Cable With Sector Shaped Multi-Strand Conductors*

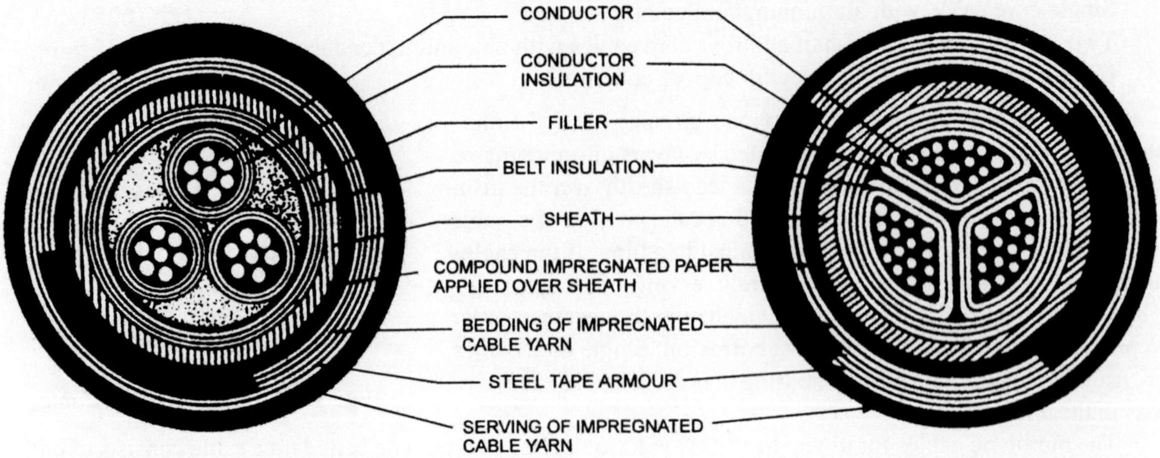

(b) Cross-section of Cable With Round Conductors

(c) Cross-section of Cable With Sector-Shaped Multistrand Conductors

Fig. 12.7 *Three-core Power Cable With Impregnated Paper Core and Belt Insulation*

perforated to facilitate the process of impregnation, the coefficient of expansion and contraction is same as that of dielectric. Additional layer of the cotton tape with fine wires of copper is wrapped round all the three cores. Over this tape, no belt insulation is provided, but the lead sheath and armouring are provided as usual. The perforations in the metallized paper sheaths assist in the complete impregnation of cable with the compound and thus the possibility of air-duct formation is eliminated. All four screens and the lead sheath are at earth potential, with the result that the electric stresses are entirely radial and the dielectric losses are therefore reduced. Another advantage of metal foil is that it increases the heat dissipating power and there are no sheath losses. All of these effects increase the current carrying capacity of the cable. These cables have been used up to 66 kV but are usually used up to 33 kV because pressurized cables have better performance than H type cables for voltages exceeding 33 kV.

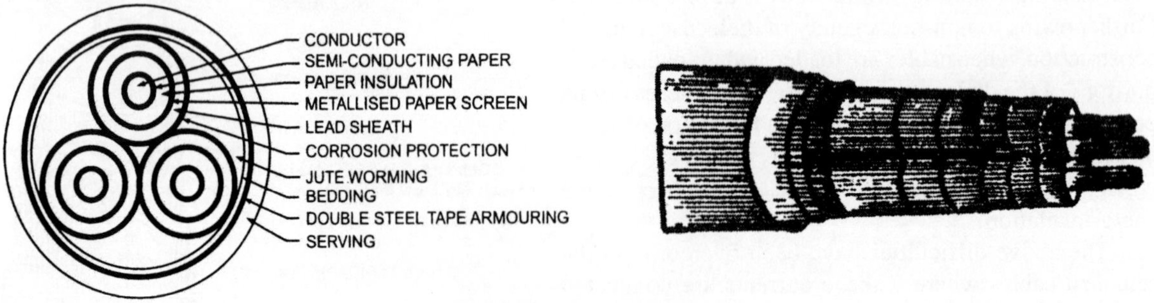

Fig. 12.8 *H-Type Cable*

(b) SL Type Cables. In SL type cables each core is first insulated with an impregnated paper and then each of them, is separately lead sheathed. Now the three cores are just equivalent to three separate cables, each having its own lead sheath. The three cables are laid up with fillers in the ordinary way, armoured and served overall with impregnated hessian tape as usual. No lead covering is provided surrounding all the three cores in addition to their individual lead sheaths. The advantages of SL type cables over H-type cables are :

(*i*) Bending of cable becomes possible owing to no overall lead sheath.

(*ii*) Less tendency for oil drainage on hilly routes owing to elimination of filler spaces containing compound.

The disadvantage of SL type cable is that the manufacturing is difficult because of thinner lead sheaths.

SL type cables can be used up to 66 kV.

(*c*) *HSL Type Cables.* Such a cable is a combination of H-type and SL type cables. In HSL type cables each conductor is insulated, sheathed with metalized paper and is then provided with lead sheath. The three cores are then laid up and provided with filler, braided, armoured and finally served.

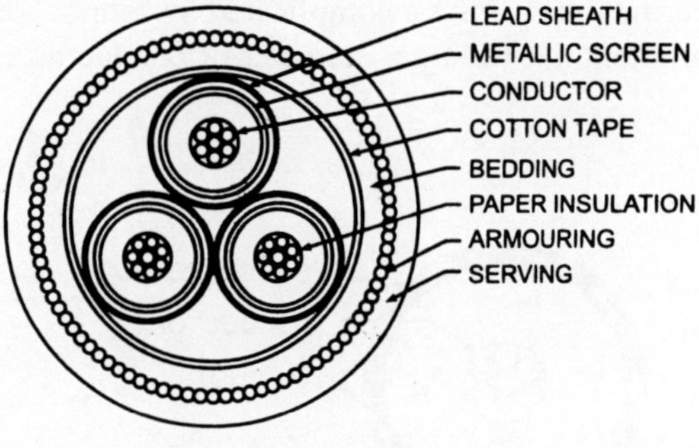

Fig. 12.9 *SL Type Cable*

Advantages of Screened Type Cables Over Belted Types. (*i*) The metal sheathed core cable has a greater core to core thickness for a given overall diameter than a plain cable so the possibility of core to core faults are reduced to a large extent.

(*ii*) Electric stresses are uniformly radial in all sections of the dielectric.

(*iii*) As there is no worming or packing so the dielectric subjected to electric stress is only paper which is quite homogeneous and therefore possibility of formation of voids within the electric field is not there.

(*iv*) The metal sheaths help in dissipation of heat, so the current carrying capacity of the cables is increased.

3. Extra High Tension Cables. The cables considered up till now are also called *solid type* and in such cables it was assumed that the dielectric is homogeneous and there are no voids in the layers. With this assumption it is necessary to stick to maximum safe dielectric stress of about 4 or 5 kV per mm and also to maximum operating temperature of 50 or 60 °C in order to ensure safety from breakdown because the dielectric is far from homogeneous and with the normal manufacturing methods, it is almost impossible to avoid voids in the layers of dielectric. The formation of these voids or vacuous spaces causes unequal voltage stresses (which may exceed the safe limits) and also temperature rise due to leakance current. Again, even if a new cable could be constructed free from voids owing to development of modern techniques, it would be impossible to avoid their formation under the influence of the movements of the compound and expansion and contraction of the paper during normal working. When these voids are subjected to the electrostatic stresses, ionization takes place, which gives rise to certain chemical action, resulting in the deterioration of the chemical compound. Thus, these voids are sometimes the primary cause of a breakdown and therefore are required to be avoided. In order to meet the increased voltage demand the extra high tension and extra super-voltage power cables, useful for 132 kV and above, have been developed. In all such cables, the voids have been eliminated by increasing the pressure of the compound and that is why such cables are also called *pressure cables.*

Pressure cables are of two types viz., (*a*) oil-filled cables, and (*b*) gas-pressure cables.

(*a*) *Oil-filled Cables.* A single core oil-filled cable is shown in Fig. 12.10. In this type of cable a channel is formed at the centre of the core by stranding the conductor wires around a hollow cylindrical steel spiral tape. This channel formed at the centre of core is filled with thin oil by means of oil reservoir and feeding tanks along its length and maintained at a pressure, not below atmospheric one at any point along the cable. This oil is the same light mineral oil of very low viscosity as used for initial impregnation. The system is so designed that when the oil gets expanded on account of increase in temperature of cable, the extra oil collects in the external reservoir, which sends it back during contraction on account of fall in temperature during light load conditions. Thus formation of voids in the dielectric is rendered impossible. Such cables are known as *single core conductor channel cables.* Its disadvantage is due to the fact that the channel is at the middle of the cable and is at full voltage with respect to

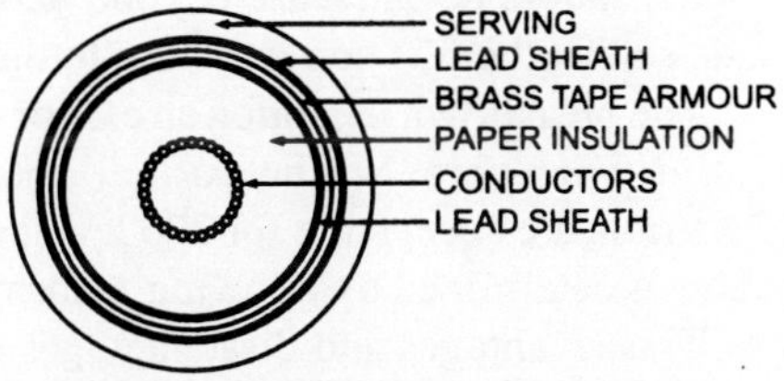

Fig. 12.10 *Single Core Conductor Channel Oil-Filled Cable*

earth, so requiring a complicated system of jointing. It is advantageous from the point of view of potential gradient due to larger diameter of conductor and its hollow construction.

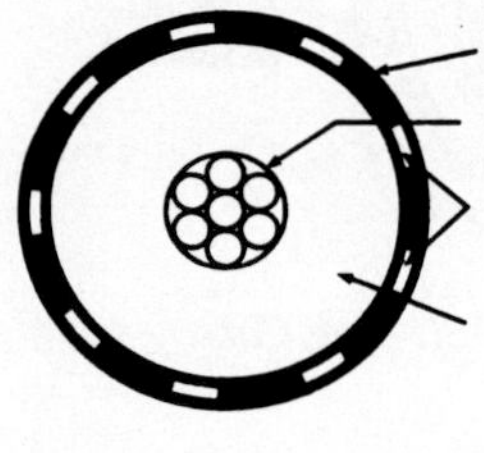

Fig. 12.11 *Single Core Sheath Channel Oil-Filled Cable*

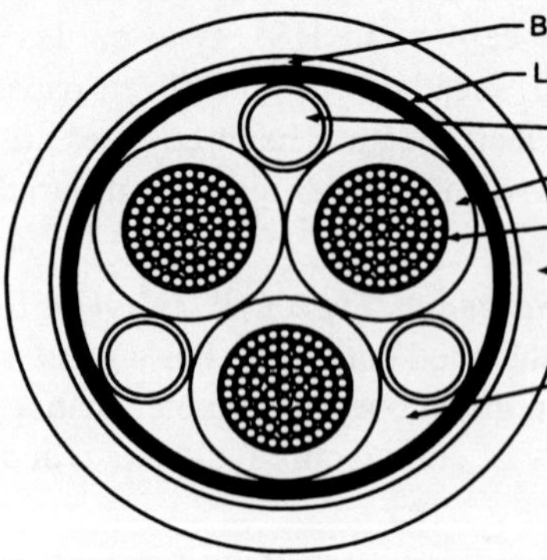

Fig. 12.12 *Three Core Filler Space Channel Oil-Filled Cable*

The other type of single core oil-filled cable is *sheath channel cable* (Fig. 12.11). In this type of cable, the conductor is made solid similar to that of solid cables and is paper insulated. In this type of cables the oil channels are produced either by grooving the sheath or by arranging spacers between dielectric and lead sheath. The latter arrangement is more sound mechanically in comparison to former one but due to high resistance to oil flow (6 to 8 times that of former type) requires large number of feeding points. In this case since the channels are at earth potential, system of joints and installation are simpler.

In 3-core oil-filled cables (Fig. 12.12) the oil ducts are accommodated in the hollow filler spaces. These channels are made of perforated metal-ribbon tubing and are at earth potential. While jointing such a cable great care is required to be taken.

Another design of 3-core oil-filled cable is flat type, as illustrated in Fig. 12.13. The flat sides are reinforced with metallic tapes and binding wires so that during increase in oil pressure, due to heating, the flat side is deformed and the cable section becomes slightly elliptical.

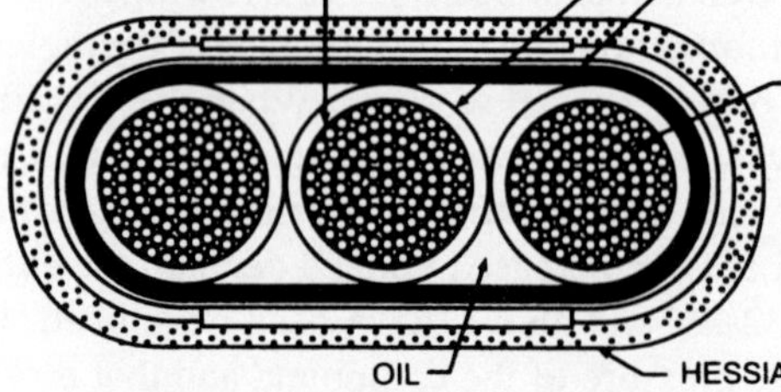

Fig. 12.13 *Flat Oil Filled Pressure Cable*

In another construction of 3-core oil filled cables use of 3-core paper insulated cable without a lead sheath is made. The cable is pulled into a steel pipe which is then filled with oil. Pumps are then used to maintain a specified oil pressure and permit it to expand and contract with the loading cycle.

Leakage of oil in oil-filled cables is a very serious problem. Automatic signalling is provided to indicate the decrease in oil pressure in any of the phases. The pressure within the cable must not go below atmospheric and a lower limit of about 21,700 N/m^2 gauge is set. The transient pressure, developed due to sudden increase of load, should not increase beyond about 8.6×10^6 N/m^2 gauge. The joints in the cable are to be properly made so that there is no restriction to the oil flow.

The breakdown inception stress for oil-filled cables is 30–40 kV/mm which is much higher than that for a solid type cable. So the normal operating stresses are raised considerably. Maximum stresses up to 13 kV/mm are acceptable for 275 kV and 400 kV oil-filled cables. The current carrying capacity of oil-filled cables is determined by allowing a normal maximum conductor temperature of 85 °C.

The advantages and disadvantages of oil-filled cables are given below:

Advantages. (*i*) Smaller overall size and smaller weight for given voltage and kVA rating due to reduction in the thickness of dielectric required.

(*ii*) No ionisation, oxidation and formation of voids.

(*iii*) More perfect impregnation.

(*iv*) Possibility of increased temperature range in service.

(*v*) Smaller thermal resistance due to decrease in dielectric thickness, so higher current rating.

(*vi*) Impregnation possible after sheathing.

(*vii*) More maximum permissible stresses.

(*viii*) Detection of fault easy, as it will come to notice immediately as oil will start leaking.

Disadvantages. (*i*) Greater cost.

(*ii*) Complicated laying of cables and maintenance.,

(*b*) *Gas-Pressure Cables.* Gas pressure cables are of two types (*i*) external pressure cables and (*ii*) gas-filled cables.

(*i*) *External Pressure Cables.* The external pressure cable, is a serious rival to the oil-filled type, and is being developed for the highest voltages. In such a cable the pressure, is applied externally and raised to such an extent that no ionisation can take place. At the same time the radial compression due to this increased pressure tend to close any voids. Also the working power factor of such a cable is improved.

The external pressure cable originally designed by Hochstadter, Vogel and Bowden is illustrated is Fig. 12.14. This cable is similar in construction to that of ordinary solid type except that it is triangular instead of circular in section and thickness of lead sheath is 75% of that of solid type cable. The triangular section reduces the weight, gives a low thermal resistance and at the same time lead sheath acts as a pressure membrane. Also there is no bedding and serving so thermal resistance is reduced. The cable is armoured with a thin metal tape so that the formation of any abnormal ties over its surface is avoided. The core may have circular or oval section but modern practice is to employ an oval section, which gives an improved copper space factor. The cable is laid in gas-tight metal pipes of somewhat larger section. The pipe is filled with nitrogen at a pressure of 12 to 15 atmospheres which continually compresses the cable radially from outside so that radial breathing of the cable occurs and any voids etc. are closed. The steel pipes used are coated with a special paint in order to avoid corrosion and it is further protected with an impregnated felt.

Fig. 12.14 *66 kV, Three-core Pressure Cable*

In comparison with a normal cable, such cables can carry 1.5 times the load current, double the operating voltage and thus transmit 3 times power. The maximum potential gradient is 10 kV/mm and the dielectric power factor at 15°C is 0.6 percent. The steel pipes provide mechanical protection to the cables. The nitrogen in steel pipes help in quenching any flame. Moreover, maintenance cost of such cables is small. The disadvantage of such cables is that cost is very high.

The above cable is pipe line type. The second type is self contained type, in which an additional reinforced lead sheath is employed, the principle being the same as that of former type. Dried nitrogen is fed into the small annular space between dielectric and sheath from gas cylinders, and finding its way along the cable maintains the whole of the dielectric at a pressure of about 14 atmospheres.

(*ii*) *Internal Pressure Cables.* Such cables are of 3-types :

1. High-pressure gas-filled cables.

2. Gas-cushion cables.

3. Impregnated pressure cables.

In high pressure gas-filled cables spaces are provided in the dielectric itself for the gas which is inert gas like nitrogen at pressure of about 12 atmospheres for super-voltage cables and about 6 atmospheres for extra high tension cables. Pressure is retained by means of a lead sheath which in the case of single core cables, has a diametral clearance of about 0.63 mm. This facilitates the axial flow of gas, which also passes along the unimpregnated strand. In the case of multi-core cables this clearance is not essential; the filler spaces and strands providing a sufficiently low resistance path for the flow of gas.

In gas-cushion cable, screened space is provided all along the length of cable, in between the lead sheath and the dielectric. The screened space is sub-divided by means of barriers and thus inert gas is stored at various points along the run of a cable. This facilitates the jointing of cable without losing gas from the entire cable, because while cutting the cable, loss of gas will be local only. The special feature of the cable is thus that no arrangement is required for the transmission of pressure to the cables from outside, so that the cable is a complete unit with its own armouring, requiring no external pipe protection whatever.

Impregnated pressure cable is similar to solid cable except that such a cable consists of mass-impregnated paper dielectric and this is maintained under a pressure of 14 atmospheres by means of nitrogen. Special reinforcement is provided to cater for the large hoop and longitudinal stresses set up. This reinforcement consists of longitudinal and circumferential metallic tapes.

The advantages of the internal pressure cables are :

(*i*) External accessories are eliminated.

(*ii*) With suitable designs the cable can be used for vertical run without any fear of drainage.

(*iii*) There is marked improvement in the power factor of the cable dielectric with the increased pressure.

Gas SF_6, because of its good thermal characteristics and high dielectric strength, is also used for insulating the cables. Cables insulated with SF_6 gas can be matched to overhead lines and can be operated corresponding to their surge impedance loading. Such cables can be employed for transmission of thousands of MVA even at ultra high voltage while the use of conventional cables is limited to 100 MVA and 500 kV.

12.10. SELECTION OF CABLES

For selection of cable of correct size and type for a particular application, the factors to be considered are as follows:

1. System Voltage. The type of operating system such as dc (2-wire or 3-wire) or ac (single phase, three-phase), earthed or unearthed and operating voltage such as 415/240 V, 11, 33, 66, 132 kV etc.

2. Current Carrying Capacity. The current rating is the most important factor, already discussed in Art 12.6.

3. Permissible Voltage Drop. The cable should be selected of such a size that the voltage drop in the cable is within permissible limits.

4. Short-circuit Rating. For selection of *lt* cables this factor is not important but for *hv* cables, this factor is the most important. It is because in high voltage systems the fault current (phase-to-phase or phase-to-earth) is very high. The cable should be so selected that it can withstand the stresses and the resulting increase in temperature caused by the maximum short circuit current produced by the phase-to-phase short-circuit up to a period of 1 second (I S : 1255 – 1967).

The last, but not the least, factor is the *economy consideration.* The cable selected satisfying the above requirements-continuous current rating, voltage grade, permissible voltage drop and short-circuit current rating (in case of *ht* cables), should also give minimum operating cost (power losses, the interest on capital cost and depreciation).

HIGHLIGHTS

1. The cable consists of three parts :
   (*a*) the *conductor or core;* the metal wire or strand of wires carrying the current :
   (*b*) the *insulation or dielectric;* a covering of insulating material to avoid leakage of current from the conductors and
   (*c*) the *protective covering* for protection of insulation from mechanical damage.
2. The function of conductor, usually known as a core, in a cable is to carry electrical current. Copper and aluminium are the materials used as conductors in power and lighting cables.
3. The insulating materials used in electric cables must possess the properties: high resistivity, high flexibility, high dielectric strength, non-inflammability, non-hygroscopic, highly resistive to moisture, acids or alkalies, capability to withstand high rupturing voltages and high temperature without much deterioration.
4. The insulating materials used in cables are: Rubber, VIR, impregnated paper, PVC, Silk and cotton.
   PVC and XLPE insulated cables are predominantly used.
5. The type of cable to be used at a particular location is determined by the mechanical considerations and the voltage at which it is required to operate. Usually the operating voltage determines the type of insulation and the cables are placed in various categories depending upon the voltage for which they are designed. According to voltage, the cables are classified as :
   (*i*) Low voltage (LT) cables for operating voltage up to 1,000 V.
   (*ii*) High voltage (HT) cables for operating voltage up to 11,000 V.
   (*iii*) Super-tension (ST) cables for operating voltage up to 33,000 V.
   (*iv*) Extra high tension (EHT) cables for operating voltage up to 66,000 V.
   (*v*) Extra super-voltage cables for operating voltage beyond 1,32,000 V.

   A cable may be a single core or multi-core depending upon the type of service for which it is intended to be used. A multi-core cable may be two-core, three-core or four core. For 3-phase, 3-wire systems either three single-core cables or three-core cables can be used depending upon the operating voltage and load demand. For 3-phase, 4-wire systems 4-core cables may be used.

EXERCISES

1. Write a short note on conducting materials used in making cables for internal wiring work and their comparison.
2. Describe the various types of wires or cables usually used in internal wiring of buildings.
3. Describe the general construction of an underground cable with a neat sketch.
4. With a neat labelled diagram, show the various parts of a high voltage single core cable.
5. Discuss the qualities the dielectrics used for cable insulation must have. Also name three dielectrics with their properties used in the manufacture of underground cables. **[Jodhpur Univ. 1978]**
6. State the classification of cables and discuss their general construction.
7. Classify the underground cables according to insulation used, voltage rating and number of cores.
   Sketch a typical 3-core cable and label the important parts.
8. Show the cross-sectional view of a three-core belted type lead sheathed paper cable; label the different components and discuss their functions.
9. Describe different types of three core cables. State specialities for use at different voltages. **[Pb. Univ. April 1982]**
10. Explain the constructional features of HSL type oil-filled power cable.
11. Mention the important factors to be considered while selecting cable.
12. Write short notes on the following :
    (*i*) General construction of underground cable.
    (*ii*) Classification of underground cables.
    (*iii*) Selection of cables.

SHORT ANSWER TYPE QUESTIONS WITH ANSWERS

Q. 1. What is difference between wire and cable ?

Ans. Single wire, may be bare or covered with insulation is called the wire while several wires stranded together is called the cable.

Q. 2. Why conductor is covered with insulating material ?

Ans. The conductor is covered with insulating material so as to prevent leakage of current from the conductor.

Q. 3. What are the essential properties, which the insulation used in cables must possess ?

Ans. The insulating materials used in cables must have the properties of high resistivity, high flexibility, high dielectric strength, non-inflammability, non-hygroscopicity, highly resistive to moisture, acids or alkalies and capability to withstand high rupturing voltages and high temperatures without much deterioration.

Q. 4. Which insulation is most widely used for covering wires/cables used in internal wiring ?

Ans. PVC.

Q. 5. Why multi-strand conductors are preferred over solid conductors ?

Ans. Multi-strand conductors are preferred over solid conductors because they provide flexibility and durability.

Q. 6. PVC cables are most widely used in internal wiring. Why ?

Ans. PVC cables are most widely used in internal wiring because PVC insulation provides better flexibility, has better insulating qualities, has no chemical effect on metal of wire and gives smaller diameter of cable.

Q. 7. Where are flexible cords used ?

Ans. Flexible cords are used as connecting wires for such purposes as from ceiling rose to lamp holder, socket outlet to portable apparatus such as radios, fans, lamps etc.

Q. 8. What is meant by voltage grading of cables ?

Ans. Voltage grading of cable specifies the safe voltage which the insulation of the cable can withstand.

Q. 9. How the size of a wire or cable is given ?

Ans. The size of a wire or cable is specified giving the number of strands used and diameter of each strand or giving the area of cross-section of conductor used.

Q. 10. How are cables specified ?

Ans. The cables are specified providing size of conductor used, type of conductor used, number of cores that cable consists of, voltage grade and type of cable.

Q. 11. For what purpose metallic sheath is provided in underground cables ?

Ans. The metallic sheath is provided around the insulation to protect it against ingress of moisture, gas or other damaging liquids (acid or alkalies) from the soil and atmosphere.

Q. 12. Why is a layer of bedding provided over the metallic sheath in underground cables ?

Ans. A layer of bedding is provided over the metallic sheath for the protection of metallic sheath against corrosion and from mechanical injury.

Q. 13. Why are copper conductors not used in power cables ?

Ans. Copper conductors are not used in power cables because of scarcity of copper.

Q. 14. What are the main requirements of insulating materials used for underground cables ?

Ans. The main requirements of the insulating materials used for underground cables are high insulation resistivity, high dielectric strength, good mechanical properties, non-hygroscopic, non-inflammable, immune to attacks by acids and alkalies.

Q. 15. Why single core cables are usually not provided with armouring ?

Ans. Single core cables are usually not provided with armouring in order to avoid excessive loss in the armour.

Q. 16. What are the factors, which are to be considered while selecting a cable for a particular service ?

Ans. The factors to be considered in selection of a cable are continuous current rating, voltage grade, permissible voltage drop and short-circuit current rating (in case of *ht* cables only) and operating cost.

❑❑❑

UNIT-VII
Electrical Wiring

Contents in This Unit

Electrical Wiring: Connectors and switches; Systems of wiring; Domestic wiring installation; Sub-circuits in domestic wiring; Simple control circuit in domestic installation; Industrial electrification.

Chapter 13: Electrical Wiring

Chapter

13 Electrical Wiring

13.1. INTRODUCTION

A network of wires connecting various accessories for distribution of electrical energy from the supplier meter board to the numerous electrical energy consuming devices such as lamps, fans and other domestic appliances through controlling and safety devices is known as a *wiring system.*

The supplier service cable feeding an installation terminates in what is usually called the *service fuse*. In an ordinary house the service fuse is called as *service cutout.* Such cutouts including service meters remain the property of the supplier and represent the furthest point of the supplier responsibility. The point at which the consumer's wiring is connected into the cutout is known as *point of commencement of supply or consumer's terminals.* From consumer terminals onwards the supply cables are entirely under the control of consumer and laid out as per his selection. A typical house-wiring circuit is shown in Fig. 13.1.

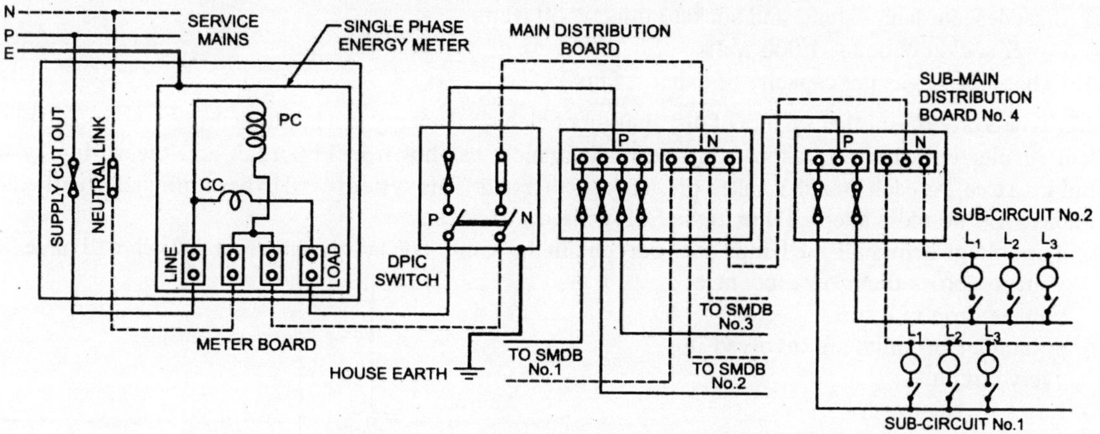

Fig. 13.1 *Typical House-Wiring Circuit*

13.2. SYSTEMS OF DISTRIBUTION OF ELECTRICAL ENERGY

Since as per recommendation of Indian Standards the maximum number of points of lights, fans and 5 A socket-outlets that can be connected in one circuit is 10 and the maximum load that can be connected in such a circuit is 800 watts, in case more load or points are required to be connected to the supply, then it is to be done by having more than one circuit.

13.2.1. Distribution Board System. In *distribution board system,* which is most commonly adopted for distribution of electrical energy in a building, the fuses of various circuits are grouped together on a distribution board, sometimes simply known as *fuse board.* Connections necessary for connecting two or more than two circuits, each consisting of ten or less number of lamps is shown in Fig. 13.2.

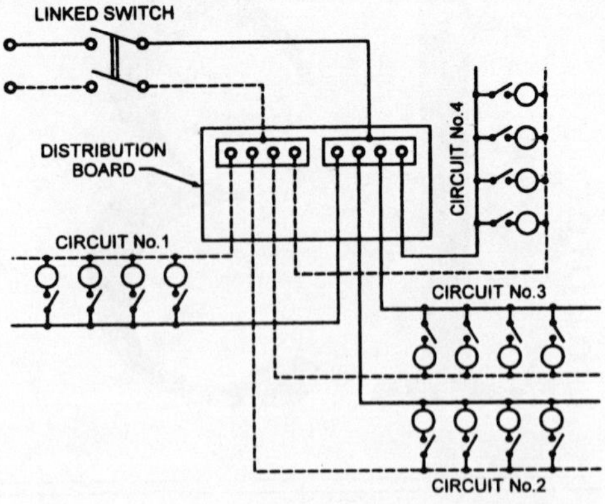

Fig. 13.2 *Distribution Board System*

The two copper strips, known as *bus-bars,* fixed in a distribution board of hard wood or metal case are connected to the supply mains through a linked switch so that the installation can be switched off as a whole from both poles of the supply if required. A fuse is inserted in the + ve or phase pole of each circuit so that each circuit is connected up through its own particular fuse. The lamps constituting each circuit need not necessarily be in the same room or even on the same floor in case of a small building and simply allocated to each circuit in such a way that the routes or "runs" for connecting them is most convenient and expenditure incurred is minimum. The distribution board shown in Fig. 13.2 has 4 ways for four circuits but there is no limit to the number of ways or circuits provided the cable feeding the board is large enough to carry the total load current. In large buildings, however, if only one distribution board were used, some of the points would be at a considerable distance from it and in such cases it is advisable to employ sub-distribution boards either to save cable or to prevent too great voltage drop at the more distant points (lamps, fans or other appliances). In such cases main distribution board controls the circuits to each sub-distribution board from which the sub-circuits are taken, as shown in Fig. 13.1. The sub-distribution boards are installed near the load.

The number of circuits and sub-circuits is decided as per number of points to be wired and load to be connected to the supply system. For determination of load of an installation the following ratings may be assumed unless the values are known or specified.

(*i*) Fluorescent lamps—50 watts
(*ii*) Incandescent lamps, fans, and socket outlets—60 watts
(*iii*) Power socket outlets—1,000 watts
(*iv*) Exhaust fans—as per capacity of exhaust fans.

13.2.2. Tree System. Another system of distribution of electrical energy in a building is the *tree system.* In this system smaller branches are taken from the main branch, as shown in Fig. 13.3, and the wiring system resembles a tree. As each branch is taken off, a fuse is inserted. This system used to be employed in early days. Nowadays it is no more adopted due to the following drawbacks :

(*i*) The voltage across all the lamps does not remain the same. The lamps in the last branch will have least voltage across them on account of voltage drop in leads.
(*ii*) A number of joints are involved in every circuit.
(*iii*) Fuses are scattered.
(*iv*) In case of occurrence of fault all the joints have to be located and if some of these joints are concealed beneath floors or roof spaces, a lot of difficulties are faced. Sometimes a number of such joints are required to be opened for testing purpose, so damage is caused to installation, conductors and building.

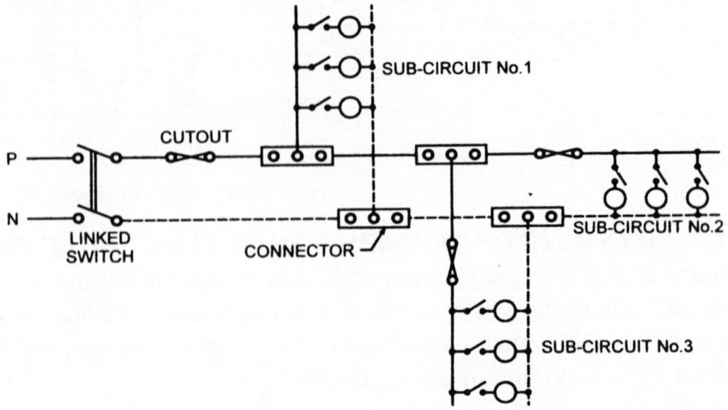

Fig. 13.3 *Tree System*

13.3. METHODS OF WIRING

There are two methods of wiring known as *joint box system* (*or Tee system*) and loop-in-system.

13.3.1. Joint Box or Tee System. In joint box system the connections to the lamps are made through joints made in joint boxes by means of suitable connectors or joints cutouts. In this method though there is a saving in the quantity of wire or cable required but the same is offset by the extra cost of joint boxes. The other disadvantage of T-connections is that the number of T-connections made in wiring system results in weakness if not properly made. Nowadays the use of this system is limited to temporary installations only as its cost is low.

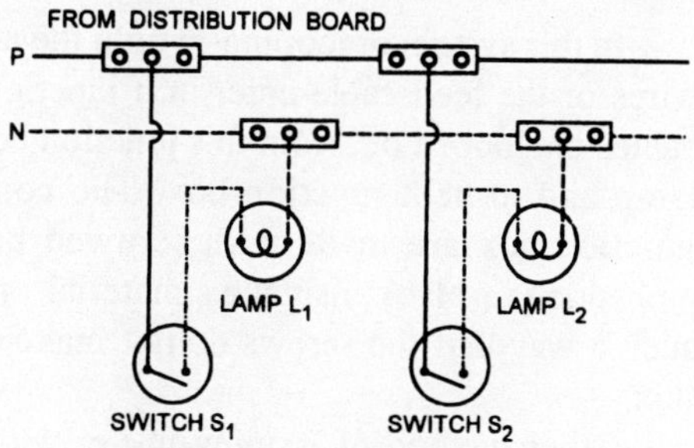

Fig. 13.4 *Joint Box System*

13.3.2. Loop-in System. This system is universally employed for connections of various lamps or other appliances in parallel. In this system when a connection is required at a light or switch, the feed conductor is looped-in by bringing it direct to the terminal and then carrying it forward again to the next point to be fed, as shown in Fig. 13.5. The switch and light feeds are carried round the circuit in a series of loops from one point to another until the last point on the circuit is reached.

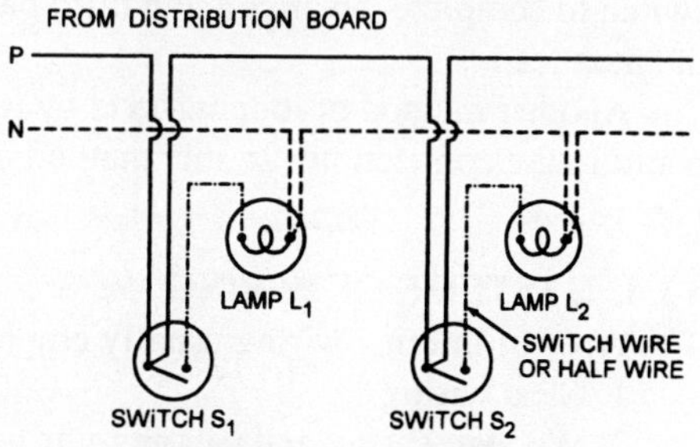

Fig. 13.5 *Loop-in System*

The phase or line conductors are looped either in switchboard or box and neutrals are looped either in switchboard or from light or fan. Line or phase should never be looped from light or fan.

In no case joint is made in the run of wire or cable.

The advantages and disadvantages of loop-in system are given below.

Advantages. 1. Joint boxes are not required.

2. In loop-in systems no joints are concealed beneath floors or in roof spaces. As they are made only at outlets so they are accessible for inspection and opening out merely by removing the fitment concerned. Hence fault location is easy.

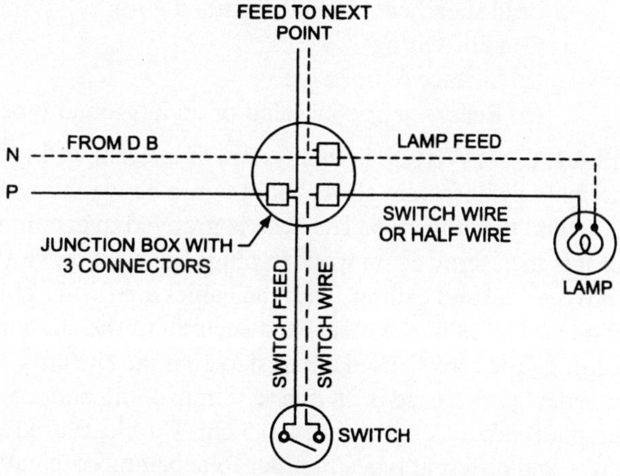

Fig. 13.6 *Looping-in With Junction-box*

Disadvantages. 1. Length of wire or cable required is more and voltage drops and copper losses are, therefore, more.

2. Looping-in in switches and lamp holders is usually difficult.

For carrying out wiring by means of twin-core metal sheathed cables, the above diagram (Fig. 13.5) is slightly modified in order to make it unnecessary for a third wire to be run down to the switch, as shown in Fig. 13.6 and use of junction box with 3 connectors is made.

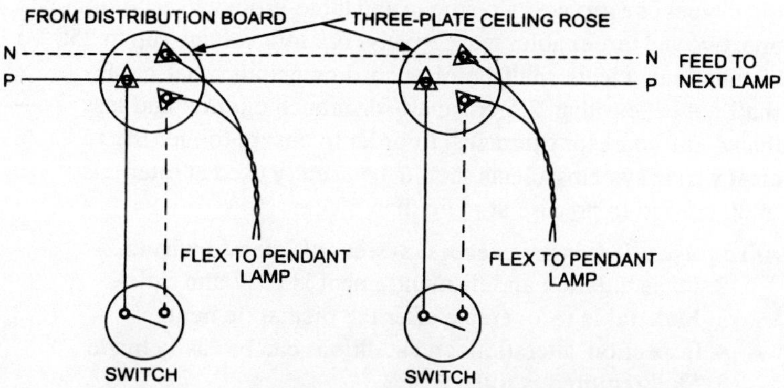

Fig. 13.7 *Looping-in From Three Plate Ceiling Rose*

In this system of looping-in both the neutral and phase wires of the feed cable enter in a junction box and three cables are looped out from this junction box to the switch, lamp and to next junction box. The connections in the junction box are made with screwed brass connectors embedded in black insulating material or in porcelain in such a way that the screws do not make contact with the box.

Other method of looping-in-i.e. with the help of 3-plate ceiling rose from which two wires are taken to the switch to complete the circuit and third pair comes out for the next feed.

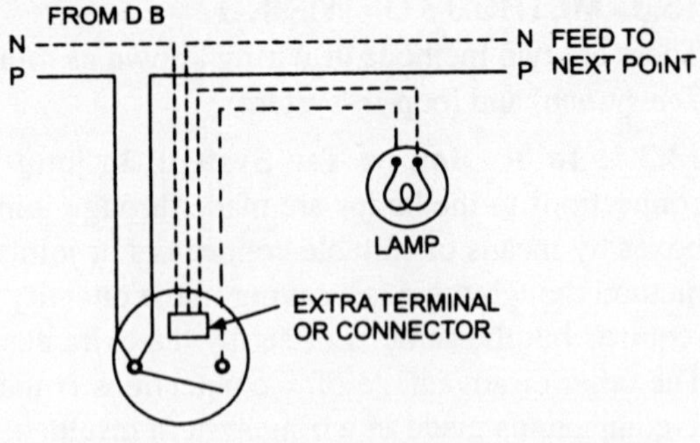

Fig. 13.8 *Looping-in Both Feeds From Switch*

Another method of looping-in is by means of a switch provided with an extra insulated terminal for the neutral wire connections. In this method of looping–in feed enters and leaves from the switch, as shown in Fig. 13.8.

13.4. SYSTEMS OF WIRING

The types of internal wiring usually employed in our country are :

1. Cleat wiring.
2. Wooden casing and capping wiring.
3. CTS or TRS or PVC sheathed wiring.
4. Lead sheathed or metal sheathed wiring.
5. Conduit wiring
   (*a*) Surface or open type
   (*b*) Recessed or concealed or underground type.

13.4.1. Cleat Wiring. In this system of internal wiring the cables used are either VIR or PVC type. The cables are held by porcelain cleats about 6 mm above the walls or ceilings. The cleats are made in two halves, one base and the other cap. The base is grooved to accommodate the cables and the cap is put over it and whole of it is then screwed on wooden plugs (gutties) previously cemented into the wall and ceiling. Thus the cables are firmly gripped between the two halves of the cleats and secured to the supporting wall and ceiling. The screws used are of size 38 mm. The cross-section of the wooden gutties used is 38 mm × 38 mm at big end, 25 mm × 25 mm at small end and length is about 6.5 cm. The cleats used are of different sizes and different types in order to accommodate cables of various sizes and different numbers of cables respectively. The cleats are of three types one groove, two groove and three-groove to accommodate one, two and three cables respectively. For low voltage (up to 250 V) installations, cleats shall be of such dimensions that cables shall not be less than 2.5 cm apart for branch circuits and less than 4 cm apart for submains. In order to ensure longer life to cleat wiring systems, cleats should be usually used at intervals of 30 cm and in no case at more than 60 cm.

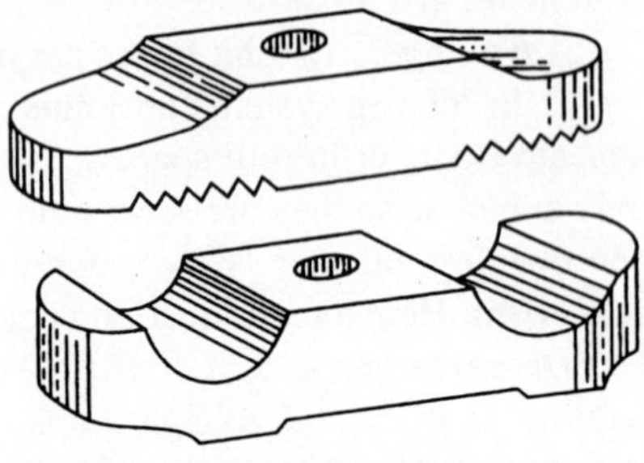

Fig. 13.9 *Cleat*

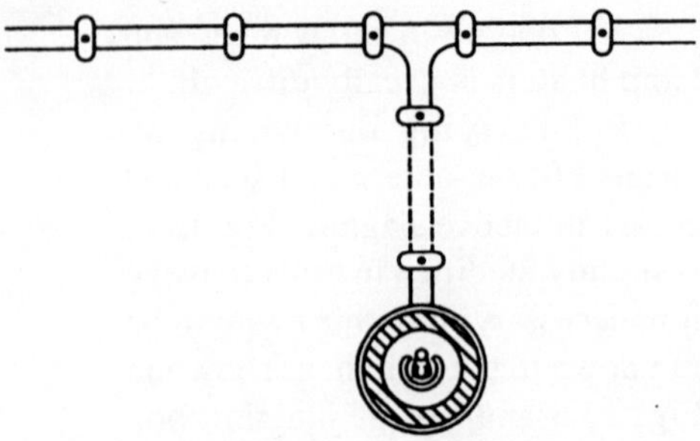

Fig. 13.10 *Cleat Wiring*

Advantages. 1. It is the cheapest system of internal wiring.

2. Its installation and dismantlement is easy and quick.
3. Material is recoverable after the dismantlement.
4. Inspection, alterations and additions can be easily made.
5. Skill required is little.

Disadvantages. 1. It is not good looking.

2. It is quite temporary and perishes quickly.

3. The wires are exposed to mechanical injury.

4. The insulation catches dampness from the atmosphere and a common salt like substance appears on the insulation which lowers the insulation resistance and causes leakage. Hence this system of wiring cannot be used in damp places unless precautions in an approved manner are adopted to effectively prevent dampness from deteriorating the insulation of conductors.

5. Oil and smoke are injurious to VIR insulation.

Fields of Applications. The wiring of this type is very suitable for temporary installations in dry places. This is also used where appearance is not so important and cost is the main consideration. This system is not suitable for use in domestic premises.

13.4.2. Wooden Casing and Capping Wiring. The cables used in this type of wiring are either VIR or PVC or any other approved insulated cables. The cables are carried through the wooden casing enclosures. The casing consists of V-shaped grooves (usually two to hold the cables of opposite polarity in different grooves) and is covered at the top by means of rectangular strip of wood, known as *capping* of same width as that of casing. The capping is screwed to the casing by means of 13 mm × 4 mm wooden screws fixed at every 15 cm on the centre fillet. To protect the casing against white ants first class seasoned teak wood varnished by shellac varnish is employed. Two or three cables of same polarity (either all phases or all neutrals) may be run in one groove and in no case the cables of opposite polarity should be run in the same groove. The casing is usually placed 3.2 mm apart from the wall or ceiling by means of porcelain distance pieces of thickness not less than 6.5 mm in order to keep the casing dry at the back. The wooden gutties on which the casing is screwed by means of 32 mm × 8 mm wooden screws are fitted into walls or ceilings at intervals not exceeding 90 cm for sizes of casing capping upto 64 mm and not exceeding 60 cm for sizes more than 64 mm. The sizes of the casing capping to be used depends upon the number and size of cables to be accommodated in any particular length of run. The Table 13.1 shows the number of cables of different sizes, which can be accommodated in each groove of different sizes of casing capping. The length in which casing capping is available, varies from 2.5 metres to 3 metres.

Fig. 13.11

Advantages. 1. Cheap in cost as compared to lead sheathed and conduit wiring system.

2. Easy to instal and rewire.

3. It provides good insulation as conductors are good distance apart.

4. Free from trouble of condensation, so advantageous in tropical countries where there might be trouble from condensation of moisture in steel conduit.

5. Easy to inspect by opening the capping.

Disadvantages. 1. This type of wiring, even being coated with paint or shellac varnish, is not damp proof so cannot be used in damp places.

2. Since there is a risk of fire, it cannot be used where there is a possibility of fire hazard.

3. Since it requires better workmanship, the labour cost is higher.

4. This type of wiring can be used only on surface and cannot be concealed in plaster.

5. It is a costly system because it needs knot free wood.

Fields of Application. This type of wiring is suitable for low voltage domestic installations in dry places and where there is no risk of fire hazard.

TABLE 13.1

Size of wood casing and capping and the number of wires that may be accommodated in one groove of the casing

Width of casing or capping in mm	*No. of grooves*	*Width of grooves in mm*	*Thickness of casing in mm*	*Thickness at the back under the groove of casing in mm*	*Thickness of outer wall in mm*	*Thickness of capping in mm*
38	2	6	16	6	7	6
44	2	6	16	6	10	10
51	2	9	19	6	10	10
64	2	13	19	10	10	10
76	2	16	25	10	10	10
89	2	16	32	10	11	13
102	2	19	32	13	13	13

Size of cable in mm²	*Number of cables that an be drawn in one groove of casing-capping of size COPPER CONDUCTORS ONLY*						
	38 mm	*44 mm*	*51 mm*	*64 mm*	*76 mm*	*89 mm*	*102 mm*
1.0	2	2	3	3	9	12	12
1.25	1	1	2	2	8	12	12
1.5	1	1	2	2	8	12	12
2.0	1	1	2	2	5	10	10
2.5	1	1	2	2	5	10	10
3.0	–	–	2	2	5	8	9
4.0	–	–	2	2	5	8	9
4.5	–	–	1	1	4	6	6
6.0	–	–	1	1	4	6	6
10.0	–	–	1	1	2	3	4
14.0	–	–	–	–	1	2	2
16.0	–	–	–	–	1	2	2
25.0	–	–	–	–	1	1	1
35.0	–	–	–	–	1	1	1
40.0	–	–	–	–	1	1	1
50.0	–	–	–	–	1	1	1

13.4.3. CTS or TRS Wiring. In this type of wiring the cables used may be single core, twin core or three core TRS cables with a circular or oval shape. Usually single core cables are preferred. TRS cables are sufficiently chemical proof, water proof, steam proof but are slightly affected by lubricating oils. TRS cables are run on well seasoned, perfectly straight and well varnished (on all four sides) teak wood batten of thickness 10 mm at least. The width of the batten depends upon the number and size of cables to be carried by it. The battens are available in width of 13, 19, 25, 31, 38, 44, 50, 56, 63, 69, and 75 mm. For guidance the number of copper conductor cables of size 3/0.736 mm or aluminium conductor cables of size 1/1.40 mm (usually used in internal wiring) that can be carried by batten of different sizes are given in Table 13.2. The wood battens are secured to the walls or ceilings by flat head wood screws to wood or other approved plugs at an interval not exceeding 75 cm. The cables are held on the wood batten by means of tinned brass link clips already fixed on the batten with brass pins and spaced at an interval of 10 cm in case of horizontal runs and 15 cm in case of vertical runs. The wiring after erection is neatly painted with two coats of oil-less non-cracking paint, as specified in IS 732 and so on.

Advantages. 1. Its installation is easy and quick and savings in labour largely compensate for the extra cost of the cable.

2. Its life is sufficiently long.
3. Within certain limits it is fire proof.
4. It can withstand the action of most chemicals such as acids and alkalies.
5. It is cheapter than other types of wiring except cleat wiring.
6. If the job is carried out with care it gives a nice appearance.

TABLE 13.2

Number of cables of size 3/0.736 mm copper conductor or 1/1.40 mm aluminium conductor single core that can be laid	*Size of batten required*	*Number and size of link clips required*
2	13 × 13 mm	1 × 38 mm
3	19 × 13 mm	1 × 50 mm
4	25 × 13 mm	2 × 38 mm
5	31 × 13 mm	1 × 38 mm and 1 × 50 mm
6	38 × 13 mm	3 × 38 mm
7	44 × 13 mm	2 × 38 mm and 1 × 50 mm
8	50 × 13 mm	1 × 38 mm and 2 × 50 mm
9	56 × 13 mm	3 × 50 mm
10	63 × 13 mm	2 × 38 mm and 2 × 50 mm
11	69 × 13 mm	1 × 38 mm and 3 × 50 mm
12	75 × 13 mm	4 × 50 mm

Disadvantages. 1. Good workmanship is required to make a sound job in TRS wiring.

2. This type of wiring cannot be recommended for use in situations open to sun or rain unless preventive steps are taken to preserve the insulation of cables.

Fields of Application. The TRS wiring is suitable for low voltage installations and is extensively used for lighting purposes everywhere i.e. in domestic, commercial or industrial buildings except workshop where it is liable to mechanical injury. It cannot be used in damp places.

This type of wiring is suitable in situations where acids and alkalies are likely to be present.

13.4.4. Lead Sheathed or Metal Sheathed Wiring. In this system of wiring, the cables used are insulated wires, TRS or PVC with an outer covering of sheath of lead-aluminium alloy containing about 95% lead. This metal sheath gives protection to the cable from mechanical injury, dampness and atmospheric corrosion. The whole lead covering is made electrically continuous and is connected to earth at the point of entry to protect against electrolytic action due to leakage current and to provide safety against the sheath becoming alive. The cables are run on well seasoned and perfectly straight teak wood batten of thickness not less than 10 mm and are fixed to it by means of link clips as in case of TRS wiring. The batten shall be well varnished with two coats at the back, and the front shall be painted with two coats of paint of colour to match the surroundings. The width of the wooden batten may vary in accordance with the number of wires it may have to carry. The batten shall be run on brick walls, stone walls or plastered walls and ceilings, steel joists or any structural steel work. The wood plugs for fixing the batten shall be of standard size as mentioned in general specifications, and be placed at a distance of 75 cm. The wood screws shall be used for fixing the batten. Only tinned brass clips of suitable size shall be used for fixing the cables on the well erected wooden batten. The greater part of the cable employed is flat twin (the cable having two insulated conductors side by side covered with red and black tape respectively and under one flat covering of lead alloy). Three-core flat cable is also used in certain cases as well as single cables under a circular sheathing of lead-alloy. Wiring with twin-core cable makes the neatest job, and therefore, one of the methods of looping-in shown in Figs. 13.6, 13.7 and 13.8 should be employed.

Advantages. 1. It provides protection against mechanical injury better than that of TRS wiring.

2. It is easy to fix and looks nice as it can be run in buildings without damaging decoration and can be painted to suit colour scheme of surroundings.

3. Its life is long if proper earth continuity is maintained throughout.

4. It can be used in damp situations provided protection against moisture effect on the ends of the cables is given.

5. It can be used in situations exposed to rain and sun provided no joint is exposed.

Disadvantages. 1. It is costlier than TRS wiring.

2. It is not suitable for places where chemical corrosion may occur.

3. In case of damage of insulation, the metal sheath becomes alive and gives shock. So, as to provide safety against electrical shock it is necessary that the sheath is properly earthed and an earth wire is run side by side with it and all pieces are properly bonded or jointed together so that not a single cover is left unearthed.

4. Skilled labour and proper supervision is required.

Fields of Application. This wiring system is suitable for low voltage (up to 250 V) installations. It may be used in places exposed to sun and rain provided no joint of any kind is exposed. It may also be used in damp places with protection against dampness coming in contact with open ends of the cable. It should not be used in places where acids and alkalies may be present.

The most common application of this system is for laying submains from pole or bracket to the electric meter laid by the supplier. This is invariably employed for this purpose, wooden batten is not used; rather lead sheathed cable is attached to the walls by means of clips fixed directly on wooden plugs of standard size.

13.4.5. Conduit Wiring. In this system of wiring steel tubes, known are *conduits,* as installed on the surface of walls by means of saddles or pipe hooks or buried under plaster and VIR or PVC cables are drawn afterwards by means of a GI wire of size of about 18 SWG. In damp situations the conduits can be spaced from the walls by means of small wooden blocks fixed below the pipes at regular intervals. In order to facilitate drawing of wires number of inspection fittings are provided along its length. The conduits should be electrically and mechanically continuous and connected to earth at some suitable point. The conduit used for this purpose is of two types namely (*i*) light gauge (or split type) conduit and heavy gauge (or screwed type) conduit. Light gauge or split conduit with a seam along its length is used for cheap work. It is not water tight or even damp proof and is not permitted on medium voltage (i.e. on voltage higher than 250 V). Screwed conduit (solid drawn or with welded seam) is used for all medium voltage (250 V to 600 V) circuits and in places where good mechanical protection and absolute protection from moisture is desired. In general the finish of the conduit is black stove enamelled, there being a smooth coat of enamel both on the inside and outside surface of the tube. Galvanised conduit is also employed, especially in damp situations when the conduit is on the surface but under ordinary conditions buried in walls it offers little, if any, advantage over good enamelled conduits.

Conduit size is stated in terms of its outer diameter. The smallest size is 12 mm seldom employed except in cheap competitive work, the next size is 16 mm and the next 19 mm after which it rises in size in 6 mm steps to 31 mm and next standard sizes are 38 mm and 50 mm. The largest standard size is 63 mm but this is not much in use.

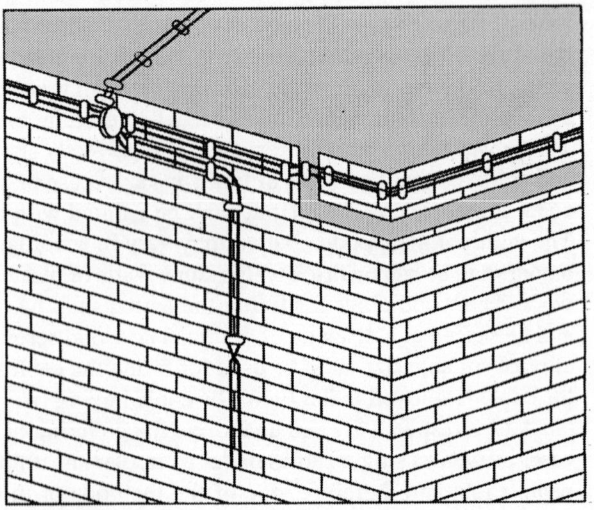

Fig. 13.12 *Illustration of Surface Conduit Wiring*

The number of cables of different wires that can be accommodated in various sizes of conduits are given in Table 13.3 (Refer I S: 732-1983).

PVC conduit pipes are also available now and are increasingly being employed in place of steel conduits. PVC conduits are cheaper in cost and the labour time saved may be as much as 25% compared to the time taken in installing steel conduit. Such conduits are resistant to acid, alkalies, oil and moisture. They can be buried in lime or cement plaster without ill effects. The main drawback of PVC conduit is its movement due to variations in temperature. Also PVC conduit is not suitable for use in locations prone to fire hazards.

TABLE 13.3

Maximum capacity of conduits for drawing-in of 1,100 V grade cables

Nominal x-sectional area in mm²	*Number and diameter of wires in mm*	*Material of conductor*	*Number of cables that can be accommodated in conduit of size*											
			19 mm		*25 mm*		*31 mm*		*38 mm*		*50 mm*		*63 mm*	
			S	*B*	*S*	*B*	*S*	*B*	*S*	*B*	*S*	*B*	*S*	*B*
1.0	1/1.12	Copper	7	5	13	10	20	14	–	–	–	–	–	–
1.25	3/0.736	Copper	7	5	12	10	20	14	–	–	–	–	–	–
1.5	1/1.40	Al	7	5	12	10	20	14	–	–	–	–	–	–
2.0	3/0.925	Copper	5	4	10	8	18	12	–	–	–	–	–	–
2.5	1/1.80	Al	6	5	10	8	18	12	–	–	–	–	–	–
3.0	7/0.736	Copper	5	4	8	6	12	10	–	–	–	–	–	–
4.0	1/2.24	Al	4	3	7	6	12	10	–	–	–	–	–	–
4.5	7/0.925	Copper	3	2	6	5	10	8	–	–	–	–	–	–
6.0	1/2.80	Al	3	2	6	5	10	8	–	–	–	–	–	–
6.75	7/1.12	Copper	2	–	5	4	8	7	–	–	–	–	–	–
10	1/3.55	Al	2	–	5	4	8	7	–	–	–	–	–	–
12	7/1.32	Copper	2	–	4	3	6	5	8	6	–	–	–	–
14	7/1.626	Copper	–	–	3	2	4	4	7	6	–	–	–	–
16	7/1.70	Al	–	–	2	–	4	3	7	6	–	–	–	–
18	19/1.12	Copper	–	–	–	–	4	3	6	5	10	7	12	8
25	7/2.24	Al	–	–	–	–	3	2	5	4	8	6	9	7
30	19/1.32	Copper	–	–	–	–	3	2	5	4	8	6	9	7
35	7/2.50	Al	–	–	–	–	2	–	4	3	7	5	8	6
40	19/1.626	Copper	–	–	–	–	–	–	3	3	6	5	8	6
50	7/3.00	Al							4	3	6	4	6	5
50	19/1.80	Copper	–	–	–	–	–	–	2	–	5	4	6	5

Note: The table shows the maximum capacity of conduit for the simultaneous drawing-in of cables. The Table applies to 1,100 V grade cables. The columns headed 'S' apply to runs of conduit which have distance not exceeding 4.25 metres between draw-in-boxes, and which do not deflect from straight by an angle of more than 15°. The columns headed B apply to runs of conduit with deflect from the straight by an angle of more than 15°.

Advantages. 1. It provides protection against mechanical damage.

2. It provides complete protection against fire due to short-circuit etc.
3. The whole system is water proof.
4. Replacement and alteration of defective wiring is easy.
5. Its life is long if the work is properly executed and
6. It is shock proof also if earthing and bonding is properly done.

Disadvantages. 1. It is very costly system of wiring.

2. Its erection is not easy and requires time.

3. Experienced and highly skilled labour is required for carrying out the job.

4. Internal condensation of moisture may cause damage to the insulation unless the system outlets are properly drained and ventilated.

Fields of Application. As this system of wiring provides protection against fire, mechanical damage and dampness so this is the only approved system of wiring for:

(*i*) places where considerable dust or fluff is present such as in textile mills, saw mills, flour mills etc.

(*ii*) damp situations.

(*iii*) in workshops for lighting and motor wirings.

(*iv*) places, where there is possibility of fire hazards such as in oil mills, varnish factories etc.

(*v*) places, where important documents are kept such as a record room.

(*vi*) residential and public buildings where appearance is the prime thing.

The recessed type conduit wiring is preferred for residential and public buildings.

PVC conduit wiring system (particularly concealed) is cheaper in cost and takes less time but does not provide protection against fire. Insurance requirements stipulate metallic conduit wiring and PVC wiring only for offices.

13.5. CHOICE OF WIRING SYSTEMS

It is certain that the success or failure of a wiring installation depends very largely on the proper selection of the type of wiring, size and position of light/fan points.

The choice of any wiring system for a particular installation should be based on technical and economic considerations, both in the context of the wiring system itself and the installation for which it is proposed. In general the following factors should be considered.

1. Safety. It is one of the most important factors to be considered. Sometimes poor workmanship may lead to dangerous results. The first and foremost consideration is safety to the person using electricity against leakage or shock. For instance, in factory where lot of fumes are produced, cleat or casing-capping wiring cannot be employed. Where there is possibility of fire hazard, conduit wiring must be used.

2. Durability. The type of wiring to be selected must be durable and it must be of proper specifications and also in accordance with assessed life and type of building. For example cleat wiring suitable for temporary buildings will definitely be unsuitable for a permanent building. The wiring should be able to withstand wear and tear due to weather and it must be capable of carrying the maximum current without overheating.

3. Appearance. The wiring must provide a good look after its installation. In case cleat or casing-capping wiring is used in a modern beautiful house it will spoil the outlook of the building. From aesthetic point of view, concealed conduit wiring system provides beautiful and good appearance but is costly. PVC wiring system also provides good appearance, and is very popular nowadays.

4. Mechanical Protection. The wiring must be protected from mechanical damage during its use.

5. Permanency. The wiring must not deteriorate unduly by action of weather, fumes, dampness etc.

6. Accessibility. In a selected wiring system there should be facilities for extension, renewal or alterations.

7. Initial Cost. The initial cost of the wiring system to be selected is one of the main points to be considered. The wiring system selected should be safe as well as economical.

8. Maintenance Cost. The wiring system selected should have, as far as possible, low maintenance cost.

The other factors, in addition to above, to be kept in view while making the choice of wiring system are load, voltage to be employed, type of building etc.

Fire hazards and insurance standards must be kept in mind in case of large factories etc.

TABLE 13.4
Comparison Between Various Systems of Wiring

S.No.	*Particulars*	*Cleat Wiring*	*Wood Casing Capping Wiring*	*TRS Wiring*	*Lead Sheathed Wiring*	*Conduit Wiring*
1.	Material required	Cleats, VIR or PVC cables, screws, gutties, blocks, boards etc.	Teak wood casing and capping, VIR or PVC cables, wooden gutties, screws, blocks, boards etc.	Teak wood batten, TRS or PVC cables, wooden gutties, screws, nails, link clips, boards etc.	Teak wood batten, lead sheath cables, wooden gutties, screws, clips, or joint clips boards, round boards etc.	Conduit pipe, VIR or PVC cables, saddles or pipe hooks, wooden gutties, screws, IC boxes, IC bends, tees etc., with IC socket and screws.
2.	Cost	low	medium	medium	costly	very costly
3.	Voltage	low (up to 250 V)	low (up to 250 V)	low (up to 250 V)	low (up to 250 V)	low or medium (up to 660 V)
4.	Life (durability)	very short	fairly long	long	long	very long
5.	Protection against fire	poor	no	fair	good	very good
6.	Mechanical protection	none	fairly good	good	fairly good	very good
7.	Dampness protection	none	poor	good	good	fairly good
8.	Appearance	not good	fair	good	fair	very good
9.	Safety	no	medium	medium	medium	high
10.	Type of labour required	semi-skilled	highly skilled	skilled	skilled	highly skilled
11.	General reliability	poor	good	good	fairly good	very good
12.	Additions or alterations to the existing wiring	very easy	difficult	easy	not very difficult	most difficult
13.	Number of points that can be wired per day by a wire man with a mate	6	4	5	4	3
14.	Fields of application	For temporary installations e.g., for functions, marriages etc.	For residential, commercial and office buildings.	For residential, commercial and office buildings and for general purposes but nowadays being replaced by PVC wiring on account of inherent advantages.	Only used for service mains etc. because of its high cost and heavy short circuit in case of leakage.	Mainly for godowns, workshops and public buildings where economy is not the prime factor.

13.6. MAIN SWITCH AND DISTRIBUTION BOARDS

As per Indian Electricity Rule 50 a suitable linked switch (a switch operating simultaneously on phase or line and neutral wires) is to be provided immediately after the meter board. This rule also stipulates that a suitable cut out must be provided just after the linked switch to protect the circuit against excessive current. The linked main switch and fuse unit may be provided as one unit or as separate units.

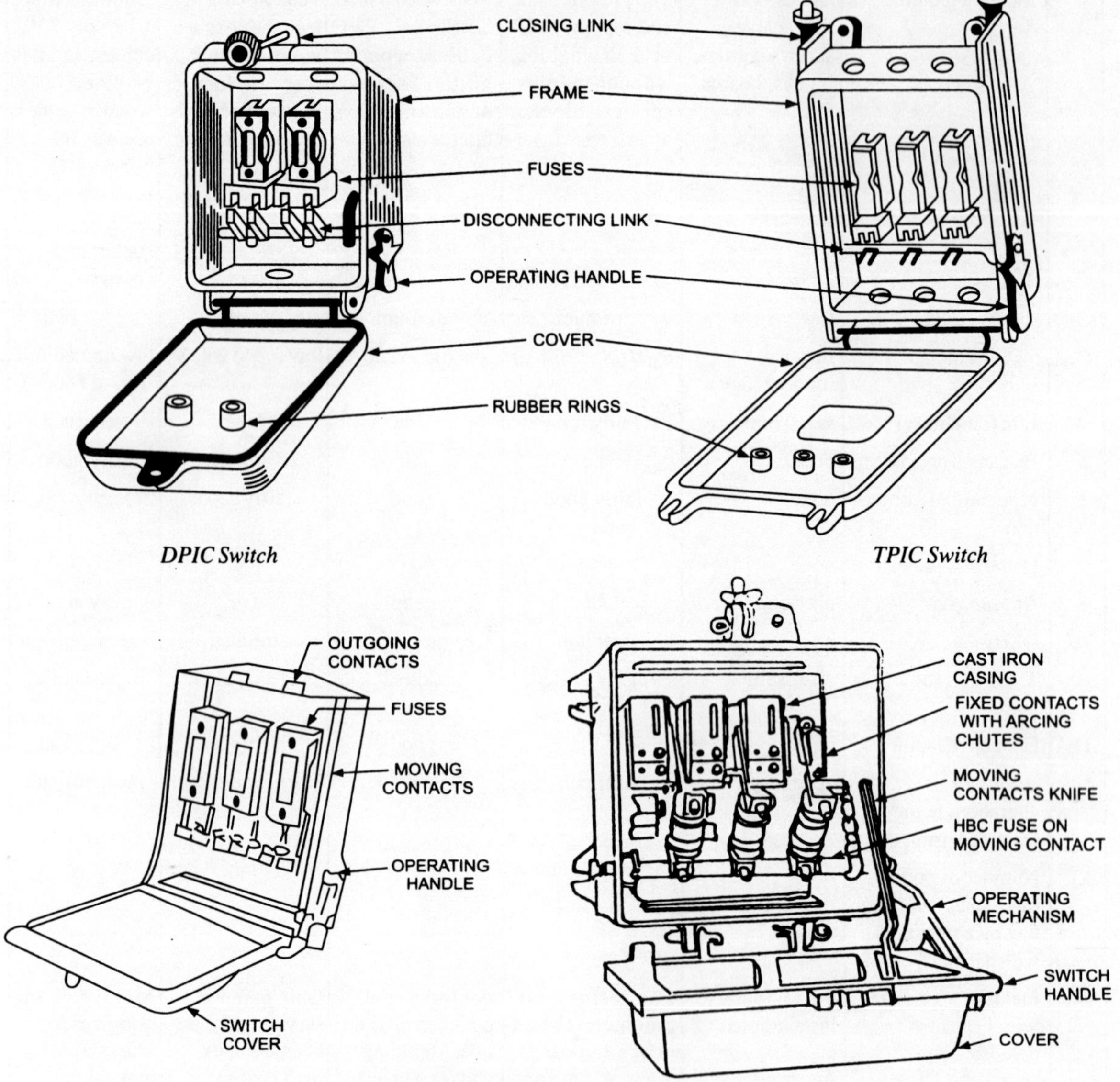

Fig. 13.13 *Switches and Switch Fuses*

Switch fuse is a combined unit and is known as an iron *clad switch*, being made of iron. It may be double pole for controlling single phase two-wire circuits or triple pole for controlling three-phase, 3-wire circuits or triple pole with neutral link for controlling 3-phase, 4-wire circuits. The respective switches are known as double pole iron clad (DPIC); *triple pole iron clad* (TPIC) and *triple pole with neutral link iron clad* (TPNIC) switchs. These switches are shown in Figs. 13.13, 13.14 and 13.15.

Since no fuse is to be provided in neutral (IE Rule 32), in DPIC switch fuses, where provision is made for fuses in both the wires, one fuse carrier is furnished with fuse element and the other with a thick copper wire.

(a) Double Pole Iron Clad Switch

(b) Triple Pole, Iron Clad Switch

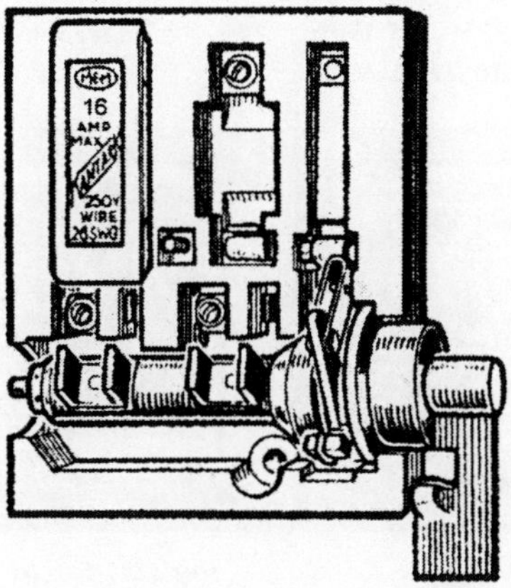

(c) 16-Amp. All-insulated Main-Switch

(d) All-insulated Switch With Moulded Cover and Base For Circuits Upto 16 Amperes Capacity

Fig. 13.14

The specifications of IC switch fuse units are given below as samples :

1. For Two-Wire DC Circuits or Single Phase AC Circuits. 240 V, 16 A, DPIC switch fuse of any make approved by IS.

2. For Three-Wire DC Circuits. 500 V, 32 A (63/100/150 or higher amperes), IS approved TPIC switch fuse.

3. For Three-Phase Balanced Load Circuits. 415 V, 32 A (63/100/150 or higher amperes), IS approved TPIC switch fuse.

4. For Three-Phase 4-Wire Circuits. 415 V, 32 A (63/100/150 or higher amperes) IS approved TPIC switch fuse with neutral link.

Switch Fuse Requirements - General

1. The switches and fuses shall be enclosed in a strong metallic enclosure.
2. The metallic enclosure shall be dust and weather proof and have a mounting arrangement on a wall or an iron frame.
3. The metallic enclosure will have an earthing terminal.

(e) Three-Pase Iron-Clad Main Switch For Power Use

Fig. 13.14

16 A 240 Volts AC SPN ISI

16 A 240 Volts AC SPN (2 Fuse)

32 A 240 Volts AC SPN ISI

32 A 240 Volts AC SPN (2 Fuse)

16 A 415 Volts AC TP ISI

16 A 415 Volts AC TP & N ISI

32 A 415 Volts AC TP ISI

32 A 415 Volts AC TP & N ISI

63 A 415 Volts AC DP

63 A 415 Volts AC TP ISI

63 A 415 Volts AC TP & N ISI

100 A 415 Volts AC TP & N ISI

100 A 415 Volts AC TP ISI

Fig. 13.15 *Air-Break Switch Fuse Combination Units [Courtsy SSK]*

4. The cover shall be interlocked with the actuating lever to prevent access to the interior whilst the lever is in the "ON" position.
5. The "ON" and "OFF" shall be clearly and indelibly marked on it.
6. The actuating lever shall be in effective electrical contact with the cover and fully insulated from the circuit.
7. Fixed contacts and other metal parts shall be nickel plated or tinned where it is desirable.

Distribution Boards. By definition, the distribution board is an assembly of parts, including one or more fuses or circuit breakers, arranged for the distribution of electrical energy to various circuits or other distribution boards known as *submain distribution boards.* The boards are usually metal-cased in sheet steel or hard-wood-cased in oak or teak. The door may be solid or glazed. The earthing terminal and locking arrangement are invariably provided. The number of ways depends upon the circuits or sub-circuits to be fed.

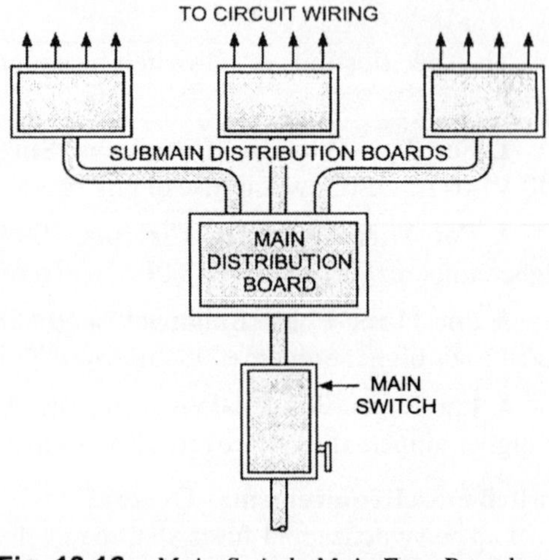

Fig. 13.16 *Main-Switch, Main Fuse Board and Submain Distribution Boards Showing The Method of Distribution of Electrical Energy To Various Floor Levels For Sectionalizing of Wiring*

2-Way 16 A 240 Volts AC | *3-Way 16 A 240 Volts AC* | *4-Way 16 A 240 Volts AC* | *6 Way 16 A 240 Volts AC* | *8-Way 16 A 240 Volts AC*

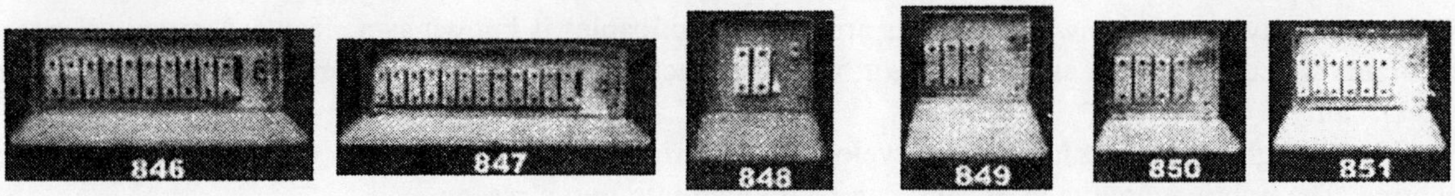

10-Way 16 A 240 V AC | *12-Way 16 A 240 V AC* | *2-Way 32 A 240 Volts AC* | *3-Way 32 A 240 Volts AC* | *4-Way 32 A 240 Volts AC* | *6-Way 32 A 240 Volts AC*

Distribution Fuse Boards [Courtesy SSK]

Separate distribution fuse boxes should be provided for light and power circuits. Some manufacturers combine linked switch and distribution boards.

Fig. 13.16 shows the main switch, main fuse board and submain distribution boards indicating the method of distribution to different floor levels to sectionalize the wiring.

Various types of fuse boards are shown in Figs. 13.17 and 13.18.

Fuse/Distribution Boards-General Requirements. Distribution boards have similar requirements as switch fuse units except that the metallic enclosure must have a locking arrangement.

(a) Six-Way Distribution Fuse Board

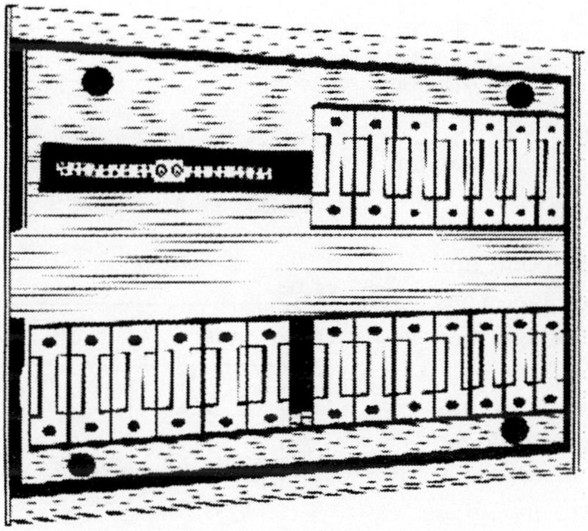

(b) Three Phase Distribution Box

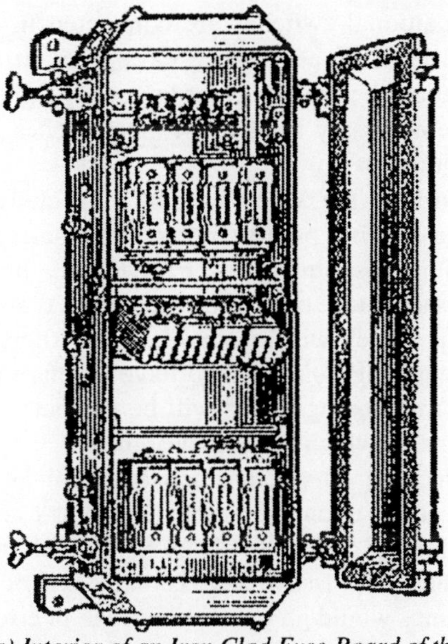

(c) Interior of an Iron Clad Fuse-Board of the Double-Pole Pattern, Showing The Fuse Racks Tilted, and the Cover Plates Which can be Drilled For Conduit Entry

The specifications of fuse boards are given below :

For Medium Size Residential Building. 6-Way, 16 A, 240 V ICDB of any make approved by IS.

For Motor Installations. 6-Way, 415 V, 300 A, Triple Pole with (or without) neutral link ICDB of any make approved by IS.

13.7. LIGHTING SUB-CIRCUITS

Electrical apparatus is connected, by cables, to the supply main and to the associated protective and controlling devices (usually fuses and switches). This arrangement of cables is known as a *circuit*. A circuit that feeds apparatus directly is known as a *sub-circuit*; when connected to a distribution board this is known as a *final sub-circuit*.

Various lighting circuits have been developed for maximum convenience and economical use of electrical lighting. The lighting circuits in general can be divided into two classes :

(a) Series and (b) Parallel circuits.

13.7.1. Series Lighting Circuits. This is the simplest arrangement. In this case if a lighting point controlled by one switch is to be wired then one cable will be taken from one terminal of the distribution board to one side of the single pole switch; from the other side of the switch to one terminal of light and then from second terminal of light to the second terminal of distribution board, as shown, Fig. 13.19. This arrangement is satisfactory if additional lighting points are not required.

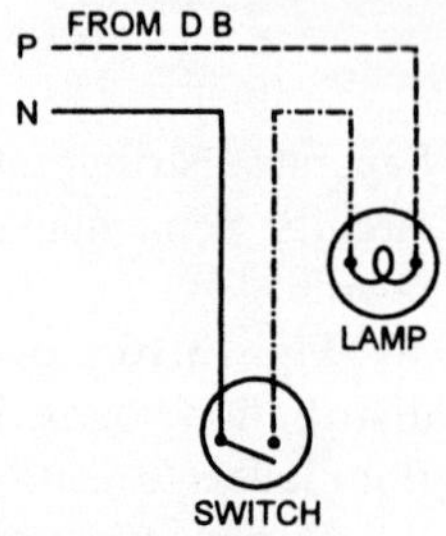

Fig. 13.19

In case more than one lighting points are to be connected in series in the same sub-circuit, they will all be controlled by one switch, as shown in Fig. 13.20.

Series circuits use is restricted to decoration purpose lighting circuits and not used for general house wiring circuits due to its following disadvantages :

(*i*) Since all the lighting points are controlled from one switch, all the lighting points will be switched on irrespective one or all are desired to be switched on.

If additional switches are connected in series there will be no advantage except creation of complications because in order to get any light at all it will be necessary to have all the switches in the 'on' position.

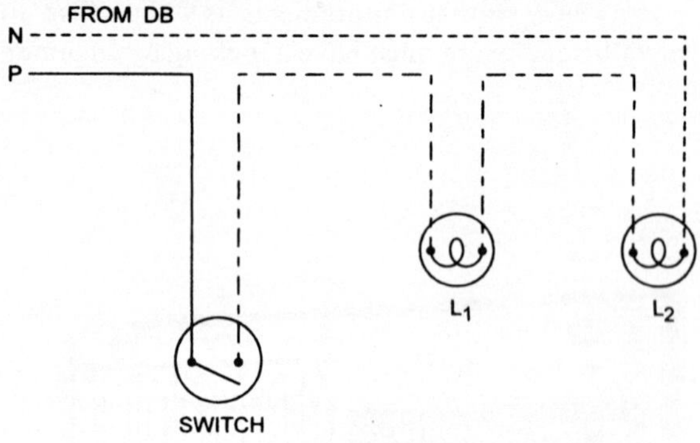

Fig. 13.20

(*ii*) No doubt the switches can be connected across the lighting points instead of in series with them, as shown in Fig. 13.21, but the disadvantage of this arrangement is that, when one switch will be closed, it will short out a lighting point and remaining points would have to share the supply voltage. Thus the remaining points will be subjected to a higher voltage than that for which they are designed.

(*iii*) Since the voltage, to which lamps will be subjected, in such circuits depends upon the number of lamps across the supply voltage, lamps of different voltage rating will be required.

(*iv*) The other disadvantage is that if one of the lamps fails, the other lamps would go out due to break in circuit.

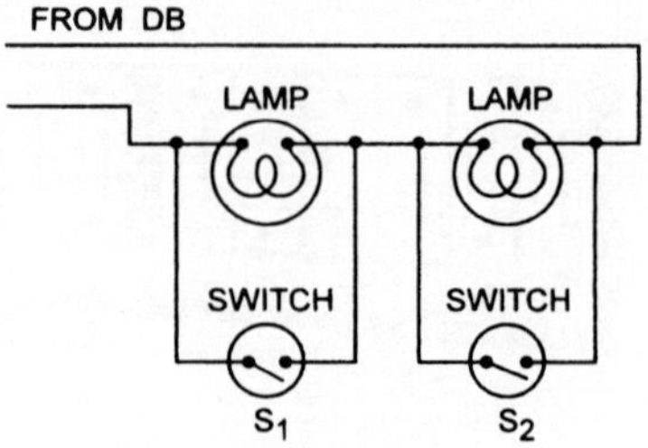

Fig. 13.21

The type of sub-circuit commonly used for house wiring is parallel sub-circuit.

13.7.2. Parallel Lighting Circuits. In this arrangement, all the lighting points are connected directly, across the distribution board terminals (across the supply) and controlled by individual switches, as will be shown in the circuits discussed in the succeeding paragraphs.

The lamps can be connected in parallel by two methods known as T-connection method and looping-in method. These methods have already been described in Art 13.3.

13.8. VARIOUS CIRCUIT DIAGRAMS

13.8.1. One Lamp Controlled From One Point. The circuit diagram of one lamp controlled from one point by one pole switch is shown in Fig. 13.22.

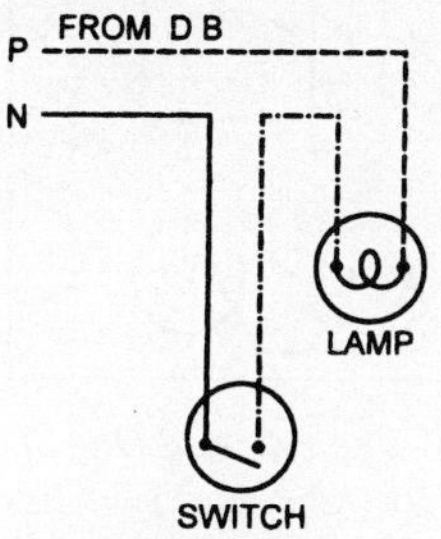

Fig. 13.22

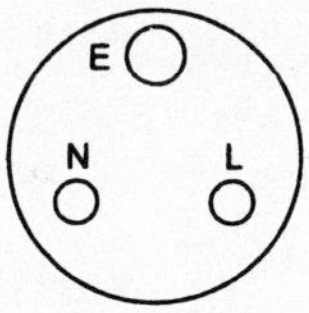

Fig. 13.23

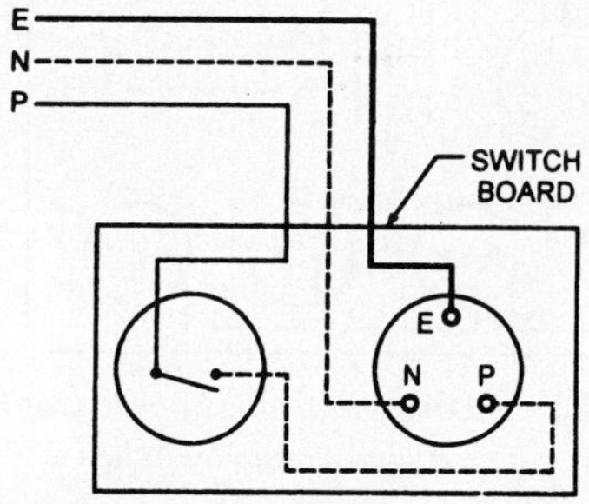

Fig. 13.24 *Connections of 3-Pin Socket-Outlet and Switch*

13.8.2. Socket Outlet Controlled By One Pole Switch. Relative positions of the sockets in a three pin sockets, when viewed from the front are shown in Fig. 13.23. The larger socket marked E is for earthing connection, the left hand socket for neutral wire and right hand socket for live wire. Connections of socket-outlet with one pole switch control is shown in Fig. 13.24.

13.8.3. Two Lamps Controlled By Individual Switches. The connections of two lamps controlled by individual switches in looping-in system are shown in Fig. 13.25.

13.8.4. Three Lamps and One Socket Outlet Controlled By Individual Switches. The connection diagram of 3 lamps and one socket outlet controlled by individual switches in looping-in system is shown in Fig. 13.26.

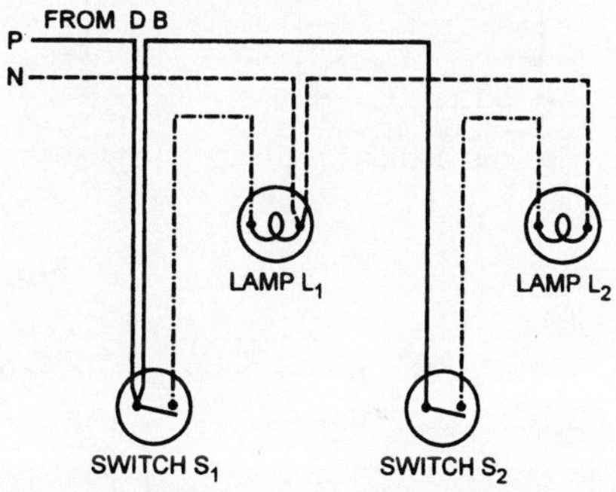

Fig. 13.25 *Two Lamps Controlled By Individual Switches*

13.8.5. One Lamp, One Fan and One Socket Outlet Controlled By Individual Switches. The connection diagram of one lamp, one fan and one socket outlet controlled by individual switches is shown in Fig. 13.27.

13.8.6. Two-Way Switching. Switch used for controlling a point or points from one position only is a one way switch. Normally it has two terminals although there is a type with an additional terminal used for looping-in. Two one-way switches cannot give complete control of a light point from two different positions, such as the stair and landing of a house as when one switch is in the 'on' position the light will remain 'on' irrespective of the position of the other switch.

For independent control from two positions, two-way switches are required. A two-way switch usually has four terminals, two of them being permanently connected together what is often known as the *bar side*. One of the bar terminals is blanked off by a screw. The live or phase wire is taken to the bar terminal of one two-way switch and bar terminal of the other switch is connected to the lamp one terminal, the neutral being looped into the other terminal of the lamp as usual. The remaining two terminals of each switch are inter-connected as shown in Fig. 13.28.

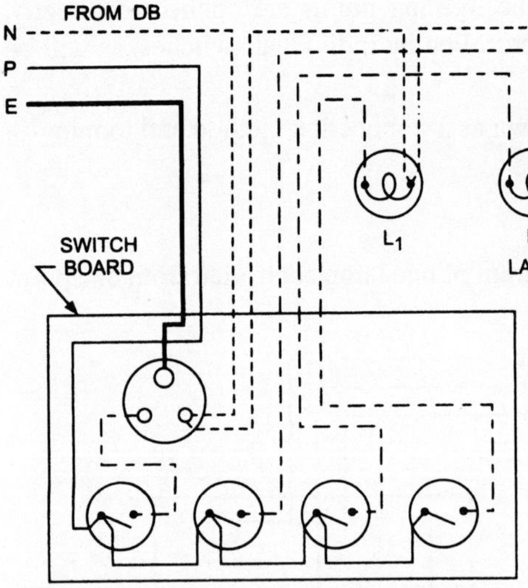

Fig. 13.26 *3 Lamps and One Socket-Outlet Controlled By Individual Switches*

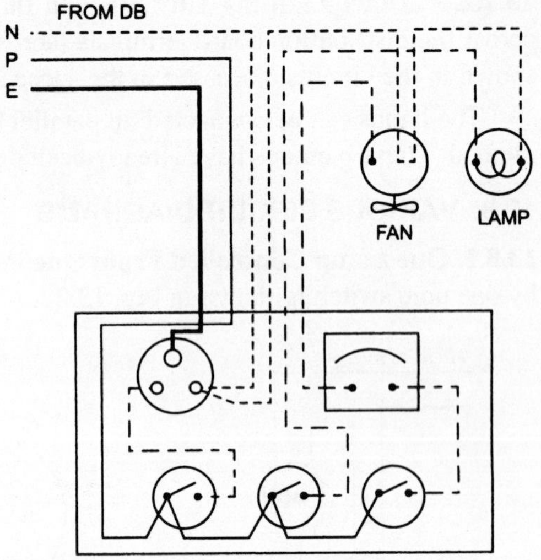

Fig. 13.27 *One Lamp, One Fan and One Socket Outlet Controlled By Individual Switches*

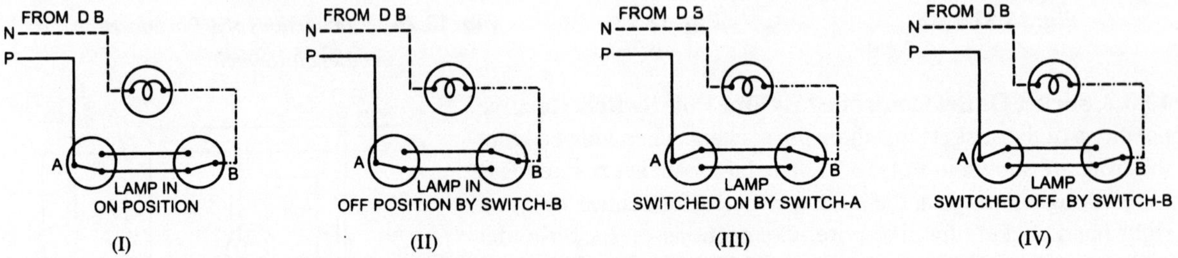

Fig. 13.28 *Stair-Case Wiring Circuits*

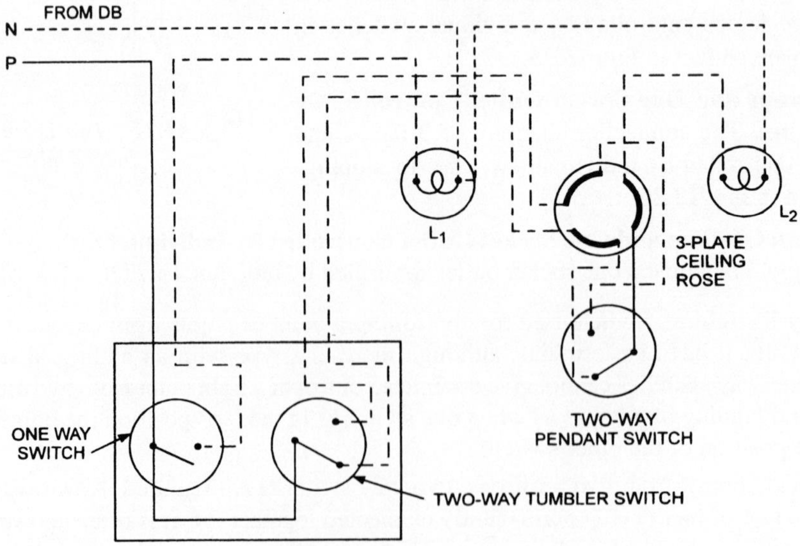

Fig. 13.29 *Bed Room Lighting Circuit*

13.8.7. Bed room Lighting. Bed room is usally provided with two lamps, one lamp near the dressing table controlled by one-way switch and another lamp near the bed controlled by two-way switches. One of these two 2-way switches should be tumbler switch and another should be pendant type suspended from a 3-plate ceiling rose. The circuit of such type lighting is shown in Fig. 13.29.

13.9. ELECTRICAL INSTALLATION FOR POWER CIRCUITS

Power circuits can be divided into two main categories.

(*i*) Power circuits for heaters, coolers, refrigerators, air-conditioners and other similar loads.

(*ii*) Power circuits for motors, generators, air-conditioning equipment/ventilation equipment, machine tools, other special purpose large loads.

The electrical installation for loads referred to at (i) above is done in the same way as for lighting and fan sub-circuit wiring. The only difference is that for lighting loads the load per circuit is restricted to 800 watts and 10 points whereas for loads referred above are restricted to 3,000 watts and two outlets on each sub-circuit (IS 4648—1968).

The electrical installation for loads referred to at (ii) above uses PVC/XLPE cables of suitable size, depending on the rating of the loads, enclosed in a rigid metallic conduit or laid underground. Flexible conduit is used to connect the end of the conduit pipe to the motor/equipment terminals. This is connected by a box connector coupling consisting of a tube with clamping arrangement for flexible conduit at one end and male threads at the other end. The male thread end is passed through the hole in the box and a lock nut is screwed over it to make connections of flexible conduit to the box. A similar coupling connects flexible conduit to the rigid conduit. Such couplings consist of clamping arrangement at one end and female threads at the other end for screwing rigid conduit into it.

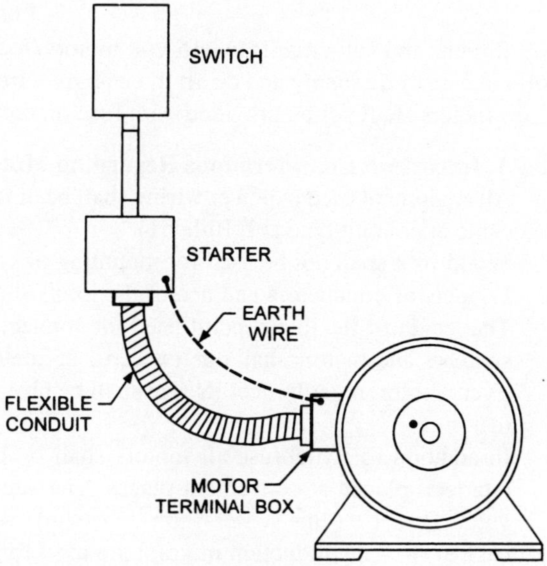

Fig. 13.30

In a large workshop or factory, power distribution is affected from a suitable location known as *distribution board* or a switchboard which normally consists of

(*i*) a set of insulated busbars.

(*ii*) each incoming/outgoing circuit controlled by a suitable means (automatic circuit breaker or linked switch and cutouts).

(*iii*) each outgoing circuit may be controlling a single 3-phase load or a group of such loads. In the latter case the outgoing circuits feed different sub-distribution boards erected at convenient locations, distributing power to various machines / equipment.

In case a distribution board is installed in an isolated place suitable means of control and isolation are to be provided at both places in accordance with Indian Electricity Rules—Rule 50 clause (a) and (c).

A line diagram showing the tapping supply from the busbar chamber to the individual machines is illustrated in Fig 13.31.

In no case looping-out on the motor circuits is to done, except when the motors are of very small capacity (such as fractional-horse power motors) and even then the same sized cable must be run from the first motor to the second and so on. Actually looping-out in case of wiring for supply to motors is hardly worthwhile and use of distribution boards and sub-distribution boards makes the job better.

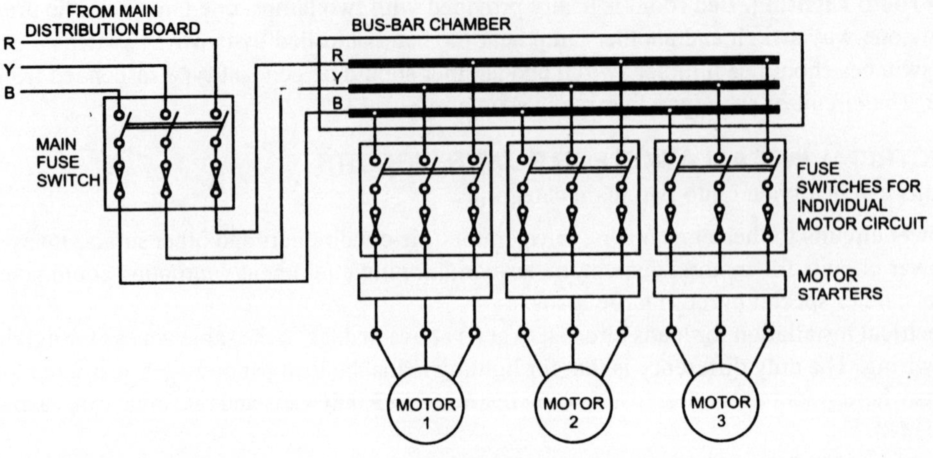

Fig. 13.31

Circuits and sub-circuits to 415 volt motors/loads must be provided with fuses on all poles i.e. on both poles in case of dc supply and on all three phase wires in case of 3-phase ac supply. Circuits supplying single phase motors shall not be provided with fuse on the neutral.

13.9.1. Important Considerations Regarding Motor Installation Wiring. These are as detailed below.

1. All equipment used in power wiring shall be of iron clad construction and wiring shall be of the armoured cable or conduit type (I E Rule 51).
2. Woodwork shall not be used for mounting of switchgear.
3. Looping of conductors and use of the joints shall not be done,
4. The length of flexible conduit used for connections between the terminal boxes of motors and starters, switches and motors shall not exceed 1.25 metres.
5. Every motor, regardless of its size shall be provided with a switchfuse placed near it. [IE Rule 50 clause (d)]
6. In addition to switchfuse all motors shall be provided with suitable means for starting and stopping (starters) placed at convenient places. The starters are used to limit the starting current to a desirable value. Direct-on-line starters, star-delta starters, auto-transformer-starters (or rotor resistance starters in case of slip-ring induction motors) are used for ac motors of rating up to 0.75 kW, above 0.75 kW and below 11 kW and above 11 kW respectively.
7. The conduit should preferably be laid in covered trenches to facilitate operator movement (safe).
8. Laying of cables must be in separate conduits for separate motors.
9. The minimum cross-section of conductor that can be used for power wiring is 2.5 mm^2 for aluminium conductor cables and 1.25 mm^2 for copper conductor cables (ISI recommendations). Hence PVC/XLPE cables of size lower than 3/0.915 mm copper or 1/1.80 mm aluminium cannot be used for power wiring.
10. The current rating of cables for supply to motor may be based on the normal full-load current of the motor but fuse rating should be based on starting current. In no case should the rating of the fuse be greater than twice the rating of the cable. So the following points must be observed:
    (*a*) For motors of capacity above 12 kW having starting current lower than twice normal full-load, the current rating of the cables is to be based on normal-full load current of motors and current-rating of fuses on motor starting current.
    (*b*) For motors of capacity below 12 kW, which have a very large starting current fuse should be of current rating to carry starting current safely and cable of current rating not lower than half the current rating of the fuse.

11. The conduit used in power wiring shall be electrically continuous throughout and connected to the frame of the motor. The frame of the motor shall be earthed by the owner by two separate and distinct connections of earth (IE Rule 61).
12. The wire used to earthing conductor shall be of copper or galvanised iron. The x-sectional area of copper earthing wire should be not smaller than half of the largest current carrying conductor used in wiring. The x-sectional area of G I wire, if used as an earthing conductor, should be such that its conductivity is not less than that of copper conductor.
13. Since supplier provides and maintains only a suitable earthed terminal at or near the point of commencement of supply at the consumer premises, the consumer is required to provide his own earthing system with an independent electrode.
14. While deciding the current rating of a main switch controlling a group of motors, starting current of one motor (highest rating) plus full-load current of remaining motors shall be considered.

HIGHLIGHTS

1. A network of wires connecting various accessories for distribution of electrical energy from the supplier meter board to the numerous electrical energy consuming devices such as lamps, fans and other domestic appliances through controlling and safety devices is known as a *wiring system.*
2. The supplier service cable feeding an installation terminates in what is usually called the *service fuse*. In an ordinary house the service fuse is called as *service cutout*. Such cutouts including service meters remain the property of the supplier and represent the furthest point of the supplier responsibility. The point at which the consumer's wiring is connected into the cutout is known as *point of commencement of supply or consumer's terminals.*
3. Since as per recommendation of Indian Standards the maximum number of points of lights, fans and 5 A socket-outlets that can be connected in one circuit is 10 and the maximum load that can be connected in such a circuit is 800 watts, in case more load or points are required to be connected to the supply, then it is to be done by having more than one circuit.
4. The number of circuits and sub-circuits is decided as per number of points to be wired and load to be connected to the supply system. For determination of load of an installation the following ratings may be assumed unless the values are known or specified.
   (*i*) Fluorescent lamps—50 watts
   (*ii*) Incandescent lamps, fans, and socket outlets—60 watts
   (*iii*) Power socket outlets—1,000 watts
   (*iv*) Exhaust fans—as per capacity of exhaust fans.
5. There are two methods of wiring viz. joint box or Tee system and loop-in system. Loop-in system is universally employed for connections of various lamps or other appliances in parallel.
6. **Cleat wiring** is the cheapest system of internal wiring and is very suitable for temporary installations in dry places.

   Wooden cashing and capping wiring is cheaper in cost as compared to lead sheathed and conduit wiring systems and is suitable for low voltage domestic installations in dry places and where there is no risk of fire hazard. This wiring system has now become obsolete.

   CTS or TRS wiring is suitable for low voltage installations and is extensively used for lighting purposes everywhere i.e. in domestic, commercial or industrial buildings except workshop where it is liable to mechanical injury. It cannot be used in damp places. This type of wiring is suitable in situations where acids and alkalies are likely to be present.

   Lead Sheathed or Metal Sheathed Wiring is suitable for low voltage (up to 250 V) installations. It may be used in places exposed to sun and rain provided no joint of any kind is exposed. It may also be used in damp places with protection against dampness coming in contact with open ends of the cable. It should not be used in places where acids and alkalies may be present.

   The most common application of this system is for laying submains from pole or bracket to the electric meter laid by the supplier. This is invariably employed for this purpose, wooden batten is not used; rather lead sheathed cable is attached to the walls by means of clips fixed directly on wooden plugs of standard size.

7. As per Indian Electricity Rule 50 a suitable linked switch (a switch operating simultaneously on phase or line and neutral wires) is to be provided immediately after the meter board. This rule also stipulates that a suitable cut out must be provided just after the linked switch to protect the circuit against excessive current. The linked main switch and fuse unit may be provided as one unit or as separate units.

   Switch fuse is a combined unit and is known as an iron *clad switch*, being made of iron. It may be double pole for controlling single phase two-wire circuits or triple pole for controlling three-phase, 3-wire circuits or triple pole with neutral link for controlling 3-phase, 4-wire circuits. The respective switches are known as double pole iron clad (DPIC); *triple pole iron clad* (TPIC) and *triple pole with neutral link iron clad* (TPNIC) switches.

8. **Distribution Boards.** By definition, the distribution board is an assembly of parts, including one or more fuses or circuit breakers, arranged for the distribution of electrical energy to various circuits or other distribution boards known as *submain distribution boards*. The boards are usually metal-cased in sheet steel or hard-wood-cased in oak or teak. The door may be solid or glazed. The earthing terminal and locking arrangement are invariably provided. The number of ways depends upon the circuits or sub-circuits to be fed. Separate distribution fuse boxes should be provided for light and power circuits. Some manufacturers combine linked switch and distribution boards.

EXERCISES

1. Draw and explain neat wiring diagram for a house installation showing connections for energy meter, main switch and distribution boards.
2. How will you determine the number of circuits required in a house wiring installation ?
3. Describe the various systems of distribution of electrical energy for internal wiring.
4. Explain the looping-in-system and cut-out system of wiring and compare them.
5. Describe in brief different types of domestic wiring systems.
6. What are the various wiring systems and state the merits and demerits of each system.
7. State various types of domestic wiring and suggest their suitability for respective installation.
8. Name various types of wiring systems. What are the advantages and disadvantages of conduit wiring ?
9. Suggest suitable wiring for any four of the following, give reasons in support of your answer :
   (*i*) Auditorium (*ii*) Radio station (*iii*) Machine shop (*iv*) Cold storage (*v*) Damp places.
10. Name the type of wiring best suited for workshops. Explain this wiring in detail and mention its advantages and disadvantages.
11. Explain what is meant by a sub-circuit. Describe the various lighting circuits.
12. Give a circuit diagram of one lamp controlled from one point by one pole switch.
13. Give a circuit diagram of one socket-outlet controlled from one point by one pole switch.
14. Give a circuit diagram of two lamps controlled by individual switches in looping–in system.
15. Draw the wiring diagram of 3 lamps and one socket-outlet controlled by individual switches in looping–in system.
16. Draw the wiring diagram, schematic diagram and single line diagram of one light, one ceiling fan with re-gulator and one 5 A 3 pin plug point each controlled by individual switches.
17. Draw schematic and wiring diagram for stair-case lighting.
18. Draw a circuit diagram for bed room lighting.
19. State some important points to be kept in view while carrying out wiring for a motor installation.
20. Write short note on "power wiring".

SHORT ANSWER TYPE QUESTIONS WITH ANSWERS

Q. 1. What is the maximum number of points of light, fan and socket-outlets that can be connected in one sub-circuit ?

Ans. Ten.

Q. 2. What is the maximum load (light and fan) that can be connected in one sub-circuit ?

Ans. 800 watts.

Q. 3. On what basis is the number of circuits/sub-circuits determined in house wiring installations ?

Ans. The number of circuits and sub-circuits is decided on the basis of total number of points to be wired and the total load to be connected to the supply system.

Q. 4. Which system is used for distribution of electrical energy in a building ?

Ans. Distribution board system.

Q. 5. Which system is used in connections of various lamps or appliances in parallel ?

Ans. Loop-in-system.

Q. 6. Which of the following wiring systems is commonly used for light/fan load in domestic and commercial buildings ?

(*a*) Cleat wiring (*b*) Casing-capping wiring (*c*) TRS wiring

(*d*) Lead sheathed wiring (*e*) PVC wiring

Ans. PVC wiring.

Q. 7. Which wiring system is employed in mechanical workshops ?

Ans. Conduit wiring system (surface or concealed).

Q. 8. What are that points which will be considered in selecting the wiring for a particular installation ?

Ans. The main points, which are considered in selection of the type of wiring for a particular installation, are safety, durability, appearance mechanical protection, permanency, accessibility and cost (initial as well as maintenance).

Q. 9. What is meant by distribution board ?

Ans. Distribution board is an assembly of parts, including one or more fuses or circuit breakers, arranged for the distribution of electrical energy to different circuits or other distribution boards known as submain distribution boards.

Q. 10. What is meant by a switch ?

Ans. A switch is a device that is used for making or breaking the electric circuit in a convenient way.

Q. 11. What is the maximum load that is usually connected in a power sub-circuit ?

Ans. 3,000 W

Q. 12. What is the maximum number of outlets, which can be connected in a power sub–circuit ?

Ans. Two

❑❑❑

UNIT-VIII
Illumination

Contents in This Unit

Illumination: Types of lamps; Fixtures and reflectors; Illumination schemes for domestic; industrial and commercial premises; Lumen requirements for different categories.

Chapter 14: Illumination

Chapter

14 Illumination

14.1. INTRODUCTION

Light is the prime factor in the human life as all activities of human beings ultimately depend upon the light. Where there is no natural light, use of artificial light is made. Artificial lighting produced electrically, on account of its cleanliness, ease of control, reliability, steady output, as well as its low cost, is playing an increasingly important part in modern every day life. Apart from its aesthetic and decorative aspects, good lighting has a strictly utilitarian value in increasing production, reducing workers fatigue, protecting their health, eyes and nervous system, and reducing accidents. The science of illumination engineering is, therefore, becoming of major importance.

Nature of Light. Light is a form of radiant energy. Various forms of incandescent bodies are the sources of light and the light emitted by such bodies depend upon the temperature of bodies. Heat energy is radiated into the medium by a body which is hotter than the medium surrounding it. The heat of the body, as seen, can be classified as red hot or white-hot. While the body is red-hot the wavelength of radiated energy will be sufficiently large and the energy available is in the form of heat. When the temperature increases the body changes from red-hot to white-hot state, the wavelength of the energy radiated becomes smaller and enters into the range of the wavelength of light. Actually the hot bodies emit heat as well as light energy, the velocity of these being equal to 3×10^8 m/s. The light waves have wavelengths varying from 0.00075 mm to 0.0004 mm. The wavelength of light is usually expressed in Angstrom; 1 Angstrom being equal to 10^{-10} metre. The ratio of the energy emitted by the body in the form of light to the total energy emitted by the body is known as the '*radiant efficiency*' of the body, which depends upon the temperature. Higher the temperature of the body ; lower the wavelength of the radiant energy and higher the radiant efficiency. The radiant efficiency will be maximum when the temperature of the body will be such that the wavelength of the shortest wave radiated by the body is 0.0004 mm (4,000 Å). It is found that maximum radiant efficiency would occur at about 6,250°C and even then it would be 20%. Since this temperature is far above the highest that has yet been obtained in practice, it is obvious that the actual efficiency of all artificial light sources, *i.e.*, those depending upon the temperature of incandescence is quite low.

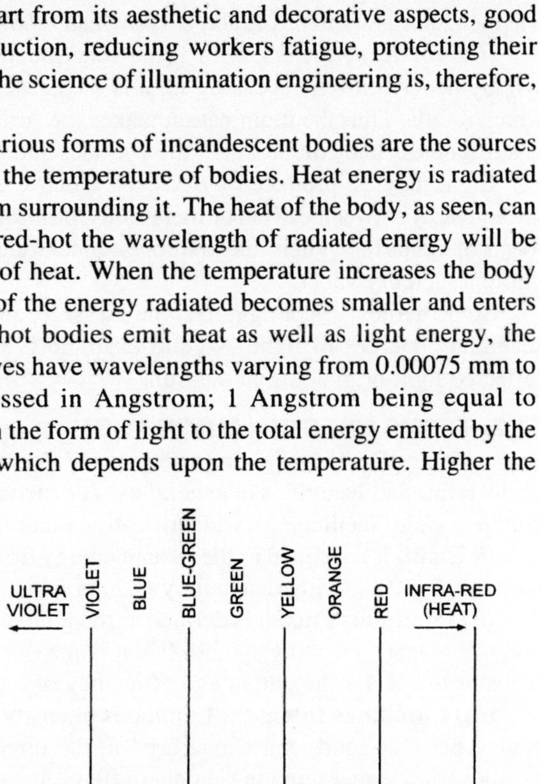

Fig. 14.1 *Wavelength and Colours*

Colour. The sensation of colour is due to the

difference in the wavelengths of the light radiations. Visible light can have a wavelength between 4,000 Å and 7,500 Å and the colour varies in the way, as shown in Figs. 14.1 and 14.2.

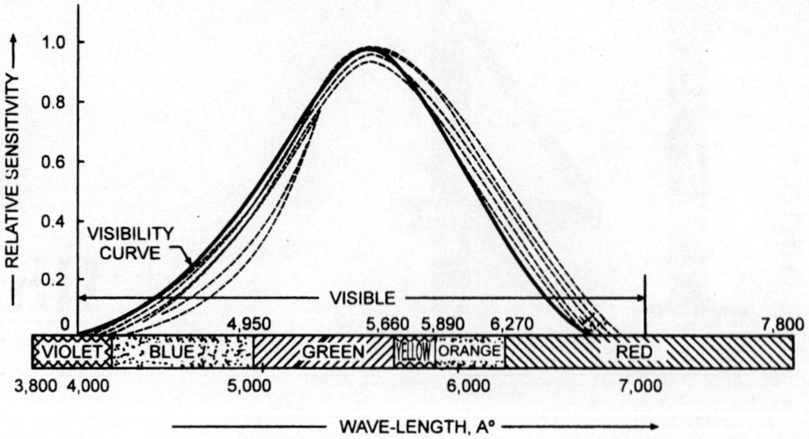

Fig. 14.2 *Eye Sensitivity Curve*

Relative Sensitivity. The sensitivity of the eye to the lights of different wavelengths varies from person to person and according to the age. The average relative sensitivity is shown in Fig. 14.2. The eye has greatest sensitivity for wavelengths of about 5,500 Å ; that is, yellow- green can be seen under poor conditions of illumination when blue or red cannot be seen under dim illumination, the sensitive curve tends to shift, as shown by the shaded region in Fig. 14.2. Therefore, violet disappears first and red remains visible. Yellow disappears last as the illumination becomes very dim. As each colour disappears, it becomes a grey shade, and finally black. The sensitivity of the eye to yellow-green radiation is taken as unity or 100% and the sensitivity to other wavelengths is expressed as a fraction or percentage of it. The relative sensitivity at any particular wavelength λ is written $K\lambda$ and is known as *relative luminosity factor.*

Illumination. differs from light very much, though generally these terms are used more or less synonymously. Strictly speaking light is the cause and illumination is the result of that light on surfaces on which it falls. Thus the illumination makes the surface look more or less bright with a certain colour and it is this brightness and colour which the eye sees and interprets as something useful, or pleasant or otherwise.

Light may be produced by passing electric current through filaments as in the incandescent lamps, through arcs between carbon or metal rods, or through suitable gases as in neon and other gas tubes. In some forms of lamps the light is due to fluorescence excited by radiation arising from the passage of electric current through mercury vapour.

Some bodies reflect light in some measure, and when illuminated from an original source they become secondary sources of light. A good example is the Moon, which illuminates the earth by means of the reflected light originating in the Sun.

14.2. TERMS USED IN ILLUMINATION

The devising of modern lighting schemes and the selection of fittings and types of lamps requires a knowledge of the terms and quantities in general use for such purposes. Therefore the following definitions are given in simple form to facilitate easy identification and reference.

(*i*) Light. It is defined as the radiant energy from a hot body which produces the visual sensation upon the human eye. It is usually denoted by Q, expressed in lumen-hours and is analogous to watt-hours.

(*ii*) Luminous Flux. It is defined as the total quantity of light energy emitted per second from a luminous body. It is represented by symbol F and is measured in lumens (or *cd-sr*). The conception of luminous flux helps us to specify the output and efficiency of a given light source.

(*iii*) Luminous Intensity. Luminous intensity in any given direction is the luminous flux emitted by the source per unit solid angle, measured in the direction in which the intensity is required. It is denoted by symbol I and is measured in candela (*cd*) or lumens per steradian.

If F is the luminous flux radiated out by source within a solid angle of ω steradian in any particular direction then $I = \frac{F}{\omega}$ lumens/steradian or candela (*cd*).

***(iv)* Lumen.** The lumen is the unit of luminous flux and is defined as the amount of luminous flux given out in a space represented by one unit of solid angle by a source having an intensity of one candle power in all directions.

$$i.e., \quad \text{Lumens} = \text{Candle power} \times \text{solid angle} = \text{CP} \times \omega$$

or total lumens given out by source of one candela is 4π lumens.

***(v)* Candle Power.** Candle power is the light radiating capacity of a source in a given direction and is defined as the number of lumens given out by the source in a unit solid angle in a given direction. It is denoted by symbol CP.

$$i.e., \text{CP} = \frac{\text{Lumens}}{\omega}$$

***(vi)* Illumination.** When the light falls upon any surface, the phenomenon is called the *illumination.* It is defined as the number of lumens, falling on the surface, per unit area. It is denoted by symbol E and is measured in lumens per square metre or lux or metre-candle.

If a flux of F lumens falls on a surface of area A, then the illumination of that surface is

$$\text{E} = \frac{\text{F}}{\text{A}} \text{ lumens/m}^2 \text{ or lux } (lx)$$

***(vii)* Lux or Metre Candle.** It is the unit of illumination and is defined as the luminous flux falling per square metre on the surface which is every where perpendicular to the rays of light from a source of one candle power and one metre away from it.

***(viii)* Lamp Efficiency.** It is defined as the ratio of the luminous flux to the power input. It is expressed in lumens per watt.

***(ix)* Specific Consumption.** It is defined as the ratio of the power input to the average candle power. It is expressed in watts per candela.

***(x)* Glare.** The size of the opening of the pupil in the human eye is controlled by its iris. If the eye is exposed to a very bright source of light, the iris automatically contracts inorder to reduce to amount of light admitted and prevent damage to retina ; this reduces the sensitivity, so that other objects within the field of vision can be only imperfectly seen. This effect is referred to 'glare' and is familiar in connection with motor-car head lights. *In other words glare may be defined as the brightness within the field of vision of such a character as to cause annoyance, discomfort, interference with vision or eye-fatigue.*

***(xi)* Space-height Ratio.** It is defined as the ratio of horizontal distance between adjacent lamps and height of their mountings.

$$i.e., \text{Space-height ratio} = \frac{\text{Horizontal distance between two adjacent lamps}}{\text{Mounting height of lamps above working plane}}$$

***(xii)* Utilisation Factor or Coefficient of Utilisation.** It is defined as the ratio of total lumens reaching the working plane to total lumens given out by the lamp.

$$i.e., \text{Utilisation factor} = \frac{\text{Total lumens reaching the working plane}}{\text{Total lumens given out by the lamp}}.$$

***(xiii)* Maintenance Factor.** Due to accumulation of dust, dirt and smoke on the lamps, they emit less light than that they emit when they are new ones and similarly the walls and ceilings ,etc., after being covered with dust, dirt and smoke do not reflect the same output of light, which is reflected when they are new.

The ratio of illumination under normal working conditions to the illumination when the things are perfectly clean is known as maintenance factor.

$$i.e., \text{Maintenance factor} = \frac{\text{Illumination under normal working conditions}}{\text{Illumination when every thing is perfectly clean}}.$$

***(xiv)* Depreciation Factor.** This is merely the inverse of the maintenance factor and is defined as the ratio of initial metre-candles to the ultimate maintained metre-candles on the working plane. Its value is more than unity.

(*xv*) Solid Angle. Plane angle is subtended at a point in a plane by two converging straight lines and its magnitude is given by

$$\theta = \frac{\text{Arc}}{\text{Radius}} \text{ radians}$$

The largest angle subtended at a point is 2π radians.

Solid angle is the angle generated by the surface passing through the point in space and the periphery of the area. Solid angle is denoted by ω, expressed in steradians and is given by the ratio of the area of the surface to the square of the distance between the area and the point.

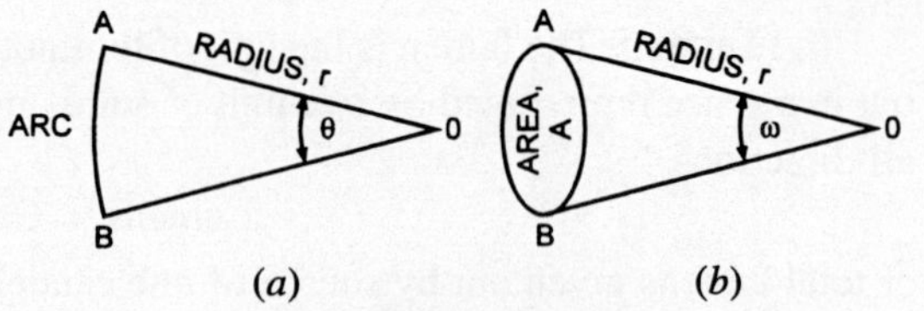

Fig. 14.3

$$i.e., \omega = \frac{\text{Area}}{(\text{Radius})^2} = \frac{A}{r^2}$$

The largest solid angle subtended at a point is that due to a sphere at its centre. If r is the radius of any sphere, its surface area is $4\pi r^2$ and the distance of its surface area from the centre is r, therefore, solid angle subtended at its centre by its surface, $\omega = \frac{4\pi r^2}{r^2} = 4\pi$ steradians.

(*xvi*) Steradian. It is the unit of solid angle and is defined as the solid angle that subtends a surface on the sphere equivalent to the square of the radius.

14.3. LAWS OF ILLUMINATION

There are two laws of illumination (*i*) Law of inverse squares and (*ii*) Lambert's cosine law.

1. Law of Inverse Squares. If a source of light which emits light equally in all directions be placed at the centre of a hollow sphere, the light will fall uniformly on the inner surface of the sphere, that is to say, each square mm of the surface will receive the same amount of light. If the sphere be replaced by one of the larger radius, the same total amount of light is spread over a larger area proportional to the square of the radius. The amount which falls upon any square mm of such a surface will therefore diminish as the radius increases, and will be inversely proportional to the square of the distance.

A similar relation holds if we have to deal with a beam of light in the form of a cone or pyramid, as shown in Fig. 14.4 (*a*). If we consider parallel surfaces which cut the pyramid at different distances from the source, the areas of these surfaces are proportional to the square of these distances, and therefore the amount of light which falls on one unit of the area of these surfaces, is inversely proportional to the square of the distance from the source. This relationship is referred to as the *law of inverse squares*.

Mathematically it can be proved as follows :

Let us consider surface area A_1 and surface area A_2 at distances r_1 and r_2 respectively from the point source S of luminous intensity I and normal to the rays, as shown in Fig. 14.4 (*a*).

Let the solid angle subtended be ω steradians

Luminous flux radiated per steradians = I

Total luminous flux radiated = $I\omega$ lumens

Illumination on the surface of area A_1

$$= \frac{I\omega}{A_1} \text{ lumens/unit area}$$

and area $A_1 = \omega r_1^2$

$\therefore$ Illumination on the surface of area A_1,

$$E_1 = \frac{I\omega}{\omega r_1^2} = \frac{I}{r_1^2} \text{ lumens per unit area}$$

Similarly illumination on the surface of area A_2,

$$E_2 = \frac{I\omega}{A_2} = \frac{I\omega}{\omega r_2^2} = \frac{I}{r_2^2} \text{ lumens per unit area.}$$

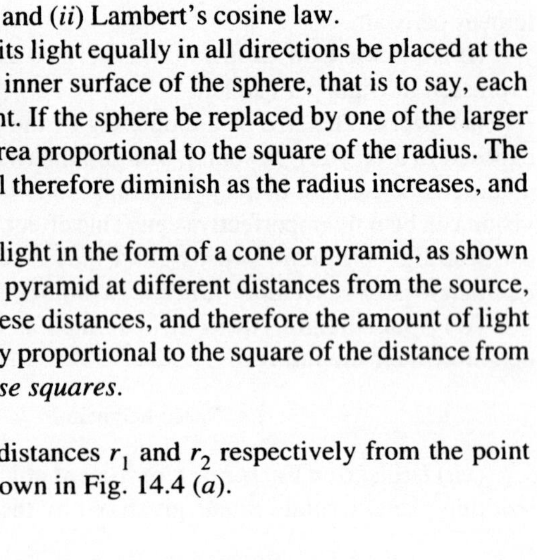

Fig. 14.4 (*a*) *Inverse Square Law*

Hence the illumination of a surface is inversely proportional to the square of the distance between the surface and the light source provided that the distance between the surface and the source is sufficiently large so that the source can be regarded as a point source.

2. Lambert's Cosine Law. Very often the illuminated surface is not normal to the direction of light as AC in Fig. 14.4 (*b*) but is inclined as AB. The area over which the light is spread is then increased in the ratio

$$\frac{AB}{AC} = \frac{1}{\cos\theta}$$

and the illumination decreases in the ratio $\frac{\cos\theta}{1}$.

The expression for the illumination then becomes

$$E = \frac{I\cos\theta}{r^2}$$

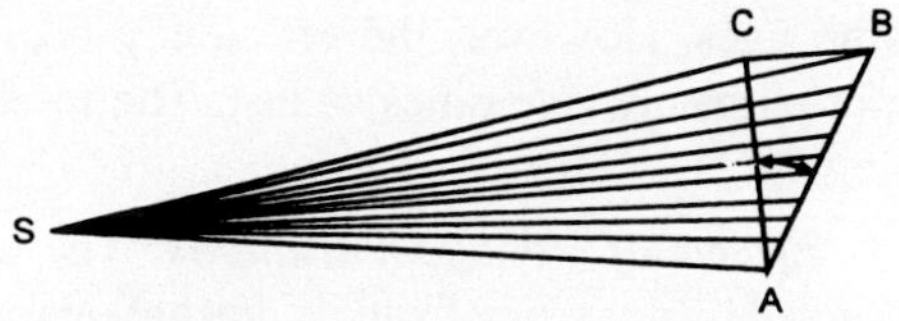

Fig. 14.4 (*b*) *Lambert's Cosine Law*

According to this law the illumination at any point on a surface is proportional to the cosine of the angle between the normal at that point and the direction of luminous flux.

14.4. ELECTRIC LAMPS

The different electric lamps/light sources in use are as detailed below.

1. Incandescent or Filament Lamps. For sometimes the arc lamps were used for general lighting purpose but as the carbon-arc lamps were complicated so these have been superseded by filament lamps for general lighting.

When an electric current is passed through a fine metallic wire, heat is produced and the temperature of the wire increases. At low temperatures the wire radiates heat energy, as the temperature of the wire increases due to heating it radiates heat as well as light energy. The higher the temperature of the wire, higher is the amount of light energy radiated. A black body when heated to 6,250°C emits the maximum energy in the visible spectrum range.

The incandescent or filament type lamp consists of a glass globe completely evacuated and a fine wire, known as *filament,* within it. The glass globe is evacuated to prevent the oxidization and convection currents of the filament and also to prevent the temperature being lowered by radiation.

The material, which can be used for the filaments of incandescent lamps, must possess the properties of high melting point, low vapour pressure, high resistivity, low temperature coefficient, ductility and sufficient mechanical strength to withstand vibrations during use.

The materials which can be used for the filaments in incandescent lamps are carbon, osmium, tantalum and tungsten. Though the carbon has a melting point of 3,500°C but its main drawback is that it starts vaporizing at very fast rate if it is operated beyond 1,800°C, moreover its temperature coefficient of resistance is negative, *i.e.*, its resistance decreases with the rise in temperature and, therefore, it takes more current from mains. Because of its low operating temperature its efficiency is quite low and is of the order of 3.5 lumens per watt. Osmium is very rare and expensive metal. The melting point of osmium is 2,600°C. Average efficiency of osmium lamp is of the order of 5 lumens per watt. The melting point of tantalum is only 2,800°C and, therefore, it cannot be operated at higher temperature to give more efficiency. The efficiency of tantalum filament lamps is about 5 lumens per watt. Nowadays tungsten is the most commonly used metal for filament due to its high melting point (3,400°C), high resistivity, low temperature coefficient (0.0051), low vapour pressure, being ductile and mechanically strong to withstand vibration during use. The hot resistance of tungsten filament is about 15 times the cold resistance and, therefore, there is a inrush of current at the switching instant. However, the inrush current is not beyond 15 times the normal current. Since the filament heats up immediately and resistance increases almost instantaneously, the inrush current attains a maximum value in 0.003 second and falls to normal operating value in 0.2 second. The average efficiency of tungsten filament lamp is about 10 lumens per watt. The light spectrum of an incandescent lamp is continuous, *i.e.*, it contains all the colours, but contains relatively excess of red and yellow radiations and less of blue and violet radiations.

Ageing Effects. The light output of an incandescent lamp decreases gradually. As a tungsten lamp ages, its light output decreases for two reasons. Evaporation of the filament tends to cause the bulb to blacken. Also, evaporation makes the filament slowly decrease in diameter, which means that the resistance of the filament increases. Therefore, an old filament draws less current and operates at a lower temperature, which reduces its light output. In turn the efficiency of the lamp (lumens output/watt input) also decreases with the passage of time. The current drawn and the power consumed by the filament decrease at the same rate as the lamp ages. However, the efficiency decreases about four times as fast, and the light output decreases approximately five times as fast. The total depreciation of light output is roughly 15% over the useful life range.

Effects of Voltage Variations. The operating characteristics of an incandescent lamp are materially affected by departure from its normal operating voltage. An increase of 5% in operating voltage increases the lumens output by 20% but shortens the life of the lamp by 50%. On the other hand, reduction of 5% in operating voltage causes a reduction of approximately 20% in lumens output but doubles the life of the lamp. The efficiency of a lamp (lumens/watt) increases with the increase in voltage owing to increase in temperature and is proportional to the square of the voltage. 1% change in applied voltage modifies lamp wattage by 1.5%, efficiency by 2%, lumens output by 3.5%. The normal life of a filament lamp is about 1,000 working hours. The variations in power consumption, lumens output, efficiency and life of incandescent lamps with the variation in voltage are shown in Fig. 14.5.

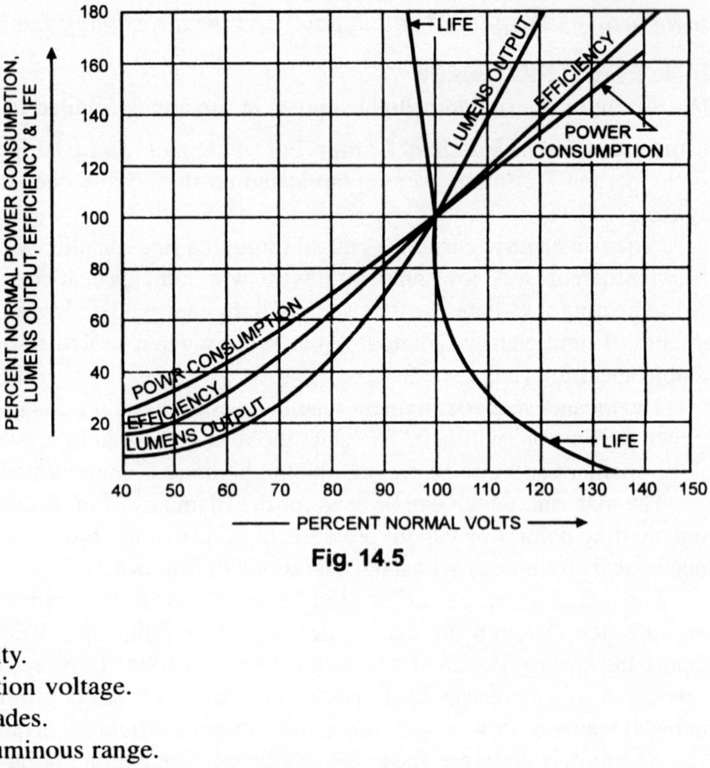

Fig. 14.5

Analytically these relationships are given as below :

Lumens output $\propto (V)^{3.55}$

Power consumption $\propto (V)^{1.55}$

Luminous efficiency $\propto (V)^2$

Life $\propto (V)^{-13}$ for vacuum lamps;

$\propto (V)^{-14}$ for gas-filled lamps

Advantages. 1. Operating power factor unity.

2. Direct operation on standard distribution voltage.
3. Availability in various shapes and shades.
4. Good radiation characteristic in the luminous range.
5. No effect of surrounding air temperature.

2. Gas-Filled Lamps. A metal filament can work in an evacuated bulb up to 2,000°C without oxidation and if it is worked beyond this temperature it vaporizes quickly and blackens the lamp. For higher efficiency it is necessary to use working temperature more than 2,000°C keeping down the evaporation, which is possible by filling the bulb with an inert gas–argon with a small percentage of nitrogen. Nitrogen is added to reduce the possibility of arcing. Krypton is the best gas for this purpose but it is so expensive that it is used only in special purpose lamps, such as miners lamps. The tungsten filament can safely burn at temperature of 2,400°C to 2,750°C according to the size of the lamp. However, due to presence of gas there is heat loss due to convection currents. This loss depends upon the surface area of filament. As such coiled coil filaments, which take much less space compared with coiled filaments are used with such lamps. A coiled coil filament is made by winding tungsten wire on a fine iron wire to form a spiral which is again wound on to a thick iron wire to form a coiled coil. (iron is later on dissolved out by acid). Helical filaments also have a slower rate of tungsten evaporation. This evaporation eventually causes bulb blackening because the tungsten vapour condenses as a black film on the inner surface of the bulb. In a gas-filled lamp, the hot gas carries the tungsten vapour upward. Therefore, a black spot forms at the top of the bulb instead of spreading over the entire inner surface, as in a high-vacuum bulb. Chemicals called '*getters*' are often placed inside the bulb to capture tungsten vapour and thereby reduce the rate of blackening. A piece of wire mesh called a *collector grid* may also be attached to each lead-in-wire to attract the particles of tungsten vapour. Efficiency of coiled-coil lamp is about 12 lumens/watt.

For low wattage lamps, however, the heat loss due to introduction of gas is more than in medium wattage lamps, so for low wattage (up to 40 watts) vacuum type lamps are used.

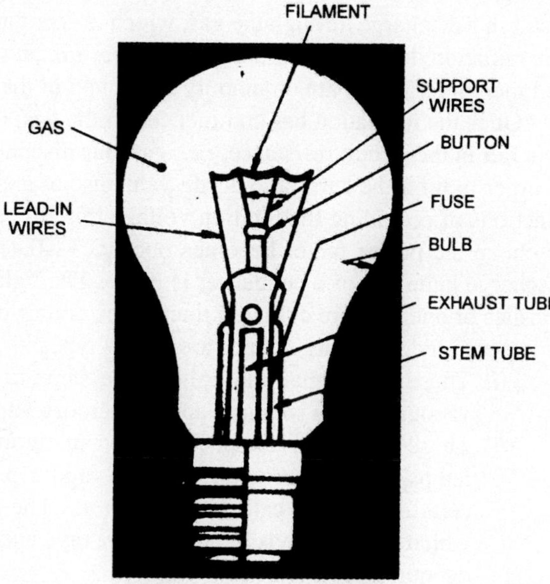

Fig. 14.6 *Gas-Filled Lamp*

3. Halogen Lamp. The halogen lamp is the latest member in the family of incandescent lamps. It possesses numerous advantages over the ordinary incandescent lamp. As already stated, the life and efficiency of an incandescent lamp fall off with use-partly due to slow evaporation of the filament and partly due to black deposit formed on the inside of the bulb. The addition of a small amount of halogen vapour to the filling gas restores part of the evaporated tungsten vapour back to the filament by means of a chemical reaction, *i.e.*, there is a sort of '*regenerative cycle*'.

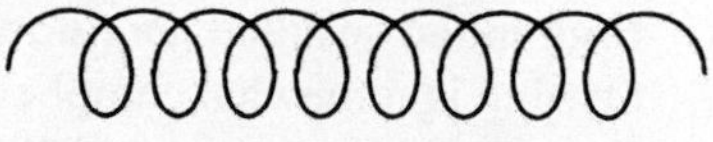

(*a*) *Coiled Filament*

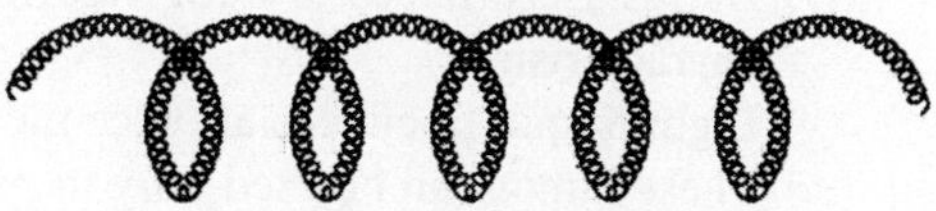

(*b*) *Coiled Coil Filament*

Fig. 14.7

Halogen lamps possess the following advantages :

(*i*) No blackening of lamp, hence no depreciation of lumens output.

(*ii*) High operating temperature with increased luminous efficiency varying from 22 lumens per watt to 33 lumens per watt.

(*iii*) Reduced dimensions of lamps–miniature size.

(*iv*) Long life—2,000 hours.

(*v*) Better colour rendition.

Halogen lamps, which are being manufactured in sizes up to 5 kW, are suitable for outdoor illumination of buildings, playing fields, large gardens, fountains, car parks, air-port runways etc. and for lighting of public halls, factories, sport-halls, photofilm and TV studios etc.

4. Gaseous Discharge Lamps. Incandescent lamp suffers from two disadvantages—low efficiency and coloured light. The gaseous discharge lamps have been developed to overcome these drawbacks.

The basic principle of a gaseous discharge lamp is illustrated in Fig. 14.8. Gases are normally poor conductors, specially at atmospheric and higher pressures, but application of suitable voltage, *called the ignition voltage*, across the two electrodes can

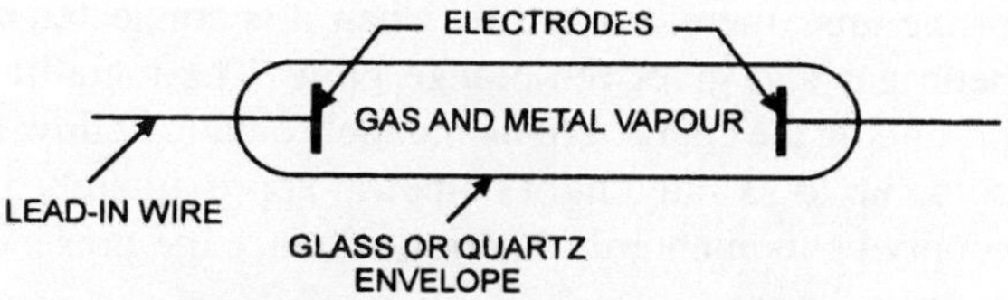

Fig. 14.8 *Gaseous Discharge Lamp*

result in a discharge through the gas, which is accompanied by electromagnetic radiation. The wavelength of this radiation depends upon the gas, its pressure, and the metal vapour used in lamp. Argon gas and sodium and mercury vapours are commonly employed in the manufacture of gaseous discharge lamps.

Once the ionisation has commenced in the gas, it has a tendency to increase continuously accompanied by a fall in the circuit resistance, *i.e.*, gaseous discharge lamp possesses a *negative resistance* characteristic. In order to limit the current to a safe value use of a choke or a ballast is made. The choke performs the dual functions of providing the ignition voltage initially, and limiting the current subsequently. Since due to use of choke the power factor becomes poor (0.3—0.4), in order to improve the power factor of the gaseous discharge lamp use of a condenser is made. The light spectrum obtained is, however, discontinuous (*i.e.*, it consists of one or more coloured lines). The colour of the light obtained depends upon the nature of the gas or vapour used. Discharge lamps are of two types:

(*i*) Those which give the light of the same colour as produced by the discharge through the gas or vapour such as sodium vapour, mercury vapour and neon gas lamps.

(*ii*) Those which use the phenomenon of fluorescence and are known as *fluorescent lamps*. In these lamps, the discharge through the vapour produces ultraviolet waves which cause fluorescence in certain materials called as *phosphors*. The inside of a fluorescent lamp is coated with a phosphor which absorbs invisible ultraviolet rays and radiates visible rays. Example is fluorescent mercury-vapour tube.

The gaseous discharge lamps are, in general, considered superior to metal filament lamps. However, they suffer from the following drawbacks:

(*i*) High initial cost and poor power factor.

(*ii*) Starting is somewhat complicated requiring starters in some cases and transformer in others.

(*iii*) These take time to attain full brilliancy.

(*iv*) Ballasts are necessary for stabilising the current since such lamps have negative resistance characteristic.

(*v*) Light output fluctuates at twice the supply frequency. The flicker causes *stroboscopic* effect.

(*vi*) These lamps can be used only in particular position.

(*a*) Sodium Vapour Discharge Lamp. Principally the sodium vapour discharge lamp consists of a bulb containing a small amount of metallic sodium, neon gas, and two sets of electrodes connected to a pin type base. The presence of neon gas serves to start the discharge and to develop enough heat to vaporize the sodium. Since long discharge paths are necessary, the discharge envelope is usually bent into U shape. The lamp operates at a temperature-like 300°C and in order to conserve the heat generated and assure the lamp operating at normal air temperatures the discharge envelope is enclosed in a special vacuum envelope designed for this purpose. The lamp must be operated horizontally, or nearly so, to keep the sodium well spread out along the tube, although some small lamps may be operated vertically, lamp cap up. Care should be taken in handling these lamps, particularly when replacing inner U-tube, for if it is broken and sodium comes in contact with moisture fire will result.

The sodium vapour lamp is only suitable for alternating current, and therefore, requires choke control. This requirement is met by operating the lamp from a stray field, step-up, tapped auto-transformer with an open-circuit secondary voltage of 470 to 480 volts. The uncorrected power factor is very low, about 0.3, and a capacitor must be used to improve the power factor to about 0.8.

When the lamp is not in operation, the sodium is usually in the form of solid deposited on the side walls of the tube, therefore, at first when it is connected across the supply mains the discharge takes place in the neon gas and gives red-orange glow. The metallic sodium gradually vaporizes and then ionizes, thereby producing the characteristic monochromatic yellow light, which makes objects appear as grey. The lamp will come up to its rated light output in approximately 15 minutes. It will restart immediately should the power supply be momentarily interrupted since the presence of vapour is quite low and the voltage is sufficient to restrike the arc.

The wiring diagram of sodium vapour lamp(s) is given in Fig. 14.9 (*c*).

The efficiency of a sodium vapour lamp under practical conditions is about 40–50 lumens/watt. The major application of this type of lamp is for highway and general outdoor lighting where colour discrimination is not required, such as street lighting, parks, rail yards, storage yards etc. Such lamps are manufactured in 45, 60, 85 and 140 watt ratings. The average life is about 3,000 hours and is not affected by voltage variations. At the end of this period the light output will be reduced by 15% due to ageing.

The lamp fails to operate when

(*i*) the filament breaks or burns out,

(*ii*) the cathode stops to emit electrons,

(*iii*) the sodium particles may concentrate on one side of the tube,

(*iv*) the lamp tube is blackened owing to sodium vapour action on the glass, in which case the output will be reduced.

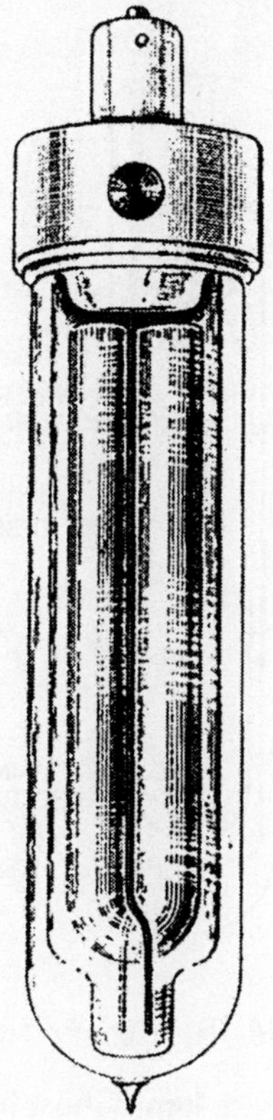

(*a*)

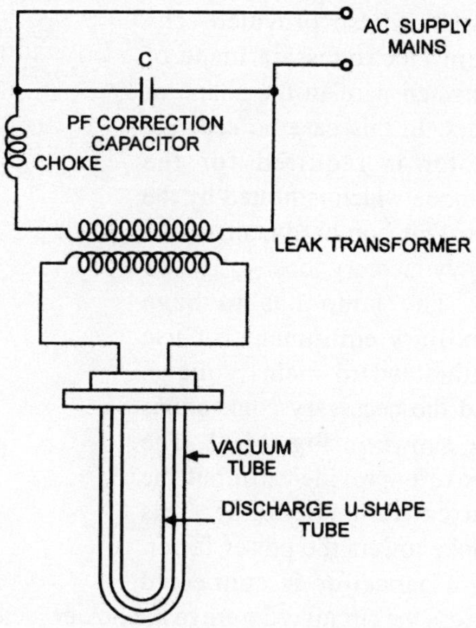

(*b*)

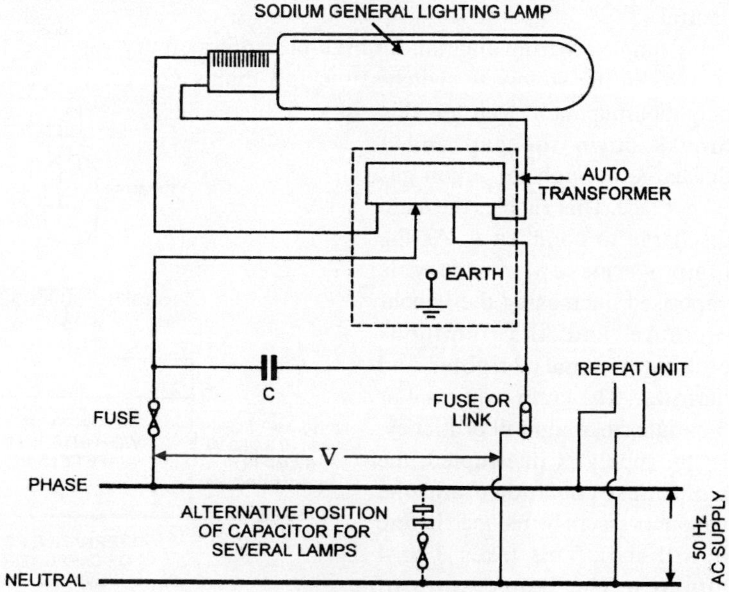

(*c*) *Wiring Diagram*

Fig. 14.9 *Sodium Vapour Discharge Lamp*

(*b*) High Pressure Mercury Vapour Discharge Lamp. The extensive use of the mercury vapour discharge lamp depends entirely upon the versatility of the mercury vapour as regards pressure, temperature, voltage and other characteristics, each change resulting in a lamp of different spectral quality and efficiency.

The mercury vapour discharge lamp is similar in construction to the sodium vapour discharge lamp. It consists of a discharge envelope enclosed in an outer bulb of ordinary glass. The discharge envelope may be of hard glass or quartz. The space between the bulb is partially or completely evacuated to prevent heat loss by convection from the inner bulb. The outer bulb absorbs harmful ultra violet rays. The inner bulb contains argon and a certain quantity of mercury. In addition to two main electrodes a starting (auxiliary) electrode connected through a high resistance (as shown in Fig. 14.10) is also provided. The main electrodes are made of tungsten wire in the shape of helix. In this case no separate heater is required for the cathode which is heated by the constant bombardment of the heavy mercury ions.

The lamp has to have auxiliary equipment for use with standard mains voltage, and the necessary connections are shown in Fig. 14.11. The choke is provided to limit the current to a safe value. This choke lowers the power factor, so a capacitor is connected across the circuit to improve the power factor. These lamps must be operated vertically, since if they are used horizontally convection will cause the discharge to touch the glass bulb, which will fail. Lamps which are intended to operate horizontally are fitted with a magnetic device which will hold the luminous column central.

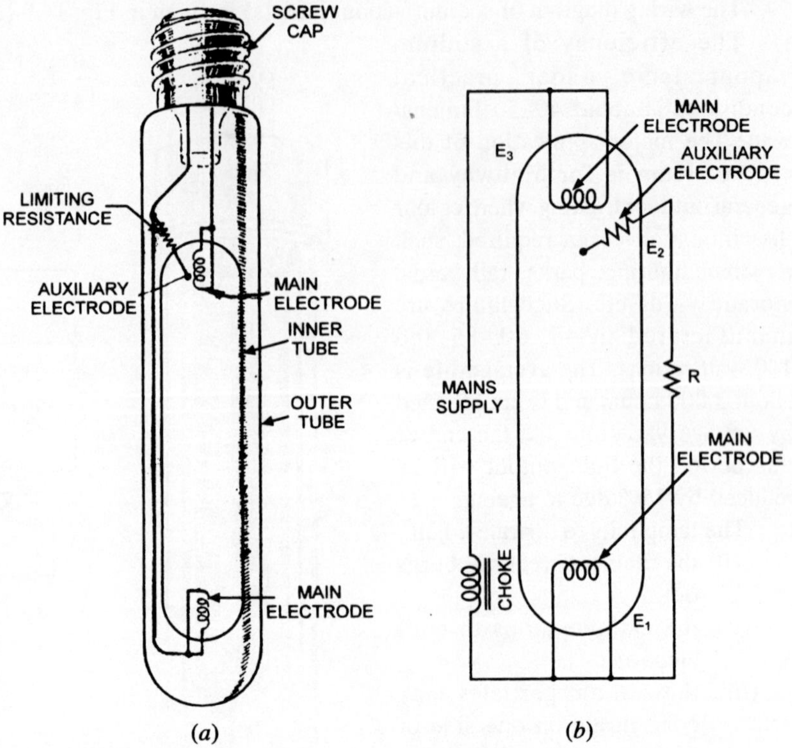

Fig. 14.10 *High Pressure Mercury Vapour Discharge Lamp*

Complete wiring diagram for high-pressure mercury vapour discharge lamp is shown in Fig. 14.11.

When the supply is switched on, full mains voltage is applied between the auxiliary electrode and neighbouring main electrode; this breaks down the gap and a discharge through the argon gas takes place. This enables the main discharge to commence. As the lamp warms up, mercury is vaporised, increasing the vapour pressure and the luminous column becomes brighter and narrower. The lamp requires 4 or 5 minutes to attain full brilliancy. If the supply is interrupted, the lamp must cool down and the vapour pressure be reduced before it will start. This takes 3 or 4 minutes. The temperature of operation inside the inner bulb is about 600°C. It gives greenish

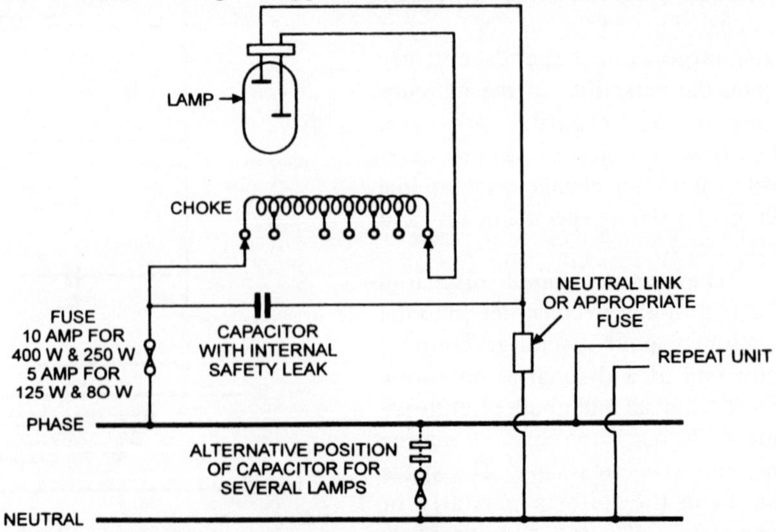

Fig. 14.11 *Complete Wiring Diagram of HPMV Lamp*

blue colour light, which causes colour distortion. The efficiency is about 30—40 lumens/watt. These lamps are manufactured in 250 and 400 watt ratings for use on 200—250 volts ac supply mains. The pressure of vapour in these lamps is 2-3 atmosphere. Lamps of this type are used for general industrial lighting, railway yards, ports, work area, shopping centres etc., where greenish-blue colour light is not objectionable.

The lamp described above is MA type. Another type, which is manufactured in 300 and 500 watt rating for use on ac as well as dc supply mains, is MAT type. This is similar to MA type except that choke is not used as a ballast. Space between two tubes instead of being evacuated consists of a tungsten filament in series with a discharge tube which acts as a ballast. When the supply is switched on, it operates as a filament lamp, its full output being given by the outer tube. At the same time the discharge or inner tube begins warming up and at a particular temperature a thermal switch operates cutting a part of the filament and thereby increasing the voltage across the discharge tube. The filament contributes a considerable portion of red rays. The combination of the rays from the filament and the blue radiations from the discharge tube produce a useful colour. As the filament acts as a resistance, the overall power factor of the lamp is about 0.95 and therefore, capacitor is not required.

Lower wattage lamps, such as 80 and 125 watts, are manufactured in a different design and using high vapour pressure of about 5-10 atmosphere. These are known as MB types. These operate in a manner similar to MA type except that resistance in series with starting electrode is large and outer bulb is of quartz, (not of ordinary glass) in order to withstand high temperature so that these lamps can be used in any position.

(*c*) Mercury Iodide Lamps. These lamps are similar in construction to high pressure mercury vapour lamps but in addition to mercury, a number of iodides are added which fill the gaps in the light spectrum, and thus, improve the colour characteristic of the light. Their efficiency is also higher (75—90 lumens/watt). A separate ignition device, in addition to the choke, is required for the mercury-iodide lamp. Such lamps are suitable for application in the fields of floodlighting, industrial lighting and public lighting.

(*d*) Neon Lamp. It is a cold cathode lamp and consists of a glass bulb filled with neon gas with a small percentage of helium. These lamps give orange pink coloured light. Electrodes are of pure iron and are spaced only few mm apart so that lamps can be made for voltages as low as 110 volts ac or 150 dc. For use on ac the electrodes are of equal size. On dc the gas glows near the negative electrode, therefore, negative electrode is made larger in size. The efficiency of neon lamp lies between 15-40 lumens/watt. Owing to discharge of the gas between the electrodes in the form of an arc, it may cause the current drawn by the lamp to increase indefinitely. This is prevented by connecting a high resistance of few thousand ohms in series and mounting it in the cap. The lamp of this type is of the size of an ordinary incandescent lamp. The power consumption is of the order of 5 watts.

Neon lamps are used as indicator lamps, night lamps, for determination of polarity of dc mains and in larger sizes as neon tubes for the purpose of advertising.

(*e*) Neon Tubes. The popularity of high-voltage neon lighting arose almost entirely from its use in advertising, for signs, or in the decorative treatment of buildings, but later the lighting field became important. The neon tube, which is used in varying lengths up to about 9 metres, may be bent into almost any desired shape during manufacture. It consists of a length of glass tubing containing two electrodes, normally cylindrical in shape, of iron, steel, or copper.

The true neon tube contains neon, but the term is now used also for tubes with fillings of other rare gases. By varying the composition of glass and adding different substances to neon gas different colours such as orange, red, yellow, green etc. are obtained. The diameters of the tube vary, and common sizes of 10, 15, 20 and 30 mm carry currents of 25, 35, 60 and 150 mA respectively. Voltage required may vary from 300 V to 1,000 V per metre of tube length and for starting the discharge a striking voltage, about 1½ times this value is required. Such voltage is obtained by making use of a step-up transformer having a high leakage reactance so that it gives a drooping characteristic. The usual operating voltage is 6,000 volts.

The tubes are mounted either on a wooden frame or a metal base. These are matched with step-up transformers by connecting suitable tappings for the rated current. Connections between letters are made by nickel wires, the glass tubings being slipped over them. The power factor of neon tubes is quite low and is improved by using capacitors. The capacitors can, however, be placed only on the low voltage side of the transformer.

A number of precautions are prescribed for supply to high voltage luminous tube sign installations under IER 1956 (Rule No. 71).

Faults. Flickering may be due to;

(*i*) Transformer secondary voltage too low; this can be rectified by adjustment of the transformer tappings.

(*ii*) Reduction of gas pressure in tube due to absorption of the gas into the electrodes, the tube may be removed and refilled by the manufacturers.

Neon tubes installed in the open requires frequent cleaning, say 4 times per year. Care must be taken that the key of the opened lock switch is removed and held by the operator while work is being done.

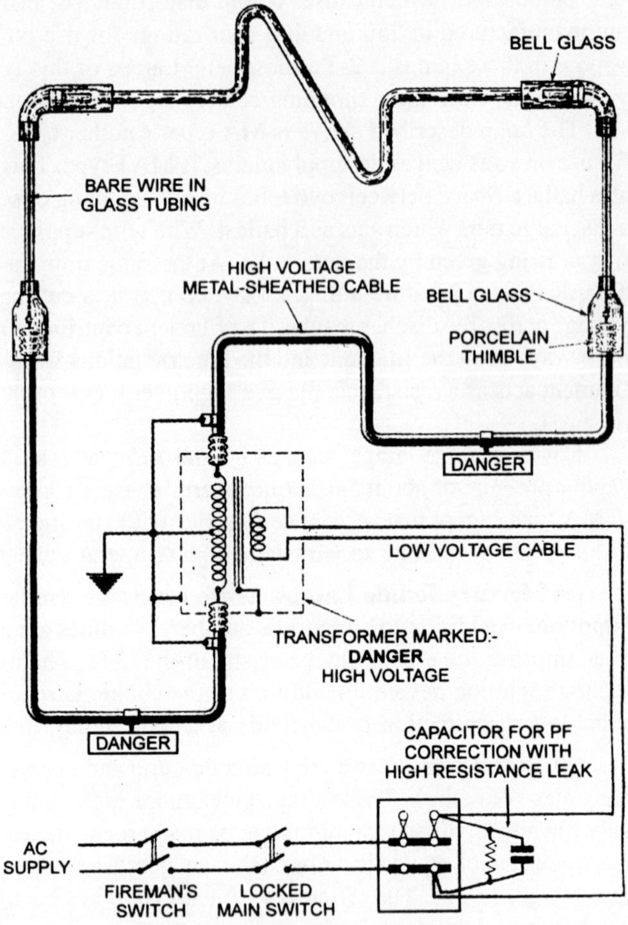

Fig. 14.12 *Neon Tube Installation*

(*f*) Fluorescent Tubes. Fluorescent lighting has a great advantage over other light sources in many applications. The tubes can be obtained in a variety of lengths, with illumination in a variety of colours. It is possible to achieve quite high lighting intensities without excessive temperature rise and, owing to the nature of light sources, the danger of glare is minimised. It must not, however, be thought that fluorescent lamps can be used indiscriminately without giving careful consideration to the type of reflector to be employed or to the correct positioning of the light source. An exposed tube in the line of vision is usually as bad as a tungsten lamp would be. The efficiency of fluorescent lamp is about 40 lumens per watt, about three times the efficiency of an equivalent tungsten filament lamp. The fluorescent tube consists of a glass tube 25 mm in diameter and 0.38 m–1.52 m in length. The inside surface of the tube is coated with a thin layer of fluorescent material in the form of a powder.

The coating materials used depend upon the colour effect desired and may consist of zinc silicate, cadmium, silicate or calcium tungstate. These organic chemicals are known as *phosphors* which transforms short-wave invisible radiations into visible light. By mixing the various powders light of any desired colour including day light can be obtained. The tube contains small quantity of argon gas at a pressure of 2.5 mm of mercury and one or two drops of mercury. It is provided with two electrodes coated with electron emissive material. A starting switch is provided in the circuit, which puts the electrodes directly across the supply mains at the time of starting, so that electrodes may get heated and emit sufficient electrons. A stabilising choke is connected in series with it, which acts as a ballast in running condition and provides a voltage impulse for starting. A capacitor is connected across the circuit to improve the power factor.

Fluorescent tubes are available in the following sizes :

Length	*Wattage*	*Length*	*Wattage*	*Length*	*Wattage*
38 cm	14 W	100 cm	25 W	152 cm	65 W
46 cm	15 W	122 cm	40 W	152 cm	80 W
61 cm	20 W				

The fluorescent lamp in common with most electrical devices may cause a certain amount of radio-interference. This interference may be caused due to one of the following factors :

1. Direct radiation from the bulb to the antenna. This effect diminishes rapidly as the radio is separated from the lamp. Thus, for example, at a radius of 2.8 metres interference from this cause is negligible.
2. Line radiation from the electric supply line to the antenna.
3. Line feedback from the lamp through the line to the radio.

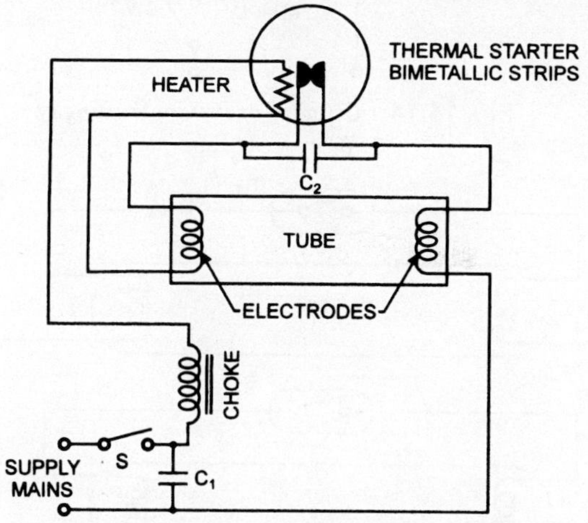

(*a*) *Incorporating a Thermal Type Starter*

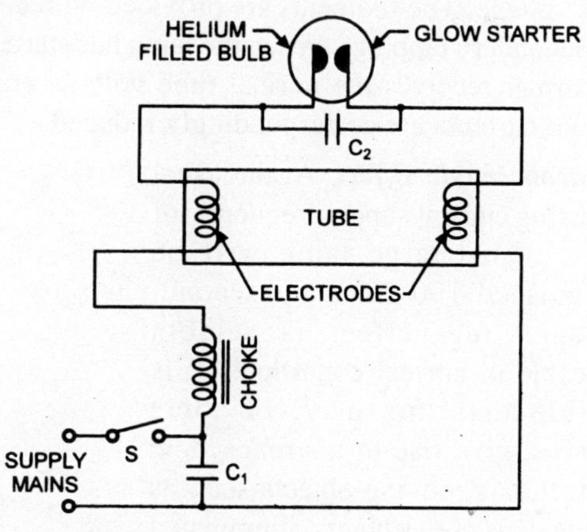

(*b*) *Incorporating a Glow Type Starter*

Fig. 14.13 *Connection Diagram For a Fluorescent Tube*

The radio-interference effect may be reduced to a minimum by connecting a small capacitor (0.05 μF) across starter terminals, as shown in Figs. 14.13 (*a*) and 14.13 (*b*).

The starting switches are of two types, namely the thermal type and the glow type.

The connections of a fluorescent tube incorporating a thermal type starter are shown in Fig. 14.13 (*a*). The thermal starter is a current operated device and consists of two metallic strips and a heater coil. The bimetallic strips are in contact with each other when the lamp is not in operation. When the supply is switched on, the two electrodes get connected in series through the thermal switch, the relatively large current rising them to incandescence. The current also flows through the heater element as a result of which bimetallic strips break contact. This causes interruption in the current flowing through the circuit, which further results in a high voltage surge across the electrodes of the tube, which is enough to strike the arc between the electrodes. This arc is then maintained by the normal lamp voltage. Bombardment of the electrode's surface by the positive mercury ions maintains their temperature so that they continue to emit electrons. Thus the tube is put in operation. The thermal switch is now generally obsolete because of its more complicated construction, greater cost and greater power loss.

The connections of a fluorescent tube incorporating a glow type starter are shown in Fig. 14.13 (*b*). The glow type starter is a voltage operated device and consists of two bimetallic electrodes enclosed in a glass bulb filled with a mixture of helium and hydrogen. Normally the contacts are open. When the supply is switched on, the potential across bimetallic electrodes causes a small glow discharge at a small current not enough to heat up the tube electrodes (filament). This discharge is enough, however, to heat the bimetallic strips of the switch causing them to bend and make contact. The result is a large current through the electrodes, their temperature being raised to incandescence and the gas in the immediate neighbourhood is ionised. After one or two seconds the bimetallic strips cool down and the contacts open. This opening of contacts in series with the choke causes a momentary high voltage, which is sufficient to start the discharge in the main tube. The starter ceases to glow as the voltage is now too low. A small capacitor is placed inside the starter to suppress arcing and radio-interference.

Nowadays starterless circuits, as instant start, rapid start etc. for such tubes are also employed (Fig. 14.14). The primary winding of an auto-transformer is connected in parallel with the tube and receives practically full mains voltage. The filaments are provided with current from secondary tappings. When the lamp has started the transformer receives the normal tube voltage and the filament currents are correspondingly reduced.

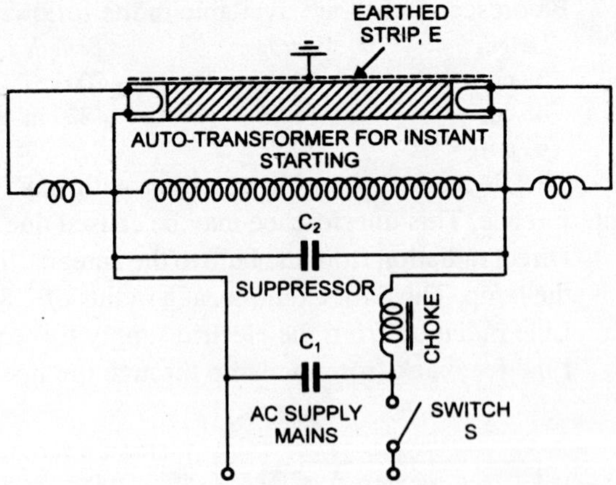

Fig. 14.14 *Circuit For Instant Starting of Fluorescent Tube*

Stroboscopic Effect. At the usual alternating current supply frequency of 50 Hz, a discharge lamp will be extinguished 100 times per second. Although this effect is seldom noticeable in normal conditions. It is possible that this may, in some instances, give rise to a stroboscopic effect, that is moving objects such as rotating parts of machinery, illuminated by this light, may appear to be moving in disunity or jerks, or rotating more slowly than their actual speed.

In actual applications, however, where this effect might cause annoyance, it can be practically eliminated in a three lamp unit by connecting each lamp on a separate phase of a 3-phase system and it can be greatly reduced in a two lamp unit by the use of a two lamp control unit, which employs a condenser in the ballast of one of the lamps. (Fig. 14.15). The current through the lamps is almost 90° out of phase and under these conditions light output of one of the lamps is at a maximum. This method has an additional advantage of giving an overall power factor of nearly unity for the unity of two lamps. In this arrangement one circuit remains at a low power factor at about 0.5 lagging, while the other circuit incorporating a series capacitor C remains at a power factor of about 0.5 leading.

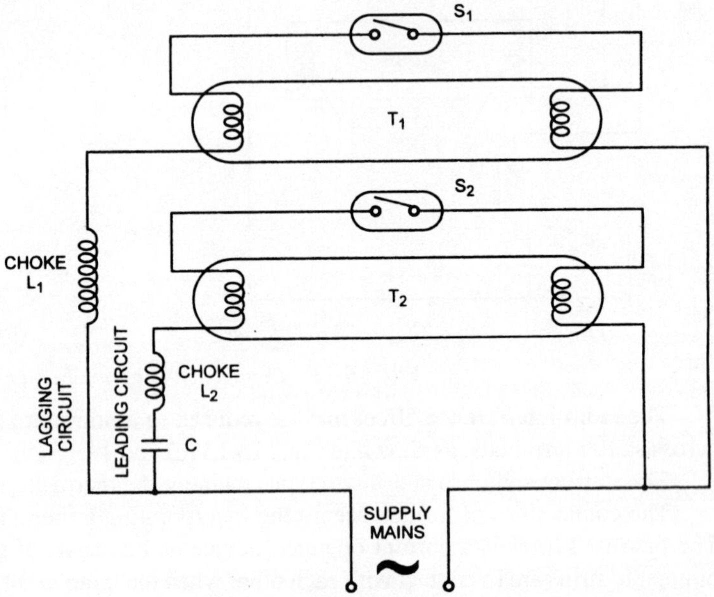

Fig. 14.15

Fluorescent Lamps For DC Supply. In the foregoing discussions, it has been assumed that the supply to the fluorescent lamps is ac. If however, the available supply is dc some special accessories and circuit modification will be required.

(*i*) The choke coil has a low impedance on dc and therefore, a ballast resistance is connected in series with the choke in order to limit the current.

(*ii*) On systems below 220 V, starting becomes less certain on dc. Only thermal type starters should be used.

(*iii*) The positive end becomes relatively dark on account of the tendency of the mercury vapour to migrate towards the negative end of the tube. In order to overcome this defect a reversing switch is included in the circuit between the supply and the fitting, as shown in Fig. 14.16.

In dc operation of fluorescent tube there is no problem of power factor correction and no stroboscopic effect. Its disadvantages are—low efficiency due to power loss in ballast series resistance, increased cost of the ballast resistance and reversing switch and less life of the tube (about 80% of that with ac operation).

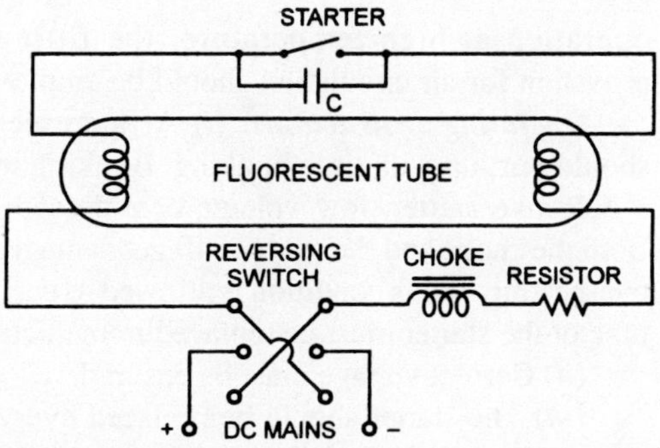

Fig. 14.16 *Operation of Fluorescent Tube in DC*

Colour of Fluorescent Light. As already mentioned the colour of fluorescent light depends upon the fluorescent powder used and the vapour pressure in the lamp. A few important variations of white colour with their applications are given below.

(i) Daylight. This is suitable for industrial applications where checking of material is carried out at high illumination levels (say 1,000 lux or more) and for normal lighting service.

(ii) Warm-White. This is mostly used for street lighting where colour rendition, due to low level of lighting (say 10 lux or less), is of little importance.

(iii) Warm-White Deluxe. It is used for illuminating rooms of large gathering such as shops, restaurants, and dwelling houses. It combines well with the light of incandescent lamps, as in show-windows etc.

(iv) White. It is useful for use in offices, drawing offices, schools and factories. It harmonizes exceptionally well with the day light, and therefore, light is a good supplement to day light.

White-Deluxe. It is very suitable for use in offices, schools and shops where colour rendition is important.

Useful Lamp Life. In general, fluorescent lamps lose their usefulness because of reduction in light output before they fail to operate. Darkening of the bulb occurs because of the effect of mercury on the fluorescent coating and because of the material given off by the electrodes. The latter specially causes darkening at the ends of the bulb late in life. The rate of depreciation in light output diminishes through out the life ; the first hundred hours produce approximately as much darkening as the following 1,000 hours. Rated output is based on conditions at 100 hours.

Frequent starting of lamps may take more life out of the electrodes than long hours of burning because momentarily there is a higher than normal voltage drop at the electrodes which causes the active material to sputter or evaporate off. If a lamp be started once a minute, for example, the hours of burning will be shorter than normal, but if it be switched on and burned continuously, its life will be longer than normal. When the active material on the electrodes is nearly exhausted, the voltage required for starting will rise and may equal or exceed the available supply. This may occur after the lamp has been started thousands of times or burned beyond its rated life. Sometimes the end of life is indicated by the lamp flashing momentarily and then going out.

The normal life of a fluorescent lamp is 7,500 hours. The average life is for three burning hours per switching operation. The actual life may vary from 5,000 to 10,000 hours depending upon the operating conditions. Light output is reduced by 15-20% after 4,000 hours of operation and it is, therefore, a good practice to replace the fluorescent lamp after 4,000-5,000 hours burning on economical grounds.

Performance Curves. Performance curves are shown in Fig. 14.17.

The effect of voltage variation in the case of fluorescent lamps is less marked as compared to the incandescent lamps. However, their life and performance are adversely affected both by low and high voltage. With increased voltage there is a greater heating of electrodes and they lose emissive material by evaporation. With reduced voltage, the current reduces causing sputtering at the electrodes shortening their life.

The best performance of fluorescent lamps is obtained at 20-25°C operating temperature. It decreases rapidly when a lamp is operated at a lower temperature or is exposed to cold wind drafts. The fluorescent lamps may be enclosed in outer envelops for their protection from cold weather in exposed locations. For operation at high temperature, the fittings with provision for air circulation should be employed.

Operating Instructions. (*i*) A fluorescent lamp should start up with little blinking. Blinking indicates a defective starter, low voltage or a defective tube. Both the choke and the starter will get damaged due to overheating, if this condition is allowed to persist. The tube or the starter must be replaced immediately.

(*ii*) Correct voltage must be ensured.

(*iii*) The starter should be replaced every time a tube is changed. A bad starter shortens the life of the tube.

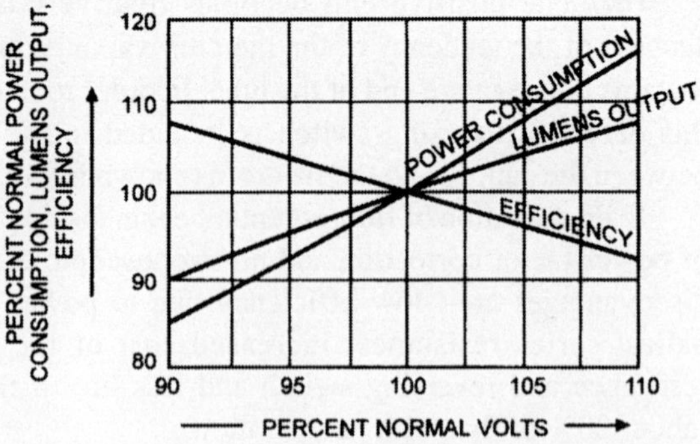

Fig. 14.17 *Performance Curves of Fluorescent Lamps*

(*iv*) Frequent switching operations should be avoided.

Merits and Demerits. As already stated the fluorescent lamps have got merits of high luminous efficiency, long life, low running cost, low glare level and less heat output. The demerits are stroboscopic effect, small wattage requiring large number of fittings and magnetic hum associated with the choke causing disturbance. The problem of noise can be solved to some extent by mounting chokes on resilient pads.

Trouble-Shooting. Various faults occurring in fluorescent tubes, their possible causes and remedies are given below :

	Symptoms	*Possible Causes*	*Remedies*
1.	Blinking on and off.	1. Low circuit voltage.	Change ballast tapping if required.
		2. Loose contact.	Check all connections.
		3. Defective starter.	Replace starter.
		4. Defective tube.	Replace tube.
		5. Wrong connections.	Check all connections.
2.	End of a tube remain lighted.	1. Short-circuited starter.	Replace starter.
3.	Failure of lamp to light up.	1. Defective tube.	Replace tube.
		2. Defective starter.	Replace starter.
		3. Defective choke.	Replace choke.
		4. Defective and loose holders.	Turn tube through 90°.
			Try another starter and check connections.
4.	Blackening of the end (early life)	1. Too low or too high voltage.	Adjust ballast tapping.
	(normal life)	1. Mercury deposits at the ends.	It will evaporate as the lamp operates.
5.	Burn out electrodes	1. Control unit in circuit or choke short-circuited.	Test with a new choke.
6.	Dark streaks along lamp	1. Mercury globules.	Rotate tube through 180°.
7.	Snaking spiralling (with glow type starter)	1. Insufficient heating due to either starter opening quickly or wrong tappings made on the ballast.	Replace starter or ballast. In case of new lamp, it cures after few days.

(*g*) Compact Fluorescent Lamps. These are the latest lamps and operate on the same principle as fluorescent tubes. Different colours can be obtained by using different fluorescent materials. They are not much larger than ordinary lamps. Overall length is reduced by bending each tube back on itself. These lamps are available in variety of shapes, colours and with different light intensity. Several reputed companies have designed the lamps to meet almost every requirement in the field of illumination offering great deal of energy saving. For instance, a 15 watt CFL offers nearly the same level of illumination as a 75 watt incandescent lamp. Lamps are available in ratings of 5, 7, 11, 15, 20 and 23 watts. These lamps are quite suitable for emergency lighting on 230 V dc except 23 W lamp.

The compact fluorescent lamp (CFL) with an electronic control gear and standard base have the following characteristics.

(*i*) Excellent colour rendering and excellent light distribution.
(*ii*) Long life—about 10 times the life of incandescent lamp.
(*iii*) High efficiency—Up to about 80% saving in energy.
(*iv*) Instant flicker free starting and flicker free operation.
(*v*) Small weight (50 to 150 gms) and are even suitable for pendant luminaires.
(*vi*) As switchable as incandescent lamps *i.e.* exceeding 500,000 switching cycles.
(*vii*) Lamp replacement intervals can be considerably prolonged by limiting the switching frequencies.

Compact fluorescent lamps (CFLs) find use in domestic installations (drawing rooms, living rooms), commercial installations (restaurants, corridors, hotels, offices) and especially where continuous economic lighting is required.

14.5. COMPARISON BETWEEN TUNGSTEN FILAMENT LAMPS AND FLUORESCENT TUBES

	Tungsten Filament Lamps	*Fluorescent Tubes*
1.	Voltage fluctuation has comparatively more effect on the light output.	Voltage fluctuation has comparatively low effect on light output as the variations in voltage are absorbed in the choke.
2.	Luminous efficiency increases with the increase in voltage of the lamp.	Luminous efficiency increases with the increase in wattage and increase in length of tube.
3.	It gives light close to natural light, therefore, objects are properly seen.	It does not give light close to natural light, therefore, colour rendering is defective.
4.	Luminous efficiency of coloured filament lamps is poor because coloured glasses are used for this purpose.	Different colour lights can be obtained by using different composition of fluorescent powder. Hence efficiency is high and better colours are obtained.
5.	Due to comparatively high working temperature heat radiations are also present.	Due to low working temperature heat radiation is low.
6.	Its brightness is more.	Its brightness is less.
7.	With the time light output is reduced.	With the time light output is gradually reduced.
8.	No stroboscopic effect.	It has objectionable stroboscopic effect.
9.	Though the life of filament lamps varies with the working voltage, however, its normal life is 1,000 working hours.	Life of fluorescent tubes is not affected so much by variations in voltage but it depends on the frequency of starting. The life of the tube is about 7,500 working hours.
10.	The initial cost per lamp is quite low.	The initial cost per tube is more.
11.	For same lumens output more lamps are required and wiring cost is more. Life of lamp is also low. Hence overall cost of maintenance is more.	For same lumens output, lesser number of tubes is required, the wiring cost is low. Since its life is comparatively more, replacement cost is low. Hence overall maintenance cost is low.

14.6. TYPES OF LIGHTING SCHEMES

The distribution of the light emitted by lamps is usually controlled to some extent by means of reflectors and translucent diffusing screens, or even lenses.

The interior lighting schemes may be classified as (*i*) direct lighting (*ii*) semi-direct lighting (*iii*) semi-indirect lighting (*iv*) indirect lighting and (*v*) general lighting.

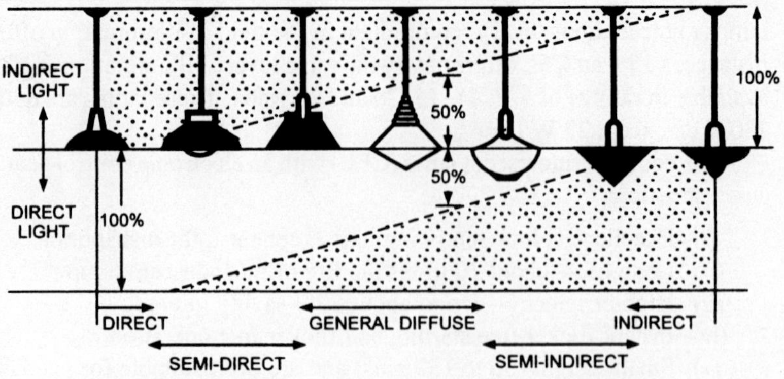

Fig. 14.18 *General Classification of Luminaires*

1. *Direct Lighting.* It is the most commonly used type of lighting scheme. In this lighting scheme more than 90 percent of total light flux is made to fall directly on the working plane with the help of deep reflectors. Though it is most efficient but causes hard shadows and glare. It is mainly used for industrial and general outdoor lighting.

2. Semi-Direct Lighting. In this lighting scheme 60 to 90 percent of the total light flux is made to fall downwards directly with the help of semi-direct reflectors, remaining light is used to illuminate the ceiling and walls. Such a lighting scheme is best suited to rooms with high ceilings where a high level of uniformly distributed illumination is desirable. Glare in such units is avoided by employing diffusing globes which not only improve the brightness towards the eye level but also improve the efficiency of the system with reference to the working plane.

3. Semi-Indirect Lighting. In this lighting scheme 60 to 90 percent of total light flux is thrown upwards to the ceiling for diffuse reflection and the rest reaches the working plane directly except for some absorption by the bowl. This lighting scheme is with soft shadows and glare free. It is mainly used for indoor light decoration purposes.

4. Indirect Lighting. In this lighting scheme more than 90 percent of total light flux is thrown upwards to the ceiling for diffuse reflection by using inverted or bowl reflectors. In such a system the ceiling acts as the light source, and the glare is reduced to minimum. The resulting illumination is softer and more diffused, the shadows are less prominent and the appearance of the room is much improved over that which results from direct lighting. It is used for decoration purposes in cinemas, theatres and hotels etc. and in workshops where large machines and other obstructions would cause troublesome shadows if direct lighting is employed.

5. General Lighting. In this scheme lamps made of diffusing glass are used which give nearly equal illumination in all directions.

14.7. LIGHTING ACCESSORIES AND FITTINGS

This section cannot pretend to give a comprehensive description of very large variety of lighting accessories and fittings available for installation work. The illustrated catalogues of manufacturers will give much information, and should be referred to by the students.

Various types of switches, ceiling rose, socket-outlets, plugs and lamp holders are described below.

1. *Switches.* A switch is used in an electric circuit as a device for making or breaking the electric circuit in a convenient way i.e. by the simple motion of a knob or handle to connect together or disconnect two terminals to which wires or cables are connected. This is the function of an ordinary single pole single way switch for switching lamps *on* and *off*.

The switches may be classified in various ways. According to the type of base material they are classified as porcelain or bakelite switches. According to colour of base they are either white or black or brown coloured switches. According to operation required they are classified as one-way switches, two-way switches, two-way centre-off switches, double pole switches etc.

(*i*) *One-Way Switches.* This type of switch consists of two terminals which can be easily seen from the back side of the switch as well, without removing the cover. The switch is always connected in series with the point (lamp, fan or socket outlet to be controlled). The switches usually available are for use up to 250 volts and 5 or 6 A. If either of the two is increased, the switch may get damaged due to sparking. These switches are fixed to the wooden supports by means of 2 nos wooden screws of 19 mm size. The cables are connected from the bottom side carefully.

(ii) Two-Way Switches. The switch of this type consists of four terminals, two of them being short circuited inside the switch. The switch of this type is usually used for stair case wiring or circuits where one point is to be controlled from two different places.

(iii) Two-Way Centre Off Switches. This switch is just like a two way switch but having three operations. In the centre it becomes off. Such switches are used when two lamps are to be operated alternately.

(iv) Double Pole Main Switches. This is a combination of two one-way switches which can be operated simultaneously as the ON-OFF terminals of both the switches are connected together by a handle made of bakelite. Such switches are used as interlinked switches when the load current is less than 5 amperes and supply voltage is under 250 volts.

In case either of the voltage or current exceeds the limits mentioned above DPIC switch is used.

(v) Push Button Switches. Such switches are used for controlling the electric bells. When the knob is pressed, the circuit is completed and bell rings and as soon as the knob is left, circuit becomes open.

(vi) Table Lamp Switches. This is a small on-off switch which is commonly used in table lamps.

(vii) Bed Switches. Such switches are used to switch off the table lamps or other lamps while going to sleep or making the lamp *on* while getting up at night. It is connected in series with one of the two flexible wires. The speciality with this switch is that fluorescent material is applied to its knob so that it may glow at night and can easily be seen in darkness. This is a pendant type switch.

The switches are of two types, known as surface switches (or tumbler switches) and flush switches (or concealed switches).

(a) Tumbler or Surface Switches. Tumbler switches are those which are fixed on mounting blocks directly fixed on the surface of the wall. Such switches project out of the surface of the wall and are in common use.

Fig. 14.19 *Surface or Tumbler Switches*

Surface switches are available in round and oblong base. Round base switches are cheap and in common use. Oblong surface switches are good in appearance, but being costly, are rarely used.

(b) Flush Switches. Flush switches, as obvious from its name are fixed flush with the wall and do not project out. These switches are used where high quality performance and appearance are required. In such switches all current carrying parts are mounted on high grade vitreous porcelain base enclosed in an iron box recessed into the wall.

Full specifications of switches controlling light, fan or socket outlet cover the type of base material (porcelain or bakelite), shape of the base (round or oblong), the colour of the base (white, black or brown), type of switch (tumbler, concealed or pendant) and the operation required (one way, two way etc).

The specifications for controlling 240 V lamps are given below.

6 A, 240 V, single way, bakelite base, round shape and black colour, tumbler switch of any make approved by I S.

Fig. 14.20 *Flush Switches*

(a) One-Way Switch
6 A, 250 V, AC

(b) Two-Way Switch
6 A, 250 V, AC

(c) Bell Push
6 A, 250 V, AC

(d) Bell Push
6 A, 250 V, AC

[Countesy SSK]

Fig. 14.21

2. *Ceiling Rose.* The ceiling rose is used to connect the pendant lamps, fans or fluorescent tubes to installation through flexible plastic or silk covered wire. It consists of two parts known as *base and cover.* The base may be of bakelite or porcelain and is fitted with two or more terminal plates separated from each other by a porcelain or bakelite bridge. The number of terminal plates fitted to ceiling rose base depends upon the number of ways of the ceiling rose i.e. two-way ceiling rose is fitted with two-terminal plates and 3-way ceiling rose is fitted with three-terminal plates so on. Fixing holes are also provided so that ceiling rose may be fixed to the round block. Each terminal plate is provided with a metallic sleeve and a binding screw on one side to connect the circuit wire and washer and clamping screw for making connections to the flexible wire on the other side. The circuit wire for connecting to ceiling rose, is opened for about 19 mm and inserted from the back so that it comes out by about 3 mm. The insulating bridge is provided with two holes through which flexible wire is passed first before making connections to the terminal plate. This is done with the object of taking weight of the lamp holder, lamp etc. from the terminal plates. Now to the threaded base the cover is screwed in clockwise direction.

The specifications of a ceiling rose are given below.

6 amperes, 240 volts, two (or three or four)-way, of any make approved by IS.

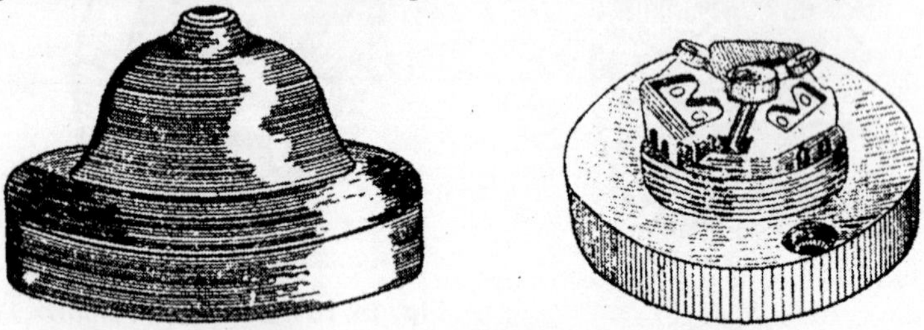

(a) Cover *(b) Base*

2-Plate Flush Type
6 A, 250 V, AC

3-Plate Flush Type
6 A, 250 V, AC

Jumbo Ceiling Rose
(2-Plate, 6 A, 250 V, AC)

[Courtesy SSK]

Fig. 14.22 *Ceiling Roses*

3. ***Socket-Outlets.*** The socket outlets are used to supply electrical connection whenever required for electrical appliances such as TVs, radios, table lamps, table fans, irons, stoves etc. Socket-outlets are of two types; two pin type and three pin type. Two pin type socket outlet has all insulated base with moulded or socket base having two terminal sleeves in which two pin plug can easily be inserted but not loosely, 2 pin socket-outlets were used for 6 amperes but nowadays these have become obsolete.

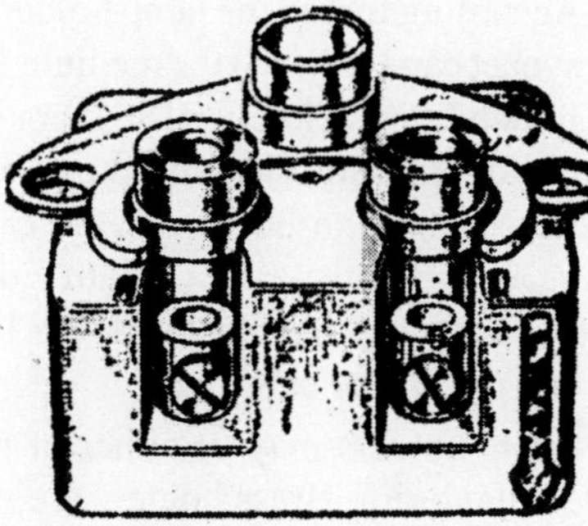

Flush Mounting (3-Pin)

The three pin socket-outlet has got three hollow terminals in which three pin plug can easily be inserted but not loosely. Two holes, being of same size, are meant for making connections to the flexible wire of the appliance and the third hole, which is bigger comparatively is meant for earth connection. Thus three holes or sleeves are for live, neutral and earth connections. The 3-pin socket outlets are also of two types (1) 6 A for table fan, table lamp, radio, TV etc. and (2) 16 A for power circuits as heater, geyser, iron etc.

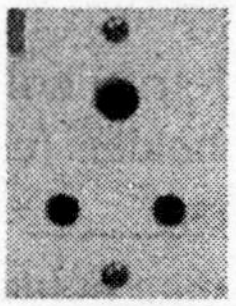

Three-Pin Socket
6 A, 250 V

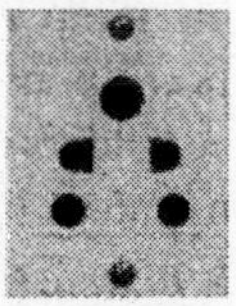

Five-Pin Socket
6 A, 250 V, AC

Fig. 14.23

The specifications of socket outlet for use for a table lamp or table fan are given below.

240 V, 6 A, three-pin (two pole and earthing pin) socket outlet of any make approved by IS.

DP Switch Socket Combined
6/16 A, 250 V, AC
With Indicator

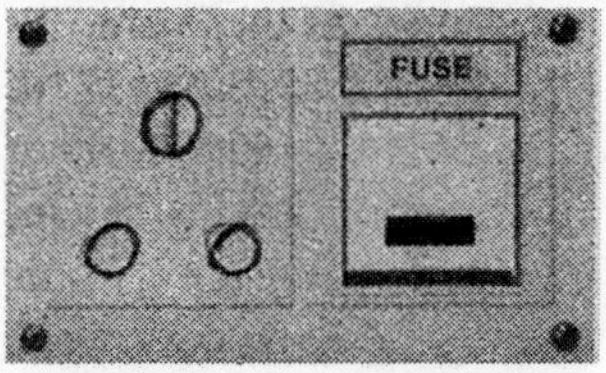

DP Switch Socket Combined
6/16 A, 250 V, AC Shuttered
With Indicator and Fuse
[Courtesy SSK]

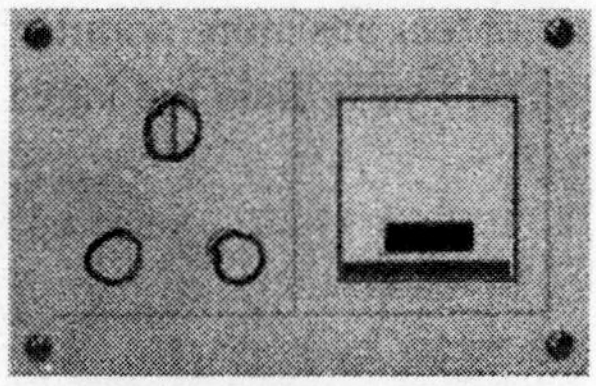

DP Switch Socket Combined
6/16 A, 250 V, AC Shuttered
With Indicator and Fuse

Fig. 14.24

4. ***Plugs.*** Plugs are used to collect the supply from the socket-out lets for electrical appliances such as table lamp, table fan, stove, iron, heater etc. Similar to socket-outlets plugs are also of two types viz two-pin and three-pin types. Two-pin plugs have become obsolete nowadays. Three-pin type plug consists of three pins usually made from brass. To the two pins which are thin and of the same size, flexible wires are connected and then covered up. To the third pin, which is thicker comparatively, earth wire form the electrical appliance is connected. The flexible wires may be twisted in the hole or there may be screws provided to keep them in place. The cover and base may be of bakelite (breakable), or of rubber or polyvinyl material (unbreakable). Similar to 3-pin socket-outlets 3-pin plugs are used for table lamps, table fans, radios etc. and 16 A plugs are used for power circuits such as stove, heater, iron etc.

The specifications of plug for use for a table lamp or a table fan are given below.

240 V, 5 A, three pin plug of any make approved by I S.

5. ***Lamp Holders.*** As the name indicates the function of lamp holder is to support the lamp and also to connect it electrically. These are designed for quick removal and replacement of the lamp. Lamp holders are of many types . A few will be described here.

According to size the lamp holders are of three types namely (*i*) Bayonet cap (BC), (*ii*) the medium Edison Screw (ES) and (*iii*) the Goliath Edison Screw (GES). For ordinary tungsten filament lamp upto 200 watts the lamp caps and thus the lamp holders are Bayonet cap; up to 300 watts the caps are Edison Screw and above 300 watts they are Goliath Edison screw. In every case where a lamp is to be installed, the lamp holder of appropriate size and type must be fitted.

Fig. 14.25 *Three Pin Plug and Cover*

Lamp holders may be either of brass or bakelite type with porcelain interior. Brass holders are more durable but may give shock if connections are poor. Though bakelite holders are not durable, but they are safer.

The lamp holders may also be classified into (*i*) batten holders (*ii*) pendant (cored-grip) holders (*iii*) angle holders (*iv*) slanting holders (*v*) bracket holders (*vi*) water tight bracket holders (*vii*) miniature lamp holders.

(*i*) *Batten Holders.* Batten holders are employed where the lamp is to be fitted to the roof or to the wall and they are directly fitted either to batten or to wooden board. Such lamp holders are bayonet type i.e. in such a lamp holder the bulb is forced in, turned slightly and left in position. Nowadays generally, bakelite batten holders are more popular than the brass ones as these are cheaper and safer than the latter.

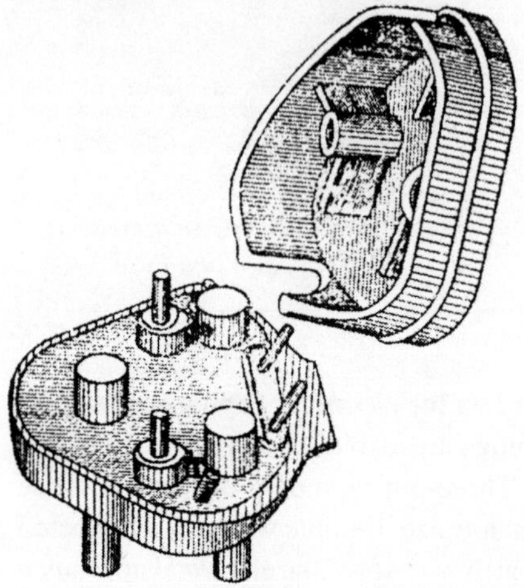

5 Amp Plug

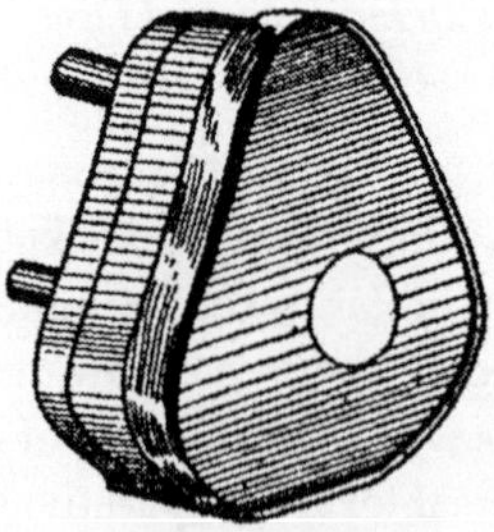

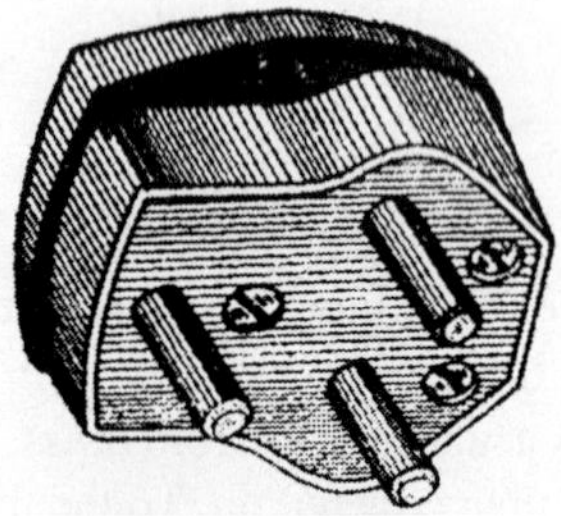

Heavy Duty Power Plug

(a) 2-Pin 6 A, 250 V, AC Plug

(b) 3-Pin Triangular With Indicator 6 A, 250 V, AC Plug

(c) 3-Pin Triangular With Indicator 16 A, 250 V, AC Plug

[Courtesy SSK]

Fig. 14.26

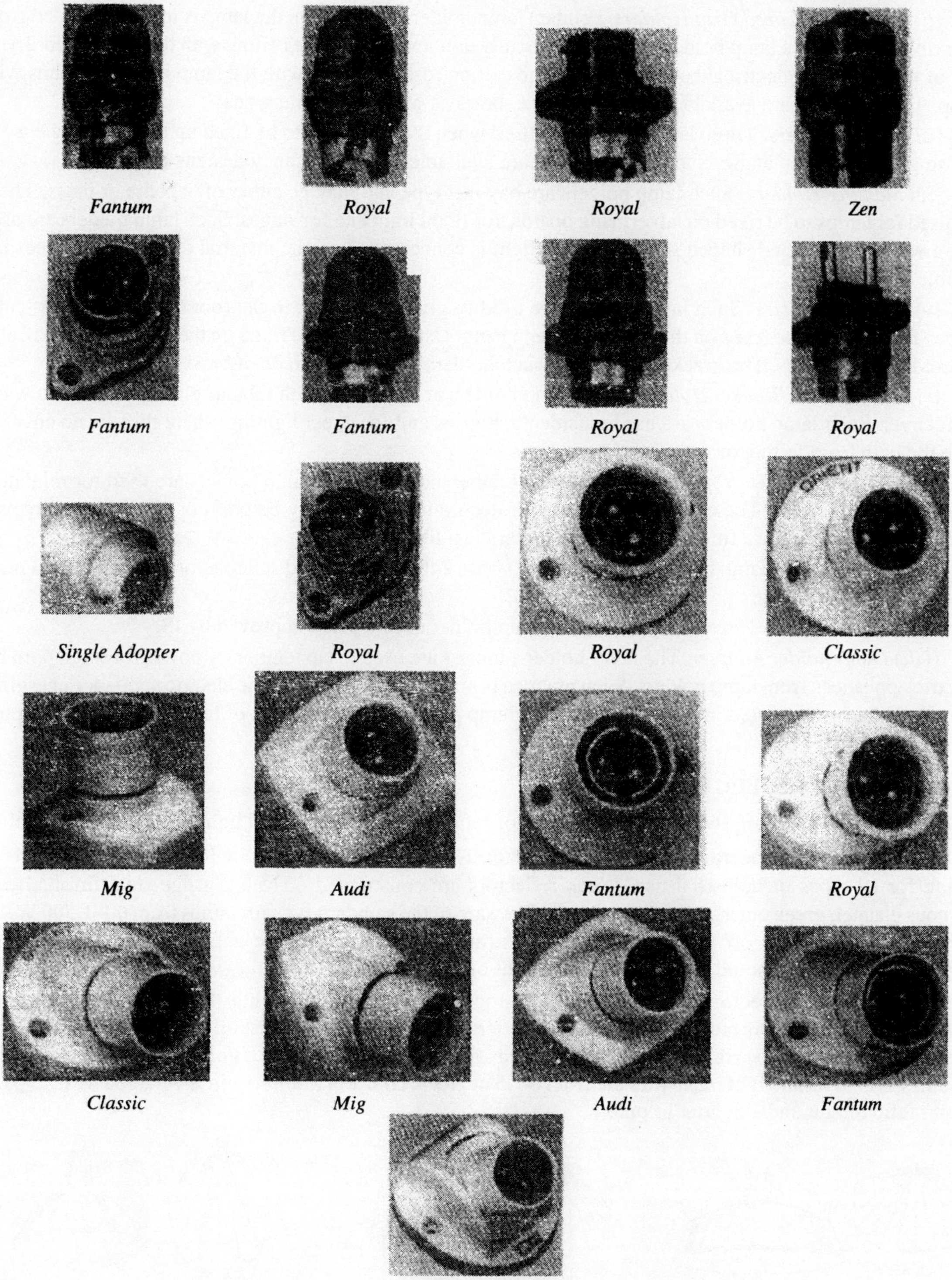

Lamp Holders and Adopters [Courtesy ORIENT]

Fig. 14.27 *Electrical Accessories and Cables*

(*ii*) *Pendant or Cored Grip Holders.* Pendant lamp holder is used when the lamp is to be suspended from a flexible cord. Such a lamp holder is hanged vertically downward from the ceiling with the flexible cord, one end of which makes electrical connection with the ceiling rose and other with the lamp holder and thus with lamp. These holders are available in both types i.e. brass as well as bakelite types.

(*iii*) *Angle Holders.* These lamp holders are used when the lamps are to be fixed up directly on the walls and to give light at an angle. Such lamp holders are available in various fancy designs and colours.

(*iv*) *Slanting Holders.* Such lamp holders are bayonet type and may be either of bakelite or brass. These are used for lamps to be fixed on advertising boards, for flood light and for stages. Such lamp holders are used along with shades (hand shaped shades) so that light is concentrated on the material displayed and does not trouble the viewer.

(*v*) *Bracket Holders.* Such lamp holders are used to give direct light in the room or above a particular place. These cannot be fixed on the roof or made to hang. Usually these are fixed on the wall. These may also be used in table lamps. The brackets used with such holders are usually made of brass.

(*vi*) *Water Tight Bracket Holders.* Such lamp holders are provided with tubular glasses fixed with water tight cover. Such lamp holders are used outside the houses and for street lighting where there is no cover to save the bulb from falling of water over it.

(*vii*) *Miniature Screw Type Lamp Holders.* As name indicates such lamp holders are used for miniature lamps like night bulbs. These can also be used for decoration. These may be brass or bakelite or porcelain type. Miniature lamps are fitted in such lamp holders just like screws.

Specification for a lamp holder for holding 100 watt, 240 volt lamp and suitable for being fixed to a brass bracket are given below:

5A, 240 V, bakelite, bayonet type bracket lamp holder of any make approved by IS.

(*viii*) *Lamp-holder Adopter.* The lamp-holder adopters are used to tap temporary power for small portable electric appliances from lamp holders. Such practice is not advised. In no case the electric appliance energized by this method is permitted in bath room or other damp places. Different types of lamp holders and adopters are shown in Fig. 14.27.

14.8. TYPES OF REFLECTORS

14.8.1. General Lighting Reflectors. The types of reflectors used for general lighting are described below.

1. Dispersive Type Reflector. A typical fitting of this type is shown in Fig. 14.28. It is considered a universal fitting for all types of industrial uses. Such reflectors are constructed of heavy gauge steel finished with vitreous enamel, green outside and white inside. They are of the standard type for lamps from 60–1,500 W, the view of filament is cut off at 70 degrees and maximum light intensities are obtained from vertical to 45 degrees. The spacing should be 1.5 times the height of lamp filament above the working plane.

2. Diffuser Type Reflector. An industrial diffusing unit which has a slightly lower efficiency than the dispersive type reflector, but the appearance is better and it is particularly useful for minimizing the glare reflected from polished surfaces. Some fittings of this type permit a little light to go upwards to relieve ceiling gloom. The maximum light intensities are from 0–45 degrees. Uniform illumination is obtained with a space-height ratio of one and a quarter to one.

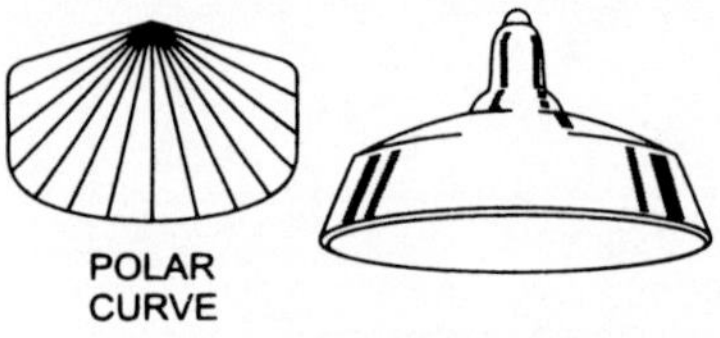

Fig. 14.28 *Dispersive Type Reflector*

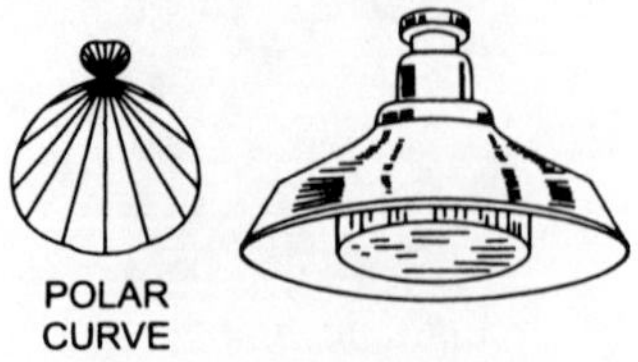

Fig. 14.29 *Diffuser Type Reflector*

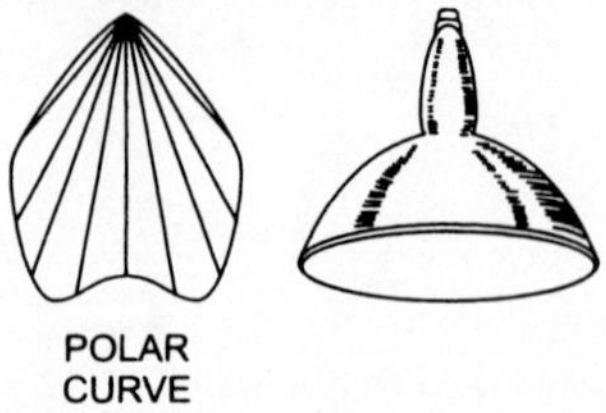

Fig. 14.30 *Concentrating Type Reflector*

3. Concentrating Type Reflectors. In this type of reflector (Fig. 14.30), the reflecting surface is designed in the shape of a deep parabola. The main distribution of light is obtained up to 45 degrees from the vertical. View of the filament is cut off at 60°. The maximum light intensities are obtained from vertical up to 25°. Uniform illumination is obtained with a space-height ratio of unity.

These reflectors are constructed in heavy-gauge steel vitreous enamelled white inside and green outside. This type of reflector is designed for shop lighting or for very high mounting heights. These reflectors are used for lamps from 60 to 1,500 watts.

4. Parabolitical and Elliptical Angle Reflectors. Such reflectors (Fig. 14.31) are constructed similarly to the fittings previously mentioned for use with 60–1,500 W lamps, the principal axis being tilted at an angle of 45° to the vertical. The maximum light intensities are obtained from 15° to 75° in front of the reflector, and view of the filament in cut off at 17° behind and 84° in front. These are usually used to augment overhead illumination, especially in the case of shop lighting, and have great scope for wall illumination. Angle reflectors are also available in elliptical types, and an example, together with the vertical elliptical reflector are shown in Fig. 14.32 with the polar curves which apply in each case.

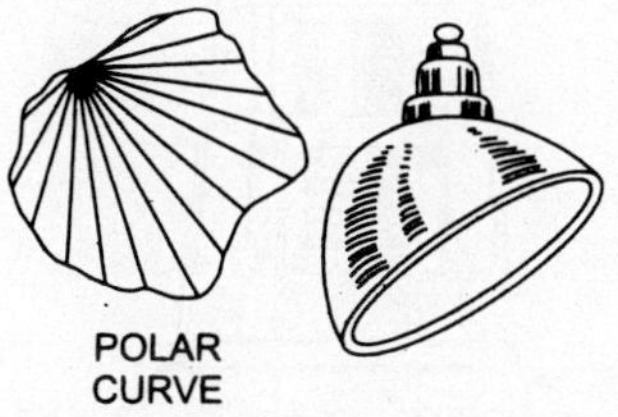

Fig. 14.31

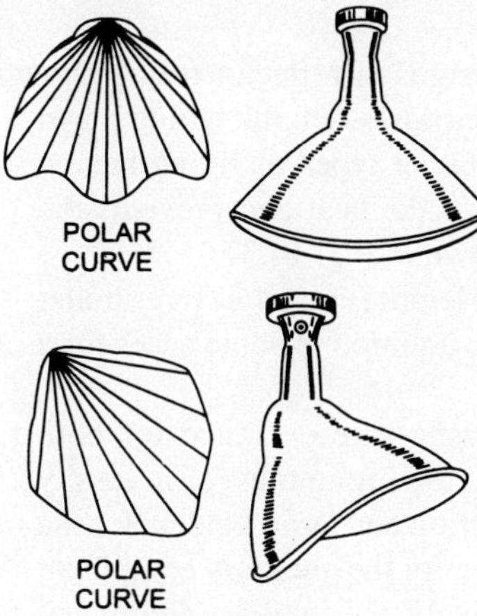

Fig. 14.32

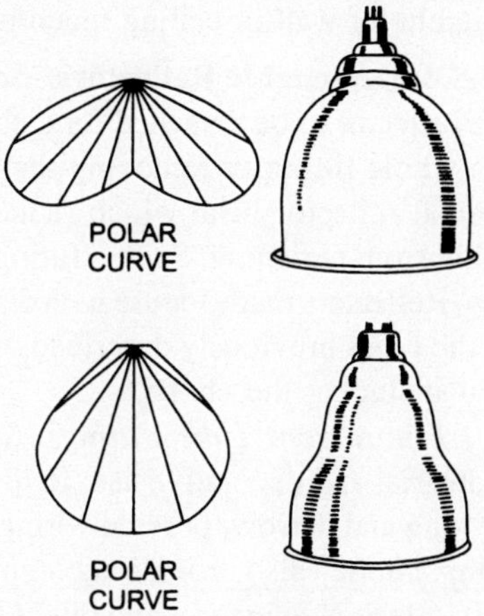

Fig. 14.33 *Extensive and Intensive Type Reflectors*

14.8.2. Local Lighting Reflectors

1. Extensive Type Reflectors. These are, also, constructed in steel with vitreous-enamel finish. The bowl is in the form of a hyperbolic curve. The maximum light intensities are obtained between the angle of 0° and 45° and the cutoff angle being 58° from the vertical, and spacing should be 1.5 times of the mounting height.

2. Intensive Type Reflectors. This is similar to extensive type, the difference lying in the shape of the reflector bowl, which is a combination of convex and concave curves.

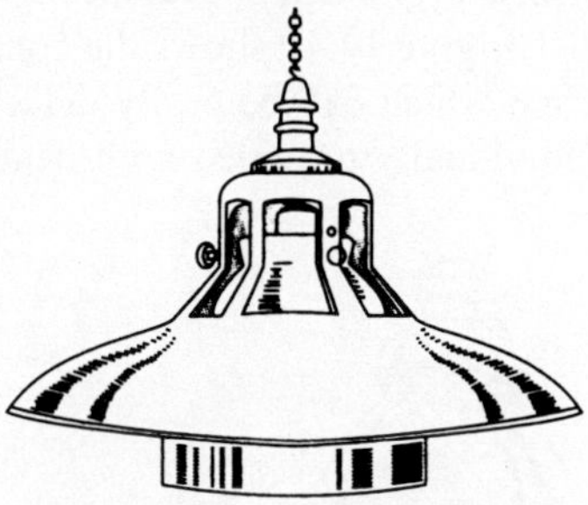

Fig. 14.34 *Benflux Reflector*

14.8.3. Other Types of Reflectors

1. Ben flux Reflector. A wide distribution fitting made in vitreous enamel with a glass diffusing shade encircling the lamp. Small apertures are left above the glass shade to ensure adequate ventilation. This also permits some light to pass upwards to provide a little ceiling light. It finds uses in both industrial and commercial applications (factories, schools and offices). Space-height ratio should be $1\frac{2}{3}$.

2. Porcelain Well-glass Reflectors. These are special reflectors manufactured in the same type as mentioned above, and are shown in Fig. 14.35. These are made for chemical works, plating shops and laundries, in fact any place where acid fumes or moisture prevail. Such fittings are for use with up to 300 W lamps.

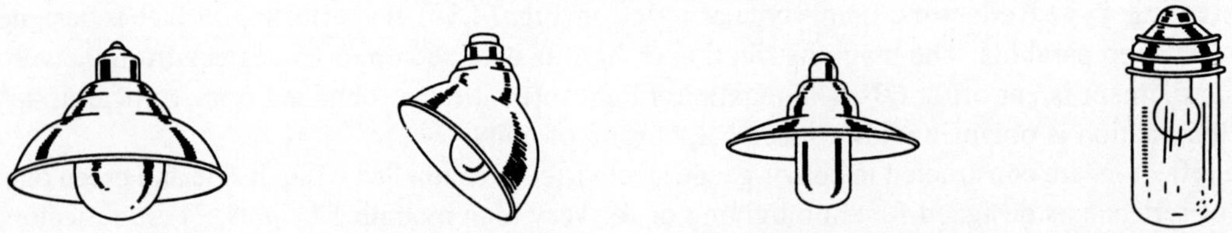

Dispersive Reflector *Parabolic Angle Reflector* *Distributing Reflector* *Well-glass Fitting*

Fig. 14.35 *Porcelain Well-glass Reflectors*

3. Bulkhead Fittings. Such fittings (shown in Fig. 14.36) are specially designed for locations where the space is so restricted that normal fittings cannot be employed, as in garages, tunnels and corridors. They provide wide lateral distributions of light by virtue of the prismatic glass fronts. The outside glass is smooth to enable the fitting to be easily cleaned. It is suitable for wall or ceiling mounting.

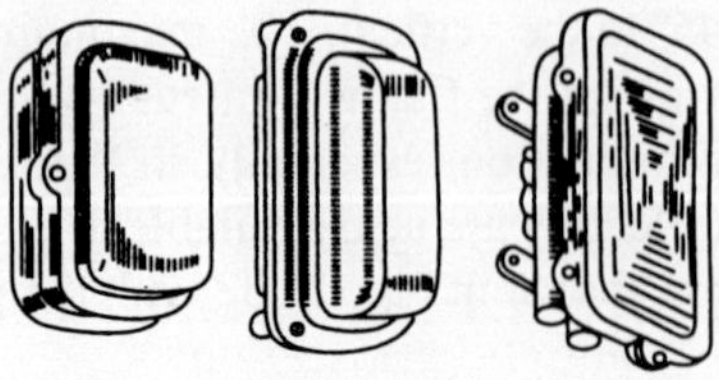

Fig. 14.36

14.8.4. Detachable Reflectors. Some designs of reflector fitting allow the reflector to be detached for cleaning purposes without taking down the whole fitting or removing the lamp. Other types of fitting have a special reflector head which radiates away the heat and prevents the cable from perishing. Such a fitting is shown in Fig. 14.37.

Reflectors made for use with discharge lamps (Fig. 14.38) are similar to the types previously described, with the addition in some cases for a unit including the choke.

Fluorescent Tube Fittings. Such fittings are designed for both industrial and decorative use. In principle, the fitting must of necessity be long and narrow, provide wiring protection and lamp holders for the lamps. It may also provide accommodation for the auxiliary equipment if it is desired to incorporate this. Alternatively, the auxiliary equipment may be housed separately, away from the fitting.

Figure 14.39 depicts a typical reflector for a 1.5 m fluorescent tube with a very clean appearance.

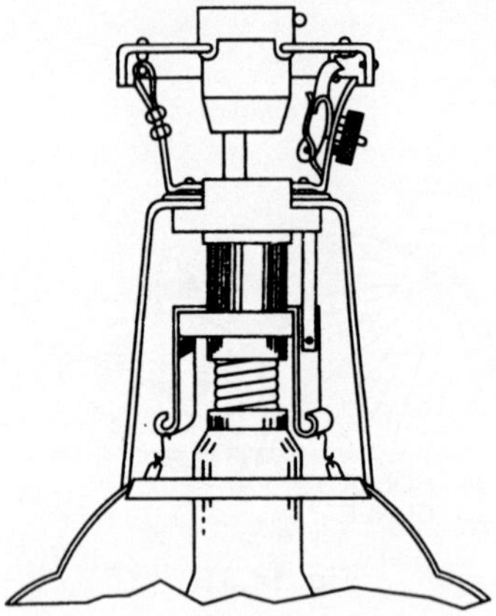

Fig. 14.37

Figure 14.40 shows the construction and it will be noticed that the auxiliary equipment is mounted on a plate which can be easily detached. All wiring and equipment are very readily accessible as all parts are individually provided with detachable covers.

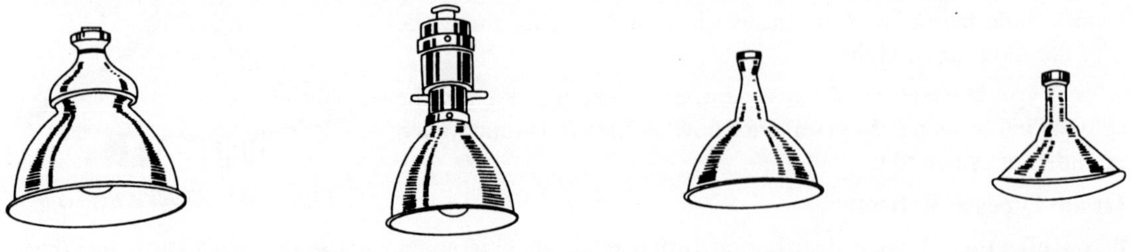

Dispersive Reflector *Dispersive Omnichoke Unit* *Concentrating Reflector* *Elliptical Reflector*

Fig. 14.38 *Discharge Lamp Reflectors*

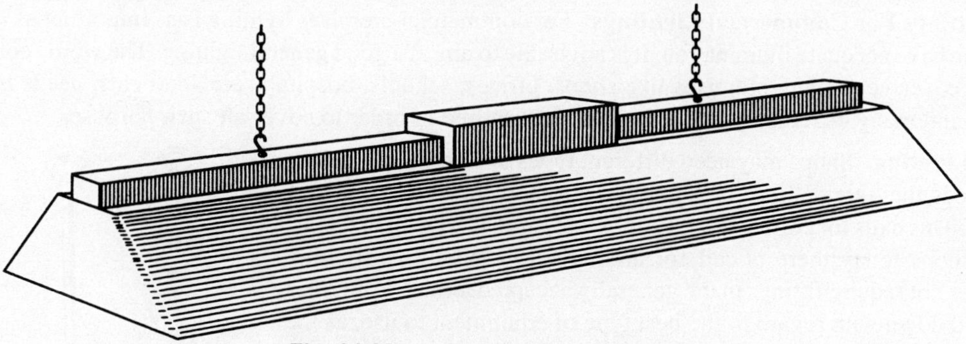

Fig. 14.39 *Fluorescent Tube Reflector*

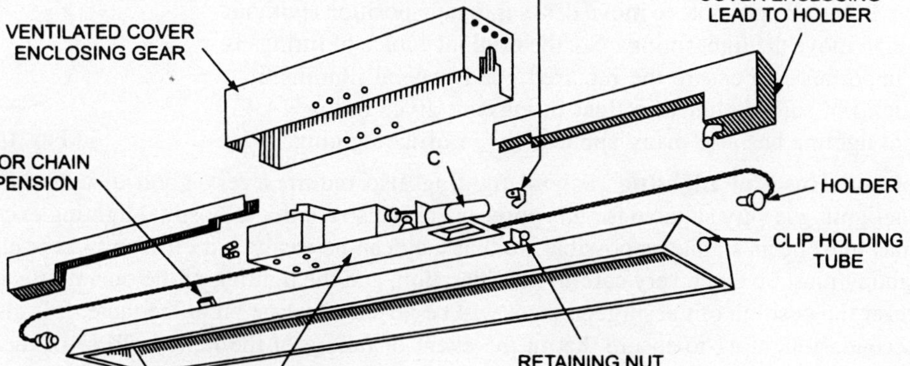

Fig. 14.40 *Tube Reflector in Detail*

14.8.5. Fittings For Domestic Lighting. It must be realized that each individual room needs a fitting depending on its size and nature of decoration. It is possible, however, to bear in mind certain points (given below) that are fairly universally applicable.

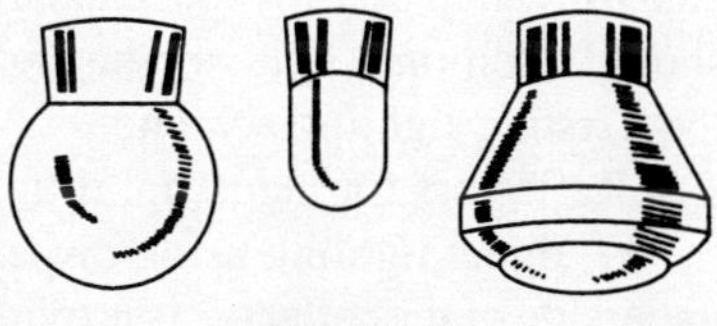

Fig. 14.41

1. Fittings for bathroom, kitchen and corridors should be of the ceiling-fitting type and be easy to clean and, in the case of bathroom and kitchen fittings, as damp-proof as possible. Figure 14.41 depicts a few typical fittings that will provide maximum possible protection.
2. The main feature to be observed about wall light fittings is that a wall fitting, unless carefully designed, tends to be dust trap. Hence, the installation should be harmonious with the decoration and be easy to clean. Figure 14.42 indicate a few wall-type fittings which are both simple in design and yet very effective from a decoration point of view.
3. Fittings for living-room, dining room, lounge and bedrooms, available in the market, are of many types, allowing plenty of choice. One important fact is that a good sense of proportion must be preserved when choosing these fittings, and the decorative aspect not allowed to overshadow the prime purpose of the fitting, which is to provide illumination.

The old idea of holding the fitting up by the flex should be avoided. Most modern fittings enclose the flex in a metal tube which takes the weight, and so prevents many possible faults.

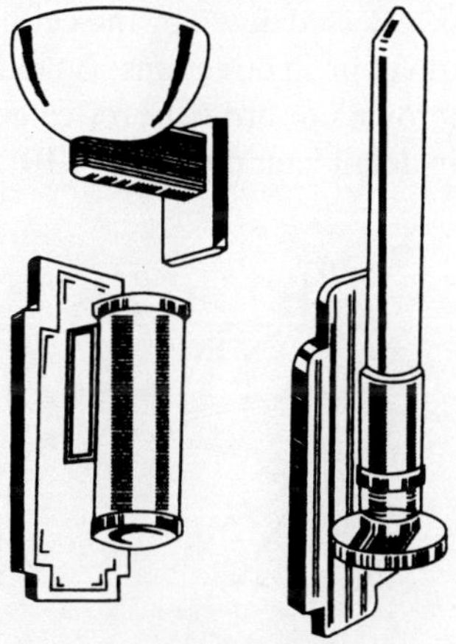

Fig. 14.42

14.8.6. Fittings For Commercial Lightings. For commercial premises lighting is as important as in factory premises to have adequate illumination. It is advisable to aim at a good general lighting. The word "commercial building" cover very many premises like shops, offices, schools, hospitals, etc., and each needs individual attention, and many different types of fittings are required in order to cover all such purposes.

1. Shop Lighting. Shops may need different types of lighting according to the goods they are selling; in some cases totally indirect lighting is advisable. This calls for cornice-type fittings and, probably, wall brackets. In most shops today there is call for architectural or built-in lighting which does not require fittings in the generally accepted term. So, no rules can be laid down with regard to the best type of equipment to use, as the lighting generally calls for special material according to the particular use to which it has to be put.

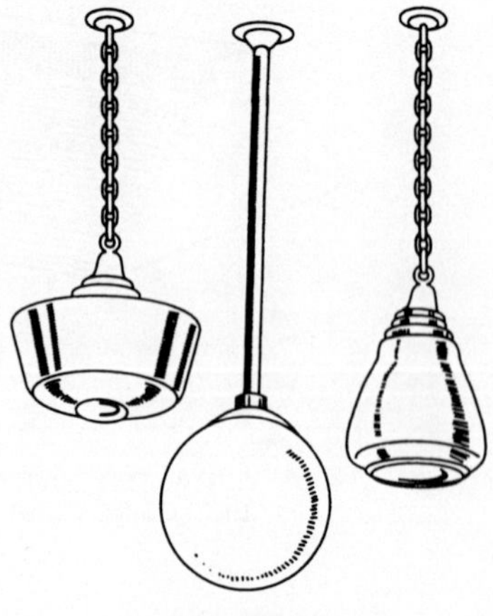

Fig. 14.43

2. Office Lighting. As nearly all work is of a clerical nature, good lighting is essential. It must be possible to move desks into new positions without having also to move the light fittings. So, the original choice of fittings is of major importance to ensure the required good general illumination. Some samples of suitable tungsten light fittings are given in Fig. 14.43. Fluorescent lighting has also many applications in office lighting.

3. School and Hospital Lighting. School lighting also requires very good distribution of light, and fluorescent lighting is very suitable for this purpose. This also applies to hospital lighting except that fittings for hospitals must be as simple as possible, so that they can be easily and efficiently kept clean. Operating theatre lighting must be given very careful consideration, and the lighting of the operating table so arranged that whatever the position of the surgeon there will be no shadow cast on to the table. It is also necessary to provide secondary lighting to ensure that in the event of failure of the lamp bulb or of the mains supply, alternative lighting will immediately be available.

4. Factory Lighting. Each branch of industry must be catered for individually with reference to the factory and workshop Act for the standard of illumination. All lighting are required to be planned, and at no time, should additional lights be haphazardly installed without due consideration as to how they will tie in with the existing lighting scheme.

Reflectors for industrial fittings are shown in Fig. 14.44, which may be broadly classified as follows:

(*i*) direct lighting using dispersive or concentrating reflectors; (*ii*) direct diffused lighting, in which the underside of the reflector is provided with an enclosed diffusing globe; (*iii*) semi-direct and enclosed fittings, from which most of the light is emitted in a downward direction, but a substantial proportion goes upward and is reflected down by the ceiling; (*iv*) general diffusing fittings, like glass spheres, in which light is allowed to travel in all directions; (*v*) indirect lighting, may be of fitting, trough or cornice form, in which all the light is thrown upward to be reflected down to the working plane from the ceiling. Factory lighting will be discussed in detail later on [Art. 14.10].

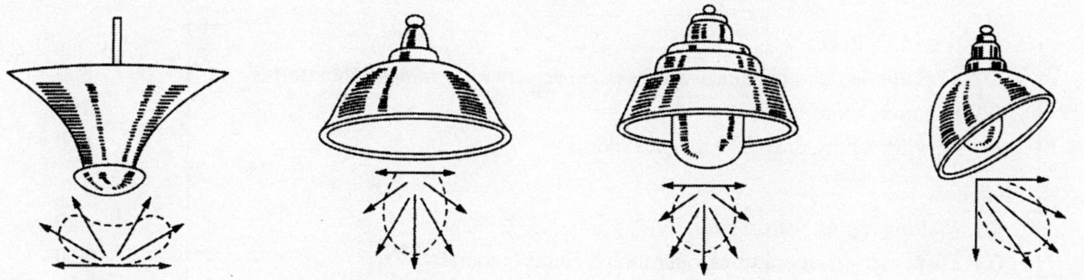

6. *Indirect* 7. *Concentrating* 8. *Vapour Proof* 9. *Angle Reflector*

Fig. 14.44 *Reflectors For Industrial Fittings*

14.9. DESIGN OF LIGHTING SCHEMES

The lighting scheme should be such that it may,

(*i*) provide adequate illumination,

(*ii*) provide light distribution all over the working plane as uniform as possible,

(*iii*) provide light of suitable colour, and

(*iv*) avoid glare and hard shadows as far as possible.

The following factors are required to be considered while designing the lighting scheme.

1. *Illumination Level.* This is the most vital factor because a sufficient illumination is the basic means where by we are able to see our surroundings, unless they are themselves light sources, since only when illuminated do the objects take on the necessary brightness. It is the task of illumination to give objects a distributed brightness. Body colours have property of reflecting light in different degrees. It is this differential brightness which gives essential perception of details. For each type of work there is a range of brightness most favourable to output, *i.e.*, which causes minimum fatigue and gives maximum output in terms of quality and quantity. Degree of illumination, to give necessary brightness to the objects depends upon (*i*) the size of the object to be seen and its distance from the observer–greater the distance of the object from observer and smaller the size of the object, greater will be the illumination required for its proper perception and (*ii*) contrast between the object and back ground—greater the contrast between the colour of the object and its background, greater will be the illumination required to distinguish the object properly. Objects which are seen for longer duration of time require more illumination than those for casual work. Similarly moving objects require more illumination than those for stationary object.

The recommended levels of illumination for different occupancies is given below:

S. No.	*Occupancy*	*Illumination Level in LUX*
1.	**Houses**	
	(*i*) Living room—general	150
	(*ii*) Living room—homework or sustained reading	300
	(*iii*) Kitchen, bedrooms, bathrooms etc.	150
2.	**Offices**	
	(*i*) Reception	150
	(*ii*) Conference room, typing rooms, general offices	300
	(*iii*) Drawing offices	400

Contd...

3.	**Schools and Colleges**	
	(*i*) Classrooms, lecture halls, workshops, library, reading tables, laboratories	300
	(*ii*) Common room, stairs	150
	(*iii*) Serving rooms, drawing halls, art rooms	600
4.	**Hospitals**	
	(*i*) Waiting rooms, wards, casualty	150
	(*ii*) Dispensaries, laboratories, operation theatres (general)	300
	(*iii*) Operation table	Special lighting
5.	**Hotels and Restaurants**	
	(*i*) Reception, dining rooms, bedrooms, lounges, stairs	150
	(*ii*) Accounts, writing desk, dressing table	300
6.	**Shops**	
	(*i*) General areas	300
	(*ii*) Stock areas	150
	(*iii*) Shop window	Special lighting
7.	**Storage Places**	
	(*i*) Loading and unloading	50
	(*ii*) General stores	150
	(*iii*) Stores of very small items	300
8.	**Play Grounds**	
	(*i*) Stadium	300
	(*ii*) Football field	200
	(*iii*) Tennis court	400
9.	Canteens	250
10.	Garage	70
11.	**Factories and Workshops**	
	(*i*) Rough work, *e.g.* frame assembly of heavy machinery	150
	(*ii*) Medium work, *e.g.* machined parts, engine assembly, vehicle body assembly	300
	(*iii*) Fine work, *e.g.* radio and telephone equipment, type writer and office equipment assembly	450
	(*iv*) Very fine work, *e.g.* assembly of very small precision mechanisms, instruments	1,500
12.	**Power Houses**	
	(*i*) Boiler house, turbine house, conveyor house, switchgear and transformer chambers	100
	(*ii*) Control rooms	300

2. *Uniformity of Illumination.* The human eye adjusts itself automatically to the brightness within the field of vision. If there is a lack of uniformity, pupil or iris of the eye has to adjust more frequently and thus fatigue is caused to the eye and productivity is reduced. It has been found that visual performance is best if the range of brightness within the field of vision is not greater than 3 : 1, which can be achieved by employing general lighting in addition to localised lighting. Apart from the consideration of causing fatigue, local lighting without using matching general lighting creates psychological feeling of loneliness, gloom and unfriendliness. The modern trend is thus towards 'localised lighting plus general lighting' and towards the adoption of "general lighting oriented towards the working surface" especially in mass production factories, offices, drawing offices, shops etc.

3. *Colour of Light.* The appearance of the body colour entirely depends upon the colour of the incident light. In general the composition of the light should be such that the colour appears natural, *i.e.*, its appearance by artificial light is not appreciably different from that by daylight. Daylight fluorescent tubes nowadays make it possible to illuminate economically even large spaces with artificial daylight giving good colour rendering and at sufficiently high level. For certain applications such as street lighting, colour of light does not matter much if different components have not to be distinguished from each other by their colours, highly efficient discharge lamps, which cause colour distortion, can be used.

4. *Shadows.* In lighting installations, formation of long and hard shadows causes fatigue to eyes and, therefore, is considered to be a shortcoming. Complete absence of shadows altogether again does not necessarily mean an ideal condition of lighting installations. Contrary, perhaps to popular opinion, a certain amount of shadow is desirable in artificial lighting as it helps to give shape to the solid objects and makes them easily recognised. Objects illuminated by shadowless light appear flat and uninteresting, contours are lost, and it is difficult for the eye to form a correct judgement of the shape of an object. However, there is one exception to this, *i.e.*, in drawing offices, where we are to see flat surfaces, shadowless light is essential otherwise shadows will hinder the work. Hard and long shadows can be avoided by (*i*) using large number of small luminaries mounted at a height not less than 2.5 metres and (*ii*) by using wide surface sources of light using globes over filament lamps or by using indirect lighting system.

5. *Glare.* It may be direct or reflected, *i.e.*, it may come directly from the light source or it may be reflected brightness such as from a desk top, nickelled machine parts, or calendered paper.

Direct glare from a source of light is the more common, and is more often a hindrance to vision. A glance at the Sun proves that an extremely bright light source causes acute eye discomfort. Light sources of far less brilliancy than the Sun, such as the filament of an incandescent lamp, or the incandescent mantle of a gas lamp, also cause discomfort by a direct glare. Reflected glare is a glare which comes to the eyes as glint or reflection of the light source in some polished surface.

Toleration of bright light sources in the intermediate vicinity is made possible by locating them at such a height as to place them above the ordinary range of vision.

Metal reflectors for industrial lighting are ordinarily provided with a skirt around the rim of the reflector.

6. *Mounting Height.* The mounting height will largely be governed by the type of the building and type of lighting scheme employed. In the case of direct lighting, in rooms of large floor area, the luminaries should be mounted as close to the ceiling as possible. Lowering them not only will make the illumination less uniform, but will also bring them more into the field of vision, thus increasing the glare, without causing an appreciable increase in the coefficient of utilisation. In the unusual case of small rooms with high ceilings, there is something to be gained by lowering the luminaries, but even here a better solution might be to use filament lamps with focusing reflectors and to mount them high.

In the case of indirect and semi-indirect lighting, it would of course be desirable to suspend the luminaries far enough down from the ceiling in order to give reasonably uniform illumination on the ceiling. In practice this is usually taken to mean that the length of the suspension tubes should be one-quarter to one-third the horizontal spacing between the rows of luminaries. Unfortunately it is often impossible to suspend them as far down as this, because it is quite generally recognized that there should be a minimum clearance of 2.5 metres between the luminaries and the floor. It follows that in rooms having low ceilings, the illumination on the ceiling is unavoidably non-uniform. This is not objectionable, because the ceiling is not a working plane.

7. *Spacing of Luminaries.* Correct spacing is of great importance to provide uniform illumination over the whole area and thus do away with comparatively dark areas which are so often found when the fittings are badly spaced.

In the case of direct and semi-direct luminaries the ratio of the horizontal spacing between rows to the height of the luminaries above the working plane depends to quite an extent on the candle power-distribution curve of the luminaire. With fluorescent luminaries it is good practice to aim at a value of unity for this ratio, and to set an upper limit of 3/4. In the case of tungsten lamps combined with focussing reflectors, the ratio of spacing to height should be about 0.6.

In the case of indirect and semi-indirect luminaries, it is good practice to aim at a horizontal spacing between rows approximately equal to a height of the ceiling above the working plane, and in no case should the horizontal spacing exceed $1\frac{1}{2}$ times this height.

In the case of fluorescent luminaries, it is a common practice to join two or more luminaries end to end so that they can share a common outlet. In fact it often works out well to use continuous rows of luminaries, especially when the specified illumination is fairly high.

8. *Colour of Surrounding Walls.* The illumination in any room depends upon the light reflected from the walls and ceilings. White walls and ceilings reflect more light as compared to coloured ones.

14.10. FACTORY LIGHTING

Adequate lighting of factories is of vital importance, as it provides improved amenities for the employees, increased production and has a definite economic value in reducing accidents with their consequent loss of time and compensation payments.

General Requirements and Types of Installations. A factory lighting installation, in common with other indoor equipments should provide an adequate illumination on the working plane and give a good distribution of light, employ simple and easily cleaned fittings and avoid glare. It is essential not only to avoid glare from the lamp itself but also reflected glare from any polished surface, which may be within the line of vision.

General Lighting. The usual scheme in factories and workshops is to mount a number of lamps at a sufficient height so that uniform distribution of light over the working plane is obtained. In large machine shops the height is governed by the necessity of keeping the lamps above the travelling crane. In such cases it is often desirable to supplement the main lighting by side lighting in order to give additional illumination on a vertical plane. Since light coloured walls and ceiling add to the effectiveness of an installation, it is necessary to get white washing or painting done.

Local Lighting. On some points fairly intense illumination is required. For this purpose local lighting can be provided by means of adjustable fittings attached to the machine or bench in question or mounted on portable floor standards. Such lamps should be mounted in deep reflectors so that glare is avoided.

Portable hand lamps attached to wall plugs by means of trailing leads are used for maintenance work and emergency lighting. Low voltage lamps of not more than 50 volts are recommended for use as portable hand lamps because such lamps have thicker filament and are, therefore, more robust than those for normal voltage and danger of shock is also avoided in these few volt lamps. The supply for such lamps can be obtained from a special low voltage distribution system running through out the factory or by means of a small transformer for each individual lamp.

Local lighting should never be employed alone, good general lighting is essential so that the dark places between the local lighting units are avoided. Dark places between the local lighting units cause fatigue to the eyes on account of its continually having to adjust itself to new conditions.

Emergency Lighting. Some lights, such as for (*i*) internal pilot lighting required for safe and speedy evacuation of personnel after main lighting circuit is off (*ii*) external pilot lighting, provided with careful shades leading to shelters required for evacuation of personnel (*iii*) for control posts, first aid centres etc. (*iv*) dials and gauges in important plants required to be watched regularly are required during an air-raid when all the factory lights are off as a matter of air-raid precaution. The circuit supplying the above emergency lights should be independently controlled. It is very desirable to provide auxiliary lighting from the source other than the main electric supply preferably from batteries or from small petrol driven generator set. If, however, emergency lighting circuits are operated from main electric supply, these should be completely separated from main lighting circuit.

14.10.1. Industrial Lighting Fittings. Reflectors for industrial purposes must be simple in design and easily cleaned. The requirements of most of the installations can be met by one of the following types of fittings.

Standard Reflectors. These reflectors are made to accommodate lamps of ratings from 40 to 1,500 watts and designed so that they give adequate and uniform illumination when they are mounted at a spacing equal to about 1.5 times their mounting height above the working plane.

Diffusing Fittings. When more diffused light is required than that given by the standard reflector a diffusing glass screen may be fixed across a standard type of reflector. Such fittings are used where highly polished articles are dealt with.

Concentrating Reflectors. A reflector with a concentrated beam is employed in large machine shops and foundaries, where the fittings are to be mounted on a considerable height above the working plane. In such places an ordinary reflector would have too wide angle of divergence and would waste a great deal of light on the walls.

Enclosed Diffusing Fittings. An opal globe completely enclosing the lamp giving a very even and well diffused light is used when light coloured walls and ceiling are there.

Angle Reflectors. Angle reflectors are used to provide illumination in a vertical plane where concentrating type reflectors are used. These can be mounted on suitable stanchions or the walls.

14.11. METHODS OF LIGHTING CALCULATIONS

A number of methods have been employed for lighting calculations, among which may be mentioned :

1. Watts per square metre method.
2. Lumen or light flux method.
3. Point to point or inverse-square law method.

1. *Watts Per Square Metre Method.* This is principally a "rule of thumb" method, very handy for rough calculation or checking. It consists in making an allowance of watts per square metre of area to be illuminated according to the illumination desired on the assumption of an average figure of overall efficiency of the system.

2. *Lumen or Light Flux Method.* This method is applicable to those cases where the sources of light are such as to produce an approximate uniform illumination over the working plane or where an average value is required.

From the size of lamp or lamps employed and from their efficiency total lumens output are determined. Multiplying the total lumens output from the source by coefficient of utilisation, the lumens received on the working plane are determined. If the lamps and surroundings are not perfectly clean, then in determination of lumens received on working plane, the depreciation factor or maintenance factor should be included, *i.e.*,

Lumens received on working plane = Number of lamps × wattage of each lamp × efficiency of each lamp in terms of lumens/watt × $\frac{\text{coefficient of utilisation}}{\text{depreciation factor}}$

Also Lumens received on working plane = Number of lamps × wattage of each lamp × efficiency of each lamp in terms of lumens per watt × coefficient of utilisation × maintenance factor.

Coefficient of Utilisation or Utilisation Factor. The whole light radiated by the lamps does not reach the working plane. The ratio of lumens reaching the working plane to the total lumens given out by the lamp or lamps, is known as *utilisation factor* or *coefficient of utilisation*. Higher the value of utilisation factor, more lumens will reach the working plane for the given lumens output of the lamps. The value of utilisation factor depends upon (*i*) the mounting height of lamps—utilisation factor decreases with the increase in mounting height of lamps (*ii*) area to be illuminated—for given height, proportion of direct light becomes more and more if floor area increases, *i.e.*, utilisation factor increases with the increase in area to be illuminated (*iii*) type of lighting—more for direct lighting and low for indirect lighting and (*iv*) colours of surroundings etc.—more for light colours and less for dark colours. Its value varies from 0.25 to 0.5 and from 0.1 to 0.25 for direct and indirect lighting schemes respectively.

Maintenance Factor. The illumination produced by a lighting installation is considerably less after a year or two of use than it was initially. The loss is due partly to the ageing of the lamps and partly to the accumulation of dust on the lamps, on the reflecting and transmitting surfaces of the fixtures and on the ceiling and walls. This fact is taken into account by including the maintenance factor, which is defined as the ratio of the ultimate maintained metre-candles on the working plane to the initial metre-candles. Its value is more if the lamp fittings are cleaned regularly, say 0.8, and less if there is much dust etc. say 0.6.

Depreciation Factor. This is merely the inverse of the maintenance factor and is defined as the ratio of the initial metre-candles to the ultimate maintained metre-candles on the working plane. Its value is more than unity.

3. *Point to Point or Inverse-Square Law Method.* This method is applicable where the illumination at a point due to one or more sources of light is required, the candle powers of the sources in the particular direction under consideration being known.

If a polar curve of lamp and its reflector giving candle powers of the lamp in different directions is known, the illumination at any point within the range of the lamp can be calculated from the inverse square law. If two and more than two lamps are illuminating the same working plane, the illumination due to each can be calculated and added. This method is not much used because of its complicated and cumbersome applications. It is employed only in some special problems, such as flood lighting, yard lighting etc.

Example 14.1

A room of size 15 × 6 metres is to be illuminated by twenty 200 W lamps. The mscp of each lamp is 250. Assume a depreciation factor 1.2 and utilisation factor 0.6. Find the average illumination produced on the floor.

Solution: Total area to be illuminated, $A = 15 \times 6 = 90 \text{ m}^2$

Total lumens given out by lamps, ϕ = Number of lamps × mscp of each lamp × 4π

$$= 20 \times 250 \times 4\pi = 62{,}832 \text{ lumens}$$

$$\text{Lumens reaching on the working plane} = \frac{\phi \times \text{utilisation factor}}{\text{depreciation factor}} = \frac{62{,}832 \times 0.6}{1.2} = 31{,}416 \text{ lumens}$$

$$\text{Illumination on the working plane, } E = \frac{\text{Lumens reaching on the working plane}}{\text{Total area to be illuminated}} = \frac{31{,}416}{90} = 350 \text{ lux.}$$

Example 14.2

An illumination on the working plane of 75 lux is required in a room 72 m × 15 m in size. The lamps are required to be hung 4 m above the work bench. Assuming a suitable space-height ratio, a utilisation factor of 0.5, a lamp efficiency of 14 lumens per watt and a candle power depreciation of 20%, estimate the number, rating and disposition of lamps.

Solution : Area to be illuminated, $A = 72 \times 15 = 1{,}080 \text{ m}^2$

Illumination required, $E = 75$ lux

Total lumens required $= A \times E = 1{,}080 \times 75 = 81{,}000$

Utilisation factor $= 0.5$

Maintenance factor $= 1 -$ candle power depreciation $= 1 - 0.2 = 0.8$

$$\text{Gross lumens required, } \phi = \frac{A \times E}{UF \times MF} = \frac{81{,}000}{0.5 \times 0.8} = 202{,}500$$

$$\text{Total wattage required} = \frac{\phi}{\text{Lamp efficiency}} = \frac{202{,}500}{14} = 14{,}464$$

80 lamps in 4 rows, each row having 20 lamps, can be used giving spacings of 3.6 metres in length and 3.75 metres in width and space-height ratios of 0.9 and 0.9375 respectively.

$$\text{Wattage of each lamp} = \frac{14{,}464}{80} = 180.8 \simeq 200 \text{ W (say)}$$ **Ans.**

Disposition of lamps is shown in Fig. 14.45.

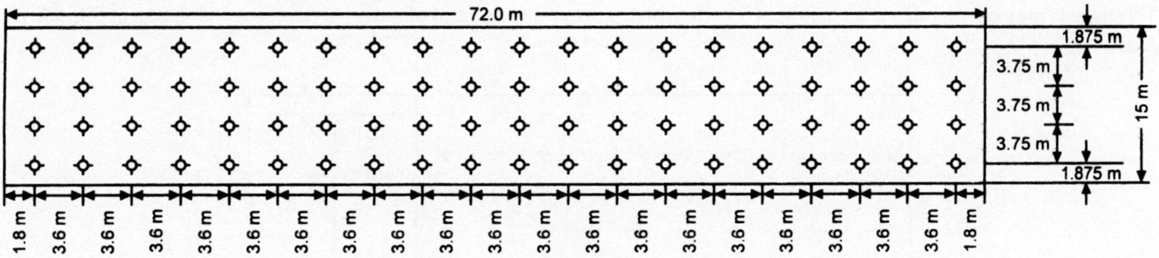

Fig. 14.45

Example 14.3

Estimate the number and wattage of lamps which would be required to illuminate a workshop space 60 × 15 metres by means of lamps mounted 5 metres above the working plane. The average illumination required is about 100 lux, coefficient of utilisation = 0.4, luminous efficiency 16 lumens per watt. Assume a space-height ratio of unity and a candle power depreciation of 20%.

Solution: Gross lumens required,

$$\phi = \frac{\text{Area to be illuminated} \times \text{average illumination required}}{\text{Coefficient of utilisation} \times \text{maintenance factor}}$$

$$= \frac{A \times E}{UF \times MF} = \frac{60 \times 15 \times 100}{0.4 \times (1 - 0.2)} = 281{,}250$$

$$\text{Total wattage required, } W = \frac{\phi}{\text{Luminous efficiency in lumens/watt}} = \frac{281{,}250}{16} = 17{,}578 \text{ watts}$$

For a space-height ratio unity, only three lamps can be mounted along the width of the room and 12 lamps can be arranged along the length of the room. Thus total number of lamps required = 12 × 3 = 36 **Ans.**

$$\text{Wattage of each lamp} = \frac{\text{Total wattage required}}{\text{Number of lamps}} = \frac{17{,}578}{36} = 488.3$$

We will take the nearest standard lamp of 500 watts. Location of the lamps is shown in Fig. 14.46.

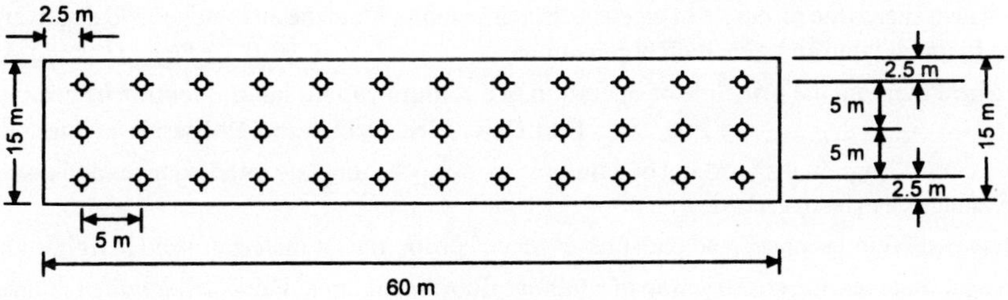

Fig. 14.46

Example 14.4

A hall 30 metres by 15 metres with a ceiling height of 5 metres is to be provided with a general illumination of 120 lumens/m². Taking a coefficient of utilisation of 0.5 and depreciation factor of 1.4, determine the number of fluorescent tubes required, their spacing, mounting height and total wattage. Take luminous efficiency of fluorescent tube as 40 lumens per watt for 80 watt tubes.

Solution: Gross lumens required,

$$\phi = \frac{A \times E \times DF}{UF} = \frac{30 \times 15 \times 120 \times 1.4}{0.5} = 151{,}200$$

$$\text{Total wattages required} = \frac{151{,}200}{40} = 3{,}780 \text{ watts} \quad \textbf{Ans.}$$

$$\text{Number of tubes required} = \frac{\text{Total wattage required}}{\text{Wattage of each tube}} = \frac{3{,}780}{80} = 47.25 \cong 48 \text{ (say)}$$

50 tubes in 5 rows, each row having 10 tubes, can be used giving spacings of 3 metres in length as well as in width. These tubes may be mounted at an height of 3 metres giving space-height ratio of unity. Arrangement is shown in Fig. 14.47. **Ans.**

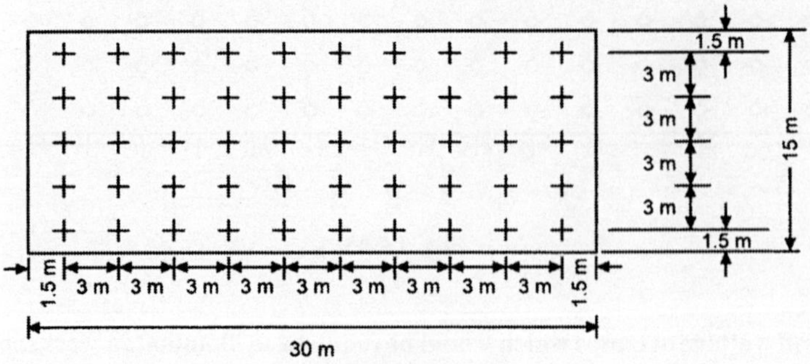

Fig. 14.47

EXERCISES

1. Explain the basic nature of light.
2. Define and explain the following terms:
   (*i*) Luminous flux (*ii*) Luminous intensity (*iii*) Illumination (*iv*) Lux or metre candle (*v*) Lamp efficiency (*vi*) Specific consumption (*vii*) Space-height ratio (*viii*) Utilisation factor (*ix*) Maintenance factor (*x*) Depreciation factor.
3. State and explain laws of illumination. **[Pb. Univ. Elec. Drives and Utilisation of Elec. Energy August 1984]**
4. Describe the construction and working of a filament lamp. Compare it with fluorescent lamp.
5. What do you understand by discharge lamp ? Explain the construction and working of high pressure mercury vapour lamp.
6. Discuss the effects of voltage variation on the life and illumination as regards tungsten filament lamp and fluorescent lamp.
7. Describe with a neat sketch the principle of electric discharge lamp. State the advantages and disadvantages of discharge lamps over the filament lamp and give their applications. **[B.H.U. Elec. Traction & Utilisation 1978]**
8. Explain with a neat diagram the principle of operation of a sodium vapour lamp. Mention its use. **[Pb. Univ. Elec. Drives and Utilisation of Elec. Energy April 1982]**
9. Describe with a circuit diagram the working of a fluorescent lamp. Enumerate its advantages and disadvantages. Explain how its disadvantages can be overcome ?
10. Distinguish between "fluorescence" and phosphorescence" properties of materials used in lighting equipments.
    Describe with neat sketches, the construction of a tubular fluorescent lamp. Draw a diagram of connections and explain the action of tube (*i*) at starting (*ii*) when operating normally.
    Compare the spectrum of such a tube lamp with that of (*i*) tungsten filament lamp (*ii*) mercury discharge lamp, and explain why the luminous efficiency of the fluorescent lamp is so much higher than that of the tungsten filament lamp. **[Gorakhpur Univ. Utilisation of Elec. Power & Traction Sept., 1983]**
11. Explain the working of a fluorescent tube with the help of the circuit diagram giving the function of various parts. How stroboscopic effect is eliminated in fluorescent tube lighting ? **[Gorakhpur Univ. Utilisation of Elec. Power & Traction 1981]**
12. Compare the merits of alternative lighting schemes using incandescent lamps, sodium vapour lamps or fluorescent lamps for industrial lighting. **[Gorakhpur Univ. Utilisation of Elec. Power & Traction 1982]**
13. What do you understand by direct, indirect, and semi-indirect lighting. **[Agra Univ. Traction and Utilisation of Elec. Power, 1982]**
14. State and describe various types of lighting schemes. **[Pb. Univ. Elec. Drives and Utilisation of Elec. Energy April, 1981]**

15. Describe different types of switches, ceiling rose, socket-outlets, plugs and lamp holders used in house wiring.

16. Describe different types of light fittings and reflectors.

17. Describe various lighting fittings used for indoor lighting giving special applications of each one of them.

18. Describe in brief the requirements of good lighting. Enumerate the factors to be considered while designing an indoor lighting scheme.

19. Explain what is meant by 'point to point method' of lighting calculations.

20. Write short notes on the following:

(*i*) Fluorescent tubes.

(*ii*) CFL.

(*iii*) Different types of reflectors.

(*iv*) Workshop lighting.

SHORT ANSWER TYPE QUESTIONS WITH ANSWERS

Q. 1. What are the advantages of artificial lighting produced electrically ?

Ans. Artificial lighting produced electrically have the advantages of cleanliness, ease of control, reliability, steady output as well as low cost.

Q. 2. What is difference between illumination and light ?

Ans. Light is the cause and illumination is the result of that light on surfaces on which it falls.

Q. 3. What is relation between lux and lumens/m^2 ?

Ans. Lux and lumens/m^2, both, are units of illumination. 1 lux = 1 lumen/m^2

Q. 4. Differentiate between lamp efficiency and specific consumption.

Ans. Lamp efficiency is the ratio of the luminous flux in lumens to the power input in watts while the specific consumption is the ratio of the power input in watts to the average candle power in candela.

Q. 5. What is difference between maintenance factor and depreciation factor ?

Ans. Maintenance factor is defined as the ratio of illumination under normal working conditions to the illumination when every thing is perfectly clean. Depreciation factor is merely inverse of the maintenance factor.

Q. 6. What are the two laws of illumination ?

Ans. (*i*) Law of inverse squares (*ii*) Lambert's cosine law.

Q. 7. Why is tungsten selected as the filament material ? **[Agra Univ. 1983]**

Ans. Nowadays tungsten is the most commonly used metal for filament due to its high melting point (3,400°C), high resistivity, low temperature coefficient (0.0051), low vapour pressure, being ductile and mechanical strong to withstand vibration during use.

Q. 8. Why does the efficiency of a filament lamp increase with the increase in operating voltage ?

Ans. The efficiency (lumens/watt) of a lamp increases with the increase in operation voltage owing to increase in temperature and is proportional to the square of the operating voltage.

Q. 9. Why sodium vapour discharge lamps are not used for general lighting ? **[Gorakhpur Univ. 1981]**

Ans. Sodium vapour discharge lamps are not used for general lighting because they have got drawback of colour discrimination.

Q. 10. Why thermal type starter for use in fluorescent tubes have become obsolete now-a-days ?

Ans. Because of more complicated construction, greater cost and greater power loss.

PROBLEMS

1. Find the illumination due to 20 bulbs of 200 W in a room of 15 m × 8 m assuming following data. Coefficient of utilisation 0.6, Depreciation factor 1.2, MSCP of 200 W lamp is 250. [**Ans.** 262 lux]
2. It is required to provide an illumination of 100 lux in a factory hall 30 m × 12 m. Assume that the depreciation factor is 0.8, the coefficient of utilisation 0.4, and the efficiency of proposed lamps 11 lumens per watt. Calculate the number of lamps and their disposition. [**Ans.** 40 lamps of 200 W each in 4 rows, each row having 10 lamps]
3. A minimum illumination of 80 lumens/m^2 is required in the factory shed of 50 m × 12 m. Calculate the number, the location and wattage of the units to be used. Assume that the depreciation factor is 0.8, coefficient of utilisation is 0.4 and the efficiency of lamp units is 14 lumens/watt. [**Ans.** 36 lamps of 300 watts in 3 rows of 12 lamps in each row]
4. A uniform illumination of 50 lux is to be obtained on the floor of a room measuring 10 m × 10 m by arranging electric lights suitably. The system to be employed should be shadowless and glare free. Calculate the number of lamps and watt rating of each lamp if the efficiency of lamp is 15 lumens/watt. Assume suitable values of coefficient of utilisation and depreciation. Indicate the layout. [**Ans.** 4 lamps; each of 250 W rating]

❑❑❑

UNIT-IX
Safety and Protection

Contents in This Unit

Safety and Protection: Safety; Electric shock; First aid for electric shock; Other hazards of electrical laboratories and safety rules; Use of multimeters; Grounding; Importance of grounding; Equipment of grounding for safety; Circuit protection devices, fuses, MCB, ELCB and relays.

Chapter 15

Safety and Protection

15.1. INTRODUCTION

This is an age of electricity. Electrical equipments are used by million of people either in their homes or in their offices/factories. The fact is that, none can live without electricity in this age. But unfortunately every year a large number of casualties occur owing to electrical accidents. In most cases, accidents occur due to lack of some basic knowledge of electricity to the users, faulty installations, poor quality equipments, violation of existing regulations and sometimes environments. Though electricity has contributed a lot towards our modern civilization, its slight mishandling may cause even fatal accidents. In short it can be said that "electricity is a good servant but bad master". Hence, we must be very much careful in handling electricity and every attempt should be made to prevent such accidents.

Accidents in an electrical installation work may occur in the following ways:

(*i*) Touching of live wires or other current carrying parts resulting in electric shock.

(*ii*) Touching of metal enclosures of electrical equipment which may have become alive resulting in electric shock.

(*iii*) Shorting of high voltage winding of a transformer to its low voltage winding causing electric shock.

(*iv*) Coming nearly the intense heat of an electric arc resulting from a fault causes burning.

(*v*) Handling of poor quality tools an improper way.

(*vi*) Improper handling of toxic and poisonous substances which are often used in the impregnation of electrical machine windings causes slow poisoning.

(*vii*) Oily floor of the workshop/laboratory premises causes slipping.

(*viii*) Use of two-pin plug may cause electric shock.

(*ix*) Careless movement inside the workshop/laboratory causes both mechanical and electrical injury.

(*x*) Violation of regulations specially regarding grounding results in electric shock.

15.2. ELECTRICAL SAFETY MEASURES

The lowest recorded voltage at which death occurred from an electric shock is 38 volts. In general, 240 volts seldom kills a fully dressed person wearing dry footwear (but may do so); but it is a fatal voltage for any one wearing damp leather shoe or perhaps with bare feet standing on a damp floor or touching earthed metal.

However, if the work is undertaken with a full sense of responsibility, proper materials and proper methods are employed as laid down in IS 732-1963 (Code of Practice for Electrical Wiring Installations), IS 4648-1968 (Guide for Electrical Layout in Residential Buildings) and Indian Electricity Rules, 1956, there is no reason why an economical and safe installation should not result.

The safety of electrical appliances and wiring is ensured, basically, three ways. Firstly, by insulation; secondly, by earthing; and thirdly, by providing a fuse wire in the circuit or sub-circuit.

Insulation. Covering of live electrical wires or other equipment preventing completely any one to come in contact with live metal is called the insulation action. For example, if right type of cable is used, there can be no danger from touching the outside of the insulating covering. Appliances of all kinds should have the live parts completely encased in porcelain or plastic materials so that even the child may not be able to insert its finger into any part that is alive.

Earthing. This is the second line of defence. It provides safety to the human beings against electric shocks. For example, in case of an appliance, whose metallic parts have been earthed, if any metallic part comes in direct contact with a bare or live wire or breakdown occurs, immediately the charge will be transferred to the earth and the user will be safe. Metallic parts of each and every appliance, except double insulated appliances, should always be earthed. With portable appliances, this is carried out by using a 3-core flexible cable and a 3-pin plug and socket outlet, the third pin of which, called the earth pin, is properly connected to earth.

Fuse. This is a third line of defence. To protect the cable, current carrying equipment and current consuming appliances connected in a circuit from damage owing to continuous flow of excessive current, that may occur due to overload, short-circuit fault or heavy earth fault, fuse of correct size and type is provided in circuit. Flow of excessive current in a circuit may also result in break out of fire hence protection against fire risk is also secured by providing the fuse in the circuit.

15.3. EFFECTS OF ELECTRIC CURRENTS ON HUMAN BODY

Bruner (1967) states that the threshold of perception of electric shock is about 1 mA. At this level a tingling sensation is felt by the subject when there is a contact with an electrified object through intact skin. With the increase in magnitude of ac, the sensation of tingling gives way to contraction of muscles. The muscular contractions increase as the current is increased and finally a value of current is reached at which the subject cannot release his grip on the current carrying conductor. The maximum current at which the subject is still capable of releasing a conductor by using muscles directly stimulated by the current is called "*let go current*". The value of this current is significant because an individual can withstand, without serious after effects, repeated exposures to his 'let go current' for at least the time required for him to release the conductor. Also, currents slightly in excess of 'let go current' would not permit the individual to release his grip from the conductor supplying current.

Based on the experiments conducted on males and females, it is generally accepted that the safe 'let go current' could be taken approximately 9 mA and 6 mA for men and women respectively.

At current levels higher than the 'let go current' the subject loses ability to control his own muscle actions and he is unable to release his grip on the electrical conductor. Such currents are very painful and hard to bear. This type of accident is called 'hold-on-type' accident, and is caused by currents in the range 20-100 mA. These currents may also cause physical injury due to powerful contraction of the skeletal muscles. However, the heart and respiratory function usually continue because of uniform spread of current through the trunk of the body.

If current contacts contact skin and passe through the trunk, at about 100 mA and above, there is a likelihood of pulling the heart into ventricular fibrillation. In this condition, the rhythmic action of the heart ceases, pumping action stops and the pulse disappears. Ventricular fibrillation is a serious cardiac emergency because once it starts, it practically never stops spontaneously. It proves *fatal* unless corrected within minutes, since the brain begins to die 2 to 4 minutes after it is robbed of its supply of oxygenated blood.

At very high currents of the order of 6 A and above, there is a danger of temporary respiratory paralysis and also of serious burns. However, if the shock duration is of only a very few seconds, there is a possibility of heart reverting to the normal rhythmic action.

The threshold of perception depends largely on the current density in the body tissues. It may vary widely depending upon the size of the current contact. At very small point contact, it is probable that even

0.3 mA current may be felt whereas a current in excess of perhaps 1 mA may not produce sensation if the contacts are somewhat larger. Similarly, depending on the size of contact, the threshold of pain may also be considerably above 1 mA, probably 10 mA if the contacts are large enough.

Besides the magnitude of current, the current duration and the relationship of current flow resistance are also important. Duration of less than 10 ms typically does not produce fibrillation whereas duration of 0.1 s or longer does. It has been found experimentally that the safe value of current in amperes (rms) which a human body can tolerate is given as

$$I = \frac{0.165}{\sqrt{t}} \text{ for } t < 3 \text{ s and } I = 9 \text{ mA for } t > 3 \text{ s.}$$

where t is the time duration in seconds of the flow of current.

The tolerable currents mentioned above are for power-frequency currents. It has been found that human body can tolerate about 5 times higher direct current. At high frequencies (3-10 kHz) still higher currents are tolerable.

15.4. TREATMENT FOR ELECTRIC SHOCK

Electric shock may be due to (*i*) contacts with parts normally alive such as ht and lt lines (*ii*) contacts with parts of the electric equipment, which are normally dead but have become alive owing to some fault, such as outer-casing of electrical machines and appareatus.

Effects of Electric shock. Victims of electric shock become unconscious and often show no sign of life (there is no breathing , the heart ceases to pump blood).

If the shock is mild, the victim will remain unconscious for a few seconds and recover with out any aid. But in case of severe shock the victim requires immediate, energetic and skilled aid, otherwise there will be a loss of human life.

In some cases of electric shock, the electric current passes through the breathing centre and causes this centre to stop sending out impulses to the breathing muscles. Consequently breathing stops abruptly. This is why artificial respiration must be applied to the patient as substitute for natural respiration.

First Aid Treatment. The victim of electric shock must be immediately freed from contact with current and given first aid.

Breaking of Contact. Cut off the supply or disconnect the victim from supply, by any means available. If the main switch is close by, switch it off at once. If not, ask some one to contact the substation supplying the area to cut off the supply temporarily in person or on telephone.

If the equipment cannot be switched off at once, break the victim's contact with the live conductor or part as follows:

In case of low voltage supply make use of a dry stick, dry board, dry rope or dry coat or other non-conductor. In no case metallic parts or wet things may be used. In case the clothes of the victim are dry or loose, he may be freed by pulling at his clothing. The victim may be pulled by his legs only if your hands are adequately protected.

For insulation from earth or from the victim's body, the rescuer should put on rubber shoes or stand on a dry board or on any other dry and non-conductor material; put on rubber gloves or wrap the hands in rubberized or dry fabric.

When the current flows to earth through the victim's body the contact may be broken by lifting the victim off ground, while taking all the precautions mentioned above.

Where necessary, the live conductor must be cut with an axe having a dry wooden handle or with an insulated cutting tool, with the rescuer reliably insulated from earth. Where several conductors are involved, they must be cut one by one.

In case of high voltage supply and if there is no alternative to break the contact of the victim from the line or equipment, short-circuit the lines wires.

Send for a doctor and if the victim is unconscious, start giving artificially respiration immediately.

Methods of Artificial Respiration. There are several methods for artificial respiration, any one may be applied.

Method 1. (*a*) Place the victim on his stomach with his face to one side, as shown in Fig. 15.1. Free the neck from clothing and see that the passage to his throat is clear. Remove all foreign materials such as false teeth, tobacco etc. if any from the mouth.

Fig. 15.1

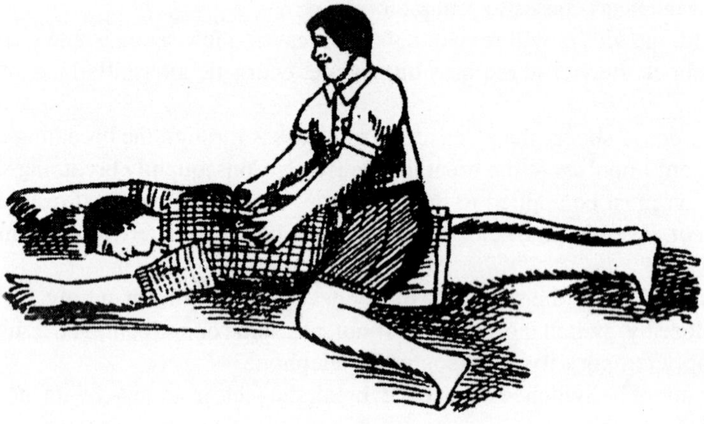

Fig. 15.2

(*b*) Kneel over the victim's back and place both the hands on his back near the lowest rib in such a way that the fingers are spread outwards while the two thumbs almost touch each other and are parallel to the spine.

(*c*) With arms straight lean forward gradually over the patient bringing the weight of your body to bear on the patient for about one second (refer to Fig. 15.1)

(*d*) Relax the pressure slowly and come to the original kneeling position without lifting your hands from the victim (refer to Fig. 15.2)

(*e*) repeat the process described in (*c*) and (*d*) at the rate of about 15 times per minute until natural breathing is established.

Method 2. In case the victim has got burns or some other injury on his chest or on front side, such as due to fall from an height, an alternative method, as described below, may be adopted for artificial respiration.

(*a*) Loosen the clothes around the chest and stomach. Place a pillow or rolled coat under the victim's shoulders so that his head falls backward. The tongue is then drawn forward.

(*b*) Kneel in the position shown in Fig. 15.3. Grasp the victim just below the elbows and draw his arms over his head until horizontal. Retain them there for about 2 seconds.

Fig. 15.3

(*c*) Next bring the patient's arms down on each side of his chest pressing inwards upon leaning on his arms so as to compress his chest. Remain in this position for about 2 seconds.

(*d*) Repeat (*b*) and (*c*) at the same rate.

Mouth To Mouth Method. This method has become more popular nowadays. In this method lay the patient on his back, be sure that his passage is clear, wipe visible foreign matter out of victim's mouth.

Fig. 15.4

Put one hand under victim's neck while with other hand tilt his head back and down as far as possible so that chin points upward.

Now place your open mouth tightly over victim's mouth and pinch his nostrils shut. Blow vigorously into victim's mouth to expand his chest. Remove your mouth to let returning air escape. Repeat inflations every 3 or 4 seconds.

To avoid direct contact it is preferable to blow through a handkerchief placed over the patient mouth and nose.

This method supplies about 10–12 times more volume of air into the patient lungs than any other method.

Important Points: 1. Do not stop artificial respiration immediately after the patient regains his respiration.

2. Do not give any drink until the patient is conscious.

3. Keep the patient warm with blankets or coats and apply hot water bottles, if available, to his feet.

4. If the patient has got burns, only the burns should be dressed properly but oil should, never be used on the burns.

5. Whatever the method of artificial respiration, you use, apply it gently and without violence.

15.5. EARTHING

Earthing means connections of the neutral point of a supply system or the non-current carrying parts of electrical apparatus, such as metallic frame work, metallic covering of cables, earth terminal of socket outlet, stay wires etc., to the general mass of earth in such a manner that at all times an immediate discharge of electrical energy takes place without danger.

Earthing is provided

1. to ensure that no current carrying conductor rises to a potential with respect to general mass of earth than its designed insulation,

2. to avoid electric shock to the human beings, and

3. to avoid risk of fire due to earth leakage current through unwanted path.

In an electric installation, if a metallic part of an electric appliance comes in direct contact with a bare or live wire (that may be due to failure of insulation or otherwise) the metal being a good conductor of electricity is charged and static charge on it will accumulate. Now if any person comes in contact with this charged metallic part, he will get a severe shock. But if the metallic parts of the appliances are earthed, the charge will be transferred to the earth immediately, as the metallic part comes in direct contact with a bare or live wire or breakdown occurs. And as the discharge takes place to earth, the impedance of path of the current is low, a large amount of current flows to earth, the instant, the current exceeds the limiting value, the fuse provided in the circuit will blow off and cut off the appliance from supply. Thus earthing of metallic parts of electrical equipment and appliances provides safety.

15.6. IS SPECIFICATIONS REGARDING EARTHING OF ELECTRICAL INSTALLATIONS

The various important specifications regarding earthing as recommended by Indian Standards are given below:

1. Distance of Earth From Building. An earthing electrode shall not be situated within a distance of 1.5 m from the building whose installation system is being earthed.

2. Size of Earth Continuity Conductor. The conductor, by means of which the metal body of an equipment or appliance is connected to the earth, is known as *earth continuity conductor* (ECC). The earth continuity can be ensured either through metal conduit, metal sheathing of metal sheathed cables or by a special earth continuity conductor. The cross-section of earth continuity conductor should not be either less than 2.9 mm^2 (14 SWG) or half of the installation conductor size.

3. Resistance of Earth. There is no hard and fast rule. The main principle regarding earth resistance is that the earth resistance should be low enough to cause flow of current sufficient to operate the protective relays or blow fuses, in the event of an earth fault. The value of earth resistance does not remain constant but changes with the weather, as it depends upon the moisture content of the soil, and is maximum during dry season. As a general rule the lower the value of earth resistance better it is but even then the following values of earth resistance (maximum permissible values) will give satisfactory results.

Large power station – 0.5 Ω
Major power station – 1.0 Ω
Small substation – 2.0 Ω
In all other cases – 5 Ω Maximum

Earth continuity inside an installation *i.e.* from the earth plate to any point in the installation – 1.0 Ω.

The sensitivity of the protective equipment, system voltage and the maximum fault current directly relate to permissible value of earth resistance. In case the earth resistance exceeds the permissible value, then in the event of earth fault, the fault current may not reach a sufficient value to operate the protective equipment (such as fuses or relays) and dangerous condition may arise.

4. The earth wire and earth electrode will be of same material.

5. The earth wire shall be taken through GI pipe of 13 mm diameter for at least 30 cm length above and below ground surface to the earth electrode to protect it against mechanical damage.

6. It is not necessary that earth wire connected to an earth electrode is run along the whole wiring system. All the earth wires run along the various sub-circuits shall be terminated and looped firmly at the main board and from main board the main earth wire shall be taken to the earth electrode. The loop earth wires used shall not be either less than 2.9 mm^2 (14 SWG) or half of the size of the sub-circuit conductor.

7. The earthing electrode shall always be placed in vertical position inside the earth or pit so that it may be in contact with all the different earth layers.

15.7. POINTS TO BE EARTHED

According to IE Rules and IEE Regulations :

(*i*) Earth pin of 3-pin lighting plug sockets and 4-pin power plug sockets should be permanently and efficiently earthed.

(*ii*) All metal casings or metallic coverings containing or protecting any electric supply line or apparatus, such as iron clad switches, iron clad distribution fuse boards, GI pipes and conduits enclosing VIR or PVC cables, the down rods of electric fans, etc; should be connected to earth.

(*iii*) The metal casings of portable apparatus such as heaters, refrigerators, hand lamps, soldering irons, electric drills etc., should be connected to earth. If any one of them is installed in a fixed position, a separate direct connection to the earth should be provided in addition to the earth wire in the connecting cable.

(*iv*) The frame of every generator, stationary motor, and as far as possible, portable motor and the metallic parts (not intended as conductors) of all transformers and any other apparatus used for regulating or controlling energy and all medium voltage energy consuming apparatus should be earthed by two separate and distinct connections with earth.

(*v*) The neutral conductor of a 3-phase, 4-wire system and the middle conductor of a 2-phase, 3-wire system should be earthed by not less than two separate and distinct connections with earth at the generating station and at the substation. It may also be earthed at one or more points along the distribution system or service line in addition to any connection with earth which may be at the consumer's premises.

(*vi*) In the case of a system comprising electric supply lines having concentric cables, the external conductor of such cables should be earthed by two separate and distinct connections with earth.

(*vii*) In a dc 3-wire system the middle conductor should be earthed at the generating station.

(*viii*) Fabricated steel transmission line towers, tubular steel or rail poles carrying overhead conductors should be earthed. For this purpose a continuous earth wire is provided and connected with earth at four points in every mile (1.61 km), the spacing between the points being as nearly equidistant as possible, alternatively, the metal work should be connected to an effective earthing device at each individual support.

(*ix*) Stay wires provided for overhead lines should be connected to earth by connecting at least one strand to the earth wire.

15.8. EARTH ELECTRODE AND EARTHING LEAD

Any wire, pipe, rod or metal plate embedded in earth for the purpose of making an effective connection with the general mass of earth is known as *earth electrode.*

The wire which connects overhead earth wire or any other apparatus to be earthed to the earth electrode is known as *earthing lead.*

According to ISI specifications :

1. The cross-section of the earthing lead, as a general rule, should not be less than half of the section of the main supply feeding the installation.

2. The cross-section of the earthing lead should, in any case, be such that it shall never fuse out even under the worst fault conditions.

3. (*i*) The cross-section of earth lead should not be less than 161 mm^2 for main connections and 64 mm^2 for branch connections in large installations.
(*ii*) Copper strip of 25 mm × 3 mm or 25 mm × 6 mm are usually employed as a ring main for connecting all the electrical apparatus to the earth.
(*iii*) Copper wire of size not less than 8 SWG (12.9 mm^2) and not more than 7/0.0915 mm (64.52 mm^2) are used for earthing in case of smaller installations.
(*iv*) The size of the continuous earth wire used with cables in domestic installations should not be either less than 2.9 mm^2 (14 SWG) or half of the installation conductor size.

15.9. METHODS OF EARTHING

The various methods of earthing are :

1. Strip or Wire Earthing. In this system of earthing strip electrodes of cross-section not less than 25 mm × 1.6 mm if of copper and 25 mm × 4 mm if of galvanised iron or steel are buried in horizontal trenches of minimum depth 0.5 metre. If round conductors are used, their cross-sectional area shall not be smaller than 3.0 mm^2 if of copper and 6 mm^2 if of galvanised iron or steel. The length of buried conductor shall be sufficient to give the required earth resistance. It shall, however, be not less than 15 metres. The electrodes shall be as widely distributed as possible, preferably in a single straight or circular trench or in a number of trenches radiating from a point. If conditions require use of more than one strip, they shall be laid either in parallel trenches or in radial trenches.

This type of earthing is used at places which have rocky soil earth bed because at such places excavation work of plate earthing is difficult.

2. Rod Earthing. In this system of earthing 12.5 mm diameter solid rods of copper or 16 mm diameter solid rods of galvanised iron or steel or hollow section 25 mm GI pipes of length not less than 2.5 metres are driven vertically into the earth either manually or by pneumatic hammer. In order to increase the embedded length of electrodes under the ground, which is sometimes necessary to reduce the earth resistance to desired value, more than one rod sections are hammered one above the other.

This system of earthing is suitable for areas which are sandy in character. This system of earthing is very cheap as no excavation work is involved.

3. Pipe Earthing. This is the most common and best system of earthing as compared to other systems suitable for the same earth and moisture conditions.

In this method of earthing, a galvanised steel and perforated pipe of approved length and diameter is placed up right in a permanently wet soil, as shown in Fig. 15.5.

The size of the pipe depends upon the current to be carried and type of soil. Usually the pipe used for this purpose is of diameter 40 mm and 2.5 metres in length for ordinary soil or of greater length in case of dry and rocky soil. The depth at which the pipe must be buried depends upon the moisture of the ground. The pipe is placed at a depth of 3.75 metres (minimum). The pipe is provided with a tapered casting at the lower end in order to facilitate the driving. The pipe at the bottom is surrounded by broken pieces of coke or charcoal for a distance of about 15 cm around the pipe. Generally alternate layers of coke and salt are used to increase the effective area of the earth and to decrease the earth resistance respectively. Another pipe of 19 mm diameter and minimum length 1.25 metres is connected at the top of GI pipe through reducing socket.

In our country in summer season the moisture in the soil decreases which causes increase in earth resistance. So a cement concrete work, as shown in Fig. 15.5, is done in order to keep the water arrangement accessible, and in summer to have an effective earth, 3 or 4 buckets of water are put through the funnel connected to 19 mm diameter pipe, which is further connected to GI pipe.

The earth wire (either GI wire or GI strip of sufficient cross-section to carry faulty current safely) is carried in a GI pipe of diameter 12 mm at a depth of about 60 cm from the ground.

Care should be taken that earth wire is well protected from mechanical injury, when it is carried over from one machine to another.

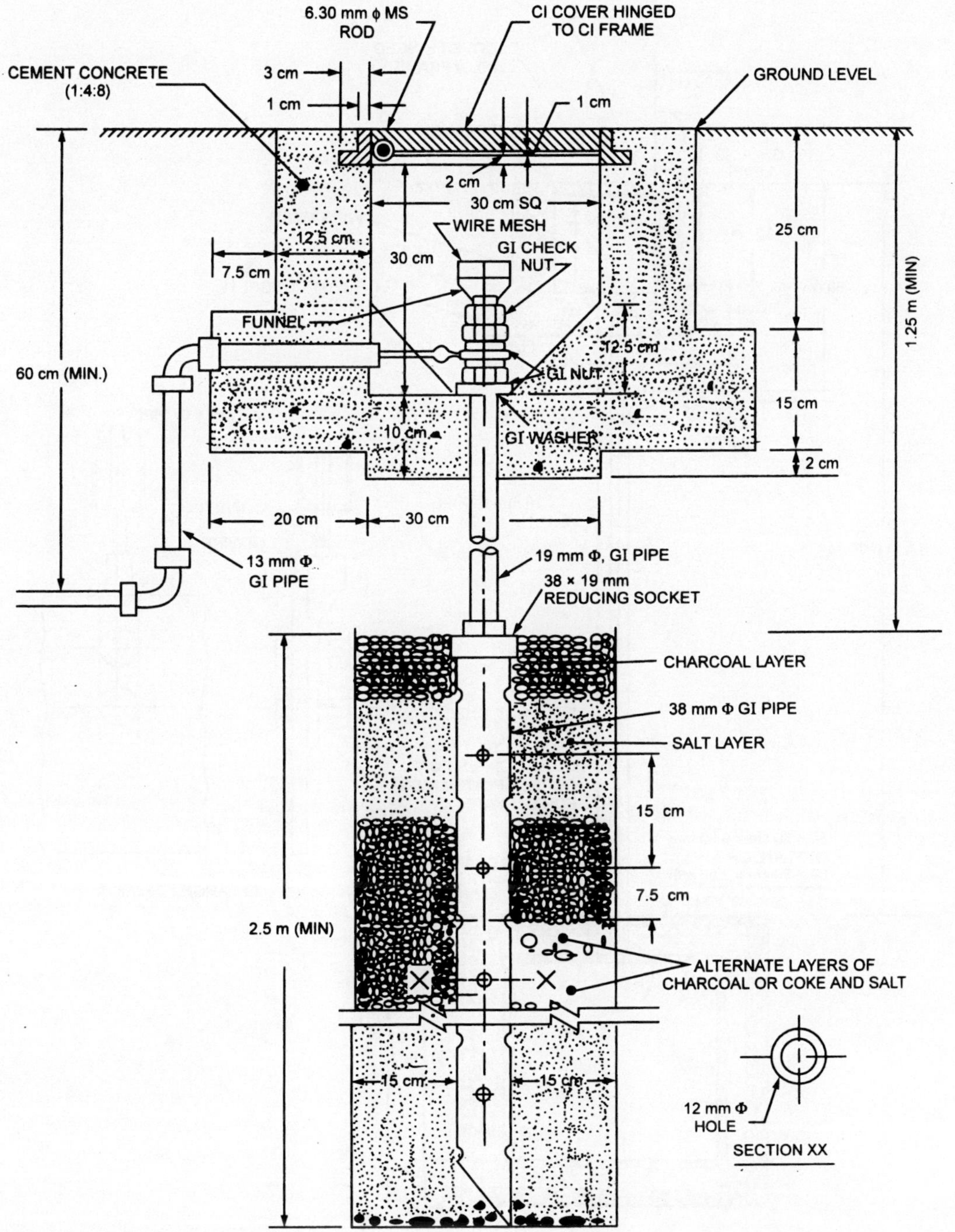

Note: Three or four buckets of water to be poured into sump every few days to keep the soil surrounding the earth pipe permanently moist.

Fig. 15.5 *A Typical Illustration of Pipe Earth Electrode*

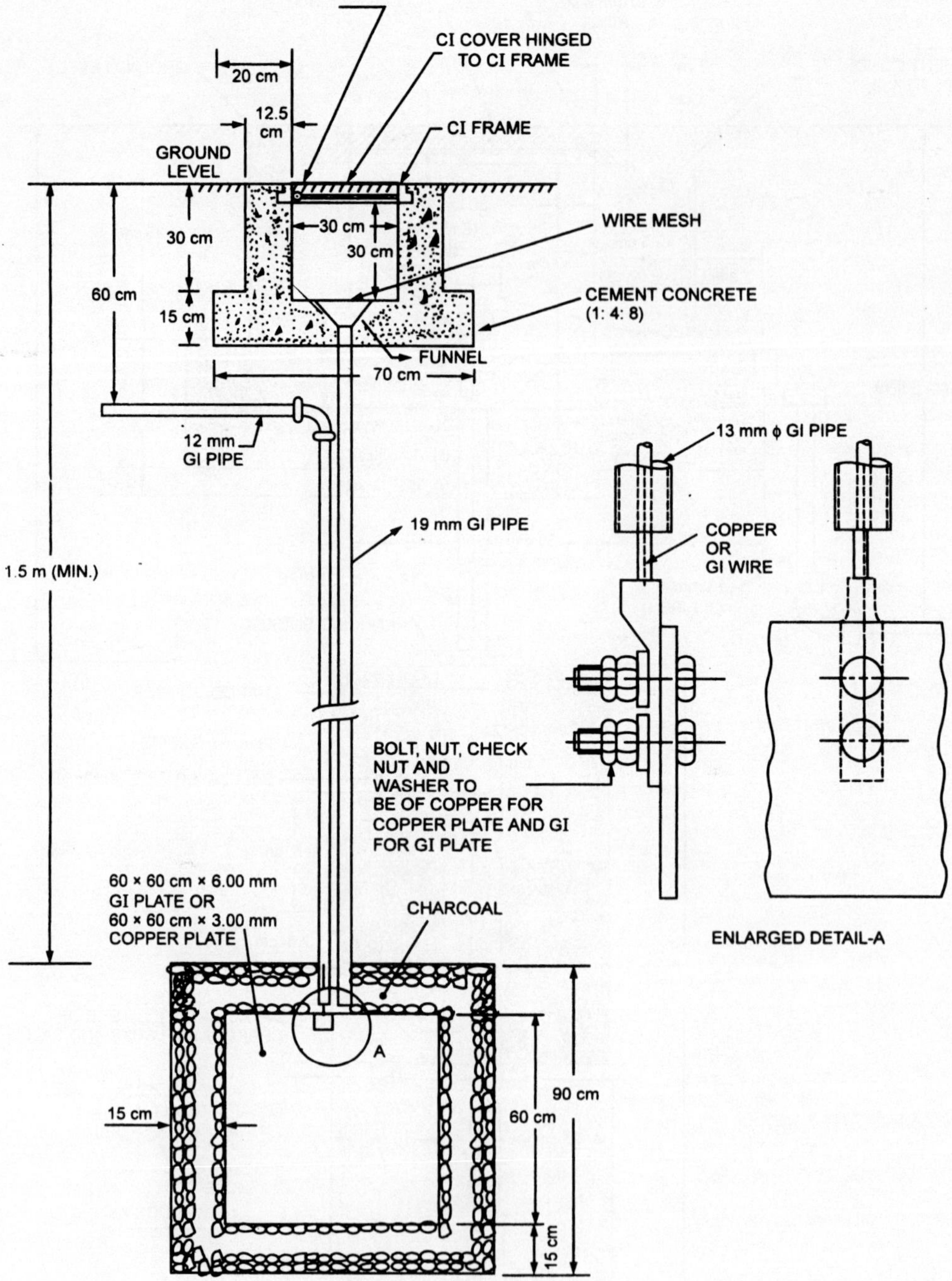

Note: Three or four buckets of water to be poured into sump every few days to keep the soil surrounding the earth pipe permanently moist.

Fig. 15.6 *A Typical Illustration of Plate Earth Electrode*

4. Plate Earthing. This is another common system of earthing. In plate earthing an earthing plate either of copper of dimensions 60 cm × 60 cm × 3 mm or of galvanised iron of dimensions 60 cm × 60 cm × 6 mm is buried into the ground with its face vertical at a depth of not less than 3 metres from ground level. The earth late is embedded in alternate layers of coke and salt for a minimum thickness of 15 cm. The earth wire (GI wire for GI plate earthing and copper wire for copper plate earthing) is securely bolted to an earth plate with the help of a bolt, nut and washer made of material of that of earth plate (made of copper in case of copper plate earthing and of galvanised iron in case of GI plate earthing).

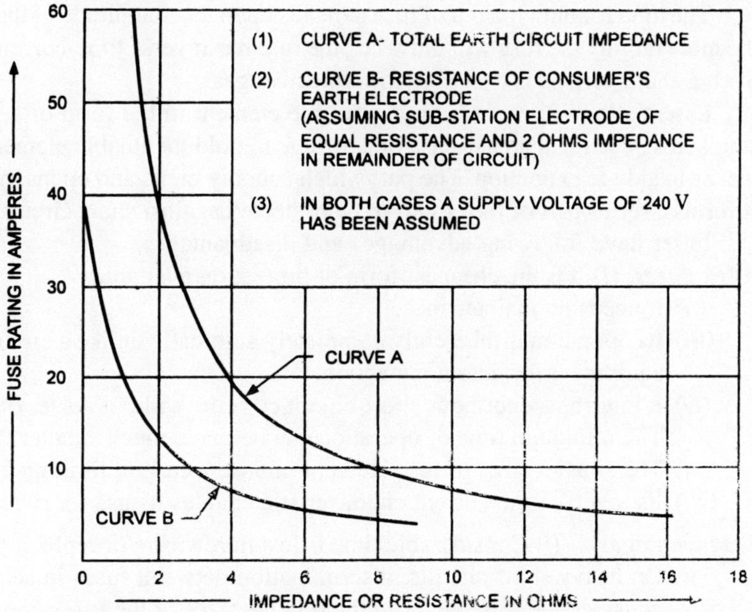

Fig. 15.7 *Recommended Earth Circuit Impedance or Resistance For Different Values of Fuse Rating*

A small masonry brick wall enclosure with a cast iron cover on top or an RCC pipe round the earth plate is provided to facilitate its identification and for carrying out periodical inspection and tests.

The copper plate and copper wire are usually not employed for grounding purposes because of their higher cost.

The recommended earth circuit impedance or resistance for different values of fuse rating are detailed in Fig. 15.7. This is to decide safety level.

15.10. FUSES

Fuse is perhaps the simplest and cheapest device used for interrupting an electrical circuit under short-circuit, or excessive overload, current magnitudes. As such, it is used for overload and/or short-circuit protection in high voltage (up to 66 kV) and low voltage (up to 400 V) installations/circuits. In high voltage circuits their use is confined to those applications where their performance characteristics are particularly suitable for current interruption.

The action of a fuse is based upon the heating effect of the electric current. In normal operating conditions, when the current flowing through the circuit is within safe limits, the heat developed in the fuse element carrying this current is readily dissipated into the surrounding air, and therefore, fuse element remains at a temperature below its melting point. However, when some fault, such as short-circuit occurs or when load connected in a circuit exceeds its capacity, the current exceeds the limiting value, the heat generated due to this excessive current cannot be dissipated fast enough and the fusible element gets heated, melts and breaks the circuit. It thus protects a machine or apparatus or an installation from damage due to excessive current.

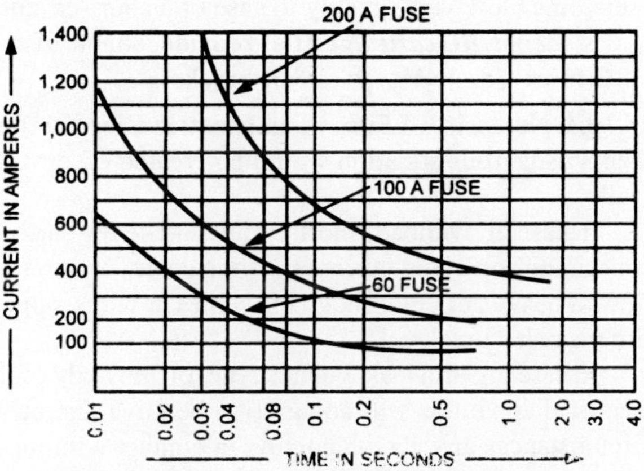

Fig. 15.8 Time-Current Characteristic

The time for blowing out of fuse depends upon the magnitude of the excessive current. Larger the current, the more rapidly the fuse will blow i.e., the fuse has inverse time-current characteristic, as shown in Fig. 15.8. Such a characteristic is desirable for protective gear.

Essentially, a fuse consists of a fusible element in the form of a metal conductor of specially selected small cross-sectional area, a case or cartridge to hold the fusible element, and in some cases, provided with a means to aid arc extinction. The part which actually melts and opens the circuit is known as the *fuse element*. It forms a series part of the circuit to be protected against short-circuit or excessive overloads.

Fuses have following advantages and disadvantages.

Advantages. (*i*) It is the cheapest form of protection available.

(*ii*) It needs no maintenance.

(*iii*) Its operation is inherently completely automatic unlike a circuit breaker which requires an elaborate equipment for automatic action.

(*iv*) It interrupts enormous short-circuit currents without noise, flame, gas or smoke.

(*v*) The minimum time of operation can be made much smaller than that with the circuit breakers.

(*vi*) The smaller sizes of fuse element impose a current limiting effect under short-circuit conditions.

(*vii*) Its inverse time-current characteristic enables its use for overload protection.

Disadvantages. (*i*) Considerable time is lost in rewiring or replacing a fuse after operation.

(*ii*) On heavy short-circuits, discrimination between fuses in series cannot be obtained unless there is considerable differences in the relative sizes of the fuses concerned.

(*iii*) The current-time characteristic of a fuse cannot always be correlated with that of the protected device.

The function of fuse wire is (*i*) to carry the normal working current safely without heating and (*ii*) to break the circuit when the current exceeds the limiting current.

Distribution circuits are protected from ground and short-circuit currents by fuses or circuit breakers so arranged as to disconnect the faulted equipment promptly from its source of supply. Fuses are used almost exclusively for the protection of cables in low-voltage light and power circuits and for transformers of rating not exceeding 200 kVA, in primary distribution systems. Circuit breakers are employed for larger amounts of power and in cases where the operation of the overload device is so frequent as to make the use of fuses impractical.

Nowadays several types of fuses are available which find extensive use in low and moderate voltage applications where frequent operations are not expected or where the use of circuit breakers is uneconomical. Most important among these is the *motor protective fuse* which may be chosen to have a suitable time-delay to permit starting inrush current to flow and also carrying of moderate motor overloads without blowing, and yet be capable of blowing early enough to avoid excessive winding temperature due to overload and at the same time blow very quickly in case of a short-circuit.

Modern HRC cartridge fuses provide reliable discrimination and accurate characteristics. In some respects HRC fuses are superior to circuit-breakers.

15.10.1. Necessity of Fuse in an Electric Circuit. If no fuse or other similar device is provided in the circuit then a dangerous situation would be created on developing of faults such as overload, short-circuit or earth faults.

In case of overload, short-circuit and heavy earth faults a heavy current will continue to flow through the consuming apparatus, current carrying cables or wires and other current carrying equipment. Due to continuous flow of heavy current through the cables or wires, apparatus etc., these will get heated up and so damaged. The fire may also break out.

In case of earth leakage fault, (i.e. on the body of the electrical apparatus becoming alive), the body of the electrical apparatus will continue to be alive and at much higher potential above that of the earth. In such circumstances any person coming in contact with the metal body of the apparatus is liable to get an electric shock, even if it is earthed.

The main function of a fuse is to blow out under a fault and isolate the faulty section from the live side. If the fuse is provided on neutral wire, in place of live wire, then in abnormal conditions though the fuse will blow out but the lamp or other apparatus still remains connected to the live wire and in case of leakage some trouble will arise and cause a considerable damage. In case the earth fault takes on the neutral wire between lamp and fuse provided in it, the fuse will blow out because the neutral wire is slightly at a higher potential with respect to earth and so the fault current flows through the neutral wire and fuse melts itself. The current will flow through the live wire, lamp, neutral wire and earth fault, even after the fuse has blown out and this may cause serious damage to the wiring, the apparatus connected or building itself.

If fuses of same capacity are provided on the phase wire and neutral, then in case of short-circuit fault, one of them will blow out first. If the fuse on neutral wire blows out first, the fuse in phase line remains intact and faulty apparatus still remains connected to the live. If some person comes in contact with the faulty apparatus, he is liable to get electric shock. In case the installation is connected to 3-phase 4-wire supply system, and fuses are provided on both live and neutral wire and fuse on neutral wire blows out then voltage of each phase to neutral will become considerably different, which is not desirable. *Hence the fuse is provided only in phase or live pole, never on neutral pole.*

15.10.2. Types of Fuses.

1. *Supply Main Fuse.* This fuse is provided by the supplier and is fixed just after the service meter and sealed by him. The seal can be broken only by authorised person of the supply authority in case of blowing out of fuse for the purpose of replacement. The rating of supply main fuse will be as per load current of the consumer.

2. *Consumer Main Fuse.* This is another fuse of rating slightly less than that of supply main fuse provided by the supplier and placed after the consumer's main switch. The rating of consumer fuse is kept slightly lower than that of supply main fuse so that in case of exceeding of current than normal current (that may occur due to overload, short-circuit or earth fault) consumer fuse, which can be replaced by him, may blow and supply main fuse may remain intact.

3. *Sub-Circuit Fuses.* The total wiring system is divided into a number of sub-circuits or branch circuits, as already explained in chapter 13. A separate fuse is provided for each branch circuit and is known as *sub-circuit or branch circuit fuse.*

4. *Point Fuses.* In good quality indoor wiring of buildings, every light and plug point is provided with its individual fuse known as *point fuse.*

15.11. FUSE UNITS

A fuse unit essentially consists of the metal fuse element or link, a set of contacts between which it is fixed and a body to support and isolate them. Many types of fuses also have some means for extinguishing the arc which appears when the fuse element melts.

The various types of fuse units, most commonly available are :

1. Round type fuse unit.
2. Kit-kat type fuse unit.
3. Cartridge type fuse unit
4. HRC (High rupturing capacity) fuse units and
5. Semiconductor fuse units

1. Round Type Fuse Unit. This type of fuse unit consists of a porcelain or bakelite box and two separated wire terminals for holding the fuse wire between them. This type of fuse is not in common use on account of its following disadvantages :

1. One of the terminals remain always energised and, therefore, for replacement of fuse either the worker will have to touch the live mains or open the main switch.

2. Appreciable arcing takes place at the instant of blowing off fuse and thus damage the terminals. After two or three arcing the fuse unit becomes unusable.

2. Rewirable or Kit-Kat Type Fuses. The most commonly used fuse in 'house wiring' and small current circuits is the semi-enclosed or rewirable fuse (also sometimes known as kit-kat type fuse). It consists of a *porcelain base* carrying the fixed contacts to which the incoming and outgoing live or phase wires are connected and a *porcelain fuse carrier* holding the fuse element, consisting of one or more strands of fuse wire, stretched between its terminals. The fuse carrier is a separate part and can be taken out or inserted in the base without risk, even without opening the main switch. If fuse holder or carrier gets damaged during use, it may be replaced without replacing the complete unit.

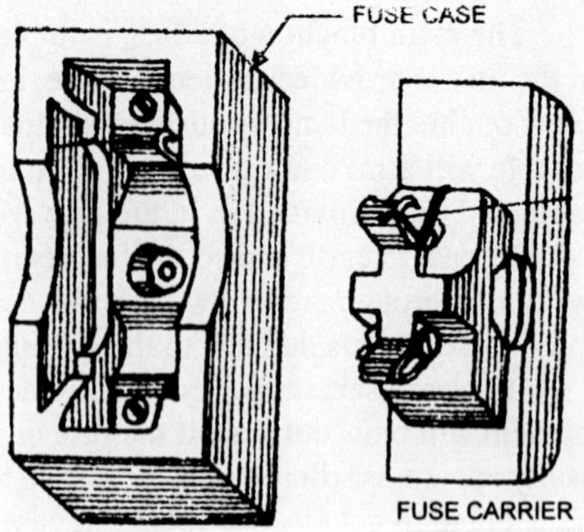

Fig. 15.9 *Rewirable or Kit-Kat Type Fuse*

The fuse wire may be of lead, tinned copper, aluminium or an alloy of tin-lead. The actual fusing current will be about twice the rated current. When two or more fuse wires are used, the wires should be kept apart and a derating factor of 0.7 to 0.8 should be used to arrive at the total fuse rating.

The specifications for rewirable fuses are covered by IS : 2086-1963. Standard ratings are 6, 16, 32, 63 and 100 A. A fuse wire of any rating not exceeding the rating of the fuse may be used in it i.e., a 80 A fuse wire can be used in a 100 A fuse, but not in the 63 A fuse.

On occurrence of fault, the fuse element blows off and the circuit is interrupted. The fuse carrier is pulled out, the blown out fuse element is replaced by new one and the supply is restored by reinserting the fuse carrier in the base.

Though such fuses have the advantages of easy removal or replacement without any danger of coming into contact with a live part and negligible replacement cost but they suffer from the following disadvantages:

(a) Unreliable Operation. The operation of the rewirable fuses is unreliable because of the following factors :

(*i*) There is a possibility of renewal by the fuse wire of wrong size.

(*ii*) The fuse wire deteriorates, over a period, due to oxidation through the continuous heating up of the element. After a relatively short period the current causes the metal to deteriorate and the fuse operates at a lower current than originally rated.

(*iii*) A fuse with normal rated capacity of say 60 A would fuse at nearly 120 amperes. Should that amount of current damage the apparatus in the circuit, obviously the size of the fuse wire must be smaller but, as its rated capacity will also be lower, the same dependence could not be placed upon it.

(*iv*) The protective capacity is uncertain i.e., no dependence can be placed upon the wire to interrupt the circuit when a given current flows. For instance although theoretically a No.17 SWG wire should interrupt a circuit when a current of 135 A is flowing, the circuit may be interrupted when a lower current flows or in other circumstances when only a much higher current flows.

(*v*) Accurate calibration of the fuse wire is impossible, as a longer fuse operates earlier than one of shorter length.

Single phasing of three-phase induction motors is a common occurrence where rewirable type fuses are employed in motor circuits.

(b) Lack of Discrimination. Due to unreliable operation, discrimination cannot be always ensured. However, it will be possible to achieve some measure of discrimination by using a fuse which has twice the rating of the next fuse ahead of it in series. For example, a fuse of 80 A will discriminate with a fuse of 40 A, but not with a fuse of 55 A.

(c) Small Time Lag. Due to the small time lag, such fuses can blow with large transient currents which are encountered during the starting of motors and switching-on operation of transformers, capacitors, fluorescent lights etc., unless fuses of sufficiently high rating are used.

(d) Low Rupturing Capacity. Rewirable fuses have limited breaking or rupturing capacity. For example, according to IS : 2086-1963 the rewirable fuse of 16 A normal current have a breaking current of 2 kA and those upto 200 A normal current have a breaking current of 4 kA.

(e) No Current-Limiting Feature.

(f) Slow Speed of Operation. No special means are employed to extinguish the arc that blows after the fuse melts. Thus arcing time is more in such fuses.

(g) Risk of Flame and Fire.

Because of simplicity and cheapness rewirable type fuses are extensively used as protective devices against overload and short-circuit on low and medium voltage circuits.

3. Cartridge Type Fuse. This is a totally enclosed type fuse unit. It essentially consists of an insulating container of bulb or tube shape and sealed at its ends with metallic cap known as *cartridge* enclosing the fuse element and filled up with powder or granular material known as filler. There are various types of materials used as filler like sand, calcium carbonate, quartz etc. There is sometimes a blow out device in the side of the tube to indicate when the fuse is blown. On overloads or short-circuits, the fusible element is heated to a high temperature causing it to vaporize. The powder in the fuse cartridge cools and condenses the vapour and quenches the arc thereby interrupting the flow of current.

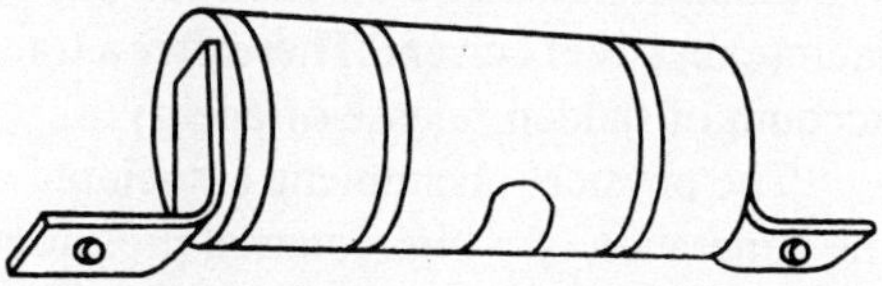

Fig. 15.10 *Cartridge Fuse*

Since it is totally enclosed it will not be possible to rewire and, therefore, the whole unit will have to be replaced, once it blows out. It provides complete security against fire risk, as it is a totally enclosed unit. The filling powder provides good insulating path and helps to extinguish the arc at the time of blowing up of fuse. This type of fuse is available up to 660 V and the current rating up to 800 A.

4. High Rupturing Capacity (HRC) Fuses. With a very heavy generating capacities of the modern power stations, extremely heavy currents would flow into the fault and the fuse clearing the fault would be required to withstand extremely high stresses in this process. A rewirable fuse may not be useful in this case and, therefore, high rupturing capacity fuses commonly known as HRC fuses, designed and developed after intensive research for use in medium and high voltage installations, are used for such duties. Their rupturing capacity is as high as 500 MVA upto 66 kV and above. The main advantages of HRC fuses are :

(*i*) They are cheap as compared with other types of circuit interrupters of same breaking capacity.
(*ii*) No maintenance is required.
(*iii*) The operation is quick and sure.
(*iv*) They do not deteriorate with time.
(*v*) They have inverse time current characteristic.
(*vi*) They are capable of clearing high as well as low currents.
(*vii*) They are quite reliable and can be selected for proper discrimination.

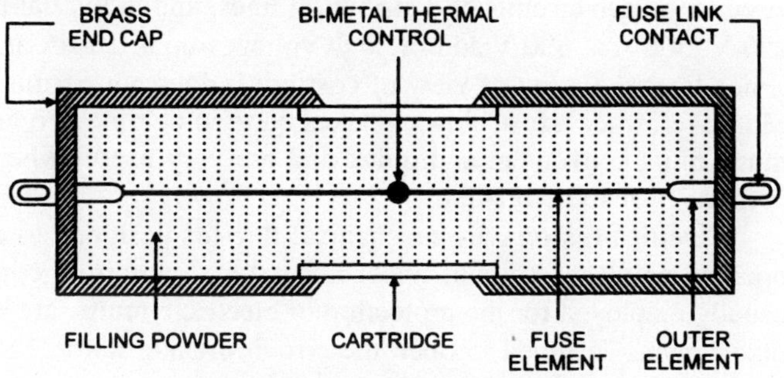

Fig. 15.11 *HRC Cartridge Fuse*

HRC fuses suffer from the following disadvantages :

(*i*) They are required to be replaced after each operation.
(*ii*) Interlocking is not possible.
(*iii*) They lack relays in complete discrimination.

Since the advantages outweigh the disadvantages to a very large extent, HRC fuses are extensively used.

(*a*) *Cartridge Type HRC Fuse.* The high rupturing capacity cartridge fuse is most popular. In its simplest form an HRC cartridge fuse consists of a heat resisting ceramic body having metal end-caps to which are welded fusible silver (or bimetallic) current-carrying elements. The complete space within the body surrounding the elements is filled with a powder which acts as an arc extinguishing agent.

The process of fusing comprises the following operations :

1. Pre-arcing operation i.e. melting of silver elements.
2. Arcing operation i.e. vaporisation of the elements.
3. Fusion of silver vapours and the filling powder, and
4. Extinction of arc under fusion process.

On the occurrence of a fault, short-circuit current flows through the fuse element and element is thus heated up to its melting point. When the melting process is completed, an arc is formed. The chemical reaction between silver vapour and filling powder tends to establish high resistance. The high resistance acts as an insulator because the fault current decreases along with the high pressure created within the fuse by the fault (excessive) current. Thereafter a transient voltage is created at the instant of fault current interruption on account of sudden release of energy.

The physical phenomena associated with the process of fusing include a sudden rise of temperature and the generation of a high internal pressure.

(*b*) *Tetra Chloride Type HRC Fuse.* It essentially consists of a glass tube filled with carbon-tetra-chloride solution and sealed at both ends with brass caps. Inside the tube a high resistance fuse wire is sealed at one end of the glass tube and the other end of the fuse wire is held by a strong phosphor bronze spiral spring fixed to the other end of the glass tube. A flexible copper wire is also similarly held between the spring and the other end of cap in parallel with the high resistance wire. On short-circuit or overload, the high resistance fuse wire melts and spring pulls back the flexible copper wire. The arc produced is extinguished by carbon-tetra-chloride vapour.

5. *Semiconductor Fuses.* These are very fast acting fuses for protection of thyristor and other electronic circuits.

15.12. PROTECTIVE DEVICES FOR WIRING INSTALLATIONS AND MOTORS

The most frequent causes for abnormal conditions are overloads and short-circuits, therefore, protection is to be provided to wiring installations and motors against overloading and short-circuiting. The devices used are fuses, relays, miniature circuit breakers (MCB), earth leakage circuit breakers.

Fuses are used for low and high voltage circuits of moderate to high capacity, where frequent operations are not expected e.g. for the protection of distribution transformers, small and medium size motors, lighting circuits, branch circuits of distribution lines, and industrial plants and commercial buildings up to 600 A at 250 V, 500 A at 600 V and in high voltages up to 100 A at 132,000 V. The protection by fuses is advantageous from the point of view of cost but it does not permit supervision and are to be replaced in the field. Moreover, when a fuse blows, it takes quite some time to renew it and restore the supply. Circuit breakers are much more convenient and quick and are very useful where slight overloads occur often or where heavy currents have to be broken.

Circuit breakers are mechanical devices designed to close or open contact members, thus closing or opening an electric circuit under normal and abnormal conditions. Automatic circuit breakers, which are usually employed for the protection of electric circuits, are equipped with a trip coil connected to a relay or other means designed to open the circuit breaker automatically under abnormal conditions, such as over-current. The automatic circuit breakers perform the following duties :

(*i*) It carries the full-load current continuously without overheating or damage.

(*ii*) It opens and closes the circuit on load.

(*iii*) It makes and breaks the normal operating current.

(*iv*) It makes and breaks the short-circuit currents of magnitude up to which it is designed for.

A circuit breaker consists of fixed and moving contacts which are touching each other under normal conditions i.e. when circuit breaker is closed. Whenever a fault occurs, the trip coils get energized, the moving contacts are pulled by some mechanism and therefore the circuit breaker is opened and the circuit is broken.

A more economical device used for protection of a motor is the externally mounted **thermal overload relay.** The construction of a thermal overload relay, used in magnetic starters, is shown diagrammatically in Fig. 15.12. The key element of a thermal overload relay is bimetallic strip and placed next to the bimetallic strip is a heater coil. When the coil gets sufficiently heated due to excessive current, the bimetallic strip flexes upwards at its outer end and releases a latch. A spring rotates the latch about its axis to actuate the tripping member which opens the normally closed contacts. The relay is manually reset after tripping by means of a reset push button.

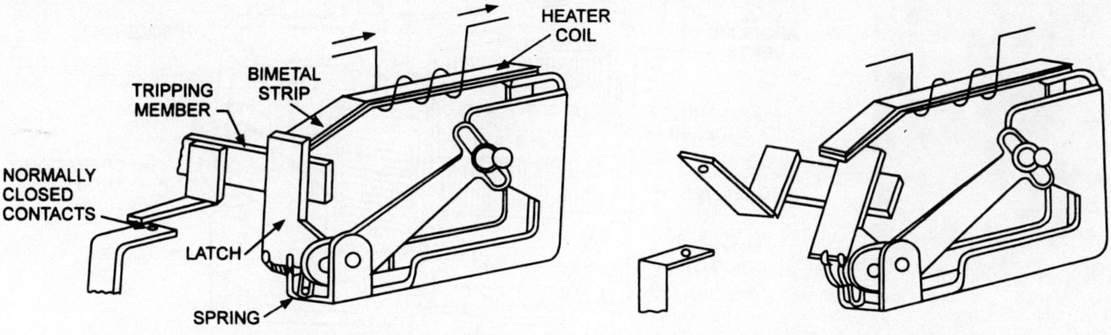

(a) Relay in Operation *(b) Relay De-energized*

Fig. 15.12 *Thermal Overload Relay*

Thermal overload relays provide protection for the motor only in the case of an overload and do not provide protection against short-circuits. This is because a short-circuit is required to be cleared instantaneously, while the heater coil takes some time to heat the metal strip to the set temperature.

Miniature circuit breaker (MCB) is a device that provides definite protection to the wiring installations and sophisticated equipment against over-currents and short-circuit faults. The outer and interior views of an MCB are shown in Fig. 15.13. Thermal operation (overload protection) is achieved with a bimetallic strip, which deflects when heated by any over-currents flowing through it. In doing so, releases the latch mechanism and causes the contacts to open. Inverse time current characteristics result, *i.e.*, greater the overload or excessive current, shorter the time required to operate the MCB. On the occurrence of a short-circuit, the rising current energizes the solenoid, operating the plunger to strike the trip lever causing immediate release of the latch mechanism. Rapidity of the magnetic solenoid operation causes instantaneous opening of contacts.

Miniature circuit breakers are available with different current ratings of 0.5, 1, 2, 2.5, 3, 4, 5, 6, 7.5, 10, 16, 20, 25, 32, 35, 40, 63, 100, 125, 160 A and voltage ratings of 240/415 V ac and upto 220 V dc. Operating time is very short (less than 5 ms). So they are very suitable for the protection of important and sophisticated equipment, such as air-conditioners, referigerators, computers etc.

Earth leakage circuit breaker (ELCB) is a device that provides protection against earth leakage. These are of two types viz, the current operated type and the voltage operated type.

Current operated earth-leakage circuit breaker is used when the product of the operating current in amperes and the earth-loop impedance in ohms does not exceed 40. Where such a circuit breaker is used, the consumer's earthing terminal is connected to a suitable earth electrode. A current-operated earth leakage applied to a 3-ϕ, 3-wire circuit is shown in Fig. 15.14. In normal conditions when there is no earth leakage the algebraic sum of the currents in the three coils of the current transformers (CTs) is zero, and no current flows through the trip coil. In case of any earth leakage, the currents are unbalanced and the trip coil is energized and thus the circuit breaker is tripped.

Voltage-operated earth leakage circuit breaker is suitable for use when the earth-loop impedance exceeds the values applicable to fuses or excess-current circuit breaker or to current-operated earth leakage circuit breaker. Such an earth-leakage trip in a 2-wire circuit is shown in Fig. 15.15. When the voltage between the earth continuity conducter (ECC) and earth electrode rises to a sufficient value, the trip coil will carry the required current to trip the circuit breaker. With such a circuit breaker the earthing lead between the trip-coil and the earth electrode must be insulated; in addition, the earth electrode must be placed outside the resistance area of any other parallel earths which may exist.

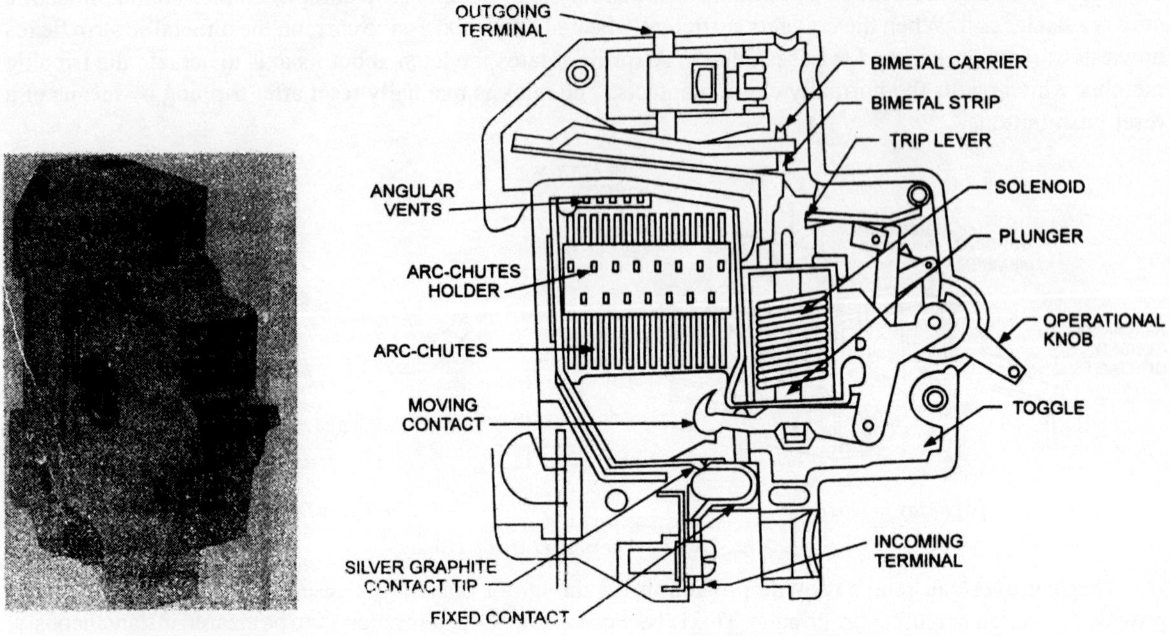

(a) Outer View *(b) Interior View*

Fig. 15.13 *Miniature Circuit Breaker (Courtesy Havells)*

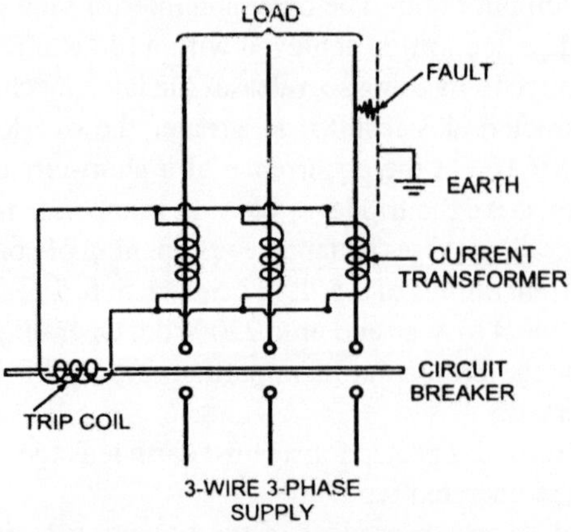

Fig. 15.14 *Current-Operated Earth-Leakage Trip*

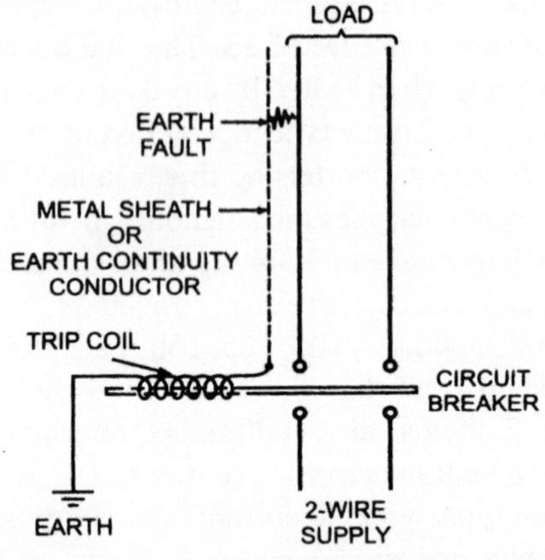

Fig. 15.15 *Voltage-Operated Earth Leakage Trip*

In both the above types of ELCB the tripping operation may be tested by means of a finger-operated test button which passes a predetermined current from the line wire through a high resistance to trip the coil and thus to earth. This test operation should be performed regularly.

Both types of earth leakage circuit breakers are arranged to work manually and may take the place of the linked switch and fuses, or the excess-current circuit breaker.

15.13. SAFETY PRACTICES

1. In all electrical works, it is essential that certain elementary safety practices are observed. It has been found that quite a large number of accidents occur due to negligence of these practices. The details of such practices are given in appendix D of IS 5216-1969.

2. In the event of fire on electrical mains or apparatus, the affected part should immediately be completely isolated from its source of supply.

Fire extinguishers, which are not insulated, should never be used in fighting fires near exposed live conductors. Only such fire extinguishers should be used on electrical mains and apparatus which are marked as suitable for this purpose. When making use of fire hose, it should be ensured that the jet of water does not come in contact with live conductors.

It is quite dangerous to throw a stream of water, wet blanket or a stream from ordinary soda-acid type extinguishers on live mains or apparatus. When found necessary to make their use, make all the neighboring mains or apparatus dead. Carbon-dioxide, carbon tetra chloride and other special types of extinguishers, sand or dry blankets be used on live conductors and static apparatus.

3. Inadequate lighting of working areas in by itself a source of danger, particularly when work is undertaken at night. No work should, therefore, be allowed in dark or poorly illuminated or ill-ventilated places.

4. Suitable safety posters in the form of DO and DON'T instructions for the guidance of working staff should be exhibited at important locations, such as generating stations, receiving stations, substations and factories. A suggested safety poster incorporating DO and DON'T instructions for general work is given in Appendix E of IS 5216-1969.

15.14. MULTIMETER OR AVOMETER

This is a multipurpose instrument as its name indicates and is used for measurement of current (dc and ac), voltage (dc and ac) and resistance. With the external source, higher resistance above (1 MΩ), inductance and capacitance can also be measured with this instrument. It is simple, compact and portable because the only power it employs is the battery for the ohm-meter.

The ammeter, the voltmeter, and the ohmmeter all use D'Arsonval movement. The difference between these instruments is the circuit in which the basic movement is employed. It is, therefore, obvious that a single instrument can be designed to perform the three measurement functions. This instrument, which contains a function switch to connect the appropriate circuits to the D'Arsonval movement, is often called a *multimeter* or *ampere-volt-ohm* (AVO) *meter* or *volt-ohm-milliammeter* (VOM).

Multimeters may be of analog type or digital type. The analog type is of the pointer and scale type i.e. it employs the basic D'Arsonval movement. However, digital multimeters (DMMs) are becoming more and more popular because of their easy readability, numerical display and improved accuracy.

The "heart" of the meter consists of a dc moving coil instrument taking very low current of the order of 1 mA for full-scale deflection. On dc a tapped universal shunt is used for current measurements and a tapped series resistance for voltage measurements. On ac a transformer feeds into a metal rectifier, the output from which goes to the dc movement. Thus the same scales are used for dc and ac measurements. The resistances are measured by applying a set voltage from an internal battery to the resistance under test and measuring the resulting current. Since the resulting current is, by Ohm's law, inversely proportional to the resistance, the scale can be calibrated directly in ohms, kilo-ohms and mega-ohms.

Selection of the particular range required is affected by rotary switch. All the accessories required for the various measurements to be carried out by the meter are housed in a meter casing, making the test-set completely self-contained.

For measurement of electrostatic capacity of condensers, inductance of low frequency choke coils, high resistance and insulation resistance of condensers an external current source is required.

For the measurement of capacitance and inductance with the help of this meter, first the meter is connected to the ac source, as shown in dotted lines [Fig. 15.16 (*a*)], and variable resistance, inserted in the primary or secondary of the transformer is adjusted so that the pointer of the meter may indicate the full scale. Then one of the testing probe is shifted to the component (L or C) to be measured putting it in series in between the current source and the probe and the meter indicates the value.

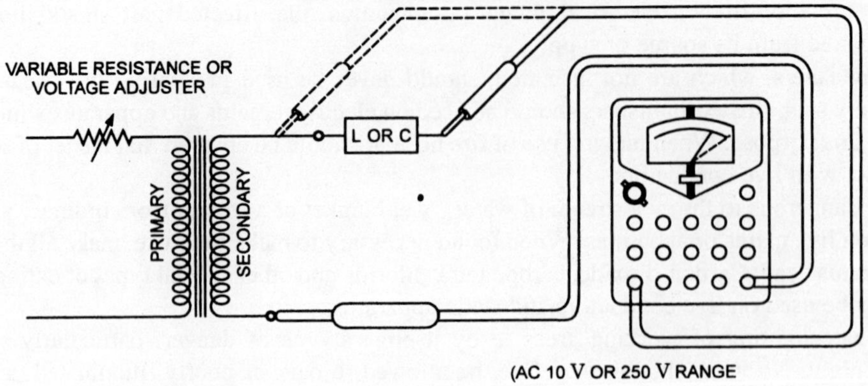

(a) Circuit For Measuring Capacitance and Inductance

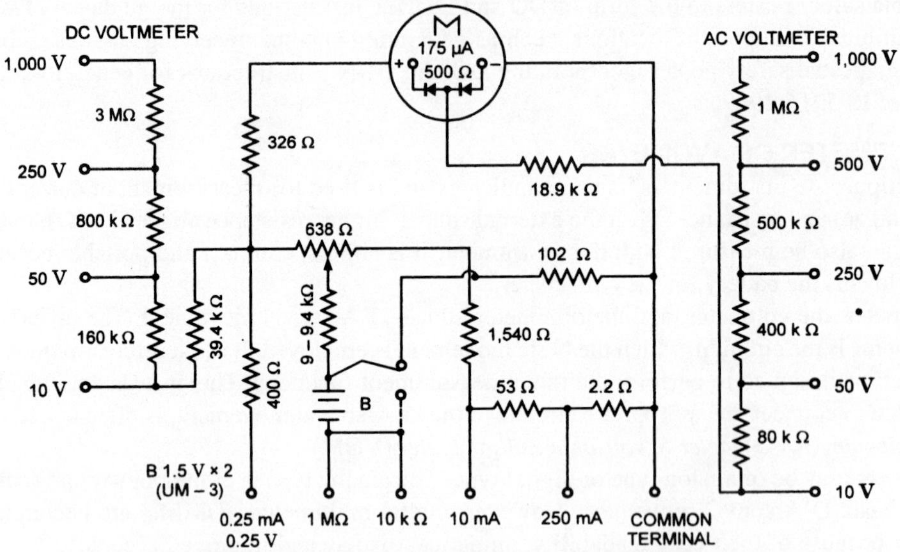

(b) Schematic Diagram of a Multimeter (AVO Meter)
(Courtesy: Atlas Radio, Electronic Industry Pvt. Ltd. Naroda, Ahmedabad-India)

Fig. 15.16

The circuit diagram of a multimeter consisting of a dc milliammeter, a dc voltmeter, an ac voltmeter and an ohmmeter is shown in Fig. 15.16 (*b*).

The circuit for measuring dc milli-amperes is given in Fig. 15.17 (*a*) and the circuit is self-explanatory. The circuit for measuring dc volts is given in Fig. 15.17 (*b*) and again the circuit is self-explanatory.

The ac voltmeter section of a multimeter is given in Fig. 15.17 (*c*). Two diodes D_1 and D_2 are employed in this circuit, forming a full-wave rectifier with the meter so connected that it receives only half of the rectified current. Diode D_1 conducts during the positive half cycle of the input waveform and makes the meter to deflect according to the average value of this half cycle. Across the meter, the other diode D_2 serves as protection. In the absence of diode D_2, the negative half cycle of the input voltage would apply a reverse voltage across diode D_1 resulting in a small leakage current in the reverse direction. The average value of the complete cycle would thus be lower than it should be for half-wave rectification. But when diode D_2 is connected, as shown in the figure, it conducts heavily and current bypasses the meter in reverse direction. The commercial multimeters often use the same scale markings for both its dc and ac voltage ranges. Since the average value of sinusoidal wave for half-wave rectification is equal to 0.45 times of its rms value, in order to obtain the same deflection on corresponding dc and ac voltage ranges, it becomes essential to lower the multiplier for the ac range proportionately.

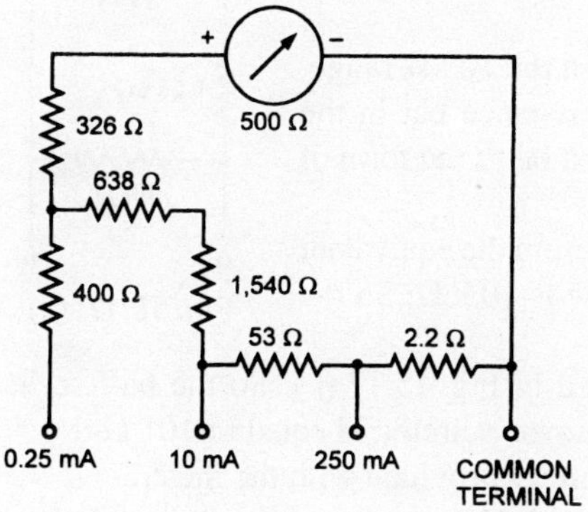

(a) DC Ammeter Section of a Multimeter

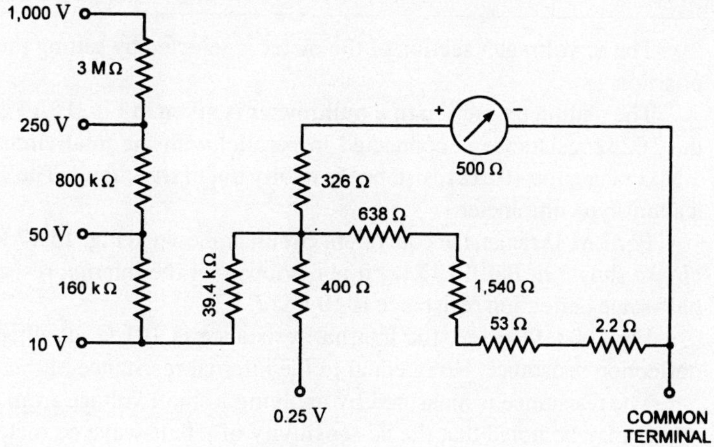

(b) DC Voltmeter Section of a Multimeter

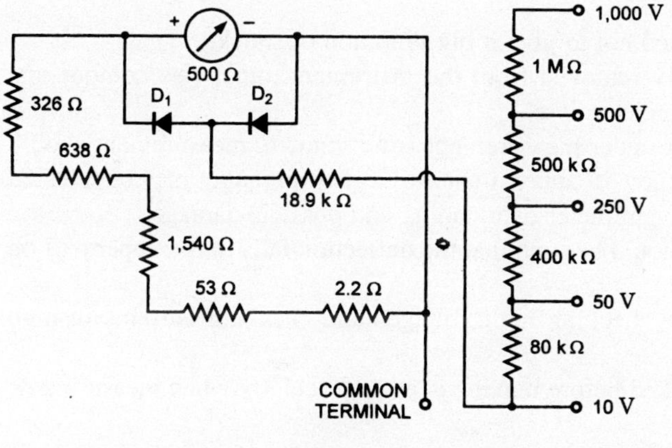

(c) AC Voltmeter Section of a Multimeter

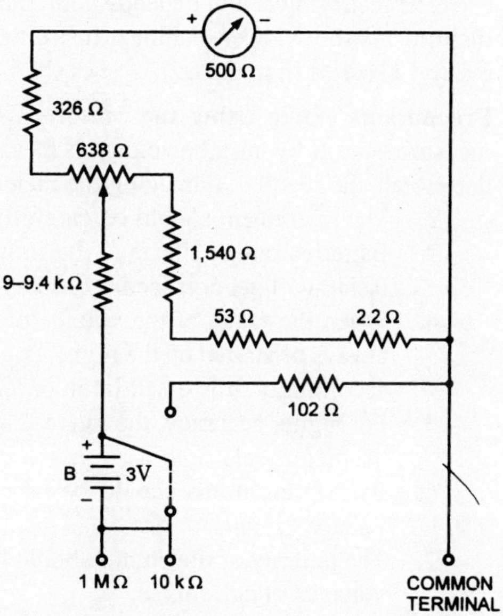

(d) Ohmmeter Section of a Multimeter

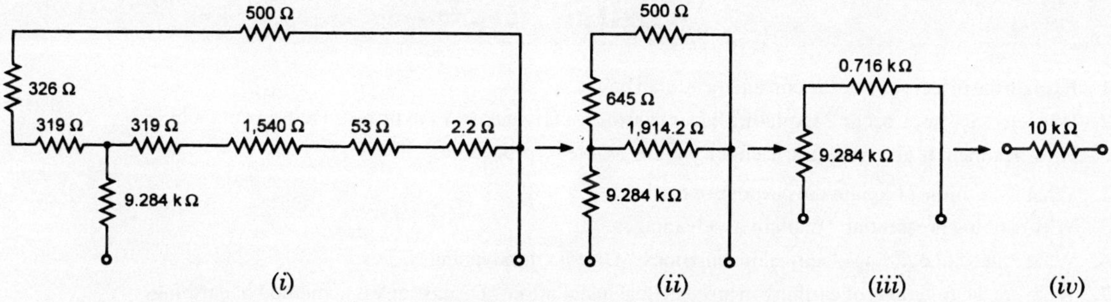

(e) Equivalent Circuit For 1 MW Range

Fig. 15.17

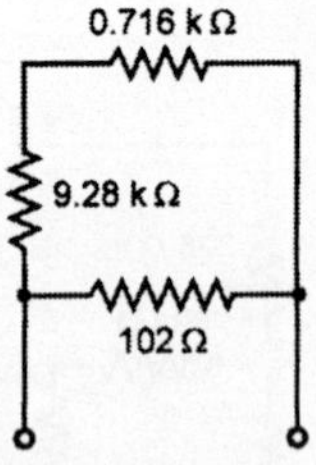

Fig. 15.17 (*f*)

The ac voltmeter section of the meter is selected by setting the ac-dc switch to the ac position.

The ohmmeter section of a multimeter is given in Fig. 15.17 (*d*). In the 10 kΩ range, the 102 Ω resistance is connected in parallel with the total circuit resistance but in the 1 MΩ range, the 102 Ω resistance is totally out of the circuit. The circuit takes the form of a shunt-type ohmmeter.

For 1 M Ω range, the equivalent circuit is shown in Fig. 15.17 (*e*). From the equivalent circuit shown in Fig. 14.17 (*e*) it is obvious that the internal resistance is 10 k Ω. So the half-scale deflection resistance is 10 k Ω.

For 100 k Ω range, the internal resistance is 101 Ω, as illustrated in Fig. 15.17 (*f*). So the half-scale deflection resistance, being equal to the internal resistance of the ohmmeter circuit, is equal to 101 Ω.

The resistance is measured by applying a small voltage from a battery provided with the meter.

It may be noted that the ac sensitivity of a half-wave ac meter is only 45 per cent of the dc sensitivity, therefore, the loading effect is greater while measuring ac voltages than that while measuring corresponding dc voltages.

The rectifying action depends upon the frequency to some extent due to capacitance effect and rectifier instruments show lower readings. Instrument indications may be in error by as much as 0.5% decrease for every 1 kHz rise in frequency.

Precautions While Using the Meter. 1. Range should be confirmed each time before carrying out any measurement. If by mistake voltage is measured on a resistance or current range, it will burn off the shunt or deteriorate the rectifier, damaging the meter itself.

2. The instrument should be carefully handled not to give it big vibration of shock.
3. Batteries out of life must be immediately removed from the instrument; otherwise components inside will get corroded by leakage of the electrolyte.
4. When the range of the voltage or current under measurement is not known, measurement should always be started on the highest range to know its approximate value. Lower ranges may be selected accordingly. It is essential in order to prevent meter overloading and possible damage.
5. For higher accuracy, the range selected should be such that the deflection falls in the upper half on the meter scale.
6. Extra precautions should be taken while checking high voltages and checking currents in high voltage circuits.
7. The polarity of the circuit should be checked before making a test, particularly when measuring dc voltages or currents.
8. The instrument must not be exposed long in high temperature and moisture.

EXERCISES

1. Explain the effects of electric current on human body.
2. Why electric shock occur ? Explain. Discuss various safety measures to be taken to prevent them.
3. How treatment is given to electric shock patient ? Give its procedure.
4. What is earthing ? Explain the purpose of earthing.
5. Why earthing is essential ? Explain its advantages.
6. What is need of earthing electrical installations ? Discuss its advantages.
7. What are the functions of earthing in an electrical installation ? Discuss any one method of earthing.
8. Give some important IS specifications regarding earthing.
9. Why is it necessary to provide earthing in an electrical installation ? State IE rules regarding points to be earthed in an electrical installation ?

10. Explain why earthing of an electrical installation and electrical equipment is important. Write the IE Rules regarding earthing of an electrical installation and equipment.
11. What do you understand by earthing ? Why is it necessary to earth electrical installation ? Draw out a neat sketch for earthing by GI pipe.
12. What do you understand by earthing ? Why is it provided ? Draw a neat sketch of pipe earthing.
13. What is fuse ? What is its purpose ? Discuss.
14. What is the function of a fuse ? Why is it connected in the phase wire ?
15. According to IE rules, a cut out (fuse) shall not be placed in the earthed neutral conductor of two-wire system. Discuss the reason to justify this rule.
16. What are the different types of fuse used ? Why is a fuse inserted in the phase wire, not in the neutral wire ? Why are HRC fuses used ?
17. Describe the various types of fuse units.
18. Why is it advisable to use HRC fuse (cartridge fuse) for motor protection instead of rewirable fuse ? Why is a fuse inserted in the live wire and not in the neutral wire ?
19. Enumerate the various protective devices used in wiring installations and motors. Explain any one of these in detail.
20. What are the various electrical quantities you can measure with a multimeter ? Can you measure capacitance with multimeter ? Justify your answer. **[U.P. Technical Univ. First Semester 2001-02]**
21. Using a suitable diagram explain the working principle of a multimeter. Also write its various applications. **[U.P. Technical Univ. First Semester 2004-05]**
22. What are the different parameters measured using multimeter ? How is resistance measured? Explain with the help of circuits. **[U.P. Technical Univ. First/Second Semester 2001-02]**
23. Which physical variable (current, voltage and resistance) does this meter inside a multimeter measure ? How is it modified to measure the other two variables ? **[U.P. Technical Univ. First/Semester 2000-01]**
24. Write short notes on the following:
    (*i*) Effects of electric current on human body.
    (*ii*) Treatment for electric shock.
    (*iii*) Earthing and its importance.
    (*iv*) MCB.
    (*v*) ELCB.
    (*vi*) Use of multimeter.

SHORT ANSWER TYPE QUESTIONS WITH ANSWERS

Q. 1. What is meant by earth continuity conductor ?

Ans. The conductor, by means of which the metal body of an equipment or an appliance is connected to the earth, is known as earth continuity conductor (ECC).

Q. 2. What are maximum permissible values of resistance of earth system for (*a*) large power stations (*b*) major power stations (*c*) small substations (*d*) in all other cases ?

Ans. Maximum permissible values of resistance of earth system are as follows :

Large power stations	– 0.5 Ω
Major power stations	– 1.0 Ω
Small substations	– 2.0 Ω
In all other cases	– 5 Ω maximum.

Q. 3. What should be the minimum size of earth continuity conductor ?

Ans. The cross-section of earth continuity conductor should not be either less than 2.9 mm^2 (14 SWG) or half of the installation conductor size.

Q. 4. What is an earth electrode ?

Ans. Any wire, pipe, rod or metal plate embedded in earth for the purpose of making an effective electrical connection with the general mass of earth is known as earth electrode.

Q. 5. What is an earthing lead ?

Ans. The wire which connects overhead earth wire or any other apparatus to be earthed to the earth electrode is known as earthing lead.

Q. 6. Why charcoal and salt is used in earthing ?

Ans. Generally alternate layers of charcoal and salt are used to increase the effective area of the earth and to reduce the earth resistance respectively.

Q. 7. What is the average resistance of a human body ?

Ans. About 1,000 Ω.

Q. 8. What is current that is considered safe for a human body ?

Ans. About 9 mA.

Q. 9. Give the applications of rod earthing.

Ans. Rod earthing is suitable for areas which are sandy in character.

Q. 10. What is meant by a fuse ?

Ans. Fuse is a device used in an electrical circuit for protecting electrical equipment against overloads and/or shor-circuits.

Q. 11. On what principle does the fuse operate ?

Ans. The action of a fuse is based upon the heating effect of the electric current. During overloads or short-circuits the fuse gets heated upto melting point due to excessive current and breaks the circuit.

Q. 12. What kind of protection is offered by a fuse ?

Ans. The fuse provides protection against short-circuits and overloads.

Q. 13. What function does the fuse perform ?

Ans. The function of a fuse wire is (*i*) to carry the normal working current safely without heating and (*ii*) to break the circuit when current exceeds the limiting current.

Q. 14. What is fuse link ?

Ans. Fuse link is that part of the fuse which needs replacement when the fuse blows out.

Q. 15. What are the purposes of using quartz sand as filler material in the construction of HRC fuse ?

Ans. The purposes of using quartz sand as filler material in the construction of HRC fuse are to absorb heat at a very high rate, to produce a substance of high resistance and avoid the oxidation of silver element.

Q. 16. Can the HRC fuse be used for both ac and dc circuits. If so under what condition ?

Ans. In general an HRC fuse meant for ac circuit may not be used for dc circuits; but however, a fuse satisfying the requirement of arc voltage can be used on both ac and dc circuits.

Q. 17. What is a multimeter ?

Ans. The instrument, which contains a function switch to connect the appropriate circuits to the d'Arsonval movement, is often called a multimeter or ampere-volt-ohm (AVO) meter or volt-ohm-milliammeter (VOM).

Q. 18. Indicate the various quantities that can be measured with a multimeter.

Ans. Multimeter is used for measurement of current (dc as well as ac), voltage (dc as well as ac) and resistance. With the external source, high resistance (exceeding 1 MΩ), inductance and capacitance can be also measured with this instrument.

Q. 19. What are the shortcomings of multimeter ?

Ans. It may be noted that the ac sensitivity of a half-wave ac meter is only 45 per cent of the dc sensitivity, therefore, the loading effect is greater while measuring ac voltages than that while measuring corresponding dc voltages.

The rectifying action depends upon the frequency to some extent due to capacitance effect and rectifier instruments show lower readings. Instrument indications may be in error by as much as 0.5 % decrease for every 1 kHz rise in frequency.

❑❑❑

Objective Types Questions

MULTIPLE CHOICE TYPE QUESTIONS

Note: Each of the following questions comprises multiple responses (two or more than two answers). Select the correct answers. In case you feel that there is more than one correct answers, select the answer which you consider the best. In any case, choose ONLY ONE answer or response for each question.

1. The condition for the validity of Ohm's law is that the
(a) temperature should remain constant.
(b) current should be proportional to voltage.
(c) resistance must be wire wound type.
(d) all of the above.

2. Two wires A and B of the same material and length l and $2l$ have radius r and $2r$ respectively. The ratio of their specific resistance will be
(a) 1 : 1 (b) 1 : 2
(c) 1 : 4 (d) 1 : 8
[U.P.S.C. I.E.S. Elec. Engineering-I, 1992]

3. If the length of a wire of resistance R is uniformly stretched to n times its original value, its new resistance is
(a) n R (b) R/n
(c) n^2 R (d) R/n^2 **[GATE 1998]**

4. The hot resistance of the filament of a bulb is higher than the cold resistance because the temperature coefficient of the filament is
(a) negative. (b) infinite.
(c) zero. (d) positive.
[U.P.S.C. I.E.S. Elec. Engineering-I, 2000]

5. Four resistances 80 Ω, 50 Ω, 25 Ω and R are connected in parallel. Current through 25 Ω resistance is 4 A. Total current of the supply is 10 A. The value of R will be
(a) 66.66 Ω (b) 40.25 Ω
(c) 36.36 Ω (d) 76.56 Ω
[U.P.S.C. I.E.S. Elec. Engineering-I, 1999]

6. All the resistances in figure shown below are 1 Ω each. The value of current 'I' is

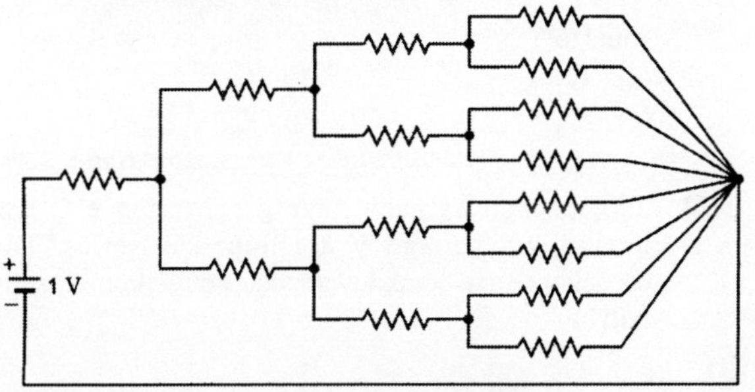

(a) $\frac{1}{15}$ A (b) $\frac{2}{15}$ A
(c) $\frac{4}{15}$ (d) $\frac{8}{15}$ A
[GATE 1992]

7. Two resistors R_1 and R_2 give combined resistance of 4.5 Ω when in series and 1 Ω when in parallel, the resistances are
(a) 2 Ω and 2.5 Ω (b) 1 Ω and 3.5 Ω
(c) 1.5 Ω and 3 Ω (d) 4 Ω and 0.5 Ω

8. For the circuit shown below the equivalent resistance will be

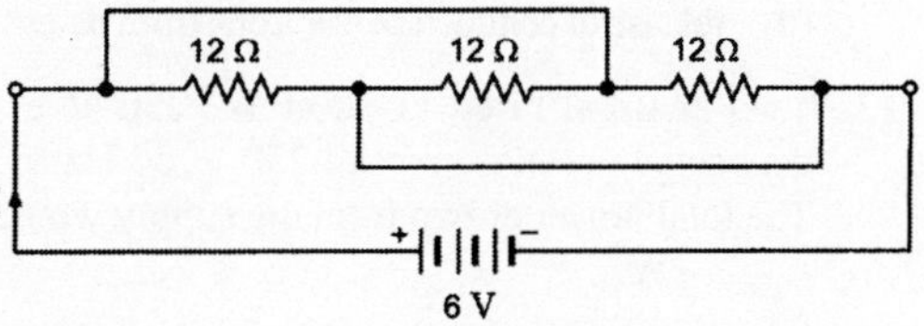

(a) 36 Ω (b) 12 Ω
(c) 4 Ω (d) 6 Ω

9. In given figure, the value of resistance R in Ω is
(a) 10
(b) 20
(c) 30
(d) 40

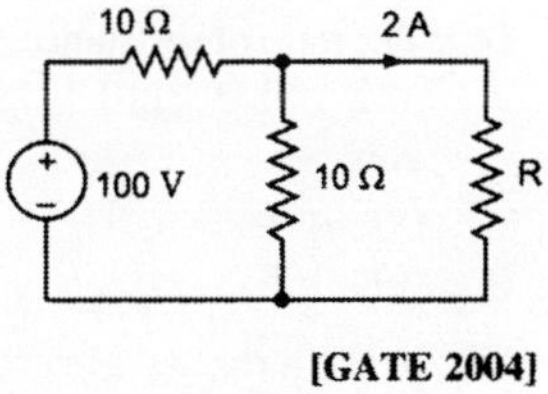

[GATE 2004]

10. In the figure given below the value of R is

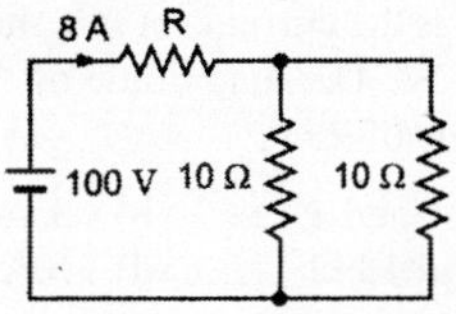

(a) 2.5 Ω (b) 5.0 Ω
(c) 7.5 Ω (d) 10.0 Ω
[GATE 2005]

Answers: 1. (a), 2. (a), 3. (c), 4. (d), 5. (c), 6. (d), 7. (c), 8. (c), 9. (b), 10. (c)

11. Consider the following circuit. What is the value of current I in the circuit shown ?

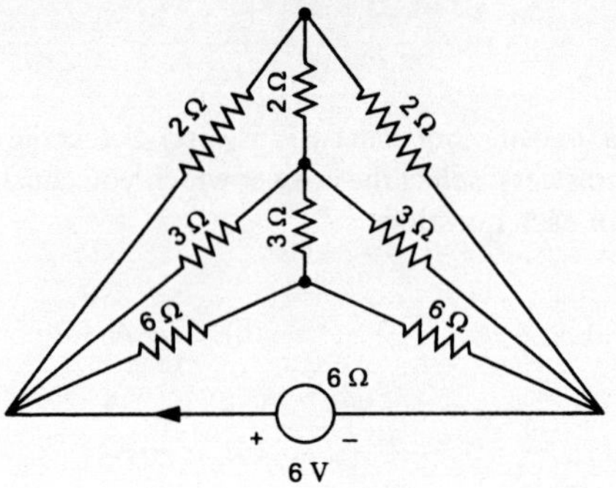

(a) 1 A (b) 2 A
(c) 3 A (d) 4 A

[U.P.S.C. I.E.S. Electrical Engineering-I, 2004]

12. When an electric current flows through a conductor, its temperature rises. This is because of
(a) mutual collisions between metal atoms.
(b) mutual collisions between conducting electrons.
(c) collisions between conduction electrons and atoms.
(d) release of conduction electrons from parent atoms.

13. Two heaters, rated at 1000 W, 250 V each are connected in series across a 250 V, 50 Hz ac mains. The total power drawn from the supply would be —
——— W.
(a) 1,000 (b) 500
(c) 250 (d) 2,000

[Delhi University 1984]

14. The ratio of resistances of a 100 W, 220 V lamp to that of a 100 W, 110 V lamp will be, at the respective voltages
(a) 4 (b) 2
(c) 1/2 (d) 1.4

15. When a resistor R is connected to a current source, it consumes a power of 18 W. When the same R is connected to a voltage source having the same magnitude as the current source, the power absorbed by R is 4.5 W. The magnitude of the current source and the value of R are
(a) $\sqrt{18}$ A and 1 Ω (b) 3 A and 2 Ω
(c) 1 A and 18 Ω (d) 6 A and 0.5 Ω

[GATE 1999]

16. Kirchhoff's laws are valid for
(a) linear circuits only.
(b) passive time invariant circuits.
(c) nonlinear circuits only.
(d) both linear and non-linear circuits.

17. In given figure R_a, R_b and R_c are 20 Ω, 10 Ω and 10 Ω respectively. The resistances R_1, R_2 and R_3 in ohms of an equivalent star-connection are

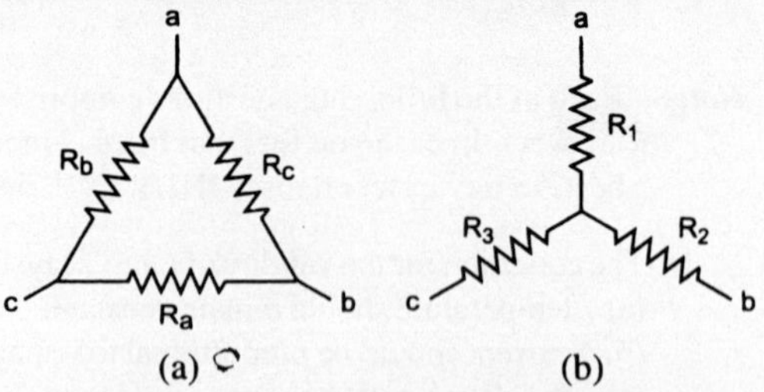

(a) 2.5, 5, 5 (b) 5, 2.5, 5
(c) 5, 5, 2.5 (d) 2.5, 5, 2.5

[GATE 2004]

18. Two charges of equal magnitude are separated by some distance. If the charges are increased by 10%; to get the same force between them, their separation must be
(a) increased by 21%. (b) increased by 10%.
(c) decreased by 10%.
(d) none of the above is correct.

19. The relation between electric intensity E, voltage applied V and the distance d between the plates of a parallel plate condenser is
(a) $E = \dfrac{V}{d}$ (b) $E = V \times d$
(c) $E = \dfrac{V}{d^2}$ (d) $E = V \times d^2$

[U.P.S.C. I.E.S. Elec. Engineering-I, 2002]

20. The electric field lines and equipotential lines
(a) are parallel to each other.
(b) are one and the same.
(c) cut each other orthogonally.
(d) can be inclined to each other at any angle.

[U.P.S.C. I.E.S. Elec. Engineering-I, 1992]

21. Six capacitors of different capacitances C_1, C_2, C_3, C_4, C_5 and C_6 are connected in series. $C_1 > C_2 > C_3 > C_4 > C_5 > C_6$. What is the total capacitance almost equal to ?
(a) C_1 (b) C_3
(c) C_4 (d) C_6

[U.P.S.C. I.E.S. Elec. Engineering-I, 2006]

22. Two capacitors each having capacitance C and breakdown voltage V are joined in series. The capacitance and breakdown voltage of the combination will be
(a) 2 C and 2 V (b) $\dfrac{C}{2}$ and $\dfrac{V}{2}$

Answers: 11. (c), 12. (c), 13. (b), 14. (a), 15. (b), 16. (a), 17. (a), 18. (b), 19. (a), 20. (c), 21. (d), 22. (d)

(c) 2 C and $\frac{V}{2}$ (d) $\frac{C}{2}$ and 2 V

23. Three capacitors each of breakdown voltage 500 V are connected in parallel. The breakdown rating of the combination will be

(a) 500 V (b) 707 V
(c) 1,000 V (d) 2,500 V

24. The number of 2 μF, 400 V capacitors needed to obtain a capacitance value of 1.5 μF rated for 1,600 V is

(a) 12 (b) 8
(c) 6 (d) 4

[U.P.S.C. I.E.S. Elec. Engineering-I, 1994]

25. The condensors of 20 μF and 40 μF capacitance are connected in series across a 90 V supply. After charging, they are removed from the supply and are connected in parallel with positive terminals connected together and similarly the negative terminals. Then the voltage across them will be

(a) 90 V (b) 60 V
(c) 40 V (d) 20 V

[U.P.S.C. I.E.S. Elec. Engineering-I, 1995]

26. If V, W, q stand for voltage, energy and charge, then V can be expressed as

(a) $V = \frac{dq}{dw}$ (b) $V = \frac{dw}{dq}$

(c) $dV = \frac{dw}{dq}$ (d) $dV = \frac{dq}{dw}$

[GATE 1996]

27. Joule/coulomb is the unit of

(a) electric field potential.
(b) potential.
(c) charge. (d) none of the above.

[U.P.S.C. I.E.S. Elec. Engineering-I, 1992]

28. The ratio of charge stored by two metallic spheres raised to the same potential is 6. The ratio of the surface areas of the sphere is

(a) 6 (b) 1/6
(c) 36 (d) $1/\sqrt{6}$

[U.P.S.C. I.E.S. Elec. Engineering-I, 1992]

29. A parallel plate capacitor has capacitance of 10 μF. If the linear dimensions of the plates are doubled and the separation between them is also doubled the value of the capacitor would be

(a) 10 μF (b) 20 μF
(c) 5 μF (d) 40 μF

30. A parallel plate capacitor has an electrode area of 100 mm^2, with a spacing of 0·1 mm between the electrodes. The dielectric between the plates is air with a permittivity of $8{\cdot}85 \times 10^{-12}$ F/m. The charge on the capacitor is 100 V. The stored energy in the capacitor is

(a) 8·85 pJ (b) 440 pJ
(c) 22·1 nJ (d) 44·3 nJ **[GATE 2003]**

31. The time constant of the network shown in figure is

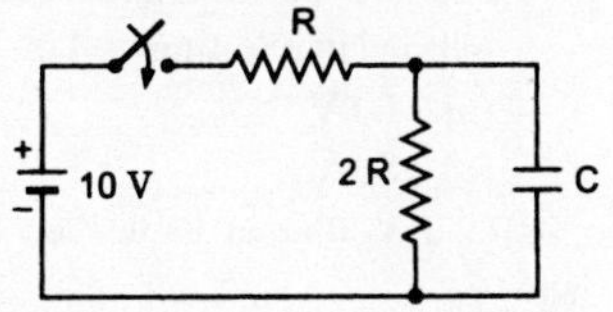

(a) 2 RC
(b) 3 RC
(c) RC/2
(d) 2 RC/3

[GATE 1992]

[**Hint** : Resistance faced by C with the source shorted is 2 R/3]

32. In the series R-C circuit shown in the figure, the voltage across C starts increasing when the dc source is switched on. The rate of increase of voltage across C at the instant just after the switch is closed (*i.e.* at $t = 0^+$), will be

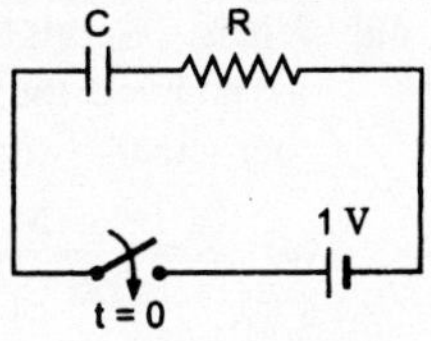

(a) zero.
(b) infinity.
(c) RC.
(d) 1/RC.

[GATE 1996]

33. Iron is a

(a) ferromagnetic. (b) ferrimagnetic.
(c) paramagentic.
(d) antiferromagnetic material.

34. Soft magnetic materials have

(a) high permeability and low coercive force.
(b) low permeability and high coercive force.
(c) high residual magnetism.
(d) low resistivity.

35. The property of a material which opposes the production of magnetic flux in it is known as

(a) mmf. (b) reluctance.
(c) permeance. (d) permittivity.

36. A coil wound over an iron ring carries certain current and establishes flux in the ring. If the area of x-section of the ring is doubled, the flux density in the core

(a) is double of the previous value.
(b) is half of the previous value.
(c) is same as the previous value.
(d) is not possible to predict.

Answers: 23. (a), 24. (a), 25. (c), 26. (b), 27. (b), 28. (c), 29. (b), 30. (d), 31. (d), 32. (d), 33. (a), 34. (a), 35. (b), 36. (c)

37. The conductance of electrical circuit is analogous in magnetic circuit by
(a) flux. (b) reluctance.
(c) permeance. (d) relative permeability.
[R.G.P.V. Bhopal Basic Elec. Engineering Jan./Feb. 2006]

38. A magnetic circuit requires 800 AT to produce a certain quantity of flux in magnetic circuit. If its excitation coil has 100 turns and 5 Ω resistance, the voltage to be applied in exciting coil is
(a) 60 V (b) 40 V
(c) 80 V (d) 8 V
[R.G.P.V. Bhopal Basic Elec. Engineering, Jan./Feb. 2006]

39. The laws of electromagnetic induction (Faraday's and Lenz's laws) are summarized in the following equation:
(a) $e = LR$ (b) $e = L\dfrac{di}{dt}$
(c) $e = -\dfrac{d\psi}{dt}$ (d) none of the above
[GATE 1998]

40. When two coils having self inductances of L_1 and L_2 are coupled through a mutual inductance M, the coefficient of coupling, K is given by
(a) $K = \dfrac{M}{\sqrt{2 L_1 L_2}}$ (b) $K = \dfrac{M}{\sqrt{L_1 L_2}}$
(c) $K = \dfrac{2M}{\sqrt{L_1 L_2}}$ (d) $K = \dfrac{L_1 L_2}{M}$
[U.P.S.C. I.E.C. Elec. Engineering-I, 1993]

41. The overall inductance of two coils connected in series, with mutual inductance aiding self inductance is L_1; with mutual inductance opposing self-inductance the overall inductance is L_2. The mutual inductance M is given by
(a) $L_1 + L_2$ (b) $L_1 - L_2$
(c) $\dfrac{1}{4}(L_1 - L_2)$ (d) $\dfrac{1}{2}(L_1 + L_2)$
[U.P.S.C. I.E.S. Elec. Engineering-I, 1993]

42. The coils having self-inductance of 10 mH and 15 mH and effective inductance of 40 mH, when connected in series aiding. What will be the equivalent inductance if we connect them in series opposing?
(a) 20 mH (b) 10 mH
(c) 5 mH (d) zero
[U.P.S.C. I.E.S. Elec. Engineering-I, 1995]

43. The magnetic field required to reduce the residual magnetisation to zero is called
(a) retentivity. (b) coercivity.
(c) hysteresis.
(d) saturation magnetisation.
[U.P.S.C. I.E.S. Elec. Engineering-I, 2005]

44. The coercive force in a ferromagnetic material is related to the

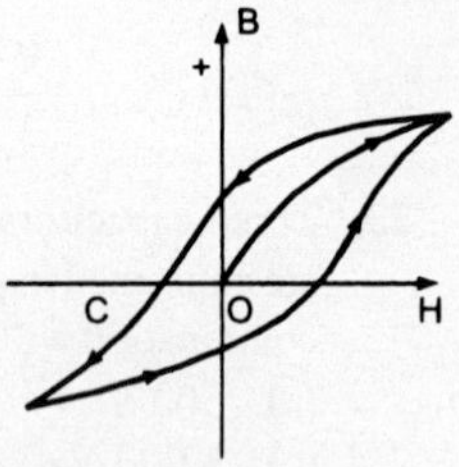

(a) area enclosed by the B-H loop.
(b) intercept on the negative H axis of the B-H loop.
(c) intercept on the positive B axis of the B-H loop.
(d) maximum value of B on the B-H loop.
[U.P.S.C. I.E.S. Elec. Engineering-I, 1993]

45. Area of hysteresis loop represents
(a) copper loss. (b) eddy curent loss.
(c) hysteresis loss.

46. When a battery of E volts is suddenly applied across an ideal inductance of L henry, the current through inductance will be
(a) zero.
(b) infinity instantaneously.
(c) increasing linearly at the rate of $\dfrac{E}{L}$ amperes/s.
(d) $\dfrac{E}{L}(1 - e^{-t/T})$.

47. A coil with a certain number of turns has a specified time constant. If the number of turns is doubled, its time consant would
(a) remain unaffected. (b) become doubled.
(c) become four fold. (d) get halved.
[U.P.S.C. I.E.S. Elec. Engineering-I, 1993]

48. A series R-L circuit with R = 100 ohm; L = 50 H, is supplied to a dc source of 100 V. The time taken for the current to rise 70% of its steady-state value is
(a) 0.3 s (b) 0.6 s
(c) 2.4 s
(d) 70% of time required to reach steady state.
[U.P.S.C. I.E.S. Elec. Engg-I, 1993]

49. A constant current of 2.8 A exists in a resistor. The rms value of current is
(a) 2.8 A (b) about 2 A
(c) 1.4 A (d) undefined.

50. The current through a resistor has a waveform as shown in figure. The reading shown by a moving coil ammeter will be _______ amperes.

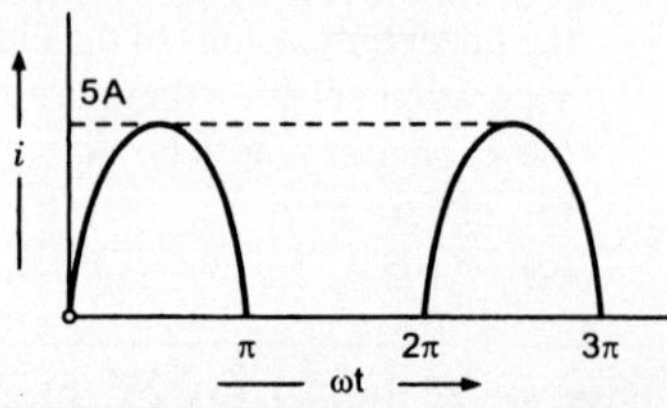

(a) $5/\sqrt{2}$
(b) $2.5/\sqrt{2}$

Answers: 37. (c), 38. (b), 39. (c), 40. (b), 41. (c), 42. (b), 43. (b), 44. (b), 45. (c), 46. (c), 47. (b), 48. (b), 49. (a), 50. (c)

(c) $5/\pi$ (d) 0

[Delhi Univ. July 1984]

51. The rms value of a half-wave rectified symmetrical square wave current of 2 A is

(a) $\sqrt{2}$ A (b) 1 A

(c) $\frac{1}{\sqrt{2}}$ A (d) $\sqrt{3}$ A **[GATE 2005]**

52. A boiler at home is switched on to the ac mains supplying power at 230 V, 50 Hz. The frequency of instantaneous power consumed is

(a) 0 Hz (b) 50 Hz.

(c) 100 Hz (d) 150 Hz **[GATE 1996]**

53. A 10 mH inductor carries a sinusoidal current of 1 A rms at a frequency of 50 Hz. The average power dissipated by the inductor is

(a) 0 W (b) 0·25 W

(c) 0·5 W (d) 1·0 W

[U.P.S.C. I.E.S. Elec. Engineering-I, 2003]

54. A circuit component that opposes the change in circuit voltage is

(a) resistance. (b) capacitance.

(c) inductance. (d) all of the above.

55. A circuit of zero lagging power factor behaves as

(a) an inductive circuit. (b) a capacitive circuit.

(c) R-L circuit. (d) R-C circuit.

56. Power loss in an electrical circuit can take place in

(a) inductance only. (b) capacitance only.

(c) inductance and resistance.

(d) resistance only.

57. In an R-L series circuit the power factor is

(a) leading. (b) lagging.

(c) zero. (d) unity.

58. The power factor of an ac circuit is equal to

(a) cosine of the phase angle.

(b) sine of the phase angle.

(c) unity for a resistive circuit.

(d) unity for a reactive circuit. **[Nagpur Univ. 1993]**

59. When a sinusoidal voltage is applied across R-L series circuit having $R = X_L$, the phase angle will be

(a) 90° (b) 45° lag

(c) 45° lead (d) 90° leading

[R.G.P.V. Bhopal Basic Elec. Engineering Jan./Feb. 2006]

60. An ac source of 200 V rms supplies active power of 600 W and reactive power of 800 VAR. The rms current drawn from the source is

(a) 10 A (b) 5 A

(c) 3·75 A (d) 2·5 A

[U.P.S.C. I.E.S. Elec. Engineering-I, 1999]

61. The voltage phasor of a circuit is $10\angle 15°$ V and the current phasor is $2\angle -45°$ A. The active and reactive powers in the circuit are

(a) 10 W and 17·32 VAR

(b) 5 W and 8·66 VAR

(c) 20 W and 60 VAR

(d) $20\sqrt{2}$ W and $10\sqrt{2}$ VAR **[GATE 1999]**

62. The source in the circuit shown is a sinusoidal source. The supply voltage across various elements are marked in the figure. The input voltage is

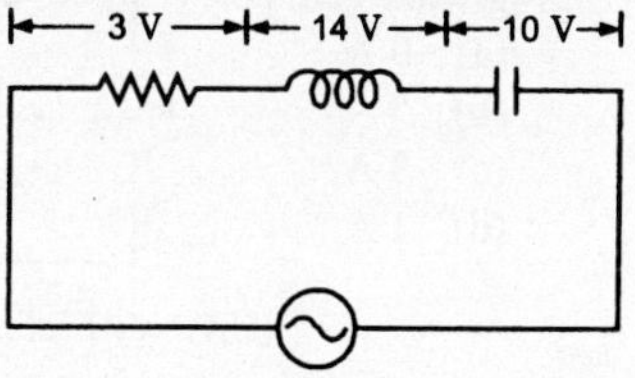

(a) 10 V

(b) 5 V

(c) 27 V

(d) 24 V

[U.P.S.C. I.E.S. Elec. Engineering-I, 1992]

63. In the circuit shown in the given figure, if the power consumed by the 5 Ω resistor is 10 W, then the pf of the circuit is

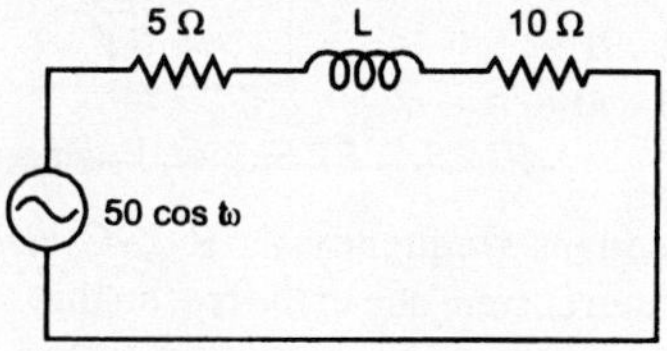

(a) 0.8 (b) 0.6

(c) 0.5 (d) zero

[U.P.S.C. I.E.S. Elec. Engineering-I, 1996]

64. If a series RLC circuit excited by a voltage $e = E \sin \omega t$ where $LC < \frac{1}{\omega^2}$

(a) current lags behind the applied voltage.

(b) current leads the applied voltage.

(c) current is in phase with the applied voltage.

(d) voltage across L and C are equal.

[U.P.S.C. I.E.S. Elec. Engineering I, 1993]

65. The current in the circuit shown

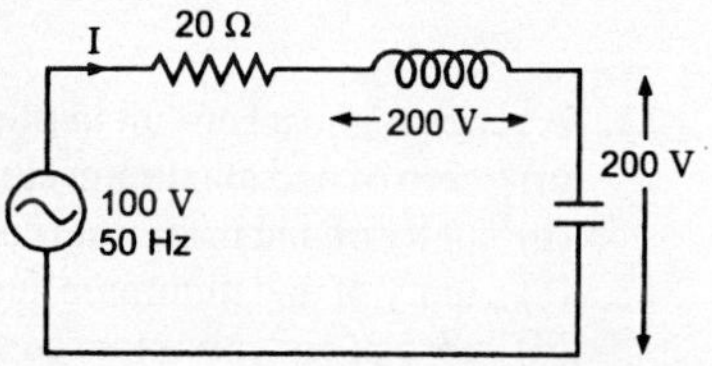

(a) 5 A

(b) 10 A

(c) 15 A

(d) 25 A

[GATE 1999]

66. In an ac circuit if voltage $V = (a + jb)$ and current $I = (c + jd)$, then the power is given by

Answers: 51. (a), 52. (c), 53. (a), 54. (b), 55. (a), 56. (d), 57. (b), 58. (a), 59. (b), 60. (b), 61. (a), 62. (b), 63. (b), 64. (b), 65. (a), 66. (b)

(a) $ac + ad$ (b) $ac + bd$
(c) $bc - ad$ (d) $bc + ad$

67. When a sinusoidal voltage is applied across R-L parallel circuit so that $R = X_L$ the phase angle will be
(a) 45° lag. (b) 45° leading.
(c) 90° lag. (d) 90° leading.
[R.G.P.V. Bhopal Basic Elec. Engineering Jan./Feb. 2006]

68. The current read by the ammeter A in the ac circuit shown in the given figure is

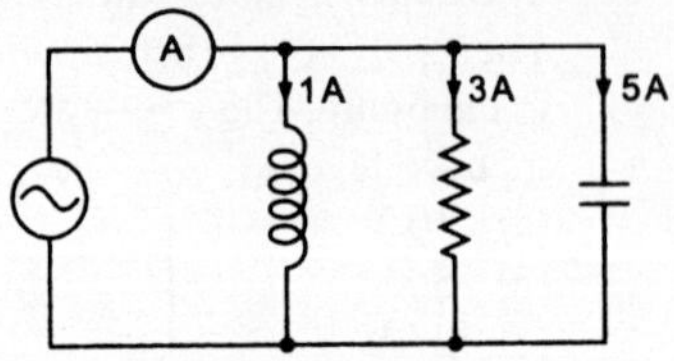

(a) 9 A
(b) 5 A
(c) 3 A
(d) 1 A
[U.P.S.C. I.E.S. Elec. Engineering- I, 1995]

69. For the given ac circuit, what is the value of I ?

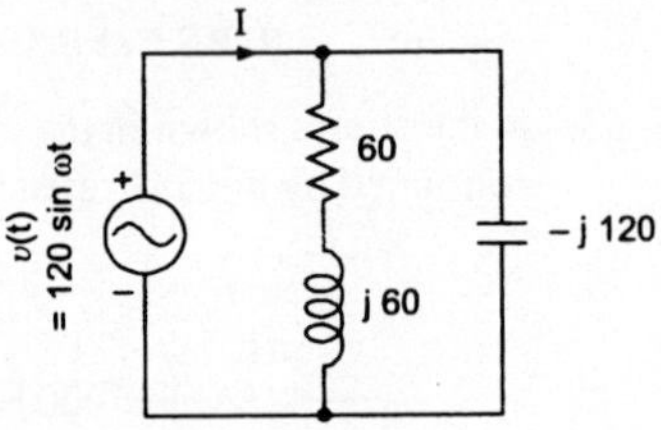

(a) $1 + j1$
(b) $1 + j0$
(c) $2 - j1$
(d) $0 + j0$
[U.P.S.C. I.E.S. Elec. Engineering-I, 2005]

70. At resonant frequency an R-L-C circuit draws maximum current due to the reason that
(a) the difference between capacitive reactance and inductive reactance is zero.
(b) the impedance is more than resistance.
(c) the voltage across the capacitor equals the applied voltage.
(d) the power factor is less than unity.
[U.P.S.C. I.E.S. Elec. Engineering-I, 2002]

71. A series R-L-C circuit will have unity power factor if operated at a frequency of
(a) 1/LC. (b) $1/\omega\sqrt{LC}$.
(c) $1/\omega^2 LC$. (d) $1/2\pi\sqrt{LC}$.
[Delhi Univ. July 1984]

72. A series-resonant circuit implies
(a) zero pf and maximum current.
(b) unity pf and maximum current.
(c) unity pf and minimum current.
(d) zero pf and minimum current.

73. In a series RLC circuit at resonance, the magnitude of voltage developed across the capacitor
(a) is always zero.
(b) can never be greater than the input voltage.
(c) can be greater than the input voltage however, it is 90° out of phase with the input voltage.
(d) can be greater than the input voltage and is in phase with the input voltage. **[GATE 2001]**

74. A coil having a resistance of 5 Ω and inductance of 0.1 H is connected in series with a condenser of capacitance 50 μF. A constant alternating voltage of 200 V is applied to the circuit. The voltage across the coil at resonance is
(a) 200 volts (b) 1,788 volts
(c) 1,800 volts (d) 2,000 volts
[U.P.S.C. I.E.S. Elec. Engineering-I, 1994]

75. In a series R-L-C circuit, the maximum voltage across the capacitor occurs at a frequency
(a) double the resonant frequency.
(b) equal to resonant frequency.
(c) $\sqrt{2}$ limes the resonant frequency.
(d) below the resonant frequency.
[U.P.S.C. I.E.S. Elec. Engineering-I, 2006]

76. For a series RLC circuit, the power factor at the lower power frequency is
(a) 0·5 lagging. (b) 0·5 leading.
(c) unity. (d) 0·707 leading.
[U.P.S.C. I.E.S. Elec. Engineering-I, 1999]

77. Q-factor of a series R-L-C circuit possessing resonant frequency of 10 Hz and bandwidth of 5 Hz is
(a) 0·5 (b) 2
(c) 2·5 (d) 50
[G.G.S.I.P. Univ. Delhi Univ. Electrical Science, May 2005]

78. The quality factor of R-L-C circuit will increase if
(a) R decreases. (b) R increases.
(c) voltage increases.

79. When Q-factor of a circuit is high, then
(a) power factor of the circuit is high.
(b) impedance of the circuit is high.
(c) bandwidth is large.
(d) none of these. **[U.P.S.C. I.E.S. Elec. Engg.-I, 1992]**

80. Resonant frequency f_r of a series R-L-C circuit is related to half power frequencies f_1 and f_2 as
(a) $f_r = \dfrac{f_1 + f_2}{2}$ (b) $f_r = \sqrt{f_1 f_2}$
(c) $f_r = f_2 - f_1$ (d) $f_r = \sqrt{f_1} + \sqrt{f_2}$

81. In a 3-phase system the emfs are
(a) 30° apart. (b) 40° apart.
(c) 90° apart. (d) 120° apart.

Answers: 67. (a), 68. (b), 69. (b), 70. (a), 71. (d), 72. (b), 73. (c), 74. (c), 75. (d), 76. (d), 77. (b), 78. (a), 79. (a), 80. (a), 81. (d)

82. Which of the following are the necessary conditions for an entire 3-Φ system to be balanced ?

1. The line voltages are equal in magnitude.
2. The phase differences between successive line voltages are equal.
3. The impedances in each of the phases are identical.

Select the correct answer using the codes given below:

Codes:

(a) 1, 2 and 3 (b) 1 and 3

(c) 1 and 2 (d) 2 and 3

[U.P.S.C. I.E.S. Elec. Engineering-I, 1996]

83. The phase sequence of the 3-phase system shown in given figure is

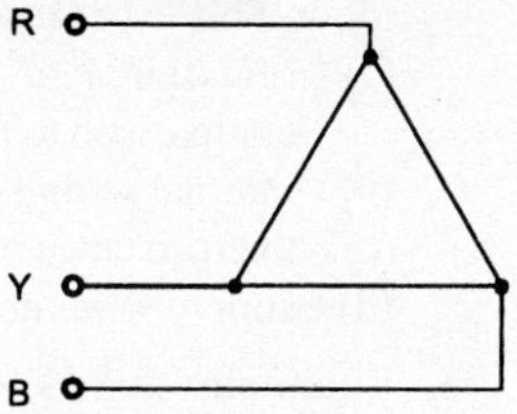

(a) RYB

(b) RBY

(c) BRY

(d) YBR **[GATE 2004]**

84. The rated voltage of a 3-phase power system is given as

(a) rms phase voltage. (b) peak phase voltage.

(c) rms line-to-line voltage.

(d) peak line-to-line voltage. **[GATE 2004]**

85. An alternator is delivering power to a balanced load at unity pf. The phase angle between the line voltage and the line current is

(a) 90° (b) 60°

(c) 30° (d) 0°

[U.P.S.C. I.E.S. Elec. Engineering-I, 1997]

86. E_{oa}, E_{ob} and E_{oc} are three phase voltages while E_{ab}, E_{bc} and E_{ca} are three line voltages of a balanced three-phase system having *a-b-c* phase sequence. In relation to E_{oc}, E_{bc} would

(a) lag by 30°. (b) lead by 30°.

(c) have the same phase.

(d) have no definite phase relationship.

[U.P.S.C. I.E.S. Elec. Engineering-I, 1997]

87. Three identical impedances are connected in delta to a 3-phase supply of 400 V. The line-current is 34.65 A and the total power taken is 14.4 kW. The resistance of the load in each phase (in ohm) is

(a) 20 (b) 16

(c) 12 (d) 10

[U.P.S.C. I.E.S. Elec. Engineering I, 1993]

88. A balanced delta connected load has an impedance of $9\angle 30°\ \Omega$ per phase. What is the impedance per phase of its equivalent star ?

(a) $27\angle 30°\ \Omega$ (b) $27\angle 90°\ \Omega$

(c) $3\angle 30°\ \Omega$ (d) $3\angle 20°\ \Omega$

[U.P.S.C. I.E.S. Elec. Engineering-I, 2004]

89. A 3-phase star-connected balanced load consumes P watts of power from a 400 V (line-to-line voltage) supply. If the same load is connected in delta across that same supply. What is the power consumption ?

(a) $\frac{P}{3}$ W (b) PW

(c) $\sqrt{3}$ PW (d) 3 PW

[U.P.S.C. I.E.S. Elec. Engineering-I, 2004]

90. A 3-phase delta-connected symmetrical load consumes P watts of power from a balanced supply. If the same load is connected in star to the same supply, then what is the power consumption ?

(a) $\frac{P}{3}$ (b) P

(c) $\sqrt{3}$ P (d) 3 P

[U.P.S.C. I.E.S. Elec. Engineering-I, 2005]

91. The power delivered to a 3-phase load can be measured by the use of 2 wattmeters only when the

(a) load is balanced. (b) load is unbalanced.

(c) 3-phase load is connected to the source through 3-wires.

(d) 3-phase load is connected to the source through 4-wires. **[U.P.S.C. I.E.S. Elec. Engineering-I, 1995]**

92. In a two wattmeter method of measuring power in a 3-phase system one of the wattmeters reads negative implying

(a) wattmeter connection is faulty.

(b) load is unbalanced.

(c) power flow is in the reverse direction.

(d) power factor is less than 0.5

93. In two wattmeter method of power measurement, one of the wattmeters will show negative reading when the load power factor angle is strictly

(a) less than 30°. (b) less than 60°.

(c) greater than 30°. (d) greater than 60°.

[U.P.S.C. I.E.S. Elec. Engineering-I, 1996]

94. While measuring power in a three phase load by two-wattmeter method, the readings of the two wattmeters will be equal and opposite when

(a) pf is unity. (b) load is balanced.

(c) phase angle is between 60° and 90°.

(d) the load is purely inductive.

[U.P.S.C. I.E.S. Elec. Engineering-I, 1995]

95. Which of the following statements is true about two-wattmeter method for power measurement in 3-phase circuit ?

Answers: 82. (a), 83. (b), 84. (c), 85. (c), 86. (b), 87. (c), 88. (c), 89. (d), 90. (a), 91. (c), 92. (d), 93. (d), 94. (d), 95. (d)

(a) Power can be measured using two wattmeter method only for star-connected 3-phase circuits.
(b) When two wattmeters show identical readings, the power factor is 0.5.
(c) When power factor is unity, one of the wattmeter reads zero.
(d) When the readings of the two wattmeters are equal but of opposite sign, the power factor is zero. **[U.P.S.C. I.E.S. Elec. Engineering-I, 1992]**

96. The ratio of the readings of two wattmeters connected to measure power in a balanced 3-phase load is 5:3 and the load is inductive. The power factor of the load is
(a) 0.917 lead. (b) 0.917 lag.
(c) 0.6 lead. (d) 0.6 lag.
[U.P.S.C. I.E.S. Elec. Engineering-I, 1994]

97. When two-wattmeter method of measurement of power is used to measure power in a balanced three-phase circuit, if the wattmeter reading is zero, then
(a) power consumed in the circuit is zero.
(b) power factor of the circuit is zero.
(c) power factor is unity.
(d) power factor is 0·5.
[U.P.S.C. I.E.S. Elec. Engineering-I, 2002]

98. If the reading of two-wattmeters are equal and positive in two-wattmeter method, the load pf in a balanced 3-phase 3-wire circuit will be
(a) zero (b) 0·5
(c) 0·866 (d) unity
[U.P.S.C. I.E.S. Elec. Engineering-I, 1999]

99. In the measurement of power on balanced load by two-wattmeter method in a 3-phase circuit, the reading of the wattmeters are 3 kW and 1 kW respectively, the latter being obtained after reversing the connections of the current coil. The pf of the load is
(a) 0·277 (b) 0·554
(c) 0·625 (d) 0·866
[U.P.S.C. I.E.S. Elec. Engineering-I, 2002]

100. W_1 and W_2 are the readings of two wattmeters used to measure power of a 3-phase balanced load. The reactive power drawn by the load is
(a) $W_1 + W_2$ (b) $W_1 - W_2$
(c) $\sqrt{3}(W_1 + W_2)$. (d) $\sqrt{3}((W_1 - W_2))$

101. The energy capacity of a storage battery is rated in
(a) kWh. (b) kW.
(c) ampere hours. (d) joules.
[U.P.S.C. I.E.S. Elec. Engineering-I, 2003]

102. As per recommendation of ISI the maximum number of points of lights, fans and socket outlets that can be connected in one sub-circuit is
(a) 8 (b) 10
(c) 15 (d) 20

103. Which of the following types of wiring is preferred for workshop lighting ?
(a) Casing-Capping wiring.
(b) Batten wiring.
(c) Concealed conduit wiring.
(d) Surface conduit wiring.

104. When a heater is connected to the power supply, the heater coil will glow but the supply wiring does not glow. This is because
(a) resistance of heater coils is very high in comparison to that of internal wiring.
(b) internal wiring is of superior material.
(c) the resistance of internal wiring is very high.
(d) supply wires are covered with insulation.

105. Luminous flux is
(a) the rate of energy radiation in the form of light waves.
(b) the part of light energy, radiated by Sun that is received on earth.
(c) measured in lux.

106. Candle power is
(a) the luminous flux emitted by the source per unit solid angle.
(b) the light radiating capacity of a source in a given direction.
(c) the unit of illumination.

107. Glare is caused due to
(a) excessive luminance.
(b) excessive lighting contrast in the field of vision.
(c) either (a) or (b).

108. Which of the following material is most commonly used for the filaments in incandscent lamps ?
(a) Tungsten. (b) Osmium.
(c) Tantalum. (d) Silver.

109. The safe operating temperature of a tungsten filament lamp is
(a) 1,000°C. (b) 2,000°C.
(c) 3,000°C. (d) 3,500°C.

110. What percentage of the input energy is radiated by filament lamps ?
(a) 60 – 70 per cent. (b) 40 – 50 per cent.
(c) 25 – 30 per cent. (d) 10 – 15 per cent.

111. An electric bulb, when broken, produces bang. It is on account of
(a) vacuum inside the bulb.

Answers: 96. (b), 97. (d), 98. (d), 99. (a), 100. (d), 101. (c), 102. (b), 103. (c), 104. (a), 105. (a), 106. (a), 107. (b), 108. (a), 109. (c), 110. (d), 111. (a),

(b) pressure inside is equal to that outside.
(c) pressure of air in the bulb.
(d) none of the above.

112. In an incandescent lamp, bird cage filament is usually used in vacuum bulb so as to
(a) increase the life span of the filament.
(b) give uniform radiations.
(c) reduce the oxidation phenomenon.
(d) reduce the convection losses.

113. Most affected parameter of a filament lamp due to variation in operating voltage is
(a) life. (b) light output.
(c) luminous efficiency.
(d) wattage.

114. The lumens output of an incandescent lamp with the variation in operating voltage is given as lumens output α V^n
where n is a constant. The value of n for tungsten is
(a) from 1.0 to 2.2 (b) from 3 to 4
(c) from 8 to 10 (d) from 10 to 12

115. Which gas is sometimes used in filament lamps ?
(a) Nitrogen. (b) Carbon dioxide.
(c) Argon. (d) Krypton.

116. The gas used in a gas-filled filament lamp is
(a) helium. (b) oxygen.
(c) nitrogen. (d) ozone.

117. In electric discharge lamps, light is produced by
(a) magnetic effect of current.
(b) heating effect of current.
(c) cathode ray emission.
(d) ionisation in a gas or vapour.

118. In electric discharge lamps for stabilization of arc
(a) a choke is connected in series with the supply.
(b) a variable resistance is connected in series with the circuit.
(c) a condenser is connected across the supply.
(d) any of the above.

119. Halogen lamps are useful for the illumination of
(a) airports.
(b) parks and large gardens.
(c) playing fields. (d) all of the above.

120. Sodium vapour lamp needs an ionisation voltage of about
(a) 5 V. (b) 20 V.
(c) 50 V. (d) 100 V.

121. The ignition voltage for a sodium lamp is about
(a) 100–125 V. (b) 200–240 V.
(c) 300–400 V. (d) 400–600 V.

122. When a sodium vapour lamp is switched on, initially the colour is
(a) red. (b) pink.
(c) yellow. (d) blue.

123. The colour of light given out by a sodium vapour discharge lamp is
(a) pink. (b) bluish green.
(c) yellow. (d) blue.

124. In a sodium vapour lamp, the discharge starts first in the
(a) argon gas. (b) neon gas.
(c) nitrogen gas. (d) krypton gas.

125. In a sodium vapour discharge lamp the neon gas
(a) changes the colour of light.
(b) prevents vapourization of filament.
(c) acts as a shield around the filament.
(d) assists in developing enough heat to vapourize the sodium.

126. The capacitor is used in auto-transformer circuit of a sodium vapour lamp in order to
(a) regulate discharge voltage.
(b) improve the circuit power factor.
(c) control lamp illumination level.
(d) protect the lamp against overvoltage.

127. The luminous efficiency of a sodium vapour lamp is .... lumens per watt.
(a) 40-50 (b) 50-100
(c) 120-200 (d) 10-12

128. High pressure mercury vapour lamps
(*a*) are similar in construction to sodium vapour lamps.
(*b*) need 4 or 5 minutes to attain full brilliancy.
(*c*) are generally used for general industrial lighting, railway yards, work areas, shoping centres etc.
(*d*) all of the above.

129. Which of the following lamps is a cold cathode lamp ?
(a) Sodium vapour lamp.
(b) Neon lamp.
(c) Mercury vapour lamp.
(d) None of the above.

130. In neon signs, neon with a mixture of mercury is used to give

Answers: 112. (b), 113. (a), 114. (b), 115. (c), 116. (c), 117. (d), 118. (a), 119. (d), 120. (a), 121. (d), 122. (b), 123. (c), 124. (b), 125. (d), 126. (b), 127. (a), 128. (d), 129. (b), 130. (b)

(a) blue colour. (b) green colour.
(c) pink colour. (d) yellow colour.

131. Neon tubes are widely used for
(a) advertising. (b) indoor lighting.
(c) road signalling. (d) airport lighting.

132. A stabilising choke is connected in the fluorescent tube circuit so as to
(a) reduce the flicker.
(b) act as a ballast in operating conditions and provide a voltage impulse for starting.
(c) act as a starter.
(d) avoid radio-interference.

133. For operation of fluorescent tube on dc supply the additional device incorporated in the tube circuit is a
(a) transformer. (b) resistor.
(c) inductor.

134. Sometimes the wheels of the rotating machinery, in fluorescent tube lighting, appear to be stationary . This is due to
(a) luminescence effect. (b) light fluctuations.
(c) stroboscopic effect. (d) all of the above.

135. The flicker effect of fluorescent lamps is more pronounced at
(a) lower voltages. (b) higher voltages.
(c) higher frequencies.
(d) lower frequencies.

136. The frequency of flickers in a fluorescent tube operating at 220 V, 50 Hz supply is
(a) 200 per second. (b) 100 per second.
(c) 50 per second. (d) 25 per second.

137. Radio-interference from a fluorescent tube can be reduced by
(a) eliminating choke.
(b) connecting a small capacitor across starter terminals.
(c) putting two tubes in parallel..
(d) any of the above.

138. In a mercury vapour lamp light, red objects appear black. This is on account of
(*a*) colour mixing.
(*b*) high wavelengths of red colour.
(c) absence of red light from lamp radiations.
(*d*) absorption of red light by the lamp.

139. Blinking of a fluorescent tube may be removed by
(a) replacement of starter.
(b) checking of all connections.
(c) change of ballast tapping.
(d) any of the above-turn by turn.

140. The cost of a fluorescent tube is more than that of an incandescent lamp on account of
(a) more components in the tube.
(b) more quantity of glass in the tube.
(c) more labour required.
(d) all of the above.

141. The disadvantage(s) of fluorescent tubes in comparison to filament lamps is/are
(a) high cost. (b) noise in choke.
(c) stroboscopic effect.
(d) all of the above.

142. Power factor is highest in case of
(a) sodium vapour lamp.
(b) mercury vapour lamp.
(c) incandescent lamp. (d) neon lamp.

143. The lamp which gives nearly monochromatic light is
(a) sodium vapour lamp.
(b) mercury vapour lamp.
(c) tungsten filament lamp.
(d) fluorescent tube.

144. The direct lighting scheme is most efficient but is liable to cause
(a) glare. (b) hard shadows.
(c) monotony. (d) both (a) and (b).

145. Total flux required in any lighting scheme depends inversely on
(a) surface area. (b) space-height ratio.
(c) illumination.
(d) coefficient of utilisation.

146. Which of the following will need the highest level of illumination ?
(a) Living room. (b) Kitchen.
(c) Proof reading. (d) Hospital wards.

147. Which of the following will need the lowest level of illumination ?
(a) Workshop. (b) Displays.
(c) Railway Platform. (d) Garage.

148. Glare is reduced by
(a) increasing the mounting height of the lamp.
(b) using diffusers.
(c) using reflectors to cut-off the light at certain angle.
(d) all of the above.

149. The lamp is provided with a reflector in order to
(a) avoid glare.
(b) provide better illumination.
(c) protect the lamp. (d) all of the above.

Answers: 131. (a), 132. (b), 133. (b), 134. (c), 135. (d), 136. (b), 137. (b), 138. (c), 139. (d), 140. (d), 141. (d), 142. (c), 143. (a), 144. (d), 145. (b), 146. (c), 147. (c), 148. (d), 149. (d)

150. Reflectors are provided with slits at the top so as to
(a) reduce heating effect.
(b) reduce colour effect.
(c) introduce chimney effect for cleaning.
(d) all of the above.

151. A fuse is normally a
(a) power limiting device.
(b) voltage limiting device.
(c) current limiting device.
(d) power factor correcting device.

152. Fuses have got advantages of
(a) cheapest type of protection.
(b) inverse time-current characteristic.
(c) no maintenance.
(d) current limiting effect under short-circuit conditions.
(e) all of the above.

153. A fuse is
(a) always connected in series with the circuit.
(b) always connected in parallel with the circuit.
(c) normally connected in series with the circuit.
(d) normally connected in parallel with the circuit.

154. A fuse is
(a) normally inserted in phase wire.
(b) normally inserted in neutral wire.
(c) never inserted in neutral wire.
(d) never inserted in phase wire.

155. Operation of fuse is based upon
(a) photo-electric effect.
(b) heating effect of electric current.
(c) magnetic effect of electric current.
(d) none of the above.

156. The fuse blows off by
(a) arcing. (b) burning.
(c) melting. (d) none of the above.

157. The primary function of a fuse is to
(a) open the circuit. (b) protect the appliance.
(c) protect the line.
(d) prevent excessive currents from flow through the circuit.

158. A fuse in a motor circuit provides protection against
(a) short-circuit. (b) overload.
(c) open circuit.
(d) short-circuit and overload.

159. Fuses normally protect a system against
(a) overcurrent. (b) undercurrent.
(c) resonance. (d) lightning surge.

160. Protection by fuses is generally not used beyond
(a) 200 A. (b) 50 A.
(c) 25 A. (d) 10 A.

161. Operating time for fuse wire is about
(a) 0.1 ms. (b) 2 ms.
(c) 0.1 s. (d) few seconds.

162. The relation between fusing current(I) and diameter(d) of fuse wire is
(a) $I \alpha d$. (b) $I \alpha^{1/d}$.
(c) $I \alpha d^{3/2}$. (d) $I \alpha d^{1/2}$.

163. Rewirable fuse has
(a) inverse time-current characteristic.
(b) linear time-current characteristic.
(c) square law time-current characteristic.
(d) none of the above.

164. The material used in liquid fuses is
(a) SF_6
(b) distilled water.
(c) carbon tetra chloride.
(d) transformer oil.

165. In comparison to rewirable fuses HRC fuses have the advantage(s) of
(a) high speed operation.
(b) high rupturing capacity.
(c) no ageing effect.
(d) all of the above.

166. In HRC fuse the time between cut-off and final current zero is called the
(a) pre-arcing time. (b) arcing time.
(c) total operating time. (d) none of these.

167. For the same rupturing capacity, the actual current to be interrupted by an HRC fuse is
(a) much more than that of any circuit breaker.
(b) much less than that of any circuit breaker.
(c) equal to that of any circuit breaker.
(d) none of the above.

168. HRC fuses provide best protection against
(a) short-circuit. (b) lightning.
(c) sparking. (d) fire.

169. Heat generated in HRC fuses is dissipated mainly through
(a) radiation. (b) convection.
(c) conduction. (d) all of these.

170. The rating of fuse is expressed in terms of
(a) amperes. (b) volts.
(c) VAR. (d) kVA.

Answers: 150. (d), 151. (c), 152. (e), 153. (a), 154. (c), 155. (b), 156. (c), 157. (d), 158. (d), 159. (a), 160. (a), 161. (b), 162. (c), 163. (a), 164. (c), 165. (d), 166. (b), 167. (b), 168. (a), 169. (a), 170. (a)

171. A circuit breaker is essentially
(a) an arc extinguisher.
(b) a current interrupting device.
(c) a power factor correcting device.
(d) a device for neutralising the effects of transients.

172. Circuit breakers usually operate under
(a) steady-state short-circuit current.
(b) sub-transient state of short-circuit current.
(c) transient state of short-circuit current.
(d) none of the above.

173. On occurence of fault on the connected circuit, a circuit breaker operates
(a) manually. (b) automatically.
(c) manually through a control switch.
(d) depending upon the circuit breaker design.

174. A circuit breaker normally operates
(a) when the power is to be supplied.
(b) when the line is to be tested.
(c) when the switch is to be put on.
(d) whenever fault occurs in the line.

175. It is fatal to touch a live wire as
(a) the voltage may cause burns to the skin.
(b) it will cause flow of current through the human body.
(c) the current may cause burns to the skin or inside the body.
(d) it may cause damage to the heart and nerve system.

176. Earthing of electrical equipment is necessary for the protection against
(a) over loading. (b) voltage fluctuation.
(c) danger of electric shock.
(d) high conductor temperature.

177. The earth wire should be
(a) good conductor of electricity.
(b) mechanically strong.
(c) both (a) and (b).
(*d*) mechanically strong but bad conductor of electricity.

178. Earth wires are made of
(a) copper. (b) aluminium.
(c) iron.
(d) galvanised stranded steel.

179. The size of the earth wire is determined on the basis of
(a) voltage of the service line.
(b) current carrying capacity of the service line.
(c) atmospheric conditions.
(d) none of the above.

180. The earth wire should not be of size smaller than
(a) 10 SWG copper. (b) 8 SWG copper.
(c) 6 SWG copper. (d) 4 SWG copper.

Answers: 171. (b), 172. (c), 173. (b), 174. (d), 175. (b), 176. (c), 177. (c), 178. (d), 179. (b), 180. (a)

TRUE/FALSE TYPE QUESTIONS

Note: Identify whether the following statements are TRUE (T) or FALSE (F)

1. One ampere means the flow of one coulomb each second.
2. Voltage applied across a circuit, acts as a force.
3. The lower the resistivity, greater the resistance.
4. Specific resistance is measured in Ω/m^3
5. The reciprocal of resistivity of a material is called its conductivity.
6. The specific resistance of copper increases with the increase in temperature.
7. The values of temperature coefficient of resistance of a given conductor are the same at different temperatures.
8. The voltage drop across each resistor is same in case of a series circuit.
9. Resistance of a tungsten filament lamp decreases with the increase in supply voltage.
10. Dielectric constant and permittivity are same.
11. Potential of every point inside a hollow charged sphere is same.
12. The charge in a capacitor is stored in dielectric field.
13. The capacitor never dissipates energy but only stores it.
14. Electric field line and equipotential line are never orthogonal.
15. The charge on an isolated conductor resides inside the conductor.

Answers: 1. (T), 2. (T), 3. (F), 4. (F), 5. (T), 6. (T), 7. (F), 8. (F), 9. (F), 10. (T), 11. (T), 12. (T), 13. (T), 14. (F), 15. (F)

16. With the insertion of an iron core into a coil, magnetic field is strengthened.
17. With the insertion of an iron core into a coil the reluctance of the iron path increases.
18. A minus sign is required to be placed before the right hand side of the expression for emf induced just to indicate the phenomenon explained by Lenz.
19. High silicon steel has comparatively higher permeability and high saturation.
20. Hysteresis loss is the amount of energy dissipated as heat in the magnetic material.
21. Eddy current losses are reduced by using thin plates of high resistivity material.
22. AC varies in magnitude and reverses in polarity.
23. Periodic time, T = supply frequency, *f*.
24. Peak-to-peak voltage is twice the peak value of voltage.
25. The phase difference between two alternating quantities is measured by the angular distance between the points where the two curves cross the base or reference line in the same direction.
26. Each 180° (or π radian) rotation of the phasor generates a complete cycle.
27. When two waves are 90° out of phase, one has its peak value when the other is at zero value.
28. An inductor opposes equally the dc as well as ac.
29. A pure inductance connected across 230 V, 50 ac supply consumes 125 watts.
30. An inductor does not offer any opposition to dc.
31. A capacitor blocks alternating current.
32. The inductors are designed to have low value of Q-factor.
33. The average power P_{av} is equal to one half of the peak power for sinusoidal waveforms.
34. The ratio of the bandwidth to the resonant frequency is called the quality factor.
35. In a series resonant circuit, lower the resistance in the circuit, more steep will be the current response.
36. In a parallel R-L circuit, as the frequency increases, impedance Z decreases.
37. In a parallel resonant circuit, the value of resistance has effect on the resonant frequency.
38. When a capacitor is connected in parallel to an inductive circuit, the phase angle increases and pf decreases.
39. In a 3-phase system the emf are 120° apart.
40. A 3-phase supply will be unbalanced when either of the three-phase voltages are unequal in magnitude or the phase angle between these phases is not equal to 120°.
41. Single phase system is more capable and reliable than 3-phase system.
42. In star-connected system

$$E_p = E_L \text{ and } I_L = \sqrt{3}\, I_P$$

43. Power output of a 3-phase system is given as

$$P = \sqrt{3}\, V_P I_P \cos \phi$$

44. In star-connected system line voltages are 30° ahead of respective phase voltages.
45. In delta-connected system line currents are 30° behind the respective phase currents.
46. When load pf is unity one of the two wattmeters reads zero.
47. During charging of a lead acid cell the specific gravity of a cell decreases.
48. Faure or pasted plates are lighter in weight and cheaper in cost than formed or plante plates.
49. The grids of positive plates are lighter than those of negative plates.
50. There is always one more positive plate than the negative plates in lead-acid batteries.
51. In a lead acid battery the positive terminal is of comparatively larger diameter than that of negative terminal.
52. To prevent local action in a battery, only distilled water is used in electrolytes.
53. Charging of sulphated battery produces very little heat.
54. A sulphated battery should be charged always for a longer period at low rate.
55. Undercharging of a lead acid battery causes increase in specific gravity of the electrolyte.
56. In constant current charging method the supply voltage is reduced as the battery gets charged.
57. In constant voltage charging, charging current increases as the battery gets charged.
58. The emf of a storage battery depends upon the size of the electrodes.
59. The plates of alkaline cells do not buckle or swell.
60. The capacity of Edison cell decreases when its electrolyte is exposed to air.

Answers: 16. (T), 17. (F), 18. (T), 19. (F), 20. (T), 21. (T), 22. (T), 23. (F), 24. (T), 25. (T), 26. (F), 27. (T), 28. (F), 29. (F), 30. (T), 31. (F), 32. (F), 33. (T), 34. (F), 35. (T), 36. (F), 37. (T), 38. (F), 39. (T), 40. (T), 41. (F), 42. (F), 43. (F), 44. (T), 45. (T), 46. (F), 47. (F), 48. (T), 49. (F), 50. (F), 51. (T), 52. (T), 53. (F), 54. (T), 55. (F), 56. (F), 57. (F), 58. (F), 59. (T), 60. (T),

61. Edison cells cannot be kept in discharged condition for a long period.

62. Short-circuits in alkaline cells do not reduce the life.

63. Loop-in-system is universally used for connections of various lamps and other appliances in parallel.

64. Conduit wiring provides protection against fire and mechanical damage.

65. Fuse is provided only in phase wire, never on neutral wire.

66. In motor wiring installations, looping of conductors and use of tee-joints is not permitted.

67. The insulation on a current carrying conductor is provided to prevent current leakage.

68. PVC conduits cannot be buried in concrete.

69. The sensation of colour is due to the difference in the wavelengths of the light radiations.

70. Solid angle subtended at a point in space is 2π steradians.

71. For the same power output low voltage lamps willl be more sturdy.

72. Filament lamps at starting will take current less than its normal operating current.

73. Incandescent filament lamps are generally of copper filaments.

74. Incandescent lamps should be used with luminaires.

75. Maintenance of discharge lamps is more cumbersome.

76. Halogen lamps are preferred for outdoor illumination of buildings.

77. Sodium vapour lamp is only suitable for operation on dc .

78. A leak transformer used in sodium vapour lamp acts as a step-down transformer initially.

79. In a sodium vapour lamp, the neon gas serves as to start the discharge and to develop sufficient heat to vapourize the sodium.

80. In mercury vapour lamp the harmful ultraviolet rays are absorbed by outer bulb.

81. Sodium vapour lamps use metallic sodium sealed in translucent aluminium oxide tubes.

82. Power factor of high pressure mercury vapour lamp is nearly unity and so do not need capacitor for power factor correction.

83. In a colour-corrected high pressure mercury vapour lamp, the outer glass envelope is coated with fluorescent powder.

84. Mercury iodide lamps are not suitable for use for flood-lighting.

85. The luminous efficiency of mercury iodided lamps ranges from 75 to 90 lumens per watt.

86. Starters used in fluorescent tubes are usually of thermal type.

87. In fluorescent tubes low voltage is required for starting.

88. Fluorescent tubes operating on dc are generally free from stroboscopic effect.

89. In a fluorescent tube starter is required for preheating the electrodes.

90. According to the Stefan Boltzman's law intensity of emitted radians is proportional to T^4 where T is the absolute temperature.

91. For the same lumens output, the operating cost of fluorescent lamp is less than the filament lamp.

92. Low voltage generally facilitates starting of fluorescent tubes.

93. With use, walls of a fluorescent tube blacken a little and dark rings appear near the ends owing to the deposition of active material from the electrodes.

94. The lumen output of a fluorescent tube is reduced, after about 4,000 hours of use, by 15-20 per cent.

95. Power factor of fluorescent lamps is high.

96. Interior lighting is designed by lumen method.

97. Lumen method takes into account the inter-reflections of light inside a room.

98. Luminaires are generally categorized as industrial, commercial or residential.

99. Lenses with prismatic patterns refract light sources to disperse the rays or to direct them most effectively.

100. Greater the size of the room lesser will be the coefficient of utilisation.

101. Maintenance factor and depreciation factor are two different names of same concept.

102. Space-height ratio is less with dispersive type reflectors than that with concentric type reflectors.

103. Smaller the space-height ratio lesser the number of lamps required by a given installation.

104. In a lighting installation selection of proper space-height ratio ensures uniformity of illumination.

Answers: 61. (F), 62. (T), 63. (T), 64. (T), 65. (T), 66. (T), 67. (T), 68. (F), 69. (T), 70. (F), 71. (T), 72. (F), 73. (F), 74. (T), 75. (T), 76. (T), 77. (F), 78. (F), 79. (T), 80. (T), 81. (T), 82. (F), 83. (T), 84. (F), 85. (T), 86. (F), 87. (F), 88. (T), 89. (T) 90. (T), 91. (T), 92. (F), 93. (T), 94. (T), 95. (F), 96. (T), 97. (T), 98. (T), 99. (T), 100. (F), 101. (T), 102. (F), 103. (F), 104. (T)

105. Ratio of light flux reaching the working plane to that emitted by the lamp(s) is called the maintenance factor.

106. Illumination level in railway yards is usually maintained at 10 to 20 lux.

107. Depreciation factor is merely the inverse of utilisation factor.

108. Maintenance factor is always more than unity.

109. The fusing current for a stranded fuse will be equal to the product of the number of strands and fusing current for one strand.

110. Breaking capacity of a fuse is the rating corresponding to the rms value of the ac component of the maximum prospective current and the system voltage.

111. The characteristics of a fuse is governed by fuse material and shape of the fuse element.

112. HRC fuses obey inverse time-current characteristic.

113. HRC fuses are rensable after blowing off.

114. A fuse has low braking capacity as compared to a circuit breaker.

115. Current rating of fuse element is always more than its fusing current.

Answers: 105. (F), 106. (T), 107. (F), 108. (F), 109. (F), 110. (T), 111. (T), 112. (T), 113. (F), 114. (T), 115. (F)

FILL IN THE BLANK TYPE QUESTIONS

Note: Fill in the blanks with appropriate word or words.

1. The rate of movement of charge in a conductor is known as ________.

2. ________ is the force that causes a current to flow.

3. Voltage across a resistance is called the ________.

4. Voltage is a form of ________ energy.

5. 1 V = 1 ________ per coulomb.

6. Electric pressure is also called the ________.

7. The ratio of voltage and current ________.

8. The curve representing Ohm's law is ________.

9. A linear resistor is one which obeys ________ law.

10. For the same dimensions of a conductor its resistance depends upon the ________ of the material.

11. Elements of electric heaters and filament lamps are made of high resistivity materials. This is done to ________ the filament length.

12. The property of a conductor due to which it induces the flow of current through the conductor is called the ________.

13. The diameter of wire is reduced to one half, keeping the length constant. The new resistance will be ________ times the original value.

14. If the diameter of the wire is halved, its current carrying capacity will become approximately ________.

15. Electrical conductivity is measured in ________.

16. With the rise in temperature the resistance of carbon ________.

17. Resistance of carbon filament lamp ________ with the increase in supply voltage.

18. Two resistors of 12 Ω and 4 Ω respectively are connected in series. A short is placed across the combination. The effective resistance will be ________.

19. A series arrangement of *n* identical resistors is changed into *n* parallel arrangement. The new equivalent resistance will become of ________ of the original resistance.

20. When one leg of a parallel circuit gets opened out, the current drawn from the supply will ________.

21. Three resistors, each of R ohms are connected to form a triangle. The resistance between any two terminals will be ________ Ω.

22. Twelve identical wires of resistance 6 Ω each are arranged to form the edges of a cube. The effective resistance between the opposite corners of the cube will be ________.

23. The values of E and I for the circuit shown in figure are ________ V and ________ A. **[GATE 1997]**

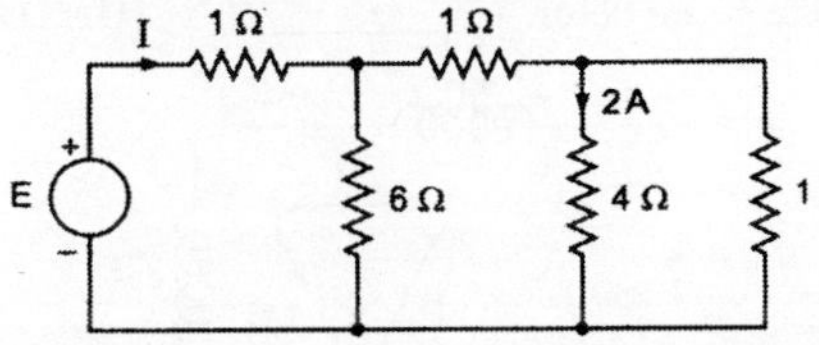

24. A 200 W, 100 V lamp is to be operated on 250 V supply. The additional resistance required to be connected in series will be ________ Ω.

25. The unit of electric intensity is ________.

26. The electric field inside a perfectly conducting media is ________.

27. A hollow conductor is at a potential V. The potential at any point inside the hollow conductor is ________.

[GATE 1994]

28. The ________ is the rate of change of potential measured in the direction of electric force.

29. ________ is a measure of the ability of a capacitor to store an electric charge.

30. The capacitance of an isolated sphere of radius 10 cm in air is equal to ________ pF. **[GATE 1997]**

31. Absolute permeability of vacuum is ________ H/m.

32. The ratio of intensity of magnetisation to the magnetizing force or intensity of magnetic field is called the ________.

33. In right hand rule when thumb points in the direction of flow of current then finger tips will point the direction of ________ of force.

34. The space between the poles of a magnet is called the ________.

35. The transfer of magnetism from one magnetic material to another is called the magnetic ________.

36. ________ force drives or tends to drive flux through a magnetic circuit.

37. The ratio of total flux (flux in the iron path) to useful flux (flux in the airgap) is known as ________ factor.

38. The product of the number of turns and flux linking with the coil is known as ________.

39. The inductance of a coil is proportional to the ________ of the number of turns, all other parameters remaining the same. **[GATE 1994]**

40. The inductance has dimensions of ________.

41. Two identical coils of negligible resistance when connected in series across a 200 V, 50 Hz source draws, a current of 10 A. When the terminals of one of the coils are reversed, the current drawn is 8 A. The coefficient of coupling between the two coils is ________. **[GATE 1997]**

42. The equivalent inductance seen at terminals A-B in figure given below is ________ H. **[GATE 1992]**

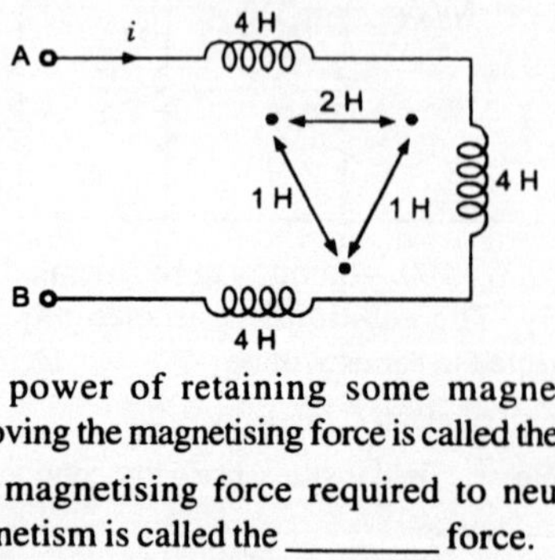

43. The power of retaining some magnetism after removing the magnetising force is called the ________.

44. The magnetising force required to neutralize the magnetism is called the ________ force.

45. The lagging of magnetic flux density B behind the magnetic field intensity is known as magnetic ________.

46. The ________ currents produce a magnetic field of their own which opposes the main magnetic field.

47. The voltage and current waveforms for an element are shown in figure. The circut element is ________ and its value is ________.

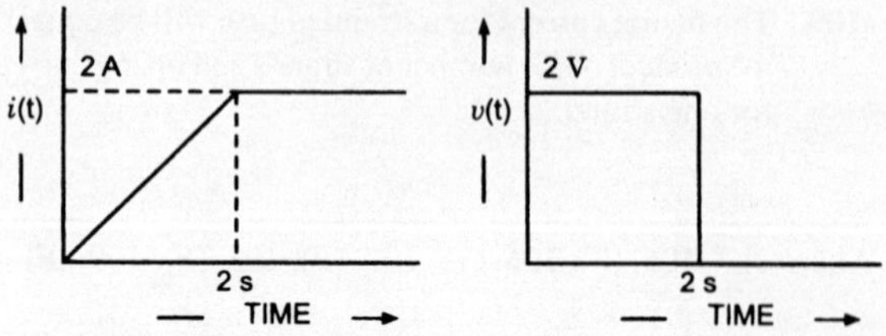

48. The form factor for half wave rectified current is________.

49. The emf induced in a coil rotating in a uniform magnetic field is maximum when the flux linkage by the coil is ________.

50. The opposition to the flow of an alternating current is called the ________.

51. In the circuit below the voltage **V**s has a phase angle of ________ with respect to V_L. **[GATE 1994]**

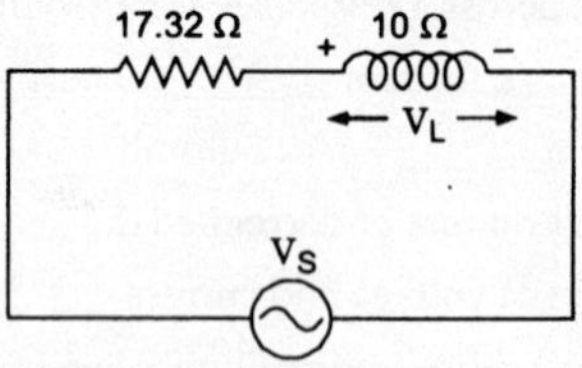

52. A series R-L-C circuit has the following parameter values R = 10 Ω, L = 0.01 H, C = 100 μF. The Q-factor of the circuit at resonance is ________.

[GATE 1995]

53. A parallel a-c circuit draws ________ current when in resonance.

54. The ability of a resonant circuit to discriminate between one particular frequency and all others is called its ________

55. Selectivities of different resonance circuits are compared in terms of their ________.

56. The ratio of bandwidth to the resonant frequency is called the ________ of the circuit.

57. A tuned circuit provides a voltage of known ________ across a known resistance.

58. If the coil employed in a tuned circuit is of low resistance, its selectivity will be ________.

59. In a parallel R-L circuit, the phase angle ________ with the decrease in frequency.

60. The frequency, at which an R-L-C network behaves as a purely resistive network, is called the ________ frequency.

61. In case of a series circuit Q factor is ________ magnification while in case of parallel circuit it is ________ magnification.

62. In a balanced 3-phase star-connected system line voltages are 30°________ of the respective phase voltages.

63. In a balanced delta-connected system line currents are 30°________ the respective phase currents.

64. In a 3 – Φ delta connection, line voltage is ________ phase voltage.

65. A delta connected load with 10Ω resistor in each phase is connected across a 400 V line to line voltage balanced 3–Φ supply. The power per phase consumed by this load is ________ kW.

66. Three resistors of 10Ω each forming a star-connected load are connected across a 3 – Φ balanced supply line to line voltage of 400 V. The total power consumed by this load is________ kW.

67. A set of 3 equal resistors, each of value R_X, connected in star across RYB of fig shown consumes the same power as the unbalanced delta-connected load shown. The value of R_X is ________Ω **[GATE 1994]**

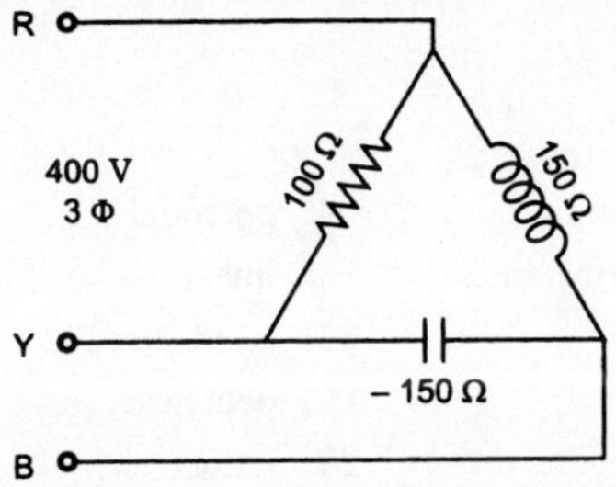

68. In a lead acid cell hydrogen is liberated at ________ plate.

69. During charging of a battery the terminal voltage decreases with the ______ in temperature.

70. In a lead acid battery porous rigid plastic sheets are inserted between plates to prevent ________ between plates.

71. The ________ of plates reduces the internal resistance and increases the capacity of the cell.

72. The supply required for charging a lead acid battery is ________.

73. Proper charging rate of battery in amperes is around ________ of its ampere-hour capacity.

74. Battery charging room is to be kept well ventilated so as to reduce the hydrogen concentration below ________ percent.

75. The best indication about the state of charging of a lead acid battery is given by ________ of electrolyte.

76. To avoid freezing in cold weather, specific gravity of the electrolyte of the lead acid cell should be kept above ________.

77. The expected life of a lead acid battery is ________ years.

78. Lead acid battery should not remain discharged normally for more than ________ hours.

79. The internal resistance of an Edison cell is about ________ times that of the lead acid cell.

80. For a given ampere hour capacity, weight of lead acid cell is ________ of that of Edison cell.

81. One Angstrom is equal to ________.

82. Human eye is most sensitive for a wavelength of ________A°.

83. The amount of light entering the eye is controlled by ________.

84. Light is a form of ____________energy.

85. Light wave travels with a velocity of ________.

86. The colour having shortest wave length is ________.

87. The colour having longest wave length is ________.

88. Steradian is defined as the solid angle that subtends a surface on the sphere equivalent to the square of the ________.

89. Light is expressed in ________.

90. ________may be defined as the brightness within the field of vision of such a character as to cause annoyance, discomfort, interference with vision or eye fatigue.

91. The unit of illumination is ________.

92. 1 lumen/m^2 is equal to ________.

93. The luminous flux is measured in ________.

94. Lumens/watt is the unit of ________.

95. The unit of solid angle is ________.

96. ________ lamp has least capacity to sustain voltage fluctuations.

97. ________ lamps emit only a faint glow and are used as indicating lamps on panels and appliances.

98. Power factor of a sodium vapour lamp is nearly ________lagging.

99. Leak transformer used in a sodium vapour lamp acts as a choke coil for limiting the ________when the lamp starts conducting.

100. The light from a sodium vapour lamp is nearly ________.

101. Neon lamps are basically __________ lamps.

102. The usual operating voltage for neon tubes is __________ volts.

103. Neon tube consists of a length of glass tubing containing two electrodes, normally cylindrical in shape, of iron, steel or __________.

104. Fluorescent lamp operating on dc supply needs __________ in addition to a starter and a choke.

105. ________ is essentially a small price of metal inserted in a circuit which melts when a predetermined value of current flows through it. (circuit breaker, fuse)

106. ________ of a fuse element is defined as the current, which the fuse wire can normally carry without overheating or melting.

107. Current rating of a fuse element depends upon the permissible ________ of the contacts of the fuse holder, fuse and upon the deterioration of fuse caused by oxidation.

108. ________ is defined as the minimum value of current at which the fuse element or fuse wire melts.

109. The fusing current depends upon the type of material used, x-sectional area, ________, size and location of the terminals, diameter of wire, type of enclosure used and type of surface.

110. If two stranded fuse wires are twisted together, then fusing current of this wire will be ________ the sum of fusing current of two wires. (equal to, less than)

111. The ratio of minimum fusing current and the current rating of the fuse element is called the ________.

112. The value of fusing factor is always ________ than 1. (more, less)

113. Breaking capacity of a fuse is the rating corresponding to the .... value of ac component of the maximum prospective current and the system voltage. (average/rms)

114. HRC fuses are capable of clearing ________ currents. (high, as well as low)

115. Operating time of in HRC fuse in the event of a fault is ________ than one cycle. (less, more)

116. The operating time of HRC fuse is ________ cycle.

117. ________ are the mechanical devices designed to close or open contact members, thus closing or opening an electrical circuit under normal and abnormal conditions.

118. It is ________ to touch a live wire as it would cause flow of current through the human body.

119. The current that is considered safe for a human body is ________.

120. Earthing of electrical equipment is necessary for the protection against danger of ________.

ANSWERS

1. electric current	**2.** EMF	**3.** voltage drop	**4.** potential
5. joule	**6.** voltage	**7.** remains constant	**8.** linear
9. Ohm's	**10.** resistivity	**11.** reduce	**12.** conductance
13. four	**14.** one-fourth	**15.** S/m	**16.** decreases
17. decreases	**18.** zero	**19.** $1/n^2$	**20.** reduce
21. 2/3 R	**22.** 5 Ω	**23.** 31, 13	**24.** 75
25. N/C	**26.** zero	**27.** zero	**28.** potential gradient
29. Capacitance	**30.** 11.11	**31.** $4\pi \times 10^{-7}$	**32.** susceptibility
33. magnetic lines	**34.** airgap	**35.** induction	**36.** Magnetomotive
37. leakage	**38.** flux linkage	**39.** square	**40.** flux per current
41. 1/9	**42.** 8	**43.** retentivity	**44.** coercive
45. hysteresis	**46.** eddy	**47.** inductor, 2 H	**48.** 1.57
49. zero	**50.** impedance	**51.** 60°	**52.** 1.0
53. minimum	**54.** selectivity	**55.** bandwidths	**56.** selectivity
57. frequency	**58.** high	**59.** increases	**60.** resonant
61. voltage, current	**62.** ahead	**63.** behind	**64.** equal to
65. 16.0	**66.** 16.0	**67.** 100	**68.** negative
69. increase	**70.** short-circuiting	**71.** interlacing	**72.** dc
73. one-eighth	**74.** three	**75.** specific gravity	**76.** 1.225
77. 2 to 5	**78.** 24	**79.** five	**80.** twice

81. 10^{-10} m	**82.** 5,500	**83.** pupil	**84.** radiant
85. 3×10^8 m/s	**86.** blue	**87.** red	**88.** radius
89. lumen-hours	**90.** Glare	**91.** lux	**92.** 1 lux
93. lumens	**94.** luminous efficiency	**95.** steradian	**96.** Incandescent
97. Glow	**98.** 0.3	**99.** current	**100.** monochromatic
101. gas discharge	**102.** 6,000	**103.** copper	**104.** a resistor
105. Fuse	**106.** Current rating	**107.** temperature rise	**108.** Fusing current
109. length	**110.** less than	**111.** fusing factor	**112.** more
113. rms	**114.** high as well as low	**115.** less	**116.** one-fourth of a
117. circuit breakers	**118.** fatal	**119.** 9 mA	**120.** electric shock

❑❑❑

INDEX

A

B

C

D

E

F

P

Q

R

S

T

U

V

W

X

□□□

Win Prizes!

ATTENTION READERS

We request you for your frank assessment regarding some of the aspects of the book given as under:

ELEMENTS OF ELECTRICAL ENGINEERING

By

J.B. Gupta

GTU

Please fill up the given space in neat capital letters.

(i) What topics of your syllabus that are important from you examination point of view are not covered in the book ?

....

....

....

....

(ii) What are the chapters wherein the treatment of the subject matter is not systematic or organised or updated ?

....

....

....

....

(iii) Have you come across misprints/mistakes/factual inaccuracies in the book? Please specify the chapters and the page numbers.

....

....

....

....

(iv) What other books on the same subject have you found/heard better that the present book? Please specify in terms of price and quality.

....

....

....

....

(v) Further suggestions and comments for the improvement of the book:

....

....

....

....

(PS: If need be please attach a separate sheet to write your views.)

Other Details :

(i) Who recommended you the book ? (Please tick in the box near the option relevant to you.)

☐ Teachers ☐ Friends ☐ Bookseller

(ii) Name of the recommending teacher, his designation and address.

....

....

....

....

(iii) Name and address of the bookseller you purchased the book form:

....

....

....

....

(iv) Name and address of your institution (Please mention the University or Board, as the case may be)

....

....

....

....

(v) Your name and complete postal address:

....

....

....

....

(vi) Course you are enrolled in:

....

....

....

....

The best assessment will be awarded each month. The award will be in the form of our publications as decided by the Editorial Board.

Please mail the filled up coupon at your earliest to:

S.K. Kataria & Sons®

4885/109, Prakash Mahal, Dr. Subhash Bhargav Lane,
Opposite Delhi Medical Association, Daryaganj, New Delhi-110002
Phone: +91-11-43551243; Mobile: +91-9871775858, 7678425209
e-mail: katariabook@gmail.com, katariabooks@yahoo.com
Website: www.skkatariaandsons.com